Central Nervous System Diseases

Contemporary Neuroscience

Central Nervous System Diseases

Innovative Animal Models from Lab to Clinic

Edited by

Dwaine F. Emerich, PhD
Director of Neuroscience, Alkermes, Inc., Cambridge, MA

Reginald L. Dean, III, MS
Department of Neuroscience, Alkermes, Inc., Cambridge, MA

Paul R. Sanberg, PhD, DSc
Univeristy of South Florida College of Medicine, Tampa, FL

Humana Press ✳ Totowa, New Jersey

© 2000 Humana Press Inc.
999 Riverview Drive, Suite 208
Totowa, New Jersey 07512

This publication is printed on acid-free paper. ∞
ANSI Z39.48-1984 (American Standards Institute) Permanence of Paper for Printed Library Materials.

Cover illustration: "Dark field photomicrograph through the forebrain of a rat dual stained for tyrosine hydroxylase and acetylcholinesterase" was provided by Jeffrey Kordower, PHD, Director, Research Center for Brain Repair, and Professor of Neurological Sciences, Rush Presbyterian Medical Center, Chicago, IL

Cover design: Patricia Cleary.

For additional copies, pricing for bulk purchases, and/or information about other Humana titles, contact Humana at the above address or at any of the following numbers: Tel.: 973-256-1699; Fax: 973-256-8341; E-mail: humana@humanapr.com

Printed in the United States of America. 10 9 8 7 6 5 4 3 2 1

Library of Congress Cataloging in Publication Data

Main entry under title:

Central nervous system diseases: innovative animal models from lab to clinic / edited by Dwaine F. Emerich, Reginald L. Dean III, Paul R. Sanberg.
        p.   cm.   -- (Contemporary neuroscience)
      Includes index.
      ISBN 0-89603-724-X (alk. paper)
      1. Central nervous system–Diseases–Animal models. I. Emerich, Dwaine F. II. Dean, Reginald L. III. Sanberg, Paul R. IV. Series.
      [DNLM: 1. Alzheimer Disease–physiopathology. 2. Aging–physiology. 3. Huntington's Disease–physiopathology. 4. Models, Biological. 5. Parkinson Disease–physiopathology. WT 155 C397 2000]
      RC361.C44   2000
      616.8'0427–dc21
      DNLM/DLC   99-25833
      for Library of Congress   CIP

To Nicole Deanna Sanberg, my daughter and inspiration

Preface

Our continued understanding of the pathophysiological underpinnings of CNS diseases as well as the development of new treatments for neurological disorders is intimately linked to the development of appropriate animal models. Developing animal models is exceedingly difficult and borrows from developments in fields as diverse as behavioral neuroscience, biochemistry, anatomy, pharmacology, and molecular biology. The contributions contained in *Central Nervous System Diseases: Innovative Animal Models from Lab to Clinic* represent the efforts from some of the leading scientists and clinicians making use of the innovations occurring in animal model development today. Given the number of researchers currently active in this and related fields, it would be inappropriate to suggest that the full range of activities were fully reflected in this single volume. Instead, experimental approaches have been chosen to illustrate rapidly developing experimental and therapeutic themes.

The chapters do not provide detailed experimental protocols. Rather the authors have been encouraged to critically examine the models available today and provide a detailed examination of their advantages, but equally as important, their limitations. With this backdrop, the contributions focus heavily on the use of animal models as a means of developing therapies for human diseases. In cases where the animal models have played important roles in developing currently approved drugs, the authors describe the contributions of the models to that process. In cases where new and innovative treatments remain in preclinical evaluations, the role that currently available models are playing in that process is also examined. In this way, the reader can appreciate the roles that animal models have played in the past, the role they play today, and the roles that they might play in the future as technological advances occur across scientific disciplines. The goal of this volume is to provide, in one source, important practical information that will prove invaluable to both experienced researchers and students.

Central Nervous System Diseases: Innovative Animal Models from Lab to Clinic is organized into five sections. The first three sections concentrate on models used in three of the most commonly studied areas. The first section contains seven chapters detailing developments in aging and Alzheimer's disease. The second section provides six chapters illustrating developments in animal models of Parkinson's disease. The third section provides five chapters describing current developments in the animal models of Huntington's disease. Within each of these first three sections, the initial chapters describe techniques that rely heavily on behavioral analyses, followed by chapters discussing models developed from infusions of neurotoxins, and finally models that have grown from developments in molecular biology. The fourth section contains three chapters describing model developments for more acute neurological conditions including traumatic brain injury and stroke. Finally, the fifth section provides three examples that concentrate on the contributions that animal models are making to the development of some of the more exciting and innovative therapeutic agents.

It is hoped that this volume will foster the continued refinement and rationalization of using animal models to help treat some of the most important neurological diseases in humans.

Dwaine F. Emerich
Reginald L. Dean, IIII
Paul R. Sanberg

Contents

Contributors

Lucy E. Annett • *Department of Experimental Psychology, University of Cambridge, Cambridge, UK*

Gary W. Arendash • *Department of Biology, University of South Florida, Tampa, FL*

Raymond T. Bartus • *Alkermes, Inc., Cambridge, MA*

Gillian P. Bates • *Medical and Molecular Genetics, Guy's Hospital, London, UK*

M. Flint Beal • *Department of Neurology, Massachusetts General Hospital, Boston, MA*

Kimberly B. Bjugstad • *Department of Psychology, University of South Florida, Tampa, FL*

Cesario V. Borlongan • *National Institutes of Health, National Institute of Drug Abuse, Intramural Research Program, Cellular Neurophysiology Branch, Baltimore, MD*

Peter J. Brasted • *MRC Cambridge Centre for Brain Repair, Cambridge, UK*

Emmanuel Brouillet • *URA CER CNRS 2210, Service Hospitalier Frédéric Joliot, DRM, DSV, CEA, Orsay Cedex, France*

Françoise Condé • *URA CER CNRS 2210, Service Hospitalier Frédéric Joliot, DRM, DSV, CEA, Orsay Cedex, France*

James M. Conner • *Department of Neurosciences, University of California San Diego, La Jolla, CA*

Lauren C. Costantini • *Neuroregeneration Laboratories, Harvard Medical School/ McLean Hospital, Belmont, MA*

Ian Creese • *Center for Molecular and Behavioral Neuroscience, Rutgers, The State University of New Jersey, Newark, NJ*

Rosalyn M. Cummings • *Department of Experimental Psychology, University of Cambridge, Cambridge, UK*

Caroline Dautry • *URA CER CNRS 2210, Service Hospitalier Frédéric Joliot, DRM, DSV, CEA, Orsay Cedex, France*

Stephen W. Davies • *Department of Anatomy and Developmental Biology, University College, London, UK*

Màté D. Döbrössy • *MRC Cambridge Centre for Brain Repair, Cambridge, UK*

Karen Duff • *Neurotransgenics Laboratory, Mayo Clinic, Jacksonville, FL*

Stephen B. Dunnett • *MRC Cambridge Centre for Brain Repair, Cambridge, UK*

Dawn M. Eagle • *MRC Cambridge Centre for Brain Repair, Cambridge, UK*

Dwaine F. Emerich • *Alkermes, Inc., Cambridge, MA*

Fred H. Gage • *Laboratory of Genetics, The Salk Institute for Biological Studies, La Jolla, CA*

WENDY R. GALPERN • *Neuroregeneration Laboratories, Harvard Medical School/McLean Hospital, Belmont, MA*

ANN-CHARLOTTE GRANHOLM • *Department of Basic Science, University of Colorado HSC, Denver, CO*

PHILIPPE HANTRAYE • *URA CER CNRS 2210, Service Hospitalier Frédéric Joliot, DRM, DSV, CEA, Orsay Cedex, France*

JASMINE M. HENDERSON • *Department of Experimental Psychology, University of Cambridge, Cambridge, UK*

JAMES G. HERDON • *Department of Anatomy and Neurobiology, Boston University School of Medicine, Boston, MA*

BARRY HOFFER • *Intermural Research Program, National Institutes of Health, National Institute of Drug Abuse, Baltimore, MD*

OLE ISACSON • *Neuroregeneration Laboratories, Harvard Medical School/McLean Hospital, Belmont, MA*

SIMRANJIT KAUR • *Center for Molecular and Behavioral Neuroscience, Rutgers, The State University of New Jersey, Newark, NJ*

A. LISA KENDALL • *Department of Experimental Psychology, University of Cambridge, Cambridge, UK*

RONALD J. KILLIANY • *Department of Anatomy and Neurobiology, Boston University School of Medicine, Boston, MA*

YOSHIHISA KITAMURA • *Department of Neurobiology, Kyoto Pharmaceutical University, Kyoto, Japan*

EFFREY H. KORDOWER • *Department of Neurological Sciences, Rush Presbyterian Medical Center, Chicago, IL*

MARK D. LINDNER • *Neurocrine Biosciences, Inc., San Diego, CA*

LAURA MANGIARINI • *Medical and Molecular Genetics, Guy's Hospital, London, UK*

VINCENT MITTOUX • *URA CER CNRS 2210, Service Hospitalier Frédéric Joliot, DRM, DSV, CEA, Orsay Cedex, France*

MARK B. MOSS • *Department of Anatomy and Neurobiology, Boston University School of Medicine, Boston, MA*

FALGUNI NATHWANI • *MRC Cambridge Centre for Brain Repair, Cambridge, UK*

MARY B. NEWMAN • *Department of Psychology, College of Medicine, University of South Florida, Tampa, FL*

HITOO NISHINO • *Department of Physiology, Nagoya City University Medical School, Mizuho-ku, Nagoya, Japan*

YASUYUKI NOMURA • *Department of Pharmacology, Graduate School of Pharmaceutical Sciences, Hokkaido University, Sapporo, Japan*

YASUNOBU OKUMA • *Department of Pharmacology, Graduate School of Pharmaceutical Sciences, Hokkaido University, Sapporo, Japan*

STÉPHANE PALFI • *URA CER CNRS 2210, Service Hospitalier Frédéric Joliot, DRM, DSV, CEA, Orsay Cedex, France*

DANIEL A. PETERSON • *Laboratory of Genetics, The Salk Institute for Biological Studies, La Jolla, CA*

JASODHARA RAY • *Laboratory of Genetics, The Salk Institute for Biological Studies, La Jolla, CA*

TREVOR W. ROBBINS • *Department of Experimental Psychology, University of Cambridge, Cambridge, UK*

BEN ROITBERG • *Department of Neurosurgery, University of Illinois Medical Center, Chicago, IL*

PAUL R. SANBERG • *Departments of Psychiatry, Neurosurgery, Pharmacology, Psychology, and Neuroscience, College of Medicine, University of South Florida, Tampa, FL*

TIMOTHY J. SCHALLERT • *Department of Psychology, University of Texas, Austin, TX*

PETER SHIN • *Department of Neurological Sciences, Rush Presbyterian Medical Center, Chicago, IL*

R. DOUG SHYTLE • *Departments of Psychiatry and Behavioral Medicine, College of Medicine, University of South Florida, Tampa, FL*

ARCHIE A. SILVER • *Departments of Psychiatry, Neurosurgery and Neuroscience, College of Medicine, University of South Florida, Tampa, FL*

RACHEL E. SMYLY • *Department of Experimental Psychology, University of Cambridge, Cambridge, UK*

JOSEPH SRAMEK • *Department of Neurological Sciences, Rush Presbyterian Medical Center, Chicago, IL*

RICHARD L. SUTTON • *Department of Surgery, Hennepin County Medical Center, Minneapolis, MN*

JENNIFER L. TILLERSON • *The University of Texas at Austin, Department of Psychology and Institute of Neuroscience, Austin, TX*

MARK H. TUSZYNSKI • *Department of Neurosciences, University of California San Diego, La Jolla, CA*

YUN WANG • *Intramural Research Program, National Institutes of Health, National Institute of Drug Abuse, Cellular Neurophysiology Branch, Baltimore, MD*

GARY L. WENK • *Division of Neural Systems, Memory and Aging, University of Arizona, Tucson, AZ*

I

AGING AND ALZHEIMER'S DISEASE

The Cholinergic Hypothesis a Generation Later
Perspectives Gained on the Use and Integration of Animal Models

Raymond T. Bartus

1. INTRODUCTION

It has been roughly 20 yr since the empirical foundation supporting the cholinergic hypothesis was first established. This evidence was generated by multiple laboratories and clinics, working simultaneously on parallel fronts. Stated in its most simple form, the cholinergic hypothesis contends that: (1) a significant dysfunction of central cholinergic neurons occurs in Alzheimer's disease (AD), with analogous, but much less severe, changes also occurring in aged humans and animals; (2) this cholinergic dysfunction contributes significantly to the cognitive impairments seen in the earlier stages of AD and advanced stages of normal aging; and (3) pharmacological enhancement of cholinergic activity can measurably improve the cognitive function of these patients. Since its inception, the cholinergic hypothesis has generated considerable attention and controversy. For example, the initial article formally articulating the hypothesis *(1)* ranks fourth among all neuroscience publications of the past 15 yr, in terms of number of citations *(2)* and yet the hypothesis has also generated a continuing stream of debate and criticism that continues to this day (e.g., see discussion of many key points in *3–7*).

In addition to stimulating discussion and experimentation, the cholinergic hypothesis has also uniquely distinguished itself in several important ways. First, the cholinergic hypothesis is the only approach that has thus far yielded approvable drugs for treating AD. Of course, the ability to generate meaningful treatments is the quintessential feature of any practical hypothesis that attempts to explain a medical problem. After years of latent but steady activity, the cholinergic hypothesis has finally begun to produce tangible benefits in this important area. Two cholinesterase inhibitors have already been approved by the FDA for treating AD and the approval of still other inhibitors is believed by many to be imminent *(8,9)*.

In addition, in contrast to the large role that serendipity played in the initial development of treatments for nearly every other central nervous system (CNS) indication (e.g., depression, schizophrenia, anxiety, Parkinson's disease), the cholinergic hypothesis provided a rational framework for developing therapies for AD. The framework was derived from the successful integration of data from biochemistry, pharmacology, novel animal models, and clinical studies. In addition, in contrast to most other new

From: Central Nervous System Diseases
Edited by: D. F. Emerich, R. L. Dean, III, and P. R. Sanberg © Humana Press Inc., Totowa, NJ

therapeutic areas today, the animal models employed by the cholinergic hypothesis have been heavily dependent on behavioral paradigms, which in turn were developed from a rich history of basic neuroscience research.

Although the relatively modest symptomatic improvement achieved thus far, as well as the existence of nonresponders, may temper some people's enthusiasm for this approach, several counterpoints offer reason for optimism. Certainly, the currently approved drugs should be viewed as merely starting points. As discussed later in this chapter, there are many reasons to anticipate that even more efficacious cholinergic therapies will become available in the future. In addition, when one considers how far the field has evolved since the time that work began on the cholinergic hypothesis, the progress achieved is truly impressive. For example, prior to the cholinergic hypothesis, no consensus yet existed for the primary symptoms of AD, or even its diagnostic criteria *(10–12)*. Indeed, the concept that selective deficits in memory occur with advanced age was still poorly accepted at that time *(13–16)*, and it was still controversial to suggest that drugs might be used to improve something as complex, conceptually narrow, and poorly understood as age-related memory deficits or AD symptoms. Certainly, earlier attempts to improve the symptoms of AD (using vasodilators, psychostimulants, and other nonspecific approaches) showed no genuine efficacy *(17–19)*. Finally, the idea of using animals to help study these problems was controversial, at best, and even heretical to some, for holdovers from earlier schools of thought *(20,21)* still argued that it was anthropomorphic to use the term "memory" when describing animal behavior (e.g., *see 15*). Thus, how could one imagine using animals to model aspects of human memory, especially involving deficits associated with human-specific diseases? Work directed toward the cholinergic hypothesis, including the animal models employed, not only contributed to the conceptualization and support of the hypothesis and its initial treatments but also to the general evolution of the neurosciences, and the specialized field of AD. In part because of this effort, enthusiasm for pharmacological treatment of AD and other neurodegenerative diseases now runs high, and the use of selective animal models for neurodegenerative diseases is no longer considered controversial.

This chapter reviews the evolution in thinking that occurred as the cholinergic hypothesis was developed, and the lessons learned along the way. In particular, special attention is given to the central role that animal models played in the endeavor and the contemporary issues that have since emerged concerning the continued use of animal models to study neurodegenerative diseases. By using the benefit of hindsight to illustrate and discuss when models proved useful and predictive and when they did not, it is hoped that the continued development and use of animal models for AD and many other neurodegenerative diseases will be facilitated.

2. A BRIEF LOOK BACKWARDS: THE SUCCESSFUL INTEGRATION OF ANIMAL MODELS AND CLINICAL RESEARCH

One of the more uniquely positive aspects of the cholinergic hypothesis is the variety of approaches that contributed to its support. How this occurred and the issues that emerged during this time offer perspectives that remain relevant to today's efforts to develop models, insight, and treatments for many of the neurodegenerative diseases.

The early foundation supporting the cholinergic hypothesis came primarily from four distinct and initially independent areas of study: animal pharmacology, biochemistry, clinical research, and the more fundamental neurosciences (*see 22* for more thorough review). By the late 1970s and early 1980s, these efforts had become well integrated, with findings in one area clearly influencing the thinking and research in others.

2.1. Early Contributions from Pharmacology Modeling

The initial suggestion that central cholinergic neurons might play an important role in the specific type of memory loss associated with advanced age was suggested by Drachman and Leavitt *(23)*. While anticholinergics had been known for years to produce amnestic syndromes in different circumstances *(24,25)*, Drachman and Leavitt noted that relatively low doses of scopolamine could produce a pattern of cognitive deficits that generally paralleled those seen in (nondemented) elderly volunteers. Soon after that seminal article, a series of publications using nonhuman primates strengthened the association of central cholinergic blockade to the pattern of memory loss seen in old age. The articles on nonhuman primate demonstrated that the deficit achieved with scopolamine closely matched the most robust and consistent cognitive deficit to occur naturally in aged monkeys whereas similar blockade of peripheral cholinergic receptors and other central neurotransmitter receptors was not able to do so (reviewed in *26*). As discussed later, these studies led directly to the first successful demonstration that age-related memory losses could be pharmacologically reduced, using the anticholinesterase physostigmine in aged monkeys *(27,28)*. During this same period of time, others demonstrated that the cognitive profile induced by cholinergic blockage also mimicked some of the cognitive deficits of AD patients *(29,30)*, while the deficit in aged primates, in turn, was shown to be operationally and conceptually similar to the severe recent episodic memory impairments that occur in AD patients *(13,31)* (*see* Fig. 1). Since these early studies, numerous preclinical and clinical laboratories have continued to compare the pattern of behavioral deficits produced by pharmacological blockade of cholinergic receptors with those occurring naturally with age and AD. Considerable controversy continues to stimulate discussion regarding the extent to which the pharmacological effects mimic those occurring naturally with age or dementia and whether cholinergic blockade provides a useful model for rational drug testing *(4,32–34)*. (There is no simple answer to the question and these issues are discussed in detail later in this chapter.)

2.2. Integration of Neurochemistry and Neuropathology

At about the same time that the initial pharmacologic studies were in full swing (i.e., the mid-1970s), three laboratories nearly simultaneously reported that the brains of AD patients exhibited a significant loss of choline acetyl transferase (ChAT) activity. Bowen et al. *(35)* first reported the decrease and found that it was correlated with the cognitive impairments of AD patients. Later that same year and early the next, Davies et al. and Perry et al. noted similarly severe decreases in ChAT activity *(36,37)*. What remained unclear was how, or indeed whether, the decrease in ChAT activity played any role in the cognitive symptoms of AD. It was widely known that ChAT is not rate

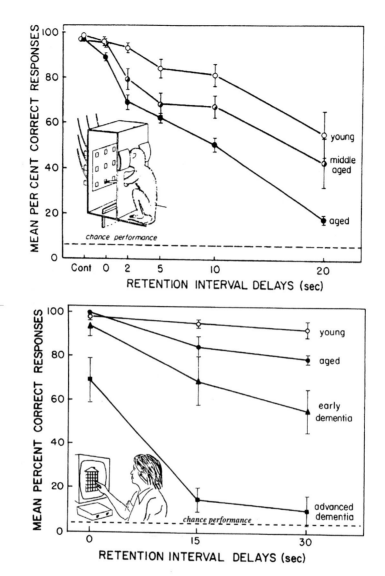

Fig. 1. Top: Performance of young (5–7 yr), middle aged (10–15 yr), and aged monkeys (>18 yr) on an automated nine-choice, subject-paced task of recent episodic memory (delayed response). The subject was required to initiate the trial by placing its face into the window, which turned on a light on one of the nine panels. A one-way viewing screen (lowered in picture to permit response, but raised during stimulus viewing and retention) prohibited the subject from responding to the stimulus until the proper time (i.e., end of "retention interval"). During the retention interval, the panel of lights was visually isolated from the monkey by the one-way screen. Note that at the short delay intervals (which were randomly interspersed), little or no age-related differences existed. However, as a longer time interval was inserted between when the stimulus was viewed and when it could be responded to (thus placing increasingly greater demands on recent memory), a progressively greater, age-related deficit was seen. In the control condition ("Cont"), the stimulus remained on during the entire trial (therefore requiring no recent, episodic memory). An additional series of control experiments *(16)* demonstrated that potential age-related differences in motivation, psychomotor coordination, sensory ability, attention, etc. could not easily account for the deficits, confirming that a

limiting for acetylcholine synthesis and that adequate acetylcholine synthesis could be maintained with losses of 90% of this enzyme *(38,39)*. Thus, on the surface, the significant, 60–70%, loss of ChAT did not necessarily imply any loss of cholinergic function, let alone a specific relationship to the memory impairments associated with AD (just as prior observations of significant losses of markers for norepinephrine and dopamine in AD brains *(40–43)* did not establish a role for these neurotransmitters in the cognitive losses of AD). However, the prior animal and human pharmacological evidence had established an important relationship between impaired cholinergic function and the conceptual and operational features of the cognitive deficits of early AD and advanced age. This knowledge, heavily dependent on data from animal models, therefore supported the empirical foundation that suggested that the loss of ChAT activity and the memory impairments in AD may not be merely coincidental.

When a significant loss of basal forebrain cholinergic neurons was soon after observed in AD brains *(44,45)*, an explanation was provided for how a reduction of ChAT activity might lead to cholinergic impairments (i.e., because the decrease actually reflected a serious and somewhat selective loss of ChAT-producing cholinergic neurons). This finding also helped further integrate the existing data from biochemical, pharmacological, and behavioral studies which, in turn, led to additional animal and human studies employing anticholinergic drugs as well as animal studies using lesions of basal forebrain cholinergic neurons to model some of the neurobehavioral deficiencies of AD. These approaches, however, remain controversial to this day, and are also the topic of discussion later in this chapter.

Biochemical determinations in aged rodent brain tissue also provided some evidence of cholinergic deficiencies (e.g., *see 46*, but inconsistencies in these effects and the relatively subtle changes reported for both muscarinic receptors and cholinergic enzymes left open the question of how much physical change in the cholinergic system actually occurs in normal aging brains. On the other hand, neurophysiological recordings of hippocampal pyramidal neurons following iontophoretic application of acetylcholine and glutamage demonstrated a robust and somewhat selective loss of cholinergic responsiveness in memory-impaired aged rats *(47,48)*. These data suggested that deficiencies in intracellular signaling of cholinoreceptive neurons are a

---

Fig. 1. *(continued)* significant age-related impairment in recent episodic memory ability was responsible for the poor performance in the aged monkeys. (Data adapted from Bartus, 1980 *[60]*.) **Bottom:** Performance of young volunteers (18–24 yr) vs healthy elderly volunteers, early AD patients, and advanced AD patients (all 57–85 yr) on a recent episodic memory task modeled after a nonhuman primate task (**above**) but designed to reflect activities of daily living. Subjects were asked to remember which room of a 25-room house, presented on a video monitor, had a light on in the window (now off). During the delay interval the subjects were required to perform a reaction time task. Note that similar (albeit modest) deficits occur in elderly volunteers, with progressively greater deficits observed in the mild to moderate and severe AD patients, respectively (similar to those observed in aged monkeys). At the 0-s delay, only the performance of the advanced AD subjects significantly differed from that of the other three groups. By 30 s post-stimulus, performance level was significantly different among all four groups. (Modified with permission from Flicker, 1984 *[13]*.)

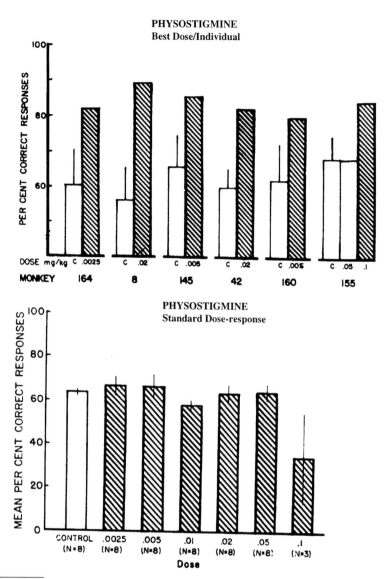

Fig. 2. Top: Performance of aged monkeys on a recent episodic memory task (i.e., a subject-paced, nine-choice delayed response task shown in **Fig. 1**) following injections of either physostigmine or vehicle. The duration of the retention interval was adjusted for each monkey so that its performance was about 60% (i.e., well above chance performance of 11%, but well below perfection). These data represent a "retest" challenge of the most effective dose of physostigmine, per individual monkey, determined by prior multiple dose-range testing in each monkey. Two out of eight monkeys showed no improvement under any dose tested and therefore are not included here. Note that all but one of the monkeys again showed a positive response to the dose previously shown to be individually effective, while that one exception did so at the next highest dose level. Note also the wide individual variation (order of magnitude difference) in the most effective dose—a phenomenon initially reported in monkeys *(27,28)*, replicated very soon after in AD patients *(51)* and since reported numerous times (*see* text, p. 9). Yet most clinical trials attempting to demonstrate efficacy of cholinomimetics do not attempt to accommodate wide individual dose variations into the protocol design. C, Mean of multiple control tests with vertical bar representing $p < .01$ confidence limits for all vehicle test days. (Data

major problem in aged brain that likely contributes to the memory losses—a hypothesis recently reiterated to explain more recent data in aged subjects *(49)*.

2.3. Early Development of Cholinergic Treatment Approaches

The initial, empirical evidence that drugs could improve age-related cognitive deficits was achieved using an animal model *(28,50)*. This experiment was performed in aged nonhuman primates using the cholinesterase inhibitor physostigmine to improve recent episodic memory deficits. In addition to demonstrating significantly improved memory performance following administration of physostigmine, this article made two other observations that still seem relevant today. First, significant variability existed in the response of individual monkeys to cholinergic drug treatment, with some subjects apparently unresponsive at any dose tested. Second, even among the positive treatment responders, significant variability in the most effective dose was seen between monkeys (e.g., *see* Fig. 2). Shortly after, similar results were reported when physostigmine was given to Alzheimer's patients *(51)*, including increased interpatient variability reminiscent of the response in aged nonhuman primates. This study in AD patients was one of a series of pioneering clinical experiments by Davis and colleagues investigating the effects of pharmacological enhancement of cholinergic activity in a variety of clinical conditions (e.g., AD, depression, schizophrenia), including tests on memory with young, healthy volunteers *(52,53)*. These initial positive effects of physostigmine in nonhuman primates and AD patients were soon replicated and extended *(54–60)*. In aged monkeys a series of studies showed that while other cholinergic drugs (such as tacrine and muscarinic agonists) could similarly improve memory performance, a variety of other agents that stimulate noncholinergic neurotransmitter receptors (e.g., dopamine, γ-aminobutyric acid, α-adrenergic) were ineffective (reviewed in *61*). Building on these data, and the broader foundation laid by the cholinergic hypothesis, oral tetrahydroaminoacridine (tacrine, THA, or Cognex®) was tested in a small group of AD patients *(62)* and shown to improve their symptoms. These results were later replicated in a multicenter trial *(63)* and in 1993 tacrine was approved by the FDA as the first treatment for the cognitive symptoms of AD. The prime mechanism of action of THA, like that of physostigmine, is the inhibition of acetylcholinesterase, thus extending the duration of synaptic acetylcholine. Although far from a miracle drug, the approval of THA was important for several reasons.

First, it provided the first clear evidence that the symptoms of AD could be significantly reduced with a drug. Second, in doing so, it provided confirmation of the link between cholinergic activity and the cognitive symptoms of AD, thus validating an important tenet of the cholinergic hypothesis. Finally, because it was generally recog-

---

Fig. 2. *(continued)* modified from Bartus, 1980 *[60]*.) **Bottom:** Data from the same study (**above**) except the replicable individual variations in peak dose were not taken into account when group means were simply computed for each dose. Note that using this analysis, it appears that no dose is effective in improving recent episodic memory in aged monkeys, when in fact the majority of monkeys exhibit consistent improvement, but under individually variable doses. Many clinical trials continue to analyze the data in this manner, potentially obscuring positive effects.

nized that tacrine was far from an ideal cholinergic drug, its success demonstrated both the need and the path to develop even better drugs.

2.4. Initial Resistance and Controversy

Despite the range of evidence supporting the cholinergic hypothesis, almost from its inception, it has been surrounded in controversy and immersed in debate. Much of the early arguments claimed it was (a) too simplistic to assume that any single neurotransmitter could be responsible for a syndrome as complicated and devastating as AD, or (b) naive to suggest that the exclusive role of cholinergic neurons is to mediate cognition; thus, it seemed unlikely to these investigators that drugs modulating cholinergic activity could ever do AD patients any good. Of course, the cholinergic hypothesis never suggested that other neurotransmitters are not involved with the deficits, only that cholinergic neurons participated in an important manner. Moreover, it was never assumed that cholinergic neurons had no other function beside cognition, only that this was one of their important functions. Finally, the cholinergic hypothesis never predicted that cholinergic drugs would likely reverse all the impairments of AD, or even completely reverse any particular symptom, only that some measurable and meaningful improvement in key symptoms associated with the recent memory impairments should be possible. These points were discussed in detail in a series of articles on the subject *(3,22,64)* and do not deserve further repetition here. However, more recent debate has focused specifically on the underlying logic of the models intended to mimic the neurobehavioral deficits of aging and AD, as well as the utility of these models for understanding the problem and searching for effective therapy. Because resolution of these issues impacts how effectively animal models might be employed for other neurodegenerative diseases, they are discussed in detail in the next section of this chapter.

3. CURRENT PERSPECTIVES ON ANIMAL MODELS ASSOCIATED WITH THE CHOLINERGIC HYPOTHESIS

The use of animal models to study AD and age-related memory loss, as well as the potential role of central cholinergic neurons, has had a controversial history. Many critics argued that animals could not be used for AD-related research because animals do not truly develop AD (in contrast to many other human diseases where animals models have proven invaluable; e.g., diabetes, hypertension, cancer, etc.). However, it was important to recognize that no model claims or intends to mirror every aspect of human disease *(65–72)*. Rather, by their nature, models are used to reproduce some aspect of the disease to study it in ways not possible in the clinical setting. Thus, the issue is not whether animals contract AD, but rather whether they share, or can be made to share, certain components of the disease to gain knowledge that may facilitate its treatment. Indeed, it is now clear that animal models played important roles from the inception of the cholinergic hypothesis through the development of the treatment approaches now shown to benefit patients. For example, as reviewed later, they helped establish an important role of the cholinergic system in the specific recent episodic memory loss that characterized the early symptoms of AD, helped in the development of cholinergic strategies and specific cholinergic drugs for treating AD, helped elucidate the age-related changes that occur in normal healthy aging, and helped establish the similarities and differences between the losses in normal aging from the more severe cognitive deficits of AD. The following text describes in greater detail the contributions made and issues generated by three different types of animal models. The individual contributions, as well as the limitations of each, is discussed so that, with the benefit of hindsight, this perspective may facilitate the further development and use of animal models to study and find treatments for neurodegenerative diseases.

3.1. Models Using Aging Non-Human Primates

Models employing aged primates played a prominent role in the early development of the cholinergic hypothesis, although not without initial resistance. Prior to the inception of the cholinergic hypothesis, there existed no consensus for whether normal aging was associated with a loss of memory, little enthusiasm for using animals to help address this issue, and widespread confusion regarding the key, early symptoms of AD and its relationship, if any, to deficits in non-AD elderly *(10–15)*. Several factors began to change this scenario in the 1970s that eventually opened the door for animal models to make contributions to the field.

3.1.1. Establishment of Age-Related Memory Deficits

One of the events that helped modify perceptions involved a series of studies with aged non-human primates, reporting that impairments in recent episodic memory were among the most robust deficits observed within a battery of tests *(27,50,73,74)*. Although others had previously reported decreases in memory ability in aged nonhuman primates *(75,76)*, the issue of whether normal age-related deficits occurred in animals and humans remained controversial until the deficit was demonstrated in an apparatus and paradigm that did not easily allow any other logical interpretation.

The importance of a well-controlled testing environment to accurately measure the intended behavioral constructs cannot be overstated; this theme will recur as other issues related to the development and use of animal models are discussed in other parts of this chapter. The apparatus used in the aforementioned primate studies was an automated, nine-choice, subject-paced device that was specifically designed to provide superior control of many nonmnemonic variables, particularly those that might confound measures of memory and other forms of cognition in aged monkeys *(73)*. It required proper visual orientation toward the stimuli for the trial to be initiated by the subject, required a minimal viewing time of the stimuli, greatly reduced the ability of the subjects to exploit extraneous (nonmnemonic) cues, and offered precise timing and control of the stimulus presentation and retention intervals. Because a correct response could be made by chance only 11.1% of the time (i.e., one out of nine), a wide range of above-chance performance provided a very sensitive measuring tool. Other features made it possible to control for potentially confounding age-related variables, such as deficits in visual acuity, psychomotor coordination, reduced appetite and motivation, etc. *(see 15,16,77* for more detailed discussion of these points). Because of these features, the series of studies performed in this apparatus unequivocally demonstrated that robust deficits in recent episodic memory occur in old monkeys *(16)*. They also demonstrated that these deficits were among the most severe and earliest manifested from among a variety of behaviors studied *(50,73,74)*. The identical deficit was later shown to exist in New World monkeys *(60)* (Fig. 1) and was remarkably indistinguishable from the major deficit induced in young monkeys given the anticholinergic agent, sco-polamine *(78)* (the latter point reinforcing recent studies in humans *(23)* (Fig. 3).

Using a similar apparatus adapted specifically for humans, the memory deficit in nonhuman primates was later shown to be operationally and conceptually similar to a robust memory deficit characterized in the early stages of AD *(13,31)* (Fig. 1). Most importantly, the conceptual and operational features of this task were similar to the loss of recent episodic memory reported in elderly humans and early-stage AD patients *(13,14,31,32,79–81)*. These and subsequent studies by other investigators (e.g., *see 82–87*) helped firmly establish recent episodic memory loss in aged primates as a con-sistent phenomenon, contributing to the current acceptance that such losses are a com-mon aspect of mammalian aging (e.g., *14,88–90*). This work also helped establish the similarity of the recent episodic memory loss that occurs in elderly monkeys, humans, and early stage AD patients, therefore providing an empirical link to use models with aged animals for studies of human cholinergic dysfunction and memory disturbances. Earlier observations of significant concentrations of amyloid plaques in aged monkey brains had provided an additional complementary pathologic link between aged mon-keys, aged humans, and AD *(91)*. Finally, the research with aged nonhuman primates helped establish age-related memory loss as a formal diagnostic entity *(14)*, while pro-viding insight into some of its biochemical sources and helping to compare it to and distinguish it from the more severe and global cognitive deficits associated with AD.

Although some investigators correctly point out that significant cognitive differ-ences exist between normal aging and AD patients *(88)*, these arguments do not mini-mize the important commonalties that do exist. Certainly, much can be learned about aging and AD by searching for and studying the differences that exist between them. At the same time, by focusing on the similarities that exist in the primary symptoms, one can exploit commonalties to develop testable hypotheses, valid animal models, and therapeutic approaches—at least for the symptoms held in common. This is what work on the cholinergic hypothesis successfully did.

3.1.2. Early Suggestion for a Frontal Lobe Dysfunction

Work with aged nonhuman primates also revealed a pattern of deficits that closely resembled that induced by selective lesions of the frontal neocortex (Table 1). These observations invoked an early hypothesis that frontal cortical dysfunction may be par-ticularly involved with the cognitive impairments of late senescence and early AD *(73)*. Thus, this hypothesis, based on work in young and aged nonhuman primates, anticipated and supported current clinical thinking that attributes important age-related losses of "executive function" to frontal cortical dysfunction *(92–94)*. Subsequent authors have since noted similar parallels between the pattern of deficits in aged mon-keys and frontal lobe dysfunction (e.g., *84,95*).

3.1.3. Drug Research with Aged Monkeys Predicted Early Clinical Results

Animal models also contributed to the development of the treatment approaches that ultimately led to success in the clinic. Not only was the initial demonstration of phar-macological improvement of age-related memory deficits achieved in aged monkeys *(27,28)*, but this same test situation generated a body of evidence that continues to have reasonable predictive value for AD clinical trials. Both the initial clinical success

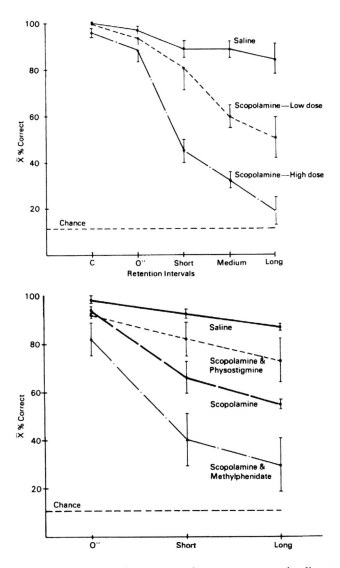

Fig. 3. Top: Mean performance of young monkeys on recent episodic memory, using the same task shown in Fig. 1. Each monkey was tested under two doses of the anticholinergic scopolamine (low dose is from .01 to .015 mg/kg; high dose is from 0.20 to 0.30 mg/kg) as well as a saline control condition. Retention intervals were adjusted per individual monkey to provide modest decline in performance under vehicle condition (maximum duration was 60 s). Note the progressive, dose-related impairment that occurs as the retention interval increases; this mimics the naturally occurring deficit found in aged monkeys. Even though accuracy on the longest retention interval was quite poor, apparent nonmemnonic effects were minimal and all monkeys easily completed the test. Similar tests using numerous other neurotransmitter modulators were not able to induce similarly selective recent memory impairments. C, Control condition where stimulus light remained on. Vertical bars represent SEM. (Data from Bartus, 1976 *[78].*) **Bottom:** Effects of combinations of scopolamine (from 0.105 to 0.02 mg/kg), physostigmine (from 0.02 to 0.03 mg/kg), and methylphenidate (0.0125 mg/kg) on performance of delayed-response task. Concurrent administration of physostigmine reliably reduced the scopolamine-induced deficit, while concurrent administration of methylphenidate potentiated the deficit, at a dose of methylphenidate that produced no measurable effects when given alone (not shown). Vertical bars represent SEM. (Data from Bartus, 1979 *[127].*)

Table 1
Similarities Between Cognitive Performance
of Aged Monkeys and Frontally Ablated Young Monkeys

Behavioral function	Aged monkeys	Frontal monkeys
Recent episodic memory (delayed response)	Severely impaired	Severely impaired
Sensitivity to interfering stimulation (intratrial)	Increased	Increased
Sensitivity to proactive interference (intertrial)	Increased	Increased
Sensory processing ability (sensory problems controlled for)	No serious impairment	No obvious impairment
Visual discrimination learning	Generally no deficit	Generally no deficit

These data suggest potentially important frontal dysfunction in mediation of some of the cognitive symptoms of age and early-stage AD. They thus offered early support for current hypotheses regarding frontal cortical dysfunction and loss of executive function in aging and dementia (e.g., *see* text for further discussion, **Subheading 3.1.2.**). (Adapted from Bartus, 1979 *[73]*).

(51,54–59) achieved with cholinesterase inhibitors as well as many negative approaches were correctly anticipated by studies with aged monkeys. Using the same test situation, the monkey data predicted a lack of efficacy with cholinergic precursor loading *(60,96)*, dopamine precursor loading *(61)*, general CNS stimulation *(17)*, neuropeptide enhancement *(97)*, and noradrenergic receptor stimulation *(98,99)*. To date, all of these results have been confirmed in AD clinical trials (Table 2).

Although stimulation of the cholinergic neurotransmitter system is the only one to produce positive effects on memory performance in the aged monkeys tested in this subject-paced, automated apparatus, Arnsten and Goldman-Rakic observed positive effects in a different test situation when adrenergic receptors were stimulated with α_2-adrenergic agonists *(100,101)*. However, attempts to replicate these observations in the subject-paced, automated apparatus described were not successful, despite testing a variety of acute doses, subchronic doses, and combinations of adrenergic and cholinergic receptor agonists *(98,99)*. Carefully controlled clinical trials also failed to achieve positive effects in AD patients, using two different adrenergic agonists, clonidine and guanfacine *(102,103)*. Nonetheless, the possibility that an age-related impairment in adrenergic frontal cortex contributes to the memory deficits in the nonhuman primates continues to intrigue and deserves further attention *(95,101,104)*. Presumably, important task differences between the two primate testing situations are responsible for the variant results. Although definitive answers are not yet possible, it seems reasonable to suggest that the task used by Arnsten and Goldman-Rakic may be more sensitive to attentional variables in the aged monkeys (as it does not impose the same observational requirements of the monkey as does the subject-paced, automated task, where no benefit was seen). Perhaps the performance benefits observed with clonidine involved some improvement at the interface between attention and mnemonic functions. Alternatively, Coull *(105)* argues for an important role of the α_2-adrenergic frontal cortical system in inhibition of distraction, providing another plausible explanation for the differences observed (as the control afforded by the automated task would also make it less sensi-

Table 2
Improvement in Recent Episodic Memory
in Aged Monkeys via Neurotransmitter Modulation

Class	Drug	Improvement
Anticholinesterase	Physostigmine	Yes
Anticholinesterase	Tetrahydroaminoacridine	Yes
Muscarinic agonist	Arecoline	Yes
Muscarinic agonist	Oxotremorine	Yes
Cholinergic and phospholipid precursor	Choline	No
Dopamine agonist	Apomorphine	No
Dopamine precursor	L-Dopa	No
GABA agonist	Muscimol	No
α-adrenergic agonist	Clonidine	No
Benzodiazepine antagonist	CGS8216	No

These drug effects were achieved in aged monkeys tested on the same recent episodic memory paradigm shown in **Fig. 1.** Note that of all neurotransmitter modulator drugs tested, only those directly affecting cholinergic system were able to improve performance on the task. *See* **Subheading 3.1.3.** for details and specific references.

tive to any beneficial drug effects on interference). Whatever the reason for the different effects, these data emphasize that variations in animal paradigms intended to measure the same constructs can produce very different results, with correspondingly differential success predicting outcomes in complicated human trials involving age-related memory deficits.

3.2. The Challenge of Using Aged Rodents to Model Age-Related Memory Impairments

Although studies of aged rodent memory have played important roles advancing biogerontology, and in elucidating many neurological mechanisms responsible for age-related behavioral deficits *(106–111)*, their role in the development of cholinergic strategies and specific treatments is less clear (except perhaps for confirming the in vivo cholinergic activity of certain lead compounds). When one looks at the nature of the work done with aged rodents, the reasons for this become more evident.

3.2.1. Do Rodent Paradigms Measure the Appropriate Constructs?

Early studies characterizing a wide range of behaviors in aged rats and mice confirmed that, like primates, aged rodents also suffer deficits on tasks intended to measure learning and memory ability *(70,112,113)*. However, a number of difficulties also became apparent, including: (1) establishing rodent paradigms of recent episodic memory (i.e., the primary form of memory loss of aged humans and early-stage AD patients), (2) achieving a high degree of experimental control over mnemonic and the multiple nonmnemonic variables that can effect outcome measures, and therefore (3) being able to conclusively argue for a specific loss of memory as responsible for the performance deficit. Although it is true that many rodent tasks have been developed that require information to be learning and/or remembered, and for which aged rodents suffer deficits, these tasks are conceptually very different from those used to measure the recent episodic memory most closely associated with cholinergic dysfunction in aged and demented AD patients and for which AD patients demonstrate significant losses early in the course of the disease.

Although severely demented AD patients exhibit deficits on nearly any cognitive task on which they are tested *(81)*, the widespread degenerative processes characteristic of end-stage AD do not help identify the key symptoms nor permit the establishment of essential causal links. Thus, an examination of the deficits in the earlier stages of the disease has proven valuable. Interestingly, memory problems are the most commonly reported complaint of cognitive decline in early AD and late stage senescence, with recent episodic memory deficits being among those affected most early and severely *(81,114)*. As discussed previously, aged nonhuman primates share this loss as a major portion of their constellation of cognitive symptoms *(73,113)* and these commonalties are not likely a coincidence. Thus, the focus on recent episodic memory in animal models is both logical and practical. Moreover, it permits animals to be used to address certain questions concerning recent memory losses in aged humans and early stage AD patients. The common features of the recent memory deficits can be operationally defined as occurring: (1) in situations where the event to be remembered is brief and discrete; (2) when little or no opportunity for practice or rehearsal exists; and (3) over a relatively rapid time frame, with retention typically decaying in minutes to hours, depending on the specific nature of the information, task, and interpolated activity. In addition, the poor retention can be shown to be independent of numerous nonmnemonic variables that might also effect accuracy. Unfortunately, when one examines tests of memory used in aged rodents, these characteristics are not usually found, making the generalization of findings in rodents to humans less certain.

3.2.2. Do Rodent Paradigms Include Sufficient Controls?

Independent of the type of memory measured and its relationship to human deficits (or lack of it), a significant limitation with rodent tasks is that they rarely include sufficient controls to permit one to conclude that the performance deficit is necessarily related to a deficiency in recent memory, per se. An important difference between the majority of rodent tasks used to study age-related memory loss and the nonhuman primate task discussed earlier, for example, is the use of several control trials within each test session under which little or no memory is required. Investigators pioneering tests of recent memory argued strongly for the importance of inserting such control trials within the test, when all other task requirements are retained, but the need for recent memory for successful performance is eliminated *(115–117)*. This control typically is accomplished by using a "zero second" delay condition, where the subject can respond immediately upon seeing the relevant stimulus, thus not having to remember what or where the stimulus was. In addition, by using variable lengths of retention intervals, differential demands are placed on the subject's memory ability. If the task is primarily measuring recent memory ability, the use of variable length retention intervals should be reflected in a temporal performance gradient (i.e., longer retention intervals yielding increasingly poorer performance). This greatly strengthens a memory interpretation of the data (*see*, e.g, Fig. 1). Because aged rodent tasks have rarely successfully incorporated this feature, one is typically forced to *assume* that the animal did not perform the task accurately because its memory was impaired. Similarly, one has to assume that performance would have been accurate if recent memory was not significantly taxed. Most often little reason exists to justify these fundamental assumptions. For example, the delayed alternation task is routinely accepted as a valid test of rodent recent memory. Yet, aged rats typically perform poorly on even the shortest retention intervals and the poor performance is accepted as evidence of a short-term memory deficit. However, it is impossible to rule out other interpretations, including deficits in their ability to distinguish between the two choices (owing to spatial sensory deficits), lower motivation to perform the task, and lack of understanding of how to optimize reward (i.e., unclear of the rules). In fact, because serious deficits occur with the shortest time intervals (when presumably relatively little memory is required), it seems logical to conclude that many nonmnemonic deficiencies are indeed contributing to the poor performance. Thus, without a proper control condition that retains all elements of the task except recent episodic memory, and where performance is near perfect, it is not justified to interpret the deficits as primarily mnemonic in nature. Similarly, when an improvement in performance is observed in such a circumstance, it is difficult to be certain the improvement is due to an effect on memory, as opposed to numerous other factors that could have been improved and that were the underlying cause of the poor performance.

The two most popular and successful tasks for evaluating age-related memory deficits in aging rats arguably are the Morris water maze and variations on radial maze tasks (e.g., eight-arm radial maze, Barnes circular platform, etc.). Both have been used to great advantage by investigators interested in studying rodent memory, in general, as well as age-related memory losses. However, it remains uncertain how suitable they are for modeling the recent, episodic memory losses associated with cholinergic dysfunction in AD patients, as well as those in human and nonhuman primates. It was long ago demonstrated that aged rats suffer serious deficiencies in spatial perception (i.e., difficulty correctly differentiating their position in space) *(118,119)*. Both the water maze and the radial maze tasks have a significant spatial component, requiring that the animal navigate in space to perform the task. Thus, this spatial deficiency places the aged rats at a distinct disadvantage before any requirement of memory is imposed. Attempts to equate for this deficiency have been few and difficult to implement. Of course, if the aged rats are unable to accurately perceive the required spatial relationships as well as their younger cohorts, how can we interpret with confidence that a deficit in ability to remember these relationships primarily reflects a loss of memory? Indeed these tasks typically demonstrate age-related deficits across all conditions, independent of the duration of the retention interval, and therefore independent of how much their memory was taxed.

The Morris water maze adds an additional confounding variable for studying aged memory in that it is both very physically demanding and presumably uses a survival instinct as its primary motivating factor (i.e., fear of drowning). This issue becomes significant when traditional probe trials are used to measure retention in this task. By design, probe trials provide no means of escape to safety and the rat is therefore required to continue swimming until rescued by the experimenter. The question that emerges is, at what point in this event is a panic response differentially induced in aged animals who are more likely to reach a state of perceived exhaustion or desperation before their younger cohorts? Similarly, to what extent might this differentially confound the intended memory measurement? Although this discussion is not meant to be critical of either these tasks or their use, it is intended to focus attention on their interpretive limitations and the need for innovative solutions when used as paradigms to model age-related recent memory. At a minimum, one needs to be able to distinguish possible memory impairments from the numerous other factors that may impair performance.

Thus, despite the importance of nonmemory control conditions and its accepted importance during the early development of animal recent memory paradigms *(115–117)*, this important point seems to have been lost in many of the more recent efforts, especially in rodent studies of age-related deficits. Yet, adopting this principle would seem to be fundamental to employing properly controlled recent memory paradigms. Omitting the nonmemory control condition is analogous to performing immunostaining without confirming the specificity of the antibody staining by running a primary deletion condition and preabsorbing with the protein against which the antibody had been raised. One simply cannot interpret what one is quantifying and what a change between groups means. Similarly, not including a nonmemory control condition in tasks intended to measure recent memory deficits seems little different than running receptor binding assays and not accounting for the nonspecific binding. One simply cannot know if what is being measured is genuine. Of course, this condition is rarely included in rodent tasks because the tasks do not accommodate it very well. Even when the best rodent memory laboratories attempt to include this control, the data do not permit an unambiguous recent memory interpretation (e.g., *120*). If performance on the shortest delay condition is not nearly perfect, with gradual decline in accuracy occurring as the retention interval increases, it is very difficult to interpret the task as one that primarily measures recent episodic memory. Thus, one cannot interpret the deficits as primarily due to impairment in recent memory. When necessary control conditions are omitted entirely, the data are even less uninterpretable *(115–117)*.

3.2.3. Issues Regarding Rodent Tasks of Episodic Memory

Even when rodent paradigms are developed that can incorporate the necessary nonmemory controls, and therefore more clearly offer information regarding recent episodic memory, they have not gained popularity among researchers. One problem likely involves the long training time required of these tasks. In addition, the tasks do not always prove suitable for studies using aged animals. For example, a variation of the radial arm maze paradigm was developed specifically to provide a test of recent episodic memory in the rat, including incorporation of the necessary "nonrecent memory" control conditions discussed earlier *(121)*. Like most variations of the eight-arm radial maze, this paradigm required that the rat first learn not to return to arms that were previously rewarded in the same session. Once this was accomplished (after weeks of training) recent episodic memory could be tested by making a subtle modification in the task. During each test session, an "information phase" limited the rat's choice to four preassigned arms. This was followed by a variable delay interval (the "memory retention" phase of the session), during which the rats were temporarily returned to their holding cage. After the delay interval expired, the rats were returned to the maze, with all eight arms available. They showed they not only remembered which arms they had been to, but that they understood the rules of the task, by now accurately going to the four remaining arms in which they had not yet been rewarded on that session (and avoiding returning to the four arms previously rewarded prior to the delay). The task can be repeated over many days or weeks (of course, randomly changing which four arms are available during the information phase), with variations in the duration of the "memory retention" phase. As the duration of the retention interval is increased, one observes a gradual decline in accuracy, reflecting a decay in recent, episodic memory. Rats subjected to ibotenic acid lesions of the nucleus basalis of Meynert (nBM) were able to accurately perform the task with short delays but exhibited selective deficits as the retention interval was increased. This provided strong evidence for a deficit in recent memory following this lesion (*121*; Fig. 4). Unfortunately, when aged rats were trained in the paradigm, they were never able to learn to accurately perform the task without any delay intervals (i.e., when little recent memory was required) (Bartus, 1986, *unpublished observations*). Thus, without excellent performance on control conditions (where little or no recent memory was required) one could not legitimately test the rats' recent, episodic memory ability. In addition, one could not even justify interpreting the problem as a deficit in recent memory for it was uncertain if the problem was one of memory, perception, motivation, etc. Thus, we were forced back to the drawing board, never to get this close to developing an effectively controlled aged rodent test for recent episodic memory again.

Despite its crude and simple nature, the passive avoidance paradigm can produce data that are more easily interpreted as reflecting losses in recent memory (as long as the paradigm is properly run). Age-related retention deficits on the one-trail passive avoidance task have been known to exist for some time (*47,112,122,123*). This deficit is both robust and consistent across rats (Fig. 5). A main advantage of the paradigm is that decreases in psychomotor activity, energy levels, etc. tend to improve retention scores on the task (by tending to keep the animal in the "learned, safe" compartment). Thus, when age-related retention deficits are observed, one can be more certain they are not due to confounding effects of these particular motoric changes which are known to occur with age. However, just as importantly for the present discussion, one can insert varying durations of retention intervals between the learning trial and the retention test, thus changing the recent memory requirements of the task. If the aged rats perform well under the short intervals (when little recent memory is required), and perform more poorly under the longer intervals (which require greater memory ability), one can conclude with more certainty that an important deficiency in recent memory has contributed to the poorer performance (*47*; Fig. 5). Note that this task also satisfies the operational and conceptual characteristics of recent episodic memory, described earlier (*see* discussion in Subheading 3.2.1.).

Interestingly, a simple and effective means of eliminating the passive avoidance deficit in aged rats is simply to arrange the procedure so that it requires multiple trials to learn the task (*124*). In the multiple-trial variation of the task, aged rats are able to learn at nearly the same rate as young rats and retention is comparable, even when measured 2 wk after training (Fig. 5). In other words, by making subtle changes in the parameters of the task, it was shifted from one primarily dependent upon recent episodic memory to one requiring other forms of learning and longer term memory. Consequently, the performance in the aged group relative to the young group changed

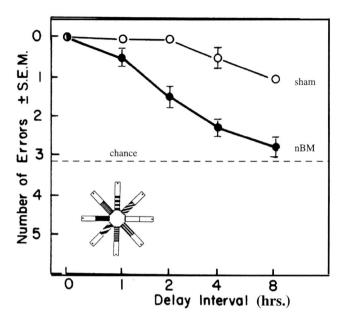

Fig. 4. Time-dependent retention gradient established in an eight-arm radial maze (schematic inserted) by rats with ibotenic acid lesions in the nucleus basalis (*filled circles*) vs sham lesions (*open circles*). The temporal performance gradient was established by allowing access to only four of the eight arms before the delay interval (selected by the experimenter and varied from session to session). Following variable delay intervals, all eight arms were available to assess the animals' ability to remember which arms had been visited earlier in the session. Delay intervals varied from 15 min to 8 h. Note that both groups performed perfectly at the shortest retention intervals and displayed increasingly poorer performance as the duration of the retention interval increased. Because performance was perfect when little or no recent episodic memory was required, these data illustrated that sufficient experimental control over nonmnemonic variables was achieved and that the animals understood the task rules. Also, because the difference between the two groups increased as the retention interval increased (thus placing greater demands on recent episodic memory), these data demonstrated an impairment in recent episodic memory following the lesion. (Data from Bartus, 1985 *[121]*.) Note the conceptual and operational similarities between this memory deficit and those reported in **Figs. 1**, **3**, and **4** for aged primates, humans, AD patients, and for monkeys given scopolamine. Unfortunately, when aged rats were tested on this paradigm, they were not able to achieve perfect performance on the shortest retention interval (Bartus, *unpublished data*), making it impossible to interpret the nature of their deficit (i.e., mnemonic or due to numerous other nonmnemonic variables). *See* text for discussion of this and other tasks used by investigators to study age-related memory deficits, as well as the challenges that exist.

accordingly, and an age-related deficit was no longer manifest. This discussion is not advocating the use of the passive avoidance paradigm over other more popular paradigms (for the passive avoidance paradigm has its own limitations). Rather, this is simply meant to illustrate that rodent tasks that share the operational and conceptual similarities to human recent, episodic memory can and should be developed. Furthermore, the selection of proper parameters is as important as the selection of the paradigm. However, achieving the necessary characteristics in a rodent task is not easy and the success achieved can be tenuous. Clearly, this is an area that deserves further concerted effort.

3.2.4. Synopsis on Rodent vs Nonhuman Primate Paradigms

There is no question that the wide variety of rodent paradigms used to study age-related deficits in learning and memory *(106–111,125)* have made valuable contributions to the field. Thus, this discussion is not meant to criticize the approaches, minimize the difficult challenges, or invalidate any individual result obtained. Rather, it is argued that, given the inherent limitations of achieving necessary control in rodent tasks of recent memory, and the differences between them and recent memory paradigms used in higher primates, caution must be exercised when interpreting data from any single rodent paradigm, while replication across many diverse paradigms would help provide some interpretive comfort.

It seems likely that these inherent limitations of rodent paradigms contribute significantly to the disagreement and confusion that exists regarding the nature of the age-related deficits observed in rodents, and the role that different neurotransmitter systems may play in the deficits. It also seems likely that this issue impacts the interpretation of data using cholinergic and other neurotransmitter blocking agents, as well as selective cholinergic lesions (discussed in subsequent sections of this chapter). Lashley demonstrated at the beginning of the century that the multiple trial tasks typically used to measure learning and memory in rodents employ widely separated brain regions involving multiple anatomical systems *(126)*. The tasks typically used to characterize age-related changes in rodents are very similar to those used by Lashley, especially when they are set up to measure learning rates and retention following multiple trial acquisition. Lashley performed a heroic series of discrete and combination brain lesions searching for the brain region(s) where memory might be stored, only to conclude facetiously that it must not be in the brain (for no brain lesions were completely successful in abolishing the ability of the rats to learn, relearn, or remember) *(126)*. Given this, what can one say when discrete neurochemical lesions fail to induce a deficit, other than that destruction of those particular neurons, in isolation, was not sufficient to cause the system to fail? It says little about whether those neurons normally participate in an important way and whether or not impairments complemented by the other changes in an aged brain, might not be responsible for very specific, but important types of functional losses. Similarly, what does it mean when aged rats are tested on such tasks and they cannot perform them? It says little about whether certain neurochemical systems or anatomical sites might be more seriously impaired than others. Similarly, it is hoped that this discussion will stimulate researchers to continue their efforts to develop tasks that gradually improve upon currently available paradigms, including implementing controls required to evaluate recent memory deficits.

In summary, the evidence accumulated over the past two decades strongly supports the concept that it is possible to use aged nonhuman primates to develop valid and reliable animal models of age-related memory impairments. Further, these models can be used to help develop mechanistic insight and treatment strategies for certain key cognitive impairments associated with AD. They should continue to be useful in predicting or verifying modest clinical observations, as well as in helping to establish rational directions for future clinical trials.

Current methodological limitations with aged rodent memory paradigms often limit their utility for studying many forms of age-related human memory deficits, and this has likely contributed to some confusion regarding the use of animal models for study-

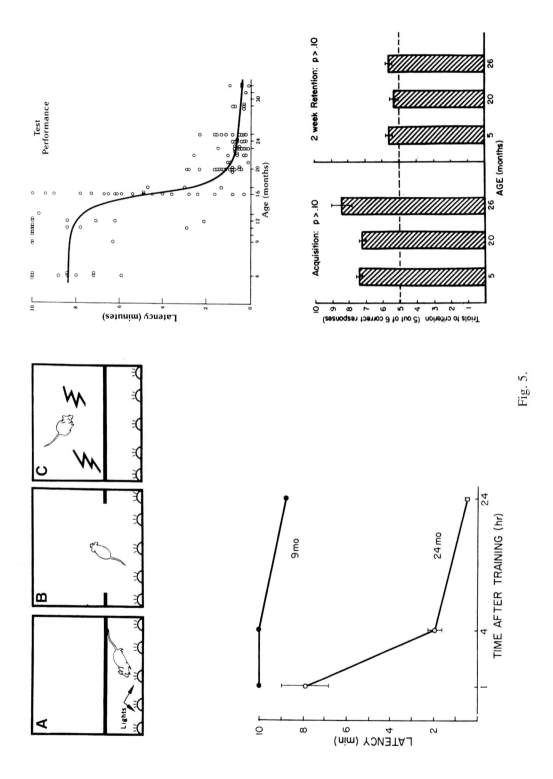

Fig. 5.

22

ing these problems *(4)*. Nonetheless, these difficulties should not discourage their continued use, or further efforts to create innovative variations so that rodent paradigms might one day possess the necessary controls and be even more reflective of human recent episodic memory.

3.3. Models Using Cholinergic Blockers

Animal models employing pharmacological blockade of cholinergic receptors proved important in the development of the cholinergic hypothesis. Following the initial observation of Drachman and Leavitt (1974) that scopolamine given to young volunteers produced a pattern of deficits on cognitive tests similar to that of elderly patients, a series of pharmacology studies in nonhuman primates strengthened the relationship between central cholinergic dysfunction and these specific age-related memory impairments, while also demonstrating apparent selectivity of the effect for central cholinergic neurons.

3.3.1. Extending the Initial Human Observations

Early studies in the same testing paradigm used to characterize cognitive deficits in aged monkeys *(16,50,73,74)* and improve performance with cholinomimetics (e.g., Figs. 1 and 2) demonstrated that the classic centrally acting cholinergic blocking drugs, scopolamine and atropine, were able to produce an amnestic effect in young monkeys that were remarkably similar to the most sensitive deficit in aged monkeys *(78)*. This scopolamine deficit in recent episodic memory in young monkeys was also shown to be dose-related, with higher doses of scopolamine producing even greater memory impairments *(78)* (Fig. 3). Finally, it appeared to be somewhat selective to CNS cholinergic neurons, in that it was attenuated by the cholinomimetic, physostigmine, but not the CNS stimulant methyphenidate (Ritalin) *(127)*. In addition, a similar deficit could not be produced with the peripherally acting anticholinergic, methylscopolamine, nor by antagonists for dopamine, β-adrenergic or nicotinic receptors *(77,78,113,128)*. Although it has been acknowledged that other neurochemical systems must also be involved in the mediation of these deficits *(1,3,22)*, few noncholinergic pharmacolo-

---

Fig. 5. *(opposite page)* **Upper left:** Schematic illustrating apparatus and steps involved with the training phase of the single-trial passive avoidance step-through paradigm. For retention tests, the animal is returned to a lighted compartment and the latency to enter darker (previously punished) compartment is measured. Higher latencies reflect increased retention of shock experience, confirmed by "no-shock" sham training condition *(not shown)*. (From Bartus, 1980 *[173]*.) **Upper right:** Robust and consistent age-related deficit on retention of one trial passive avoidance task. No age-related differences in latency existing on the training trial *(data not shown; all scores < 30 s)*, while control tests indicated no age-related difference in shock sensitivity. (From Bartus, 1985 *[71]*.) **Lower left:** Time-dependent, age-related decline in retention of single trial passive avoidance paradigm. Together with other control tests, these data support the interpretation that impairments in recent, episodic memory contribute to the age difference seen on this task. (From Lippa, 1980 *[47]*.) **Lower right:** Equivalent acquisition and retention of passive avoidance task between young and aged rats when the task was modified to require several trials to learn (i.e., successfully avoiding shock). Under these conditions, the operational definition of recent, episodic memory no longer applies, and an age-related deficit is no longer manifest. (From Bartus, 1982 *[124]*.)

gical agents have yet been identified in this testing situation that produce similar specific, age-related effects on recent memory performance (and this work has been ongoing for more than two decades now). The two most notable exceptions are diazepam and tetrahydrocannabinol (THC) *(128)*; they not only produce similar memory impairment in nonhuman primates (mimicking those of aged monkeys), but are also know to be amnestic when administered to humans (*see* Table 3). Interestingly, both drugs interact directly and indirectly with cholinergic neurons *(129–132)*.

Work is continuing with other neurotransmitter systems to establish similar relationships. For example, age-related changes in both noradrenaline and dopamine have been implicated in memory impairments observed in aged monkeys *(95,100,101,104)*. Additional work will determine the extent to which these systems play similarly important roles in the memory deficits of aging and dementia and the extent to which these hypotheses might lead to effective treatment. Alternatively, some have argued for an important role of these neurotransmitter systems in mediating attention and distraction (e.g., *105*) which could easily influence memory scores in test environments not specifically designed to reduce the impact of these effects on performance.

In summary, animal models involving pharmacologic blockade of neurotransmitter receptors can and have helped differentiate which of the many biochemical changes observed in AD brains were likely associated with the serious episodic memory loss associated with AD. Many of the same paradigms have been employed to help differentiate among alternative treatment approaches, providing a direct link between the earliest empirical activity surrounding the cholinergic hypothesis and the present success with cholinesterase inhibitors in the clinic.

3.3.2. Contrasts Between Early Cholinergic Pharmacology and Recent Models of Age-Related Memory Loss

Research on the role of cholinergic activity in learning and memory dates back to the early 1960s *(133–135)* where deficits were reported when anticholinergic agents were given to rats. The subsequent, pioneering work of Deutsch *(136,137)* is often correctly recognized for focusing attention on the cholinergic system as an important mediator of learning and memory. However, time has shown that the impact of this work on our understanding of the role of cholinergic neurons in age-related memory deficits, AD, or their treatment, has been minimal, providing an illustration for the importance of the particular behavioral paradigms used in animal models. One of Deutsch's key observations was that both improvement and impairment could be achieved with the identical dose of a cholinomimetic, simply by changing the time interval between when a learned event occurred and when the drug was administered. Certainly, this is not a phenomenon seen in models of aging recent memory, nor in any human clinical trials. This peculiar phenomenon likely is due to the particular tasks employed by Deutsch. Not only were they relatively complex, but like most rodent tasks of learning and memory, they do not permit substantial control over many nonmnemonic variables. Thus, while what is measured may indeed involve how much an animal learns or remembers, the effects of a drug might change the outcome measurement by affecting a wide variety of nonmnemonic variables (such as motivation, motor skills, attention, sensory ability, etc.), whose sensitivity to cholinergic manipulation may vary differentially over time. A related point is that the particular multiple

Table 3
Ability of Drugs in Young Monkeys to Induce Selective Recent Memory
Impairments Analogous to Those of Aged Monkeys

Drug	Class	Memory deficit induced?	Amnesic effect in clinic?
Scopolamine	Centrally acting muscarinic antagonist	Yes	Yes
Atropine	Centrally acting muscarinic antagonist	Yes	Yes
Methscopolamine	Peripherally acting muscarinic antagonist	No	No
Mecamylamine	Centrally acting nicotinic antagonist	No	No
Haloperidol	Centrally acting dopaminergic antagonist	No	No
Propanolol	Centrally acting β-adrenergic antagonist	No	No
Amitriptyline	Tricyclic antidepressant (with strong anticholinergic properties)	Yes	Yes
Desipramine	Tricyclic antidepressant (with weak anticholinergic properties)	No	No
Diazepam	Anxiolytic (with affinity for all benzodiazepine receptors)	Yes	Yes
CL 218,872	Anxiolytic (with preferential affinity for type 1 benzodiazepine receptor)	Weak	Unknown
Tetrahydrocannabinol	Psychoactive cannabinoid	Yes	Yes
Pentobarbital	Sedative hypnotic	Weak	Weak
Methylphenidate	CNS stimulant (pharmacologically similar to amphetamine)	No	No
Cinanserin	Serotonin antagonist	No	No
Methysergide	Serotonin antagonist	No	No

These data were collected in the same apparatus used to collect shown data in Figs. 2 and 3 and Table 1. Note that in this paradigm, all drugs inducing robust memory deficits either directly or indirectly impair cholinergic function (*see* text for details). Also note similarity between drugs inducing memory deficit in monkeys on this task and reported amnestic properties of drugs in humans. *See* text for more detailed discussion and references.

trial learning/memory paradigms employed by Deutsch in these early studies appear to measure behaviors not clearly related to the primary aged and AD memory deficits characterized by others a decade later (e.g., *13,73,113*). That is, Deutsch's tasks do not share many conceptual or operational similarities with AD deficits noted in more recently used paradigms, nor have they been shown to reflect memory loss in aged animals (e.g., *see 71*). On the other hand, Deutsch's pioneering studies did draw early attention to the cholinergic system as a mediator of learning and memory (just as others had similarly done for catecholamines, various neuropeptides, glutamate, and seroto-nin, e.g., *123,139*). This attention, coupled with prior clinical experience regarding the amnestic properties of cholinergic blocking drugs, made it logical to use cholinergic drugs as tools to study possible biochemical roles in the age-related cognitive deficits. Importantly, a major difference emerged between the evidence for the role the cholin-ergic system plays, vs that of many other systems also implicated in mediating learning and memory. That is, numerous studies have since shown a particularly important role of the cholinergic system in mediating some of the specific types of losses that occur with age and AD, while a similarly strong relationship has yet to be established for other neurotransmitters. This is not to suggest that other neurotransmitter systems do not play important roles in mediating many other forms of learning and memory, or that at some point in time they may not also be shown to contribute to the specific types of deficits associated with AD and advanced aging. Nonetheless, for the time being, at least, only cholinergic drugs have been shown conclusively to improve cognitive symptoms in AD.

3.3.3. Do Anticholinergics Provide Useful Models for Aging or AD?

It would be a great oversimplification to ignore the fact that important differences exist between the deficits induced by scopolamine and certain cognitive symptoms in AD. Many authors have reported that the pattern of cognitive deficits induced by scopolamine does not match perfectly those observed with AD. Indeed, the lack of unity between cholinergic blockade and AD has often been interpreted as evidence against the cholinergic hypothesis (e.g., *4,32,34,140*). Yet, variations between the two clinical profiles should surprise no one, for many functional differences exist between blocking postsynaptic receptors, versus the constellation of cholinergic deficits and noncholinergic deficits that likely exist in normal aging and especially AD. For this reason, the issue should not be how exactly one situation matches the other, but rather whether sufficient similarities exist to support a plausible cause–effect relationship for some of the important cognitive symptoms. The fact that an overlap in key symptoms can be achieved by artificially blocking postsynaptic muscarinic receptors provides circumstantial evidence supporting an important cholinergic role, despite the lack of unity with all the symptoms. The fact that cognition in aged animals and humans has been shown to be even more sensitive to cholinergic blockade *(32,141,142*; Bartus, *unpublished observation*), and that the pattern in elderly humans given anticholinergics trends toward a pattern seen in AD patients *(32)*, offers further support for the hypothesis. Unity of effects across such widely divergent subject populations and conditions would be most surprising and is not required as support for the hypothesis. Perhaps a better question for clinical studies would be to ask whether blocking other neurotransmitter systems, alone and in combination with cholinergic receptor blockers, produces deficits that more closely mimic those of aging and AD. No study has yet produced convincing evidence for the involvement of any system other than the cholinergic, although it seems unlikely that this will persist indefinitely.

Fibiger *(4)* states that it is a "logical fallacy" to "attempt to characterize the effects of systematically administered anticholinergic agents within a unitary theoretical framework," (p. 221) such as a key role in age-related memory because cholinergic neurons innervate widespread areas of the brain, and because of differences in the role and level of cholinergic neurons in different brain regions. He further argues that under some conditions, cholinergic activity in one brain region might be more important or more easily affected than under other conditions. Of course, this latter suggestion is the most likely reason why some paradigms have been able to elucidate the relationship between cholinergic function and specific forms of memory and others have not. It also offers a plausible explanation for why some paradigms have produced data more predictive of the beneficial effects of cholinergic agents than have other paradigms. Rather than demonstrating the inherent illogical nature of the approach, these issues simply point to the difficulty involved with conducting research with animals intended to model complex human diseases heavily dependent upon behavioral considerations. The fact that (1) a dose-related, selective impairment in recent memory was achieved with central cholinergic blockade that closely mimics that occurring naturally with age (and this deficit, in turn, was shown to model a robust impairment in AD patients); (2) the fact that no other neurotransmitter system has been able to consistently induce a similar impairment, including blocking cholinergic receptors in the periphery; and (3) the fact that cholinomimetics significantly reduced the anticholinergic memory impairment, all provide mutually corroborating support for the hypothesis that cholinergic neurons play an important role in the memory deficit. Of course, these data also accurately predicted that cholinergic drugs should be able to measurably reduce the cognitive impairments in AD patients. Taken together, the data also emphasize the importance of designing and implementing a well-controlled test paradigm for evaluating complex constructs such as memory. They also demonstrate that animal models can be helpful in understanding and developing treatments for serious neurodegenerative diseases in humans.

3.4. Attempts to Develop Models Using Cholinergic Brain Lesions

Paradigms employing brain lesions, typically in rats, have been used to study the neuropathological underpinnings of the cognitive deficits of AD, as well as to test potential treatments for the symptoms of AD. The rationale behind this approach rests with reports of relatively selective loss of cholinergic neurons in the basal forebrain (nBM) of AD patients *(44,45)*. Although the lesion approach has generated substantial controversy for modeling AD, it has nonetheless yielded much new and interesting information on the potential function of cholinergic and noncholinergic neurons emanating from the basal forebrain and the possible role played by the loss of these neurons in AD.

3.4.1. Questions from the Start

Early work testing a range of behaviors in rats following excitotoxic injections into the nBM region established that relatively robust deficits in memory tasks could be achieved relative to other behaviors *(79)*. However, it was also soon demonstrated that the relationship between recent, episodic memory impairments (using an eight-arm radial maze) and the loss of cholinergic neurons and markers was not simple, for with extensive overtraining, the rats were able to gradually recover the memory ability on the task, even though cortical cholinergic deficiencies persisted *(121)*. In addition, this training-dependent recovery of function was shown to generalize to a completely novel task (retention of passive avoidance) even though no specific training on the latter task occurred, and the cholinergic deficits continued to persist *(143)*. Later studies by Dunnett et al. *(144,145)* and then Wenk et al. *(146)* clearly demonstrated that the loss of memory they characterized following these lesions did not correlate well with variations in the loss of cholinergic markers induced by different neural toxins. Still later use of a cholinergic-specific toxin (IgG-saporin) demonstrated that selective destruction of cholinergic neurons in the nBM failed to produce clear evidence of memory impairments *(147,148)*. The implication has been that degeneration of cholinergic neurons in this brain region in AD must not contribute to any of the impairments that characterize the disease.

3.4.2. Do Basal Forebrain Cholinergic Neurons Play a Role in AD Memory Deficits?

Do the lesion data truly demonstrate that loss of basal forebrain cholinergic neurons in AD play no role in the cognitive deficits associated with the disease, as some *(4)* have suggested. Although these lesion data, alone, do not permit a definitive answer, the weight of all the evidence, considered collectively, does not easily support that conclusion. First of all, no one has ever questioned whether the behavioral paradigms employed in the majority of lesion studies adequately reflects any of the key symptoms of AD. It would be insightful to determine whether tasks primarily measuring recent episodic memory might be more sensitive to selective loss of cholinergic basal forebrain neurons than those used thus far. Second, although it is true that the earliest studies created lesions of the nBM that included noncholinergic neurons, and the deficits obtained therefore cannot be interpreted as due exclusively to loss of cholinergic function, the data do not argue that the loss of cholinergic neurons did not play some important role. To justify such an interpretation would require a lesion that selectively *spared* cholinergic neurons while destroying noncholinergic neurons in the area; this has yet to be done. Similarly, the lack of a clear memory deficit with more recent lesions selective to cholinergic neurons does not argue that the neurons may not play an important contributory role. Gallagher and Colombo *(89)* correctly point out these lesion data merely demonstrate that destruction of these particular cholinergic neurons may be necessary but not sufficient to achieve a significant loss of memory ability and that other cholinergic systems (e.g., striatal) may need to be affected to generate the expected deficits. Of course, it was long ago suggested that to create more accurate and possibly more predictive models of age-related memory deficits, it might be necessary to combine lesions of basal forebrain structures with other brain areas also degenerated in AD *(69)*; it was also suggested that subjecting aged animals (with their constellation of natural brain changes) to such lesions might be even more effective *(69)*.

Notwithstanding the uncertainties created by the lesion data, parallel research with aged rats given exogenous nerve growth factor (NGF) helps support an important role of basal forebrain cholinergic neurons in age-related memory loss. Given that the two most profound events induced by exogenous NGF administered to aged rats involve an enhancement in the vitality of these neurons and a decrease in their memory deficits *(149,150)*, it seems difficult to avoid the conclusion that an important link between the two exists. Similarly, transplanting neuronal stem cells into the nBM that have been genetically modified to produce NGF yields similar benefits to aged rats *(151)*. Finally, recent evidence from NGF knockout mice that exhibit both cholinergic deficiencies and memory deficits further substantiate the link *(152)*. As in aged rats, both the cholinergic deficiencies and the memory deficits are improved when these rats are given exogenous NGF.

Perry et al. *(6)* make an interesting observation regarding the poor correlation between loss of cholinergic markers following brain lesions in rats and the memory deficits that follow. They note that despite the established role of the cholinergic system in mediating memory impairments in AD patients (and the proven ability of cholinergic drugs to reduce these impairments), there exists no apparent correlation between the number of cholinergic neurons in the nBM and the presence or severity of dementia *(6)*. Thus, these data in AD patients warn against drawing overly simplistic interpretations of the lack of a correlation between neurochemical indices and behavior when acute, discrete lesions are performed in young rats.

3.4.3. On the Utility of Lesion Models

It was suggested long ago that lesion studies in animals have significant limitations, and they should not be expected to unravel the complex mysteries of human Alzheimer's and other dementing diseases *(69)*. Nonetheless, it was further suggested that information gained from these studies, and new thinking generated by the questions raised, should ultimately contribute to a more complete understanding of many important aspects of the disease *(69)*. Indeed this has begun to occur. For example, the lesion studies with rodents, when considered within the broader context of other data, suggest that loss of cholinergic neurons of the nBM likely contribute to, but are not solely responsible for, the cognitive declines of AD *(89)*. Moreover, it is now suspected that this brain region may be involved with the mediation of attention in cognitive tasks more so than memory *(89,153,154)*. Interestingly, some of the positive trials with recent cholinesterase inhibitors have noted that some of the improvement in AD patients may occur through enhanced attention *(155,156)*. Thus, although animal lesion studies cannot address many questions related to AD, they have shown that they can address some, and in so doing, offer unique perspectives that can stimulate new thinking and generate further empirical testing. Thus, lesion studies can provide a valuable tool in the scientific pursuit to understanding some of the nuances of AD as well as in our continuing quest to improve its treatment.

4. CURRENT PERSPECTIVES AND FUTURE DIRECTIONS FOR CHOLINERGIC THERAPY FOR AD

Although the FDA approval of the initial cholinomimetics for AD provided independent, objective validation of the cholinergic hypothesis, and the likely approval of additional drugs offers hope for even greater efficacy, few are yet satisfied with these treatments. Rather, these initial successes establish that this and related approaches can indeed improve the symptoms of AD, offering hope that further rational work will result in continual improvements. These improvements will not only include superior cholinomimetics, but will likely also include innovative drug combinations. The following is intended to provide an assessment of the current cholinomimetic approaches, some future directions for treatment approaches, and the role that animal models have and should continue to play.

4.1. Cholinesterase Inhibitors: Current First-Line Therapy for AD

Although tacrine was the first cholinesterase inhibitor approved by the FDA for AD (in 1993), it was actually an old drug that was selected for testing because it could be given orally and had an acceptably long half-life (offering major practical advantages over the form of physostigmine available at the time). It was not until late in 1996 that the first, rationally designed, new cholinergic was approved by the FDA—donepezil. Donepezil (Aricept) almost immediately established itself as superior to tacrine, if not in terms of greater efficacy, certainly in terms of a greater proportion of responders and substantially fewer side effects *(157)*. Since that time, several other cholinesterase inhibitors have progressed through the development pipe line, and based on positive, placebo-controlled studies published in peer-reviewed journals, many are expected to achieve FDA approval in the foreseeable future *(8,9)* (*see* Table 4). Most of these inhibitors were designed or are claimed to offer certain advantages over physostigmine, tacrine, and perhaps donepezil, ranging from greater CNS vs peripheral nervous system (PNS) activity, longer duration of action, specificity for specific brain regions, or effects in multiple, biochemical pathways. Thus, they are claimed to correct many of the deficiencies noted long ago for the initial cholinergic agents *(1,3,26,28,64)*. Of course, it remains to be seen whether these modified pharmacological profiles will translate into clearly superior therapeutic effects or even whether the improved profiles that are claimed for the emerging drugs will hold up to further scientific scrutiny.

4.2. Perspectives on the Modest Effects of Cholinergic Therapy

Although one would hope that even greater and more consistent effects might eventually be achieved with future drugs, the FDA's approval of these drugs nonetheless represents clear confirmation for the main tenets of the cholinergic hypothesis. Certainly, the standards regarding *a priori* predictions and scientific rigor demanded by the FDA for drug approval far exceed any peer review process more generally applied to scientific concepts and data. Moreover, even the relatively subtle clinical response reported to date offers reason for measured optimism. First, history teaches us that as experience is gained with any new therapeutic target and treatment approach, and refinements in analog drugs are made, the treatment effects very often improve. In the case of AD therapy, the development of more effective cholinergic drugs seems a likely possibility. Moreover, the concept of attempting to restore the appropriate balance between multiple neurotransmitter systems, initially proposed at the inception of the cholinergic hypothesis *(1)*, has yet to be systematically explored.

Finally, as more insight is gained into the nature of the subpopulations of responders and the wide individual variations in most effective dose, one might also expect additional improvements in treatment responses to be achieved. The fact that few controlled studies have attempted to account for this variable represents an interesting irony in our efforts to develop effective treatments for AD. Both the initial animal studies *(27,28,60)*, as well as the seminal studies in AD patients *(51)*, noted subgroups of responders and nonresponders, as well as individual variations in peak dose–response. This observation has since been confirmed numerous times *(29,51,56,58)*. Despite this, most studies designed for regulatory approval have opted to test a more limited dose range. Moreover, individual patients are typically either arbitrarily assigned to a particular dose, or else are tested to establish their maximum tolerated dose (rather than peak efficacious dose). This is especially puzzling because a detailed meta-analysis performed more than a decade ago concluded that "cholinergic enhancement of memory function in AD showed that such therapies do have an effect *if dosages are individualized*" *(158*, p. 237, emphasis added). More recent reviews have continued to note this consistent aspect of AD patients response to cholinergic drugs (e.g., *159,160*). Some have attempted to link the identity of the subgroups to genetic markers (such as *apoE4* alleles) *(161)*, whereas others have suggested that more prosaic pharmacokinetic/ bioavailability reasons exist *(162,163)*. In reality, because this phenomenon is seen consistently in both aged New World and Old World monkeys, as

Table 4
Survey of Drugs for AD Modulating Neurotransmitter Function

Drug	Mechanism	Company	Status
Cognex (tacrine; THA)	Cholinesterase inhib.	Warner-Lamber/ Parke-Davis	FDA Approval—1993
Aricept (donepizil)	Cholinesterase inhib.	Eisai/Pfizer	FDA Approval—1996
Exelon (ENA713; revastigmine)	Cholinesterase inhib.	Novartis	NDA filed
Metrifonate	Cholinesterase inhib. (irrev.)	Bayer	NDA filed; on hold (toxic.)
Synapton (CR Physostigmine)	Cholinesterase inhib.	Forrest	NDA filed; initial disapproval
Galanthamine (Reminyl)	Cholinesterase inhib.	Hoechst/Janssen	Phase III; NDA filing anticipat
Eptastigmine (MF-201)	Cholinesterase inhib.	Mediolanum Pharm.	Phase III
Xanomeline	Muscarinic agonist	Lilly	Phase II/III—skin patch
Memric (SB202026)	Muscarinic agonist	SmithKline Beechem	PhIII—missed primary endpts.
AF102B	Muscarinic agonist	Forrest	Status uncertain
Milameline	Muscarinic agonist	Hoechst & Parke-Davis	Phase III; poor Rx index
Eldepryl (selegiline)	MAO inhibitor	Somerset	Failed efficacy—PhIII
Propentofylline (Viviq)	Adenosine and NGF modulator	Hoechst	Phase III
Clonidine	α_2-Adrenergic agonist		No apparent efficacy
Guanficine	α_2-Adrenergic agonist		No apparent efficacy

Informal survey of drugs that modulate neurotransmitter function that are in, or have completed, pivotal trials for AD (ca. late 1998).

well as diverse groups of AD patients, it seems unlikely that any single variable is totally responsible.

Although some have argued that a probe treatment (with tacrine) can distinguish responders from non-responders *(160)* prior to entry into double-blind studies, others have failed to show such a predictive benefit (with physostigmine) *(164)*. Whatever the source(s) of the individual variability, the opportunity to enrich test populations for purposes of regulatory approval merits further effort in future clinical trials. Alternatively, investigators run the risk that outcomes will be less positive than what might otherwise be possible.

Of course, it is worth noting that supporters of the cholinergic hypothesis never argued for "cholinergic replacement to reverse completely the symptoms of AD, let alone halt its insidious attack on the mind" *(22)* (p. 344). Nonetheless, it has been argued that significant improvements in quality of life, as well as reductions in the cost of caring for patients, could be achieved if the symptoms of AD could be reduced sufficiently so that greater self-care is possible and the need for expensive institutionalization delayed *(165)*. Indeed, a recent survey estimated that each AD case collectively costs families and the health care system nearly $200,000—much of it associated with long-term care and eventual institutionalization *(166)*. Thus, even the modest efficacy achieved thus far with cholinergic drugs can produce significant benefit to AD patients, their families, and the health care system. It is also particularly encouraging that some investigators have not only noted improvements in cognitive and global status with these early cholinesterase inhibitors, but also an apparent retardation in the progression of the disease *(8,167)*. Importantly, even these modest effects are far superior to those achieved with other palliative pharmacological approaches to AD to date. Certainly, they stand in contrast to the consistent failures seen with attempts to similarly modulate other neurotransmitter systems, using noradrenergic agonists, monoamine oxidase inhibitors, etc. *(102,103,168,169)*.

4.3. Muscarinic Agonists: Will Their Potential Overcome Their Problems?

Muscarinic agonists differ from cholinesterase inhibitors in that they directly stimulate the postsynaptic cholinergic receptor, thus mimicking the effects of endogenous acetylcholine released from the synaptic bouton. A potential advantage of this class of drugs is that they do not require the production and release of endogenous acetylcholine to positively modulate cholinergic neurons. A potential disadvantage is that they activate postsynaptic receptors independent of when the neuron normally would (and possibly should) be firing. Many do not consider this an important limitation, as cholinergic neurons are believed to function by modulating or amplifying afferent inputs (e.g., *7*) and tonic stimulation might therefore be quite effective. Early studies with aged monkeys *(60)* and AD patients *(56)* suggested that muscarinic cholinergic agonists might improve these memory impairments. However, several muscarinic agonists have been tested in AD clinical trials, and although some preliminary reports of efficacy were generated *(9)*, further development work on many of these compounds seems to have been terminated (miscellaneous press releases; Table 4). Thus, either the therapeutic effects were not sufficient and/or the cholinergic side effects induced were too severe. This may represent an example where both preclinical models, as well as closely monitored Phase I/II trials, may not be sufficiently sensitive to a serious impediment to widespread treatment (i.e., serious side effects). Only time and additional research will provide the answer.

The likelihood that cholinergic side effects can be sufficiently reduced remains uncertain. Acetylcholine serves as the neurotransmitter for all voluntary movement and is the ultimate and penultimate mediator of the parasympathetic and sympathetic nervous systems, respectively. Central cholinergic activity has been associated with a wide range of behaviors and functions, including various gastrointestinal functions, sleep, depression, nociception, water intake, temperature regulation, and arousal (e.g., *see 26*). Thus, achieving a robust therapeutic effect on memory, without inducing significant side effects, will likely always represent a formidable task. It would not be too surprising if the enhanced tonic stimulation that results from cholinergic agonists produces such robust cholinergic side effects that their routine use as a palliative treatment for AD is precluded. In fact, in those instances in which cholinergic agonists have failed to induce significant cholinergic side effects, one has to question whether the lack of efficacy might not be related to the doses being too low to effectively amplify the activity of central cholinergic neurons.

4.4. Cholinergic Precursors: Classic Example of a "False Start"

From the inception of the cholinergic hypothesis, clinicians were eager to test the concept that cholinomimetics might be effective in the treatment of the memory deficits in AD patients. Because cholinergic precursors (e.g., choline and its normal dietary source, lecithin) have a wide safety margin and relatively lax government regulations regarding their use, the vast majority of the earliest clinical studies used this approach. The approach was based on prior biochemical research in animals, suggesting that the availability of choline was the rate-limiting step in the synthesis of acetylcholine *(38,170).* However, the clinical studies that followed produced a string of disappointing outcomes *(see 171).* Ironically, early studies with both aged rodents *(71,172)* and nonhuman primates *(60)* predicted that once age-related deficits were manifest, precursor loading would not measurably reduce the impairments, and an early review of the clinical literature *(1)* indicated that this approach was not able to demonstrate even preliminary evidence of efficacy. It was suggested that certain assumptions inherent in the rationale underlying the precursor approach did not likely apply to geriatric and AD patients, and in fact had never been adequately tested or confirmed *(171).* For example, important age-related and Alzheimer's-related changes in cholinergic neurons had been characterized, involving decreased high-affinity choline uptake, reduced acetyl coenzyme A production (the other necessary constituent for acetylcholine synthesis), and serious presynaptic neuronal degeneration. Any of these deficiencies would impair the ability of exogenous cholinergic choline or lecithin to enhance acetylcholine activity. Thus, it is not surprising that clinical trials in AD patients based on this approach were not successful. These experiments, which failed to reduce existing age-related deficits, stand in contrast to other preclinical studies demonstrating a potential prophylactic benefit of chronic precursor loading for age-related cognitive defects, providing that *intervention occurs prior* to the time the deficits become manifest and is maintained for a substantial portion of the subject's life span *(71,173).*

Despite the clear flaws in the cholinergic precursor approach to treating AD symptoms, the series of clinical failures were interpreted by some as convincing evidence that the entire cholinergic hypothesis was flawed. In reality, this incident merely provides a good example of how well-intended but ill-conceived clinical strategies can develop from over enthusiastic and selective attention to incomplete preclinical studies. In this case, in vivo animal models that had begun to show some predictive validity and failed to improve geriatric memory with precursor loading *(see* Table 2) were ignored in favor of data generated primarily from in vitro biochemical paradigms always restricted to brain tissue from young healthy rodents. Moreover, the latter studies never tested, let alone confirmed that the phenomenon would translate to improved function in aged, let alone AD brain tissue. With the benefit of hindsight, it is easy to see how significant loss of time and clinical resources could have been avoided if preclinical investigators had been more circumspect and self-critical about their data and clinicians had been more sophisticated in discriminating conflicting data from among different animal studies and in critically applying those data to the design of clinical strategies. This episode illustrates the difficulty of applying preclinical data to clinical strategies and the need for preclinical and clinical researchers to maintain an ongoing dialog as efforts to develop future treatments continue.

4.5. What's Next?

The FDA approval of cholinesterase inhibitors represents a crucial but incomplete step in our efforts to treat Alzheimer's patients. On the basis of the FDA's sanction of these drugs, it is logical to conclude that the concept of cholinergic improvement of AD symptoms, in principle, has been proven. The question remaining is to what extent might further advances in this field produce even greater and more consistent palliative effects? Several approaches for improving upon existing treatments are apparent.

4.5.1. New Chemistry

Rational modifications in the chemical structures of existing cholinesterase inhibitors represents a reasonable goal toward improving their biological profile, with greater selectivity to the brain likely to be of particular benefit. Greater CNS selectivity seems particularly important because the prevalence of peripheral side effects (especially gastrointestinal) has proven to be a serious impediment to the successful development and general use of current cholinergic agents in AD. It is surprising that more progress has not been made in this area, for the major deficiencies in available cholinergic drugs were described and discussed well over a decade ago *(3,22,26)*. In fact, of the six cholinesterase inhibitors currently enjoying the most attention (Table 4), all but two are actually very old compounds, resurrected from the generic, chemical warehouses (the two newly synthesized exceptions to this point are Aricept® and Exelon®). Thus, the likelihood that dedicated medicinal or combinatorial chemistry can generate new chemical entities that, through design or serendipity, have significant advantages over existing drugs seems quite plausible; certainly, the pharmaceutical industry has proven many times that it can successfully perform the type of incremental improvements required of this approach.

An analogous approach to improving the therapeutic index is the design of cholinergic prodrugs that would remain relatively inactive until they have had time to penetrate the blood–brain barrier. Blocking the active pharmacophore with a lipid moiety that is cleaved after the drug has had time to distribute to tissues (including the CNS) is an example of such an approach. Perhaps the recent and imminent FDA approvals for the first generation cholinergic treatments, coupled with the recognition that considerable room for further improvement exists, will stimulate further interest and effort within the pharmaceutical industry.

4.5.2. Manage Side Effects

Better management of the cholinergic side effects induced by current drugs provides another means of improving therapeutic outcome. For example, peripheral antagonism of cholinergic receptors (through anticholinergics that do not cross the blood–brain barrier) might significantly reduce the peripheral side effects, thus allowing more effective (i.e., higher) doses of cholinomimetics to be administered. Similarly, drugs given concomitantly with the cholinesterase inhibitors (such as ondansetron to reduce nausea, or peripheral opiates for diarrhea) might also allow higher, more effective cholinergic doses to be given. An interestingly innovative approach toward improving the pharmacokinetics and thus perhaps reducing the gastrointestinal side effects of the muscarinic agonist zonomeline has been attributed to Lilly. Following completion of clinical trials, where significant side effects limited the tolerability of the drug, they reportedly will continue developing the drug in the form of a skin patch, thus bypassing the gastrointestinal system *(174,175)*. This decision implies that Lilly likely observed sufficient signs of efficacy with the muscarinic agonist to merit this additional effort and cost. However, as discussed earlier, it remains unclear whether the problem of cholinergic side effects will preclude the successful development of muscarinic agonists; creative development efforts such as this should eventually help provide the answer.

4.5.3. Employ Polypharmacy Approach to Enhance Efficacy

There seems little doubt that the memory deficits that occur in AD (as well as those associated with normal aging) include deficiencies in neurotransmitters beside the cholinergic system. Indeed, the likelihood that other neurotransmitter systems also contribute to the loss of memory in both aging and Alzheimer's disease was suggested at the inception of the cholinergic hypothesis *(1)*. Although practical considerations dictated that the initial efforts to gain FDA approval for treating AD symptoms be limited to single-agent therapies (e.g., *see 165*), it is now feasible (and less complicated) to consider combining the approved drugs with others having different pharmacological profiles. Certainly, combinations of drugs intended to enhance neurotransmitter function beyond the cholinergic system would be interesting. Similarly, preliminary studies suggest that estrogen might improve AD patients, possibly reducing the risk of AD *(176,177)*. Because it has also been suggested that estrogen works, in part, through modulating cholinergic neurons *(176,178)*, the combination of estrogen and a cholinomimetic might provide an opportunity for pharmacological synergism. Finally, the basal forebrain cholinergic neurons that are particularly degenerated in AD patients (and that have been implicated in mediating some of the AD cognitive loss) are responsive to exogenous nerve growth factor *(179)*. Because NGF improves the viability and function of these neurons, it has been suggested that NGF might provide a therapy for AD pathology *(179)*. Certainly, if this hypothesis holds true, then combining NGF with an effective cholinomimetic may be even more effective, for somewhat healthier cholinergic neurons should be even more responsive to the positive effects of cholinergic modulation. Presumably, as basic research continues on the biochemical substrates of episodic recent memory, as well as the specific neurochemical deficiencies that contribute to the losses in AD, additional treatment opportunities will present themselves.

5. PARTING PERSPECTIVES

The initial clinical success achieved by the cholinergic hypothesis is the product of decades of dedicated work done by numerous independent groups. Although still not universally accepted, the cholinergic hypothesis has generated a wide range of empirical support, bound together in part by a wealth of data from animal models. Despite the logical basis for many of these models, it was warned long ago that the "value of any model or approach will depend not on the inherent logic of the principles that guided its development, but on its ability to make meaningful predictions about the clinical condition it was designed to study" *(68)*. Of course, the quintessential prediction of the cholinergic hypothesis is that appropriate cholinergic agents should provide meaningful and measurable symptomatic relief to many AD patients. Thus, the FDA's approval of the initial two cholinesterase inhibitors provides objective, independent confirmation of the most important tenets of the hypothesis, thereby cementing the components of the foundation upon which the hypothesis now firmly rests.

A question that remains to be answered is how much more effective other, more rationally designed cholinergic drugs might prove to be. Although the answer remains uncertain, the discussion in the previous sections suggests there is reason to be optimistic. There is also reason to believe that other treatment approaches may eventually be used, either alone or in combination with cholinergic agents to provide additional therapeutic opportunities. There seems to be little doubt that, in the final analysis, other neurotransmitter systems will be identified that contribute to the loss of cognitive function in aged or demented patients. For example, recent studies with estrogen, vitamin E, α-amino-3-hydroxy-5-methyl-4-isoxazole propionic acid (AMPA) agonists, and antiinflammatory agents all seem to offer preliminary promise, although they have yet to satisfy the rigorous standards and scrutiny of the FDA, as has the cholinergic hypothesis *(8,168,180)*. In the end, the eventual success of other approaches will only add to the contributions already established by the cholinergic hypothesis, as the seminal approach that led the way. History teaches us that the first step is very often the most difficult to achieve and that subsequent efforts are facilitated by the prior success. Despite its limitations, the cholinergic hypothesis has provided the initial (and so far only) treatment approach with genuine, positive effects in AD patients since the disease was first characterized more than a century ago.

Although the ultimate treatment goal in AD certainly involves the progression or pathogenesis of the disease, it will likely be many years, if not decades, before such treatments are routine. Despite the intense effort of many talented investigators, none of the progress achieved to date has yet been realized via the development of a novel, approved treatment. Moreover, it is difficult to know when such a breakthrough will occur. Thus, palliative treatment will likely be the only recourse for AD patients for some significant period of time. More importantly, however, even after effective pathogenic treatment becomes available, the slow progression of the disease and the difficulty of early diagnosis will necessarily dictate that many patients will incur some neurobehavioral dysfunction prior to beginning pathogenic treatment. Presumably, this dysfunction will continue to require palliative treatment, even after the underlying pathology has been arrested. For these reasons, it seems safe to anticipate the continuing need for palliative treatments well into the future—perhaps until the disease can be prevented.

When early work supporting the cholinergic hypothesis was still being integrated, it was argued that the hypothesis "necessarily suffers from a certain degree of inherent oversimplification and possible myopia. Yet, advantages gained by the research activities it helps stimulate and direct must not be underestimated. It seems clear that the opportunity it provides for increasing our understanding of age-related memory problems and for developing an effective means of reducing the severity of the cognitive symptoms of aged brain and AD provide adequate testimony for the continued value of the hypothesis as an important heuristic tool" *(3)* (p. 441). Given the initial success achieved in treating AD, as well as the knowledge gained regarding the memory losses associated with AD, the effects of selective brain lesions and drugs, and increased sophistication in the development and use of animal models, this statement seems to be as true today as it was a generation ago.

ACKNOWLEDGMENTS

The author thanks Alexis Perkins for excellent assistance in numerous aspects of preparing this review and the senior Editor, Dr. Dwaine Emerich, for effective arm-twisting in achieving my commitment to write it. Finally, I wish to express gratitude to all my friends and colleagues who, over the years, contributed to the ideas and concepts discussed in this chapter, either directly by helping me with my experiments and by reporting their own, or indirectly by the thinking they inspired.

REFERENCES

1. Bartus, R., Dean, R., Beer, B., and Lippa, A. (1982) The cholinergic hypothesis of geriatric memory dysfunction. *Science* **217,** 408–417.
2. *Web of Science™ Citation Indexes* (1988) Institute for Scientific Information, Philadelphia, PA.
3. Bartus, R. (1986) Cholinergic treatment for age-related memory disturbances: dead or barely coming of age?, in *Treatment Development Strategies for Alzheimer's Disease* (Crook, T., ed.), Powley Associacites, Madison, CT, pp. 421–450.
4. Fibiger, H. (1991) Cholinergic mechanisms in learning, memory and dementia: a review of recent evidence. *TINS* **14,** 220–223.
5. Ridley, R., Baker, H., and Fine, A. (1991) Cholinergic controversies. *TINS* **14,** 482–483.
6. Perry, E., Irving, D., and Perry, R. (1991) Cholinergic controversies. *TINS* **14,** 483–484.
7. Sarter, M., Bruno, J., and Dudchenko, P. (1991) Cholinergic controversies. *TINS* **14,** 484.
8. Davis, K. (1998) Future therapeutic approaches to Alzheimer's disease. *J. Clin. Psychiatry* **59 (Suppl. 11),** 14–16.
9. Tariot, P. and Schneider, L. (1996) Contemporary treatment approaches to Alzheimer's disease. *Consult. Pharm.* **11 (Suppl. E),** 16–24.
10. Wisniewski, H. and Terry, R. (1976) Neuropathology of the aging brain, in *Aging: Neurobiology of Aging* (Terry, R. and Gershon, S. eds.), Raven Press, New York, pp. 265–280.
11. Roth, M. (1978) Diagnosis of senile dementia and related forms of dementia, in *Alzheimer's Disease: Senile Dementia and Related Disorders* (Katzman, R., ed.), Raven Press, New York, pp. 25–39.
12. Constantinidis, J. (1978) Is Alzheimer's disease a major form of senile dementia? Clinical, anatomical, and genetic data, in *Alzheimer's Disease: Senile Dementia and Related Disorders* (Katzman, R., ed.), Raven Press, New York, pp. 15–25.
13. Flicker, C., Bartus, R., Crook, T., and Ferris, S. (1984) Effects of aging and dementia upon recent visuospatial memory. *Neurobiol. Aging* **5,** 275–283.
14. Crook, T., Bartus, R., Ferris, S., Whitehouse, P., Cohen, G., and Gershon, S. (1986) Age-associated memory impairment: proposed diagnostic criteria and measure of clinical change. *Dev. Neuropsychol.* **2,** 261–276.
15. Bartus, R., Fleming, D., and Johnson, H. (1995) Aging in the rhesus monkey: debilitating effects on short term memory. *J. NIH Res.* **7,** 69–77.
16. Bartus, R., Fleming, D., and Johnson, H. (1978) Aging in the rhesus monkey: debilitating effects on short-term memory. *J. Gerontol.* **33,** 858–971.
17. Bartus, R. (1979) Psychostimulants of the central nervous system: effects on short term memory in young vs. aged monkeys. *J. Am. Geriatr. Soc.* **27,** 11–19.
18. Sathananthan, G. and Gershon, S. (1975) Cerebral vasodilators: a review, in *Aging* (Gershon, S. and Raskin, A., eds.), Raven Press, New York, pp. 155–168.

19. Lehmann, H. and Ban, T. (1975) Central nervous system stimulants and anabolic substances in geropsychiatric therapy, in *Aging* (Gershon, S. and Raskin, A., eds.), Raven Press, New York, pp. 179–197.
20. Spence, K. (1960) *Behavior Theory and Learning*, Prentice-Hall, Englewood Cliffs, NJ.
21. Watson, J. (1967) *Behavior: An Introduction to Comparative Psychology*, New York: Holt, Rinehart, and Winston.
22. Bartus, R., Dean, R., Pontecorvo, M., and Flicker, C. (1985) The cholinergic hypothesis: an historical overview, current perspective, and future directions, in *Annals of the New York Academy of Sciences: Memory Dysfunction: Integration of Animal and Human Research from Clinical and Preclinical Perspectives* (Olton, D., ed.), New York Academy of Sciences, New York.
23. Drachman, D. and Leavitt, J. (1974) Human memory and the cholinergic system: A relationship to aging? *Arch. Neurol.* **30,** 113–121.
24. Hollister, L. (1968) *Chemical Psychoses.* Charles C. Thomas, Springfield, IL.
25. Longo, V. (1966) Behavioral and electroencephalographic effects of atropine and related compounds. *Pharmacol. Rev.* **18,** 965–996.
26. Bartus, R., Dean, R., and Flicker, C. (1987) Cholinergic psychopharmacology: an integration of human and animal research in memory, in *Psychopharmacology: a Third Generation of Progress* (Usdin, E., ed.), Raven Press, New York, pp. 219–232.
27. Bartus, R. (1978) Aging in the Rhesus monkey: specific behavioral impairments and the effects of pharmacological intervention, in *Recent Advances in Gerontology*, (Orimo, H., ed.), Elsevier, Tokyo, pp. 225–228.
28. Bartus, R. (1979) Physostigmine and memory: effects in young and aged non-human primates. *Science* **270,** 1087–1089.
29. Davis, K., Mohs, R., Davis, B., Levy, M., Rosenberg, G., Horrath, T., DeNigris, Y., Ross, A., Decker, P., and Rothpearl, A. (1981) Cholinomimetic agents and human memory: clinical studies in Alzheimer's disease and scopolamine dementia, in *Strategies for the Development of an Effective Treatment for Senile Dementia*, (Crook, T. and Gershon, R., eds.), Mark Powley Associates, New Caanan, CT.
30. Smith, C. and Swash, M. (1978) Possible biochemical basis of memory disorder in Alzheimer's disease. *Ann. Neurol.* **3,** 471–473.
31. Flicker, C., Dean, R., Bartus, R., Ferris, S., and Crook, T. (1985) Animal and human memory dysfunctions associated with aging, cholinergic lesions and senile dementia, in *Annals of the New York Academy of Sciences: Memory Dysfunctions: Integration of Animal and Human Research from Clinical and Preclinical Perspectives* (Olton, D., ed.), New York Academy of Sciences, New York, pp. 515–517.
32. Flicker, C., Ferris, S., and Serby, M. (1992) Hypersensitivity to scopolamine in the elderly. *Psychopharmacology (Ber.)* **107,** 437–441.
33. Sunderland, T., Tariot, P., Murphy, D., Weingartner, H., Mueller, E., and Cohen, R. (1985) Scopolamine challenges in Alzheimer's disease. *Psychopharmacology* **87,** 247–49.
34. Litvan, I., Sirigu, A., Toothman, J., and Grafman, J. (1995) What can preservation of autobiographical memory after muscarinic blockade tell us about the scopolamine model of demenita? *Neurology* **45,** 387–389.
35. Bowen, D., Smith, C., White, P., and Davison, A. (1976) Neurotransmitter-related enzymes and indices of hypoxia in senile dementia and other abiotrophies. *Brain* **99,** 459–496.
36. Davies, P. and Maloney, J. (1976) Selective loss of central cholinergic neurons in Alzheimer's disease. *Lancet* **ii,** 1403.
37. Perry, E., Perry, R., Blessed, G., and Tomlinson, B. (1977) Necropsy evidence of central cholinergic deficits in senile dementia. *Lancet* **i,** 189.
38. Haubrich, D. and Chippendale, T. (1977) Regulation of acetylcholine synthesis in nervous tissue. *Life Sci.* **20,** 1465–1478.

39. Jenden, D., Jope, R., and Weiler, M. (1976) Regulation of acetylcholine synthesis: does cytoplasmic acetylcholine control high affinity choline uptake? *Science* **194,** 635–637.
40. Arai, H. and Kosaka, K. (1984) Changes of biogenic amines and their metabolites in post-mortem brains from patients with Alzheimer-type dementia. *J. Neurochem.* **43,** 388–393.
41. Cross, A., Crow, T., Johnson, J., Joseph, M., Perry, E., Perry, R., Blessed, G., and Tomilson, B. (1983) Monoamine metabolism in senile dementia of Alzheimer type. *J. Neurol. Sci.* **60,** 383–392.
42. Cross, A., Crow, T., Perry, E., Pery, R., Blessed, G., and Tomilson, B. (1981) Reduced dopamine-beta-hydroxylase activity in Alzheimer's disease. *Br. Med. J.* **282,** 93–94.
43. Soininen, H., Macdonald, E., Rekonen, M., and Riekkinen, P. (1981) Homovanillic acid and 5-hydroxyindolacetic acid levels in cerebrospinal fluid of patients with senile dementia of Alzheimer type. *Acta Neurol. Scand.* **64,** 101–107.
44. Whitehouse, P., Price, D., Clark, A., Coyle, J., and Delong, M. (1981) Alzheimer's disease: evidence for selective loss of cholinergic neurons in the nucleus basalis. *Ann. Neurol.* **10,** 122–126.
45. Whitehouse, P., Price, D., Struble, R., Clark, A., Coyle, J., and Delong, M. (1981) Alzheimer's disease and senile dementia: loss of neurons in the basal forebrain. *Science* **215,** 1237–1239.
46. Enna, S., Samorajski, T., and Beer, B., eds. (1980) *Aging: Brain Neurotransmitters and Receptors in Aging and Age-Related Disorders*, Vol. 17, Raven Press, New York.
47. Lippa, A., Pelham, R., Beer, B., Critchett, D., Rean, R., and Bartus, R. (1980) Brain cholinergic dysfunction and memory in aged rats. *Neurobiol. Aging* **1,** 13–19.
48. Lippa, A., Critchett, D., Ehlert, F., Yamamura, H., Enna, S., and Bartus, R. (1981) Age-related alterations in neurotransmitter receptors: an electrophysiological and biochemical analysis. *Neurobiol. Aging* **2,** 3–8.
49. Gallagher, M. (1998) Effects of aging on cognition and hippocampal/cortical systems. *Soc. Neurosci. Abstr.* **24,** 129.
50. Bartus, R., Dean, R., and Fleming, D. (1979) Aging in the rhesus monkey: effects on visual discrimination and reversal learning. *J. Gerontol.* **34,** 209–219.
51. Davis, K., Mohs, R., and Tinklenberg, J. (1979) Enhancement of memory by physostigmine. *N. Engl. J. Med.* **301,** 946.
52. Davis, K. and Berger, P. (1978) Pharmacological investigations of the cholinergic imbalance hypothesis of movement disorders and psychosis. *Biol. Psychiatry* **13,** 23–49.
53. Davis, K. and Berger, P., eds. (1979) *Brain Acetylcholine and Neuropsychiatric Disease*, Plenum Press, New York.
54. Davis, K., Mohs, R., Davis, B., Rosenberg, G., Horvath, T., and DeNigris, Y. (1981) Cholinomimetic agents in human memory: preliminary observations in Alzheimer's disease, in *Cholinergic Mechanisms: Phylogenetic Aspects, Central and Peripheral Synapses, and Clinical Significance*, (Pepeu, G. and Ladinsky, H., eds.), Plenum Press, New York, pp. 929-936.
55. Davis, K. and Mohs, R. (1982) Enhancement of memory processes in Alzheimer's disease with multiple-dose intravenous physostigmine. *Am. J. Psychiatry* **139,** 1421–1424.
56. Christie, J., Shering, A., Ferguson, J., and Glen, A. (1981) Physostigmine and arecoline: effects of intravenous infusions in Alzheimer's presenile dementia. *Br. J. Psychiatry* **138,** 46–50.
57. Thal, L., Masur, D., and Sharpless, N. (1986) Acute and chronic effects of oral physostigmine and lecithin in Alzheimer's disease. *Prog. Neuropsychopharmacol. Biol. Psychiatry* **10,** 617–636.
58. Mohs, R., Davis, B., and Johns, C. (1985) Oral physostigmine treatment of patients with Alzheimer's disease. *Am. J. Psychiatry* **142,** 28–33.
59. Stern, Y., Sano, M., and Mayeux, R. (1988) Long-term administration of oral physostigmine in Alzheimer's disease. *Neurology* **38,** 1837–1841.

60. Bartus, R., Dean, R., and Beer, B. (1980) Recent memory deficits in aged cebus monkeys and facilitation with central cholinimimetics. *Neurobiol. Aging* **1**, 145–152.

61. Bartus, R., Dean, R., and Beer, B. (1983) An evaluation of drugs for improving memory in aged monkeys: implication for clinical trials in humans. *Psychopharmacol. Bull.* **19**, 168–184.

62. Summers, W., Majovski, L., Marsh, G., Tachiki, K., and Kling, A. (1986) Oral tetra-hydroaminoacridine in long-term treatment of senile dementia. *N. Engl. J. Med.* **315**, 1241–1245.

63. Gracon, S. (1996) Evaluation of tacrine hydrochloride (Cognex) in two parallel-group studies. *Acta Neurol. Scand. Suppl.* ***165, 114–122.***

64. Bartus, R. (1990) Drugs to treat age-related neurodegenerative problems: the final frontier of medical science? *J. Am. Geriatr. Soc.* **38**, 680–695.

65. Bartus, R. (1988) The need for common perspectives in the development and use of animal models for age-related cognitive neurodegenerative disorders. *Neurobiol. Aging* **9**, 445–451.

66. Bartus, R. and Dean, R. (1987) Animal models for age-related memory disturbances, in *Animal Models of Dementia: A Synaptic Neurochemical Perspective* (Coyle, J., ed.), Alan R. Liss, New York, pp. 67–79.

67. Bartus, R. and Dean, R. (1987) On possible relationships between Alzheimer's disease, age-related memory loss and the development of animal models, in *Alzheimer's Disease: Problems, Prospects, Perspectives* (Altman, H., ed.), Plenum Press, New York, pp. 129–139.

68. Bartus, R. (1986) Commentary on animal models of Alzheimer's and Parkinson's disease. *Integrat. Psychiatry* **4**, 74–75.

69. Bartus, R., Flicker, C., Dean, R., Fisher, S., Pontecorvo, M., and Figueiredo, J. (1986) Behavioral and biochemical effects of nucleus basalis magnocellularis lesions: implications and possible relevance to understanding or treating Alzheimer's disease, in *Progress in Brain Research* (Van Haaren, Z., ed.), Elsevier, Amsterdam, pp. 345–361.

70. Bartus, R., Flicker, C., and Dean, R. (1983) Logical principles in the development of animal models of age-related memory loss, in *Clinical and Preclinical Assessment in Geriatric Psychopharmacology* (Crook, T., ed.), Mark Powley Associates, New Caanan, CT, pp. 263–299.

71. Bartus, R. and Dean, R. (1985) Developing and utilizing animals models in the search for an efffective treatment for age-related memory disturbances, in *Normal Aging, Alzheimer's Disease, and Senile Dementia* (Gottfries, C., ed.), University of Brussels Press, Brussels, pp. 231–267.

72. Bartus, R., Flicker, C., Dean, R., and Beer, B. (1983) Behavioral and pharmacological studies using animal models of aging: implications for studying and treating dementia of Alzheimer's type. *Banbury Report 1515: Biological Aspects of Alzheimer's Disease* **1515**, 207–218.

73. Bartus, R. (1979) The effects of aging on visual memory, sensory processing and discrimination learning in a non-human primate, in *Aging: Sensory Systems and Communication in the Elderly* (Ordy, J., ed.), Raven Press, New York, pp. 85–113.

74. Bartus, R. and Dean, R. (1979) Recent memory in aged non-human primates: hypersensitivity to visual interference during retention. *Exp. Aging Res.* **5**, 385–400.

75. Riopelle, A. and Rogers, C. (1965) Age changes in chimpanzees, in *Behavior of Non-human Primates* (Schrier, A., ed.), Academic Press, New York, pp. 449–462.

76. Medin, D., O'Neill, P., Smeltz, E., and Davis, R. (1973) Age differences in retention of concurrent discrimination problems in monkeys. *J. Gerontol.* **28**, 63–67.

77. Bartus, R. (1978) Short-term memory in the Rhesus monkey: effects of dopamine blockade via acute haloperidol administration. *Pharmacol. Biochem. Behav.* **9**, 353–357.

78. Bartus, R. and Johnson, H. (1976) Short-term memory in the Rhesus monkey: disruption from the anticholinergic scopolamine. *Pharmacol. Biochem. Behav.* **5**, 39–46.
79. Flicker, C., Dean, R., Watkins, D., Fisher, S., and Bartus, R. (1983) Behavioral and neurochemical effects following neurotoxic lesions of a major cholinergic input to the cerebral cortex in the rat. *Pharmacol. Biochem. Behav.* **18**, 973–981.
80. Flicker, C., Ferris, S., Crook, T., and Bartus, R. (1987) A visual recognition memory test for the assesment of cognitive function in aging and dementia. *Exp. Aging Res.* **13**, 127–132.
81. Flicker, C., Ferris, S., Crook, T., Bartus, R., and Reisburg, B. (1985) Cognitive function in normal aging and early dementia, in *Senile Dementia of the Alzheimer Type* (Traber, J., ed.), Springer-Verlag, Berlin, pp. 1–16.
82. Bachevalier, J. (1993) Behavioral changes in aged rhesus monkeys. *Neurobiol. Aging. 14, 619.*
83. Moss, M. (1993) The longitudinal assessment of recognition memory in aged rhesus monkeys. *Neurobiol. Aging* **14**, 635.
84. Rapp, P. (1993) Neuropsychological analysis of learning and memory in the aged nonhuman primate. *Neurobiol. Aging* **14**, 627.
85. Voytko, M. (1993) Cognitive changes during normal aging in monkeys assessed with an automated test apparatus. *Neurobiol. Aging* **14**, 643.
86. Rapp, P. (1988) Toward a nonhuman primate model of age-dependent cognitive dysfunction. *Neurobiol. Aging* **9**, 503–505.
87. Moss, M., Rosene, D., and Peters, A. (1988) Effects of aging on visual recognition memory in the rhesus monkey. *Neurobiol. Aging* **9**, 495–502.
88. Albert, M. (1997) The ageing brain: normal and abnormal memory. *Philos. Trans. R. Soc. Lond.* **352**, 1703–1709.
89. Gallagher, M. and Colombo, P. (1995) Aging: the cholinergic hypothesis of cognitive decline. *Curr. Opin. Neurobiol.* **5**, 161–168.
90. Luszcz, M. and Bryan, J. (1999) Toward understanding age-related memory loss in late adulthood. *Gerontology* **45**, 2–9.
91. Wisniewski, H., Ghetti, B., and Terry, R. (1973) Neuritic (senile) plaques and filimentous changes in aged rhesus monkeys. *J. Neuropathol. Exp. Neurol.* **32**, 566–584.
92. Goldman-Rakic, P. (1996) The prefrontal landscape: implications of functional architecture for understanding human mentation and the central executive. *Philos. Trans. R. Soc. Lond.* **351**, 1445–1453.
93. Hanninen, T., Hallikainen, M., Koivisto, K., Partanen, K., Laakso, M., Riekkinen, P., and Soininen, H. (1997) Decline of frontal lobe functions in subjects with age-associated memory impairment. *Neurology* **48**, 148–153.
94. Rapp, P. and Heindel, W. (1994) Memory systems on normal and pathological aging. *Curr. Opin. Neurol.* **7**, 294–298.
95. Arnsten, A. (1993) Postsynaptic alpha-2 receptor stimulation improves memory in aged monkeys: indirect effects of yohimbine versus direct effects of clonidine. *Neurobiol. Aging* **14**, 597–603.
96. Bartus, R., Crook, T., and Dean, R. (1987) Current progress in treating age-related memory problems: a perspective from animal preclinical and human clinical research, in *Geriatric Clinical Pharmacology* (Wood, G. and Strong, R., eds.), Raven Press, New York, pp. 71–94.
97. Bartus, R., Dean, R., and Beer, B. (1982) Neuropeptide effects on memory in aged monkeys. *Neurobiol. Aging* **3**, 61–68.
98. Bartus, R. and Dean, R. (1988) Effects of cholinergic and adrenergic enhancing drugs on memory in aged monkeys, in *Current Research in Alzheimer Therapy* (Giacobini, E. and Becker, R., eds.), Taylor & Francis, New York, pp. 179–190.
99. Bartus, R. and Dean, R. (1988) Lack of effect of clonidine on memory test in aged cebus monkeys. *Neurobiol. Aging* **9**, 409–411.

100. Arnsten, A., Cai, J., Steere, J., and Goldman-Rakic, P. (1995) Dopamine D2 receptor mechanisms contribute to age-related cognitive decline: the effects of quinpirole on memory and motor performance in monkeys. *J. Neurosci.* **15,** 3429–3439.

101. Arnsten, A. and Goldman-Rakic, P. (1985) Alpha-2-adrenergic mechanisms in prefrontal cortex associated with cognitive decline in aged nonhuman primates. *Science* **230,** 1273–1275.

102. Mohr, E., Schlegel, J., Fabbrini, G., Williams, J., Mouradian, M., Mann, U., Claus, J., Fedio, P., and Chase, T. (1989) Clonidine treatment in Alzheimer's disease. *Arch. Neurol.* **46,** 376–378.

103. Schlegel, J., Mohr, E., Willaims, J., Mann, U., Gearing, M., and Chase, T. (1989) Guanfacine treatment of Alzheimer's disease. *Clin. Neuropharmacol.* **12,** 124–128.

104. Arnsten, A. and Goldman-Rakic. (1985) Catecholamines and cognitive decline in aged nonhuman primates, in *Annals of the New York Academy of Science* (Olton, D., ed.), New York Academy of Sciences, New York, pp. 218–241.

105. Coull, J. (1994) Pharmacological manipulations of the alpha 2-noradrenergic system. Effects on cognition. *Drugs Aging* **5,** 116–126.

106. Solomon, P., Beal, M., and Pendlebury, W. (1988) Age-related disruption of classical conditioning: a model systems approach to memory disorders. *Neurobiol. Aging* **9,** 535–547.

107. Gallagher, M. and Pelleymounter, M. (1988) Spatial learning deficits in old rats: a model for memory decline in the aged. *Neurobiol. Aging* **9,** 548–557.

108. Kesner, R. (1988) Reevaluation of the contribution of the basal forebrain cholinergic system to memory. *Neurobiol. Aging* **9,** 609–617.

109. Barnes, C. (1988) Aging and the physiology of spatial learning. *Neurobiol. Aging* **9,** 563–569.

110. Landfield, P. (1988) Hippocampal neurobiological mechanisms of age-related memory dysfunction. *Neurobiol. Aging* **9,** 571–581.

111. Olton, D. (1988) Aging and spatial cognition. *Neurobiol. Aging* **9,** 569–571.

112. Dean, R., Scozzafava, J., Goas, J., Beer, B., and Bartus, R. (1981) Age-related differences in behavior across the life span of the C57BL/6J mouse. *Exp. Aging Res.* **7,** 427–451.

113. Bartus, R. (1980) Cholinergic drug effects on memory and cognition in animals, in *Aging in the 1980's: Psychological Issues* (Poon, L., ed.), American Psychological Association, Washington, DC.

114. Semple, S., Smith, C., and Swash, M. (1982) The Alzheimer's disease syndrome, in *Alzheimer's Disease: A Report of Progress in Research* (Corkin, S., ed.), Raven Press, New York, pp. 93–107.

115. Jarrard, L. and Moise, S. (1971) Short-term memory in the monkey, in *Cognitive Processes of Nonhuman Primates* (Jarrard, L., ed.), Academic Press, New York, pp. 1–24.

116. Medin, D. and Davis, R. (1974) Memory, in *Behavior of Nonhuman Primates: Modern Research Trends* (Schreier, A. M. and Stollnitz, F., eds.), Academic Press, New York.

117. Fletcher, H. and Davis, J. K. (1965) The delayed response problem, in *Behavior of Nonhuman Primates* (Schrier, A. M., ed.), Academic Press, New York, pp. 129–165.

118. Barnes, C., McNaughton, B., and O'Keefe, J. (1983) Loss of place specificity in hippocampal complex spike cells of the senescent rat. *Neurobiol. Aging* **4,** 113–119.

119. Barnes, C., Nadel, L., and Honig, W. (1980) Spatial memory deficit in senescent rats. *Can. J. Psychol.* **34,** 29–39.

120. Shen, J., Barnes, C., Wenk, G., and McNaughton, B. (1996) Differential effects of selective immunotoxic lesions of medial septal cholinergic cells on spatial working and reference memory. *Behav. Neurosci.* **110,** 1181–1186.

121. Bartus, R., Flicker, C., Dean, R., Pontecorvo, M., Figueriedo, J., and Fisher, S. (1985) Selective memory loss following nucleus basalis lesions: long term behavioral recovery despite persistent cholinergic deficiencies. *Pharmacol. Biochem. Behav.* **23,** 125–135.

122. Gold, P. and McGauch, J. (1975) Changes in learning and memory during aging, in *Advances in Behavioral Biology* (Ordy, J., ed.), Plenum Press, New York, pp. 145–148.

123. Kubanis, P. and Zornetzer, S. (1981) Age-related behavioral and neurobiological changes: a review with an emphasis on memory. *Behav. Neural. Biol.* **31,** 115–172.

124. Bartus, R. (1982) The one-trial passive avoidance procedure as a possible marker of aging, in *Biological Markers of Aging* (Schneider, E. and Reff, M., eds.), National Institutes of Health, Washington, DC, pp. 67–77.

125. Ingram, D., Spangler, E., Iijima, S., Ikari, H., Kuo, H., Greig, N., and London, E. (1994) Rodent models of memory dysfunction in Alzheimer's disease and normal aging: moving beyond the cholinergic hypothesis. *Life Sci.* **55,** 2037–2049.

126. Lashley, K. (1929, reprinted 1963) *Brain Mechanisms and Intelligence: A Quantitative Study of Injuries to the Brain,* Dover, New York.

127. Bartus, R. (1979) Evidence for a direct cholinergic imvolvement in the scopolamine-induced amnesia in monkeys: effects of concurrent administration of physostigmine and methylphenidate with scopolamine. *Pharmacol. Biochem. Behav.* **9,** 833–836.

128. Dean, R. and Bartus, R. (1982) Drug induced memory impairments in non-human primates. *Soc. Neurosci. Abstr.* **8,** 322.

129. Askew, W., Kimball, A., and Ho, B. (1974) Effect of tetrahydrocannabinols on brain acetylcholine. *Brain Res.* **69,** 375–378.

130. Domino, E. (1981) Cannabinoids and the cholinergic system. *J. Clin. Pharmacol.* **21,** 249s–255s.

131. Ladinsky, H., Consolo, C., Bellantuono, C., and Garattini, S. (1981) Interaction of benzodiazepines with known and putative neurotransmitters in the brain, in *Handbook of Biological Psychiatry* (Praag, M. V., ed.), Marcel Dekker, New York.

132. Clemmesen, L., Mikkelsen, P., and Lund, T. (1984) Assessments of the anticholinergic effects of antidepressants in a single-dose cross over study of salivation and plasma levels. *Psychopharmacology* **82,** 348–354.

133. Herz, A. (1960) Die Gedentung der Gahung für die wirkung von scopolamin und ahnlichen. *Z. Biol.* **112,** 104–112.

134. Meyers, B. and Domino, E. (1964) The effect of cholinergic blocking drugs on spontaneous alteration in rats. *Arch. Int. Pharmacodyn.* **150,** 3–4.

135. Meyers, B., Roberts, D., Ricputi, R., and Domino, E. (1964) Some effects of muscarinc cholinergic blocking drugs on behavior and the electrocorticogram. *Psychopharmacology* **5,** 289–300.

136. Deutsch, J., Hamburg, M., and Dahl, H. (1966) Anticholinesterase-induced amnesia and its temporal aspects. *Science* **151,** 221–223.

137. Deutsch, J. (1971) The cholinergic synapse and the site of memory. *Science* **174,** 788–794.

138. Bartus, R. (1993) General overview: past contributions and future opportunities using aged nonhuman primates. *Neurobiol. Aging* **14,** 711–714.

139. Rigter, H. and Crabbe, J. (1979) Modulation of memory by pituitary hormones and related peptides. *Vitam. Horm.* **37,** 153–241.

140. Beatty, W., Butters, N., and Janowsky, D. (1986) Patterns of memory failure after scopolamine treatment: implications for cholinergic hypotheses of dementia. *Behav. Neural. Biol.* **45,** 196–211.

141. Molchan, S., Martinez, R., Hill, J., Weingartner, H., Thompson, K., Vitiello, B., and Sunderland, T. (1992) Increased cognitive sensitivity to scopolamine with age and a perspective on the scopolamine model. *Brain Res. Rev.* **17,** 215–226.

142. Newhouse, P. (1990) Cholinergic studies in dementia and depression, in *New Directions in Understanding Dementia and Alzheimer's Disease* (Zandi, T., ed.), Plenum Press, New York, pp. 65–76.

143. Bartus, R., Pontecorvo, M., Flicker, C., Dean, R., and Figuerido, J. (1986) Behavioral recovery following bilateral lesions of the nucleus basalis does not occur spontaneously. *Pharmacol. Biochem. Behav.* **24**, 1287–1292.

144. Dunnett, S., Whishaw, I., Jones, G., and Bunch, S. (1987) Behavioal, biochemical, and histochemical effects of different neurotoxic amino acids injected into nucleus basalis magnocellularis of rats. *Neuroscience* **20**, 653–669.

145. Dunnett, S., Everitt, B., and Robbins, T. (1991) The basal forebrain-cortical cholinergic system: interpreting the functional consequences of excitotoxic lesions. *Trends Neurosci.* **14**, 494–501.

146. Wenk, G., Stoehr, J., Quintana, G., Mobley, S., and Wiley, R. (1994) Behavioral, biochemical, histological, and electrophysiological effects of 192 IgG-saporin injections into the basal forebrain of rats. *J. of Neurosci.* **14**, 5986–5995.

147. Wenk, G., Harrington, C., Tucker, D., Rance, N., and Walker, L. (1992) Basal forebrain neurons and memory: a biochemical, histological, and behavioral study of differential vulnerability to ibotenate quisqualate. *Behav. Neurosci.* **106**, 909–923.

148. Baxter, M. and Gallagher, M. (1997) Cognitive effects of selective loss of basal forebrain cholinergic neurons—implications for cholinergic therapies of Alzheimer's disease, in *Pharmacological Treatment of Alzheimer's Disease* (Brioni, J. and Decker, M., eds.), Wiley-Liss, New York.

149. Fischer, W., Sirevaag, A., Wiegand, S., Lindsay, R., and Bjorklund, A. (1994) Reversal of spatial memory impairments in aged rats by nerve growth factor and neurotrophins 3 and 4/5 but not by brain-derived neurotrophic factor. *Proc. Natl. Acad. Sci. USA* **91**, 8607–8611.

150. Fischer, W., Chen, K., Gage, F., and Bjorklund, A. (1992) Progressive decline in spatial learning and integrity of forebrain cholinergic neurons during aging. *Neurobiol. Aging* **13**, 2–23.

151. Serrano-Martinez, A. (1996) Long-term functional recovery from age-induced spatial memory impairments by nerve growth factor gene transfer to the rat basal forebrain. *Proc. Natl. Acad. Sci. USA* **93**, 6355–6360.

152. Chen, K., Nishimura, M., Armanini, M., Crowley, C., Spenser, S., and Philips, H. (1997) Disruption of a single allele of the nerve growth factor gene results in atrophy of basal forebrain cholinergic neurons and memory deficits. *J. Neurosci.* **17**, 7288–7296.

153. Voytko, M. (1996) Cognitive functions of the basal forebrain cholinergic system in monkeys: memory or attention? *Behav. Brain Res.* **75**, 13–25.

154. Voytko, M., Olton, D., Richardson, R., Gorman, L., Tobin, J., and Price, D. (1994) Basal forebrain lesions in monkeys disrupt attention but not learning and memory. *J. Neurosci.* **14**, 167–186.

155. Lawrence, A. and Sahakian, B. (1995) Alzheimer's disease, attention, and the cholinergic system. *Alzheimer Dis. Assoc. Disord.* **9 (Suppl. 2),** 43–49.

156. Alhainon, K., Helkala, E., and Riekkinin, P. (1993) Psychometric discrimination of tetrahydroaminoacridine responders in Alzheimer patients. *Dementia* **4**, 54–58.

157. Rogers, S., Farlow, M., Doody, R., Mohs, R., and Friedhoff, L. (1998) A 24-week double-blind placebo-controlled trial of donepezil in patients with AD. *Neurology* **50**, 136–145.

158. Jorm, A. (1986) Effects of cholinergic enhancement therapies on memory function in Alzheimer's disease: a meta-analysis of the literature. *Austr. N. Zeal. J. Psychiatry* **20**, 237–240.

159. Eagger, S. and Harvey, R. (1995) Clinical heterogeneity: responders to cholinergic therapy. *Alzheimer Dis. Assoc. Disord.* **9**, 37–42.

160. Alhainen, K. and Riekkinen, J. (1993) Discrimination of Alzheimer's patients responding to cholinesterase inhibitor therapy. *Acta Neurol. Scand.* **Suppl. 149**, 16–21.

161. Farlow, M., Lahiri, D., and Poirer, J. (1996) Apolipoprotein E genotype and gender influence response to tacrine therapy. *Ann. NY Acad. Sci.* **802,** 101–110.

162. Grothe, D., Piscitelli, S., and Dukoff, R. (1998) Penetration of tacrine into cerebrospinal fluid in patients with Alzheimer's disease. *J. Clin. Psychopharmacol.* **18,** 78–81.

163. Lou, G., Montgomery, P., and Sitar, D. (1996) Bioavailability and pharmacokinetic disposition of tacrine in elderly patients with Alzheimer's disease. *J. Psychiatry Neurosci.* **21,** 334–339.

164. Stern, Y., Sano, M., and Mayeux, R. (1987) Effects of oral physostigmine in Alzheimer's disease. *Ann. Neurol.* **22,** 306–310.

165. Bartus, R. (1986) Drugs to treat age-related cognitive disorders: on the threshold of a new era in the pharmaceutical industry?, in *Treatment Development Strategies for Alzheimer's Disease* (Crook, T., ed.), Powley Associates, Madison, CT.

166. Schumock, G. (1998) Economic considerations in the treatment and management of Alzheimer's disease. *Am. J. Health Syst. Pharmacol.* **55 (Suppl. 2),** S17–21.

167. Giacobini, E. (1994) Therapy for Alzheimer's disease. Symptomatic or neuroprotective? *Mol. Neurobiol.* **9,** 115–118.

168. Schneider, L. and Tariot, P. (1994) Emerging drugs for Alzheimer's disease: mechanisms of action and prospects for cognitive enhancing medications. *Med. Clin. North Am.* **78,** 911–935.

169. Somerset, P., *Somerset Pharmaceuticals Reports Results of Clinical Trial in Alzheimer's Disease* in *PR Newswire*, Nov. 3, 1997.

170. Cohen, E. and Wurtman, R. (1975) Brain acetylcholine: increases after systemic choline administration. *Life Sci.* **16,** 1095–1102.

171. Bartus, R., Dean, R., and Beer, B. (1984) Cholinergic precursor therapy for geriatric cognition: it's past, present, and a question of its future, in *Nutrition in Gerontology* (Ordy, J., ed.), Raven Press, New York, pp. 191–226.

172. Bartus, R. T., Dean, R. L., Sherman, K. S., Friedman, E., and Beer, B. (1981) Profound effects of combining choline and piracetam on memory enhancement and cholinergic functions in aged rats. *Neurobiology of Aging* **2,** 105–111.

173. Bartus, R., Dean, R., Goas, J., and Lippa, A. (1980) Age-related changes in passive avoidance retention and modulation with chronic dietary choline. *Science* **209,** 301–303.

174. Schwarz, R. (1997) Muscarinic receptor agonists under development for the treatment of Alzheimer's disease. *Alzheimer's Dis. ID Res. Alert* **2,** 5459–5465.

175. Pang, D. (1998) Alzheimer's disease. *AAPS News* **1,** 11–15.

176. Yaffe, K., Sawaya, G., and Lieberburg, I. (1998) Estrogen therapy in post-menopausal women: effects on cognitive function and dementia. *JAMA* **279,** 688–695.

177. Henderson, V., Paganini-Hill, A., Emanuel, C., Dunn, M., and Buckwalter, J. (1994) Estrogen replacement therpay in older women. Comparisons between Alzheimer's disease cases and nondemented control subjects. *Arch. Neurol.* **51,** 896–900.

178. Gibbs, R. (1994) Estrogen and nerve growth factor-related systems in brain. Effects on basal forebrain cholinergic neurons and implications for learning and memory processes and aging. *Ann. NY Acad. Sci.* **743,** 165–196.

179. Hefti, F., Hartikka, J., and Knusel, B. (1989) Function of neurotrophic factors in the adult and aging brain and their possible use in the treatment of neurodegenerative diseases. *Neurobiol. Aging* **10,** 515–534.

180. Ingvar, M., Ambros-Ingerson, J., Davis, M., Granger, R., Kessler, M., Rogers, G., Scher, R., and Lynch, G. (1997) Enhancement by an ampakine of memory encoding in humans. *Exp. Neurol.* **146,** 553–559.

2

Patterns of Cognitive Decline
in the Aged Rhesus Monkey

Mark B. Moss, Ronald J. Killiany, and James G. Herndon

1. INTRODUCTION

It is only in the last 25 yr that significant strides have been taken toward the development of primate models of normal human aging. Together with advances in the fields of molecular biology, neuroimaging, and behavioral neuroscience, these models have emerged as potentially important components in our understanding of the age-related cognitive decline in humans (1). This is not to deny that studies using other species, particularly rodents, have provided a wealth of information on the neurobiological substrates of aging. However, the rich behavioral repertoire of the monkey, particularly its remarkable ability to perform a variety of complex short-term memory tasks, many of which have been administered to normal humans in geriatric research studies, has made this species particularly suited to assess cognitive function at a level that is not possible in nonprimates.

This chapter reviews, from a human neuropsychological perspective, earlier findings and recent advances in our understanding of cognitive decline that occurs in the aging monkey. In large part, we rely on data obtained in our multidisciplinary research program exploring the neural substrates of cognitive decline in the aged monkey.

2. LIFE SPAN OF THE RHESUS MONKEY

A study from our program examining the survival rates of 763 monkeys kept at two locations at Yerkes Regional Primate Research Center (2) showed that the maximum life-span of the rhesus monkey extends into the fourth decade, but that only 50% of monkeys reached 16 yr of age, and only about 25% reached 25 yr of age. The typical adult life span of the rhesus monkey may be considered to extend from 5 up to 30 yr of age. On this basis, we have concluded that monkeys over 20 yr of age can be considered old. Within the aged range, taking into account the estimated ratio of monkey to human years of age of approx 1:3, we consider monkeys 19–23 yr of age as "early" senescent animals, those from 24 to 28 yr of age as "advanced" aged animals, and those reaching 29 yr of age or older as "the oldest of the old."

From: Central Nervous System Diseases
Edited by: D. F. Emerich, R. L. Dean, III, and P. R. Sanberg © Humana Press Inc., Totowa, NJ

3. COGNITIVE FUNCTION

3.1. Attention

The domain of attention incorporates an integrated set of cognitive skills that provide a basis for higher cortical functioning. It encompasses a number of interrelated abilities such as orienting, alerting, selection, and vigilance *(3–5)*. Studies of attention date back to the late 1800s *(6)* and have helped to bridge the gap between purely behavioral and neuroanatomical work *(7)*. Since the 1960s, however a vast majority of the studies focusing on attentional processes in the nonhuman have been neurophysiological in nature rather than neuropsychological.

Several aspects of attention have been shown to be preserved in normal human aging *(8,9)*. However, few studies have assessed directly attentional functioning in aged monkeys. Davis *(10)* demonstrated that aged monkeys were more likely to evidence distraction in initial learning, and several studies, by virtue of unimpaired performance on simple two-choice discrimination learning tasks, have inferred normal attentional processing in aged monkeys. Recently, adapting attentional tasks from those used in humans by Posner *(11,12)* and by Baxter and Voytko *(13)* showed that aged monkeys were as proficient as young adult monkeys at shifting attention to a peripherally cued spatial location. Thus, initial studies suggest that attentional processes in aged monkeys, as in aged humans, are spared.

3.2. Learning and Memory

3.2.1. Simple Discrimination Learning

Learning in monkeys has traditionally been assessed by the use of simple two-choice discrimination tasks. In a typical paradigm, monkeys are rewarded for choosing the designated correct stimulus of a stimulus pair (objects, colors, patterns, or stimulus position) presented simultaneously over a series of trials to a learning criterion of 90% accuracy. In most studies, it has been shown that aged monkeys, as a group, do not evidence any significant difficulty in simple discrimination learning *(14–19)*, although in two studies *(20;* Moss et al., *unpublished data)* it was found that a group of aged monkeys evidenced significant impairment in learning object discriminations, suggesting that simple discrimination learning is preserved in aging until the end of the life-span.

3.2.2. Concurrent Discrimination Learning

Concurrent discrimination is another aspect of learning that has been assessed both in monkeys *(21–23)* and in humans *(24)*. Concurrent discrimination tasks assess the ability to learn associations between reward and a given stimulus objects in the presence of proactive and retroactive interference. Bartus and Dean *(25)* showed that aged monkeys were impaired on a concurrent discrimination task, leading to the suggestion that aged monkeys are particularly sensitive to interference. In one form of the task, 20 pairs of objects are presented in succession daily until the animal performs all 20 discriminations with no errors in one session *(16)*. A significant difference was found in performance of this task between a group of oldest old monkeys and a group of young adult control animals. In contrast, a group of early senescent and advanced aged monkeys were, as a group, unimpaired on the task. Recent findings from our program (Moss et al., *unpublished data)* support the findings of Bachevalier et al. *(16)* and suggest that the capacity to simultaneously learn multiple object reward associations declines with age.

3.2.3. Delayed Nonmatching-to-Sample

Normal human aging is associated with a decline in several domains of cognitive function among which short-term memory is perhaps the earliest and most prominent *(26)*. Several aspects of short-term memory commonly assessed in aged humans are recall of word lists, of aurally presented short narratives, and of visually presented designs on visual or spatial locations. In some instances, recognition paradigms, in which the relevant stimuli are presented in a "yes–no" or forced-choice format, are used in conjunction with recall tasks. In monkeys, because of the difficulty in creating a true free recall test condition, memory assessment is limited primarily to recognition paradigms. One test commonly used to assess visual recognition memory in the monkey is the delayed nonmatching-to-sample (DNMS) task *(16,17,27–30)*. This is a benchmark visual recognition memory task that has been used by several laboratories to assess memory in monkeys and resembles, in many respects, clinical tests that are used to assess memory function in patients with a variety of neurological disorders *(26)*. As a group, aged monkeys evidence impairment on the acquisition and performance of this task, although several studies have reported normal to near-normal performance by individual aged monkeys *16,27–30)*. Nevertheless, several studies spanning the last 10 years have collectively revealed a significant relationship between performance on the DNMS task and aging across the entire adult age range in monkeys.

On closer inspection of the performance of the early senescent monkey on the memory conditions of the DNMS task in our study, monkeys showed a significant impairment when delays greater than 30 s were interposed between the sample and recognition trials or when list of items were used. At first glance, this appeared at variance with the study by Presty et al. *(30)*, who did not find an overall impairment in the group of early senescent monkeys using a combined performance measure. However, it is plausible that this difference is due to methodological differences in the studies. Taken together, the data from the studies by Presty et al. *(30)* and ours lead to two conclusions. First, the impairment on the performance conditions of the DNMS by early senescent monkeys is relatively mild and is evident only under conditions of relatively high memory demands. Second, as pointed out by Rapp and Amaral *(27)*, there does not appear to be any significant relationship between rate of acquisition and accuracy on the performance conditions on DNMS. This suggests that the two phases of the task may require different cognitive abilities and that they may, in turn, be differentially affected by age.

The pattern of impairments observed in advanced aged monkeys (25–29 yr of age) differed from the one observed in early senescent monkeys *(31)*. Only one of the six monkeys in the advanced aged group reached learning criterion as efficiently as young adult monkeys, a percentage significantly lower than that seen in our study with early senescent monkeys *(31)*. On the delay condition of the DNMS, the accuracy of advanced age animals was significantly worse than that of the young adult animals, evidencing about a 15% decrement in performance when delays were extended from 30 to 300 s. Similarly, the accuracy of performance for the advanced aged group declined significantly relative to young adult animals when lists of items were presented.

With regard to the behavioral assessment of the oldest of the old rhesus monkeys, precious few data are available. In a large cohort of aged monkeys studied by Bachevalier et al. *(16)*, the oldest group, consisting of four monkeys aged 28–31 yr (originally assessed 2 yr earlier at 26–29 yr of age), evidenced a pattern of deficits that differed from the other two aged groups (one of middle age, 16–19 yr and the other in early senescence, 21–25 yr). In another study assessing performance of aged monkeys ranging in age from 22 to 33 yr on the DNMS task, Rapp and Amaral *(27)* noted that a 33-yr-old monkey, the oldest monkey in the study, was the only animal that could not reach learning criterion, even within 3000 trials, or compared to the 200–400 trials required by the other monkeys in the study to master their task.

Recently, we had the opportunity to assess a group of five monkeys between the ages of 30 and 35. As a group, and individually, these oldest old monkeys were markedly impaired relative to young adults on acquisition of the DNMS task. Monkeys in this oldest group required a mean number of trials twice that required by monkeys of advanced age. These findings suggest that survival beyond advanced age may be associated with a marked and precipitous decline in the ability to learn new operational cognitive rules, a pattern that is beginning to emerge in studies of normal human aging *(32)*.

In contrast to their severe impairment on acquisition of DNMS, the oldest old monkeys evidenced only a moderate (15–20%) decline in accuracy on the memory component of the DNMS task. This finding suggests that oldest old monkeys may experience a more rapid rate of forgetting than any of the younger animals.

Based on our findings and on those of other investigators *(27,30)*, we believe that in the monkey there is a significant but mild deficit in recognition memory associated with aging. This deficit begins to appear at a relatively early age (about the third decade of life) and appears to plateau sometime around the middle of the third decade of life. No additional decline in recognition memory appears to take place during the oldest old age of the process of successful aging. The ability to learn new strategies, on the other hand, appears to be affected at a different rate and course. This deficit does not appear to become significant until more advanced age (about the middle of the third decade of life) and declines in performance appear to continue into the oldest old age.

3.2.4. Delayed Recognition Span

The DNMS task relies upon a two-alternative, forced choice paradigm in which the monkey is required to discriminate, after a delay, which one of two objects was previously presented. However, it is limited to two-choice recognition. In an effort to characterize recognition memory within spatial as well as nonspatial stimulus domains, as well as to assess the memory capacity or "load" of aged monkeys, we developed a task that uses the same win-shift strategy as the DNMS, the delayed recognition span task (DRST, *[33,34]*). This task employs a nonmatching paradigm but requires the monkey to identify a novel stimulus among an increasing array of previously presented stimuli using either spatial or nonspatial stimulus cues. The task has been used with monkeys to assess memory function following damage to the hippocampus in infants *(34,35)* or adults *(36)*, and, in humans, has been used to differentiate patterns of memory dysfunction in a variety of neurologic disorders *(37–39)*. In addition, it has been used for assessing memory function in normal human aging *(40)*. The task can be administered using different classes of stimulus material to help characterize recognition memory deficits across several stimulus domains.

Relative to young adults, early senescent monkeys were significantly impaired on both the spatial and object component of the DRST. Moreover, a linear relationship was observed between age and degree of impairment. In contrast to their performance on the spatial and object conditions of the DRST, early senescent monkeys, as a group, were not significantly impaired on the color condition. This finding was somewhat surprising because one would have expected similar results in both nonspatial conditions (i.e., color and object). The results obtained suggest that different mechanisms may be involved in processing color information.

Monkeys of advanced age administered the DRST obtained lower recognition spans across both spatial and color stimulus classes than did young adult monkeys. On average, the monkeys in the advanced aged group retained only two thirds of the amount of information in memory as that retained by the monkeys in the young adult group *(41)*.

We found that the oldest old monkeys were impaired relative to young adults on the spatial and object conditions of the DRST. However, their mean span did not differ significantly from that of monkeys of advanced age. These findings would suggest that age-related decline in recognition span may not change beyond advanced age.

One aspect of the DRST to be considered is the series of repeated trials that were embedded within each condition of the task. Both repeated and unique sequences were included in the delayed recognition span test as a means of assessing the ability to learn and recall different types of information under the stimulus conditions within the same task. Hebb *(42)* found that in humans, the recall of an arbitrary sequence of highly familiar information, such as digits, improved when the same series was embedded repeatedly within presentations of random sequences. Similarly, we found that both the aged and the young adult monkeys in our program performed better on an embedded repeated sequence than they did on the unique sequences although the performance level of the aged animals was still significantly lower than that of young adults. These findings suggest that aged monkeys, like young adults, can benefit from repetition, even though the improvement is modest.

We have also had the opportunity to longitudinally assess recognition memory function in a different cohort of young and aged monkeys. We tested three young adult (5–7 yr) and four monkeys of advanced age (23–26 yr) 4 yr following initial assessment on the spatial recognition span test. The pattern of findings in the initial assessment revealed that the aged animals obtained significantly lower recognition span than that of the young adults. Four years later, two of the three animals in the young adult group showed a slight decline in performance, whereas one showed a slight improvement. In the aged group, a more striking dichotomous performance was noted. As in the young adult group, two animals showed very slight decline, and the remaining two showed marked decline which, in one case, was severe. We were intrigued by two aspects of the findings. First, of the seven animals tested, only one showed a severe decline in function over the 4 yr interval. Second, and perhaps more important, was the fact that there was minimal loss of function in two of the aged animals, including one that was 25 yr of age at the initial testing.

An error analysis revealed that the two markedly impaired aged monkeys, unlike monkeys that did not evidence a significant decline over the 4 yr, tended to make a greater proportion of their errors by selecting the most recently presented disk. Whether this effect represents enhanced perseverative tendencies, or is related more to serial order effects, remains to be evaluated.

In summary, longitudinal assessment of recognition memory in the rhesus monkey reveals that in aged animals, unlike in young adults, performance is quite variable. Some animals clearly evidence marked decline while others remain virtually unchanged, even after a period of 4 yr. It appears therefore, that like aged humans, aged nonhuman primates are characterized by marked individual differences. Although it is important to understand the neurobiological basis for the marked decline observed in some animals, it is perhaps more important to understand why some animals appear to age successfully.

In view of these differential age effects on learning and memory function, it is tempting to conclude that age-related cognitive declines in learning and memory function result from two separate processes. Thus, the capacity for short-term storage of information, although impaired with aging, reaches asymptotic decline in advanced age. In contrast, it appears that the capacity for learning new rules may undergo progressive decline throughout the latter part of the life-span. Whether this decline in learning new rules extends to other forms of learning (i.e., associative learning) remains to be determined. Further, it is unclear whether the differential effects of aging on new learning and memory load or storage can be functionally dissociated in the pathogenesis of normal aging.

3.2.5. Delayed Response

The delayed response task (DR) is a classic behavioral task known to be sensitive to frontal cortical system damage, as first demonstrated in the 1930s with Jacobson's *(43)* studies in the chimpanzee. Over the last 50 yr, several investigators have confirmed and extended these early findings in a variety of species, including the monkey *(44–49)*. Battig et al. *(48)* and later Goldman and Rosvold *(50)* demonstrated an impairment on DR in monkeys with lesions of the caudate nucleus. Similarly, an impairment on DR was found following lesions of the hippocampal formation *(51)*, but as reported in the same study, did not extend to lesions of the amygdala. Perhaps the most robust effect on DR performance is that following damage to the prefrontal cortex. Goldman and Rosvold *(44)* behaviorally fractionated the frontal cortical regions and identified the dorsolateral surface as a critical locus. This includes the same region in which Peters et al. *(52)* have identified a significant correlation between thinning of layer 1 cortex and cognitive impairment in aged monkeys. In humans, the DR task has been used to elucidate the nature of memory dysfunction in patients with alcoholic Korsakoff's syndrome *(53)* and schizophrenia *(54–56)* along with the patient's relatives *(57)*. Of particular interest, deficits on the DR task have been recently reported in intermediate aged adults *(53)* but little has been reported in more advanced aged subjects.

In aged monkeys, the DR task has been used by several investigators *(15,28,58–63)* to characterize memory dysfunction. Taken together, the findings indicate that aged monkeys evidence increasing degrees of impairment with longer delays, although in many instances they are impaired even on delays as short as 5 s *(58,59)*. Further, aged monkeys appear to be disproportionately affected by increased delays, a pattern that is in contrast to that seen on the DNMS task (e.g., *28–30*). This suggests that the DR task may be more sensitive to age-related cognitive decline raising the possibility that frontal cortical systems may be disproportionately vulnerable to the aging process.

3.3. Executive System Function

Although there exists an impressive literature on age-related decline in memory with normal aging and age-related disease, the number of studies aimed at the assessment of executive function is more limited. Executive system functions are a set of integrative cognitive skills that are vital to performing higher level activities of daily living. They encompass abilities such as cognitive flexibility, cognitive tracking, set maintenance, abstraction, divided attention, insight, and social judgment. Cognitive flexibility refers to the capacity to change a response pattern with changing reinforcement contingencies. Cognitive tracking refers to the capacity to keep track of correctly identify correct response–reward associations among frequently alternating contingencies. Set maintenance refers to the ability to maintain a given mode of responding. Abstraction refers to the capacity to identify a common element among stimuli that appear to differ along several dimensions. Divided attention is a term used to describe the capacity to attend to two or more stimulus sources simultaneously. In large measure, many of these functions share the requirement of inhibiting interfering responses.

In normal aged humans, a variety of studies have been reported showing impairments on tasks of executive system function *(64–66)*. One of the most consistent findings is the decrement in performance on tasks of divided attention *(68–71)*. Studies in normal adults have also reported age-related deficits in abstraction *(72,73)*.

Evidence from the neuropsychological, neuropathological, and neurological literature suggests that the executive functions are mediated by the prefrontal cortical areas *(74–78)*. However, the relationship of the age-related decline in this cognitive domain to specific neuropathological and neurochemical alterations remains unclear.

In nonhuman primates, studies of frontal lobe function date back to the 1930s, with Jacobson's *(43)* classical studies of the delayed response task in the chimpanzee. Over the last 50 yr, several investigators have confirmed and extended these early findings in a variety of species, including the monkey *(45–47,49,50)*. Three tests that were found to be sensitive, but not specific to lateral prefrontal cortical damage, were the delayed response task, a test of visuospatial memory *(48)*, the conditional position response test, a task of conditional learning *(44)*, and place or spatial reversal learning, a test of cognitive flexibility *(79,80)*. Of these three tasks, the reversal learning test appears to be the most sensitive in discriminating between various types of cortical damage *(79)*.

Only a limited number of studies have been performed in aged monkeys using reversal learning tasks and have generated what appear to be contradictory findings. Bartus et al. *(19)* reported an age-related impairment on both pattern and color reversals, while Rapp *(18)* reported no impairment on either a pattern or object reversal. We reported, in a preliminary study *(14)*, an age-related impairment on spatial but not object reversals.

While these results seem confusing at first glance, two factors stand out that may account for these findings. The first is task complexity. Bartus et al. *(19)* used more difficult tasks than Rapp *(18)* and Lai et al. *(14)*. The second factor is the age of the monkeys being studied. Bartus et al. *(19)* examined monkeys that were 18–22 yr of age, Rapp *(18)* examined monkeys without known birth dates that were thought to be about 23–27 yr of age, while Lai et al. *(14)* examined monkeys with known birth dates that were 20–28 yr of age. Because it is quite clear from the work of Bachevalier et al. *(16)*, and from the previous work we have presented in this chapter, that monkeys over the age of 19 are not a homogeneous group, it may be that individual differences account for differences in the results on tests of executive function.

We have compared the performance of early senescent, advanced age, and oldest old monkeys to young adult animals on spatial and object reversals *(14*, and Moss et al., *unpublished data)*. For spatial reversals, in increasing severity, early senescent, advanced aged, and oldest old monkeys were impaired on the task. Early senescent monkeys were impaired only on the first reversal, and the oldest old were impaired on all three reversals. For object reversal learning, only the monkeys in the oldest old group evidenced a significant impairment relative to young adults.

Taken together with results of previous reports, these findings suggest that aging is associated with a progressive impairment in executive system function in the nonhuman primate, a finding that parallels that of normal human aging.

3.4. Motor Skills

Motor skills are a complex set of activities encompassing voluntary and involuntary regulation, integration, proprioception, dexterity, and strength of both gross and fine muscle groups. They are evident both actively in all responses of an organism and passively in the posture or appearance of the organism.

Motor skills govern a person's capacity to adapt, change, control, and respond to his or her surroundings. Impairments of motor skills have the tendency to interfere with, or alter everyday life. These types of impairments have been described in a variety of ways, such as paralysis, apraxia, ataxia, motor weakness, and motor slowing *(66)*.

In normal aged humans, studies have been conducted on a wide variety of motor skills, ranging from the ability to track movement with a finger *(81)* to functional abilities such as brushing teeth *(82)*, to even highly complex skills such as playing golf *(83)*. Overall, these studies and many others have found a decline in both speed and accuracy with age *(84–88)*.

Many investigators have also noted an increased amount of variability in the data of the aged human subjects. This may in part be due to the varying degree of physical fitness. It has been established that healthy adults, particularly those deemed fit in terms of cardiovascular health, have faster reaction times than those who were sedentary or unfit *(89–92)*. It has even been reported that some aged individuals show little to no loss of motor function when compared to the performance of young adult subjects; however, in a cross-sectional study of this type *(93)*, it is possible that aged subjects may have been inadvertently selected in a biased fashion which would have tended to exclude impaired individuals.

With respect to aged monkeys, only a limited number of studies have been performed. Two of these studies measured reaction time while the monkeys were performing a memory task *(10,16)* and found that monkeys, like human subjects, evidenced a significant increase in reaction time with age. A third study *(13)*, designed to assess the most expeditious response made by the monkey, failed to find an effect of aging. These results suggest that there may be an overall motor slowing with age in the monkey, but that the capacity to muster a rapid response on demand remains relatively intact.

Our studies of motor system function in the aged monkey have focused on tasks of fine motor control and motor skill. For example, we assessed the performance of young adult monkeys, early senescent monkeys, advanced aged monkeys, and oldest old monkeys on both types of tasks *(94)*. We used the task developed by Kuypers *(95)* to assess fine motor control. The apparatus for this task is a testing board with a series of wells of decreasing size. The monkey is simply required to obtain the food reward from each well. We have also used a motor skill task *(96)* to assess the performance of our aged monkeys. This task required the monkey to guide a lifesaver candy along a bent wire to obtain the candy as a reward within 1 min. Response latency, errors, and success rate were recorded. Overall, we have not observed any significant group differences in performance on either task. However, a few individual monkeys in the oldest old group performed worse on both tasks than any of the younger monkeys in terms of response speed and number of errors.

A more thorough assessment of motor skill was conducted by another group of investigators *(97)*. They used the same lifesaver paradigm but utilized more difficult combinations of bends in the wire. They found no significant differences in performance on relatively easy portions of the task. However, they found a significant group difference in performance on this task with monkeys in the early senescent, advanced aged, and oldest old aged ranges requiring significantly more time to solve the difficult portion of the task than monkeys of young adult age.

Taken together, the results of these studies suggest that in the monkey, as in the human, there is an age-related decline in response speed and accuracy. This decline can be mitigated with exercise, practice, and optimal general health. However, as the tasks become more complex, the deleterious effects of age become more pronounced.

4. NEUROBIOLOGICAL CONSIDERATIONS

The neurobiological basis of age-related cognitive dysfunction is not yet well understood, but has become an increasing focus of study in recent years. By relating the pattern of behavioral performance with findings from neurobiological studies of aged monkeys, several authors have identified a variety of changes that may play a role. Structures or neural systems that have been implicated in these deficits include the medial temporal lobe/hippocampal system *(27,29)* and prefrontal cortices *(14,52,98,99)*.

With regard to the hippocampal formation, previous studies have identified error patterns on memory tasks in aged monkeys that parallel those seen in lesion studies in young adult monkeys. Rapp and Amaral *(27)* identified a subset of animals that were disproportionately impaired on the delay and list conditions of the delayed nonmatching-to-sample task, a pattern that is consistent with medial temporal lobe damage in young monkeys *(16,51,100–106)*. In an earlier study *(29)*, we identified a pattern of errors on the list condition of the DNMS by aged monkeys that is also characteristic of monkeys with hippocampal damage *(105)*.

It is tempting to consider aging as a mild functional lesion of the hippocampus, a notion that has been advanced by several investigators in the field *(107,109)*. In this regard, however, it is of interest that neuron loss does not appear to occur in the hippocampal formation in the aged monkey. At least two groups *(104,110)* have provided initial evidence that no age-related cell loss occurs in the hippocampal CA1 subfield or subiculum of the monkey, and recent evidence suggests only a selective and mild loss of synapses in the dentate gyrus *(111)*. Similarly, Amaral *(112)* has reported no cell loss in the hippocampus or entorhinal cortex. Yet West *(113)* has reported an age-related neuronal loss in subfield CA4 and the subiculum in the hippocampus of normal aged humans. Thus, it remains to be determined to what extent the hippocampus undergoes age-related cell loss in primates.

It is possible that more subtle "sublethal" changes may be occurring in the hippocampus that do not fully manifest the same degree of impairment as that found following near total removal of this structure. For example, recent work in our laboratory has shown that the hippocampus in monkeys undergoes volumetric loss with age and, moreover, this change appears related to cognitive decline *(114)*. When combined with the finding of Tigges et al. *(111)* that the number of synapses in the hippocampus do not change with age, one is left to conclude the volumetric loss is attributable to other changes in cell morphology such as shrinkage, dendritic regression or, as has been suggested recently, damage to and loss of white matter *(115)*.

The growing body of evidence of impairment on tasks of spatial memory and those requiring retention of temporal order *(28)* implicates the prefrontal cortex as another major neuropathological locus in aged monkeys, a view advanced earlier by Bartus and his colleagues *(25,98)*. Adding support to this notion are findings that aged monkeys are impaired on tasks of executive system function *(14,116)*, a cognitive domain that includes abilities likely mediated by prefrontal cortical areas *(100,117)*. Similarly, the impaired performance that aged monkeys show on delayed response *(28,58,59, 61–63,117)* and reversal learning *(15)* tasks have been historically associated with prefrontal cortical damage both in monkeys and humans *(44,48,118,119)*. Consistent with this view, an analysis of the errors made by early and advanced aged monkeys demonstrated significantly more perseverative errors on the DRST and reversal tasks among early and advanced aged monkeys than among young adults. However, the extent to which frontal cortical system pathology may underlie the deficits observed in aged monkeys remains unclear.

Studies are converging to suggest that there is no significant age-related loss of neurons in area 46 of the prefrontal cortex *(52)* and area 4 of the frontal cortex *(120)* as well as in other cortical regions *(109,110,112,121,122)*. Although neurons in area 46 of the prefrontal cortex do not undergo age-related cell loss, the white matter in this region has been shown to undergo marked alteration with age *(52)*. Moreover, a recent study from these same investigators revealed a loss of synapses and a reduction in thickness of layer I in area 46 in a group of aged rhesus monkeys as compared to group of young adults *(99)*.

It has been noted that when there is formation of amyloid plaques with aging, there is a greater likelihood of them developing in the prefrontal and temporal cortices than in more caudal regions of the brain *(108)*. Nevertheless, it appears there is no relationship between cognitive impairment and plaque density *(123)*, suggesting that age-related accumulation of amyloid in plaques is not a significant pathogenic feature of normal aging in the monkey. Support for this supposition has recently been obtained in a study of amyloid plaques in the monkey which indicates that the monkey mainly has the less toxic Aβ 1–40 in contrast to humans where the more toxic Aβ 1–42 predominates *(124)*.

Despite evidence of preserved number of neurons in prefrontal cortex, reduction in endogenous concentrations of dopamine in this area has been reported *(117)*. Loss of serotonin binding sites has also been reported in both the frontal cortex and hippocampus with age *(125)*. This is particularly interesting, as neuronal loss has been demonstrated in the brainstem nucleus raphe dorsalis and nucleus centralis superior of aged monkeys *(126)*, both of which are serotonergic and project mainly to the forebrain. Additional evidence of subcortical neuron loss within the cholinergic basal forebrain nuclei is conflicting. Rosene *(109)* has reported an age-related decline in the number of AChE-positive neurons in all three major basal forebrain nuclei, the medial septum, diagonal band, and the nucleus basalis, and Stroessner-Johnson et al. *(127)* reported age-related loss of ChAT-positive neurons in the medial septum. However, Voytko et al. *(128)* has recently reported no age-related cell loss in the nucleus basalis using an antibody to the p75 low-affinity nerve growth factor receptor. Clearly, the possible contribution of neuron loss in cortically projecting subcortical sites to age-related cognitive decline needs further investigation.

One intriguing change that has recently been noted in both humans and monkeys is an apparent loss of white matter, but not gray matter, that occurs with age *(15, 115,129)*. This is supported by a recent neuropathological study of area 46 in aged monkeys that confirms the presence of myelin degeneration in the deep layers of cortex and underlying white matter *(52)*. There has also been some preliminary indication that loss of white matter volume determined by MRI may correlate positively and significantly with cognitive decline *(15)*. Given the nature of the cognitive deficits found with aging, it is likely that there is a disruption of the functional integrity of prefrontal cortical systems and medial temporal lobe structures such as the hippocampus. This is made more intriguing in the context of myelin degeneration which would compromise communication within and between both areas.

In conclusion, the results of the present investigation add to the growing body of evidence that rhesus monkeys undergo cognitive declines that begin at a relatively early age, a finding that is in parallel with observations made in normal human aging *(26)*. Initial findings from anatomical studies raise the possibility that the bases for this decline may be related more to a deterioration in the efficiency and quality of cortical information processing due to dystrophic myelin or altered neuronal function, rather than to cortical neuronal cell loss.

5. CONCLUSIONS

In conclusion, evidence has accumulated to show that aged monkeys, like normal aged humans, evidence age-related cognitive decline and that the pattern of this decline changes with age. Further, as is the case in humans, the changes that occur with normal aging do not appear to be universal. Some monkeys appear to age quite successfully, while others of the same chronological age appear to age unsuccessfully. One of the primary challenges of research on the effects of normal aging is to determine the neurobiological basis of these differing patterns of age-related change.

ACKNOWLEDGMENTS

This research was supported by NIH Grants PO1-AG00001 and R55-AG12610 from the National Institute on Aging and by NIH Grant RR-00165 from the National Center for Research Resources to the Yerkes Regional Primate Research Center. Boston University School of Medicine and the Yerkes Regional Primate Research Center are both fully accredited by the American Association for Accreditation of Animal Laboratory Care. The authors wish to thank our colleagues, Drs. Douglas Rosene, Alan Peters, Tom Kemper, Julie Sandell, Johannes Tigges, Carmella Abraham, and Bradley Hyman, for their contributions to this body of work, and Jessica Berman, Beverly Duryea, Claudia Fitzgerald, and Tara Moore for their valuable assistance with this project.

REFERENCES

1. Peters, A., Rosene, D. L., Moss, M. B., Kemper, T. L., Abraham, C. R., Tigges, J., and Albert, M. S. (1996) Neurobiological bases of age-related cognitive decline in the rhesus monkey. *J. Neuropathol. Exp. Neurol.* **55,** 861–874.
2. Tigges, J., Gordon, T. P., McClure, H. M., Hall, E. C., and Peters, A. (1988) Survival rate and life span of rhesus monkeys at the Yerkes Regional Primate Research Center. *Am. J. Primatol.* **15,** 263–273.
3. Posner, M. I. and Peterson, S. E. (1990) The attention system of the human brain, in *Annual Review of Neuroscience*, Vol. 13, Annual Reviews Inc, Palo Alto, CA, pp. 25–42.
4. Parasuraman, R. and Davies, R. (1984) *Varieties of Attention*, Academic Press, New York.
5. Hasher, L. and Zacks, R. (1979) Automatic and effortful processes in memory. *J. Exp. Psychol.* **108,** 356–388.
6. James, W. (1890) *Principles of Psychology*, Vol. 1, Holt, Rhinehart, and Winston, New York.
7. Sperry, R. L. (1988) Psychology's mentalist paradigm and the religion/science tension. *Am. Psychol.* **43,** 607–613.
8. Greenwood, P. M., Parasuraman, R., and Hazby, J. V. (1993) Changes in visuospatial attention over the adult lifespan. *Neuropsychologia* **31,** 471–485.
9. Hartly, A. A., Kieley, J. M., and Slabach, E. H. (1990) Age differences and similarities in the effects of cues and prompts. *J. Exp. Psychol. Hum. Percept. Perform.* **16,** 523–537.
10. Davis, R. T. (1978) Old monkey behavior. *Exp. Gerontol.* **13,** 237–250.
11. Posner, M. I. (1988) Structures and functions of selective attention, in *Master Lectures in Clinical Neuropsychology* (Boll, T. and Bryant, B., eds.), American Psychological Association, Washington, DC, pp. 173–202.
12. Posner, M. I. and Cohen, Y. (1984) Components of performance, in *Attention and Performance* (Bouma, H. and Bowhuis, D. eds), Lawrence L. Erlbaum, Hillsdale, NJ, pp. 531–556.
13. Baxter, M. G. and Voytko, M. L. (1996) Spatial orienting of attention in adult and aged rhesus monkeys. *Behav. Neurosci.* **110,** 898–904.
14. Lai, Z. C., Moss, M. B., Killiany, R. J., Rosene, D. L., and Herndon, J. G., (1995) Executive system dysfunction in the aged monkey: spatial and object reversal learning. *Neurobiol. Aging* **16,** 947–995.
15. Lai, Z. C., Rosene, D. L., Killiany, R. J., Pugliese, D., Albert, M. S., and Moss, M. B. (1995) Age-related changes in the brain of the rhesus monkey: MRI changes in white matter but not gray matter. *Soc. Neurosci. Abstr.* **21,** 156.

16. Bachevalier, J., Landis, L. S., Walker, L. C., Brickson, M., Mishkin, M., Price, D. L., and Cork, L. C. (1991) Aged monkeys exhibit behavioral deficits indicative of widespread cerebral dysfunction. *Neurobiol. Aging* **12**, 99–111.
17. Arnsten, A. F. T. and Goldman-Rakic, P. S. (1990) Analysis of alpha-2 adrenergic agonist effects on the delayed non-matching to sample performance of aged rhesus monkeys. *Neurobiol. Aging* **11**, 583–590.
18. Rapp, P. R. (1990) Visual discrimination and reversal learning in the aged monkey *(Macaca mulatta). Behav. Neurosci.* **104**, 876–888.
19. Bartus, R. T., Dean, R. L., and Fleming D. L. (1979) Aging in the rhesus monkey: effects on visual discrimination learning and reversal learning. *J. Gerontol.* **34**, 209–219.
20. Voytko, M. L. (1993) Cognitive changes during normal aging in monkeys assessed with an automated test apparatus. *Neurobiol. Aging* **14**, 643–644.
21. Malamut, B. L, Saunders, R. C., and Mishkin, M. (1984) Monkeys with combined amygdalo-hippocampal lesions succeed in object discrimination learning despite 24-hour intertrial intervals. *Behav. Neurosci.* **98**, 759–769.
22. Moss, M. B., Mahut, H., and Zola-Morgan, S. (1981) Concurrent discrimination learning of monkeys after hippocampal, entorhinal or fornix lesions. *J. Neurosci.* **1**, 227–240.
23. Cowey, A. and Gross, C. G. (1970) Effects of foveal prestriate and inferotemporal lesionas on visual discrimination by rhesus monkeys. *Exp. Brain Res.* **11**, 128–144.
24. Oscar-Berman, M. and Zola-Morgan, S. (1980) Comparative neuropsychology and Korsakoff's syndrome. II - Two-choice visual discrimination learning. *Neuropsychologia* **18**, 513–525,
25. Bartus, R. T. and Dean, R. L. (1979) Recent memory in aged non-human primates: hypersensitivity to visual interference during retention. *Exp. Aging Res.* **5**, 385–400.
26. Albert, M. S. and Moss, M. B. (1996) Neuropsychology of aging: findings in humans and monkeys, in *Handbook of the Biology of Aging*, 4th edit. (Schneider, E. L., Rowe, J. W.; eds. Morrison, J. H.; Section ed.), Academic Press, San Diego, pp. 217–233.
27. Rapp, P. R. and Amaral, D. G. (1991) Recognition memory deficits in a subpopulation of aged monkeys resemble the effects of medial temporal lobe damage. *Neurobiol. Aging* **12**, 481–486.
28. Rapp, P. R. and Amaral, D. G. (1989) Evidence for a task-dependent memory dysfunction in the aged monkey. *J. Neurosci.* **9**, 3568–3576.
29. Moss, M. B., Rosene, D. L., and Peters, A. (1988) Effects of aging on visual recognition memory in the rhesus monkey. *Neurobiol. Aging* **9**, 495–502.
30. Presty, S. K., Bachevalier. J., Walker, L. C., Struble, R. G., Price, D. L., Mishkin, M., and Cork, L. C. (1987) Age differences in recognition memory of the rhesus monkey *(Macaca mulatta). Neurobiol. Aging* **8**, 435–440.
31. Killiany, R. J., Moss, M. B., Rosene, D. L., Herndon, J., and Lai, Z. C., Recognition memory function in early senescent rhesus monkeys, *Psychobiology*, in press.
32. Howieson, D. B., Holm, L. A., Kaye, J. A., Oken, B. S., and Howieson, J. (1993) Neurologic function in the optimally healthy oldest old. *Neurology* **43**, 1882–1886.
33. Moss, M. B. (1983) Assessment of memory in amnesic and dementia patients: adaptation of behavioral tests used with non-human primates. *INS Bull.* **3**, 10–17.
34. Rehbein, L. and Mahut, H. (1983) Long-term deficits in associative and spatial recognition memory after early hippocampal damage in monkeys. *Soc. Neurosci. Abstr.* **9**, 639.
35. Killiany, R. J. and Mahut, H. (1991) Ontogenetic development of recognition memory span in rhesus macaques. *IBRO Abstr*, **165**.
36. Beason, L., Moss, M. B., and Rosene, D. L. (1990) Effects of entorhinal, parahippocampal, or basal forebrain lesions on recognition memory in the monkey. *Soc. Neurosci. Abstr.* **16**.
37. Lange, K. W., Robbins, T. W., Marsden, C. D., James, M., Owen, A. M., and Paul., G. M. (1992) L-dopa withdrawl in Parkinson's disease selectively impairs cognitive performance in tests sensitive to frontal lobe function. *Psychopharmacology*, **107**, 394–400.

38. Salmon, D. P., Granholm, E., McCullough, D., Butters, N., and Grant, I. (1989) Recognition memory span in mild and moderately demented patients with Alzheimer's disease. *J. Clin. Exp. Neuropsychiatry* **4**, 429–443.

39. Moss, M. B., Albert, M. S., Butters, N., and Payne, M. (1986) Differential patterns of memory loss among patients with Alzheimer's disease, Huntington's disease, and alcoholic Korsakoff's syndrome. *Arch. Neurol.* **43**, 239–246.

40. Inouye, S. K., Albert, M. S., Mohs, R., Sun, K., and Berkman, L. F. (1993) Cognitive performance in a high-functioning community-dwelling elderly population. *J. Gerontol.* **48**, 146–151.

41. Moss, M. B., Killiany, R. J., Lai, Z. C., Rosene, D. L., and Herndon, J. G. (1997) Recognition memory span in rhesus monkeys of advanced age. *Neurobiol. Aging* **18**, 13–19.

42. Hebb, D. O. (1961) Distinctive features of learning in the higher animal, in *Brain Mechanisms and Learning* (Delafresnay, J. F., ed.), Blackwell, Oxford, pp. 58–72.

43. Jacobson, C. F. (1935) Functions of the frontal association are in primates. *Arch. Neurol. Psychiatry* **33**, 558–569.

44. Goldman, P. S. and Rosvold, H. E. (1970) Localization of function within the dorsolateral prefrontal cortex of the rhesus monkey. *Exp Neurol.* **27**, 291–304.

45. Iversen, S. D. (1970) Interference and inferotemporal memory deficits. *Brain Res.* **19**, 277–289.

46. Butter, C. M., Mishkin, M., and Mirsky, A. F., (1968) Emotional responses toward humans in monkeys with selective frontal lesions. *Physiol. Behav.* **3**, 213–215.

47. Gross, C. G. (1963) Comparison of the effects of partial and total lateral frontal lesions on test performance by monkeys. *J. Comp. Physiol. Psychol.* **56**, 41–47.

48. Battig, K., Rosvold, H. E., and Mishkin, M. (1960) Comparison of the effects of frontal and caudate lesions on delayed response and alternation in monkeys. *J. Comp. Physiol. Psychol.* **53**, 400–440.

49. Mishkin, M. (1957) Effects of small frontal lesions on delayed alternation in monkeys. *J. Neurophysiol*, **20**, 615–622.

50. Goldman, P. S. and Rosvold, H. E. (1972) The effects of selective caudate lesions in infant and juvenile rhesus monkeys. *Brain Res.* **43**, 53–66.

51. Zola-Morgan, S., Squire, L. R., and Amaral, D. G. (1989) Lesions of the hippocampal formation but not lesions of the fornix or the mammillary nuclei produce long-lasting memory impairment in monkeys. *J. Neurosci.* **9**, 898–913.

52. Peters, A., Gruia-Leahu, D., Moss, M. B., and McNally, K. (1994) The effects of aging on area 46 of the frontal cortex of the rhesus monkey. *Cereb. Cortex* **6**, 621–635.

53. Oscar-Berman, M., Hutner, N., and Bonner, T. R. (1992) Visual and auditory spatial and nonspatial delayed-response performance by Korsakoff and non-Korsakoff alcoholic and aging individuals. *Behav. Neurosci.* **106**, 613–622.

54. Park, D., Holzman, P. S., and Lenzenweger, M. F. (1995a) Individual differences in spatial working memory in relation to schizophrenia. *J. Abnor. Psychol.* **104**, 355–363.

55. Park, S. and Holzman, P. S. (1992) Schizopherenics show spatial working memory deficits. *Arch. Gen. Psychiatry* **49**, 975–982.

56. Raine, A., Lencz, T., Reynolds, G. P., Harrison, G., Sheard, C., Medley, I., Reynolds, L. M., and Cooper, J. E. (1992) An evaluation of structural and functional prefrontal deficits in schizophrenia: MRI and neuropsychological measures. *Psychiary Res.* **45**, 123–137.

57. Park, D., Holzman, P. S., and Goldman-Rakic, P. S. (1995b) Spatial working memory deficits in the relatives of schizophrenics. *Arch. Gen. Psychiatry* **52**, 821–828.

58. Walker, L. C., Kitt, C. A., Struble, R. G., Wagster, M. V., Price, D. L., and Cork, L. C. (1988) The neural basis of memory decline in aged monkeys. *Neurobiol. Aging* **9**, 657–666.

59. Arnsten, A. F. T. and Goldman-Rakic, P. S. (1985) a-2 adrenergic mechanisms in prefrontal cortex associated with cognitive decline in aged nonhuman primates. *Science* **230,** 1273–1276.
60. Davis, R. T., Bennet, C. L., and Weisenburger, R. P. (1982) Repeated measurements of forgetting by rhesus monkeys *(Macaca mulatta). Percept. Mot. Skills* **55,** 703–709.
61. Marriott, J. G. and Abelson, J. S. (1980) Age differences in short-term memory of test-sophisticated rhesus monkeys. *Age* **3,** 7–9.
62. Bartus, R. T., Fleming, D., and Johnson, H. R. (1978) Aging in the rhesus monkey: Debilitating effects on short-term memory. *J. Gerontol.* **34,** 209–219.
63. Medin, D. L., and Davis., R. T. (1974), in *Behavior of Non-Human Primates* (Shrier, A. M. and Stollmitz, F., eds.), Academic Press, New York, pp. 1–47.
64. Albert, M. S.,Wolfe J., and Lafleche, G. (1990) Differences in abstraction ability with age. *Psychol. Aging* **5,** 94–100.
65. Albert, M. S., Duffy, F. H., and Naeser, M. A. (1987) Nonlinear changes in cognition and their neurophysiological correlates. *Can. J. Psychiatry* **41,** 141–157.
66. Lezak, M. D. (1995) *Neuropsychological Assessment,* 3rd edit., Oxford University Press, New York.
67. Talland, G. A. (1961) Effects of aging on the formation of sequential and spatial concepts. *Percept. Mot. Skills* **13,** 210.
68. Craik, F. I. M. and Simon, E. (1980) Age differences in memory: the role of attention and depth of processing, in *New Directions in Memory and Aging* (Poon, L. W., Fozard, J. L., Cermak, L. S., and Thompson, L. W., eds.), Lawrence L. Erlbaum, Hillsdale, NJ.
69. Craik, F. I. M. (1971) Age differences in recognition memory. *Q. J. Exp. Psychol.* **23,** 316–323.
70. Brinley, J. and Fichter, J. (1970) Performance deficits in the elderly in relation to memory load and set. *Gerontology* **25,** 30–35.
71. Broadbent, D. E. and Gregory, M. (1965) Some confirmatory results on age differences in memory for simultaneous stimulation. *Br. J. Psychol.* **56,** 77–80.
72. Mack, J. L. and Carlson, N. J. (1978) Conceptual deficits and aging: the category test. *Percept. Mot. Skills* **46,** 123–128.
73. Bromley, D. (1957) Effects of age on intellectual output. *J. Gerontol.* **12,** 318–323.
74. Shimamura, A. P., Janowsky, J. S., and Squire, L. R. (1991) What is the role of frontal lobe damage in memory disorders? in *Frontal Lobe Function and Dysfunction* (Levin, H. S., Eisenberg, H. M., and Benton, A. L., eds.), Oxford University Press, Oxford, pp. 173–198.
75. Stuss, D. T. (1991) Interference effects on memory function in posteukotomy patients: an attentional perspective, in *Frontal Lobe Function and Dysfunction* (Levin, H. S., Eisenberg, H. M., and Benton, A. L., eds.), Oxford University Press, Oxford, pp. 157–172.
76. Albert, M. S. (1988) Cognitive function, in *Geriatric Neuropsychology* (Albert, M. S. and Moss, M. B., eds.), The Guilford Press, New York, pp. 33–53.
77. Damasio, A. R. (1985) The frontal lobes, in *Clinical Neuropsychology* (Heilman, K. M. and Valenstein, E., eds.), Oxford University Press, New York, pp 339–376.
78. Stuss, D. T. and Benson, D. F. (1986) *The Frontal Lobes,* Raven Press, New York.
79. Pohl, W. G. (1973) Dissociation of spatial discrimination deficits following frontal and parietal lesions in monkey. *J. Comp. Physiol. Psychol.* **82,** 227–239.
80. Goldman, P. S., Rosvold, H. E., Vest, B., and Galkin, T. W. (1971) Analysis of the delayed-alternation deficit produced by dorsolateral prefrontal lesions in the rhesus monkey. *J. Comp. Physiol. Psychol.* **77,** 212–220.
81. Carey, J. R., Bogard, C. L., King, B. A., and Suman, V. J. (1994) Finger-movement tracking scores in healthy subjects. *Percpt. Mot. Skills* **79,** 563–576.
82. Felder, R., Reveal, M., Lemon, S., and Brown, C. (1994) Testing toothbrushing ability of elderly patients. *Spec. Care Dentist* **14,** 153–157.

83. Over, R. and Thomas, P. (1995) Age and skilled psychomotor performance: A comparison of younger and older golfers. *Int. J. Aging Hum. Dev.* **41,** 1–12.
84. Rabbit, P. M. A. (1979) How old and young subjects monitor and control responses for accuracy and speed. *Br. J. Psychol.* **70,** 305–311.
85. Fozard, J. L. and Thomas, J. C. (1975) Psychology of aging: basic findings and some psychiatric applications, in *Modern Perspectives in the Psychiatry of Old Age* (Howells, J. G., ed.), Brunner/Mazel, New York.
86. Loveless, N. E. and Sanford, A. J. (1974) Effects of age on contingent negative variation and preparatroy set in a reaction time task. *J. Gerontol.* **29,** 52–63.
87. Surwillo, W. W. (1968) Timing of behavior in senescence and the role of the central nervous stystem, in *Aging and Behavior* (Talland, G. A., ed.), Academic Press, New York.
88. Birren, J. E. and Botwinick, J. (1955) Speed of response as a function of perceptual difficulty and age. *J. Gerontol.* **10,** 433–436.
89. Light, K. C. (1978) Effects of mild cardiovascular and cerebrovascular disorders on serial reaction time performance. *Exp. Aging Res.* **4,** 3–22.
90. Spirduso, W. W. (1975) Reaction and movement time as a function of age and physical activity level. *J. Gerontol.* **35,** 850–865.
91. Abrahams, J. P. and Birren, B. E. (1973) Reaction time as a function of age and behavioral predisposition to coronary heart disease. *J. Gerontol.* **28,** 471–478.
92. Botwinick, J. and Thompson, L. W. (1968) Age differences in reation time: an artifact? *Gerontologist* **8,** 25–28.
93. Rowe, J. W. and Kahn, R. (1987) Human aging: useful vs. successful. *Science* **237,** 143–149.
94. Ronald, P. (1994) Motor skill learning, performance and handedness in aged rhesus monkeys. Unpublished Master's Thesis, Department of Anatomy, Boston University School of Medicine.
95. Lawrence, D. G. and Kuypers, H. G. J. M. (1968) The functional organization of the motor system in the monkey. I. The effects of bilateral pyramidal lesions. *Brain* **91,** 1–18.
96. Davis, R. T., McDowell, A. A., Deter, C. W., and Stelle, J. P. (1954) Performance of rhesus monkeys on selected laboratory tasks presented before and after a large single dose of whole-body X radiation. *Exp. Gerontol.* **4,** 20–26.
97. Brickson, M., Bachevalier, J., Watermeier, L. S., Walker, L. C., Struble, R. G., Price, D. L., Mihkin, M., and Cork, L. C. (1987) Performance of aged rhesus monkeys on a visuospatial task. *Soc. Neurosci. Abstr.* **13,** 1627.
98. Bartus, R. T. (1979) Effects of aging on visual memory, sensory processing and discrimination learning in a nonhuman primate, in *Sensory Systems and Communication in the Elderly.* (Ordy, J. M. and Brizzee, K., eds.), Raven Press, New York, pp. 137–145.
99. Peters, A. et al. (1998) The effect of aging on layer 1 in area 46 of prefrontal cortex, *Cereb. Cortex* **8,** 671–684.
100. Bachevalier, J. and Mishkin M. (1986) Visual recognition impairment follows ventromedial but not dorsolateral prefrontal lesions in monkeys. *Behav. Brain Res.* **20,** 249–261.
101. Murray, E. A. and Mishkin, M. (1986) Visual recognition in monkeys following rhinal cortical ablations combined with either amygdalectomy or hippocampectomy. *J. Neurosci.* **6,** 1991–2003.
102. Zola-Morgan, S. and Squire, L. R. (1985) Medial temporal lesions in monkeys impair memory on a variety of tasks sensitive to human amnesia. *Behav. Neurosci.* **99,** 22–38.
103. Zola-Morgan, S. and Squire, L. R. (1986) Memory impairment in monkeys following lesions limited to the hippocampus. *Behav. Neurosci.* **100,** 155–160.
104. Murray, E. A. and Mishkin, M. (1984) Severe tactual as well as visual memory deficits follow combined removal of the amygdala and hippocampus in monkeys. *J. Neurosci.* **4,** 2565–2580.

105. Mahut, H., Zola-Morgan, S., and Moss, M. B. (1982) Hippocampal resections impair associative learning and recognition memory in the monkey. *J. Neurosci.* **2,** 1214–1229.
106. Mishkin, M. (1978) Memory in monkeys severely impaired by combined but not by separate removal of amygdala and hippocampus. *Nature* **273,** 297–298.
107. Poon, L. W. (1985) Differences in human memory with aging, in *Handbook of the Psychology of Aging* (Dirren, J. E. and Schaie, K. W., eds.), Van Nostrand Reinhold, New York, pp. 427–462.
108. Struble, R. J., Price, D. L., Jr., Cork, L. C., and Price, D. L. (1985) Senile plaques in cortex of aged normal monkeys. *Brain Res.* **361,** 267–275.
109. Rosene, D. L. (1993) Comparing age-related changes in the basal forebrain and hippocampus of the rhesus monkey. *Neurobiol. Aging* **14,** 669–670.
110. West, M. J., Amaral, D. G., and Rapp, P. R. (1993) Preserved hippocampal cell number in aged monkeys with recognition memory deficits. *Soc. Neuroci. Abstr.* **19,** 599.
111. Tigges, J., Herndon, J. G., and Rosene, D. L. (1999) Preservation into old age of synaptic number and size in the supragranular layer of the dentate gyrus in rhesus monkeys. *Anat. Rec.,* in press.
112. Amaral, D. G. (1993) Morphological analyses of the brains of behaviorally characterized aged nonhuman primates. *Neurobiol. Aging* **14,** 671–672.
113. West, M. J. (1993) Regionally specific loss of neurons in the aging human hippocampus. *Neurobiol. Aging* **14,** 287–293.
114. Rosene, D. L, Moss, M. B., Berman, J. D., Williams, K. A., and Killiany, R. J. (1999) Preservation of neuron number and size in the hippocampal formation of the aging rhesus monkey. *Neurobiol. Aging,* submitted.
115. Rosene, D. L., Lai, Z. C., Killiany, R. J., Moss, M. B., Jolesz, F., and Albert, M. S. (1999) Age related loss of white matter with preservation of gray matter in the forebrain of the rhesus monkey—an MRI study. *Neurobiol. Aging,* submitted.
116. Bachevalier, J. (1993) Behavioral changes in aged rhesus monkeys. *Neurobiol. Aging* **14,** 619–621.
117. Goldman-Rakic, P. S. and Brown, R. M. (1981) Regional changes in monoamines in cerebral cortex and subcortical structures of aging rhesus monkeys. *Neuroscience* **6,** 177–178.
118. Oscar-Berman, M., McNamara, P., and Freedman, M. (1991) Delayed-response tasks: Parallels between experimental ablation studies and findings in patients with frontal lesions, in *Frontal Lobe Function and Dysfunction* (Levin, H. S., Eisenberg, H. M., and Benton, A. L., eds.), Oxford University Press, New York, pp. 230–255.
119. Milner, B. (1963) Effects of different brain lesions on card sorting. *Arch. Neurol.* **9,** 90–100.
120. Tigges, J., Herndon, J. G., and Peters, A. (1990) Neuronal population of area 4 during life span of the rhesus monkey. *Neurobiol. Aging* **11,** 201–208.
121. Peters, A., and Sethares, C. (1993) Aging and the Meynert cells in rhesus monkey primary visual cortex. *Anat. Rec.* **236,** 721–729.
122. Vincent, A. E., Peters, A., and Tigges, J. (1989) Effects of aging on the neurons within area 17 of rhesus monkey cerebral cortex. *Anat. Rec.* **223,** 329–341.
123. Sloane, J., Pietropaolo, M., Rosene, D. L., Moss, M. B., Peters, A., Kemper, T. L., and Abraham, C. R., In Press, Lack of correlation between plaque burden and cognition in the aged monkey.
124. Gearing, M., Tigges, J., Mori, H., and Mirra, S. S. (1996) Alpha beta is a major form of Beta-amyloid in nonhuman primates. *Neurobiol. Aging* **17,** 903–908.
125. Marcusson, J. O., Morgan, D. G., Winblad, B., and Finch, C. E. (1984) Serotonin-2 binding sites in human frontal cortex and hippocampus. Selective loss of S-2A sites with age. *Brain Res.* **311,** 51–56.
126. Kemper, T. L. (1993) The relationship of cerebral cortical changes to nuclei in the brainstem. *Neurobiol. Aging* **14,** 659–660.

127. Stroessner-Johnson, H. M., Rapp, P. R., and Amaral, D. G. (1992) Cholinergic cell loss and hypertrophy in the medial septal nucleus of the behaviorally characterized aged rhesus monkey. *J. Neurosci.* **12,** 1936–1944.
128. Voytko, M. L., Sukhov, R. C., Walker, L. C., Breckler, S. J., Price, D. L., and Koliatson V. E. (1995) Neuronal number and size are preserved in the nucleus basalis of aged rhesus monkeys. *Dementia* **6,** 131–141.
129. Guttman, C., Jolesz, F., Kikinis, R., Killiany, R. J., Moss, M. B., Sandor, T., and Albert, M. S. (1998) White matter and gray matter differences with age. *Neurology* **50,** 1283–1295.

3

Cholinergic Lesions as a Model of Alzheimer's Disease

Effects of Nerve Growth Factor

James M. Conner and Mark H. Tuszynski

1. INTRODUCTION

Alzheimer's disease (AD) is an age-related neurodegenerative disorder that currently affects 4 million people in the United States alone. It is a progressively debilitating condition that, in its later stages, often leaves victims without the cognitive abilities to care for themselves. The estimated direct and indirect costs of diagnosis, treatment, and long-term institutional care for AD patients is currently estimated to be in excess of $100 billion annually *(1)*. The emotional and social impact of the disease is immeasurable. With advances in general healthcare contributing to greater longevity, 8 million Americans are expected to develop AD by the first quarter of the 21st century unless parallel advances are made in the treatment of AD to alter the devastating course of this disease.

An obstacle to finding effective therapies for AD is the complexity and variability in the presentation of the disease. Pathologically, the presence of extracellular amyloid plaques and intraneuronal neurofibrillary tangles are considered a hallmark of AD *(2)*. However, plaques can occur in aging individuals without cognitive impairment, in densities within the criteria for diagnosis of AD *(3)*. In addition, the presence of neurofibrillary tangles is neither an ubiquitous *(4,5)* nor unique *(6–8)* feature of AD pathology. Moreover, the distribution and extent of lesions in AD pathology (including plaques, tangles, as well as cellular and synaptic loss) can vary widely across individuals. For example, whereas a loss of basal forebrain cholinergic neurons, particularly within the Ch4 region, is found in most AD cases, the degree of cholinergic cell loss in this region may vary substantially between individuals (from 30% to 95% *[9]*). For other ascending brain systems, such as serotonergic, dopaminergic, and noradrenergic cell populations, cell loss may vary from almost nonexistent (0–5%) to moderately severe (40–80%) *(9)*.

Variability in the pathology of AD gives rise to differences in the types and extent of cognitive deficits that occur in patients. For instance, some individuals with AD may have prominent language impairments while others have predominantly visuospatial deficiencies *(10)*, likely reflecting primary pathological damage to either the left or

From: Central Nervous System Diseases
Edited by: D. F. Emerich, R. L. Dean, III, and P. R. Sanberg © Humana Press Inc., Totowa, NJ

right cerebral hemisphere, respectively. In addition to the pathological variability in AD, differences in the etiology of the disease may exist. Supporting the idea that AD may derive from multiple origins are the roughly 20% of patients with a genetic predisposition to the disease and the smaller proportion of cases inherited in a strictly Mendelian fashion, whereby multiple defective genes have been implicated in the disease process including *presenilin-1*, *presenilin-2*, *APOE4*, and the amyloid precursor protein (see *[11]*). Just as no single gene is responsible for all identified forms of familial AD, it is unlikely that a single causative agent will be found accounting for all remaining "spontaneous" forms of AD. Evidence for such variability in AD etiology and pathology suggests that a single therapeutic approach is unlikely to be beneficial for all AD patients and that eventual therapies may need to target the multiple brain systems that degenerate in AD, or address a variety of different genetic abnormalities.

Often, a crucial step in making rapid progress toward finding therapies for a given disease is the identification of animal models accurately reproducing the etiology of the disease, as well as the disease's clinical and pathological manifestations. In the case of AD, good animal models have been difficult to come by. The full spectrum of biochemical and pathological abnormalities characteristic of AD have not been found to occur spontaneously in any animal species other than the human, and the complexity of the disease has made it difficult to generate animals with a full range of experimentally induced AD pathological alterations. Thus, animal models necessarily become directed toward reproducing specific subsets of abnormalities associated with AD. This chapter, and many of those that follow, point out particular animal models of AD and discuss how these models may be used in seeking clinical treatments for this disease.

2. THE INVOLVEMENT OF THE CHOLINERGIC SYSTEM IN AD

Many neuronal systems undergo degeneration in AD *(9)*, and the loss of a single system is unlikely to account for the broad spectrum of cognitive abnormalities seen in the disease. Various lines of evidence, however, have implicated the specific impairment of the basal forebrain cholinergic system as a primary component of age-related memory disturbances. Taken together, these arguments have come to make up what is known as the "cholinergic hypothesis." In brief, evidence supporting the cholinergic hypothesis are as follows: (1) drugs that block cholinergic function, via blockade of muscarinic cholinergic receptors in the brain, disrupt cognitive functions and produce memory and attention deficits similar to those seen in AD *(12,13)*; drugs that potentiate cholinergic function, such as acetylcholinesterase (AChE) inhibitors, can enhance cognitive performance and can reverse behavioral deficits induced by anticholinergics *(14,15)*; (2) cholinergic activity within the cortex and hippocampus of AD patients is markedly reduced when compared to age-matched controls *(16,17)*, yet the expression of muscarinic cholinergic receptors is not altered, indicating that the defective component of cholinergic transmission lies in the presynaptic element; and (3) further evidence that cholinergic dysfunction in AD arises from defects in the presynaptic element comes from studies demonstrating a marked reduction in the number of cholinergic cell bodies in the basal forebrain, the sole source of afferent cholinergic innervation to the cortex and hippocampal formation *(18–22)*. In addition, the degeneration of the basal forebrain system is one of the earliest and most consistent changes observed in the brains of AD patients, and the extent of cholinergic degeneration, especially within the nucleus basalis-to-cortex system, correlates with the clinical severity of the disease *(23)*, the density of amyloid plaques *(24,25)*, and the loss of synapses *(26)*. Based upon this evidence, animal models of central cholinergic damage can be useful for identifying compounds that augment cholinergic function, perhaps leading to a partial amelioration of the cognitive deficits associated with AD.

3. THERAPEUTIC MODULATION OF BASAL FOREBRAIN CHOLINERGIC FUNCTION

To the extent that dysfunction of the basal forebrain cholinergic system does contribute to the overall cognitive decline in AD, it can be predicted that reversing this dysfunction will ameliorate, at least partially, the cognitive impairment. To date, multiple strategies for enhancing cholinergic function within the CNS have been explored including: (1) pharmacological modulation of postsynaptic cholinergic receptors *(27–29)*; (2) pharmacological modulation of acetylcholine production, secretion, reuptake, or degradation *(30–33)*; (3) grafting acetylcholine-producing cells into the CNS environment (either fetal basal forebrain cholinergic neurons or genetically modified fibroblasts) *(34,35)*; and (4) pharmacological modulation of *endogenous* basal forebrain cholinergic neuron functional performance *(36–39)*.

When designing strategies for modulating cholinergic function it is important to bear in mind the nature of cholinergic communication between the basal forebrain and its cortical and hippocampal targets. The notion that a loss of basal forebrain cholinergic neurons can be functionally compensated for by systemically delivering muscarinic agonists or cholinesterase inhibitors presumes that cholinergic transmission takes places largely by tonically stimulating postsynaptic receptors. Experimental evidence would indicate that this simplified view of "tonic" cholinergic transmission is inaccurate (see *[40]* and following paragraph) and may account for the lack of a more substantial impact from the administration of systemic cholinergic enhancers on cognitive dysfunction in AD.

The activation of the basal forebrain system and target release of acetylcholine is not uniform, but appears to be both *stimulus-linked* and *phasic*. Activities involving arousal or attention, such as goal-directed, motivated behaviors, or exposure to novel stimuli, are known to specifically activate the basal forebrain system *(41–43)*. In addition, the synaptic communication between the basal forebrain and its targets is often phasic, and the temporal aspect of this synaptic activity may be essential for proper cholinergic function (such as bursting to induce theta activity in the hippocampus) *(44–48)*. Lastly, the actions of acetylcholine within basal forebrain targets is highly dependent upon the topography of transmitter release, both on a macroscopic and a microscopic level. Studies have indicated that acetylcholine release may differ significantly across cortical regions depending upon the nature of the behavior an animal is engaged in *(49)*. In addition, the postsynaptic actions of acetylcholine may be either stimulatory or inhibitory, depending upon the type of postsynaptic cell the neurotransmitter acts *(50)*, or the location on the postsynaptic cell where the acetylcholine acts (dendritic vs somal) *(51)*. Bearing in mind the complexities of the basal forebrain cholinergic system, it may be assumed that strategies restoring normal cholinergic function by modulating endogenous systems will eventually result in greater clinical benefits.

4. ANIMAL MODELS

It is in the context of identifying strategies for manipulating endogenous cholinergic function that animal models of basal forebrain degeneration have proven most useful. Two main categories of basal forebrain cholinergic degeneration models are generally used: (1) *acute lesion models* in which degeneration of basal forebrain cell bodies or their axons is caused by direct mechanical or chemical damage, and (2) *spontaneous degeneration models* in which cholinergic neurons undergo atrophy or death as a result of target ablation or aging. Both models have unique advantages and disadvantages, and each has played an important role in investigating possible therapeutic strategies for treating AD.

4.1. Acute Cholinergic Degeneration Models

Models of *acute* basal forebrain cholinergic degeneration include the well-characterized fimbria–fornix transection, and various electrolytic and chemotoxic ablation paradigms. These models all involve direct damage to the basal forebrain neurons or their processes and are most often characterized by a rapid onset of neuronal atrophy, lasting between a few days and a few weeks. Varying degrees of cell death have been reported with acute lesions of the basal forebrain cholinergic system, although obtaining experimental confirmation of cell death has been a controversial issue *(52–54)*. Furthermore, in many of these lesion models there is an accompanying loss of cholinergic phenotype in surviving basal forebrain cholinergic neurons, a loss of cholinergic function within basal forebrain cholinergic targets, and evidence for axonal and dendritic atrophy, all features reminiscent of basal forebrain cholinergic cell pathology in AD. Thus, acute lesion models may serve as an excellent tool for identifying compounds capable of providing neuroprotective actions for degenerating basal forebrain cholinergic neurons and restoring or enhancing cholinergic function.

4.1.1. Fimbria–Fornix Transection

One of the earliest models of lesion-induced CNS neuronal degeneration to be investigated was the fimbria–fornix transection (FFT). In this model, cholinergic cell bodies within the medial septum are physically separated from their hippocampal innervation target by an acute transection of their axons, thereby interrupting anterograde and retrograde communication between the septal neurons and their hippocampal target. Axotomy of the septo-hippocampal pathway is most often achieved by either a knife cut *(55)* or an aspirative lesion *(56)*. The aspirative lesion is thought to provide a more complete and reproducible lesion, although it results in more extensive damage to the overlying cortical and callosal structures. The fortuitous anatomy of the septo-hippocampal pathway, in which the axons of medial septal neurons course in a tight bundle (the fornix) to innervate the hippocampus, has made this an ideal system to manipulate experimentally. FFT lesions produce a near complete cholinergic denervation of the hippocampal target (with the exception of a minor cholinergic projection through a ventral pathway), and provide a highly reproducible model of neuronal degeneration among axotomized basal forebrain cholinergic neurons. Typically, FFT lesions result in a loss of choline acetyltransferase (ChAT)-immunoreactive cell labeling (a specific marker of cholinergic neurons) of approx 80% magnitude *(57,58)* *(see*

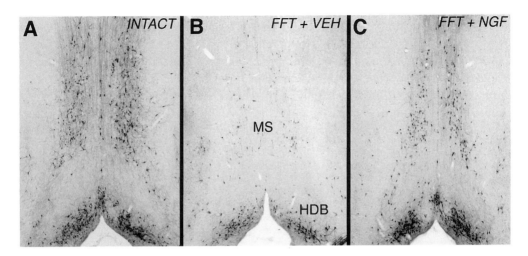

Fig. 1. NGF protects basal forebrain cholinegic neurons against lesion-induced degeneration. Within the basal forebrain of an intact adult rat (**A**), numerous cholinergic neurons within the medial septum (MS) and horizontal limb of the diagonal band (HDB) can be identified using choline acetyltransferase immunohistochemisry. Following a bilateral fimbria–fornix transection (**B**), many of the cholinergic neurons within the MS and HDB can no longer be detected, and infusion of vehicle alone fails to prevent this degeneration. However, an infusion of NGF (12 μg/d for 14 d) unilaterally within the striatum can prevent cholinergic degeneration bilaterally within the medial septum and diagonal band (**C**).

Fig. 1A,B). Studies monitoring basal forebrain cell loss using another marker of cholinergic neurons, the p75 low-affinity neurotrophin receptor, also demonstrate an 80% loss of cellular immunolabeling *(58,59)*. However, only a 65% loss of basal forebrain neurons was observed in the initial investigations documenting this phenomenon, where AChE was used as a marker for the cholinergic cell population *(55,60,61)*. In addition to a loss of cholinergic cell bodies following axotomy, the remaining neurons within the medial septum shrink significantly (by about 20% compare to unlesioned controls) *(55,58)*.

Pioneering studies using the FFT lesion model of basal forebrain cholinergic degeneration first demonstrated that nerve growth factor (NGF) could protect against this lesion-induced degeneration of basal forebrain cholinergic neurons in the medial septum, restoring ChAT and p75 cell counts to normal intact levels *(56,61,61)* *(see* Fig. 1C). NGF administration also prevented lesion-induced cellular atrophy, and even caused slight cellular hypertrophy in some cases *(55,58)*. The observation that NGF could protect against lesion-induced degeneration was not only an important breakthrough in developing possible therapies for treating AD, but also caused an explosion of interest in the field of trophic factor therapy for neurodegenerative diseases. In the years that followed, other trophic factors (including ciliary neurotrophic factor, brain-derived neurotrophic factor, neurotrophin-4/5, and basic and acidic fibroblast growth factor) were found capable of protecting basal forebrain cholinergic neurons from FFT-induced neuronal degeneration *(63–68)*, although none have demonstrated the specificity and potency of NGF.

The majority of studies using the FFT lesion model have been carried out in rats, but the occurrence of lesion-induced degeneration of basal forebrain cholinergic neurons, and the ability of NGF to prevent this degeneration, has also been confirmed in primates with fornix lesions *(69–71)*. Lesion-induced cellular degeneration in primates occurs in 50–75% of cholinergic neurons (compared to 80% degeneration in rats), and can be reversed almost completely with NGF administration. As was noted in rats, cellular hypertrophy of basal forebrain cholinergic neurons was also seen following NGF infusions into lesioned primates. The success of these latter primate studies has served as a strong preclinical basis for considering NGF therapy for the clinical treatment of AD.

In addition to identifying molecules with neuroprotective qualities, the FFT model has also been important for elucidating possible roles for neurotrophic factors in CNS regenerative events. NGF has been shown to be capable of (1) stimulating supranormal levels of ChAT synthesis in surviving basal forebrain cholinergic neurons *(58,71,72)*; (2) enhancing neuritic sprouting from axotomized basal forebrain cholinergic neurons *(70,73,74)*; (3) restoring appropriate cholinergic fiber innervation to basal forebrain targets *(75–77)*, with possible restoration of appropriate cholinergic synaptic physiology *(78)*; and (4) partially restoring appropriate behavioral performance following septo-hippocampal regeneration *(77)*. The ability of NGF to contribute significantly to many critical CNS regenerative events, eventually leading to a beneficial functional outcome in animal models of neurodegeneration, further suggests that NGF may be an important clinical tool for treating cholinergic degeneration in AD.

4.1.2. Other Models of Acute Basal Forebrain Degeneration

A second basal forebrain cholinergic cell group of interest in AD is the population of neurons within the nucleus basalis, providing the sole source of cholinergic innervation to the cortex. Unlike the septo-hippocampal system, the rather diffuse anatomy of the basalis-cortical system has made it difficult to develop reproducible transection-lesion paradigms. However, alternate methods for lesioning this cell population have been explored and these studies have provided additional support for a role of trophic factors in preventing lesion-induced degeneration of this cell population.

Electrolytic ablations of either the nucleus basalis or medial septal regions have been used to study cholinergic degeneration *(79)*. In these models, a wire electrode is stereotaxically placed within the region of interest and a defined current is used to destroy tissue in the vicinity of the electrode. By controlling the intensity of electrical current passing through the tissue, and the time the current is passed, these lesions provide accurate and relatively reproducible destruction of basal forebrain cholinergic cell populations. However, electrolytic lesions are nonspecific and will destroy all cell populations within the lesioned area, as well as fibers passing through the region. In addition, electrolytic lesions are most often irreversible and cannot be used to evaluate compounds with potential neuroprotective qualities.

Chemotoxic lesions, employing a variety of agents, have also been widely used to target destruction of both nucleus basalis and medial septal cholinergic neurons. The specificity of chemotoxic lesions can vary widely depending upon the particular compound used. Various glutamate analogs, including quisqualate, ibotenic acid, and α-amino-3-hydroxy-5-methyl-4-isoxazolepropionic acid (AMPA), have been used to induce lesions in the basal forebrain, although the specificity of these agents for cholinergic neurons has been questioned *(80)*. Compounds such as the choline analog AF64A initially appeared to exhibit lesion specificity for the cholinergic system, but later studies suggested that cell loss resulting from the use of this compound could vary substantially *(81,82)*. More recently, an immunotoxin directed against the p75 neurotrophin receptor, which is found exclusively in association with cholinergic neurons in the basal forebrain, has been developed and used with relatively good success for producing reliable and specific lesions of the basal forebrain cholinergic systems *(83,84)*. High doses of immunotoxin are capable of reducing ChAT activity in the cortex by 80–90% and hippocampal activity by almost 95% *(85)*. By varying the dose of immunotoxin given, it is possible to produce graded lesions of the basal forebrain cholinergic system *(86)*, with accompanying graded behavioral impairments on a number of relevant tasks.

Importantly, lesions targeting cholinergic cell bodies do not afford the same opportunity for exploring neuronal protection as axonal transection paradigms because, in many cases, the neuronal cell bodies are destroyed nearly instantaneously by the chemotoxic or electrolytic lesion. However, chemotoxic lesions have been used successfully to examine the ability of agents, such as NGF, to reverse *partial* cholinergic cell damage and to restore cholinergic function by modulating remaining unlesioned cholinergic neurons *(87,88)*.

Just as animal models of cholinergic degeneration are simplified approximations of actual AD pathology, *acute* models of degeneration are likely to be simplified approximations of chronic neurodegeneration, such as those that occur in association with AD. One distinct drawback of acute degeneration models is the confounding influence of the lesion upon the cell population is being investigated. The ability of NGF to stimulate ChAT synthesis *(72)*, induce cellular hypertrophy *(58)*, and elicit axonal sprouting *(73,89)* is greater in *lesioned* basal forebrain cholinergic neurons than in *intact* controls, indicating that the lesion may be an important component in determining neuronal responsiveness to a trophic factor stimulus. Another potential caveat in most studies using models of acute basal forebrain cholinergic degeneration is that putative neuroprotective agents have routinely been administered beginning at the time of the lesion. Thus, although these studies may provide evidence for the ability of a factor to prevent subsequent atrophy from taking place (making them potential therapeutic tools for preventing additional degeneration in AD), they do not address the important issue of whether these factors can reverse existing neuronal dysfunction. This is potentially an important issue in chronic neurodegenerative diseases where extensive cell damage may already exist at the time of diagnosis. Studies looking at the effects of delayed NGF treatment upon acutely transected basal forebrain cholinergic neurons *(52,58)* have addressed this issue and indicated that delayed NGF treatment (for up to 3 mo postlesion) was capable of partially reversing preexisting cellular atrophy in FFT-lesioned medial septal neurons. Similar studies in rats receiving ibotenic acid lesions of the nucleus basalis have also demonstrated that a delayed NGF treatment could partially reverse the lesion-induced loss of cortical ChAT activity *(85)*. Lastly, it remains unknown whether the cascade of events leading to cell death in an acute degeneration model is similar to that taking place in chronic neurodegenerative diseases. However, evidence from studies thus far has indicated that NGF is effective at preventing cell death mediated by either a necrotic or apoptotic mechanism (see review [90]).

4.2. Spontaneous Cholinergic Degeneration Models

More recently, models of *spontaneous* basal forebrain cholinergic degeneration have been developed that overcome some of the shortfalls of acute lesion models and more accurately reflect the type of cellular changes seen in chronic neurodegenerative diseases such as AD. Following excitotoxic destruction of the hippocampal formation (91,92) or ischemia-induced damage within the cortex (50,93), basal forebrain cholinergic neurons within either the medial septum or nucleus basalis magnocellularis, respectively, undergo neuronal degeneration. Although neither of these models is associated with a reduction in cholinergic cell numbers, significant cellular atrophy does occur. In both target ablation models, the extent of cholinergic cellular atrophy is correlated with the magnitude of damage to the target tissue, and losses of 30–40% are generally induced (50,91). Neurodegeneration in these models occurs over a more protracted timeframe than is seen following acute lesions, and thus more closely resembles the chronic neurodegeneration observed in AD. The finding that NGF can protect against neuronal degeneration in these animal models by reversing cholinergic cell atrophy (94,95) demonstrates that basal forebrain neurons remain responsive to the neuroprotective actions of NGF even when the neurons have not been damaged directly.

4.2.1. Age-Related Basal Forebrain Degeneration

Perhaps one of the most interesting animal models of basal forebrain cholinergic degeneration for evaluating potential neuroprotective agents and identifying strategies for manipulating endogenous cholinergic function in the intact basal forebrain system is the *aged animal*. In rats, just as in humans, aging has been associated with a deterioration of cognitive function. Basal forebrain cholinergic degeneration has also been reported to occur to varying extents in aged rodents. For example, a loss of immunolabeling for cholinergic neuronal markers has been reported to occur over a range from 0% to 40% in different studies of aged rats. This variability in magnitude may be accounted for in part by differences in rodent species, neuronal markers (ChAT, p75, or AChE), and quantification methodologies used in the different studies (see [96]). However, the most thorough analysis of cholinergic cell loss in aging rodents, employing unbiased stereological techniques, has indicated that a 30% loss of cholinergic neurons does occur within the nucleus basalis magnocellularis (97). Whether remaining cholinergic neurons undergo cell atrophy (98) or hypertrophy (99) is also unclear, and unbiased stereological techniques for assessing cell size in aged rodents have not yet been published. Despite these study-to-study differences, cognitive impairment is regularly demonstrated in aged rodents, and on tasks such as spatial navigation, the degree of cognitive impairment correlates with reductions in cholinergic neuronal numbers or size (98,100,101). Limited studies in primates have also shown age-associated alterations in basal forebrain cholinergic neuronal morphology (102). Our own recent studies using unbiased stereological methods have confirmed that an age-related loss of immunolabeling occurs in basal forebrain cholinergic neurons in nonhuman primates (103).

A number of studies have indicated that NGF administration to cognitively impaired aged animals is capable of reversing cholinergic cell shrinkage in basal forebrain cholinergic neurons and ameliorating cognitive deficits *(36,104–108)*. However, in some studies NGF administration did not reverse the age-associated cell loss, suggesting that a subpopulation of basal forebrain cholinergic neurons were either dead, or were no longer responsive to the actions of exogenous NGF *(36)*. Yet these data from animal models of spontaneous age-related degeneration are important, not only because they demonstrate NGF can modulate the performance of basal forebrain cholinergic neurons that have not been acutely damaged, but also that NGF can reverse preexisting cellular degeneration.

An issue that remains to be elucidated is the mechanism of neuronal degeneration in animal models of aging and the relation of this mechanism to the neuronal degeneration that occurs in AD. Indeed, many studies of neuronal loss in AD suggest that patterns of cellular degeneration in normal aging and in AD fundamentally differ, and that AD is not simply an accelerated form of cellular damage found during normal aging. Thus, it is uncertain that NGF, which reduces the extent of cholinergic neuronal atrophy in animals models of aging, will also protect cholinergic neurons in AD. Despite these uncertainties, a demonstration that NGF therapy also prevents cholinergic neuronal degeneration in aged primates would provide further justification for trials of NGF therapy in AD. Age-associated cognitive impairment occurs in primates (*see [109]*), although the extent to which degeneration of basal forebrain cholinergic system contributes to this cognitive dysfunction is unknown *(110)*. Recent preliminary studies in aging primates have indicated that exogenous NGF delivery by ex vivo gene transfer may be capable of reversing the age-related loss of p75 receptor-positive cholinergic neurons within the Ch4 region *(103)*, although it remains unknown whether this recovery is associated with improved cognitive function.

5. CONCLUSIONS

It is a difficult task to generate animal models simulating a disease as complex as AD. Thus, nearly any animal model will serve only as an approximation of the actual disease, and results obtained from a particular model must be considered carefully before extrapolating their relevance to the human disease. In the case of animal models of lesion-induced basal forebrain cholinergic degeneration, it is clear that these models are most relevant for investigating a particular aspect of AD pathology, namely the disease-related dysfunction of the basal forebrain cholinergic system. Even with this caveat clearly established, models of lesion-induced damage of central cholinergic systems remain relevant for investigating AD. Extensive evidence suggests that dysfunction of the basal forebrain cholinergic system is a consistent and significant component in AD pathology and is likely to contribute significantly to disease-related cognitive impairment. For this reason, it is worthwhile continuing to pursue therapeutic strategies targeting the basal forebrain system with the hope that reversing cholinergic damage may partially restore normal cognitive function in AD.

Beyond the direct clinical relevance of targeting therapeutic strategies to ameliorate cholinergic dysfunction in AD, there are several reasons that animal models of central cholinergic damage remain important. To date, animal models of central cholinergic dysfunction are some of the best characterized models of central neurodegeneration and they continue to serve as excellent tools for studying various fundamental questions pertaining to neurodegenerative disease in general (including mechanisms of cellular atrophy and conditions necessary for trophic factor responsiveness). These models also continue to serve as useful test systems for evaluating new treatment strategies and novel drug delivery systems for introducing trophic factors into the CNS. Findings obtained from investigations of models of basal forebrain cholinergic degeneration will be transferable to other models of neurodegeneration and thus will be applicable to other CNS neurodegenerative diseases.

ACKNOWLEDGMENTS

Supported by the National Institute of Health (AG10435, NS37083) and Veterans Affairs.

REFERENCES

1. Diaz, B. R. and Yamazaki, R. S. (1998) Advances and challenges in the prevention of Alzheimer's disease. *Pharmaceut. Res.* **15,** 386–398.
2. Terry, R. D., Masliah, E., and Hansen, L. A. (1994) Structural basis of the cognitive alterations in Alzheimer's disease, *Alzheimer's Disease* (Terry, R. D., Katzman, R., and Bick, K. L., eds.), Raven Press, New York, pp. 179–196.
3. Cikson, D. W., Crystal, H. A., Mattiace, L. A., Masur, D. M., Blau, A. D., Davies, P., Yen, S. H., and Aronson, M. K. (1992). Identification of normal and pathological aging in prospectively studied nondemented elderly humans. *Neurobiol. Aging* **13,** 179–189.
4. Katzman, R., Terry, R., Deteresa, R., Brown, T., Davies, P., Fuld, P., Renbing, X., and Peck, A. (1988) Clinical, pathological, and neurochemical changes in dementia: a subgroup with preserved mental status and numerous neocortical plaques. *Ann. Neurol.* **23,** 138–144.
5. Terry, R. D., Hansen, L. A., Deteresa, R., Davies, P., Tobias, H., and Katzman, R. (1987) Senile dementia of the Alzheimer's type without neocortical neurofibrillary tangle. *J. Neuropathol. Exp. Neurol.* **46,** 262–268.
6. Greenfield, J. G. and Bosanquet, F. D. (1953) Brainstem lesions in parkinsonism. *J. Neurol. Neurosurg. Psychiatry* **16,** 213–226.
7. Mandybur, T. I., Nagpaul, A. S., Pappas, Z., and Niklowitz, W. J. (1977) Alzheimer neurofibrillary change in subacute sclerosing panencephalitis. *Ann. Neurol.* **1,** 103–107.
8. Popovitch, E. R., Wisniewski, H. M., Barcikowska, M., Silverman, W., Bancher, C., Sersen, E., and Wen, G. Y. (1990) Alzheimer's neuropathology in non-Down's syndrome mentally retarded adults. *Acta Neuropathol.* **80,** 362–367.
9. Geula, C. and Mesulam, M-M. (1994) Cholinergic systems and related neuropathological predilection patters in Alzheimer's disease, *Alzheimer's Disease* (Terry, R. D., Katzman, R., and Bick, K. L., eds.), Raven Press, New York, pp. 263–291.
10. Berg, L., and Morris, J. C. (1994) Diagnosis, in *Alzheimer's Disease* (Terry R. D., Katzman, R., and Bick, K. L., eds.), Raven Press, New York, pp. 9–25.
11. Tanzi, R. E., Kovacs, D. M., Kim, T-W., Moir, R. D., Guenette, S. Y., and Wasco, W. (1996) The gene defects responsible for familial Alzheimer's disease. *Neurobiol. Dis.* **3,** 159–168.
12. Drachman, D. A. and Leavitt, J. (1974) Human memory and the cholinergic system: a relationship to aging? *Arch. Neurol.* **30,** 113–121.
13. Drachman, D. A., Noffsinger, D., Sahakian, B. J., Kurdziel, S., and Fleming, P. (1980)

Aging, memory, and the cholinergic system: a study of dichotic listening. *Neurobiol. Aging* **1**, 39–43.

14. Davis, K. L., Mohs, R. C., Tinklenberg, J. R., Pfefferbaum, A., Hollister, L. E., and Kopell, B. S. (1978) Physostigmine: improvement of long-term memory processes in normal humans. *Science* **201**, 272–274.

15. Mewaldt, S. P. and Ghoneim, M. M. (1979) The effects and interactions of scopolamine, physostigmine and methamphetamine on human memory. *Pharmacol. Biochem. Behav.* **10**, 205–210.

16. Bartus, R. T., Dean, R. L., Beer, B., and Lippa, A. S. (1982) The cholinergic hypothesis of geriatric memory dysfunction. *Science* **217**, 408–417.

17. Coyle, J. T., Price, P. H., and Delong, M. R. (1983) Alzheimer's disease: a disorder of cortical cholinergic innervation. *Science* **219**, 1184–1189.

18. Arendt, T., Bigl, V., Arendt, A., and Tennstedt, A. (1983) Loss of neurons in the nucleus basalis of Meynert in Alzheimer's disease, paralysis agitans and Korsakoff's disease. *Acta Neuropathol.* **61**, 101–108.

19. Iraizoz, I., De Lacalle, S., and Gonzalo, L. M. (1991) Cell loss and nuclear hypertrophy in topographical subdivisions of the nucleus basalis of Meynert in Alzheimer's disease. *Neuroscience* **41**, 33–40.

20. Vogels, O. J. M., Broere, C. A. J., Ter Laak, H. J., Ten Donkelaar, H. J., Nieuwenhuys, R., and Shulte, B. P. M. (1990) Cell loss and shrinkage in the nucleus basalis of Meynert complex in Alzheimer's disease. *Neurobiol. Aging* **11**, 3–13.

21. Whitehouse, P. J., Price, D. J., Clark, A., Coyle, J. T., and Delong, M. (1981) Alzheimer's disease: Evidence for selective loss of cholinergic neurons in the nucleus basalis. *Ann. Neurol.* **10**, 122–126.

22. Whitehouse, P. J., Price, D. L., Struble, R. G., Clark, A. W., Coyle, J. T., and Delong, M. R. (1982) Alzheimer's disease and senile dementia: loss of neurons in the basal forebrain. *Science* **215**, 1237–1238.

23. Wilcock, G. K., Esiri, M. M., Bowen, D. M., and Smith, C. C. T. (1982) Alzheimer's disease. Correlation of cortical choline acetyltransferase activity with the severity of dementia and histological abnormalities. *J. Neurol. Sci.* **57**, 407–417.

24. Arendt, T., Bigl, V., Tennstedt, A., and Arendt, A. (1984) Correlation between cortical plaque count and neuronal loss in the nucleus basalis in Alzheimer's disease. *Neurosci. Lett.* **48**, 81–85.

25. Perry, E. K., Tomlinson, B. E., Blessed, G., Bergman, K., Gibson, P. H., and Perry, R. H. (1978) Correlation of cholinergic abnormalities with senile plaques and mental test scores in senile dementia. *Br. Med. J.* **2**, 1457–1459.

26. Masliah, E., Terry, R. D., Alford, M., Detersa, R., and Hansen, L. A. (1991) Cortical and subcortical patterns of synaptophysin-like immunoreactivity in Alzheimer's disease. *Am. J. Pathol.* **138**, 235–246.

27. Avery, E. E., Baker, L. D., and Asthana, S. (1997) Potential role of muscarinic agonists in Alzheimer's disease. *Drugs Aging* **11**, 450–459.

28. Freedman, S. B., Harley, E. A., Marwood, R. S., and Patel, S. (1990) In vivo characterization of novel efficacious muscarinic receptor agonists. *Eur. J. Pharmacol.* **187**, 193–199.

29. Soncrant, T. T., Raffaele, K. C., Asthana, S., Berardi, A., Morris, P. P., and Haxby, J. V. (1993) Memory improvement without toxicity during chronic, low dose intravenous arecoline in Alzheimer's disease. *Psychopharmacology* **112**, 421–427.

30. Feurerstein, T. J. and Seeger, W. (1997) Modulation of acetylcholine release in human cortical slices: possible implications for Alzheimer's disease. *Pharmacol. Ther.* **74**, 333–347.

31. Rosenberg, G. S. and Davis, K. L. (1982) The use of cholinergic precursors in neuropsychiatric disease. *Am. J. Clin. Nutr.* **36**, 709–720.

32. Taylor, P. (1998) Development of acetylcholinesterase inhibitors in the therapy of Alzheimer's disease. *Neurology* **51,** S30–35.
33. Thal, L. J. (1994) Clinical trials in Alzheimer's disease, in *Alzheimer's Disease* (Terry, R. D., Katzman, R., and Bick, K. L. eds.), Raven Press, New York, pp. 431–444.
34. Dunnet, S. (1991) Cholinergic grafts, memory and aging. *TINS* **14,** 371–376.
35. Fisher, L. J., Schinstine, M., Salvaterra, P., Dekkar, A. J., Thai, L., and Gage, F. H. (1993) In vivo production and release of acetylcholine from primary fibroblasts genetically modified to express choline acetyltransferase. *J. Neurochem.* **61,** 1323–1332.
36. Fischer, W., Wictorin, K., Bjorklund, A., Williams, L. R., Varon, S., and Gage, F. H. (1987) Amelioration of cholinergic neuron atrophy and spatial memory impairment in aged rats by nerve growth factor. *Nature* **329,** 65–68.
37. Hefti, F. (1994) Neurotrophic factor therapy for nervous system degenerative diseases. *J. Neurobiol.* **25,** 1418–1435.
38. Simpkins, J. W., Green, P. S., Gridley, K. E., Singh, M., De Fiebre, N. C., and Rajakumar, G. (1997) Role of estrogen replacement therapy in memory enhancement and the prevention of neuronal loss associated with Alzheimer's disease. *Am. J. Med.* **103,** 19S–25S.
39. Tuszynski, M. H., Kordower, J. H., Mufson, E., and Gage, F. H. (1998) Neurotrophic factors, gene therapy and Alzheimer's disease, in *CNS Regeneration: Basic Science and Clinical Applications* (Tuszynski, M. H. and Kordower, J. H., eds.), Academic Press, San Diego.
40. Sillito, A. M. and Murphy, P. C. (1987) The cholinergic modulation of cortical function, in *Cerebral Cortex*, Vol. 6 (Jones, E. G. and Peters, A., eds.), Plenum Press, New York, pp. 161–185.
41. Richardson, R. T. and Delong, M. R. (1986) Nucleus basalis of Meynert neuronal activity during a delayed response task in the monkey. *Brain Res.* **399,** 364–368.
42. Rigdon, G. C. and Pirch, J. H. (1986) Nucleus basalis involvement in conditioned neuronal responses in rat frontal cortex. *J. Neurosci.* **6,** 2535–2542.
43. Wilson, F. A. W. and Rolls, E. T. (1990) Learning and memory is reflected in the response of reinforcement-related neurons in the primate basal forebrain. *J. Neurosci.* **10,** 1254–1267.
44. Bland, B. H. (1986) The physiology and pharmacology of hippocampal formation theta rhythms. *Prog. Neurobiol.* **26,** 1–54.
45. Huerta, P. T. and Lisman, J. E. (1993) Heightened synaptic plasticity of hippocampal CA1 neurons during a cholinergically induced rhythmic state. *Nature* **364,** 723–725.
46. Jerusalinsky, D., Edgar, K., and Izquierdo, I. (1997) Cholinergic neurotransmission and synaptic plasticity concerning memory processing. *Neurochem. Res.* **22,** 507–515.
47. Stewart, M. and Fox, S. E. (1990) Do septal neurons pace the hippocampal theta rhythm? *TINS* **13,** 163–168.
48. Winson, J. (1978) Loss of hippocampal theta rhythm result in spatial memory deficit in rat. *Science* **201,** 160–163.
49. Butt, A. E., Testylier, G., and Dykes, R. W. (1997) Acetylcholine release in rat frontal and somatosensory cortex is enhanced during tactile discrimination learning. *Psychobiology* **25,** 18–33.
50. Sofroniew, M. V., Pearson, R. C. A., Eckenstein, F., Cuello, A. C., and Powell, T. P. S. (1983) Retrograde changes in cholinergic neurons in the basal forebrain of the rat following cortical damage. *Brain Res.* **289,** 370–374.
51. Valentino, R. J. and Dingledine, R. (1981) Presynaptic inhibitory effect of acetylcholine in the hippocampus. *J. Neurosci.* **1,** 784–792.
52. Hagg, T., Manthorpe, M., Vahlsing, H. L., and Varon, S. (1988) Delayed treatment with nerve growth factor reverses the apparent loss of cholinergic neurons after acute brain damage. *Exp. Neurol.* **101,** 303–312.

53. Naumann, T., Straube, A., and Frotshcer, M. (1997) Recovery of ChAT immunoreactivity in axotomized rat cholinergic septal neurons despite reduced NGF receptor expression. *Eur. J. Neurosci.* **9,** 1340–1349.

54. Tuszynski, M. H., Armstrong, D. M., and Gage, F. H. (1990) Basal forebrain cell loss following fimbria/fornix transection. *Brain Res.* **508,** 241–248.

55. Gage, F. H., Armstrong, D. M., Williams, L. R., and Varon, S. (1988) Morphological responses of axotomized septal neurons to nerve growth factor. *J. Comp. Neurol.* **269,** 147–155.

56. Williams, L. R., Varon, S., Peterson, G. M., Wictorin, K., Fisher, W., Bjorklund, A., and Gage, F. H. (1986) Continuous infusion of nerve growth factor prevents forebrain neuronal death after fimbria-fornix transection. *Proc. Natl. Acad. Sci. USA* **83,** 9231–9235.

57. Gage, F. H., Wictorin, K., Fishcer, W., Williams, L. R., Varon, S., and Bjorklund, A. (1986) Retrograde cell changes in medial septum and diagonal band following fimbria-fornix transection: quantitative temporal analysis. *Neuroscience* **19,** 241–255.

58. Hagg, T., Fass-Holmes, B., Vahlsing, H. L., Manthorpe, M., Conner, J. M., and Varon, S. (1989) Nerve growth factor (NGF) reverses axotomy-induced decreases in choline acetyltransferase, NGF receptor and size of medial septum cholinergic neurons. *Brain Res.* **505,** 29–38.

59. Batchelor, P. E., Armstrong, D. M., Blaker, S. N., and Gage, F. H. (1989) Nerve growth factor receptor and choline acetyltransferase colocalization in neurons within the rat forebrain: response to fimbria fornix transection. *J. Comp. Neurol.* **284,** 187–204.

60. Armstrong, D. M., Terry, R. D., Detersa, R. M., Bruce, G., Hersh, L. B., and Gage, F. H. (1987) Response of septal cholinergic neurons to axotomy. *J. Comp. Neurol.* **264,** 421–436.

61. Hefti, F. (1986) Nerve growth factor promotes survival of septal cholinergic neurons after fimbrial transections. *J. Neurosci.* **6,** 2155–2162.

62. Kromer, L. F. (1987) Nerve growth factor treatment after brain injury prevents neuronal death. *Science* **235,** 214–216.

63. Alderson, R. F., Wiegand, S. J., Anderson, K. D., Cai, N., Cho, J. Y., Lindsay, R. M., and Altar, C. A. (1996) Neurotrophin-4/5 maintains the cholinergic phenotype of axotomized septal neurons. *Eur. J. Neurosci.* **8,** 282–290.

64. Figueiredo, B. C., Piccardo, P., Maysinger, D., Clarke, P. B., and Cuello, A. C. (1993) Effects of acidic fibroblast growth factor on cholinergic neurons of nucleus basalis magnocellularis and in spatial memory task following cortical devascularization. *Neuroscience* **56,** 955–963.

65. Hagg, T., Quon, D., Higaki, J., and Varon, S. (1992) Ciliary neurotrophic factor prevents neuronal degeneration and promotes low affinity NGF receptor expression in the adult rat CNS. *Neuron* **8,** 145–158.

66. Morse, J. K., Wiegand, S. J., Anderson, K., You, Y., Cai, N., Carnahan, J., Miller, J., Distefano, P., Altar, C. A., Lindsay, R. M., and Alderson, R. F. (1993) Brain-derived neurotrophic factor (BDNF) prevents the degeneration of medial septal cholinergic neurons following fimbria transection. *J. Neurosci.* **13,** 4146–4156.

67. Otto, D., Frotscher, M., and Unsicker, K. (1989) Basic fibroblast growth factor and nerve growth factor administered in gel foam rescue medial septal neurons after fimbria fornix transection. *J. Neurosci. Res.* **22,** 83–91.

68. Tooyama, I., Sasaki, K., Oomura, Y., Li, A. J., and Kimura, H. (1997) Effect of acidic fibroblast growth factor on basal forebrain cholinergic neurons in senescence-accelerated mice. *Exp. Gerontol.* **32,** 171–179.

69. Koliatsos, V. E., Nauta, H. J., Clatterbuck, R. E., Holtzman, D. M., Mobley, W. C., and Price, D. L. (1990) Mouse nerve growth factor prevents degeneration of axotomized basal forebrain cholinergic neurons in the monkey. *J. Neurosci.* **10,** 3801–3813.

70. Kordower, J. H., Winn, S. R., Liu, Y-T., Mufson, E. J., Sladek, J. R., Hammang, J. P., Baetge, E. E., and Emerich, D. F. (1994) The aged monkey basal forebrain: rescue and sprouting of axotomized basal forebrain neurons after grafts of encapsulated cells secreting human nerve growth factor. *Proc. Natl. Acad. Sci. USA* **91,** 10,898–10,902.

71. Tuszynski, M. H., U, H-S., Amaral, D. G., and Gage, F. H. (1990) Nerve growth factor infusions in primate brain reduces lesion-induced cholinergic neuronal degeneration. *J. Neurosci.* **10,** 3604–3614.

72. Williams, L. R., Jodelis, K. S., and Donald, M. R. (1989) Axotomy-dependent stimulation of choline acetyltransferase activity by exogenous mouse nerve growth factor in adult rat basal forebrain. *Brain Res.* **488,** 243–266.

73. Hagg, T. and Varon, S. (1993) Neurotropism of nerve growth factor for adult rat septal cholinergic axons in vivo. *Exp. Neurol.* **119,** 37–45.

74. Yunshao, H., Zhibin, Y., Yaoming, G., Guobi, K., and Yici, C. (1992) Nerve growth factor promotes collateral sprouting of cholinergic fibers in the septohippocampal cholinergic system of aged rats with fimbria transection. *Brain Res.* **586,** 27–35.

75. Hagg, T., Vahlsing, H. L., Manthorpe, M., and Varon, S. (1990) Septo-hippocampal cholinergic axonal regeneration through peripheral nerve bridges: quantification and temporal development. *Exp. Neurol.* **109,** 153–163.

76. Hagg, T., Vahlsing, H. L., Manthorpe, M., and Varon, S. (1990) Nerve growth factor infusion into the denervated adult rat hippocampal formation promotes its cholinergic reinnervation. *J. Neurosci.* **10,** 3087–3092.

77. Tuszynski, M. H. and Gage, F. H. (1995) Bridging grafts and transient nerve growth factor infusions promote long-term central nervous system rescue and partial functional recovery. *Proc. Natl. Acad. Sci. USA* **92,** 4621–4625.

78. Tuszynski, M. H., Buzsaki, G., and Gage, F. H. (1990) Nerve growth factor infusions combined with fetal hippocampal grafts enhance reconstruction of the lesioned septohippocampal projection. *Neuroscience* **36,** 33–44.

79. Kesner, R. P., Crutcher, K. A., and Measom, M. O. (1986) Medial septum and nucleus basalis magnocellularis lesions produce order memory deficits in rats which mimic symptomatology of Alzheimer's disease. *Neurobiol. Aging* **7,** 287–295.

80. Dunnett, S. B., Everitt, B. J., and Robbins, T. W. (1991) The basal forebrain-cortical cholinergic system: interpreting the functional consequences of excitotoxic lesions. *TINS* **14,** 494–501.

81. Loren, S. A., Kindel, G., Dong, X. W., Lee, J. M., and Hanin, I. (1991) Septal choline acetyltransferase immunoreactive neurons: dose dependent effects of AF64A. *Brain Res. Bull.* **26,** 965–971.

82. McGurk, S. R., Hartgraves, S. L., Kelly, P. H., Gordon, M. N., and Butcher, L. L. (1987) Is ethylcholine mustard aziridinium ion a specific cholinergic neurotoxin? *Neuroscience* **22,** 215–224.

83. Schliebs, R., Robner, S., and Bigl, V. (1996) Immunolesion by 192IgG-saporin of the rat basal forebrain cholinergic system: a useful tool to produce cortical cholinergic dysfunction. *Prog. Brain Res.* **109,** 253–264.

84. Wiley, R. G., Berbos, T. G., Deckwerth, T. L., Johnson, E. M., and Lappi, D. A. (1995) Destruction of the cholinergic basal forebrain using immunotoxin to rat NGF receptor: modeling the cholinergic degeneration of Alzheimer's disease. *J. Neurol. Sci.* **128,** 157–166.

85. Dekker, A. J. and Thal, L. J. (1992) Effect of delayed treatment with nerve growth factor on choline acetyltransferase activity in the cortex of rats with lesions of the nucleus basalis magnocellularis: dose requirements. *Brain Res.* **584,** 55–63.

86. Waite, J. J., Chen, A. D., Wardlow, M. L., Wiley, R. G., Lappi, D. A., and Thal, L. J.

(1995) 192 Immunoglobulin G-saporin produces graded behavioral and biochemical changes accompanying the loss of cholinergic neurons of the basal forebrain and cerebellar Purkinje cells. *Neuroscience* **65,** 463–476.

87. Robner, S., Yu, J., Pizzo, D., Werrback-Perez, K., Schliebs, R., Bigl, V., and Perez-Polo, J. R. (1996) Effects of intraventricular transplantation of NGF-secreting cells on cholinergic basal forebrain neurons after partial immunolesion. *J. Neurosci. Res.* **45,** 40–56.

88. Winkler, J. and Thal, L. J. (1995) Effects of nerve growth factor treatment on rats with lesions of the nucleus basalis magnocellularis produced by ibotenic acid, quisqualic acid and AMPA. *Exp. Neurol.* **136,** 234–250.

89. Conner, J. M., Fass-Holmes, B., and Varon, S. (1994) Changes in nerve growth factor immunoreactivity following entorhinal cortex lesion—a possible molecular mechanism regulating cholinergic sprouting. *J. Comp. Neurol.* **345,** 409–418.

90. Pettmann, B. and Henderson, C. E. (1998) Neuronal cell death. *Neuron* **20,** 633–647.

91. Sofroniew, M. V., Cooper, J. D., Svendsen, C. N., Crossman, P., Ip, N. Y., Lindsay, R. M., Zafra, F., and Lindholm, D. (1993) Atrophy but not death of adult septal cholinergic neurons after ablation of target capacity to produce mRNAs for NGF, BDNF, and NT-3. *J. Neurosci.* **13,** 5263–5276.

92. Sofroniew, M. V., Galletly, N. P., Isacson, O., and Svendsen, C. N. (1990) Survival of adult basal forebrain cholinergic neurons after loss of their target neurons. *Science* **247,** 338–342.

93. Funnell, W. R., Maysinger, D., and Cuello, A. C. (1990) Three-dimensional reconstruction and quantitative evaluation of devascularizing cortical lesions in the rat. *J. Neurosci. Methods* **35,** 147–156.

94. Garofalo, L., Ribeiro-Da Silva, A., and Cuello, A. C. (1992) Nerve growth factor-induced synaptogenesis and hypertrophy of cortical cholinergic terminals. *Proc. Natl. Acad. Sci. USA* **89,** 2639–2643.

95. Liberini, P., Pioro, E. P., Maysinger, D., Ervin, F. R., and Cuello, A. C. (1993) Long-term protective effects of human recombinant nerve growth factor and monosialoganglioside GM1 treatment on primate nucleus basalis cholinergic neurons after neocortical infarction. *Neuroscience* **53,** 625–637.

96. Sarter, M. and Bruno, J. P. (1998) Age-related changes in rodent cortical acetylcholine and cognition: main effects of age versus age as an intervening variable. *Brain Res. Rev.* **27,** 143–156.

97. Smith, M. L. and Booze, R. M. (1995) Cholinergic and GABAergic neurons in the nucleus basalis region of young and aged rats. *Neuroscience* **67,** 679–688.

98. Fischer, W., Gage, F. H., and Bjorklund, A. (1989) Degenerative changes in forebrain cholinergic nuclei correlate with cognitive impairments in aged rats. *Eur. J. Neurosci.* **1,** 34–45.

99. Armstrong, D. M., Scheffield, R., Buzsaki, G., Chen, K. S., Hersh, L. B., Nearing, B., and Gage, F. H. (1993) Morphologic alterations of choline acetyltransferase-positive neurons in the basal forebrain of aged behaviorally characterized Fisher 344 rats. *Neurobiol. Aging* **14,** 457–470.

100. Fischer, W., Chen, K. S., Gage, F. H., and Bjorklund, A. (1992) Progressive decline in spatial learning and integrity of forebrain cholinergic neurons in rats during aging. *Neurobiol. Aging 13,* 9–23.

101. Koh, S., Chang, P., Collier, T. J., and Loy, R. (1989) Loss of NGF receptor immunoreactivity in basal forebrain neurons of aged rats: correlation with spatial memory impairments. *Brain Res.* **498,** 397–404.

102. Stroessner-Johnson, H. M., Rapp, P. R., and Amaral, D. G. (1992) Cholinergic cell loss and hypertrophy in the medial septal nucleus of behaviorally characterized aged rhesus monkeys. *J. Neurosci.* **12,** 1936–1944.

103. Smith, D. E., McKay, H. L., Roberts, J. A., and Tuszynski, M. H. (1998) Intraparenchymal delivery of NGF by ex vivo gene transfer reverses age-related loss of expression of p75NTR in basal forebrain cholinergic neurons. *Soc. Neurosci. Abstr.* **24,** 541.

104. Backman, C., Rose, G. M., Hoffer, B. J., Henry, M. A., Bartus, R. T., Friden, P., and Granholm, A-C. (1996) Systemic administration of a nerve growth factor conjugate reverses age-related cognitive dysfunction and prevents cholinergic neuron atrophy. *J. Neurosci.* **16,** 5437–5442.

105. Chen, K. S. and Gage, F. H. (1995) Somatic gene transfer of NGF to the aged brain: behavioral and morphological amelioration. *J. Neurosci.* **15,** 2819–2825.

106. Lindner, M. D., Kearns, C. E., Winn, S. R., Frydel, B., and Emerich, D. F. (1996) Effects of intraventricular encapsulated hNGF-secreting fibroblasts in aged rats. *Cell Transplant.* **5,** 205–223.

107. Markowska, A. L., Koliatsos, V. E., Breckler, S. J., Price, D. L., and Olton, D. S. (1994) Human nerve growth factor improves spatial memory in aged but not young rats. *J. Neurosci.* **14,** 4815–4823.

108. Martinez-Serrano, A., Fischer, W., and Bjorklund, A. (1995) Reversal of age-dependent cognitive impairments and cholinergic neuron atrophy by NGF-secreting neural progenitors grafted to the basal forebrain. *Neuron* **15,** 473–484.

109. Price, D. L., Martin, L. J., Sisodia, S. S., Walker, L. C., Voytko, M. L., Wagster, M. V., Cork, L. C., and Koliatsos, V. E. (1994) The aged nonhuman primate: a model for the behavioral and brain abnormalities occurring in aged humans, in *Alzheimer's Disease* (Terry, R. D., Katzman, R., and Bick, K. L., eds.), Raven Press, New York, pp. 231–245.

110. Voytko, M. L. (1996) Cognitive functions of the basal forebrain cholinergic system in monkeys: memory or attention? *Behav. Brain Res.* **75,** 13–25.

4

The Immunolesioned Animal as a Model of Transmitter Dysfunction

Gary L. Wenk

1. INTRODUCTION

Central nervous system (CNS) diseases can affect multiple neural systems and produce widespread changes throughout the brain, or they can selectively involve relatively fewer neurotransmitter systems within discrete brain regions, such as Alzheimer's diseases (AD, *1*). Considerable effort has been given to the design of a useful animal model of AD *(2,3)*. This chapter reviews recent progress in the development of animal models of selective transmitter dysfunction and focuses upon the introduction of highly selective toxins that can act as biochemical scalpels to target specific neurons within the brain. The first section discusses the recent attempts to produce selective cholinergic system dysfunction as an animal model of AD, followed by a summary of the results of studies that have used novel "immunotoxins" to reproduce a specific element of the pathology associated with AD. This is followed by a section devoted to recent investigations using immunotoxins directed against either noradrenergic or substance P-receptive neurons.

2. ANIMAL MODELS OF CHOLINERGIC SYSTEM DYSFUNCTION

Animal models have become a critical component in our recent endeavors to understand the brain mechanisms involved in normal cognitive processes, the pathological bases of cognitive impairments, and the therapeutic interventions that might one day be used to alleviate these impairments. Their use in the study of AD has been particularly important because this disease will have a profound impact on society as the population ages, and, in addition, it promises to be an area of major research for many years. In this chapter, special emphasis is placed upon an analysis of animal models of the cholinergic hypofunction that is associated with AD for two reasons. First, one of the major sites of pathological degeneration in humans with AD is in the basal forebrain cholinergic system, the nucleus basalis of Meynert *(4–6)*. Consequently, experimental lesions in animals have been placed in analogous anatomical sites in rats, including the nucleus basalis magnocellularis (NBM) and medial septal area (MSA). The NBM contains large cholinergic neurons that are diffusely distributed throughout the ventral

From: Central Nervous System Diseases
Edited by: D. F. Emerich, R. L. Dean, III, and P. R. Sanberg © Humana Press Inc., Totowa, NJ

pallidum and substantia innominata *(7–9)*. These NBM neurons provide a topographically organized cholinergic input to the entire neocortical mantle, olfactory bulbs, and
amygdala *(10,11)*. In contrast, the cholinergic neurons within the MSA provide an input
into the hippocampus via the fornix *(10)*. Second, a selective immunotoxin for the
cholinergic forebrain neurons was the first such toxin that has been widely used to
study a specific disease of the CNS, that is, AD. Because the major behavioral difficulties of AD patients is associated with memory and attention problems *(12–14)*, many of
the behavioral tests for animals have assessed memory or attentional ability *(15–17)*.
Thus studies of the behavioral impairments exhibited by animals with selected lesions
of these forebrain regions have often used diagnostic test batteries designed to determine the extent to which the behavioral changes seen following lesions are caused by a
selective mnemonic or attentional impairment.

2.1. Lesion Analyses Using Nonselective Neurotoxins

Behavioral deficits associated with lesions produced by injections of excitatory
amino acid agonists or AF64A into the NBM and MSA have been demonstrated in a
variety of tasks *(18–20;* for review *see 2,21,22)*. Performance decrements produced by
these lesions were initially interpreted as being the result of impairments in learning
and memory abilities *(2,23)*. The possibility that forebrain cholinergic systems might
not always play a role in the neural processes that underlie memory alone was initially
demonstrated using a selective inhibitor of choline acetyltransferase (ChAT), the rate-
limiting enzyme in the production of acetylcholine *(24)*. The near total loss of ChAT
activity throughout the brain, and substantial decline in acetylcholine levels, did not
impair memory *(24)*. The possibility that cholinergic neurons might not always be
required for learning and memory *per se* was first shown in an elegant study that
injected a variety of different excitatory amino acid agonists into the basal forebrain
(25). This study found that the degree of destruction of basal forebrain cholinergic
neurons did not correlate with the degree of impairment in a memory task. These results
were later confirmed and extended by other laboratories *(22,26–31)*. Unfortunately,
because these studies relied on nonspecific excitatory amino acid neurotoxins, such
as ibotenic or kainic acid or α-amino-3-hydroxy-5-methyl-4-isoxazole propionic
acid (AMPA), to produce the discrete NBM lesions, the precise role of basal forebrain
cholinergic neurons in the memory or attention deficits could not be proven conclusively *(32)*.

2.2. Lesion Analyses Using a Selective Immunotoxin

Recently, it became possible to produce a selective and discrete lesion of basal forebrain cholinergic neurons by injection of the immunotoxin 192 IgG-saporin *(33–36)*.
This compound consists of a molecule of saporin (a 30-kDa protein isolated from
Saponaria officinalis), a ribosome-inactivating toxin *(37)*, combined with a monoclonal
antibody (192 IgG) for the low-affinity nerve growth factor (NGF) receptor (i.e., p75
protein). The monoclonal antibody binds to the low-affinity NGF receptor and is then
internalized and retrogradely transported to the cell body *(38)*. The saporin molecule is
subsequently released into the cytoplasm to inhibit protein synthesis and induce cell
death. There is a high degree of correspondence between cellular cholinergic phenotype and expression of the low-affinity NGF receptor in the basal forebrain *(33)*. Injec-

tions of the 192 IgG-saporin immunotoxin produced long-lasting and apparently selective depletions in cholinergic markers throughout the cortex, hippocampus, and olfactory bulbs *(33,34,39–49)*. Although the toxin is selective for p75-expressing neurons, two studies identified short- and long-term changes in noncholinergic neurotransmitter markers, that is an increase in dopamine and its metabolites within the olfactory bulbs *(50)* and a decrease in norepinephrine levels within the hippocampus *(51)*. However, another study did not confirm these changes in catecholamine function *(52)*. Such variations in the levels of biomarkers of noncholinergic systems may represent compensatory alterations by other neural following the removal of cholinergic afferents in specific brain regions.

Since its introduction, the 192 IgG-saporin immunotoxin has been used to investigate the influence of the removal of forebrain cholinergic neurons upon the following: seizure-induced loss of somatostatin within the hippocampus *(53)*; the expression, density, and location of muscarinic receptors *(45,54–56)*, the frequency and duration of slow-wave and REM sleep *(57,58)*; the regional production of NGF *(59–61)*; and the synaptic density within the olfactory bulbs, a major site of cholinergic innervation *(62)*. Virtually all of the cholinergic cells within the basal forebrain are vulnerable to this toxin, with the exception of those that send an efferent projection to the amygdala *(63)*. The sparing of cholinergic inputs to the basolateral amygdala has been suggested as a possible explanation for the intact memory abilities in rats following injection of 192 IgG-saporin into the basal forebrain *(41,63)*.

Because early investigators injected large amounts of the immunotoxin directly into the ventricles, most of the cholinergic neurons throughout the basal forebrain, including cells within the NBM, MSA, and vertical and horizontal limbs of the diagonal band, were effectively destroyed *(42,47,51,52,57)*. A few studies have injected smaller quantities of the toxin in an attempt to produce more discrete lesions of the NBM or MSA without damaging cholinergic neurons within nearby structures, that is, the caudate nucleus. Although the 192 IgG-saporin injections significantly and reliably decreased biomarkers of cholinergic function throughout the neocortex and hippocampus, its effect upon behavior has been quite variable.

In the Morris water maze task, injection of 192 IgG-saporin directly into the MSA did not impair performance in young rats *(39,40,48,64,65)*. Injection into the NBM impaired spatial but not navigational performance in an early study *(40)*; however, later studies did not find significant impairments in spatial navigation ability in the water maze task *(39,48,65)*. Surprisingly, injection of 192 IgG-saporin directly into both the NBM and MSA decreased cholinergic cell number in the both the NBM and MSA but did not impair performance *(65)*. In general, the effects have been dose dependent, that is, injection of large amounts *(40,51,52,66)* but not small amounts *(67)*, of the immunotoxin into the ventricles impaired performance in the Morris water maze task.

In the radial arm maze task, injection of 192 IgG-saporin directly into the MSA produced no impairment in delay-dependent performance *(68)*, a mild impairment during the acquisition phase of training *(65,69)*, and a significant impairment following injection of high doses of the immunotoxin but only with long interchoice delays *(70)*. Injection into the NBM alone, or a combined injection into both NBM and MSA, produced a slight impairment in working memory but no impairment in reference memory *(65)*.

In an operant, delayed matching-to-position task, performance was impaired by an injection of large amounts of the immunotoxin into the ventricles *(71–73)*, a combined injection into both the NBM and MSA *(43)*, or a single injection into the MSA *(48)*. In all of these studies cortical and/or hippocampal cholinergic markers were significantly decreased. In contrast, injection of a large amount of the immunotoxin into the ventricles had no effect upon object discrimination ability *(74)*.

Injection of 192 IgG-saporin into the ventricles impaired performance in an inhibitory avoidance task, although this impairment might have been due to the hyperkinesis produced by extensive cerebellar cell loss *(52)*. Injection directly into the NBM either slightly impaired *(48,75)* or had no effect upon *(49)* inhibitory avoidance learning. Injection of the immunotoxin into the MSA did not impair inhibitory avoidance performance or the development latent inhibition in a conditioned taste aversion task *(76)*. Finally, injection of the immunotoxin into the ventricles had no effect on performance in a delayed alternation test of spatial working memory *(49,67)*.

The interpretation of the results of studies that injected large amounts (>1.0 μg) of the immunotoxin into the ventricles is somewhat limited owing to the dose-dependent loss of Purkinje cells in the cerebellum and the potential impairment in motoric abilities *(40–42,52,66,71,77)* or implicit memories of sensorimotor learning *(77)*.

The observation that patients with AD have difficulty shifting visuospatial attention *(12–14)* was the rationale for the design of recent studies that have used this immunotoxin to investigate whether NBM cholinergic neurons regulate attention. Attention is most often defined in operational terms, that is, in relation to the behavioral task being used to investigate it. Frequently, subjects are required to choose between equally salient sources of stimulation or equally valid ways of interpreting stimulus information. Attention has many different components, including, but not limited to, vigilance, arousal, stimulus discrimination ability, and expectancy, which can be expressed in either a sustained or divided manner. The results of these investigations are discussed below.

Injection of the immunotoxin into the NBM impaired an animal's ability to increase the associability of a cue when it was an inconsistent predictor of another cue, but not its ability to demonstrate latent inhibition, that is, to decrease the associability of a cue when it was extensively preexposed prior to conditioning *(15)*. NBM lesions also impaired a rat's ability to detect visual signals of variable length and discriminate these signals from nonsignal events *(16)*. The loss of both MSA and NBM cholinergic cells impaired a rat's ability to discriminate between independent sensory stimuli and conferred a response bias to reward-related visual stimuli *(17)*. These results suggest that basal forebrain cholinergic neurons are involved with the control of shifting attention to potentially relevant, and brief, sensory stimuli that predict a biologically relevant event.

Electrophysiological studies have also used this immunotoxin to define the role of NBM and MSA cholinergic cells in cortical desynchrony, hippocampal theta and kindling activity, and in the postsynaptic action of acetylcholine within the hippocampus *(78)*. Injection of 192 IgG-saporin into the NBM significantly reduced the magnitude of a reward-related, negative slow potential recorded over the frontal cortex (Stoehr and Wenk, *unpublished findings*). These slow potentials are analogous to the contingent negative variation seen in humans, an electrical phenomenon that may represent specific psychological processes, such as expectancy, attention, or arousal *(79)*. Injection of the immunotoxin into the MSA reduced the power of theta activity in the hippocampus, but did not alter its main frequency *(57,80)* and facilitated kindling within the hippocampus *(81)*.

Primate models of AD have also investigated the effects of basal forebrain lesions produced by injection of nonselective excitatory amino acids *(21,82,83)*. Because the 192 IgG antibody used in the studies described previously was raised against rat p75, it does not bind to and destroy primate forebrain cholinergic cells. Recently an alternative immunotoxin was produced using a monoclonal antibody, ME 20.4, raised against the human p75 neurotrophin receptor *(84)*. This antibody binds to cholinergic neurons in the basal forebrain of primates *(85)*, and, when coupled with saporin, is an effective and selective cholinergic immunotoxin *(86)*. Injection of the ME 20.4 immunotoxin into the NBM of marmosets (*Callithrix jacchus*) impaired the acquisition and retention of simple visual discriminations *(86)*, an effect similar to that seen in lesioned rats.

3. LESIONS OF NONCHOLINERGIC SYSTEMS

3.1. An Immunotoxin for Noradrenergic Neurons

Several methods have been used to destroy central noradrenergic neurons, including (N2-chloroethyl)-N-ethyl-2-bromobenzyzlamine (DSP-4, *77,87*), 6-hydroxydopamine *(88)*, and 1-methyl-4-phenyl-1,2,3,6-tetrahydropyridine (MPTP, *89*). Although all are effective neurotoxins, the latter two also destroy dopaminergic neurons and all three have differential actions upon noradrenergic neurons within the locus coeruleus and lateral tegmental area *(88,89)*. Recently an effective immunotoxin that selectively targets noradrenergic neurons was introduced *(90–92)*. This immunotoxin consists of a monoclonal antibody to the noradrenergic cell-specific enzyme dopamine β-hydroxylase (DBH) that has been conjugated by disulfide bond to saporin. DBH is a suitable target because, following the release of norepinephrine, the membrane-bound form of this intravesicular enzyme is exposed to the extracellular environment before being recycled by endocytosis and retrogradely transported to the cell body *(93)*. The monoclonal antibody binds to the membrane-bound DBH, is endocytosed, and the saporin molecule is released into the cytoplasm to inhibit protein synthesis and induce cell death. The intraventricular administration of this immunotoxin dose-dependently destroyed noradrenergic cells within the brain stem without injury to the nearby dopaminergic neurons in the substantia nigra and ventral tegmental area, or the serotonergic neurons in the raphe nuclei, or the basal forebrain cholinergic neurons *(92)*. The intravenous administration of this immunotoxin was also able to produce an irreversible sympathectomy in both adult and neonatal rats *(90,91)*.

3.2. An Immunotoxin for Neurokinin 1 (Substance P) Receptors

A similar approach was taken to selectively target neurons that express neurokinin-1 receptors in the brain. After substance P binds to its G-protein-coupled neurokinin-1 receptor it is rapidly internalized into the neuron *(94)*. The conjugate of substance P bond to saporin produced an immunotoxin that is selective for neurons that express neurokinin-1 receptors that are postsynaptic to neurons that release substance P *(95)*. This immunotoxin has been used successfully to destroy lamina I spinal cord neurons that express the substance P receptor *(96)*. Substance P is released in the spinal cord onto neurokinin-1 receptors following noxious stimulation. This immunotoxin was recently used to reveal the important role that substance P plays in hyperalgesia *(96)*.

4. SUMMARY

The studies discussed in this chapter demonstrate that the immunolesioning method can be used to destroy a specific population of neurons, provided that certain conditions exist, that is, that a cell exhibit specific functional properties that make it selectively vulnerable. For example, the cell should express a specific molecule, for example, a protein receptor, that can be identified by an appropriate antibody. The specificity of this molecule for a particular neurotransmitter system will define and limit the selectivity of the immunotoxin. The loss of noncholinergic cerebellar Purkinje cells following the injection of 192 IgG-saporin into the ventricles is an example of the type of nonspecific cell loss that can occur when many different cell types express a common cell surface molecule, that is, the p75 protein. This molecule–antibody complex must then be internalized so that the cellular toxin, for example, saporin, can gain access to the cytoplasm and ultimately inhibit critical metabolic processes. The three immunotoxins discussed, that is, for cholinergic, noradrenergic, and substance P-receptive neurons, are only the first members of what is certain to become a large family of immunotoxins that can act as biochemical scalpels that will be used to study animal models of transmitter dysfunction.

ACKNOWLEDGMENT

The preparation of this chapter was supported by the Alzheimer's Association, IIRG-95-004.

REFERENCES

1. Whitehouse, P. J., Price, D. L., Clark, A. W., Coyle, J. T., and DeLong, M. R. (1981) Alzheimer Disease: Evidence for selective loss of cholinergic neurons in the nucleus basalis. *Ann. Neurol.* **10,** 122–126.
2. Olton, D. S. and Wenk, G. L. (1987) Dementia: animal models of the cognitive impairments produced by degeneration of the basal forebrain cholinergic system, in *Psychopharmacology: The Third Generation of Progress* (Meltzer, H. Y., ed.), Raven Press, New York, pp. 941–953.
3. Wenk, G. L. and Olton, D. S. (1987). Basal forebrain cholinergic neurons and Alzheimer's disease, in *Animal Models of Dementia: A Synaptic Neurochemical Perspective,* Vol. 33: *Neurology and Neurobiology* (Coyle, J. Y., ed.), Alan R. Liss, New York, pp. 81–101.
4. Allen, S. J., Dawbarn, D., and Wilcock, G. K. (1988) Morphometric immunochemical analysis of neurons in the nucleus basalis of Meynert in Alzheimer's disease. *Brain Res.* **454,** 275–281.
5. Coyle, J. T., Price, D. L., and DeLong, M. R. (1983) Alzheimer's disease: a disorder of cortical cholinergic innervation. *Science* **219,** 1184–1190.
6. McGeer, P. L., McGeer, E. G., Suzuki, J., Dolman, C. E., and Nagai, T. (1984) Aging, Alzheimer's disease and the cholinergic system of the basal forebrain. *Neurology* **34,** 741–745.
7. Fibiger, H. C. (1982) The organization and some projections of cholinergic neurons of the mammalian forebrain. *Brain Res. Rev.* **4,** 327–388.
8. Sofroniew, M. V., Eckenstein, F., Thoenen, H., and Cuello, A. C. (1982) Topography of choline acetyltransferase-containing neurons in the forebrain of the rat. *Neurosci. Lett.* **33,** 7–12.

9. Wenk, H., Bigl, V., and Meyer, U. (1980) Cholinergic projections from magnocellular nuclei of the basal forebrain to cortical areas in rats. *Brain Res. Rev.* **2**, 295–316.

10. Lamour, Y., Dutar, P., and Jobert A. (1982) Topographic organization of basal forebrain neurons projecting to the rat cerebral cortex. *Neurosci. Lett.* **34**, 117–122.

11. Lehmann, J., Nagy, J. I., Atmadja, S., and Fibiger, H. C. (1980) The nucleus basalis magnocellularis: The origin of a cholinergic projection to the neocortex of the rat. *Neuroscience* **5**, 1161–1174.

12. Freed, D. M., Corkin, S., Growdon, J. H., and Nissen, M. J. (1989) Selective attention in Alzheimer's disease: characterizing cognitive subgroups of patients. *Neuropsychology* **27**, 325–339.

13. Parasuraman, R. and Haxby, J. V. (1993) Attention and brain function in Alzheimer's disease: a review. *Neuropsychology* **7**, 242–272.

14. Scinto, L. F. M., Daffner, K. R., Castro, L., Weintraub, S., Vavrik, M., and Mesulam, M.-M. (1994) Impairment of spatially directed attention in patients with probable Alzheimer's disease as measured by eye movements. *Arch. Neurol.* **51**, 682–688.

15. Chiba, A. A., Bucci, D. J., Holland, P. C., and Gallagher, M. (1995) Basal forebrain cholinergic lesions disrupt increments but not decrements in conditioned stimulus processing. *J. Neurosci.* **15**, 7315–7322

16. McGaughy, J., Kaiser, T., and Sarter, M. (1996) Behavioral vigilance following infusions of 192 IgG-saporin into the basal forebrain: selectivity of the behavioral impairment and relation to cortical AChE-positive fiber density. *Behav. Neurosci.* **110**, 247–265.

17. Stoehr, J. D., Mobley, S. L., Roice, D. D., Brooks, R., Baker, L. Wiley, R. G., and Wenk, G. L. (1997) The effects of selective cholinergic basal forebrain lesions and aging upon expectancy in the rat. *Neurobiol. Learn. Mem.* **67**, 214–227.

18. Dunnett, S. B., Everitt, B. J., and Robbins, T. W. (1991) The basal forebrain-cortical cholinergic system: interpreting the function consequences of excitotoxic lesions. *Trends Neurosci.* **14**, 494–501.

19. Muir, J. L., Everitt, B. J., and Robbins, T. W. (1994) AMPA-induced excitotoxic lesions of the basal forebrain: a significant role for the cortical cholinergic system in attentional function. *J. Neurosci.* **14**, 2313–2326.

20. Muir, J. L., Page, K. J., Sirinathsinghji, D. J. S., Robbins, T. W., and Everitt, B. J. (1993) Excitotoxic lesions of the basal forebrain cholinergic neurons: effects on learning, memory and attention. *Behav. Brain Res.* **57**, 123–131.

21. Wenk, G. L. (1997) The nucleus basalis magnocellularis cholinergic system: 100 years of progress. *Neurobiol. Learn. Mem.* **67**, 85–95.

22. Wenk, G. L., Harrington, C. A., Tucker, D. A., Rance, N. E., and Walker, L. C. (1992) Basal forebrain neurons and memory: a biochemical, histological and behavioral study of differential vulnerability to ibotenate and quisqualate. *Behav. Neurosci.* **106**, 909–923.

23. Flicker, C., Dean, R., Watkins, D. L., Fisher, S. K., and Bartus, R. T. (1983) Behavioral and neurochemical effect following neurotoxic lesions of a major cholinergic input to the cerebral cortex in the rat. *Pharmacol. Biochem. Behav.* **18**, 973–981.

24. Wenk, G. L., Sweeney, J., Hughey, D., Carson, J., and Olton, D. (1986) Cholinergic function and memory: extensive inhibition of choline acetyltransferase fails to impair radial maze performance in rats. *Pharmacol. Biochem. Behav.* **25**, 521–526.

25. Dunnett, S. B., Whishaw, I. Q., Jones, G. H., and Bunch, S. T. (1987) Behavioral, biochemical and histochemical effects of different neurotoxic amino acids injected into nucleus basalis magnocellularis of rats. *Neuroscience* **20**, 653–669.

26. Etherington, R., Mittleman, G., and Robbins, T. W. (1987) Comparative effects of nucleus basalis and fimbria-fornix lesions on delayed matching and alternation tests of memory. *Neurosci. Res. Commun.* **1**, 135–143.

27. Page, K. J., Everitt, B. J., Robbins, T. W., Marston, H. M., and Wilkinson, L. S. (1991) Dissociable effects on spatial maze and passive avoidance acquisition and retention

following AMPA- and ibotenic acid-induced excitotoxic lesions of the basal forebrain in rats: differential dependence on cholinergic neuronal loss. *Neuroscience* **43**, 457–472.

28. Riekkinen, M., Riekkinen, P., and Riekkinen, P., Jr. (1991) Comparison of quisqualic and ibotenic acid nucleus basalis magnocellularis lesions on water-maze and passive avoidance performance. *Brain Res. Bull.* **27**, 119–123.

29. Robbins, T. W., Everitt, B. J., Marston, H. M., Wilkinson, J., Jones, G. H., and Page, K. J. (1989) Comparative effects of ibotenic acid- and quisqualic acid-induced lesions of the substantia innominata on attentional function in the rat: further implications for the role of the cholinergic neurons of the nucleus basalis in cognitive processes. *Behav. Brain Res.* **35**, 221–240.

30. Robbins, T. W., Everitt, B. J., Ryan, C. N., Marston, H. M., Jones, G. H., and Page, K. J. (1989) Comparative effects of quisqualic and ibotenic acid-induced lesions of the substantia innominata and globus pallidus on the acquisition of a conditional visual discrimination: differential effects on cholinergic mechanisms. *Neuroscience* **28**, 337–352.

31. Wenk, G. L., Markowska, A., and Olton, D. S. (1989) Basal forebrain lesions and memory: alterations in neurotensin, not acetylcholine, may cause amnesia. *Behav. Neurosci.* **103**, 765–769.

32. Fibiger, H. C. (1991) Cholinergic mechanisms in learning, memory and dementia: a review of recent evidence. *Trends Neurosci.* **14**, 220–223.

33. Book, A. A., Wiley, R. G., and Schweitzer, J. B. (1994) 192 IgG-saporin: I. Specific lethality for cholinergic neurons in the basal forebrain of the rat. *J. Neuropathol. Exp. Neurol.* **53**, 95–102.

34. Book, A. A., Wiley, R. G., and Schweitzer, J. B. (1995) 192 IgG-saporin. 2. Neuropathology in the rat brain. *Acta Neuropathol.* **89**, 519–526.

35. Wiley, R. G., Oeltmann, T. N., and Lappi, D. A. (1991) Immunolesioning: selective destruction of neurons using immunotoxin to rat NGF receptor. *Brain Res.* **562**, 149–153.

36. Wiley, R. G., Berbos, T. G., Deckwerth, T. L., Johnson, E. M., Jr., and Lappi, D. A. (1995) Destruction of the cholinergic basal forebrain using immunotoxin to rat NGF receptor: modeling the cholinergic degeneration of Alzheimer's disease. *J. Neurol. Sci.* **128**, 157–166.

37. Barthelemy, I., Martineau, D., Ong, M., Matsunami, R., Ling, N., Benatti, L., Cavallaro, U., Soria, M., and Lappi, D. A. (1993) Expression of saporin, a ribosome-inactivating protein from the plant *Saponaria officinalis,* in *Escherichia coli. J. Biol. Chem.* **268**, 6541–6548.

38. Ohtake, T., Heckers, S., Wiley, R. G., Lappi, D. A., Mesulam, M. M., and Geula, C. (1997) Retrograde degeneration and colchicine protection of basal forebrain cholinergic neurons following hippocampal injections of an immunotoxin against the P75 nerve growth factor receptor. *Neuroscience* **78**, 123–133.

39. Baxter, M. G., Bucci, D. J., Gorman, L. K., Wiley, R. G., and Gallagher, M. (1995) Selective immunotoxic lesions of the basal forebrain cholinergic cells: effects on learning and memory in rats. *Behav. Neurosci.* **109**, 714–722.

40. Berger-Sweeney, J., Heckers, S., Mesulam, M.-M., Wiley, R. G., Lappi, D. A., and Sharma, M. (1994) Differential effects on spatial navigation of immunotoxin-induced cholinergic lesions of the medial septal area and nucleus basalis magnocellularis. *J. Neurosci.* **14**, 4507–4519.

41. Heckers, S., Ohtake, T., Wiley, R. G., Lappi, D. A., Geula, C., and Mesulam, M. (1994) Complete and selective cholinergic denervation of rat neocortex and hippocampus but not amygdala by an immunotoxin against the p75 NGF receptor. *J. Neurosci.* **14**, 1271–1289.

42. Nilsson, O. G., Leanza, G., Rosenblad, C., Lappi, D. A., Wiley, R. G., and Björklund, A. (1992) Spatial learning impairments in rats with selective immunolesion of the forebrain cholinergic system. *NeuroReport* **3**, 1005–1008.

43. Robinson, J. K., Wenk, G. L., Wiley, R. G., Lappi, D. A., and Crawley, J. N. (1996) 192-IgG-saporin immunotoxin and ibotenic acid lesions of nucleus basalis and medial septum produce comparable deficits on delayed nonmatching to position in rats. *Psychobiology* **24,** 179–186.

44. Rossner, S., Hartig, W., Schliebs, R., Bruckner, G., Brauer, K., Perez-Polo, J. R., Wiley, R. G., and Bigl, V. (1995) 192 IgG-saporin immunotoxin-induced loss of cholinergic cells differentially activates microglia in rat basal forebrain nuclei. *J. Neurosci. Res.* **41,** 335–346.

45. Rossner, S., Schliebs, R., Perez-Polo, J. R., Wiley, R. G., and Bigl, V. (1995) Differential changes in cholinergic markers from selected brain regions after specific immunolesion of the rat cholinergic basal forebrain system. *J. Neurosci. Res.* **40,** 31–43.

46. Rossner, S., Yu, J., Pizzo, D., Werrbach-Perez, K., Schliebs, R., Bigl, V., and Perez-Polo, J. R. (1996) Effects of intraventricular transplantation of NGF-secreting cells on cholinergic basal forebrain neurons after partial immunolesion. *J. Neurosci. Res.* **45,** 40–56.

47. Singh, V. and Schweitzer, J. B. (1995) Loss of p75 nerve growth factor receptor mRNA containing neurons in rat forebrain after intraventricular IgG 192-saporin administration. *Neurosci. Lett.* **194,** 117–120.

48. Torres, E. M., Perry, T. A., Blokland, A., Wilkinson, L. S., Wiley, R. G., Lappi, D. A., and Dunnett, S. B. (1994) Behavioral, histochemical and biochemical consequences of selective immunolesions in discrete regions of the basal forebrain cholinergic system. *Neuroscience* **63,** 95–122.

49. Wenk, G. L., Stoehr, J. D., Quintana, G., Mobley, S., and Wiley, R. G. (1994) Behavioral, biochemical, histological, and electrophysiological effects of 192 IgG-saporin injections into the basal forebrain of rats. *J. Neurosci.* **14,** 5986–5995.

50. Waite, J. J., Wardlow, M. L., Chen, A. C., Lappi, D. A., Wiley, R. G., and Thal, L. J. (1994) Time course of cholinergic and monoaminergic changes in rat brain after immunolesioning with 192 IgG-saporin. *Neurosci. Lett.* **169,** 154–158.

51. Walsh, T. J., Kelly, R. M., Dougherty, K. D., Stackman, R. W., Wiley, R. G., and Kutscher, C. L. (1995) Behavioral and neurobiological alterations induced by the immunotoxin 192-IgG-saporin: cholinergic and non-cholinergic effects following i.c.v. injection. *Brain Res.* **702,** 233–245.

52. Waite, J. J., Chen, A. D., Wardlow, M. L., Wiley, R. G., Lappi, D. A., and Thal, L. J. (1995) 192 Immunoglobulin G-saporin produces graded behavioral and biochemical changes accompanying the loss of cholinergic neurons of the basal forebrain and cerebellar purkinje cells. *Neuroscience* **65,** 463–476.

53. Jolkkonen, J., Kähkönen, K., and Pitkänen, A. (1997) Cholinergic deafferentation exacerbates seizure-induced loss of somatostatin-immunoreactive neurons in the rat hippocampus. *Neuroscience* **80,** 401–411.

54. Levey, A. I., Edmunds, S. M., Hersch, S. M., Wiley, R. G., and Heilman, C. J. (1995) Light and electron microscopic study of m2 muscarinic acetylcholine receptor in the basal forebrain of the rat. *J. Comp. Neurol.* **351,** 339–356.

55. Levey, A. I., Edmunds, S. M., Koliatsos, V., Wiley, R. G., and Heilman, C. J. (1995) Expression of m1-m4 muscarinic acetylcholine receptor proteins in rat hippocampus and regulation by cholinergic innervation. *J. Neurosci.* **15,** 4077–4092.

56. Roβner, S., Schliebs, R., Härtig, W., and Bigl, V. (1995) 192 IgG-saporin-induced selective lesion of cholinergic basal forebrain system: neurochemical effects on cholinergic neurotransmission in rat cerebral cortex and hippocampus. *Brain Res. Bull.* **38,** 371–381.

57. Bassant, M. H., Apartis, E., Jazat-Poindessous, F. R., Wiley, R. G., and Lamour, Y. A. (1995) Selective immunolesion of the basal forebrain cholinergic neurons: effects on hippocampal activity during sleep and wakefulness in the rat. *Neurodegeneration* **4,** 61–70.

58. Kapás, L., Obál, F., Jr., Book, A. A., Schweitzer, J. B., Wiley, R. G., and Krueger, J. M. (1996) The effects of immunolesions of nerve growth factor-receptive neurons by 192IgG-saporin on sleep. *Brain Res.* **712**, 53–59.

59. Rossner, S., Schliebs, R., Hartig, W., Perez-Polo, J. R., and Bigl, V. (1997) Selective induction of c-Jun and NGF in reactive astrocytes after cholinergic degenerations in rat basal forebrain. *NeuroReport* **8**, 2199–2202.

60. Roβner, S., Wörtwein, G., Gu, Z., Yu, J., Schliebs, R., Bigl, V., and Perez-Polo, J. R. (1997) Cholinergic control of nerve growth factor in adult rats: evidence from cortical cholinergic deafferentation and chronic drug treatment. *J. Neurochem.* **69**, 947–953.

61. Yu, J., Wiley, R. G., and Perez-Polo, R. J. (1996) Altered NGF protein levels in different brain areas after immunolesion. *J. Neurosci. Res.* **43**, 213–223.

62. Pallera, A. M., Schweitzer, J. B., Book, A. A., and Wiley, R. G. (1994) 192 9IgG-saporin causes a major loss of synaptic content in rat olfactory bulb. *Exp. Neurol.* **127**, 265–277.

63. Heckers, S. and Mesulam, M. (1994) Two types of cholinergic projections to the rat amygdala. *Neuroscience* **2**, 383–397.

64. Bannon, A. W., Curzon, P., Gunther, K. L., and Decker, M. W. (1996) Effects of intra-septal injection of 192-IgG-saporin in mature and aged Long–Evans rats. *Brain Res.* **718**, 25–36.

65. Dornan, W. A., McCampbell, A. R., Tinkler, G. P., Hickman, L. J., Bannon, A. W., Decker, M. W., and Gunther, K. L. (1996) Comparison of site-specific injections into the basal forebrain on water maze and radial arm maze performance in the male rat after immuno-lesioning with 192 IgG saporin. *Behav. Brain Res.* **82**, 93–101.

66. Leanza, G., Nilsson, O. G., Wiley, R. G., and Bjorklund, A. (1995) Selective lesioning of the basal forebrain cholinergic system by intraventricular 192 IgG-saporin: behavioural, biochemical and stereological studies in the rat. *Eur. J. Neurosci.* **7**, 329–343.

67. Pappas, B. A., Davidson, C. M., Fortin, T., Nallathamby, S., Park G. A., Mohr, E., and Wiley, R. G. (1996) 192 IgG-saporin lesion of basal forebrain cholinergic neurons in neo-natal rats. *Brain Res.* **96**, 52–61.

68. McMahan, R. W., Sobel, T. J., and Baxter, M. G. (1997) Selective immunolesions of hippocampal cholinergic input fail to impair spatial working memory. *Hippocampus* **7**, 130–136.

69. Shen, J., Barnes, C. A., Wenk, G. L., and McNaughton, B. L. (1996) Differential effects of selective immunotoxic lesions of medial septal cholinergic cells on spatial reference and working memory. *Behav. Neurosci.* **110**, 1181–1186.

70. Walsh, T. J., Herzog, C. D., Gandhi, C., Stackman, R. W., and Wiley, R. G. (1996) Injec-tion of IgG 192-saporin into the medial septum produces cholinergic hypofunction and dose-dependent working memory deficits. *Brain Res.* **726**, 69–79.

71. Leanza, G., Muir, J., Nilsson, O. G., Wiley, R. G., Dunnett, S. B., and Bjorklund, A. (1996) Selective immunolesioning of the basal forebrain cholinergic system disrupts short-term memory in rats. *Eur. J. Neurosci.* **8**, 1535–1544.

72. McDonald, M. P., Wenk, G. L., and Crawley, J. N. (1997) Analysis of galanin and the galanin antagonist M40 on delayed non-matching-to-position performance in rats lesioned with the cholinergic immunotoxin 192 IgG-saporin. *Behav. Neurosci.* **111**, 552–563.

73. Steckler, T., Keith, A. B., Wiley, R. G., and Sahgal, A. (1995) Cholinergic lesions by 192 IgG-saporin and short-term recognition memory: role of the septohippocampal projection. *Neuroscience* **66**, 101–114.

74. Vnek, N., Kromer, L. F., Wiley, R. G., and Rothblat, L. A. (1996) The basal forebrain cholinergic system and object memory in the rat. *Brain Res.* **710**, 265–270.

75. Zhang, Z. J., Berbos, T. G., Wrenn, C. C., and Wiley, R. G. (1996) Loss of nucleus basalis magnocellularis, but not septal, cholinergic neurons correlates with passive avoidance impairment in rats treated with 192-saporin. *Neurosci. Lett.* **203**, 214–218.

76. Dougherty, K. D., Salat, D., and Walsh, T. J. (1996) Intraseptal injection of the cholinergic immunotoxin 192-IgG saporin fails to disrupt latent inhibition in a conditioned taste aversion paradigm. *Brain Res.* **736,** 260–269.

77. Glickstein, M. and Yeo, C. (1990) The cerebellum and motor learning. *J. Cognit. Neurosci.* **2,** 69–80.

78. Jouvenceau, A., Billard, J. M., Wiley, R. G., Lamour, Y., and Dutar, P. (1994) Cholinergic denervation of the rat hippocampus by 192-IgG-saporin: electrophysiological evidence. *NeuroReport* **5,** 1781–1784.

79. Tecce, J. J. (1972) Contingent negative variation (CNV) and psychological processes in man. *Psychol. Bull.* **77,** 73–108.

80. Lee, M. G., Chrobak, J. J., Sik, A., Wiley, R. G., and Buzsaki, G. (1994) Hippocampal theta activity following selective lesion of the septal cholinergic system. *Neuroscience* **62,** 1033–1047.

81. Kokaia, M., Ferencz, I., Leanza, G., Elmer, E., Metsis, M., Kokaia, Z., Wiley, R. G., and Lindvall, O. (1996) Immunolesioning of basal forebrain cholinergic neurons facilitates hippocampal kindling and perturbs neurotrophin messenger RNA regulation *Neuroscience* **70,** 313–327.

82. Voytko, M. L., Olton, D. S., Richardson, R. T., Gorman, L. K., Tobin, J. T., and Price, D. L. (1994) Basal forebrain lesions in monkeys disrupt attention but not learning and memory. *J. Neurosci.* **14,** 167–186.

83. Wenk, G. L. (1993) A primate model of Alzheimer's disease. *Behav. Brain Res.* **57,** 117–122.

84. Ross, A. H., Grob, P., Bothwell, M., Elder, D. E., Ernst, C. S., Marano, N., Ghrist, B. F. D., Slemp, C. C., Herlyn, M., Atkinson, B., and Koprowski, H. (1984) Characterisation of nerve growth factors in neural crest tumors using monoclonal antibodies. *Proc. Natl. Acad. Sci. USA* **81,** 6681–6685.

85. Maclean, C. J., Baker, H. F., Fine, A., and Ridley, R. M. (1996) The distribution of p75 neurotrophin receptor-immunoreactive cells in the forebrain of the common marmoset (*Callithrix jacchus*). *Brain Res.* **43,** 197–208.

86. Fine, A., Hoyle, C., Maclean, C. J., Levatte, T. L., Baker, H. F., and Ridley, R. M. (1997) Learning impairments following injection of a selective cholinergic immunotoxin, ME20.4 IgG-saporin, into the basal nucleus of Meynert in monkeys. *Neuroscience* **81,** 331–343.

87. Wenk, G. L., Hughey, D., Boundy, V., Kim, A., Walker, L., and Olton, D. S. (1987) Neurotransmitters and memory: the role of cholinergic, serotonergic and noradrenergic systems. *Behav. Neurosci.* **101,** 325–332.

88. Kostrzewa, R. M. and Jacobowitz, D. M. (1974) Pharmacological actions of 6-OHDA. *Pharmacol. Rev.* **26,** 199–288.

89. Hallman, H., Lange, J., Olson, L., Stromberg, L., and Jonsson, G. (1985) Neurochemical and histochemical characterization of neurotoxic effects of MPTP on brain catecholamine neurons in the mouse. *J. Neurochem.* **44,** 117–127.

90. Picklo, M. J., Wiley, R. G., Lappi, D. A., and Robertson, D. (1994) Noradrenergic lesioning with an anti-dopamine beta-hydroxylase immunotoxin. *Brain Res.* **666,** 195–200.

91. Picklo, M. J., Wiley, R. G., Lonce, S., Lappi, D. A., and Robertson, D. (1995) Anti-dopamine beta-hydroxylase immunotoxin-induced sympathectomy in adult rats. *J. Pharmacol. Exp. Ther.* **275,** 1003–1010.

92. Wrenn, C. C., Picklo, M. J., Lappi, D. A., Robertson, D., and Wiley, R. G. (1996) Central noradrenergic lesioning using anti-DBH-saporin: anatomical findings. *Brain Res.* **740,** 175–184.

93. Strudelska, D. R. and Brimijoin, S. (1989) Partial isolation of two classes of dopamine β-hydroxylase-containing particles undergoing rapid axonal transport in rat sciatic nerve. *J. Neurochem.* **53,** 623–631.

94. Mantyh, P. W., Allen, C. J., Ghilardi, J. R., Rogers, S. D., Mantyh, C. R., Liu, H., Basbaum, A. I., Vigna, S. R., and Maggio, J. E. (1995) Rapid endocytosis of a G-protein-coupled receptor: substance P evoked internalization of its receptor in the rat striatum in vivo. *Proc. Natl. Acad. Sci. USA* **92,** 2622–2626.
95. Wiley, R. G. and Lappi, D. A. (1997) Destruction of neurokinin-1 receptor expressing cells in vitro and in vivo using substance P-saporin in rats. *Neurosci. Lett.* **230,** 97–100.
96. Mantyh, P. W., Rogers, S. D., Honore, P., Allen, B. J., Ghilardi, J. R., Li, J., Daughters, R. S., Lappi, D. A., Wiley, R. G., and Simone, D. A. (1997) Inhibition of hyperalgesia by ablation of lamina I spinal neurons expressing the substance P receptor. *Science* **278,** 275–279.

5

An Intracerebral Tumor Necrosis Factor-α Infusion Model for Inflammation in Alzheimer's Disease

Kimberly B. Bjugstad and Gary W. Arendash

1. INTRODUCTION

Alzheimer's disease (AD) is a progressive neurodegenerative disorder of insidious onset, characterized by extensive short- and long-term memory loss in association with a variety of cognitive disabilities. AD is most commonly seen after the age of 60 yr, although it often occurs earlier in genetically predisposed individuals. An estimated 5–10% of individuals over the age of 65 have the severe dementia of AD and as many as 50% of individuals over the age of 85 have AD. The disease is characterized by a profound "cerebral atrophy," manifested by cortical atrophy (thinning) and ventricular enlargement—both of which are correlated with the extent of dementia *(1–3)*. The cerebral atrophy of AD affects specific parts of the brain (i.e., cerebral cortex, hippocampus, entorhinal cortex) and is due to loss of neurons and synapses therein *(1,4)*. Recent evidence suggests that at least some of this neuronal loss in AD is due to apoptosis (programmed cell death), as apoptotic cell bodies are much more numerous in AD brains compared to age-matched controls *(5,6)*.

The two characteristic neuropathologic lesions in AD are neurofibrillary tangle-containing neurons and neuritic plaques—both of which are present in normal aged brains, but to a much greater extent in AD brains. Neurofibrillary tangles, which have yet to be directly linked to the AD process, are comprised mainly of abnormally phosphorylated τ protein. Neuritic plaques, the density of which has been correlated with degree of dementia, are comprised of a central core of β-amyloid (a 40–43-amino acid peptide), surrounded by degenerative nerve terminals, reactive astrocytes, and "activated" microglia. Although the cause(s) of AD is (are) currently unknown, mounting evidence provides a compelling case for involvement of two interrelated mechanisms within the AD brain—inflammation and free-radical-induced oxidative stress *(7–9)*. Along this line, studies clearly suggest that inflammatory cytokines released by activated glial cells (particularly astrocytes and microglia associated with neuritic plaques) participate in a local inflammatory cascade that promotes β-amyloid deposition, free radical formation, and resultant cell death in AD.

In this chapter, we first indicate supportive evidence for involvement of brain inflammation and associated free radical mechanisms in the AD process. We then pro-

From: Central Nervous System Diseases
Edited by: D. F. Emerich, R. L. Dean, III, and P. R. Sanberg © Humana Press Inc., Totowa, NJ

pose tumor necrosis factor-α (TNF-α) as a key inflammatory cytokine in AD and describe our initial findings involving a TNF-α intracerebral infusion model of brain inflammation that is relevant to AD. It is suggested that such an in vivo model could be most useful in testing the utility of new antiinflammatory and/or antioxidant drugs with the potential to prevent or treat AD.

2. EVIDENCE THAT ALZHEIMER'S DISEASE INVOLVES CEREBRAL INFLAMMATION

It has been suggested that the etiology or at least the progression of AD involves an inflammatory response by the immune system. Several studies have indicated that there is an inverse relationship between AD and the use of antiinflammatories, especially nonsteroidal antiinflammatory drugs (NSAIDs) *(10,11)*. The inverse relationship between AD and NSAID use is maintained even when other variables such as education, age, gender, and other medications were controlled for *(12)*. Direct evidence of this relationship was seen in a 1-yr longitudinal study that found that AD patients on an antiinflammatory regime had a higher level of functioning than nontreated AD patients *(13)*.

Histologic analyses of AD brains and cerebrospinal fluid (CSF) sampling adds support for a relationship between AD and inflammation, while also suggesting that AD involves a "local" or central nervous system (CNS) immune activation rather than the activation of the peripheral immune system. In this context, immunohistochemical analysis of AD brains showed an enhanced staining for immune markers specific for microglia, the brain's resident immune cells, but failed to find significant staining for peripheral immune markers such as immunoglobulins or T-cell subsets *(14)*. Also, CSF samples taken from AD patients showed an increase in antibrain antibodies for microglia, a result that appears to be specific for AD compared to other forms of dementia *(15)*. If, as those data suggest, AD is a local inflammatory disease, then this may account for the discordant results from studies searching for elevated immune markers (i.e., TNF-α) in the periphery via sera from AD patients *(16,17)*.

The hallmark characteristic of AD is the presence of neuritic plaques. Closely associated with the β-amyloid core of these plaques are activated microglia cells. These microglia are heavily stained for the inflammatory cytokines TNF-α, interleukin-1 (IL-1), and interleukin-6 (IL-6), as well as for activated complement proteins, and adhesion molecules that facilitate recruitment and positioning of immune cells *(7,18)*. One key question is whether this inflammatory reaction is an end-stage of neuritic plaques or whether it is contributory to their development.

β-Amyloid, the peptide of 40–42 amino acids forming the core of neuritic plaques, has been found to be neurotoxic when administered in "aggregated" form, both in cell culture experiments and following intracerebral infusion *(19–22)*. These data support the yet unproven hypothesis that aggregation of β-amyloid and its ensuing deposition are key steps that precede neuronal loss in AD. Other in vitro studies have indicated that "aggregated" β-amyloid can enhance production of oxygen free radicals in CNS homogenates *(23–25)*. A synergistic relationship may, in fact, exist between β-amyloid and cerebral inflammation in AD, with at least one final common pathway being enhanced free radical production and resultant neuronal loss/dysfunction *(see* Fig. 1).

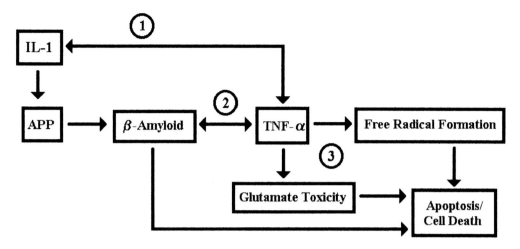

Fig. 1. Flow diagram showing the possible interactions between TNF-α, β-amyloid, and the resulting apoptosis/cell death. Path 1 shows how TNF-α can influence β-amyloid production through its influence on other cytokines that increase APP synthesis, specifically IL-1. Path 2 shows a more synergetic relationship between TNF-α and β-amyloid, with β-amyloid inducing TNF-α release via microglia and TNF-α inducing β-amyloid formation via ACT. Path 3 shows that elevated TNF-α can result in apoptosis and cell death, through the induction of free radical formation and glutamate toxicity.

Along this line, there is evidence that β-amyloid induces an inflammatory cytokine reaction in the brain. Specifically, β-amyloid has been shown to induce the release of TNF-α, IL-1, IL-6, and reactive oxygen species from glial cells *(26–29)*.

There is also evidence to suggest that inflammatory cytokines may be involved in the formation and deposition of β-amyloid into neuritic plaques. IL-1 can enhance amyloid precursor protein (APP) synthesis, the protein from which β-amyloid is cleaved *(30,31)*. In addition, IL-1 and TNF-α can up-regulate the transcription and synthesis of α_1-antichymotrypsin (ACT), an acute phase protein involved in β-amyloid fibril formation *(32)*. Furthermore, complement proteins and complement regulatory proteins are present in neuritic plaques *(33)*. β-Amyloid has the potential to bind C1q, the first protein in a series of proteins that must be combined to form the complement *(30)*. Because C1q has six binding sites, it has the potential to bring multiple β-amyloid proteins together *(34)*. It is noteworthy that brain areas with a low incidence of plaque formation (i.e., cerebellum) have low levels of C1q, whereas areas with high plaque formation (i.e., cortex) display high levels of C1q *(34)*.

The aforementioned studies suggest that activated glia, inflammatory cytokines, complement proteins, and β-amyloid are all part of a complex series of interactions in AD. This series of inflammatory–β-amyloid interactions could be self-propagating and causative to major neuropathologic aspects of AD, such as neuritic plaques, activated microglia within neuritic plaques, and activation of complement proteins within the brain. Whether brain inflammation or β-amyloid is ultimately determined to be "primary" to the AD disease process may be of lesser importance than the apparent synergy between these two processes, which we (and others) propose to result in a positive feedback, self-propagating pathogenic process. Thus, even if cytokines are a result rather than causative factor in AD, they are probably still directly involved in the disease process and can certainly exacerbate disease progression.

3. TNF-α AS A KEY CYTOKINE IN AD

There are three cytokines that are considered to be the major mediators of inflammation: IL-1, IL-6, and TNF-α *(35)*. Of these, increasing evidence indicates that TNF-α may play the most important role in the pathogenesis of AD. TNF-α is a 26-kDa inflammatory cytokine released by monocytes, macrophages, microglia, and astrocytes, as well as the peripheral immune natural killer (NK) cells, T cells, and B cells *(8,36,37)*. TNF-α is thought to be a primary mediator of inflammation, as it can induce the release of other cytokines (IL-1, IL-6, and IL-8) and complement proteins, as well as the production of free radicals *(38–40,80)*. In addition, the transcription and synthesis of TNF-α can in turn be influenced by IL-1, IL-2, complement proteins, interferon-γ (IFN-γ), and lipopolysaccharide (LPS), creating a self-perpetuating feedback loop *(37)*.

TNF-α has two receptors that it can bind to—one that is approx 55–60 kDa in size and thus referred to as p55 or TNF-R1; the second is 75–80 kDa and has been referred to as p75 or TNF-R2 *(37,41,42)*. These receptors appear to have diametrical opposed properties. The p55 receptor appears to be involved with cytotoxic properties of TNF-α *(23,37)*. For example, transgenic mice that fail to express the p55 receptor appear immune to septic shock resulting from LPS-induced elevations in TNF-α *(43)*. Conversely, the p75 receptor may play a protective role after injury. Blocking the p75 receptors in cultured human neuronal cells significantly increases the cell injury/death after exposure to hypoxic conditions or following treatment with β-amyloid *(44)*. There is some evidence that TNF-α had neuroprotective effects against β-amyloid toxicity *(26)*, although such neuroprotective properties of TNF-α are usually observed in embryonic or immature neuronal cultures *(26,45)*. Along this line, one study using perinatal and adult rodent tissue found that TNF-α could induce neurotoxicity in adult tissue but had no effect in the perinatal tissue *(46)*. In Alzheimer's patients the density of both TNF-α receptor subtypes is increased compared to age-matched controls, but no change in receptor affinities occurs *(47)*.

The concept of TNF-α as the main inflammatory cytokine in AD comes from two lines of evidence: (1) the connection between TNF-α and β-amyloid, and (2) the connection between TNF-α and neuronal loss via free radical production and/or glutamate toxicity. Figure 1 illustrates a conceivable way in which TNF-α and β-amyloid may interact to produce neuronal loss and ultimately the progression of AD.

The connection between TNF-α and β-amyloid is both physiologic and anatomic. Activated microglia cells (the primary source of TNF-α) are consistently found surrounding the amyloid core of neuritic plaques. Immunohistochemical staining of these cells shows staining for IL-1 and lesser staining for IL-6—but the greatest staining is for TNF-α *(7)*. In microglia, β-amyloid stimulates TNF-α release *(29,42,48)* and, in the presence of IFN-γ, β-amyloid-induced release of TNF-α is enhanced. β-Amyloid can also enhance LPS-induced release of TNF-α release from astrocytes *(27)*. Furthermore, β-amyloid-induced production of nitric oxide (NO•) from microglia is mediated by TNF-α *(29)*. As stated earlier, ACT, a protease inhibitor thought to regulate β-amyloid fibril formation *(49)*, can be up-regulated by TNF-α and IL-1, but not by IL-6 *(50,51)*. ACT tends to be found in astrocytes closely associated with plaques

(49,52). Moreover, an increase in ACT is found in AD brains compared to brains from the normal aged *(52)*.

The relationship, then, between β-amyloid and TNF-α involves a mutual, interactive enhancement not seen with other cytokines. Of the primary mediators of inflammation (IL-1, IL-6, or TNF-α), TNF-α has the highest staining around microglia associated with the amyloid core of neuritic plaques, β-amyloid induces TNF-α release from microglia, and finally TNF-α can affect β-amyloid fibril formation through its effect on ACT. The result of this synergistic relationship between β-amyloid and TNF-α could be neuronal loss. β-Amyloid-induced apoptosis was studied in a variety of human tissues, including neuroblastoma cell lines, and it was found that when TNF-α was added with β-amyloid, an increase in apoptotic cell death was seen in these cultures *(53)*. Also, consistent with this notion are in vitro and in vivo studies showing the neurotoxicity of β-amyloid (*see* Subheading 2.), as well as in vitro studies demonstrating that TNF-α is toxic to neurons and oligodendrocytes *(44,54–56)*. Indeed, TNF-α alone can initiate apoptosis in neuronal cell cultures *(46,57,58)*, with apoptotic cells beginning to appear within 48–72 h after exposure to TNF-α *(55,56)*.

The process of TNF-α induced cell death appear to be mediated, at least in part, by enhanced free radical production and a resultant increase in cellular oxidative damage to DNA, proteins, and lipids *(81,82)*. In this context, TNF-α has been shown to induce formation of reactive oxygen species (i.e., superoxide radical, hydroxyl radical, and hydrogen peroxide) and to cause DNA damage *(59,60,82)*. In addition, TNF-α mediates the increased microglial production of NO• induced by β-amyloid *(29)*. Consistent with involvement of free radicals/oxidative damage in the mechanism of cytotoxicity of TNF-α are studies demonstrating that antioxidants and iron chelators can prevent TNF-α-induced cytotoxicity. For example, the DNA damage and cell death induced by TNF-α in L929 cells (an immortalized mouse fibroblast cell line) could be prevented by the iron chelator desferoxamine or the oxygen radical scavenger butylated hydroxyanisole (BHA) *(59,82)*. In addition, in a human neuroblastoma cell line (SK-N-MC), the oxygen radical scavenger *N*-acetylcysteine (NAC) reduced the number of TNF-α-induced apoptotic cells *(56)*.

In addition to free radical/oxidative damage mechanisms, there is evidence that TNF-α induces cytotoxicity through an interrelated mechanism, namely, excessive glutamate transmission (glutamate toxicity). Glutamate toxicity has long been proposed to play a role in the pathogenesis of AD *(61)*. In human neuronal cell cultures, TNF-α selectivity enhances glutamate toxicity in a dose- and time-dependent manner, although neither IL-1 or IL-6 have any such enhancement *(62)*. In human CNS cell cultures, the TNF-α-induced enhancement of glutamate toxicity appears to involve a number of processes, including: (1) enhancement of glutamate binding to postsynaptic *N*-methyl-D-aspartate (NMDA) receptors, (2) decreased uptake of glutamate by neurons and glia, and/or (3) a suppression of glutamine synthetase levels *(62,63)*. Singularly, or in combination, these three potential TNF-α effects would be expected to increase glutamate transmission postsynaptically, resulting in increased intraneuronal Ca^{2+} levels and stimulation of free radical formation through activation of Ca^{2+}-dependent enzymes *(9)*. Thus, TNF-α has the potential to increase free radical formation not only directly, but also indirectly via enhanced glutamate transmission (Fig. 1). Parenthetically, the elevated intraneuronal Ca^{2+} levels in the above scenario would also be expected to induce neurotoxicity through activation of Ca^{2+}-dependent endonucleases, proteases, and lipases.

4. AN INTRACEREBRAL TNF-A INFUSION MODEL FOR BRAIN INFLAMMATION

Most of the evidence concerning AD and inflammation presented thus far has involved in vitro work. Few studies have extended the findings of this in vitro work to in vivo models, although recently developed transgenic mouse models that overexpress mutant/fragmental APP are beginning to yield promising data supportive of an inflammatory response induced by elevated brain β-amyloid levels. Most notable among these new transgenic mouse lines is the "Swedish Mutation" (APP_{sw}) wherein the overexpressed human APP contains two mutant amino acid substitutions found in a large Swedish family with early-onset AD (64). These mice develop an age-related increase in brain β-amyloid levels, neuritic plaques in cortex and hippocampus, and cognitive impairment (64). Moreoever, APP_{sw} mice develop a significant microglial response, consisting of increased numbers of activated, enlarged microglia in and around neuritic plaques (65). A second APP transgenic mouse model (the PDAPP transgenic), which overexpresses human APP containing a single mutant amino acid substitution, also develops age-related β-amyloid deposition. This deposition is associated with an extensive astrogliosis (66). Despite widespread β-amyloid deposition and characteristics of an inflammatory response in both of these transgenic models, however, neither has been shown to result in neuronal loss through 18–21 mo of age. Involving a different approach is yet a third transgenic model for AD that involves overexpression of the C-terminal 104 amino acids of APP—the "C-100" transgenic model (67). Although β-amyloid deposition in this model is not as robust as in the previously mentioned models, cognitive deficits are apparent, as is reduced maintenance of long-term potentiation (LTP). Moreoever, an astrogliosis and microgliosis occur in hippocampus and cortex, along with a cell loss (nonserologic counts) in the CA1 region of hippocampus. Thus, initial studies involving APP transgenics have linked β-amyloid formation with certain aspects of a brain inflammatory response. Also providing such a link are rat studies involving intrahippocampal infusions of LPS (68), wherein deficits in active avoidance learning were observed and prevented with nonsteroidal antiinflammatory drug pretreatment.

Other than the foregoing studies, very little in vivo work has explored the involvement of brain inflammation in AD—particularly the involvement of individual proinflammatory cytokines. As documented in the previous section, TNF-α could be a key cytokine in the development of an inflammatory response in AD. Consistent with this notion are the interactions of TNF-α with β-amyloid and the neurotoxic effects of TNF-α, the later of which appears to involve free radical mechanisms, glutamate toxicity, and apoptosis. Therefore, an inflammatory-based animal model to discretely examine the impact of elevated TNF-α levels in the brain could provide valuable information regarding the neuropathologic, neurotoxic, and behavioral impact that this key inflammatory cytokine may have in AD. It is in that context that we have recently developed an intraventricular TNF-α infusion model for brain inflammation that has relevance to

AD, as well as other neurodegenerative diseases wherein an inflammatory response is observed.

4.1 General Protocol

To study the physiologic, behavioral, and neuropathological consequences of elevated brain TNF-α levels, young adult male Sprague–Dawley rats were prepared for long-term (7 d) intracerebroventricular infusions of TNF-α. Stainless steel cannulas were permanently implanted bilaterally into each lateral ventricle. Four to five days following intracerebroventricular surgery, animals were weighed and divided into two equal weight groups. On the first day of treatment, each animal was placed into a restraining device to receive bilateral intracerebroventricular infusions of either 50 ng/μL of human recombinant TNF-α (*n* = 13) or 1 μL of isotonic saline (*n* = 13). Each infusion was done over a 1-min time period. Intracerebroventricular infusions were given daily for 7 d (d 1–7), during which time animals were also weighed daily.

Following the 7 d of intracerebroventricular treatment, approximately half of the animals in each treatment group were tested in a standard Morris water maze *(69)*. These animals were tested daily for 7 d to evaluate the rate of learning/acquisition (d 8–14). On the day following completion of acquisition testing, each animal was given a single probe trial to evaluate the level of memory retention. The Morris water maze consists of a circular pool divided into four quadrants and filled with water. Along the pool walls are black visual cues and multiple visual cues were present around the pool. During acquisition testing, an escape platform was placed in one of the quadrants. The top of the escape platform is just below the water line of the pool so that it cannot be seen by the swimming rat. The rat must use the visual cues in and around the pool walls to find and escape onto the submerged platform. For each trial, the time from placement into the pool to finding the platform was recorded as "latency" and is used as a measure of learning. Four trials were done daily for acquisition, with the animal placed at a different location in the maze to initiate each trial. An average daily latency was calculated and used for statistical analysis. For the retention probe trial, the platform was removed. Each animal was given 60 s to "search" for the escape platform, and the percentage of time spent searching in each quadrant was recorded. A higher percentage of swim time spent in the quadrant that formerly contained the platform was used as a measure of memory retention. During behavior testing, animals were weighed every other day.

On the day following the retention trial, animals were anesthetized with sodium pentobarbital and intracardially perfused with 4% neutral buffered formalin. Brains were subsequently removed from the cranium and stored at 20°C until histologic processing. Thionin-stained brain sections were used to determine cannula locations in each animal, as well as ventricular area, neocortical thickness, neostriatal area, and dorsal hippocampal volume. Only those brains in which the cannulas clearly penetrated the lateral ventricles were used.

The remaining animals (i.e., those that were not behaviorally tested) were killed 3 h after the last intracerebroventricular infusion on d 7. Again, animals were anesthetized

WEIGHTS

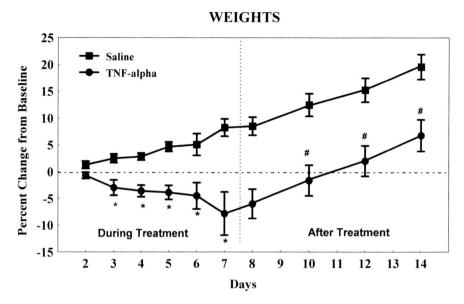

Fig. 2. Effects of intracerebroventricular infusions of TNF-α on body weight during 7 d of treatment (d 2–7) and for the 7 d following treatment (d 8–14). Weights reflect the percent change from d 1 baseline weights. *Significantly different from the saline control group (p < .05). #Significantly different from weights at d 7, the end of the treatment period (*p* < .05).

and intracardially perfused with neutral buffered formalin. Brains were stored for no more than 24 h before a 3-mm thick coronal slice of the cannula site was taken and paraffin embedded. From this slice, 6-μm sections were taken at the deepest point of cannula penetration and processed for *in situ* end labeling (ISEL) of apoptotic cells. Quantification of parenchymal apoptotic cells adjacent to the infusion site and located in the neocortex, basal ganglia, or septum was done for two sections from each animal and averaged together. The total number of apoptotic cells as well as the number of apoptotic cells in each specified area were determined for each infusion site.

4.2. TNF-α-Induced Weight Changes

During the week of intracerebroventricular treatment, TNF-α infused animals had a progressive weight loss compared to saline-infused animals, which gained weight during this time (*see* Fig. 2). By d 7 of treatment, TNF-α-treated animals showed a 7.8% loss in weight from their d 1 weights, while the weight of saline controls increased by 8.3%. Weight loss is a common side effect of elevated TNF-α levels and occurs independently of other cytokines *(70,71,80)*. When TNF-α treatment stopped, animals began to gain weight at a rate equal to the saline controls, so that by d 14 TNF-α-treated animals were 6.8% heavier than their original d 1 weights. Although multiple mechanisms probably contribute to the weight loss often associated with AD, the weight loss presently observed through intracerebral TNF-α infusion suggests that elevated brain levels of TNF-α could be contributory to the weight loss seen in many AD patients *(72,73)*.

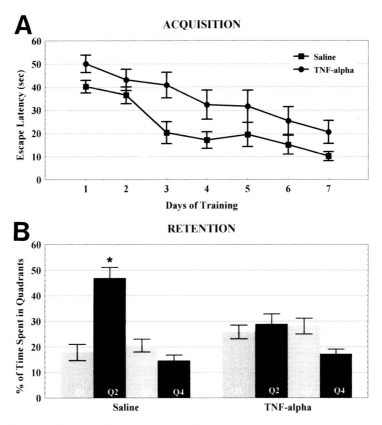

Fig. 3. Effects of intracerebroventricular infusion of TNF-α on water maze performance. **(A)** Water maze acquisition during the 7 d after treatment. Daily escape latencies were calculated for each animal from the mean of four daily trials. **(B)** Water maze retention on the day following the completion of acquisition testing. Percentage of time spent in each of the four quadrants (Q1, Q2, Q3, and Q4) was recorded. Q2 was the quadrant that previously contained the escape platform during acquisition. *Significantly different from the other quadrants and from the other group's Q2 quadrant ($p < .05$).

4.3. TNF-α-Induced Cognitive Deficits

During acquisition testing, animals that had previously been given daily intracerebroventricular injections of TNF-α had consistently longer escape latencies than saline controls (Fig. 3A). Because there was no significant difference between both groups' swim speed, the longer escape latencies were attributed to the effects of TNF-α on cognitive performance. Because both groups did display statistically equal latencies (learning) by d 7, a memory retention probe trial was run.

The cognitive deficits displayed by the TNF-α-infused animals during acquisition testing were also evident during the memory retention probe trial. Animals infused with TNF-α showed no quadrant preference and spent approximately equal percentages of time searching in all quadrants for the escape platform (*see* Fig. 3B). By contrast, saline-infused animals spent significantly more time in the quadrant (Q2) formerly containing the platform than any other quadrant, indicating that they remembered the location of the platform.

It is clear by these results that inflammation, in this case initiated by increased brain levels of TNF-α, induce cognitive deficits. These results represent the first evidence that elevated brain levels of TNF-α can induce cognitive impairment both in learning and memory. One other model of central nervous system (CNS) inflammation has also shown cognitive deficits. Glial fibrillary acidic protein (GFAP)–IL-6 transgenic mice have an additional gene for IL-6 inserted into a gene that codes for the astrocyte protein GFAP, allowing an overexpression of IL-6 specifically in the brain. These transgenic mice show a cognitive deficit in avoidance learning, which increases with age *(74)*. The reason we have not presented IL-6 as a primary mediator in the etiology of AD is that IL-6 seems to be more associated with normal aging rather than a disease state. Serum levels of IL-6, which may or may not reflect CNS levels of IL-6, are typically below detection in young adult humans and primates; with age, however, IL-6 levels increase to detectable levels *(14,19)*. In fact, in rhesus monkeys the levels of IL-6 are correlated with age *(75)*, suggesting that increases in IL-6 are a normal consequence of aging.

4.4. TNF-α-Induced Neuropathologic Changes

Neuropathologic analysis of TNF-α- and saline-infused animals revealed that intracerebroventricular infusions of TNF-α induced an enlargement of the lateral ventricles exceeding any enlargement due to mechanical damage of intracerebroventricular saline infusion (*see* Fig. 4). The ventricles from TNF-α-infused animals had a larger area than those from saline controls (*see* Table 1). Irrespective of intracerebroventricular treatment, a significant correlation ($r = 0.63$ and 0.55 for acquisition and memory retention, respectively) was found between ventricle size and cognitive performance—animals with larger ventricles had longer escape latencies during acquisition and spent less time searching in the former platform quadrant during memory retention. As stated in the Introduction to this chapter, ventricular enlargement is a common characteristic of AD *(3,76,77)*. Moreover, the ventricular enlargement seen in AD patients is correlated with several tests of cognitive performance, including the Mini Mental State Exam *(3,77)*.

No differences were found between the two treatment groups for neocortical thickness, neostriatal area, or dorsal hippocampal volume (Table 1). Interestingly, however, neocortical thickness was correlated with cognitive performance when animals in both treatment groups were considered ($r = 0.51$ and 0.58 for acquisition and memory retention, respectively). Animals with thicker cortices had shorter acquisition latencies and spent more time in the former platform quadrant than animals with thinner cortices. Even though we did not find a TNF-α-induced effect on cortical thickness, a correlation between cortical atrophy and cognitive performance is clearly documented in AD patients *(3)*. Thus, while our TNF-α infusion model does not produce the gross cortical atrophy of AD, the combined results from all animals in this study do provide strong evidence for cortical involvement in cognitive functions of the rat.

4.5. Apoptosis

Following 7 d of intracerebroventricular treatment, animals infused with TNF-α had signigicantly more apoptotic cells near lateral ventricular infusion sites compared to

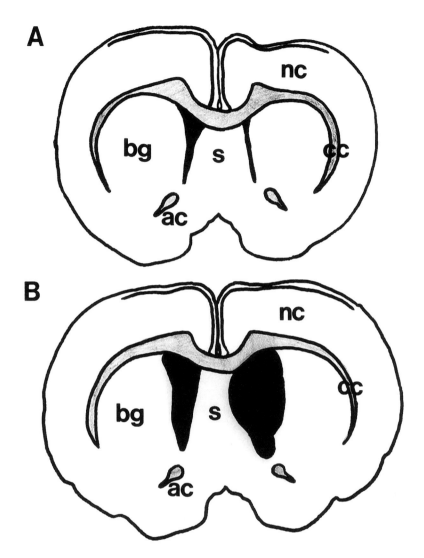

Fig. 4. Lateral ventricular enlargement due to intracerebroventricular infusions of TNF-α. **(A)** A coronal section of a saline-infused control animal shows minimal ventricular enlargement (*blackened area*), due to mechanical damage. **(B)** A coronal section of a TNF-α-infused animal shows greater ventricular enlargement than the saline-infused animal. *Abbreviations:* ac, anterior commissure; bg, basal ganglia; cc, corpus callosum; nc, neocortex; s, septum.

saline-infused controls. Figure 5 shows two diagrams of brain sections taken at the same coronal level from a saline-infused and a TNF-α-infused animal to show the typical location and density of apoptotic cells. It is apparent that the intracerebroventricular infusion process itself induced mechanical damage, resulting in a certain level of apoptotic cell death along the ventricular wall and in cortical areas through which the cannulas penetrated. TNF-α induced a significant 19.5% increase in apoptosis over and above that induced by mechanical damage. Although this TNF-α-induced increase was significant for the total number of cells in the neocortex, corpus callosum, septum, and basal ganglion, significant region-specific increases were not observed. Figure 6

Table 1
**Effects of Intracerebroventricular Infusions of TNF-α on Ventricular Area,
Neocortical Thickness, Neostriatal Area, and Dorsal Hippocampal Volume**

	Saline (intracerebroventricular infusions)	TNF-α (intracerebroventricular infusions)
Ventricular area (mm^2)	4.1 ± 1.3	12.8 ± 4.9*
Neocortical thickness (mm)	1.63 ± 0.08	1.63 ± 0.04
Neostriatal area (mm^2)	62.1 ± 1.8	59.9 ± 1.7
Dorsal hippocampal volume (mm^3)	11.2 ± 0.8	10.9 ± 0.5

*Significantly different ($p < .05$) from saline-infused controls. $N = 6$–7 for each group.

shows photomicrographs of parenchymal basal ganglia tissue adjacent to the endothelial lining of the lateral ventricle for a saline-infused (Fig. 6A) and a TNF-α-infused (Fig. 6B) animal. A higher magnification view of several apoptotic cells from the TNF-α-infused animal is presented in Fig. 6C.

Prior to the present results, several in vitro studies had deomonstrated that TNF-α can directly induce apoptosis in neuronal cell cultures *(46,57,58)* and that free radicals/ oxidative processes are involved in this apoptotic cell death *(56)*. Our present finding of increased brain apoptosis induced by TNF-α infusions provides the first in vivo evidence that TNF-α induces apoptosis. Moreover, these results suggest that increased TNF-α secretion by activated microglia in AD brains could play an important role in the increased cell death via apoptosis that is seen in AD *(5,6)*. The fact that brain apoptosis in AD is not correlated with amyloid deposition *(5)* suggests that the presence of β-amyloid may not be sufficient to explain all the cell loss seen in AD brains and that CNS inflammation may play a greater role in cell loss than previously thought.

5. GENERAL DISCUSSION AND CONCLUSIONS

This chapter first presented the increasing evidence that AD is an inflammatory-based disease, involving strong interactions with free radical/oxidative damage mechanisms. In that context, a key inflammatory cytokine in AD appears to be TNF-α because: (1) activated microglia associated with neuritic plaques stain most intensely for TNF-α compared to other cytokines *(7)*, (2) β-amyloid induces TNF-α release from microglia *(29,42,48)*, (3) TNF-α induces free radical formation and causes oxidative damage *(59,60,82)*, and (4) neuronal death in AD is thought to occur primarily through apoptosis, a process that is induced by TNF-α *(46,56–58)*. Indeed, we proposed that there is a mutual, interactive enhancement in AD between β-amyloid and TNF-α, resulting in neurodegenerative effects (Fig. 1). Given the evidence for inflammatory/ oxidative damage in AD *(7–9)*, it is not at all surprising that the onset and/or the progression of AD can be delayed by treatment with NSAIDs or with the antioxidant vitamin E *(10,11,13,78)*.

Also in this chapter, we have presented a new animal model pertinent to AD that is based on brain inflammation, namely, rats given long-term intracerebroventricular infusions of TNF-α. We present evidence that such chronic infusions of TNF-α

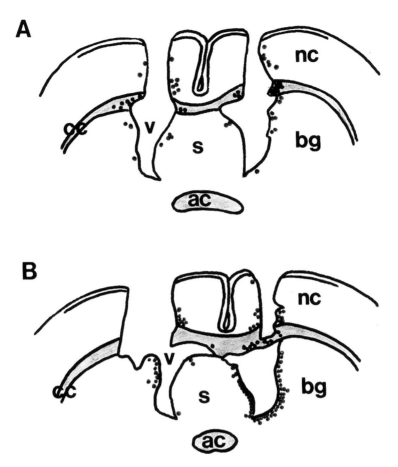

Fig. 5. The distribution of apoptotic cells around the ventricular infusion sites following seven days of intracerebroventricular infusions. **(A)** A saline-infused control animal and **(B)** a TNF-α-infused animal. Each dot represents one apoptotic cell found at that location. Note the higher number of apoptotic sites in the TNF-α-infused brain, particularly for the right side. *Abbreviations:* ac, anterior commissure; bg, basal ganglia; cc, corpus callosum; nc, neocortex; s, septum; v, ventricle.

induce cognitive impairment, ventricular enlargement, apoptosis, and weight loss—all of which characterize AD. Therefore, this animal model should be useful in testing the efficacy of antiinflammatory/antioxidant-based drugs with a potential to treat AD. Indeed, we have recently utilized our TNF-α infusion model in demonstrating that a novel synthetic antioxidant (CPI-1189) can prevent the cognitive impairment, ventricular enlargement, apoptosis, and weight loss induced by intracerebroventricular TNF-α infusions *(79)*. Thus, antioxidant/antiinflammatory-based therapies for AD can be screened in the TNF-α model presented in this chapter.

As with any new animal model for a CNS disease, however, a more extensive characterization of the TNF-α infusion model is certainly desirable, as are improvements/modifications to the model so as to make it as relevant as possible to the AD process. In that context, the following characterizations/improvements, many of which we are

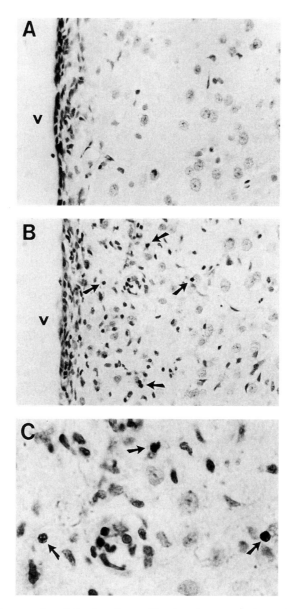

Fig. 6. Photomicrographs of the basal ganglia showing parenchymal apoptotic cells. (**A**) The basal ganglia along the ventricle wall of a saline-infused animal. (**B**) The basal ganglia along the ventricle wall of a TNF-α-infused animal. *Arrows* in (B) point to four apoptotic cells as indicated by the ISEL staining method for apoptosis. (**C**) A higher magnification view of the upper three apoptotic cells seen in B. *Abbreviations:* v, ventricle. Magnification for (A) and (B) is 400x and for (C) is 1000x.

currently studying, are desirable to more fully explore the potential of intracerebral TNF-α infusion as a model for brain inflammation in AD:

1. The fact that AD is a disease of aging clearly directs future studies to use "aged" rats for TNF-α infusion, rather than young adult rats.
2. Any inflammatory cell response induced by TNF-α infusions needs to be examined through immunostaining for activated microglia and astrocytes. This is particularly

important in view of the interrelationship between TNF-α and β-amyloid (Fig. 1) and, therefore, the possibility that TNF-α might encourage at least some β-amyloid fibrillary formation and/or glial cell accumulation near intracerebral infusion sites. Although it is unlikely that infused TNF-α causes neuritic plaque formation, it is not at all clear if neuritic plaques (or neurofibrillary tangles for that matter) are even involved in the pathogenesis of AD.

3. Because in vitro studies have shown that TNF-α directly induces free radical formation, various brain areas near intracerebroventricular infusion sites should be evaluated for TNF-α effects on free radical levels and oxidative damage.

4. Additional time points following initiation of long-term TNF-α infusion should be characterized—particularly time points earlier than those selected thus far (i.e., 7 and 14 d). This is of particular importance for characterizing the time course of TNF-α-induced apoptosis, as neuronal cell culture studies show that apoptotic cells begin appearing within 48–72 h of TNF-α exposure *(55,56)*.

5. In view of neuronal loss as a primary neuropathologic feature of AD, it would be desirable to quantitate any neuronal losses in neocortical and hippocampal areas resulting from TNF-α infusions.

In summary, this chapter introduced a novel rodent model for brain inflammation in AD that involves long-term intracerebral infusion of the inflammatory cytokine TNF-α. Results thus far obtained from the model have provided initial evidence linking brain inflammation, cognitive impairment, and apopotosis—all of which are characteristic of AD. Thus, the model is supportive of TNF-α as a key neuropathologic factor in AD. In the context that testing of new antiinflammatory and antioxidant drugs to treat AD will require prior testing of drug utility in appropriate animal models, the present TNF-α infusion model would appear to offer a number of desirable advantages.

In a broader perspective, it should be noted that the study of cytokine involvement and inflammation in AD is in its infancy. Although ever increasing evidence is supportive of brain inflammation as a key component of the AD process, it is also apparent that AD is multifactoral in pathogenesis. Nonetheless, inflammatory cytokines such as TNF-α are likely to be key players in the AD process, making it probable that further research in this area will advance our understanding of AD and lead to effective therapies therein.

ACKNOWLEDGMENTS

Support for the original research presented in this chapter was provided by Centaur Pharmaceuticals, Inc. (Sunnyvale, CA). We would also like to personally thank Drs. W. G. Flitter and W. A. Garland for their involvement in the conception and design of the studies.

REFERENCES

1. Hansen, L. A., DeTeres, R., Davies, P., and Terry, R. D. (1988) Neocortical morphometry, lesion counts, and choline acetyltransferase levels in the age spectrum of Alzheimer's disease. *Neurology* **38,** 48–54.
2. Katzman, R. and Jackson, J. E. (1991) Alzheimer Disease: Basic and clinical advances. *J. Am. Geriatr. Soc.* **39,** 516–525.
3. Slansky, I., Herholz, K., Pietrzyk, U., Kessler, J., Grond, M., Mielke, R., and Heiss, W. D. (1995) Cognitive impairment in Alzheimer's disease correlates with ventricular width and atrophy-corrected cortical glucose metabolism. *Neuroradiology* **37,** 270–277.

 4. DeKosky, S. T. and Scheff, S. W. (1990) Synapse loss in frontal cortex biopsies in Alzheimer's disease: correlation with cognitive severity. *Ann. Neurol.* **27**, 457–464.
 5. Lassman, H., Bancher, C., Breitschopf, H., Wegial, J., Bobinski, M., Fellinger, K., and Wisniewski, H. M. (1995) Cell death in Alzheimer's disease evaluated by DNA fragmentation in situ. *Acta Neuropathol.* **89**, 35–41.
 6. Smale, G., Nichols, N. R., Brady, D. R., Finch, C. E., and Horton, W. E. (1995) Evidence for apoptotic cell death in Alzheimer's Disease. *Exp. Neurol.* **133**, 225–230.
 7. Dickson, D. W., Lee, S. C., Mattiace, L. A., Yen, S. C., and Brosnan, C. (1993) Microglia and cytokines in neurological disease, with special reference to AIDS and Alzheimer's disease. *Glia* **7**, 75–83.
 8. Mrak, R. E., Sheng, J. G., and Griffin, W. S. T. (1995) Glial cytokines in Alzheimer's disease: review and pathogenic implications. *Hum. Pathol.* **26**, 816–823.
 9. Olanow, C. W. and Arendash, G. W. (1994) Metals and free radicals in neurodegeneration. *Curr. Opin. Neurol.* **7**, 548–558.
10. Breitner, J. C. (1996) The role of anti-inflammatory drugs in the prevention and treatment of Alzheimer's disease. *Ann. Rev. Med.* **47**, 401–411.
11. McGeer, P. L., Schulzer, M., and McGeer, E. G. (1996) Arthritis and anti-inflammatory agents as possible protective factors for Alzheimer's disease: a review of 17 epidemiologic studies. *Neurology* **47**, 425–432.
12. Andersen, K., Launer, L. J., Oh, A., Hoes, A. W., Breteler, M. M., and Hofman, A. (1995) Do nonsteroidal anti-inflammatory drugs decrease the risk for Alzheimer's disease? The Rotterdam study. *Neurology* **45**, 1441–1445.
13. Rich, J. B., Rasmusson, D. X., Folstein, M. F., Carson, K. A., Kawas, C., and Brandt, J. (1995) Nonsteroidal anti-inflammatory drugs in Alzheimer's disease. *Neurology* **45**, 51–57.
14. Eikelenboom, P., Rozemuller, J. M., Kraal, G., Stam, F. C., McBride, P. A., Bruce, M. E., and Fraser, H. (1991) Cerebral amyloid plaques in Alzheimer's disease but not in scrapie-affected mice are closely associated with a local inflammatory process. *Virchows Arch. B Cell. Pathol.* **60**, 329–336.
15. Dahlstrom, A., McRae, A., Polinsky, R., Nee, L., Sadasivan, B., and Ling, E. A. (1994) Alzheimer's disease cerebrospinal fluid antibodies display selectivity for microglia. *Mol. Neurobiol.* **9**, 41–54.
16. Cacabelos, R., Alvarez, X. A., Franco-Maside, A., Fernandez-Novoa, L., and Caamano, J. (1994) Serum tumor necrosis factor (TNF) in Alzheimer's disease and multi-infarct dementia. *Methods Find. Exp. Clin. Pharmacol.* **16**, 29–35.
17. Fillit, H., Ding, W., Buee, L., Kalman, J., Altstiel, L., Lawlar, B., and Wolf-Klein, G. (1991) Elevated circulating tumor necrosis factor levels in Alzheimer's disease. *Neurosci. Lett.* **129**, 318–320.
18. Arvin, B., Neville, L. F., Barone, F. C., and Feuerstein, G. Z. (1996) The role of inflammation and cytokines in brain injury. *Neurosci. Biobehav. Rev.* **20**, 445–452.
19. Emre, M., Geula, C., Ransil, B. J., and Mesulam, M. M. (1992) The acute neurotoxicity and effects upon cholinergic axons of intracerebrally injected β-amyloid in the rat brain. *Neurobiol. Aging* **13**, 553–559.
20. Malouf, A. T. (1992) Effect of β-amyloid peptides on neurons in hippocampal slice cultures. *Neurobiol. Aging* **13**, 543–551.
21. Pike, C. J., Walencewicz, A. J., Calabe, C. G., and Cotman, C. W. (1991) In vitro aging of β-amyloid protein causes peptide aggregation and neurotoxicity. *Brain Res.* **563**, 311–314.
22. Rush, D. K., Aschimies, S., and Merriman, M. C. (1992) Intracerebral β-amyloid produces tissue damage: is it neurotoxic? *Neurobiol. Aging* **13**, 591–594.
23. Behl, C., Davis, J. B., Lesley, R., and Schubert, D. (1994) Hydrogen peroxide mediates β-amyloid protein toxicity. *Cell* **77**, 817–827.
24. Harris, M. E., Hensley, K., Butterfield, D. A., Leedle, R. A., and Carney, J. M. (1995)

Direct evidence of oxidative injury produced by the Alzheimer's β-amyloid peptide (1–40) in cultured hippocampal neurons. *Exp. Neurol.* **131**, 193–202.

25. Manelli, A. M. and Puttfarcken, P. S. (1995) β-Amyloid induced toxicity in rat hippocampal cells: in vitro evidence for the involvement of free radicals. *Brain Res. Bull.* **38**, 569–576.

26. Barger, S. W., Horster, D., Furukawa, K., Goodman, Y., Krieglstein, J., and Mattson, M. P. (1995) Tumor necrosis factor α and β protect neurons against β-amyloid-peptide toxicity: evidence for involvement of a κβ-binding factor and attenuation of peroxide and Ca^{2+} accumulation. *PNAS* **92**, 9328–9332.

27. Forloni, G., Mangiarotti, F., Angeretti, N., Lucca, E., and De Simoni, M. G. (1997) β-Amyloid fragments potentiate IL-6 and TNF-α secretion by LPS in astrocytes but not in microglia. *Cytokine* **9**, 759–762.

28. McRae, A., Dahlstrom, A., and Ling, E. A. (1997) Microglial in neurodegenerative disorders: emphasis on Alzheimer's disease. *Gerontology* **43**, 95–108.

29. Meda, L., Cassatella, M. A., Szendrel, G. I., Otvos, L. Jr., Baron, P., Villalba, M., Ferrari, D., and Rossl, F. (1995) Activation of microglial cells by β-amyloid protein and interferon-γ. *Nature* **374**, 647–650.

30. Eikelenboom, P., Zhan, S. S., VanGool, W. A., and Allsop, D. (1994) Inflammatory mechanisms in Alzheimer's disease. *Trends Pharmacol. Sci.* **15**, 447–450.

31. Sheng, J. G., Ito, K., Skinner, R. D., Mrak, R. E., Rovnaghi, C. R., Van Eldic, L. J., and Griffin, W. S. (1996) In vivo and in vitro evidence supporting a role for the inflammatory cytokine interleukin-1 as a driving force in Alzheimer's pathogenesis. *Neurobiol. Aging* **17**, 761–766.

32. Kisilevsky, R. (1994) Inflammation-associated amyloidogenesis. *Mol. Neurobiol.* **8**, 65–66.

33. Zhann, S. S., Veerhuis, R., Kamphorst, W., and Eikelenboom, P. (1995) Distribution of β-amyloid associated proteins in plaques in Alzheimer's disease and in non-demented elderly. *Neurodegeneration* **4**, 291–297.

34. Rogers, J. (1995) Inflammation as a pathogenic mechanism in Alzheimer's disease. *Arzneimittelforschung* **45**, 439–442.

35. Shiosaki, K. and Puttfarcken, P. (1996) Emerging opportunities in neuroinflammatory mechanisms of neurodegeneration, in *Annual Reports in Medicinal Chemistry*, Vol. 30 (Bristol, J. A., ed.), Academic Press, New York, pp. 31–40.

36. Streit, W. J. and Kincaid-Colton, C. A. (1995) The brain's immune system. *Sci. Am.* **11**, 54–61.

37. Tracey, K. J. and Cerami, A. (1993) Tumor necrosis factor: an update review of its biology. *Crit. Care Med.* **21**, S415–S422.

38. Benigni, F., Faggioni, R., Sironi, M., Fantuzzi, G., Vandenabeele, P., Takahashi, N., Sacco, S., Fiers, W., Buurman, W. A., and Ghezzi, P. (1996) TNF receptor p55 plays a major role in centrally mediated increases of serum IL-6 and corticosterone after intracerebroventricular injection of TNF. *J. Immunol.* **157**, 5563–5568.

39. Greenspan, H. C., and Aruoma, O. I. (1994) Oxidative stress and apoptosis in HIV infection: a role for plant derived metabolites with synergistic antioxidant activity. *Immunol. Today* **15**, 209–213.

40. Pace, G. W. and Leaf, C. D. (1995) The role of oxidative stress in HIV disease. *Free Radic. Biol. Med.* **19**, 523–528.

41. Beutler, B. and Grau, G. E. (1993) Tumor necrosis factor in the pathogenesis of infectious diseases. *Crit. Care Med.* **21**, S423–S435.

42. Mattson, M. P., Barger, S. W., Furukawa, K., Bruce, A. J., Wyss-Coray, T., Mark, C. J., and Mucke, L. (1997) Cellular signaling roles of TGF-β, TNF-α, and bAPP in brain injury responses and Alzheimer's disease. *Brain Res. Rev.* **23**, 47–61.

43. Pfeffer, K., Matsuyama, T., Kundig, T. M., Wakeham, A., Kishihara, K., Shahimian, A., Wiegmann, K., Ohashi, P. S., Kronke, M., and Mak, T. W. (1993) Mice deficient for the 55 kd tumor necrosis factor receptor are resistant to endotoxic shock, yet succumb to *L. monecytogenes* infection. *Cell* **73,** 457–467.

44. Shen, Y., Li, R., and Shiosaki, K. (1997) Inhibition of p75 tumor necrosis factor receptor by antisense oligonucleotides increase hypoxic injury and β-amyloid toxicity in human neuronal cell lines. *J. Biol. Chem.* **272,** 3550–3553.

45. Cheng, B., Christakos, S., and Mattson, M. P. (1994) Tumor necrosis factors protect neurons against metabolic-excitotoxic insults and promote maintenance of calcium homeostasis. *Neuron* **12,** 139–153.

46. Yin, D., Kendo, S., Barnett, G. H., Morimura, T., and Takeuchi, J. (1995) Tumor necrosis factor-alpha induces p53-dependent apoptosis in rat glioma cells. *Neurosurgery* **37,** 758–762.

47. Bongioanni, P., Romano, M. R., Sposito, R., Castagna, M., Boccardi, B., and Borgna, M. (1997) T-cell tumor necrosis factor α receptor binding in demented patients. *J. Neurol.* **244,** 418–425.

48. Klegeris, A., Walker, D. G., and McGeer, P. L. (1997) Interaction of Alzheimer β-amyloid peptide with the human monocyte cell line THP-1 results in a protein kinase C-dependent secretion of tumor necrosis factor α. *Brain Res.* **747,** 114–121.

49. Abraham, C. R., Shirahama, T., and Potter, H. (1990) Alpha 1-antichymotrypsin is associated solely with amyloid deposits containing the beta-protein. Amyloid and cell localization of alpha 1-antichymotrypsin. Neurobiol. Aging **11,** 123–129.

50. Lieb, K., Fiebich, B. L., Schaller, H., Berger, M., and Bauer, J. (1996) Interleukin-1 beta and tumor necrosis factor-alpha induce expression of alpha 1-antichymotrypsin in human astrocytoma cells by activation of nuclear factor-kappa β. *J. Neurochem.* **67,** 2039–2044.

51. Machein, U., Lieb, K., Hull, M., and Fiebich, B. L. (1995) IL-1β and TNFα, but not IL-6, induce α_1-antichymotrypsin expression in the human astrocytoma cell line U373 MG. *NeuroReport* **6,** 2283–2286.

52. Ishiguro, K., Shoji, M., Yamaguchi, H., Matsubara, E., Ikeda, M., Kawarabayashi, T., Harigaya, Y., Okamoto, K., and Hirai, S. (1993) Differential expression of alpha 1-antichymotrypsin in the aged human brain. *Virchows Arch. B Cell. Pathol. Mol. Pathol.* **64,** 221–227.

53. Blasko, I., Schmitt, T. L., Steiner, E., Trieb, K., and Grubeck-Loebenstein, B. (1997) Tumor necrosis factor alpha augments amyloid beta protein (25-35) induced apoptosis in human cells. *Neurosci. Lett.* **238,** 17–20.

54. Gelbard, H. A., Dzenko, K. A., DiLoreto, D., del Cerro, M., and Epstein, L. G. (1993) Neurotoxic effects of tumor necrosis factor alpha in primary human neuronal cultures are mediated by activation of the glutamate AMPA receptor subtype: implication for AIDS neuropathogenesis. *Dev. Neurosci.* **15,** 417–422.

55. Selmaj, K., Raine, C. S., Farooq, M., Norton, W. T., and Brosnan, C. F. (1991) Cytokine cytotoxicity against oligodendrocytes: apoptosis induced by lymphotoxin. *J. Immunol.* **147,** 1522–1529.

56. Talley, A. K., Dewheurst, S., Perry, S. W., Dollard, S. C., Gummuluru, S., Fine, S. M., New, D., Epstein, L. G., Gendelman, H. E., and Gelbard, H. A. (1995) Tumor necrosis factor alpha-induced apoptosis in human neuronal cells: protection by the antioxidant *N*-acetylcysteine and the genes *bcl-1* and *crm* A. *Mol. Cell. Biol.* **15,** 2359–2366.

57. Baier-Bitterlich, G., Fuchs, D., Murr, C., Reibnegger, G., Werner-Femayer, G., Sgonc, R., Bock, G., Dierich, M. P., and Wachter, H. (1995) Effects of neopterim and 7,8-dihydroneopterin on tumor necrosis factor-α induced programmed cell death. *FEBS Lett.* **364,** 234–238.

58. Larrick, J. W. and Wright, S. C. (1990) Cytotoxic mechanism of tumor necrosis factor-α. *FASEB J.* **4,** 3215–3223.

59. Goosens, V., Grooten, J., DeVos, K., and Fiers, W. (1995) Direct evidence for tumor necrosis factor-induced mitochondrial reactive oxygen intermediates and their involvement in cytotoxicity. *Proc. Natl. Acad. Sci. USA* **92,** 8115–8119.

60. Hennet, T., Richter, C., and Peterhans, E. (1993) Tomur necrosis factor-α induces superoxide anion generation in mitochondria of L929 cells. *Biochem. J.* **289,** 587–592.

61. Maragos, W. F., Greenamyre, T., Penney, J. B. Jr., and Young, A. B. (1987) Glutamate dysfunction in Alzheimer's disease: an hypothesis. *Trends Neurosci.* **10,** 65–68.

62. Chao, C. C., and Hu, S. (1994) Tumor necrosis factor-alpha potentiates glutamate neurotoxicity in human fetal brain cell cultures. *Dev. Neurosci.* **16,** 172–179.

63. Fine, S. M., Angel, R. A., Perry, S. W., Epstein, L. G., Rothstein, J. D., Dewhurst, S., and Gelbard, H. A. (1996) Tumor necrosis factor α inhibits glutamate uptake by primary human astrocytes. Implications for pathogenesis of HIV-1 dementia. *J. Biol. Chem.* **271,** 15,303–15,306.

64. Hsiao, K., Chapman, P., Nisen, S., Eckman, C., Harigaya, Y., Younkin, S., Yang, F., and Cole, G. (1996) Correlative memory deficits, Aβ elevation, and amyloid plaques in transgenic mice. *Science* **274,** 99–102.

65. Frautschy, S. A., Yang, F., Irrizarry, M., Hyman, B., Saido, T. C., Hsiao, K., and Cole, G. M. (1998) Microglial response to amyloid plaques in APP$_{sw}$ transgenic mice. *Am. J. Pathol.* **152,** 307–317.

66. Irizarry, M. C., Soriano, F., McNamrar, M., Page, K. J., Schenk, D., Games, D., and Hyman, B. T. (1997) A-beta deposition is associated with neuropil changes, but not with overt neuronal loss in the human amyloid precursor protein V717F (PDAPP) transgenic mouse. *J. Neurosci.* **15,** 7053–7059.

67. Nalbantoglu, J., Tirado-santiago, G., Lahsaini, A., Poirier, J., Gonclaves, O., Verge, G., Momli, F., Weiner, S.A., Massicotte, G., Julien, J. P., and Shapiro, M. L. (1997) Impaired learning and LTP in mice expressing the carboxyterminus of the Alzheimer amyloid precursor protein. *Nature* **387,** 500–505.

68. Ma, T. C. and Zhu, X. Z. (1997) Suppression of lipopolysaccharide-induced impairment of active avoidance and interleukin-6-induced increase of prostaglandin E2 release in rats by indometacin. *Arnzeimittelforschung* **47,** 595–597.

69. Socci, D. J., Sanberg, P. R., and Arendash, G. W. (1995) Nicotine enhances Morris water maze performance of young and aged rats. *Neurobiol. Aging* **16,** 857–860.

70. Bluthe, R. M., Pawlowski, M., Suarez, S., Parnet, P., Pittman, Q., Kelly, K. W., and Dantzer, R. (1994) Synergy between tumor necrosis factor alpha and interleukin-1 in the induction of sickness behavior in mice. *Psychoneuroendocrinology* **19,** 197–207.

71. Fantino, M. and Wieteska, L. (1993) Evidence for a direct central anorectic effect of tumor-necrosis-factor-alpha in the rat. *Physiol. Behav.* **53,** 477–483.

72. Grundman, M., Corey-Bloom, J., Jernigan, T., Archibald, S., and Thal, C. J. (1996) Low body weight in Alzheimer's disease is associated with medial temporal cortex atrophy. *Neurology* **46,** 1585–1591.

73. White, H., Pieper, C., Schmader, K., and Fillenbaum, G. (1996) Weight changes in Alzheimer's disease. *J. Am. Geriatr. Soc.* **44,** 265–272.

74. Heyser, C. J., Masliah, E., Samimi, A., Campbell, I. L., and Gold, L. H. (1997) Progressive decline in avoidance learning paralleled by inflammatory neurodegeneration in transgenic mice expressing interleukin 6 in the brain. *Proc. Natl. Acad. Sci. USA* **94,** 1500–1505.

75. Ershler, W. B., Sun, W. H., Binkley, N., Gravenstein, S., Volk, M. J., Kamoske, G., Klopp, R. G., Roecker, E. B., Daynes, R. A., and Weindruch, R. (1993) Interleukin-6 and aging: blood levels and mononuclear cell production increase with advancing age and in vitro production is modifiable by dietary restriction. *Lymphokine Cytokine Res.* **12,** 225–230.

76. Forstl, H., Zerfass, R., Geiger-Kabisch, C., Sattel, H., Besthorn, C., and Hentschel, F. (1995) Brain atrophy in normal ageing and Alzheimer's disease. Volumetric discrimination and clinical correlations. *Br. J. Psychiatry* **167,** 739–746.
77. Murphy, D. G., DeCarli, C. D., Daly, E., Gillette, J. A., McIntosh, A. R., Haxby, J. V., Teichberg, D., Shapiro, M. B., Rapaport, S. I., and Horwitz, B. (1993) Volumetric magnetic resonance imaging in men with dementia of the Alzheimer's type: correlation with disease severity. *Biol. Psychiatry* **34,** 612–621.
78. Sano, M., Ernesto, C., Thomas, R. G., Klauber, M. R., Schafer, K., Grundamn, M., Woodbury, P., Growdon, J., Cotman, C. W., Pfeiffer, E., Schneider, L. S., and Thal, L. J. (1997) A controlled trial of selegiline, alpha-tocopherol, or both as treatment for Alzheimer's Disease. The Alzheimer's Disease Cooperative Study. *N. Engl. J. Med.* **336,** 1216–1222.
79. Bjugstad, K. B., Flitter, W. D., Garland, W. A., Su, G. C., and Arendash, G. W. (1998) Preventative actions of a synthetic antioxidant in a novel animal model of AIDS dementia. *Brain Res.* **795,** 349–357.
80. Strieter, R. M., Kunkel, S. L., and Bone, R. C. (1993) Role of tumor necrosis factor-α in disease states and inflammation. *Crit. Care Med.* **21,** S447–S463.
81. Halliwell, B. (1992) Oxygen radicals as key mediators in neurological disease: fact or fiction? *Ann. Neurol.* **32,** S10–S15.
82. Shoji, Y., Uedono, Y., Ishikura, H., Takeyama, N., and Tanaka, T. (1995) DNA damage induced by tumour necrosis factor-α in L929 cell is mediated by mitochondrial oxygen radical formation. *Immunology* **84,** 543–548.

6

The Senescence-Accelerated Mouse as a Possible Animal Model of Senile Dementia

Yasuyuki Nomura, Yasunobu Okuma, and Yoshihisa Kitamura

1. INTRODUCTION

Recent demographic trends toward a markedly aging population have led to concerns about more people developing age-associated diseases, in particular senile dementia. Therefore, it is important to clarify the basic mechanism of age-related changes in brain functions such as learning and cognitive deficiency and to develop safe and effective means of prevention and treatment of age-associated diseases. To do so, a useful animal model of age-associated diseases is essential. As one alternative model, Takeda et al. *(1,2)* developed the senescence-accelerated mouse (SAM) as a murine model of accelerated aging. SAM strains have a shortened life-span and develop early manifestations of senescence, including decreased activity, alopecia, lack of hair glossiness, skin coarseness, periophthalmic lesions, increased lordokyphosis, and systemic senile amyloidosis. The SAMP8 strain of senescence-related prone mice (SAMP) shows an age-related deterioration in learning ability compared with the control strain SAMR1, senescence-resistant mice (SAMR) *(3–5)*. We studied neurochemical changes in the SAMP8 brain compared to the SAMR1 brain during aging *(6,7)* and the effectiveness of several drugs in preventing age-related changes in the SAMP8 brain. In this chapter, we report the neurochemical and pharmacological findings of a study of SAMP8 and SAMR1 mice and discuss the significance and limitations of SAMP8 in basic studies of the aging mechanism and the screening and development of novel cognitive enhancers.

2. NEUROCHEMICAL CHANGES IN THE SAMP8 BRAIN

SAMP8 mice seem to be useful for studying the mechanism of brain aging and memory deficiency in humans. We studied neurochemical changes in aging related to learning and cognition. We examined: (1) the glutamatergic and cholinergic system, (2) protein kinase C (PKC), (3) anti-anxiety-related receptors, (4) glial cells, and (5) amyloid precursor proteins (APPs) in the cerebral cortex and hippocampus of SAMP8 and control SAMR1 mice. Our results are shown in Table 1.

From: Central Nervous System Diseases
Edited by: D. F. Emerich, R. L. Dean, III, and P. R. Sanberg © Humana Press Inc., Totowa, NJ

Table 1
Neuro- and Histochemical Alterations in SAMP8 Brain

Neuro- and histochemical parameters	SAMP8	Comments
Content of glutamate and glutamine	↑	
Release of ACh and NA	↓	
NMDA receptor/channels	↓	Learning and memory ↓
mACh receptors	↓	
PKC	↓	
5-HT$_{1A}$ receptors	CC ↑ HP ↓	Anxiety ↓
Central BDZ receptors	CC ↑ HP ↓	
Anti-GFAP antibody immunostaining	↑	Gliosis ↑
Peripheral BDZ receptors	↑	
Expression of APP mRNA	↑	Aging ↑
C-terminal fragment of APP	↑	

mACh, muscarinic acetylcholine; NA, noradrenaline; BDZ, benzodiazepine; GFAP, glial fibrillary acidic protein; APP, amyloid precursor protein; CC, cerebral cortex; HP, hippocampus; ↑, increase; ↓, decrease in SAMP8 vs SAMR1.

2.1. Changes in Glutamatergic Functions in SAMP8 Mice

The amounts of glutamate (Glu) and glutamine (Gln) were higher in the hippocampal and cerebral cortex of SAMP8 mice at a younger age than in the SAMR1 mice, whereas the amounts of aspartate (Asp) and alanine (Ala) were lower. This suggests that the metabolic pathway from α-ketoglutarate to Glu predominates, probably because of higher transaminase or lower Glu dehydrogenase (GDH) activity in SAMP8 mice. Patients with olivopontocerebellar atrophy (OPCA) are GDH deficient, and their plasma Glu levels are higher than in healthy controls *(8)*. These patients are characterized by a variable loss of neurons in the inferior olivary nuclei, the ventral part of the pons, and in the cerebellar cortex *(9)*. In addition, the local accumulation of an excitatory amino acid is known to cause neuronal degeneration, resulting from the neurotoxic effects of high concentrations of Glu and its analogs in Glu receptive areas *(10)*. Several of the observed age-related pathomorphological changes in the SAM brain *(11)*, and the learning and memory dysfunction in young SAMP8 mice *(3)*, may be caused by changes in Glu and Gln metabolism in the SAMP8 brain. The depolarization-stimuli evoked release of endogenous Glu from slices of the hippocampus and cerebral cortex of SAMR1 brains is observed to decline with increasing age. Stimulus-induced release decreased at an earlier age in SAMP8 mice, but the release of amino acids that are either neurotransmitters (Glu, Asp, and γ-aminobutyric acid [GABA]) or nontransmitters (Gln, Ala, and taurine) in response to high levels of K^+ increases in older SAMP8 mice *(12,13)*. KCl at a concentration of 50 m*M* is a drastic stimulus that is known to cause depolarization. Drastic stimuli may cause cytosolic amino acids to leak from presynaptic sites in SAMP8 mice more than 9 mo old. Thus, the nerve terminals from which amino acid transmitters are released seem to become increasingly fragile in the hippocampus and cerebral cortex of aging SAMP8 mice.

2.2. Changes in NMDA-Induced Acetylcholine and Noradrenaline Release in SAMP8 Mice

N-Methyl-D-aspartate (NMDA) induced Ca^{2+}-dependent release of [^{3}H]acetyl-choline ([^{3}H]ACh) and [^{3}H]noradrenaline ([^{3}H]NA) in brain slices *(14,15)*. The release of [^{3}H]ACh and [^{3}H]NA was inhibited by AP5, phencyclidine, MK-801, and Mg^{2+}, suggesting that NMDA-evoked release occurs via NMDA receptor/channels. However, quisqualate (QA)- and kainate (KA)-induced release occurs even in the presence of Mg^{2+}. Thus, neurotransmitter release seems to be evoked by NMDA, QA, and KA through different mechanisms. In SAMP8 brains, the NMDA-evoked release of [^{3}H]ACh and [^{3}H]NA was markedly lower than in SAMR1 brains *(14,15)*. This suggests that a significant loss of ACh- and NA-containing neurons occurs in the SAMP8 brain.

2.3. Changes in Neurotransmitter Receptors in the SAMP8 Brain

The B_{max} value of [^{3}H]MK-801 binding to NMDA receptor/channels in the SAMP8 cerebral cortex was also lower than that in the SAMR1 mice *(13,16)*. NMDA receptor/channels are found on the soma of cholinergic neurons in the striatum and on both the soma and terminals of adrenergic neurons in the hippocampus. These modulate the release of ACh and NA from the respective terminal *(17)*. NMDA receptor activity in these tissues seems to be specifically deficient in SAMP8 neurons. The B_{max} of [^{3}H]quinuclidinyl benzoate ([^{3}H]QNB) and [^{3}H]pirezepine binding activity (to M_1 ACh receptors) decreased in the hippocampus *(16)*, and the Bmax of [^{3}H]rauwolscine binding (to α_2-adrenoceptors) increased in the cerebral cortex *(18)*. These alterations in cholinergic and noradrenergic activity may be related to the learning and memory dysfunction of SAMP8 mice.

2.4. Changes in PKC and Nitric Oxide Synthase in SAMP8 Mice

Recently, PKC and calmodulin-dependent protein kinase II (CaMKII) were identified as key enzymes in long-term potentiation (LTP) in the hippocampus. In fact, an exciting study has just reported that PKC-γ *(19)* and α-CaMKII *(20)* knockout mice have learning and memory impairment. We determined that PKC and CaMKII positively regulate NMDA receptor/channels *(21,22)*. Therefore, we examined the binding of [^{3}H]phorbol-12 and 13-dibutyrate (PDBu) (for PKC) to both the cytosol and membrane fractions of the hippocampus in SAM *(16)*. The binding activities in both fractions were lower in SAMP8 than in SAMR1 mice. The reduced PKC levels or activity in the hippocampus could underlie the learning and memory dysfunction in SAMP8 mice. The roles of CaM and CaMKII levels in the SAMP8 brain are now under investigation. In addition, activation of NMDA receptor/channels induces production of nitric oxide (NO). NO is also possibly involved in LTP in the hippocampus and LTD in the cerebellum. NO is produced from L-arginine by NO synthase (NOS). There are three isozymes of NOS: neuronal (nNOS), endothelial (eNOS), and inducible (iNOS). All three types of NOS exist in the brain. Because nNOS knockout mice have residual NOS activity in the brain *(23)*, it is possible that eNOS or iNOS may play a key role in LTP and LTD. In preliminary experiments, we found that the binding activity of [^{3}H]N^G-nitro-L-arginine (NNA) (for NOS) was decreased in the cytosol fractions of the cerebral cortex and cerebellum of SAMP8 mice.

2.5. Changes in Anti-Anxiety-Related Receptors in SAMP8 Mice

Miyamoto et al. *(24)* suggested that as well as having learning and memory dysfunction, SAMP8 mice are less anxious. Central type (ω_1 and ω_2) benzodiazepine (BDZ) receptor agonists (i.e., diazepam and clonazepam), and 1A type of 5-hydroxytryptamine (5-HT_{1A}) receptor agonists (i.e., buspirone) have anti-anxiety effects. We examined the binding of [³H]flunitrazepam (for BDZ receptors) and [³H]8-hydroxy-2-(di-*n*-propylamino)tetralin ([³H]8-OH-DPAT) (for 5-HT_{1A} receptors). In 12-mo-old SAMP8 mice, the binding activities of both [³H]flunitrazepam and [³H]8-OH-DPAT were decreased in the hippocampus, but increased in the cerebral cortex *(16)*. Because little is known about changes in the levels of endogenous BDZ and 5-HT receptor agonists, it is not clear whether the changes seen in the BDZ and 5-HT_{1A} receptors of SAMP8 mice are the causes or the effects of the reduction in anxiety.

2.6. Accelerated Gliosis and Abnormal Formation of APP-Like Proteins

[³H]PK-11195 (for ω_3-BDZ receptors) is known to selectively bind to glial cells in the brain and this binding activity is useful as a neurochemical marker of gliosis. In patients with Alzheimer's disease, [³H]PK-11195 binding is increased in the brain *(25)* and is an index of neuronal damage *(26)*. In addition, this binding is observed in primarily cultured astrocytes but not in neuronal cells, and increases in parallel with astrocyte cell growth. In 2-mo-old mice, the binding activity in the cerebral cortex was higher in the SAMP8 strain than in SAMR1. This was followed by a steep increase in both strains until the age of 6 mo *(16)*. After the age of 6 mo, the binding continued to increase with age in SAMP8 mice, but did not change in SAMR1 mice. In addition, marked immunoreactivity of antibodies against glial fibrillary acidic protein (GFAP) is observed in the entorhinal cortex and brain stem of SAMP8 mice *(27)*. In Northern blotting experiments, we found a slight increase in APP mRNA with age. In contrast, immunoblot analysis showed that with age there was an increased reaction of a 27-kDa protein with antibody against the C-terminal of APP, but no marked changes in the reaction of 90–130 kDa APP (probably APP_{695}) in SAMP8 mice. Another preparation of anti-APP antibody (22C11) immunoreacted with neuron-like cells in the hippocampus and cerebral cortex of both SAMR1 and SAMP8 mice. However, this antibody also reacted with several granular structures in the cerebral cortex and reactive astrocyte-like cells surrounding sites of spongy degeneration in the brain stem reticular formation of aged SAMP8 mice *(27)*. These observations suggest that neuronal degeneration induces accelerated gliosis and the abnormal formation of APP-like proteins. These events also seem to be related to the learning and memory dysfunction see in SAMP8 mice.

3. EFFECTS OF DRUGS ON THE CHANGES IN SAMP8 MICE

Next, we looked at the prophylactic and therapeutic effects several drugs exert in SAMP8 mice. We investigated the effects of: (1) bifemelane, a clinically effective nootropic, on mACh receptors in the hippocampus of SAMP8; (2) acidic fibroblast growth factor (aFGF) on learning and the reduction of mACh- and NMDA-receptors in the brain of SAMP8 mice; (3) facteur thymique sérique (FTS), a nonapeptide isolated from the thymus *(28),* on superoxide dismutase (SOD) activity and malondialdehyde content in SAMP8 mice, and (4) Dan-shen methanol extract and its major ingredient,

Table 2
Effects of Drugs on Learning/Cognitive Deficiencies in SAMP8

Drugs	Functions improved	Neurochemical effects	References
Bifemelane	—	mAChR ↑	*(31)*
aFGF	Spatial cognition	mAChR ↑	*(37,38)*
		NMDAR ↑	
FTS	—	Cu,Zn SOD ↑	*(45)*
		Malondialdehyde ↓	
Dan-shen extract	Spatial cognition	NMDAR ↑	*(46)*
Lithospermate B Ca/Mg salt	Spatial cognition	PKC ↑	*(47)*

mAChR, muscarinic acetylcholine receptors; NMDAR, NMDA receptors; SOD, superoxide dismutase; PKC, protein kinase C; ↑, increase; ↓, decrease by treatment vs control.

lithospermate B Ca/Mg salt, on cognitive deficiency in SAMP8 mice. The findings are summarized in Table 2.

3.1. Effects of Bifemelane on ACh Receptors

Bifemelane hydrochloride, 4-(*o*-benzylphenoxy)-*N*-methylbutylamine, a nootropic used for cognitive and emotional disturbances related to cerebrovascular disease, improves scopolamine-induced memory deficits in rats *(29)* and enhances ACh release evoked by high potassium levels *(30),* suggesting the possible involvement of the cerebral cholinergic system in the action of bifemelane. We examined the effect of bifemelane on specific [^{3}H]QNB binding to mACh receptors in the brain of SAMP8 mice. Single (10 mg/kg, i.p.) and repeated (10 mg/kg/day, i.p. for 10 d) administration of bifemelane induced an increase in the B_{max} of [^{3}H]QNB binding in the hippocampus *(31)*. Bifemelane probably exerts its pharmacological effects through activation of the cholinergic system in the hippocampus of SAM.

3.2. Effects of aFGF on Learning and ACh and NMDA Receptors

aFGF is known to (1) control food intake, (2) depolarize hypothalamic neurons, and (3) affect protection and differentiation in neurons *(32–36)*. The effects of aFGF on learning and memory loss in SAMP8 mice were examined with the passive avoidance response test and Morris's water maze task. aFGF was administered in a dose of 7 µg/kg, s.c. to male SAMP8 mice once a week from 3 wk to 9 mo after birth. We found that aFGF prevented the memory and learning deficits that occur in SAMP8 mice. The treatment prevented the reduction in the number of cerebral cortical mACh and NMDA receptors. The aFGF treatment also suppresses the immunohistochemical reduction in choline acetyltransferase in the septum of SAMP8 mice *(37,38)*. Taken together, aFGF prevents the degeneration of ACh-containing neurons during development and aging in SAMP8 mice.

3.3. Effects of FTS on SOD and Malondialdehyde

Several studies indicate that FTS exerts a number of immunobiological effects. It (1) functions in the activation and differentiation in T cells *(39),* (2) suppresses experimental allergic encephalomyelitis (EAE) *(40),* (3) improves rheumatoid arthritis *(41)* and cellular immunity *(42),* (4) stimulates spontaneous DNA synthesis in thymocytes

(43), and (5) increases the neurotransmitter levels in the brain *(44)*. With age, FTS levels in a number of tissues decline progressively until it disappears completely. We examined the effect of FTS on SOD activity and malondialdehyde content in SAMP8 mice. FTS (0.1–1 mg/kg/d, s.c. for 21 d) enhanced activity of Mn-SOD and Cu,Zn-SOD in the kidneys of SAMP8 mice and Cu,Zn-SOD activity in the brains of both SAMP8 and SAMR1 mice. FTS decreased the malondialdehyde content in both the brains and kidneys of SAMP8 mice *(45)*. We propose that FTS inhibits lipid peroxidation, and is possibly effective as an anti-aging drug.

3.4. Effects of Dan-shen Methanol Extract and Lithospermate B Ca/Mg Salt on Cognitive Deficiency

Insufficient blood flow to the brain causes memory and learning disturbances. Dan-shen (*Salviae miltiorrhizae radix*) enhances the blood flow in some peripheral organs, but the mechanism of its action is not well understood. We examined the effects of Dan-shen methanol extract (DME) and its major ingredient, lithospermate B Ca/Mg salt, on the spatial learning ability of SAMP8 mice by means of Morris's water maze task. DME dissolved in water was administered orally for 3 wk at a dose of 500 mg/kg/ d. The treatment with DME improved spatial learning and emotional function in SAMP8 mice and increased the number of NMDA receptors in the cerebral cortex *(46)*. Lithospermate B Ca/Mg salt (60 mg/kg/d, p.o. for 3 wk) also improved spatial learning, but not emotional function, in SAMP8 mice. Lithospermate B treatment appears to act by increasing the amount of PKC in the cytosol fraction of the hippo-campus *(47)*.

4. POSSIBLE SIGNIFICANCE AND FUTURE STUDIES OF SAM

The recent results of the behavioral, neurochemical, and pharmacological studies discussed previously suggest that SAMP8 mice are useful for studying the fundamental mechanism of brain aging in humans, and are a pertinent animal model of senile dementias, such as Alzheimer's disease and ischemic dementia. The abnormal APP metabolism in the SAMP8 brain suggests that the strain seems to develop Alzheimer-type dementia. However, the increased Glu content in the brain suggests that the dementia may be ischemic dementia, as ischemia leads to an increase in the Glu level in the extracellular fluid of the brain, followed by neuronal damage by excitotoxic Glu. It is important to determine whether SAMP8 is an animal model of Alzheimer's type dementia or cerebral ischemia-induced dementia, both behaviorally and neuro-chemically. The genotype(s) responsible for the learning/memory dysfunction and the accelerated aging phenotypes have never been identified. We have made a consider-able effort to isolate the genes involved in the functional/phenotype type aging in the SAMP8 brain using molecular biological techniques, but have not yet succeeded. It will also be of interest to prepare transgenic/targeted gene-disruption animals as models of dementia and age-associated diseases, once these genes have been successfully isolated.

SAMP8 animals are useful for evaluating the prophylactic and therapeutic effects of drugs for dementia and other age-associated diseases. We have to find and establish simple, practical criteria and methods for screening the effectiveness of medicines using SAMP8 mice and tissues. We believe that SAMP8 mice have several merits in the development of novel drugs, especially prophylactics for age-associated diseases, as

SAMP8 is a spontaneously manifested loss of normal function during aging. Another accelerated aging strain, SAMP10, also seems to be a possible model of age-related learning deficiency. SAM P10 mice develop atrophy of the forebrain. We found that SAMP8 mice had decreased emotional function using the forced swimming test, as well as cognitive deficiency using Morris's water maze test. The relative merits of SAMP10 mice compared to those of SAMP8 mice as models of age-associated diseases will be clarified in future.

ACKNOWLEDGMENTS

We thank Dr. T. Takeda and Dr. M. Hosokawa (Kyoto University) for originally providing the SAMP8, SAMP10, and SAMR1 mice. We also thank Dr. Y. Oomura (Nippon-Zoki Pharmaceutical Company), Dr. K. Sasaki (Toyama University), Dr. T. Namba, Dr. M. Hattori, and Dr. S. Kadota (Toyama Medical and Pharmaceutical University) for assistance with the pharmacological studies. This work was supported in part by a Grant-in-Aid from the Ministry of Education, Science, Sports and Culture, Japan.

REFERENCES

1. Takeda, T., Hosokawa, M., Takeshita, S., Irino, M., Higuchi, K., Matsusita, T., Tomita, Y., Yasuhira, K., Hashimoto, H., Shimizu, K., Ishii, M., and Yamamura, T. (1981) A new murine model of accelerated senescence. *Mech. Aging Dev.* **17,** 183–194.
2. Takeda, T., Hosokawa, M., and Higuchi, K. (1991) Senescence-accelerated mouse (SAM): novel murine model of accelerated senescence. *J. Am. Geriatr. Soc.* **39,** 911–919.
3. Miyamoto, M., Kiyota, Y., Yamazaki, N., Nagoaka, A., Matsuo, Y., and Takeda, T. (1986) Age related changes in learning and memory in the senescence-accelerated mouse (SAM). *Physiol. Behav.* **38,** 399–406.
4. Ohta, A., Hirano, T., Yagi, H., Tanaka, S., Hosokawa, M., and Takeda, T. (1989) Behavioral characteristics of the SAM-P/8 mice strain sidman active avoidance task. *Brain Res.* **498,** 195–198.
5. Yagi, H., Katoh, S., Akiguchi, I., and Takeda, T. (1988) Age-related deterioration of ability of acquisition in memory and learning in senescence accelerated mouse: SAM-P/8 as an animal model of disturbances in recent memory. *Brain Res.* **474,** 86–93.
6. Nomura, Y., Kitamura, Y., Zhao, X.-H., Ohnuki, T., Takei, M., Yamanaka, Y., and Nishiya, T. (1994) Neurochemical studies on aging in SAM brain, in *The SAM Model of Senescence* (Takeda, T., ed.), Excerpta Medica, Amsterdam/Tokyo, pp. 83–88.
7. Nomura, Y., Yamanaka, Y., Kitamura, Y., Arima, T., Ohnuki, T., Nagashima, K., Ihara, Y., Sasaki, K., and Oomura, Y. (1996) Senescence accelerated mouse: neurochemical studies on aging. *Ann. NY Acad. Sci.* **786,** 410–418.
8. Plaitakis, A., Berl, S., and Yahr, M. D. (1984) Neurological disorders associated with deficiency of glutamate dehydrogenase. *Ann. Neurol.* **15,** 144–153.
9. Kanazawa, I., Kwak, S., Sasaki, H,. Mizusawa, H., Muramoto, O., Yoshizawa, K., Nukina, N., Kitamura, K., Kurisaki, H., and Sugita, K. (1985) Studies on neurotransmitter markers and neuronal cell density in the cerebellar system in olivopontocerebellar atrophy and cortical cerebellar atrophy. *J. Neurol. Sci.* **71,** 193–208.
10. Olney, J. W. (1990) Excitotoxic amino acids and neuropsychiatric disorders. *Annu. Rev. Pharmacol. Toxicol.* **30,** 47–71.
11. Sugiyama, H., Akiyama, H., Akiguchi, I., Kameyama, M., and Takeda, T. (1987) Loss of dendritic spines in hippocampal CA1 pyramidal cells in senescence accelerated mouse (SAM)—a quantitative Golgi study. *Clin. Neurol.* **27,** 841–845.

12. Nomura, Y., Kitamura, Y., and Zhao, X.-H. (1991) Aging in glutamatergic system and NMDA receptor-ion channels in brain of senescence-accelerated mouse, in *NMDA Receptor Related Agents: Biochemistry, Pharmacology and Behavior* (Kameyama, T., Nabeshima, T., and Domino, E. F., eds.), NPP Books, Ann Arbor, pp. 287–298.

13. Kitamura, Y., Zhao, X.-H., Ohnuki, T., Takei, M., and Nomura, Y. (1992) Age-related changes in transmitter glutamate and NMDA receptor/channels in the brain of senescence-accelerated mouse. *Neurosci. Lett.* **137,** 169–172.

14. Zhao, X.-H., Kitamura, Y., and Nomura, Y. (1992) Age-related changes in NMDA-induced [^{3}H]acetylcholine release from brain slices of senescence-accelerated mouse. *Int. J. Dev. Neurosci.* **10,** 121–129.

15. Zhao, X.-H. and Nomura, Y. (1990) Age-related changes in uptake and release of L-[^{3}H]noradrenaline in brain slices of senescence accelerated mouse. *Int. J. Dev. Neurosci.* **81,** 267–272.

16. Nomura, Y., Kitamura, Y., Ohnuki, T., Arima, T., Yamanaka, Y., Sasaki, K., and Oomura, Y. (1997) Alterations in acetylcholine, NMDA, benzodiazepine receptors and protein kinase C in the brain of the senescence-accelerated mouse: an animal model useful for studies on cognitive enhancers. *Behav. Brain. Res.* **83,** 51–55.

17. Snell, L. D. and Johnson, K. M. (1986) Characterization of the inhibition of excitatory amino acid-induced neurotransmitter release in the rat striatum by phencyclidine-like drugs. *J. Pharmacol. Exp. Ther.* **238,** 938–946.

18. Kitamura, Y., Zhao, X.-H., Ohnuki, T., and Nomura, Y. (1989) Ligand-binding characteristics of [^{3}H]QNB, [^{3}H]prazosin, [^{3}H]rauwolscine, [^{3}H]TCP and [^{3}H]nitrendipine to cerebral cortical and hippocampal membranes of senescence accererlated mouse. *Neurosci. Lett.* **106,** 334–338.

19. Abeliovich, A., Paylor, R., Chen, C., Kim, J. J., Wehner, J. M., and Tonegawa, S. (1993) PKC gamma mutant mice exhibit mild deficits in spatial and contextual learning. *Cell* **75,** 1263–1271.

20. Silva, A. J., Paylor, R., Wehner, J. M., and Tonegawa, S. (1992) Impaired spatial learning in alpha-calcium-calmodulin kinase II mutant mice. *Science* **257,** 206–211.

21. Urushihara, H., Tohda, M., and Nomura, Y. (1992) Selective potentiation of NMDA-induced currents by protein kinase C in *Xenopus* oocytes injected with rat brain RNA. *J. Biol. Chem.* **267,** 11697–11700.

22. Kitamura, Y., Miyazaki, A., Yamanaka, T., and Nomura, Y. (1993) Stimulatory effects of protein kinase C and calmodulin kinase II on NMDA receptor/channels in the postsynaptic density of rat brain. *J. Neurochem.* **61,** 100–109.

23. Huang, P. L., Dawson, T. M., Bredt, D. S., Snyder, S. H., and Fishman M. C. (1993) Targeted disruption of the neuronal nitric oxide synthase gene. *Cell* **75,** 1273–1286.

24. Miyamoto, M., Kiyota, Y., Nishiyama, M., and Nagaoka, A. (1992) Senescence-accelerated mouse (SAM): age-related reduced anxiety-like behavior in the SAM-P/8 strain. *Physiol. Behav.* **51,** 979–985.

25. Diorio, D., Welner, S. A., Butterworth, R. F., Meaney, M. J., and Suranyi-Cadotte, B. E. (1991) Peripheral benzodiazepine binding sites in Alzheimer's disease frontal and temporal cortex. *Neurobiol. Aging* **12,** 255–258.

26. Benavides, J., Fage, D., Carter, C., and Scatton, B. (1987) Peripheral type benzodiazepine binding sites are a sensitive indirect index of neuronal damage. *Brain. Res.* **421,** 167–172.

27. Kitamura, Y., Yamanaka, Y., Nagashima, K., and Nomura, Y. (1994) The age-related increase in markers of astrocytes and amyloid precussor proteins in the brain of senescence-accelerated mouse (SAM), in *The SAM Model of Senescence* (Takeda, T., ed.), Excerpta Medica, Amsterdam/Tokyo, pp. 359–362.

28. Dardenne, M., Pleau, J. M., Blouquit, J. Y., and Bach, J. F. (1980) Chracterization of facteur thymique séique (FTS) in the thymus II. Direct demonstration of the presence of FTS in thymosin fraction V. *Clin. Exp. Immunol.* **42,** 477–482.
29. Tode, A., Egawa, M., and Nagai, R. (1983) Effect of 4-(*o*-benzylphenoxy)-*N*-methyl-butylamine hydrochloride (MCI-2016) on the scopolamine-induced deficit of spontaneous alternation behavior in rats. *Jpn. J. Pharmacol.* **33,** 775–784.
30. Saito, K., Honda, S., Egawa, M., and Tobe, A. (1985) Effecs of bifemelane hydrochloride (MCI-2016) on acetylcholine release from cortical and hippocampal slices of rats. *Jpn. J. Pharmacol.* **39,** 410–414.
31. Kitamura, Y., Ohnuki, T., and Nomura, Y. (1991) Effects of bifemelane, a brain function improver, on muscarinic receptors in the CNS of senescence-accelerated mouse. *Jpn. J. Pharmacol.* **56,** 231–235.
32. Hanai, K., Oomura, Y., Kai, Y., Nishikawa, K., Shimizu, N., Morita, H., and Plata-Salaman. C. R. (1989) Central action of acidic fibroblast growth factor in feeding regulation. *Am. J. Physiol.* **256,** R217–223.
33. Oomura, Y., Sasaki, K., Suzuki, K., Muto, T., Li, A. J., Ogita, Z., Hanai, K., Tooyama, I., Kimura, H., and Yanaihara, N. (1992) A new brain glucosensor and its physiological significance. *Am. J. Clin. Nutr.* **55 (Suppl. 1),** 278S–282S.
34. Sasaki, K., Oomura, Y., Muto, T., Suzuki, K., Hanai, K., Tooyama, I., Kimura, H., and Yanaihara, N. (1991) Effects of fibroblast growth factors and platelet-derived growth factor on food intake in rats. *Brain Res. Bull.* **27,** 327–332.
35. Gospodarowicz, D., Neufeld, G., and Schweigerer, L. (1986) Fibroblast growth factor. *Mol. Cell. Endocrinol.* **46,** 187–204.
36. Neufeld, G., Gospodarowicz, D., Dodge, L., and Fujii, D. K. (1987) Heparin modulation of the neurotropic effects of acidic and basic fibroblast growth factors and nerve growth factor on PC12 cells. *J. Cell. Physiol.* **131,** 131–140.
37. Oomura, Y., Sasaki, A., Li, A., Kimura, H., Tooyama, I., Hanai, K., Nomura, Y., Kitamura, Yanaihara, N., and Yago, H. (1994) FGF facilitation of learning and memory and protection from memory loss in senescence accelerated mice, in *The SAM Model of Senescence* (Takeda, T., ed.), Excerpta Medica, Amsterdam/Tokyo, pp. 415–418.
38. Oomura, Y., Sasaki, K., Li, A., Yoshii, H., Fukata, Y., Yago, H., Kimura, H., Tooyama, K., Hanai, K., Nomura, Y., Kitamura, K., and Yanaihara, N. (1996) Protection against impairment of memory and immunoreactivity in senescence-accelerated mice by acidic fibroblast frowth factor. *Ann. NY Acad. Sci.* **786,** 337–347.
39. Kaufman, D. B. (1980) Maturational effects thymic hormones on human helper and suppressor T cells: effects of FTS (factuer thymique serique) and thymosin. *Clin. Exp. Immunol.* **39,** 722–727.
40. Nagai, Y., Osanai, T., and Sakakibara, K. (1982) Intensive suppression of experimental allergic encephalomyelitis (EAE) by serum thymic factor and therapeutic implication for multiple sclerosis. *Jpn. J. Exp. Med.* **52,** 213–219.
41. Amor, B., Dougados, M., Mery, C., Gery, A. D., Choay, J., Dardenne, M., and Bach, J. F. (1984) Thymuline (FTS) in rheumatoid arthritis. *Arthritis Rheum.* **27,** 117–118.
42. Bordigoni, P., Faure, G., Bene, M. C., Dardenne, M., Bach, J. F., Duheille, J., and Olive, D. (1982) Improvement of cellular immunity and lgA production in immunodeficient children after treatment with synthetic serum thymic factor (FTS). *Lancet* **ii,** 293–297.
43. Blazsek, I. and Lenfant, M. (1983) The stimulatory effect of serum thymic factor (FTS) on spontaneous DNA synthesis of mouse thymocytes. *Cell Tissue Kinet.* **16,** 247–257.
44. Vécsei, L., Faludi, M., and Najbauer, J. (1987) The effect of "facteur thymique sérique" (FTS) on catecholamine and serotonin neurotransmission in discrete brain regions of mice. *Acta Physiol. Hungarica* **69,** 129–132.

45. Zhao, X.-H., Awaya, A., Kobayashi, H. C., and Nomura, Y. (1990) Effects of repeated administration of facteur thymique serique (FTS) on biochemical changes related to aging in senescence-accelerated mouse (SAM). *Jpn. J. Pharmacol.* **53,** 311–319.
46. Arima, T., Baba, I., Hori, H., Kitamura, Y., Namba, T., Hattori, M., Kadota, S., and Nomura, Y. (1996) Ameliorating effects of dan-shen methanol extract on cognitive deficiencies in senescence-accelerated mouse. *Kor. J. Gerontol.* **6,** 14–21.
47. Arima, T., Matsuno, J., Hori, H., Kitamura, Y., Namba, T., Hattori, M., Kadota, S., and Nomura, Y. (1997) Ameliorating effects of calcium/magnesium lithospermate B on cognitive deficiencies in senescence-accelerated mouse. *Kor. J. Gerontol.* **7,** 17–24.

7

Transgenic Mice Overexpressing Presenilin cDNAs

Phenotype and Utility in the Modeling of Alzheimer's Disease

Karen Duff

1. INTRODUCTION

Genetic analysis has implicated four genes in the etiology of Alzheimer's disease (AD): the amyloid precursor protein (APP) gene, the presenilin 1 and presenilin 2 (PS1 and PS2) genes, and the apolipoprotein E gene (reviewed in *1*). Transgenic mice with mutations in these genes replicate some of the features of the disease *(1–8)* and while molecular analysis of the effects of the mutations in these genes has implicated APP metabolism and the production of Aβ1–42(43) as important in the initiation of the pathogenesis of the disease *(6–9)*, the relationship between the presenilins (PSs) and APP remains obscure, although a direct association seems probable *(10)*.

2. TRANSGENIC MODELS OVEREXPRESSING AD-RELATED GENES

2.1. APP Mice

Four transgenic APP lines have been described that show some of the pathology (essentially, neuritic plaques) of AD. The most robust of these include the Athena/ Exemplar mouse (PDAPP) *(2)*, the Hsiao mouse (Tg2576) *(3)*, and the Novartis mouse *(5)*. The PDAPP mouse uses an APP minigene containing the APP I717F mutation with expression driven by the platelet-derived growth factor (PDGF) promoter. The Hsiao mouse uses an APP695 cDNA with the K/M670/1N/L mutation, driven by the human prion promoter, whereas the Novartis mouse uses an APP751 K/M670/1N/L construct under the control of a thy-1 promoter. These mice produce high levels of total Aβ (in excess of 20 pm/g wet wt of tissue) and all develop amyloid deposits in the cortex and hippocampus between 6 mo and 1 yr of age. None develop tangles or PHF-paired helical filaments (PHF), although some abnormally phosphorylated τ immunoreactivity is seen. The two mice with the highest amyloid load (PDAPP and Tg2576) do not show extensive cell loss *(11,12)* but claims of cell loss in the Novartis mouse have been made. The Tg2576 mouse has been reported to show behavioral abnormalities suggestive of age-related hippocampal dysfunction *(3)* but the PDAPP mouse has been shown to be cognitively impaired from a young age, with no further impairment developing with increased age and amyloid burden *(13)*. It therefore seems that these

From: Central Nervous System Diseases
Edited by: D. F. Emerich, R. L. Dean, III, and P. R. Sanberg © Humana Press Inc., Totowa, NJ

mice are reliable models of amyloid deposition, but cannot yet be considered to be models for human AD.

2.2. Presenilin Mice

Although the functions of the presenilins are not understood, the identification of two *C. elegans* isologues of PS1 (reviewed in *14*) have provided some clues. The first isologue is *spe-4*. Mutations in this gene disrupt spermatogenesis through disruption of protein trafficking in the Golgi. The second isologue is *sel-12,* mutations in which produce an egg laying defect, probably through disruption of the Notch signaling pathway. A functional relationship between *sel-12* and PS1 is illustrated by the observation that PS1 can rescue *sel-12* mutant *C. elegans* (reviewed in *14*), and further support for a role of the PS in Notch signaling comes from the observation that PS1 knockout mice show developmental abnormalities similar to those seen in mice in which components of the Notch system have been knocked out *(15,16)*. A role for the presenilins (particularly PS2) in apoptosis has also been proposed *(17,18)*.

Presenilin biology has been examined in vivo through the creation of transgenic mice. Mice from different laboratories *(6–8)* differed in the promoter and strain of mice used but were similar in achieving high levels of protein production (one- to threefold over endogenous) in neuronal regions of the brain. Full-length PS1 (at 46 kDa) was seen only in mice producing large amounts of human protein, as expected from studies of transfected cells that show that the protein is rapidly turned over *(19)*. The holoprotein undergoes endoproteolytic cleavage in a highly regulated manner, generating terminal derivatives that accumulate to saturable levels at a 1:1 stoichiometry *(19)*. This was also seen in the mouse brain where the transgene derived protein was processed correctly into 18-kDa N-terminal and 28-kDa C-terminal fragments. The human fragments have slightly retarded mobility on a gel compared to the endogenous mouse PS1 derivatives, and when the levels of both mouse and human derivatives were directly compared, it was found that the presence of the human protein led to diminished accumulation of the mouse PS1 fragments *(20)*. Both transgenic mice and transfected cells overexpressing an exon 9 deletion (ΔE9) cDNA do not show cleavage of the mutant protein but, interestingly, the mice still show a reduction in the amount of mouse PS1 fragments *(20)*. The reason for this diminution in the mouse cleavage products is as yet unclear. Perhaps the most interesting observation to have come from the mice so far is that the fragments derived from mutant PS1 accumulate to a greater degree than fragments from wild-type *(20)*. Moreover, the uncleaved ΔE9 mutant cDNA also accumulates to a greater degree than the fragments derived from wild-type human PS1 *(20)*. These observations together suggest that the mutations in PS1 somehow affect the protein's metabolism but how that relates to the pathogenesis of AD remains to be seen.

Perhaps the greatest insight into how mutations in PS1 might cause Alzheimer's disease has come from the analysis of Aβ levels in mice overexpressing mutant and wild-type PS1 transgenes. A report by Scheuner et al. *(9)* on Aβ levels in fibroblasts from AD patients with PS1 mutations first indicated a link between PS1 and APP processing. Reports have now been published that describe the same effect in mouse brain tissue in all three sets of PS1 transgenic mice. The first report *(6)* examined the levels of mouse Aβ1–40 and 1–42(43) using a sandwich Elisa system. The two subsequent reports *(7,8)* examined the level of human Aβ derived from a mutant human APP trangene that had been crossed into the PS1 mice. All three reports showed essentially the same results: overexpressing mutant PS1 in the brains of transgenic mice led to the elevation of Aβ1–42, but not Aβ1–40. This effect was a direct result of the mutation and not overexpression of the human protein as the overexpression of wild-type PS1 did not have any significant effect on Aβ levels. The physiological significance of the specific elevation of Aβ1–42(43) appears to be the observation that AD patients with PS1 mutations show plaques composed primarily of Aβ1–42(43) *(21)*. The fact that AD causing mutations in APP and PS1 both led to an elevation in Aβ strongly suggests that this event is a significant and possibly crucial mechanism underlying the etiology of AD.

So far, the elevation of Aβ1–42 is the only AD-related phenotype reported for the mutant PS1 transgenic mice. Mice have been examined for AD-related changes in histopathology at 1 yr of age, but no obvious changes have been observed. The mice are currently undergoing intensive analysis in several laboratories for any changes in behavior, electrophysiology, pathology, and biochemistry but the results are likely to be subtle, especially if the AD phenotype in PS1-linked cases is related solely to Aβ levels and not to a defect in the PS1 protein itself.

To date, there are no published reports of PS2 mutant mice, but the effect of PS2 transgene expression on Aβ levels is expected to be the same in mice as has been reported for transfected cells *(22)*. There are also no reports of targeted (knockin) mutant PS1/2 mice although they are currently under construction in several laboratories.

3. MODULATING AND ENHANCING THE PHENOTYPE: TRANSGENIC CROSSES

One of the great advantages of transgenic animals is that different mice can be mated together so that the effect of transgene interaction on the disease phenotype can be observed. This type of cross has been performed with the mutant APP transgenic line, Tg2576 *(3)* and a mutant PS1 line, PDPS1 *(6)*. The Tg2576 line has been shown to develop Aβ deposits in the brain between 9 and 12 mo of age, and deposition is temporally linked to Aβ elevation. The PDPS1 line shows an approx 1.5 fold elevation of Aβ1–42(43) from birth. When these two animals are mated together, Aβ containing deposits are observed in the cortex and hippocampus of doubly transgenic progeny as early as 12 wk of age *(23* and Duff et al., *unpublished observation)*. It would appear, therefore, that marginal increases in Aβ1–42(43) levels, evoked by the PS1 mutant transgene, can accelerate the deposition process by several months (also shown in a cross by Borchelt et al., *24*). As the mice have a severe Aβ pathology for an extended time relative to singly transgenic APP mice, other features of the disease such as τ abnormalities or major cell loss may become apparent as the animals age. It may, however, be necessary to cross them with yet more transgenic animals such as human τ overexpressing mice *(25)* to try and generate a more complete phenotype.

4. PS/APP MICE AS A MODEL SYSTEM FOR USE IN DRUG DEVELOPMENT

How can the transgenic models that have been created be used for drug development? If amyloid is at the root of AD as suggested by the effect of mutations in both the presenilins and APP on Aβ levels, then drugs that modulate amyloid should be effective against the disease. Many pharmaceutical companies are currently developing therapeutic agents designed to alleviate Aβ elevation and/or deposition which they hope will reverse the cognitive impairment that is so devastating in AD. These agents mainly fall into two categories: secretase inhibitor types of drug designed to lower the amounts of amyloid in the cell such that the concentrations do not reach levels required for fibrillarization, or drugs designed to reduce the aggregation potential of the fibrils themselves. Although the transgenic mice that have been created so far cannot be considered to be accurate models of human AD, they are excellent models of the deposition process. The doubly transgenic (Tg2576/PDPS1) animals routinely form deposits by 12 wk of age and are an ideal resource for the study of drugs designed to either reduce Aβ levels or break down deposits. Although it is hoped that modulating amyloid in the human brain will halt or retard the progression of the disease, one issue that seems critical is whether AD patients who already have some amyloidosis can be treated in the same way. The question of whether amyloid plaques can be resolved once they have formed is therefore an important question in potential treatment strategies. We, and others, aim to assess whether amyloid deposits can be resolved through the use of inducible transgenes. If APP (or PS1 in an APP/PS cross) can be switched off after deposits have formed, will the brain be able to remove the amyloid and if so, at what stage in the disease's progression is this effective? Inducible transgenes for use in the brain are currently under development (*see 26* for review).

Although there is good evidence that amyloid deposition lies at the root of AD etiogenesis, it seems likely that AD is a multistep disease, with several parts of the phenotype contributing to disease progression. This can be most clearly seen in AD patients with a huge amyloid burden who do not show signs of cognitive impairment *(27),* and in transgenic animals that have accumulated amyloid but do not show cell loss *(11,12).* The identification of a secondary AD type phenotype in the mice other than amyloid deposition would help to test the efficacy of drugs that are unlikely to affect the deposition process. This could include drugs against the inflammatory response such as nonsteroidal antiinflammatory drugs (NSAIDs) and antioxidants such as vitamin E. In an attempt to examine how the cell responds to an environment of elevated and deposited amyloid, we are examining our PS/APP mice for signs of cell disturbance such as τ phosphorylation, oxidative damage, inflammatory/immune response, and morphological changes at the electron microscopic level. Once a secondary marker has been identified, we will be able to test the effects of a wider range of drugs on these downstream events. Unfortunately, until robust behavioral changes can be shown in the PS/APP animals, we will not be able to investigate the effects of drug treatment on cognitive ability, but the difference between human and mouse cognition might make this an inappropriate area of study in terms of drug efficacy.

5. OVERVIEW

Although each type of transgenic mouse may not recreate the expected disease phenotype, the potential to cross different animals together to study both the biochemical and physiological consequence of gene interaction makes the continued generation of transgenic animals carrying Alzheimer's associated genes a valid pursuit. We have extended and enhanced the phenotypes of two separate mice such that the doubly transgenic PS/APP mouse now shows real utility for the testing of compounds designed to modulate Aβ levels.

REFERENCES

1. Hardy, J. (1997) Amyloid, the presenilins and Alzheimer's disease. *Trends Neurosci.* **20,** 154–159.
2. Games D., Adams, D., Alessandrini, R., Barbour, R., Berthelette, P., Blackwell, C., et al. (1995) Alzheimer-type neuropathology in transgenic mice overexpressing the V717F b-amyloid precursor protein. *Nature* **373,** 523–527.
3. Hsiao, K., Chapman, P., Nilsen, S., Eckman, C., Harigaya, Y., Younkin, S., Yang, F., and Cole, G. (1996) Correlative memory deficits, Aβ elevation and amyloid plaques in transgenic mice. *Science* **274,** 99–102.
4. Nalbantoglu, J., Tirado-Santiago, G., Lahsaini, A., Poirier, J., Goncalves, O., Verge, G., Momoli, F., Welner, S. A., Massicotte, G., Julien, J. P., and Shapiro, M. L. (1997) Impaired learning and LTP in mice expressing the carboxy terminus of the Alzheimer amyloid precursor protein. *Nature* **387,** 500–505.
5. Sturchler-Pierrat, C., Abramowski, D., Duke, M., Wiederhold, K. H., Mistl, C., Rothacher, S., Ledermann, B., Burki, K., Frey, P., Paganetti, P. A., Waridel, C., Calhoun, M. E., Jucker, M., Probst, A., Staufenbiel, M., and Sommer, B. (1997) Two amyloid precursor protein transgenic mouse models with Alzheimer disease-like pathology. *Proc. Natl. Acad. Sci. USA* **94,** 13,287–13,292.
6. Duff, K., Eckman, C., Zehr, C., Yu, X., Prada, C. M., Perez-tur, J., Hutton, M., Buee, L., Harigaya, Y., Yager, D., Morgan, D., Gordon, M. N., Holcomb, L., Refolo, L., Zenk, B., Hardy, J., and Younkin, S. (1996) Increased amyloid-b42(43) in brains of mice expressing mutant presenilin 1. *Nature* **383,** 710–713.
7. Borchelt, D. R., Thinakaran, G., Eckman, C. B., Lee, M. K., Davenport, F., Ratovitsky, T., Prada, C. M., Kim, G., Seekins, S., Yager, D., Slunt, H. H., Wang, R., Seeger, M., Levey, A. I., Gandy, S. E., Copeland, N. G., Jenkins, N. A., Price, D. L., Younkin, S. G., and Sisodia, S. S. (1996) Familial Alzheimer's disease-linked presenilin 1 variants elevate Abeta1-42/1-40 ratio in vitro and in vivo. *Neuron* **17,** 1005–1013
8. Citron, M., Westaway, D., Xia, W., Carlson, G., Diehl, T., Levesque, G., Johnson-Wood, K., Lee, M., Seubert, P., Davis, A., Kholodenko, D., Motter, R., Sherrington, R., Perry, B., Yao, H., Strome, R., Lieberburg, I., Rommens, J., Kim, S., Schenk, D., Fraser, P., St George Hyslop, P., and Selkoe, D. J. (1997) Mutant presenilins of Alzheimer's disease increase production of 42-residue amyloid beta-protein in both transfected cells and transgenic mice. *Nat. Med.,* 67–72.
9. Scheuner, D., Eckman, C., Jensen, M., Song, X., Citron, M., Suzuki, N., Bird, T. D., Hardy, J., Hutton, M., Kukull, W., Larson, E., Levy-Lahad, E., Viitanen, M., Peskind, E., Poorkaj, P., Schellenberg, G., Tanzi, R., Wasco, W., Lannfelt, L., Selkoe, D., and Younkin, S. (1996) Secreted amyloid beta-protein similar to that in the senile plaques of Alzheimer's disease is increased in vivo by the presenilin 1 and 2 and APP mutations linked to familial Alzheimer's disease. *Nat. Med.* **2,** 864–870.
10. Weidemann, A., Paliga, K., Durrwang, U., Czech, C., Evin, G., Masters, C. L., and Beyreuther, K. (1997) Formation of stable complexes between two Alzheimer's disease gene products: presenilin-2 and b-amyloid precursor protein. *Nat. Med.* **3,** 328–332.
11. Irizarry, M. C., McNamara, M., Fedorchak, K., Hsiao, K., and Hyman, B. T. (1997) APPSw transgenic mice develop age-related A beta deposits and neuropil abnormalities, but no neuronal loss in CA1. *J. Neuropathol. Exp. Neurol.* **56,** 965–973.
12. Irizarry, M. C., Soriano, F., McNamara, M., Page, K. J., Schenk, D., Games, D., and Hyman, B. T. (1997) Abeta deposition is associated with neuropil changes, but not with

overt neuronal loss in the human amyloid precursor protein V717F (PDAPP) transgenic mouse. *J. Neurosci.* **15;17,** 7053–7059.

13. Justice, A. and Motter, R. (1997) Behavioral characterisation of PDAPP transgenic Alzheimer mice. Abs tr. 636.6, Society for Neurosciences, New Orleans.

14. Haass, C. (1997) Presenilins: genes for life and death. *Neuron* **18,** 687–690.

15. Wong, P. C., Zheng, H., Chen, H., Becher, M. W., Sirinathsinghji, D. J., Trumbauer, M. E., Chen, H. Y., Price, D. L., Van der Ploeg, L. H., and Sisodia, S. S. (1997) Presenilin 1 is required for Notch1 and Dll1 expression in the paraxial mesoderm. *Nature* **387,** 288–292.

16. Conlon, R. A., Reaume, A. G., and Rossant, J. (1995) Notch1 is required for the coordinate segmentation of somites. *Development* **121,** 1533–1545.

17. Wolozin, B., Iwasaki, K., Vito, P., Ganjei, J. K., Lacana, E., Sunderland, T., Zhao, B., Kusiak J., Wasco, W., and D'Adamio, L. (1996) Participation of presenilin 2 in apoptosis: enhanced basal activity conferred by an Alzheimer mutation. *Science.* **274,** 1710–1713.

18. Vito, P., Lacana, E., and D'Adamio, L. (1996) Interfering with apoptosis: $Ca^{(2+)}$-binding protein ALG-2 and Alzheimer's disease gene ALG-3. *Science.* **271,** 521–525.

19. Thinakaran, G., Borchelt, D. R., Lee, M. K., Slunt, H. H., Spitzer, L., Kim, G., et al. (1996) Endoproteolysis of presenilin 1 and accumulation of processed derivatives in vivo. *Neuron* **17,** 181–190.

20. Lee, M. K., Borchelt, D. R., Kim, G., Thinakaran, G., Slunt, H. H., Ratovitski, T., Martin, L. J., Kittur, A., Gandy, S., Levey, A. I., Jenkins, N., Copeland, N., Price, D. L., and Sisodia, S. S. (1997) Hyperaccumulation of FAD-linked presenilin 1 variants in vivo. *Nat. Med.* **3,** 756–760.

21. Mann, D. M., Iwatsubo, T., Cairns, N. J., Lantos, P. L., Nochlin, D., Sumi, S. M., Bird, T. D., Poorkaj, P., Hardy, J., Hutton, M., Prihar, G., Crook, R., Rossor, M. N., and Haltia, M. (1996) Amyloid-b protein (Aβ) deposition in chromosome 14-linked Alzheimer's disease: predominance of Aβ42(43). *Ann. Neurol.* **40,** 149–156.

22. Tomita, T., Maruyama, K., Saido, T. C., Kume, H., Shinozaki, K., Tokuhiro, S., Capell, A., Walter, J., Grunberg, J., Haass, C., Iwatsubo, T., and Obata, K. (1997) The presenilin 2 mutation (N141I) linked to familial Alzheimer disease (Volga German families) increases the secretion of amyloid b protein ending at the 42nd (or 43rd) residue. *Proc. Natl. Acad. Sci. USA* **94,** 2025–2030.

23. Holcomb, L., Gordon, M. N., McGowan, E., Yu, X., Benkovic, S., Jantzen, P., Wright, K., Saad, I., Mueller, R., Morgan, D., Sanders, S., Zehr, C., O'Campo, K., Hardy, J., Prada, C. M., Eckman, C., Younkin, S., Hsiao, K., and Duff, K. (1998) Accelerated Alzheimer-type phenotype in transgenic mice carrying both mutant amyloid precursor protein and presenilin 1 transgenes. *Nat. Med.* **4,** 97–100.

24. Borchelt, D. R., Ratovitski, T., van Lare, J., Lee, M. K., Gonzales, V., Jenkins, N. A., Copeland, N. G., Price, D. L., and Sisodia, S. S. (1997) Accelerated amyloid deposition in the brains of transgenic mice coexpressing mutant presenilin 1 and amyloid precursor proteins. *Neuron* **19,** 939–945.

25. Gotz, J., Probst, A., Spillantini, M. G., Schafer, T., Jakes, R., Burki, K., and Goedert, M. (1995) Somatodendritic localization and hyperphosphorylation of tau protein in transgenic mice expressing the longest human brain tau isoform. *EMBO J.* **14,** 1304–1313.

26. Saez, E., No, D., West, A., and Evans, R. M. (1997) Inducible gene expression in mammalian cells and transgenic mice. *Curr. Opin. Biotechnol.* **8,** 608–616.

27. Snowdon, D. A. (1997) Aging and Alzheimer's disease: lessons from the Nun Study. *Gerontologist* **3,** 150–156.

II

PARKINSON'S DISEASE

8

Intervention Strategies for Degeneration of Dopamine Neurons in Parkinsonism

Optimizing Behavioral Assessment of Outcome

Timothy Schallert and Jennifer L. Tillerson

1. INTRODUCTION

Parkinsonism has a complex and variable neuropathology and etiology *(1,2)*, but the major common feature is progressive loss of dopamine (DA) neurons in the substantia nigra leading to motor impairment. Although understanding the initial causes and molecular mechanisms for degeneration of the nigrostriatal system would be invaluable for eventual treatment, in the meantime effective approaches to preventing the pathological and disabling clinical signs of the disorder may come from early detection and interventions that have quite general neuroprotective properties. Behavior-based diagnostic assessment aimed at detecting threshold level neuropathology would be critical, both in clinical practice and preclinically.

Many types of animal models of Parkinson's disease are available. Selecting a clinically predictive one is crucial, but has always been difficult. Most investigators agree that appropriate behavioral methods for assessing the functional effects of interventions are as essential as neurochemical and anatomical assays of the integrity of nigrostriatal DA neurons. However, no single test can be counted on with certainty to provide an optimal screening tool because this would require first discovering reliably successful interventions and then backtesting them on various animal models. Until this happens, the interim choice of animal models should be based on sensitivity and clear functional similarities to the neurological characteristics of Parkinsonism. Although some researchers have argued that only primate models are useful, there is no consensus or current basis to suggest that one species is more predictive than another in the transition from research to patient *(3)*.

For this chapter we were asked simply to use our experience in developing rat models to guide us in writing a selective overview of behavioral tests developed in our laboratory that we believe are most useful and promising for determining the efficacy of early therapeutic interventions in Parkinson's disease, and to provide references and enough detail about the methods to permit other investigators to use them.

Featured first and highlighted is a test of spontaneous forelimb use, which is highly reliable, easy to score, does not require DAergic drugs, and can detect a wide range of

From: Central Nervous System Diseases
Edited by: D. F. Emerich, R. L. Dean, III, and P. R. Sanberg © Humana Press Inc., Totowa, NJ

depletion both acutely and chronically. This test has been modified from a motor test of forelimb asymmetry, described first by Schallert and Lindner *(4)*, who used it to evaluate the effects of rescuing SNr neurons from transsynaptic degeneration following unilateral loss of intrinsic neurons in the neostriatum. It has since been shown to be very useful for evaluating the effects of neurotrophic interventions after partial loss of DA neurons caused by unilateral infusions of the DA neurotoxin 6-hydroxydopamine (6-OHDA) into the striatum or nigrostriatal projections, which are well-established models of important neurochemical characteristics of Parkinson's disease *(5–15)*. The test has also been adopted for models of unilateral degeneration of neurons intrinsic to the striatum, for sensory motor cortex injury, for hemispinal damage, and for other sensorimotor regions *(4,16–18)*.

Weight shifting movements initiated by the forelimbs are examined without experimenter handling, and because the movements observed are those typically used by the animal in its home cage, repeated testing does not influence the asymmetry score. In rats, the forelimbs are used to initiate movements that require weight shifting, much like legs are used by humans when they walk. Difficulty in initiating steps from a standing posture is one of the primary signs of extensive degeneration of DAergic neurons in the substantia nigra *(2,19,20)*. Thus, this test has clinical relevance. It has been used successfully to determine the effects of gene therapy and other interventions having clinical promise (e.g., *21–24*). This chapter describes our most recent version of the forelimb asymmetry test in detail and includes data validating its utility. Additional useful tests are also summarized.

2. TEST OF FORELIMB USE ASYMMETRY FOR WEIGHT SHIFTING DURING VERTICAL EXPLORATION

Use of the forelimbs for postural support can be analyzed by videotaping rats in a transparent cylinder (20 cm in diameter and 30 cm high) for 10 min (*see* Fig. 1). The cylinder encourages use of the forelimbs for vertical exploration and for landing after a rearing movement, although the home cage can be used *(4,15,17,18,23,25–32)*. All scoring is done blind to the condition of the animal.

2.1. Behaviors

The following behaviors are videotaped and scored in terms of forelimb-use asymmetry during vertical movements along the wall of the cylinder and for landings after a rear: (1) independent use of the left or right forelimb for contacting the wall of the cylinder during a full rear, to initiate a weight-shifting movement or to regain center of gravity while moving laterally in a vertical posture along the wall; (2) independent use of the left or right forelimb to land after a rear; (3) simultaneous, or near-simultaneous, use of both the left and right forelimbs for contacting the wall of the cylinder during a full rear and for lateral movements along the wall; and (4) simultaneous, or near-simultaneous, use of both the left and right forelimbs for landing after a rear.

2.2. Cylinder Requirements

The cylinder can be made of Plexiglas or glass, and should be kept clean for viewing. The dimensions of the cylinder may vary according to the size of the animals being tested. It should be high enough that the animal cannot reach the top edge by rearing,

Fig. 1. Rat in a cylinder used to assess forelimb asymmetry for weight shifting during exploration of vertical surfaces and for landing after a rear.

and it should be wide enough to permit a 2 cm space between the tip of the snout and the base of the tail when the animal is standing with all four limbs on the ground. A large mirror is placed behind the cylinder at an angle such that the forelimbs can be recorded whenever the animal turns away from the camera. To facilitate finding a movement sequence for a particular animal when searching the videotape for scoring, the tilter function of the camera or a small label depicting the animal's identification number should be placed to the side of the cylinder so that it can be seen during the entire videotaped sequence without obstructing the view of the behavior of the animal. However, to maintain blind ratings, no indication of the specific treatment (e.g., 6-OHDA vs sham; hemisphere infused; intervention vs sham procedure) should be apparent to the raters of the videotapes.

2.3. Recording Requirements

The speed setting on the video camera should be "standard play." The animal should be given two trials of 5 min each per week, although the amount of time should be long enough to obtain at least 20 vertical movements along the wall of the enclosure and at least 10 landing movements. More movements may increase the reliability of the score, but generally a minimum of 10 movements on the wall and 10 landing movements are sufficient for any one trial.

The light level should be low, or red light can be used, to encourage movement. Testing the animals during the dark part of the light/dark cycle increases the number of movements. Alternatively, wood shavings from the animal's own home cage can be placed at the bottom of the cylinder. Ensuring that the animals are well handled and familiar with the cylinder prior to surgery increases rearing movements. Because the extent of exposure to the cylinder does not affect the degree of limb use asymmetry, and because most animals are more active during the first few minutes of a trial, more trials of shorter duration may be carried out to obtain a sufficient number of movements for reliable measurement.

2.4. Scoring

A VCR with slow motion and, preferably, frame-by-frame capabilities should be used. If a movement is ambiguous (i.e., the rater cannot determine readily whether one limb is being used independently or simultaneously), that particular movement should not be scored. Each behavior (e.g., walls and lands) should be expressed in terms of (1) percent use of the ipsilateral (nonimpaired) forelimb relative to the total number of ipsilateral, contralateral, and simultaneous (both) limb use observations; (2) percent use of the contralateral (impaired) forelimb relative to the total number of ipsilateral, contralateral, and simultaneous (both) limb use observations; and (3) percent simultaneous (both) limb use relative to the total number of ipsilateral, contralateral, or simultaneous (both) limb use observations.

During a rear, the first limb to contact the wall with clear weight support (without the other limb contacting the wall within 0.4 s) is scored as an independent wall placement for that limb. After the first limb contacts the wall, a delayed placement of the other limb on the wall while the first limb remains anchored on the wall is counted as an additional movement and scored as simultaneous (both). For example, if a 6-OHDA-treated animal places its ipsilateral limb on the wall, followed by delayed contact with both forelimbs, the animal would receive a score of "one ipsilateral" and "one both" for that sequence. If only one forelimb contacts the wall, all lateral movements thereafter are each scored as independent movements of that limb until the other forelimb contacts the wall with weight support, at which point one "both" is scored. If the rat continues to explore the wall laterally in a rearing posture while alternating both limbs on the wall (wall stepping) a "both" is recorded, and every additional combination of two-limb movements (wall stepping) would receive a "both" score. If one limb remains stationary but in contact with the wall while the other makes small adjusting steps, this is scored only as one "both." Thus, both paws must be removed from the vertical surface before another movement can be scored. If the animal removes both forelimbs from the wall during a rear and then immediately resumes wall exploration, the movements are again scored as independent (left or right) or simultaneous (both) as described previously.

Landings are scored in the nearly same fashion as wall movements. After a rear in which the wall is contacted at least once, the first limb to contact the ground with clear weight support (rear termination) is scored as a land for that limb. If both limbs land with clear weight support at the same time (or within 0.4 s of each other), this is scored as a "both" land. However, if after landing the rat continues to move along the ground (stepping), these movements are not scored because unilateral dopamine depletion does not impair use of the contralateral forelimb for regaining center of gravity during horizontal movements on the ground unless the speed of weight shifting is extreme, as when the experimenter imposes rapid weight shifting (*12,33*; see Subheading 6.)

Preoperatively, if an animal shows a preference for use of a given forelimb, the DA-depleting neurotoxin can be placed in the opposite hemisphere so that the postoperative effects are not masked or reduced by any endogenous asymmetry. Note that by doing this, some sham-operated animals will have a small asymmetry bias opposite to that of animals with unilateral DA depletion (i.e., the limb contralateral to the assigned sham-operated hemisphere may remain the preferred limb for use in vertical exploration). Animals without a preoperative preference should be randomly assigned to a lesion side. Occasionally an animal will be so inactive prior to surgery that it fails to rear and make at least 20 exploratory movements along the wall and at least 10 landing movements in a session. In this case, it is recommended that the animal not be included in the study, as it is possible that this inactivity will continue postoperatively.

2.5. Analysis

Separate analyses of wall-associated use vs landing-associated use should be carried out because, after preclinical intervention, the two types of movements may show different recovery rates or endpoints. Movements involving both forelimbs, as opposed to the ipsilateral forelimb, may be particularly sensitive to some beneficial interventions.

Limb use ratios are calculated as median or mean ± SEM "ipsilateral" limb use (ipsi / ipsi + contra + both), median or mean ± SEM, "contralateral" limb use (contra / ipsi+ contra + both), and median or mean ± SEM "both" limb use (both/ ipsi+ contra + both) for wall and again for landing movements (multiply by 100 to obtain percentages).

However, wall-associated ratios and landing ratios can be averaged together for scores that reflect equal contributions from asymmetries in wall movements and landings (*see* Figs. 2 vs 3). This corrects for variability in the absolute number of landing movements vs wall movements among animals or between groups.

To obtain a single overall limb use asymmetry score, the percent contra is subtracted from the percent ipsi, for wall behavior as well as for landing. Alternatively, the ipsilateral limb use ratio, averaged across wall and landing (*see* earlier), can be used to reflect ipsilateral asymmetry. Either of these two methods provides adequate weighting among ipsilateral, contralateral, and both limb uses and will produce scores that represent the overall asymmetry, which is highly correlated with degree of interstriatal DA differences (23). Thus, for example, animals that exclusively use their ipsilateral forelimb, without independent use of the contralateral forelimb and without use of the contralateral forelimb in combination simultaneously with ipsilateral forelimb use, will show the maximal limb use bias (100%) and are likely to have sustained DA depletion >95%. However, animals that fail to use the contralateral forelimb independent of the ipsilateral forelimb, yet use the contralateral forelimb in tandem with the ipsilateral forelimb (both), will receive a less severe limb use asymmetry score and are likely to have sustained a less severe DA depletion.

2.6. Recent Data

Rats were treated with various doses of 6-OHDA or vehicle (sham) unilaterally infused into the medial forebrain bundle or at four sites in the rostral striatum, which yielded a range of DA depletion (assayed by HPLC with electrochemical detection) from nonsevere (41–79%) to severe (80–99%), with a few animals showing very severe depletions (96–99%). Rats with severe to very severe depletions displayed contralateral rotational behavior in response to apomorphine. Overall forelimb use asymmetry was correlated significantly with the degree of striatal DA loss ($r = .924$; 23).

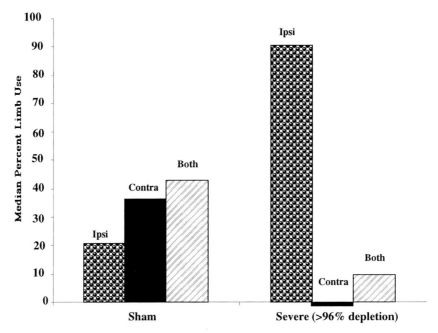

Fig. 2. Median percent limb use for ipsilateral, contralateral, and both movements for sham and very severely depleted (>96% DA depletion) animals at 7 d post 6-OHDA. Limb use was calculated as (1) ipsilateral limb use (ipsi wall/ ipsi wall + contra wall + both wall) + (ipsi land/ ipsi land + contra land + both land)/2, (2) contralateral limb use (contra wall/ ipsi wall + contra wall + both wall) + (contra land/ ipsi land + contra land + both land)/2, and (3) both limb use (both wall/ ipsi wall + contra wall + both wall) + (both land/ ipsi land + contra land + both land)/2. Rats with > 96% DA depletion preferentially used their ipsilateral forelimb for all movements.

As shown in Figs. 2 and 3, rats with depletions >96% used their ipsilateral forelimb almost exclusively for vertical exploration along the walls of the cylindrical enclosure and for landing after a rear. Accordingly, the asymmetry score for use of the ipsilateral forelimb (Fig. 2) and the overall asymmetry score (Fig. 3) were both about 90%. Sham-operated animals used both limbs, either independently (the forelimb ipsilateral or contralateral to the assigned sham hemisphere) or simultaneously (both; Fig. 2), which is reflected in the absence of a positive overall asymmetry score (Fig. 3).

Rats sustaining DA depletion <96% also showed preferential use of the ipsilateral forelimb, but co-used the contralateral and ipsilateral forelimbs almost as often as sham animals did (Fig. 4). As a group, rats sustaining 30–70% DA depletion used the contralateral forelimb independent of the ipsilateral forelimb more often than rats with >80% depletion (mostly for landing; Fig. 4), which led to a significantly lower overall asymmetry score (Fig. 5). Nevertheless, both groups showed significant overall asymmetries.

Recovery from asymmetry of forelimb use was not apparent, even in moderately depleted rats up to 40 d after 6-OHDA infusion. Thus the forelimb asymmetries found when the animals were tested during the first 2 wk postsurgery (Fig. 6) were comparable to that found when the animals were tested during d 21–40 postsurgery (Fig. 7). It

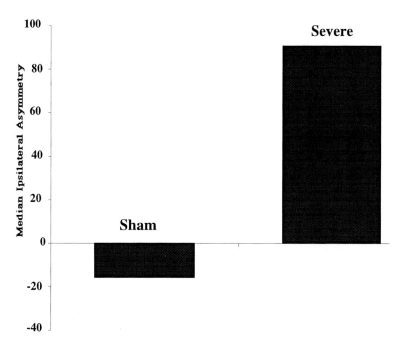

Fig. 3. Median overall asymmetries for sham and very severely depleted (>96%) rats. Severely depleted animals displayed an extreme ipsilateral limb use bias (>90%). *See* text for method of determining overall forelimb asymmetry score.

is possible that either no recovery occurred or degeneration continued at a rate sufficient to mask ongoing plasticity.

Figures 8 and 9 show wall-associated movements without landings included. In animals that have >80% DA depletion, little or no use of the contralateral forelimb occurs along the walls of the enclosure. That is, the ipsilateral forelimb is preferentially used and is not co-used with the contralateral forelimb, in contrast to landing movements during which the contralateral forelimb may assist the ipsilateral forelimb for postural support. Therefore, wall-associated overall movement scores are more extreme (compare with Fig. 5) and may represent a more sensitive index of the adverse effects of DA depletion on the capacity to initiate weight shifting.

3. BILATERAL TACTILE STIMULATION TEST

There are two phases to this test, which was developed by Schallert et al. *(34–41)* to measure forelimb somatic–sensorimotor asymmetries. Like the limb-use asymmetry test, this test can be used repeatedly without being subject to the influence of practice *(38,43,44).*

3.1. Phase One: Detection of Sensorimotor Asymmetry

The first phase of the bilateral-tactile stimulation test indicates the presence or absence of an asymmetry in forelimb tactile sensation. The animal is briefly (5–10 s) removed from its home cage and small adhesive backed patches (113.1 mm^2), which can be obtained commercially from Avery, are applied to the radial aspect of the wrist

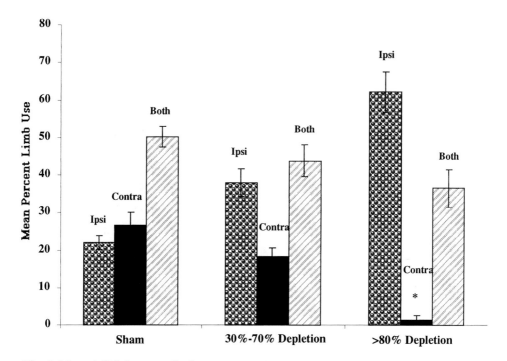

Fig. 4. Mean ± SEM percent limb use averaged across wall and landing behavior for ipsilateral, contralateral, and both limbs in tandem for animals with 30–70% depletion or > 80% depletion. Note that animals with <96% DA depletion use the contralateral forelimb substantially in tandem with the ipsilateral forelimb (both), but rarely independently unless the depletion is moderate. *See* Fig. 2 caption and text for forelimb asymmetry calculations. *Significantly different from ipsilateral.

of both forelimbs. The rat is returned to its home cage and the order (left vs right) in which it contacts the stimuli and the order in which it removes the stimuli are recorded, along with the latencies to do so. Four to five trials are conducted. Each trial is terminated upon removal of both stimuli, or after 2 min has passed.

In rats with severe unilateral DA depletion, unlike after cortical injury, the latency to remove the contralateral stimulus is typically very much delayed relative to the latency to contact the contralateral stimulus (reflecting a movement initiation deficit). In contrast to the measure of order of stimulus removal, removal of the contralateral stimulus is altered by both recovery mechanisms and experience. Moreover, during the first week or so after severe DA depletion, the animal may neglect the contralateral stimulus altogether, although it may make paw shaking movements reflecting sensitivity but a lower level of sensorimotor integration *(34)*. If animals are not housed singly, the other animals in that cage should be removed during testing. If the floor contains sawdust, all but a thin layer should be removed. Otherwise, the forelimbs may sink into the sawdust and displace the adhesive stimuli. Before surgery, all animals should be well handled and given at least 10 trial of practice removing the stimuli, each separated by at least 2 h. It is not necessary to score this practice behavior. An additional four or five preoperative trials should be scored to determine preoperative bias. After each trial, both preopertively and postoperatively, an interval of at least 5 min should elapse before

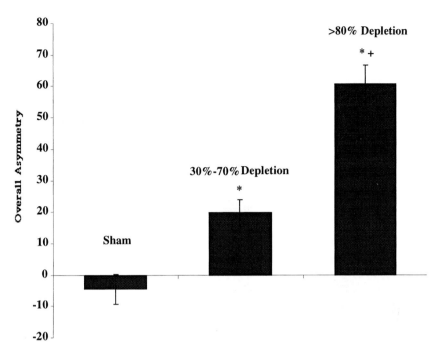

Fig. 5. Mean ± SEM overall asymmetry score for sham animals and animals with moderate or severe DA depletion averaged across d 7–40 postsurgery and averaged for wall and landing movements. *See* text for method of determining overall forelimb asymmetry score. *Significantly different from sham. +Significantly different from 30–70% depletion.

the next trial to avoid habituation. To save time, in a testing session all animals in the study can be given one trial each before beginning a second trial.

Any animal that fails to consistently remove both stimuli within 20 s prior to surgery should be excluded from the study. Preoperatively, if an animal shows an 80% preference for stimuli placed on a given forelimb, the lesion can be placed in the opposite hemisphere so that the postoperative effects are not masked by any endogenous asymmetry. Animals without a preoperative preference would be randomly assigned to a lesion side. However, because the limb-use asymmetry test described above is easier to adopt and is *chronically* more sensitive to nonsevere levels of DA depletion than the adhesive removal test, it is recommended that the hemisphere selected for DA depletion should be based on endogenous asymmetry in the limb-use test.

3.2. Phase Two: Magnitude of Sensorimotor Asymmetry

If the animal contacts and removes stimuli from only one side in 80–100% of the trials, a second phase of the test, somatosensory neutralization, is administered to assess the magnitude of the asymmetry. In the somatosensory neutralization test, the size of the stimulus on the limb ipsilateral to the response bias is decreased from trial to trial by 14.1 mm^2 while the size of the stimulus contralateral to the response bias is increased by 14.1 mm^2. The ratio of the size of the ipsilateral relative to the size of the contralateral stimulus necessary to reverse or neutralize the "contact" response bias is recorded as the magnitude of asymmetry. As in the other behavioral tests, experimenters should

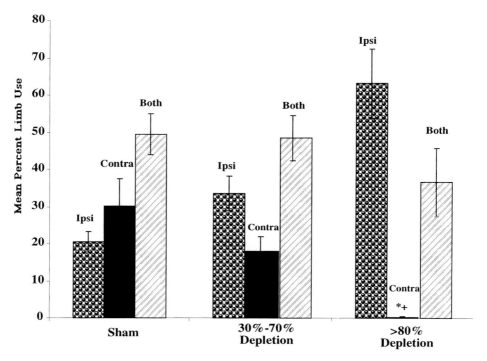

Fig. 6. Mean ± SEM percent forelimb use 7–14 d after unilateral 6-OHDA in rats with 0, 30–70% or >80% DA depletion. *Significantly different from ipsilateral. +Significantly different from 30–70% contralateral.

be blind to the condition of the animal. The magnitude of asymmetry, which can be reduced to zero or reversed with low doses of direct acting DA agonists, is more sensitive to differences in sensory asymmetries between animals than the standard adhesive removal test and should be used always to test the effects of any intervention (34,44).

4. PLACING TEST

Asymmetries in forelimb placing behaviors are assessed using the vibrissae-elicited forelimb placing and extinction placing tests. In both tests, animals are held by their torsos with their forelimbs hanging freely. Each forelimb is tested independently by orienting one side of the animal toward a countertop and moving the animal slowly and laterally toward the edge of the countertop until the vibrissae of that side make contact with the edge. Intact rats typically place the forelimb (both ipsilateral and contralateral forelimbs) quickly onto the edge of the counter. Lesioned animals, on the other hand, typically place the limb ipsilateral to the lesion but fail to place the contralateral limb reliably. Ten trials of each forelimb are performed in a balanced order. The forelimb extinction test is identical to the nonextinction placing test except that a light stimulus is applied to the opposite vibrissae (opposite to the vibrissae contacted by the countertop) by the experimenter's finger. In lesioned animals, this stimulus exacerbates placing asymmetries, possibly owing to sensory input impinging on the vibrissae ipsilateral to the lesion. Once again, 10 trials for each forelimb are carried out. The number of successful placing reactions for each forelimb are recorded (described in

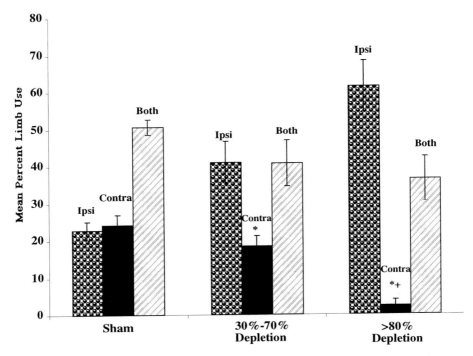

Fig. 7. Mean ± SEM percent forelimb use 21–40 d after unilateral 6-OHDA in rats with 0, 30–70% or >80% DA depletion. Note that in these animals no recovery occurred relative to forelimb use 7–14 d postsurgery, even in moderately depleted animals (compare data in Fig. 6). *Significantly different from ipsilateral. +Significantly different from 30–70% contralateral.

18,26,28). The experimenter should be blind to the condition of the animals. Both types of placing asymmetries are recorded as the percent unsuccessful contralateral placements.

5. SINGLE-LIMB AKINESIA TEST

To test forelimb akinesia, stepping movements and direction of steps made by the ipsilateral and contralateral forelimbs are assessed using an isolated forelimb test *(23,26,33,45).* In this test, the rat is held by its torso with its hindquarters and one forelimb lifted above the surface of a table so that the weight of its body is supported by one forelimb alone.

To minimize the influence of postural asymmetry and head turning, the rat's weight is centered over the isolated limb, and its head and forequarters are gently oriented forward by the experimenter's thumb and index finger. A normal, well-tamed animal held in this way typically steps about readily when either limb is isolated, but a rat with a very severe unilateral 6-OHDA lesion steps only, or primarily, when the ipsilateral limb is isolated. Moderate DA depletions yield transient hypokinesia in which the contralateral limb makes fewer steps than the ipsilateral forelimb. The latency to initiate a step is recorded. Also, the number of steps and direction relative to the damaged hemisphere made during the 30–60-s trial is recorded for each forelimb. If a directional step is less than about 30° from midline, it is counted as forward. Stepping movements are

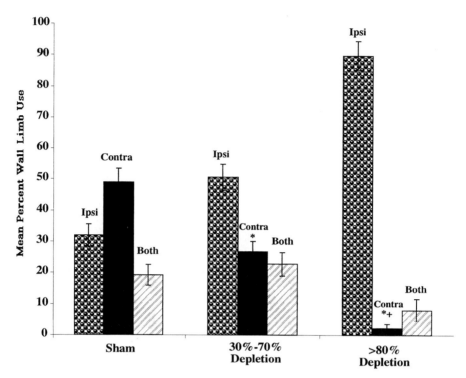

Fig. 8. Mean ± SEM percent use of the forelimb for wall-only behavior, calculated for ipsi-lateral, contralateral, and both movements. Note that in contrast to wall + landing scores, rats with >80% DA depletion used the ipsilateral forelimb almost exclusively (compare with Fig. 4, >80% "both"). *Significantly different from ipsilateral. +Significantly different from 30–70% contralateral.

defined as animal-initiated steps that substantially shift the animal's weight to a new location. Dyskinetic, in-place, or treading movements (observed frequently as a response to low doses of apomorphine in severely depleted animals) should be noted and rated on a 0–4 scale, but should not counted as steps. The order in which the limbs are tested should be varied randomly (the experimenter again must be blind as to the intervention or surgical condition of the rats).

Data in *severely* injured animals ($n = 8$ per group) indicate that when one forelimb is examined independent of the other in its capacity to move spontaneously while the other limb is immobilized by the experimenter, only the ipsilateral forelimb can reli-ably initiate motor function involving clear shift of weight (*see* Table 1).

DA receptor agonists, such as apomorphine or amphetamine, can increase the num-ber of steps made by the contralateral forelimb only if the depletion is not very severe. However, if the depletion is very severe (>97% depletion), akinesia is not reversed substantially. DA agonists increase ipsilateral forelimb stepping (*see* Table 2). How-ever, the direction of the stepping in this limb is contraversive in response to apomor-phine, which together with the hypokinesia in the contralateral forelimb, contributes substantially to the tight contraversive turning behavior observed in response to this

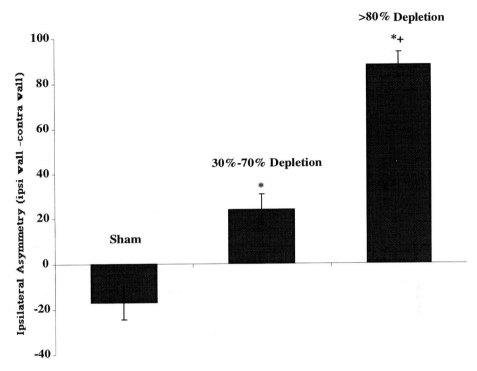

Fig. 9. Mean ± SEM overall asymmetry scores for wall-only movements 7–40 d postsurgery in rats with 0, 30–70%, and >80% DA depletion. Note that because the >80% depleted animals rarely used the contralateral forelimb even in tandem with the ipsilateral forelimb ("both"), the overall asymmetry score for wall behavior is almost 90%, compared with the 60% ipsilateral asymmetry score when landing movements are included together with wall movements (*see* Fig. 5). *Significantly different from sham. +Significantly different from 30–70% depletion.

drug *(33,46)*. Indirect acting DA agonists, such as amphetamine, increase stepping only in the ipsilateral forelimb and the direction of the stepping is either forward or ipsiversive, which leads to wider circling during exposure to amphetamine.

6. BRACING (ADJUSTING STEPS) TEST

An additional test (bracing test) can be conducted to assess the capacity to adjust stepping in response to experimenter-imposed lateral movement *(12,26,33,45)*. The animal is held as described previously in the single-limb test, but is moved by the experimenter slowly sideways or in the forward direction. The number and size of adjusting steps is recorded as the animal is moved 1 m in 5-, 10-, and 15-s periods. A ratio of the number of steps initiated by the contralateral limb compared with the number of steps initiated by the ipsilateral limb is calculated for each speed and direction (moving ipsilateral to the paw side or contralateral to the paw side). Previous work has indicated that at slow speeds stepping can occur but faster speeds cause 6-OHDA-treated animals to brace or drag the affected forelimb rather than make catchup steps; however, direct DA agonists permit adequate stepping even at faster speeds *(12,26,45)*.

Table 1
Median Steps per Minute in Forelimb Akinesia Test

	Contralateral forelimb		Ipsilateral forelimb	
	6-OHDA	Sham	6-OHDA	Sham
	1.5	18.5	21.3	20.4

Table 2
Median Steps per Minute in Forelimb Akinesia Test
During Apomorphine (0.01 mg/kg s.c.) or Amphetamine (2 mg/kg)

	Contralateral forelimb		Ipsilateral forelimb	
	6-OHDA	Sham	6-OHDA	Sham
Apomorphine	3	40	50	51
Amphetamine	2	48	42	41

6.1. Clinical Relevance

The single-limb akinesia test is a model of movement initiation involving weight shifts in Parkinson's disease. Initiation of weight shift is sensitive to direct DA receptor agonists in both humans and rats when the degeneration of DA neurons is not severe, and becomes less sensitive to DA agonists when the degeneration is very severe. Dyskinesias of the impaired forelimb occur more reliably than movement initiation with increasing doses of DA agonists.

The bracing test is a model of postural balance that assesses capacity to regain postural stability and center of gravity when rapid weight shifts are imposed (as in the "standing pull test" used by neurologists in Parkinson's disease). Deficits in the bracing test are ameliorated by DA agonists even at severe levels of degeneration. In end-stage Parkinsonism the effects of direct-acting DA receptor agonists on the capacity to rapidly move the supporting limbs in response to an imposed weight shift (including that associated with falling) may be one of the reliable benefits of drug therapy. Indeed, some patients report that to initiate walking, they let themselves begin to fall; however, if their akinesia is severe, this procedure can be dangerous unless they are under the influence of a DA agonist, such as L-Dopa.

7. ORIENTING MOVEMENT TIME AND DISENGAGE TESTS

The animals must be tested in their home cage for head orienting in response to perioral tactile stimulation (47,48,49,50). A wooden probe (23 cm long, 2 mm in diameter) is used to stimulate the vibrissae on each side of the body. If the cages are not comprised of wire mesh, two holes must be drilled in each wall 3 cm from the floor.

7.1. Orienting Movement Time

Stimulation is presented unilaterally to the right and left side at 2-s intervals until the animal either orients to the probe or 300 s elapses. Timing should begin when the probe first contacts the perioral region. The rat's latency to respond to the stimulus is recorded to the nearest second. A detailed comparison can be made between the type of orienting movements in the contralateral vs ipsilateral directions, as the topography of the lateral head movement can be different on the affected side *(47)*. The normal response is a lateral movement toward the probe. Animals with moderate to severe DA depletion may learn to use a rotatory movement of the head, followed by a ventral tuck, or other unusual strategy to orient to the probe when the contralateral vibrissae are stimulated. Ipsilaterally directed head movements in response to contralateral stimulation should not be recorded as a positive orienting response, even if they resulted in contact with the probe. A rapid 360° ipsilateral turn in response to contralateral vibrissae touch would permit such contact if the experimenter cannot remove the probe from the cage quickly enough, but this is an abnormal, misdirected movement and thus should be discounted.

7.2. Disengage Deficit

Responsiveness to perioral stimulation can be assessed concurrent with or in the absence of eating, drinking, or grooming behavior *(47,48)*. To encourage eating without food deprivation, a piece of chocolate (3.4 g) is placed in the home cage. When the animal begins eating the chocolate, the perioral stimulation test is carried out as described previously (orienting movement time). During eating, drinking, or grooming, orienting to contralateral, but not ipsilateral, stimulation is suppressed in animals with unilateral, moderate DA depletion. That is, these animals fail to disengage from eating behavior only when stimulated on the contralateral side. However, if the depletion is not severe, orienting occurs to contralateral stimulation in the absence of ongoing consummatory behaviors. The disengage test has been used to examine the success of fetal tissue grafts in DA-depleted rats *(52)*.

8. DRUG-INDUCED TURNING

The most widely adopted method for screening potentially beneficial therapies is the DA-agonist induced turning test, which measures the number and direction of horizontal circling movements (also called the rotational test; *10,46,51*). Lateral turning behavior is easy to quantify, and the differential circling response to indirect vs direct receptor agonists is quite useful for distinguishing between severe and nonsevere cases of DA depletion after unilateral nigrostriatal injury. The method is not as sensitive to partial DA depletion as other tests such as the forelimb asymmetry test described previously. However, extensive research indicates that transplanted tissue or trophic factors can reduce amphetamine-induced ipsiversive turning or apomorphine-induced contraversive turning and these interventions often are found to be associated with increased DA in the striatum *(10)*.

One concern is that apomorphine-induced turning can be reduced by intervention-induced damage to the intrinsic striatal neurons, yet this effect might be overlooked and viewed as a positive outcome. Moreover, increased striatal DA may not be the only major target for an intervention, or even the most critical. Indeed, increasing intrastriatal DA may not necessarily improve the condition of Parkinson's disease patients in a substantially meaningful way or without adverse effects, especially if the DA is uncoupled from behaviorally relevant neural systems *(see 3,33,45,52,66)*. Nevertheless, unless there are concerns about DA neurotoxicity or specific reasons for avoiding exposure to DA agonists, this test could be included routinely in any battery of tests used to screen interventions.

9. OTHER TESTS

In addition to the tests reviewed in this chapter that do not require pharmacological challenges, there are other nondrug tests used to examine the behavior of rats with unilateral striatal DA depletion. For example, Whishaw and others have devised ingenious methods for examining forelimb reaching behavior in rats that are sensitive to DA depletion caused by unilateral 6-OHDA *(42,53–56)*.

Spirduso et al. developed an extremely sensitive test of high-speed reaction time that could be adapted to examine the reactive capacity of each forelimb independently *(57)*. The animals were trained to hold a lever down and attend to specific cues that signal oncoming mild footshock, which could be prevented by releasing the lever as fast as possible. The interval between the cue and the footshock was reduced progressively to obtain the reactive capacity of each animal. Slings were fitted to immobilize one or the other forelimb, which permitted assessment of reactive capacity in each limb separately. This test can detect a range of 30–99% DA depletion with considerable accuracy. The clinical relevance is obvious. However, during the early laborious shaping phases of the task it is imperative to use highly skilled experimenters with good instincts about adjusting the cue-shock interval from trial to trial to adapt to individual differences among the animals in their responses to success or failure. Others have since described a similar reaction time test using food as reinforcement *(58)*, but it is unclear whether either test is more advantageous.

Quick screening tests for large interhemispheric differences in DA depletion include various postural asymmetry tests that rely on the animal's negative geotaxis combined with directional responsivity. Among these are a slanted grid test *(34,59,60)* and a test in which the animal is hung upside down and the direction of body angle is determined *(34)*. This latter test recently has been called the elevated body swing test, which has been likened to hemiparkinsonian postural asymmetry and other disorders of the basal ganglia *(61)*. Thigmotaxic scanning tests developed by Huston and his colleagues *(10,62,63)*, like the placing test described previously, exploit asymmetries in vibrissae sensitivity to tactile stimulation *(9,34,47,64,65)*.

Lindner et al. *(66)* describe an automated fixed ratio lever press task that appears to have considerable utility in detecting bilateral DA depletion without severely depleting DA. The task requires the animal to make movements as fast as possible during a 30-min period. By requiring the animal to make more bar presses for each food pellet, the level of difficulty can be gradually increased, which increases the sensitivity of the task.

10. CONCLUSION AND FUTURE DIRECTIONS

Why or how DA neurons degenerate in Parkinson's disease are important issues for research, but it is possible that the neurons could be rescued without yet answering these questions, for example, by an intervention that imparts a general resistance to cell death. The bane of any biotechnology company is sponsoring a disappointing clinical trial after years of expense developing a potentially therapeutic agent that appears to work well in their animal model. A worthy goal, therefore, would be to find preclinical tests that are highly likely to reverse the motor symptoms of Parkinson's disease or halt the progression of Parkinsonian symptoms. We recommend that a slow unilateral nigrostriatal degeneration model be used to test the effects of the intervention using a battery of tests, including the forelimb asymmetry test which is simple to carry out and score, is stable chronically, is very sensitive even to partial DA depletion, and does not require drug challenges. Moreover, because Parkinson's disease is an aging related disorder and aging–lesion interactions may be relevant *(20,22,35,67)*, it is important to ensure that the intervention has efficacy in aging animals as well. Finally, consideration should be given to both preoperative *(68)* and postoperative *(8,17,23,68)* experiential events that might dramatically enhance or detract from the efficacy of the intervention. A major influence on the ability of the intervention to improve anatomical and functional outcome in both the animal model and the patient may be time-sensitive motor rehabilitation manipulations. Thus, during ongoing degeneration of nigrostriatal neurons, but not at later time points, there may be a window of opportunity to take advantage of the compensatory capacity of injury- and use-dependent endogenous trophic factor expression *(18,23,69)*. A well-designed regimen of combining neurotrophic factors with physical therapy that specifically targets the impaired motor system may modulate and enhance neural events linked to interventive mechanisms.

ACKNOWLEDGMENTS

We thank Michael Zigmond, Sandy Castro, Silke Morin, Leigh Humm, and Shiela Fleming for their help.

REFERENCES

1. Conford, M. E., Chang, L., and Miller, B. L. (1995) The neuropathology of Parkinsonism: an overview. *Brain Cognit.* **28,** 321–341.
2. Hornykiewicz, O. (1975) Parkinson's disease and its chemotherapy. *Biochem. Pharmacol.* **24,** 1061–1065.
3. Schallert, T. (1995) Models of neurological defects and defects in neurological models. *Brain Behav. Sci.* **18,** 68–69.
4. Schallert, T. and Lindner, M. D. (1990) Rescuing neurons from trans-synaptic degeneration after brain damage: helpful, harmful or neutral in recovery of function? *Canad. J. Psychol.* **44,** 276–292.
5. Schallert, T. and Wilcox, R. E. (1985) Neurotransmitter-selective brain lesions, in *Neuromethods (Series 1: Neurochemistry), General Neurochemical Techniques* (Boulton, A. A. and Baker, G. B. eds.), Humana Press, Totowa, NJ, pp. 343–387.
6. Anden, N. E., Dahlstrom, A., Fuxe, K., and Larsson, K. (1966) Functional role of the nigro-neostriatal dopamine neurons. *Acta Pharmacol. Toxicol.* **24,** 263–274.
7. Ungerstedt, U. (1968) 6-Hydroxy-dopamine induced degeneration of central monoamine neurons. *Eur. J. Pharmacol.* **5,** 107–110.
8. Zigmond, J. M. J., Abercrombie, E. D., Berger, T. W., Grace, A. A., and Stricker, E. M. (1993) Compensatory responses to partial loss of dopaminergic neurons: studies with 6-hydroxydopamine, in *Current Concepts in Parkinson's Disease Research* edited by (Schneider, J. and Gupta, M., eds.), Hogrefe & Huber.
9. Marshall, J. F. (1979) Somatosensory inattention after dopamine-depleting intracerebral 6-OHDA injections: spontaneous recovery and pharmacological control. *Brain Res.* **177,** 311–324.

10. Schwarting, R. K. and Huston, J. P. (1996) The unilateral 6-hydroxydopamine lesion model in behavioral brain research. Analysis of functional deficits, recovery, and treatments. *Prog. Neurobiol.* **50,** 275–331.
11. Schallert, T., Whishaw, I. Q., Ramirez, V. D., and Teitelbaum, P. (1978) Compulsive, abnormal walking caused by anticholinergics in akinetic, 6-hydroxydopamine-treated rats. *Science* **199,** 1461–1463.
12. Schallert, T., De Ryck, M., Whishaw, I. Q., Ramirez, V. D., and Teitelbaum, P. (1979) Excessive bracing reactions and their control by atropine and L-DOPA in an animal analog of Parkinsonism. *Exp. Neurol.* **64,** 33–43.
13. Ichitani, Y., Okamura, H., Masamoto, Y., Nagatsu, I., and Abata, Y. (1991) Degeneration of nigral dopamine neurons after 6-hydroxydopamine injection to the rat striatum. *Brain Res.* **549,** 350–353.
14. Dunnett, S. B., Bjorklund, A., Stenevi, U., and Iversen, S. D. (1981) Behavioral recovery following transplantation of substantia nigra in rats subjected to 6-OHDA lesions of the nigrostriatal pathway. I. Unilateral lesions. *Brain Res.* **215,** 147–161.
15. Lee, C. S., Sauer, H., and Björklund, A. (1996) Dopaminergic neuronal degeneration and motor impairments following lesion by intrastriatal 6-hydroxydopamine in the rat. *Neuroscience* **72,** 641–653.
16. Liu, Y., Kim, D., Himes, B. T., Chow, S. Y., Schallert, T., Murray, M., Tessler, A., and Fischer, I. (1999) Transplants of fibroblasts genetically modified to express BDNF promote regeneration of adult rat rubrospinal axons and recovery of forelimb function. *J. Neurosci.* **19,** 4370–4387.
17. Jones, T. A. and Schallert, T. (1992) Overgrowth and pruning of dendrites in adult rats recovering from neocortical damage. *Brain Res.* **581,** 156–160.18.
18. Jones, T. A. and Schallert, T. (1994) Use-dependent growth of pyramidal neurons after neocortical damage. *J. Neurosci.* **14,** 2140–2152.
19. Dunnett, S. B. and Iversen, S. D. (1992) The functional role of mesotelencephalic dopamine systems. *Biol. Rev.* **67,** 491–518.
20. Langston, J. W. (1990) Predicting Parkinson's disease. *Neurology* **40,** 70–74.
21. Choi-Lundberg, D. L., Lin, Q., Schallert, T., Crippens, D., Davidson, B. L., Chang, Y. N., Chiang, Y. L., Qian, J., Bardwaj, L., and Bohn, M. C. (1998) Behavioral and cellular protection of rat dopaminergic neurons by an adenoviral vector encoding glial cell line-derived neurotrophic factor. *Exp. Neurol.* **154,** 261–275.
22. Conner, B., Kozlowski, D. A., Schallert, T., Tillerson, J. L., Davidson, B. L., and Bohn, M. C. (1999) The differential effects of adenoviral vector mediated glial cell line-derived neurotgrophic factor (GDNF) in the striatum vs substantia nigra of the aged parkinsonian rat. *J. Neurosci.*, submitted.
23. Tillerson, J. L., Castro, S., Zigmond, M. J., and Schallert, T. (1998) Motor rehabilitation of forelimb use in unilateral 6-hydroxydopamine (6-OHDA) rat model of Parkinson's disease. *Neurosci. Abstr.* **672.18** 1720.
24. Johnston, R. E. and Becker, J. B. (1997) Intranigral grafts of fetal ventral mesencephalic tissue in adult 6-hydroxydopamine-lesioned rats can induce behavioral recovery. *Cell Transplant.* **6,** 267–276.
25. Schallert, T. and Jones, T. A. (1993) "Exuberant" neuronal growth after brain damage in adult rats: The essential role of behavioral experience. *J. Neural Transplant. Plast.* **4,** 193–198.
26. Lindner, M. D., Winn, S. R., Baetge, E. E., Hammang, J. P., Gentile, F. T., Doherty, E., McDermott, P. E., Frydel, B., Ullman, M. D., Schallert, T., and Emerich, D. F. (1995) Implantation of encapsulated catecholamine and GDNF-producing cells in rats with unilateral dopamine depletions and Parkinsonian symptoms. *Exp. Neurol.* **132,** 62–76.
27. Schallert, T., Kozlowski, D. A., Humm, J. L., and Cocke, R. R. (1997) Use-dependent

events in recovery of function, in *Advances in Neurology: Brain Plasticity* (Freund, H.-J., Sabel, B. A., and Witte, O. W., eds.), vol. 70, Lippincott-Raven, Philadelphia.

28. Kozlowski, D. A., James D. C., and Schallert, T. (1996) Use-dependent exaggeration of neuronal injury following unilateral sensorimotor cortex lesions. *J. Neurosci.* **16,** 4776–4786.

29. Kawamata, T., Dietrich, W. D., Schallert, T., Gotts, J. E., Cocke, R. R., Benowitz, L. I., and Finklestein, S. P. (1997) Intracisternal basic fibroblast growth factor (bFGF) enhances functional recovery and upregulates the expression of a molecular marker of neuronal sprouting following focal cerebral infarction. *Proc. Nat. Acad. Sci. USA* **94,** 8179–8184.

30. Humm, J. L., Kozlowski, D. A., James, D. C., Gotts, J. E., and Schallert, T. (1998) Use-dependent exacerbation of brain damage occurs during an early post-lesion vulnerable period. *Brain Res.* **783,** 286–292.

31. Humm, J. L., Kozlowski, D. A., Bland, S. T., James, D. C., and Schallert, T. (1999) Progressive expansion of brain injury by extreme behavioral pressure: is glutamate involved? *Exp. Neurol.* **157,** 349–358.

32. Schallert, T. and Kozlowski, D. A. (1998) Brain damage and plasticity: use-related enhanced neural growth and overuse-related exaggeration of injury, in *Cerebrovascular Disease* (Ginsberg, M. D. and Bogousslavsky, J., eds.), Blackwell Science, New York, pp. 611–619.

33. Schallert, T., Norton, D., and Jones, T. A. (1992) A clinically relevant unilateral rat model of parkinsonian akinesia. *J. Neural Transplant. Plast.* **3,** 332–333.

34. Schallert, T., Upchurch, M., Lobaugh, N., Farrar, S. B., Spiruso, W. W., Gilliam, P., Vaughn, D., and Wilcox, R. E. (1982) Tactile extinction: distinguishing between sensorimotor and motor symmetries in rats with unilateral nigrostriatal damage. *Pharmacol. Biochem. Behav.* **1,** 455–462.

35. Schallert, T., Upchurch, M., Wilcox, R. E., and Vaughn, D. M. (1983) Posture-independent sensorimotor analysis of inter-hemispheric receptor asymmetries in neostriatum. *Pharmacol. Biochem. Behav.* **18,** 753–759.

36. Schallert, T., Hernandez, T. D., and Barth, T. M. (1986) Recovery of function after brain damage: severe and chronic disruption by diazepam. *Brain Res.* **379,** 104–111.

37. Schallert, T. (1988) Aging-dependent emergence of sensorimotor dysfunction in rats recovered from dopamine depletion sustained early in life, in *Annals of the New York Academy of Science: Central Determinants of Age-Related Decline in Motor Function* (Joseph, J. A., ed.), New York Academy of Sciences, New York, pp. 108–120.

38. Schallert, T. and Whishaw, I. Q. (1984) Bilateral cutaneous stimulation of the somatosensory system in hemidecorticate rats. *Behav. Neurosci.* **98,** 518–540.

39. Schallert, T. and Whishaw, I. Q. (1985) Neonatal hemidecortication and bilateral cutaneous stimulation in rats. *Dev. Psychobiol.* **18,** 501–514.

40. Hernandez, T. D. and Schallert, T. (1988) Seizures and recovery from experimental brain damage. *Exp. Neurol.* **102,** 318–324.

41. Barth, T. M., Jones, T., and Schallert, T. (1990) Functional subdivisions of the rat sensorimotor cortex. *Behav Brain Res.* **39,** 73–95.

42. Whishaw, I. G., O'Connor, W. T., and Dunnett, S. B. (1986) The contributions of motor cortex, nigrostriatal dopamine and caudate-putamen to skilled forelimb use in the rat. *Brain* **199,** 805–843.

43. Markgraf, C. G., Green, E., Hurwitz, B. E., Morikawa, E., Dietrich, W. D., McCabe, P. M., Ginsberg, M. D., and Schneiderman, N. (1992) Sensorimotor and cognitive consequences of middle cerebral artery occlusion in rats. *Brain Res.* **575,** 238–246.

44. Dunnett, S. B., Hernandez, T. D., Summerfield, A., Jones, G. H., and Arbuthnott, G. (1988) Graft-derived recovery from 6-OHDA lesions: specificity of ventral mesencephalic graft tissues. *Exp. Brain. Res.* **71,** 411–424.

45. Olsson, M., Nikkhah, G., Bentlage, C., and Bjorklund, A. (1995) Forelimb akinesia in the rat Parkinson model: differential effects of dopamine agonists and nigral transplants as assessed by a new stepping test. *J. Neurosci.* **15,** 3863–3875.
46. Ungerstedt, U. and Arbuthnott, G. (1970) Quantitative recording of rotational behavior in rats after 6-OHDA lesions of the nigrostriatal dopamine system. *Brain Res.* **24,** 485–493.
47. Schallert, T. and Hall, S. (1988) 'Disengage' sensorimotor deficit following apparent recovery from unilateral dopamine depletion. *Behav. Brain Res.* **30,** 15–24.
48. Hall, S. and Schallert, T. (1988) Striatal dopamine and the interface between orienting and ingestive functions. *Physio. Behav.* **44,** 469–471.
49. Schallert, T., Petrie, B. F., and Whishaw, I. Q. (1989) Neonatal dopamine depletion: spared and unspared sensorimotor and attentional disorders and effects of further depletion in adulthood. *Psychobiology* **17,** 386–396.
50. Hall, S., Rutledge, J. N., and Schallert, T. (1992) MRI, brain iron and experimental Parkinson's disease. *J. Neurol. Sci.* **113,** 1–11.
51. Mandel, R. J., Yurek, D. M., and Randall, P. K. (1990) Behavioral demonstration of a reciprocal interaction between dopamine receptor subtypes in the mouse striatum: possible involvement of the striato-nigral pathway. *Brain Res. Bull.* **25,** 285–292.
52. Nikkhah, G., Duan, W.-M., Knappe, U., Jodicke, A., and Bjorklund, A. (1993) Restoration of complex sensorimotor behavior and skilled forelimb use by a modified nigral cell suspension transplantation approach in the rat Parkinson model. *Neuroscience* **56,** 33–43.
53. Miklyaeva, E. I. and Whishaw, I. Q. (1996) Hemiparkinson analogue rats display active support in good limbs versus passive support in bad limbs on a skilled reaching task of variable height. *Behav. Neurosci.* **110,** 117–125.
54. Montoya, C. P., Campbell, H. L., Pemberton, K. D., Dunnett, S. B. (1991) The 'staircase test': a measure of independent forelimb reaching and grasping abilities in rats. *J. Neurosci. Methods* **36,** 2–3.
55. Barneoud, P., Parmentier, S., Mazadier, M., Miquet, J. M., Boireau, A., Dubedat, P., and Blanchard, J. D. (1995) Effects of complete and partial lesions of the dopaminergic mesotelencephalic system on skilled forelimb use in the rat. *Neuroscience* **67,** 837–848.
56. Sabol, K. E., Neill, D. B., Wages, S. A., Church, W. H., and Justice, J. B. (1985) Dopamine depletion in a striatal subregion dirupts performance of a skilled motor task in the rat. *Brain Res.* **335,** 33–43.
57. Spirduso, W. W., Gilliam, P. E., Schallert, T., Upchurch, M., Vaughn, D. M., and Wilcox, R. E. (1985) Reactive capacity: a sensitive behavioral marker of movement initiation and nigro-striatal dopamine function. *Brain Res.* **335,** 45–54.
58. Amalric, M. and Koob, G. F. (1987) Depletion of dopamine in the caudate nucleus but not in nucleus accumbens impairs reaction-time performance in rats. *J. Neurosci.* **7,** 2129–2134.
59. Hoyman, L., Weese, G. D., and Frommer, G. P. (1978) Tactile discrimination performance deficits following neglect-producing unilateral lateral hypothalamic lesions in the rat. *Physiol. Behav.* **22,** 139–147.
60. Marshall, J. F. (1980) Basal ganglia dopaminergic control of sensorimotor functions related to motivated behavior, in *Neural Mechanisms of Goal-Directed Behavior and Learning* (Thompson, R. F., Hicks, L. H., and Shvyrkov, V. B., eds.), Academic Press, New York.
61. Borlongan, C. V. and Sanberg, P. R. (1996) Asymmetrical motor behavior in animal models of human diseases: the elevated body swing test, in *Motor Activity and Movement Disorders.* (Sanberg, P. R., Ossenkopp, K. P., and Kavaliers, M., eds.), Humana Press, Totowa, NJ.
62. Fornaguera, J., Carey, R. J., Huston, J. P., and Schwarting, R. K. W. (1994) Behavioral asymmetries and recovery in rats with different degrees of unilateral striatal dopamine depletion. *Brain Res.* **664,** 178–188.

63. Huston, J. P., Steiner, H., Weiler, H.-T., Morgan, S., and Schwarting, R. K. W. (1990) The basal ganglia-orofacial system: studies on neurobehavioral plasticity and sensory-motor turning. *Neurosci. Biobehav. Rev.* **14,** 433–446.

64. Rutledge, J. N., Hilal, S. K., Schallert, T., Silver, A. J., Defendini, R. D., and Fahn, S. (1987) Magnetic resonance imaging of Parkinsonisms, in *Recent Developments in Parkinson's Disease,* Vol. II. (Fahn, S., Marsden, C. D., Goldstein, M., and Calne, D. B., eds.), S. Macmillan Healthcare Information, Florham Park, NJ.

65. Marshall, J. F., Richardson, J. S., and Teitelbaum, P. (1974) Nigrostriatal bundle damage and the lateral hypothalamic syndrome. *J. Comp. Physiol. Psychol.* **87,** 808–830.

66. Lindner, M. D., Plone, M. A., Jonathan, F. M., Blaney, T. J., Salamone, J. D., and Emerich, D. F. (1997) Rats with partial striatal dopamine depletions exhibit robust and long-lasting behavioral deficits in a simple fixed-ratio bar-pressing. *Behav. Brain Res.* **85,** 25–40.

67. Fearnley, J. M. and Lees, A. J. 1991. Ageing and Parkinson's disease: substantia nigra regional selectivity. *Brain* **114,** 2283–2301.

68. Schallert, T. (1989) Preoperative intermittent feeding or drinking regimens enhance post-lesion sensorimotor function, in *Preoperative Events: Their Events on Behavior Following Brain Damage* (Schulkin, J. ed.), Lawrence Erlbaum Associates, Mahwah, NJ, pp. 1–20.

69. Schallert, T., Humm, J. L., Bland, S., Kolb, B., Aronowski, J., and Grotta, J. (1999) Activity-dependent growth factor expression and related neuronal events in recovery of function after brain injury. *Proceedings of the Princeton Conference on Cerebraovascular Disorders,* in press.

9

Development of Behavioral Outcome Measures for Preclinical Parkinson's Research

Mark D. Lindner

1. INTRODUCTION

The discovery and development of novel therapeutics for central nervous system (CNS) diseases is dependent on contributions from a wide range of different fields of expertise—from initial reductionistic, molecular efforts to identify a specific target mechanism and to produce or discover therapeutic agents with the desired mechanism of action, to a more molar or systems approach assessing the potential efficacy/safety of those therapeutic agents at the behavioral level in animal models. In keeping with the goals and objectives of this book, this chapter describes how behavioral pharmacologists participate in the process of preclinical research and development. Although the examples included in this chapter are studies that were conducted to develop novel treatments for Parkinson's disease, they are intended to illustrate the general approach or process that should be relevant for many CNS indications with behavioral endpoints.

First, preclinical scientists should take their direction, and pursue development of novel treatments, based on the current needs of the patient population. Although this may seem obvious, it is easy to stray from what should be the focus of preclinical discovery and development, and preclinical scientists can easily find that they have become engrossed in the pursuit of questions that have tremendous personal appeal but are not directly related to the primary objective. For that reason, it is important to ensure explicitly and repeatedly that the focus is maintained on the primary objectives of discovering and developing novel therapeutics.

Selection of a therapeutic target should be based on the needs of the patient population, and the needs of the patient population are perhaps best articulated by the clinicians devoted to treating particular patient populations. For that reason preclinical scientists should pay close attention to the clinical literature. In addition, the clinical relevance and predictive validity of the preclinical models is dependent on the degree to which the models simulate the clinical condition, which is another reason why it is important to develop familiarity with the details of the clinical condition.

Finally, the most efficient research process will capitalize on the work that has already been conducted, and the tools that are already available. For that reason, preclinical scientists should be familiar with the preclinical literature and the models that

From: Central Nervous System Diseases
Edited by: D. F. Emerich, R. L. Dean, III, and P. R. Sanberg © Humana Press Inc., Totowa, NJ

are available. One of the first steps of the preclinical behavioral pharmacologist is often to use initial screens to reduce the number of potential therapeutics for further testing. These initial screens are selected for their ability to rapidly screen large numbers of compounds, but they usually have little clinical relevance and poor predictive validity.

More sophisticated models and more clinically relevant behavioral measures provide more power and better predictive validity. Typically, continued development and validation of the preclinical models and the assessment of potential efficacy/safety of novel treatments are complementary. To address the questions that arise during the process of conducting preclinical research to develop novel therapeutic agents, it is often desirable to further develop and improve the existing models. The improved models that result from those efforts help to close the gap between the clinical condition and the existing models. This usually increases the predictive validity of the preclinical assessments and also often increases the sensitivity and power of the preclinical model.

In addition to addressing the question of whether novel agents are active and should be pursued into the clinic or not, preclinical research may also provide useful information about how to proceed with a clinical trial to maximize the potential for demonstrating therapeutic efficacy. For example, preclinical studies often provide information about how to administer therapeutic agents and what types of behavioral endpoints are most likely to be improved. Preclinical research may also indicate which subpopulation of patients or which stage of the disease might be most amenable to the therapeutic effects of a particular novel therapeutic.

Again, this chapter is intended to illustrate the process of preclinical research at the behavioral level and the kind of information that can be gleaned from this process, by presenting a series of studies conducted to develop novel treatments for Parkinson's disease. It is hoped that the process outlined in this chapter will be useful even for those who are pursuing other indications.

2. CLINICAL CONDITION AND THERAPEUTIC TARGET

Initially, details of the clinical condition should be examined, and unmet needs of the patient population should be identified. Parkinson's disease is characterized by behavioral deficits related to the progressive degeneration of nigrostriatal dopaminergic neurons (*1–8*). Some beneficial effects can be obtained for a few years during the initial phases of the disease by supplementing or replacing the declining levels of striatal dopamine. The dopamine precursor L-Dopa and the dopamine agonist apomorphine are both effective replacement therapies that improve behavioral function in Parkinson's patients (*1,9–15*).

Although systemic administration of pharmacological doses of L-Dopa produce some beneficial effects, it is difficult to maintain constant blood levels and impossible to confine delivery to the basal ganglia with systemic L-Dopa. These limitations in the control of levels and specificity of delivery sites may be related to the side effects seen with systemic L-Dopa: the "on–off" phenomenon (sudden unexpected periods of akinesia), dyskinesias, and L-Dopa-induced psychoses (*10,16,17*). If constant levels of L-Dopa could be delivered continuously and selectively to the basal ganglia, these side effects and adverse effects might be reduced. Further, if dose selection was not constrained by side effects, greater reductions in the behavioral deficits might be possible.

3. SELECTION OF TARGET MECHANISM

Having identified specific problems in the treatment of a population of patients, the next step is to identify a therapeutic mechanism to pursue. One way to achieve continuous, nonfluctuating, site-specific delivery of therapeutic substances may be to implant dopamine and/or L-Dopa-producing cells into the target site—the striatum or substantia nigra. In fact, transplants of fetal ventral mesencephalic tissue have exhibited therapeutic potential in animal models of Parkinson's disease *(18–23)* and in Parkinson's patients *(24–27,28)*. Adrenal chromaffin transplants have also shown some potential for treatment of Parkinson's disease in animal models *(29–31)* and in clinical trials *(32,33)*. In addition to fetal tissues and adrenal chromaffin cells, PC12 cells, established as a clonal line of rat adrenal pheochromocytoma cells, constitutively release dopamine and L-Dopa *(34)*.

The use of cell lines such as PC12 cells and other genetically engineered xenogeneic cells to produce dopamine and L-Dopa would preclude the need to harvest cells from fetal or adult hosts, as such cell lines can be produced and maintained in tissue cultures. Such cell lines could also be banked, cloned, and purified to reduce the risk of viral contamination or the presence of other adventitious agents. In fact, implants of PC12 cells have already been shown to have therapeutic potential in animal models of Parkinson's disease *(29,35,36)*, but without immunosuppression of the host, the number of surviving cells gradually decreased or the grafts were completely rejected *(35,36)*, and with immunosuppression of the host, some of the implants developed into tumors *(29)*. The problem of potential tumorigenicity and/or host rejection can be addressed by encapsulating the cells in a semipermeable membrane, and numerous studies have shown that PC12 cells encapsulated in semipermeable membrane continue to survive and deliver therapeutic agents in animal models of Parkinson's disease for long periods of time without immunosuppression of the host *(37–43)*.

4. INITIAL PRECLINICAL MODEL AND TEST FOR "ACTIVITY"

Having identified a clinical target and a target therapeutic, whenever possible the initial test for potential efficacy is conducted in screens that are already available. Dopamine depletions can be produced experimentally by infusing 6-hydroxydopamine (6-OHDA). Behavioral deficits are not detectable unless severe (>90%) depletions are produced, but rats with severe bilateral depletions are adipsic and aphagic and die *(44,45)*. Rats with unilateral dopamine depletions live but do not exhibit any apparent behavioral deficits upon gross observation. Therefore, severe unilateral dopamine depletions are produced in young rats. Denervation supersensitivity develops in the depleted striatum, and systemic dopamine agonists can be administered to produce asymmetric stereotypies that cause the rats to turn involuntarily in one direction or to rotate in circles *(46,47)*. The number of rotations can be easily quantified and is used as a measure of drug "activity." Therapeutic effects of transplants are evident by the degree of attenuation of denervation supersensitivity and reductions in the amount of drug-induced rotations.

Polymer-encapsulated PC12 cells have been shown to reduce drug-induced rotations in numerous studies *(42,43,48,49)*, and rats with striatal implants of polymer-encapsulated PC12 cells exhibit substantial decreases in the number of drug-induced

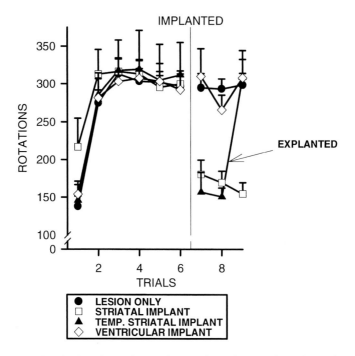

Fig. 1. Apomorphine-induced rotations. The number of contralateral rotations was the same for all groups before surgery. After surgery, the number of apomorphine-induced contralateral rotations was substantially reduced, but only in rats implanted with PC12 cell-loaded devices in the striatum, and only as long as the devices remained implanted.

rotations for as long as 6 mo after implantation *(42)*. The specificity of the effect is supported by the fact that it is dependent on catecholamine diffusion through the denervated striatal parenchyma. Striatal implants of polymer-membrane devices not loaded with PC12 cells do not affect the number of rotations *(43,49)*, and reductions in apomorphine-induced rotations are evident only in rats implanted with encapsulated PC12-cells in the denervated striatum, not in rats with cell-loaded devices implanted in the lateral ventricles (Fig. 1), and the effect lasts only as long as the devices remain in place in the striatum *(48)*.

5. RATIONALE FOR MORE CLINICALLY RELEVANT MEASURES

The development of the rotometry test measuring drug-induced rotations in rats with severe unilateral dopamine depletions was a tremendous contribution to the literature and has been invaluable since its initial introduction into the field *(46,47,50)*. However, over the years it has become apparent that, although this test continues to be of tremendous value, there are certain weaknesses related to its use. Complete reliance on changes in drug-induced rotations to assess the potential efficacy of novel treatments has been criticized because of this measure's lack of clinical relevance and specificity (e.g., effects that are not therapeutic also reduce the number of drug-induced rotations) *(50–54)*.

Although behavioral deficits are not evident upon gross observation of rats with severe unilateral dopamine depletions more sophisticated, more clinically relevant measures of Parkinsonian symptoms have been developed and are available. With the use of more sophisticated testing procedures, it has been demonstrated that rats with dopamine depletions produced by medial forebrain bundle (MFB) or nigrostriatal infusions of 6-OHDA exhibit behavioral deficits similar to those reported in Parkinson's patients *(11,49,55–70),* namely: akinesia/bradykinesia, rigidity, tremor, paradoxical kinesia (temporary restoration of motor function activated by intense environmental stimulation), aphagia, and sensorimotor neglect.

Having demonstrated that polymer-encapsulated PC12 cells implanted in the striatum exhibit therapeutic potential in the rotometry test, it was important to assess their therapeutic potential with these more sophisticated, more clinically relevant behavioral tests of parkinsonian symptoms. A battery of tests for Parkinsonian symptoms were conducted after striatal implantations of polymer-encapsulated PC12 cells or baby hamster kidney cells transfected to produce glial cell line derived neurotrophic factor (GDNF) in rats with severe dopamine depletions. These tests included measures of the ability to initiate and control guided movements in response to environmental stimuli as described in the following subsections *(49).*

5.1. Akinesia

The experimenter held the rat so that it was standing on one forelimb. The experimenter did not move the rat, but it was allowed to move on its own. The number of steps taken with the forelimb the rat was standing on was recorded during a 30-second trial for each forelimb as described previously *(72).*

5.2. Rigidity

Rats were placed on a smooth stainless-steel surface and tested as previously reported *(70).* The experimenter placed one hand along the side of the rat and gently pushed it laterally across the width of the cart (90 cm) at approx 20 cm/s. The number of steps taken with the forelimb on the side in which the rat was moving was recorded. Rats were moved two times in each direction in each test session.

5.3. Sensorimotor Neglect

This test assesses the rat's ability to make directed forelimb movements in response to sensory stimuli *(80).* Rats were held so that their limbs were hanging unsupported. They were raised up to the side of a table so that their whiskers made contact with the top surface. They were raised with the length of their body parallel to the edge of the table top on 10 trials, and on another 10 trials they were raised with their bodies perpendicular to the edge of the tabletop. Each forelimb was tested for 10 trials in response to unilateral tactile stimulation of the vibrissae, and another 10 trials with the more salient bilateral tactile stimulation of the vibrissae.

5.4. Postural/Locomotor Forelimb Use

The amount that each forelimb was used independently and spontaneously was measured by observing the rat in a small Plexiglas cage (17 cm × 25 cm × 22 cm). The cumulative time, in seconds, that each forepaw was used independently for postural support or locomotion, was recorded during a 5-min period. Forelimb use was cumu-

Table 1
Lesion and Treatment Effect Sizes for Parkinsonian Symptoms

Behavioral test	Unit of measure	Side		Effect size (ω^2)[a]	
		Ipsilateral	Contralateral	Lesion	PC12 or BHK-GDNF cells
Akinesia	Spontaneous steps in 30-s trial	17.17 ± 0.71	2.74 ± 0.26	0.94	0.11
Rigidity/ bracing	Steps taken while being gently pushed laterally for 90 cm	8.08 ± 0.14	4.86 ± 0.24	0.82	0.00
Sensori- motor neglect	Forelimb placed on tabletop in response to tactile stimulation of the vibrissae (10 trials)	8.30 ± 0.30	0.79 ± 0.25	0.86	0.00
Postural/ locomotor forelimb use	Seconds of independent forelimb use in 5-min trial	43.08 ± 2.84	12.57 ± 1.20	0.61	0.00

[a]ω^2 Effect size (proportion of variance accounted for).

lated during times when only one forepaw was in contact with the floor or walls of the cage *(81,82)*.

6. DESIRE FOR ADDITIONAL MODEL DEVELOPMENT/VALIDATION

The behavioral tests of Parkinsonian symptoms were very sensitive to the effects of the dopamine depletions, but failed to detect any therapeutic potential related to either polymer-encapsulated PC12 cells or BHK-GDNF cells *(49)*. The magnitude of the lesion effect was very large and accounted for a tremendous proportion of the variance in the data, but the magnitude of the treatment effects were not statistically significant (Table 1). There are several ways to interpret the fact that the polymer-encapsulated cells failed to produce detectable, therapeutic effects in these tests of Parkinsonian symptoms. First, it is possible that these "more clinically relevant" measures of Parkinsonian symptoms are not valid. Although several studies have reported that these measures are responsive to drugs that are effective in the clinical setting: L-Dopa *(70,83)* and apomorphine *(61)*, a full dose–response curve has not been conducted under carefully controlled, blinded conditions. Second, it is possible that these tests are simply not sensitive enough to detect effects produced by devices with such low output of therapeutic agents. In other words, these tests might not be valid and/or might not be sufficiently sensitive.

One way to test the potential validity of these measures was to assess the effects of a range of doses of oral Sinemet. If these measures of Parkinsonian symptoms are valid, they should be responsive to Sinemet. In addition, using these tests to examine shifts in the dose–response curve to oral Sinemet might increase their sensitivity to therapeutic effects of transplants. However, assessing potential shifts in the dose–response curve to oral Sinemet is of little value unless the dose-response curve to potential adverse effects can also be quantified. While the therapeutic gold standard, L-Dopa:carbidopa (Sinemet) effectively attenuates Parkinsonian symptoms, the beneficial effects of this drug are limited by the dyskinesias that occur at higher doses. The range of effective doses, from the minimum dose that produces beneficial effects to the dose that produces intolerable dyskinesias, is referred to as the "therapeutic window." As the disease progresses, the threshold dose that produces undesirable dyskinesias decreases, closing the therapeutic window and making it progressively more difficult to obtain beneficial effects with the drug, approaching the point at which intolerable dyskinesias are produced at doses below the minimum effective dose *(84)*. Therefore, increasing the sensitivity to the therapeutic effects of Sinemet is of little value unless it can be done so that the therapeutic window can be widened—without increasing the sensitivity to the adverse effects of oral Sinemet. Because novel treatments such as transplants, growth factors, and gene therapies would almost certainly be tested and used in conjunction with Sinemet and because such novel treatments might have some impact on the therapeutic window for Sinemet *(85)*, it would be extremely valuable to assess, preclinically, the effects of novel treatments not just on the beneficial effects of Sinemet, but on the therapeutic window for Sinemet.

7. MODEL VALIDATION AND DEVELOPMENT

To validate and possibly further develop the model, another study was conducted to assess the validity of the "more clinically relevant" behavioral measures of Parkinsonian symptoms, and to determine if these behavioral tests could be used to quantify the therapeutic window to oral Sinemet in rats. The results of this study demonstrate that Parkinsonian symptoms are responsive to oral Sinemet (Fig. 2). Furthermore, stereotypies/dyskinesias can also be quantified in rats (Fig. 3), and these two sets of measurements can be combined to quantify the therapeutic window to oral Sinemet in rats (Fig. 4).

8. IMPROVED MODEL INCREASES POWER
FOR ASSESSING THERAPEUTIC POTENTIAL

Having validated that the measures of Parkinsonian symptoms are responsive to oral Sinemet and that they can be used to quantify the therapeutic window to oral Sinemet in rats, measures of the therapeutic window are now available to test further the therapeutic potential of striatal implants of polymer-encapsulated PC12 cells. Because any transplantation procedure would have to be used as an adjunct to oral Sinemet, the ability to examine transplants as an adjunct to oral Sinemet in preclinical studies provides additional power that was not available before. Furthermore, cell-based delivery of dopamine and L-Dopa may not produce direct effects, but may widen the therapeutic window for oral Sinemet. Therefore, the ability to quantify the therapeutic window to oral Sinemet not only increases the ability to assess therapeutic potential, but it also increases predictive validity of the results and increases the sensitivity to potential therapeutic effects.

The next study assessed the effects of striatal implants of polymer-encapsulated

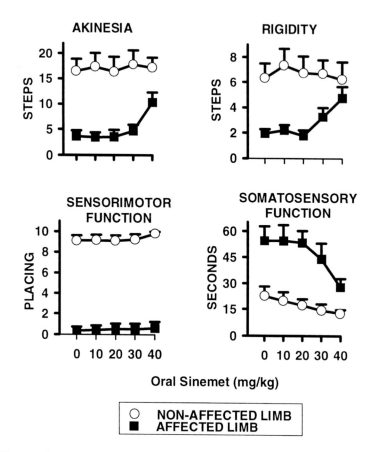

Fig. 2. Sinemet dose–response curve on neurological measures of Parkinsonian symptoms. Performance is shown for both the affected limb (contralateral to the lesion) and the nonaffected limb (ipsilateral to the lesion). The Akinesia task assesses the number of steps initiated with each forepaw during a 30-s trial; each forelimb is assessed independently. The test for Rigidity assesses the number of steps taken with each forelimb when the rats is pushed sideways over a distance of 90 cm. The test for sensorimotor function requires the rat to place its paw on the surface of a table in response to stimulation of its whiskers. The test of somatosensory function assesses the latency to remove adhesive stimuli attached to the radial surface of each forepaw.

PC12 cells on the therapeutic window of oral Sinemet *(86)*. While the measures of Parkinsonian symptoms were being validated and it was being determined whether the therapeutic window to oral Sinemet could be quantified in rats, the cell development group worked to increase catecholamine output from the encapsulated cell devices. Catecholamine output was successfully increased in the rodent-sized devices (7 mm in length, 1 mm in diameter). The devices with more output had approximately five times more output than the previous study *(49)*, and the devices with lower output still had as much output as in the previous study (Fig. 5).

Four tests of Parkinsonian symptoms were conducted: (1) For the test of akinesia, the experimenter held the rat so that it was standing with most of its body weight resting on one forelimb, and allowed to initiate steps with that forelimb. The number of steps taken with the forelimb the rat was standing on was recorded separately during

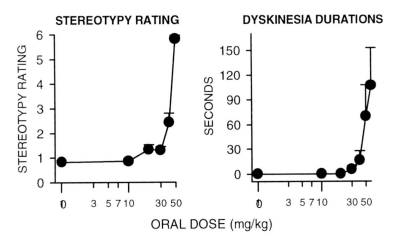

Fig. 3. Sinemet dose–response curve on Stereotypy/Dyskinesia Measures. The stereotypy rating scale is adapted from reports in the literature (0–50 mg/kg). The measure of dyskinesia durations quantifies behaviors that are perhaps more legitimate indices of dyskinesias, not affected by changes in wakefulness and simple activity level (0–60 mg/kg). Both measures revealed a nonlinear function between the dose of oral Sinemet and the degree of stereotypies and dyskinesias.

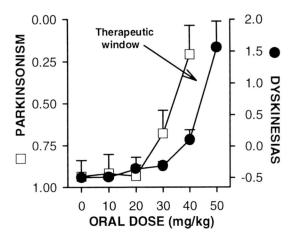

Fig. 4. The therapeutic window of oral Sinemet. Measures of parkinsonism were standardized (Z scores, *see* text) for both limbs. Although performance of the affected limb improved, it did not reach the level of the nonaffected limb. The performance on the nonaffected limb (not shown) was stable at approx –0.73 ± 0.15 across all doses of Sinemet. Note that anti-Parkinsonian effects begin to emerge at 30–40 mg/kg, but that the rats become untestable owing to severe dyskinesias at 50 mg/kg.

30-s trials for each forelimb. (2) To assess rigidity or bracing behavior, rats were placed on a smooth stainless-steel surface. The experimenter placed one hand along the side of the rat and gently pushed the animal laterally 90 cm at approx 20 cm/s. The number of steps taken with the forelimb on the side in which the rat was moving was recorded. Each trial included moving the rat two times in each direction. (3) A test designed to assess ability to respond to and remove focal somatosensory stimuli was also included.

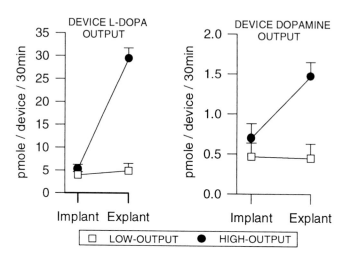

Fig. 5. Device dopamine and L-Dopa output before implantation and after explantation. Devices were arbitrarily divided into two groups based on postexplant catecholamine output. Rats implanted with the 50% of the devices that had more catecholamine output after explantation were grouped together as the "High-Output" group, and the rats implanted with the 50% of the devices that had less catecholamine output after explantation were grouped together as the "Low-Output" group.

Pairs of circular adhesive papers (with surface areas of 113.1 mm^2) were attached to the distal-radial areas of each forelimb. The forepaws were held away from the animal's mouth while it was returned to its home cage. The latencies to remove the stimuli with the mouth were recorded for each forelimb. The maximum cutoff was 2 min. The rats received five trials per session with a 1–2 min intertrial interval. (4) The forelimb placing test assesses ability to make directed forelimb movements in response to sensory stimuli. Rats were held so that their limbs were hanging unsupported. They were then raised to the side of a table so that their whiskers made contact with the top surface with the length of their body parallel to the edge of the table top on 10 trials for each forelimb. Rats normally place their forelimb on the tabletop almost every time.

The rats implanted with low-output devices were significantly improved after the higher doses of oral Sinemet, 24 and 36 mg/kg. High-output devices produced a significant improvement on their own, without Sinemet, and with every dose of Sinemet (Fig. 6). In addition, Sinemet-induced dyskinesias were not increased by encapsulated PC12 cells, which suggests that polymer-encapsulated PC12 cells increase the therapeutic window for oral Sinemet (Fig. 7).

9. RESULT OF MODEL DEVELOPMENT

Results with the initial screen using drug-induced rotations suggested that polymer-encapsulated PC12 cells have therapeutic potential. This was encouraging, but the rotometry model has little clinical relevance and poor predictive validity, so it was desirable to assess the therapeutic potential of polymer-encapsulated PC12 cells further with more clinically relevant measures of Parkinsonian symptoms. Polymer-encapsulated PC12 cells failed to produce detectable therapeutic effects in these tests

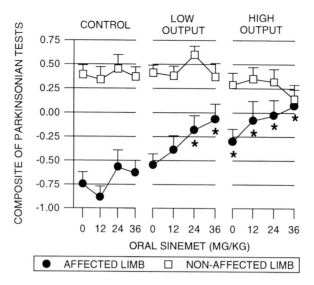

Fig. 6. Composite Parkinsonian score produced by combining four behavioral measures of Parkinsonian symptoms. Both oral Sinemet and encapsulated PC12 cell devices attenuated the degree of Parkinsonism in the affected forelimb. Asterisks mark all points at which performance with the affected limb was significantly improved relative to the control group with vehicle only.

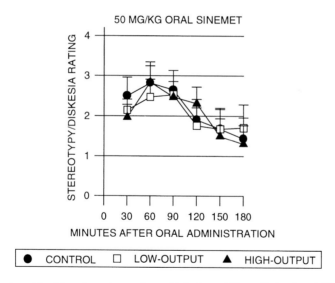

Fig. 7. Stereotypy/dyskinesia ratings after a high dose (50 mg/kg) of oral Sinemet. Encapsulated PC12 cell devices did not significantly increase the adverse effects produced by high doses of oral Sinemet.

of Parkinsonian symptoms, which could have been due to a lack of validity or sensitivity of these tests. Additional work conducted to test the validity and possibly increase the sensitivity of these tests supported the validity of these measures, and demonstrated that they could be combined with measures of stereotypies/dyskinesias to quantify the therapeutic window and thus increase the sensitivity of these measures. A final study of the effects of striatal implants of polymer-encapsulated PC12 cells demonstrated that these devices produce therapeutic effects if device output is high enough, and that they widen the therapeutic window to oral Sinemet.

In other words, increasing the clinical relevance and sensitivity of the model increased the ability to assess the therapeutic potential of novel therapeutic agents, which was not possible before the model had been further developed. The more relevant and sensitive model could then be used to address additional questions about the therapeutic potential of this novel treatment. Furthermore, in addition to demonstrating that cell-based delivery of L-Dopa and dopamine from polymer-encapsulated striatal devices have therapeutic potential, these results also demonstrate that these devices might exhibit more robust effects if they were used as an adjunct to oral Sinemet. These results not only provide information about the therapeutic potential of these devices, but also suggest ways to maximize the probability of identifying their therapeutic potential in a clinical trial.

REFERENCES

1. Hornykiewicz, O. (1975) Parkinson's disease and its chemotherapy. *Biochem. Pharmacol.* **24,** 1061–1065.
2. Hughes, A. J., Daniel, S. E., Blankson, S., and Lees, A. J. (1993) A clinicopathologic study of 100 cases of Parkinson's disease. *Arch. Neurol.* **50,** 140–148.
3. Lloyd, K. G. (1977) CNS compensation to dopamine neuron loss in Parkinson's disease. *Adv. Exp. Med. Biol.* **90,** 255–266.
4. Marsden, C. D. (1982) The mysterious motor function of the basal ganglia: The Robert Warternberg Lecture. *Neurology* **32,** 514–539.
5. Riederer, P. and Wuketich, S. (1976) Time course of nigrostriatal degeneration in Parkinson's disease. *J. Neural Transm.* **38,** 277–301.
6. Rogers, M. W. and Chan, C. W. Y. (1988) Motor planning is impaired in Parkinson's disease. *Brain Res.* **438,** 271–276.
7. Scatton, B., Rouquier, L., Javoy-Agid, F., and Agid, Y. (1982) Dopamine deficiency in the cerebral cortex in Parkinson disease. *Neurology* **32,** 1039–1040.
8. Schwab, R. S. and Zieper, I. (1965) Effects of mood, motivation, stress and alertness on the performance in Parkinson's disease. *Psychiatr. Neurol.* **150,** 345–357.
9. Baroni, A., Benvenuti, F., Fantini, L., Pantaleo, T., and Urbani, F. (1984) Human ballistic arm abduction movements: effects of L-Dopa treatment in Parkinson's disease. *Neurology* **34,** 868–876.
10. Delaney, P. and Fermaglich, J. (1976) Parkinsonism and levodopa: a five-year experience. *J. Clin. Pharmacol.* **16,** 652–659.
11. Fairley, P. C. and Marshall, J. F. (1986) Dopamine in the lateral caudate-putamen of the rat is essential for somatosensory orientation. *Behav. Neurosci.* **100,** 652–663.
12. Kempster, P. A., Frankel, J. P., Stern, G. M., and Lees, A. J. (1990) Comparison of motor response to apomorphine and levodopa in Parkinson's disease. *J. Neurol. Neurosurg. Psychiatry* **53,** 1004–1007.
13. Lees, A. J. (1993) Dopamine agonists in Parkinson's disease: a look at apomorphine. *Fundam. Clin. Pharmacol.* **7,** 121–128.
14. Muenter, M. D. (1982) Initial treatment of Parkinson's disease. *Clin. Neuropharmacol.* **5,** S2–S12.
15. Schelosky, L. and Poewe, W. (1993) Current strategies in the drug treatment of advanced Parkinson's disease—new modes of dopamine substitution. *Fundam. Clin. Pharmacol.* **7,** 121–128.

16. Birkmayer, W. and Riederer, P. (1975) Responsibility of extrastriatal areas for the appearance of psychotic symptoms (clinical and biochemical human post-mortem findings). *J. Neural Transm.* **37,** 175–182.
17. Luquin, M. R., Scipioni, O., Vaamonde, J., Gershanik, O., and Obeso, J. A. (1992) Levodopa-induced dyskinesias in Parkinson's disease: clinical and pharmacological classification. *Movement Disord.* **7,** 117–124.
18. Bjorklund, A., Dunnett, S. B., Stenevi, U., Lewis, M. E., and Iversen, S. D. (1980) Reinnervation of the denervated striatum by substantia nigra transplants: functional consequences as revealed by pharmacological and sensorimotor testing. *Brain Res.* **199,** 307–333.
19. Dunnett, S. B., Bjorklund, A., Schmidt, R. H., Stenevi, U., and Iversen, S. D. (1983) Intracerebral grafting of neuronal cell suspensions IV. Behavioural recovery in rats with unilateral 6-OHDA lesions following implantation of nigral cell suspensions in different forebrain sites. *Acta Physiol. Scand.* **522 (Suppl.),** 29–37.
20. Dunnett, S. B., Bjorklund, A., Stenevi, U., and Iversen, S. D. (1981) Grafts of embryonic nigra reinnervating the ventrolateral striatum ameliorate sensorimotor impairments and akinesia in rats with 6-OHDA lesions of the nigrostriatal pathway. *Brain Res.* **229,** 209–217.
21. Dunnett, S. B., Bjorklund, A., Stenevi, U., and Iversen, S. D. (1981) Behavioural recovery following transplantion of substantia nigra in rats subjected to 6-OHDA lesions of the nigrostriatal pathway. I. Unilateral lesions. *Brain Res.* **215,** 147–161.
22. Perlow, M. J., Freed, W. J., Hoffer, B. J., Seiger, A., Olson, L., and Wyatt, R. J. (1979) Brain grafts reduce motor abnormalities produced by destruction of nigrostriatal dopamine system. *Science* **204,** 643–647.
23. Stenevi, U., Bjorklund, A., and Dunnett, S. B. (1980) Functional reinnervation of the denervated neostriatum by nigral transplants. *Peptides* **1 (Suppl. 1),** 111–116.
24. Lindvall, O., Brundin, P., Widner, H., Rehncrona, S., Gustavii, B., Frackowiak, R., Leenders, K. L., Sawle, G., Rothwell, J. C., and Marsden, C. D. (1990) Grafts of fetal dopamine neurons survive and improve motor function in Parkinson's disease. *Science* **247,** 574–577.
25. Lindvall, O., Rehncrona, S., Brundin, P., Gustavii, B., Astedt, B., Widner, H., Lindholm, T., Bjorklund, A., Leenders, K. L., and Rothwell, J. C. (1989) Human fetal dopamine neurons grafted into the striatum in two patients with severe Parkinson's disease. A detailed account of methodology and a 6-month follow-up. *Arch. Neurol.* **46,** 615–631.
26. Lindvall, O., Sawle, G., Widner, H., Rothwell, J. C., Bjorklund, A., Brooks, D., Brundin, P., Frackowiak, R., Marsden, C. D., Odin, P., and Rehncrona, S. (1994) Evidence for long-term survival and function of dopaminergic grafts in progressive parkinson's disease. *Ann. Neurol.* **35,** 172–180.
27. Lindvall, O., Widner, H., Rehncrona, S., Brundin, P., Odin, P., Gustavii, B., Frackowiak, R., Leenders, K. L., Sawle, G., and Rothwell, J. C. (1992) Transplantation of fetal dopamine neurons in Parkinson's disease: one-year clinical and neurophysiological observations in two patients with putaminal implants. *Ann. Neurol.* **31,** 155–165.
28. Zabek, M., Mazurowski, W., Dymecki, J., Stelmachow, J., and Zawada, E. (1994) A long term follow-up of fetal dopaminergic neurons transplantation into the brain of three parkinsonian patients. *Restor. Neurol. Neurosci.* **6,** 97–106.
29. Bing, G., Notter, M. F. D., Hansen, J. T., and Gash, D. M. (1988) Comparison of adrenal medullary, carotid body and PC12 cell grafts in 6-OHDA lesioned rats. *Brain Res. Bull.* **20,** 399–406.
30. Date, I., Felten, S. Y., and Felten, D. L. (1990) Cografts of adrenal medulla with peripheral nerve enhance the survivability of transplanted adrenal chromaffin cells and recovery of the host nigrostriatal dopaminergic system in MPTP-treated young adult mice. *Brain Res.* **537,** 33–39.

31. Stromberg, I., Herrera-Marschitz, M., Ungerstedt, U., Ebendal, T., and Olson, L. (1985) Chronic implants of chromaffin tissue into the dopamine-denervated striatum. Effects of NGF on graft survival, fiber growth and rotational behavior. *Exp. Brain Res.* **60,** 335–349.

32. Ahlskog, J. E., Kelly, P. J., van Heerden, J. A., Stoddard, S.L., Tyce, G. M., Windebank, A. J., Bailey, P. A., Bell, G. N., Blexrud, M. D., and Carmichael, S. W. (1990) Adrenal medullary transplantation into the brain for treatment of Parkinson's disease: clinical outcome and neurochemical studies. *Mayo Clin. Proc.* **65,** 305–328.

33. Goetz, C. G., Tanner, C. M., Penn, R. D., Stebbins, G. T., Gilley, D. W., Shannon, K. M., Klawans, H. L., Comella, C. L., Wilson, R. S., and Witt, T. (1990) Adrenal medullary transplant to the striatum of patients with advanced Parkinson's disease: 1-year motor and psychomotor data. *Neurology* **40,** 273–276.

34. Greene, L. A. and Tischler, A. S. (1976) Establishment of a noradrenergic clonal line of rat adrenal pheochromocytoma cells which respond to nerve growth factor. *Proc. Natl. Acad. Sci. USA* **73,** 2424–2428.

35. Freed, W. J., Patel-Vaidya, U., and Geller, H. M. (1986) Properties of PC12 pheochromocytoma cells transplanted to the adult rat brain. *Exp. Brain Res.* **63,** 557–566 (Abstr.).

36. Hefti, F., Hartikka, J., and Schlumpf, M. (1985) Implantation of PC12 cells into the corpus striatum of rats with lesions of the dopaminergic nigrostriatal neurons. *Brain Res.* **348,** 283–288.

37. Aebischer, P., Goddard, M., Signore, A. P., and Timpson, R. L. (1994) Functional recovery in hemiparkinsonian primates transplanted with polymer-encapsulated PC12 cells. *Exp. Neurol.* **126,** 151–158.

38. Aebischer, P., Tresco, P. A., Winn, S. R., Greene, L. A., and Jaeger, C. B. (1991) Long-term cross-species brain transplantation of a polymer-encapsulated dopamine-secreting cell line. *Exp. Neurol.* **111,** 269–275.

39. Emerich, D. F., McDermott, P. E., Krueger, P. M., Frydel, B., Sanberg, P. R., and Winn, S. R. (1993) Polymer-encapsulated PC12 cells promote recovery of motor function in aged rats. *Exp. Neurol.* **122,** 37–47.

40. Emerich, D. F., McDermott, P. E., Krueger, P. M., and Winn, S. R. (1994) Intrastriatal implants of polymer-encapsulated PC12-cells—effects on motor function in aged rats. *Prog. Neuropsychopharmacol. Biol. Psychiatry* **18,** 935–946.

41. Tresco, P. A., Winn, S. R., and Aebischer, P. (1992) Polymer encapsulated neurotransmitter secreting cells: potential treatment for Parkinson's disease. *ASAIO J.* **38,** 17–23.

42. Tresco, P. A., Winn, S. R., Tan, S., Jaeger, C. B., Greene, L. A., and Aebischer, P. (1992) Polymer-encapsulated PC12 cells: long-term survival and associated reduction in lesion-induced rotational behavior. *Cell Transplant.* **1,** 255–264.

43. Winn, S. R., Tresco, P. A., Zielinski, B., Greene, L. A., Jaeger, C. B., and Aebischer, P. (1991) Behavioral recovery following intrastriatal implantation of microencapsulated PC12 cells. *Exp. Neurol.* **113,** 322–329.

44. Heikkila, R. E., Shapiro, B. S., and Duvoisin, R. C. (1981) The relationship between loss of dopamine nerve terminals, striatal [^{3}H]spiroperidol binding and rotational behavior in unilaterally 6-hydroxydopamine-lesioned rats. *Brain Res.* **211,** 285–292.

45. Ungerstedt, U. (1971) Adipsia and aphagia after 6-hydroxydopamine induced degeneration of the nigro-striatal dopamine system. *Acta Physiol. Scand. Suppl.* **367,** 95–122.

46. Creese, I., Burt, D. R., and Snyder, S. H. (1977) Dopamine receptor binding enhancement accompanies lesion-induced behavioral supersensitivity. *Science* **197,** 596–598.

47. Ungerstedt, U., Ljungberg, T., and Greg, G. (1974) Behavioral, physiological, and neurochemical changes after 6-hydroxydopamine-induced degeneration of the nigro-striatal dopamine neurons. *Adv. Neurol.* **5,** 421–426.

48. Emerich, D. F., Winn, S. R., and Lindner, M. D. (1996) Continued presence of intrastriatal but not intraventricular polymer-encapsulated PC12 cells is required for alleviation of behavioral deficits in Parkinsonian rodents. *Cell Transplant.* **5,** 589–596.

49. Lindner, M. D., Plone, M. A., Mullins, T. D., Winn, S. R., Chandonait, S. E., Stott, J. A., Blaney, T. J., Sherman, S. S., and Emerich, D. F. (1997) Somatic delivery of catecholamines in the striatum attenuate parkinsonian symptoms and widen the therapeutic window of oral Sinemet in rats. *Exp. Neurol.* **145,** 130–140.

50. Marshall, J. F. and Ungerstedt, U. (1977) Striatal efferent fibers play a role in maintaining rotational behavior in the rat. *Science* **198,** 62–64.

51. Barker, R. and Dunnett, S. B. (1994) Ibotenic acid lesions of the striatum reduce drug-induced rotation in the 6-hydroxydopamine-lesioned rat. *Exp. Brain Res.* **101,** 365–374.

52. Isacson, O. (1995) Behavioral effects and gene delivery in a rat model of Parkinson's disease. *Science* **269,** 856.

53. Reading, P. J. and Dunnett, S. B. (1994) 6-Hydroxydopamine lesions of nigrostriatal neurons as an animal model of Parkinson's disease, in *Toxin-Induced Models of Neurological Disorders* (Woodruff, M. L., et al., eds.), Plenum Press, New York, pp. 89–119.

54. Robinson, T. E. and Becker, J. B. (1983) The rotational behavior model: asymmetry in the effects of unilateral 6-OHDA lesions of the substantia nigra in rats. *Brain Res.* **264,** 127–131.

55. Cousins, M. S., Carriero, D. L., and Salamone, J. D. (1997) Tremulous jaw movements induced by the acetylcholinesterase inhibitor tacrine: effects of antiparkinsonian drugs. *Eur. J. Pharmacol.* **322,** 137–145.

56. Dunnett, S. B., Whishaw, I. Q., Rogers, D. C., and Jones, G. H. (1987) Dopamine-rich grafts ameliorate whole body motor asymmetry and sensory neglect but not independent limb use in rats with 6-hydroxydopamine lesions. *Brain Res.* **415,** 63–78.

57. Finn, M., Jassen, A., Baskin, P., and Salamone, J. D. (1997) Tremulous characteristics of the vacuous jaw movements induced by pilocarpine and ventrolateral striatal dopamine depletions. *Pharmacol. Biochem. Behav.* **57,** 243–249.

58. Hall, S. and Schallert, T. (1988) Striatal dopamine and the interface between orienting and ingestive functions. *Physiol. Behav.* **44,** 469–471.

59. Marshall, J. F. (1979) Somatosensory inattention after dopamine-depleting intracerebral 6-OHDA injections: spontaneous recovery and pharmacological control. *Brain Res.* **177,** 311–324.

60. Marshall, J. F., Berrios, N., and Sawyer, S. (1980) Neostriatal dopamine and sensory inattention. *J. Comp. Physiol. Psychol.* **94,** 833–846.

61. Marshall, J. F. and Gotthelf, T. (1979) Sensory inattention in rats with 6-hydroxydopamine-induced degeneration of ascending dopaminergic neurons: apomorphine-induced reversal of deficits. *Exp. Neurol.* **65,** 398–411.

62. Marshall, J. F. and Teitelbaum, P. (1974) Further analysis of sensory inattention following lateral hypothalamic damage in rats. *J. Comp. Physiol. Psychol.* **86,** 375–395.

63. Marshall, J. F. and Ungerstedt, U. (1976) Apomorphine-induced restoration of drinking to thirst challenges in 6-hydroxydopamine-treated rats. *Physiol. Behav.* **17,** 817–822.

64. Miklyaeva, E. I., Castaneda, E., and Whishaw, I. Q. (1994) Skilled reaching deficits in unilateral dopamine-depleted rats: impairments in movement and posture and compensatory adjustments. *J. Neurosci.* **14,** 7148–7158.

65. Miklyaeva, E. I., Martens, D. J., and Whishaw, I. Q. (1995) Impairments and compensatory adjustments in spontaneous movement after unilateral dopamine depletion in rats. *Brain Res.* **681,** 23–40.

66. Miklyaeva, E. I. and Whishaw, I. Q. (1996) HemiParkinson analogue rats display active support in good limbs versus passive support in bad limbs on a skilled reaching task of variable height. *Behav. Neurosci.* **110,** 117–125.

67. Miklyaeva, E. I., Woodward, N. C., Nikiforov, E. G., Tompkins, G. J., Klassen, F., Ioffe, M. E., and Whishaw, I. Q. (1997) The ground reaction forces of postural adjustments during skilled reaching in unilateral dopamine-depleted hemiparkinson rats. *Behav. Brain Res.* **88,** 143–152.

68. Salamone, J. and Baskin, P. (1996) Vacuous jaw movements induced by acute reserpine and low-dose apomorphine: possible model of parkinsonian tremor. *Pharmacol. Biochem. Behav.* **53,** 179–183.

69. Schallert, T. (1988) Aging-dependent emergence of sensorimotor dysfunction in rats recovered from dopamine depletion sustained early in life. *Ann. NY Acad. Sci.* **515,** 108–120.

70. Schallert, T., De Ryck, M., Whishaw, I. Q., Ramirez, V. D., and Teitelbaum, P. (1979) Excessive bracing reactions and their control by atropine and L-Dopa in an animal analog of Parkinsonism. *Exp. Neurol.* **64,** 33–43.

71. Schallert, T. and Hall, S. (1988) 'Disengage' sensorimotor deficit following apparent recovery from unilateral dopamine depletion. *Behav. Brain Res.* **30,** 15–24.

72. Schallert, T., Jones, T. A., and Norton, D. (1992) A clinically relevant unilateral rat model of Parkinsonian akinesia. *J. Neural Transplant. Plast.* **3,** 332–333.

73. Schallert, T., Upchurch, M., Lobaugh, N., Farrar, S. B., Spirduso, W. W., Gilliam, P., Vaughn, D., and Wilcox, R. E. (1982) Tactile extinction: distinguishing between sensorimotor and motor asymmetries in rats with unilateral nigrostriatal damage. *Pharmacol. Biochem. Behav.* **16,** 455–462.

74. Schallert, T., Upchurch, M., Wilcox, R. E., and Vaughn, D. M. (1983) Posture-independent sensorimotor analysis of inter-hemispheric receptor asymmetries in neostriatum. *Pharmacol. Biochem. Behav.* **18,** 753–759.

75. Spirduso, W. W., Gilliam, P. E., Schallert, T., Upchurch, M., Vaughn, D. M., and Wilcox, R. E. (1985) Reactive capacity: a sensitive behavioral marker of movement initiation and nigrostriatal dopamine function. *Brain Res.* **335,** 45–54.

76. Whishaw, I. Q., Coles, B. L. K., Pellis, S. M., and Miklyaeva, E. I. (1997) Impairments and compensation in mouth and limb use in free feeding after unilateral dopamine depletions in a rat analog of human Parkinson's disease. *Behav. Brain Res.* **84,** 167–177.

77. Whishaw, I. Q., Funk, D. R., Hawryluk, S. J., and Karbashewski, E. D. (1987) Absence of sparing of spatial navigation, skilled forelimb and tongue use and limb posture in the rat after neonatal dopamine depletion. *Physiol. Behav.* **40,** 247–253.

78. Whishaw, I. Q., Gorny, B., Trannguyen, L. T. L., Castaneda, E., Miklyaeva, E. I., and Pellis, S. M. (1994) Making two movements at once—impairments of movement, posture, and their integration underlie the adult skilled reaching deficit of neonatally dopamine-depleted rats. *Behav. Brain Res.* **61,** 65–77.

79. Whishaw, I. Q., O'Connor, W. T., and Dunnett, S. B. (1986) The contributions of motor cortex, nigrostriatal dopamine and caudate-putamen to skilled forelimb use in the rat. *Brain* **109,** 805–843.

80. Lindner, M. D., Plone, M. A., Francis, J. M., and Emerich, D. F. (1996) Validation of a rodent model of Parkinson's disease: evidence of a therapeutic window for oral Sinemet. *Brain Res. Bull.* **39,** 367–372.

81. Lindner, M. D. (1990) Behavioral consequences of preventing trans-synaptic degeneration with intraventricular muscimol after striatal damage. Unpublished dissertation, 135 pp.

82. Schallert, T. and Lindner, M. D. (1990) Rescuing neurons from trans-synaptic degeneration after brain damage: helpful, harmful, or neutral in recovery of function? *Can. J. Psychol.* **44,** 276–292.

83. Olsson, M., Nikkhah, G., Bentlage, C., and Bjorklund, A. (1995) Forelimb akinesia in the rat Parkinson model: differential effects of dopamine agonists and nigral transplants as assessed by a new stepping test. *J. Neurosci.* **15,** 3863–3875.
84. Mouradian, M. M., Heuser, I. J. E., Baronti, F., Fabbrini, G., Juncos, J. L., and Chase, T. N. (1989) Pathogenesis of dyskinesias in Parkinson's disease. *Ann. Neurol.* **25,** 523–526.
85. Langston, J. W., Widner, H., Goetz, C. G., Brooks, D., Fahn, S., Freeman, T., and Watts, R. (1992) Core assessment program for intracerebral transplantations (CAPIT). *Movement Disord.* **7,** 2–13.
86. Lindner, M. D., Winn, S. R., Baetge, E. E., Hammang, J. P., Gentile, F. T., Doherty, E., McDermott, P. E., Frydel, B., Ullman, M. D., Schallert, T., and Emerich, D. F. (1995) Implantation of encapsulated catecholamine and GDNF-producing cells in rats with unilateral dopamine depletions and parkinsonian symptoms. *Exp. Neurol.* **132,** 62–76.

10

Behavioral Assessment in the Unilateral Dopamine-Depleted Marmoset

L. E. Annett, R. E. Smyly, J. M. Henderson, R. M. Cummings, A. L. Kendall, and S. B. Dunnett

1. INTRODUCTION

A number of radically different approaches to the treatment of Parkinson's disease are currently under investigation in laboratories and clinics throughout the world. The approaches differ greatly in the extent to which they attempt to repair the neuronal circuitry and replace the neurotransmitters destroyed as a result of the disease process. For example, a strategy currently popular in the clinic involves the destruction or disabling of overactive brain areas, whereas alternative approaches include transplantation of embryonic neurons or genetic manipulation of remaining intrinsic neurons so that they come to release the neurotransmitter dopamine *(1–3)*. Despite fundamental differences between these approaches there are no clear predictions as to the extent by which the functional recovery that may be achieved in each case may differ. Comparisons across patient groups may not be straightforward, as often only one experimental treatment is employed within a single research center and some treatment strategies have not yet, or have only recently, reached the stage of clinical trials. An objective of our current research program is to compare systematically within a single paradigm a number of these potential novel treatments for Parkinson's disease. The experimental model comprises unilateral stereotaxic 6-hydroxydopamine (6-OHDA) lesions in the nigrostriatal bundle of the small New World monkey, the common marmoset (*Callithrix jacchus*), together with a battery of behavioral tests designed to provide quantitative measures of lateralized dopaminergic dysfunction *(4,5)*.

The purpose of this chapter is to present a critical evaluation of unilateral dopamine depletion in the marmoset as a model for the investigation of prospective treatments for Parkinson's disease. Some of the benefits and the disadvantages of the model are illustrated with reference to our work exploring three different treatment strategies: embryonic neural grafts, direct infusions of a dopamine agonist, and lesions of the subthalamic nucleus. A particular advantage of the model is that behavioral deficits produced by the unilateral 6-OHDA lesion are profound and long-lasting and can be used in the functional assessment of treatment effects, while avoiding some of the difficulties of work-

From: Central Nervous System Diseases
Edited by: D. F. Emerich, R. L. Dean, III, and P. R. Sanberg © Humana Press Inc., Totowa, NJ

ing with severely parkinsonian bilaterally lesioned monkeys. The range of behavioral tests employed reveals differences in the profile of treatment effects that might not be apparent from a limited set of behavioral measures.

2. WHY USE MONKEYS AND WHY MARMOSETS?

Experimental studies employing animal models of neurodegenerative disease play a critical role in the development of techniques first devised in the test tube or cell culture dish. Application of a new technique in vivo may reveal potential difficulties that might not otherwise become apparent before clinical application. For treatment strategies that aim to manipulate the functional interplay between neuronal pathways, for example, using drugs or direct interventions in the brain with stimulators or lesions, refining the procedure necessarily has to be done in vivo. For many experimental questions rodent models will be most appropriate for the further development of a procedure toward clinical application. It is important that monkeys are used only for those experimental questions that cannot be addressed in rodents, or in the case of novel treatment strategies developed in rodents when there is doubt as to whether the treatment will also be effective in primates, or as to how to translate the procedures into the primate brain. Indeed in the United Kingdom, primates may be used legally only when there is no suitable rodent alternative. Examples of specific issues for which primate studies are necessary for the advancement of neural transplantation techniques toward clinical application include the following.

(1) The organization and size of the target structures in the basal ganglia differ considerably between primates and rodents. For example, the monkey and the human (but not the rat) striatum is differentiated into discrete caudate nucleus and putamen. A key issue for neural transplantation in patients is whether it is necessary to distribute a limited amount of embryonic tissue across as many striatal sites as possible, or whether symptomatic relief can be achieved through grafts placements at a few critical sites at which reinnervation is essential for producing functional recovery *(2)*. This issue cannot be addressed easily in rodents because grafts often grow so that they come to occupy most of the dennervated striatum. Obtaining complete reinnervation of the larger striatum is more problematic in primates. Investigation of the functional importance of graft location within the striatum has only just begun in monkeys *(6),* and needs to be extended to include further graft sites and behavioral tests to determine which sites, or combination of sites, are likely to bring the greatest benefits to patients.

(2) Cell survival following transplantation into the brain may differ fundamentally between primates and rodents. We have consistently found that embryonic marmoset tissue, especially nigral tissue, does not dissociate as easily into a cell suspension prior to implantation as embryonic rat tissue prepared by the same methods. The uneven suspension may contribute to the considerable between-subject variability in graft survival in marmosets *(7)*. The variable graft survival and functional recovery observed in our marmoset studies mirrors the substantial variability in the improvements noted in one of the largest series of patients with nigral grafts published to date *(8)*. Recent developments in the use of neuroprotective and growth factor strategies in rodents show promise for increasing the numbers of neurons that reliably survive transplantation *(9,10)*. It is important that these strategies are now employed in primates to investigate whether the problem of inconsistent graft survival can be resolved.

(3) Behavioral studies of graft function in rodents rely heavily on rotation, but patients with Parkinson's disease don't rotate. The complex behavioral repertoire of monkeys provides greater scope for determining the extent of the behavioral recovery produced by grafts, and for comparing the effects of alternative treatment strategies across a range of behavioral measures. Subtle but important differences between treatment effects may become apparent that would not be evident from limited behavioral tests in rodents. Our marmoset models emphasize the assessment of skilled reaching ability, which is more obviously relevant to the motoric impairments experienced by patients *(4,5)*. By employing tests of both reaching and rotation within the same animals, the marmoset model provides a real bridge between rodent and clinical studies.

The advantages of using marmosets rather than Old World monkey species to model parkinsonism are numerous. Home Office regulations in the United Kingdom require the use of New World in preference to Old World monkeys when the needs of the project can be accomplished by use of the former type of monkey. Their small size (adults normally weigh between 300 and 400 g) makes them relatively easy to handle, which is an important consideration for studies aiming to make detailed behavioral assessments. With the appropriate care and attention to husbandry they are successful laboratory primates *(11)*. Stereotaxic surgery can be conducted with a high degree of accuracy based on atlas coordinates, combined in some situations with skull or other landmarks, avoiding the need for scans to locate target brain structures *(12)*. An important reason for using marmosets in neural transplantation research is their excellent breeding record in captivity. A stable breeding pair of marmosets will normally produce twins every 5 mo. The reliable procurement of embryos at predictable stages of development is crucial for successful graft survival *(7,13)*. Interestingly, the incidence of three embryos *in utero* is considerably higher than would be predicted from the ratio of twin to triplet births, suggesting that reabsorption of one or more embryos may be a natural mechanism by which marmosets regulate the number of their offspring *(14)*. As well as the unilateral lesion model which is the focus of the present chapter, marmosets have been used successfully in the bilateral 1-methyl-4-phenyl-1,2,3,6-tetrahydropyridine (MPTP) model of parkinsonism *(15,16)*, to assess neural grafts as a potential treatment for Huntington's disease *(17)*, in a model of stroke *(18)*, and in studies of cholinergic, hippocampal, and frontal lobe function employing cognitive tests *(19,20)*.

3. THE UNILATEAL 6-OHDA LESION

A key issue in this discussion of the benefits and disadvantages of using the unilateral 6-OHDA lesion in marmosets is what constitutes a "good" primate model of parkinsonism. It may be considered that the best models are those in which the experimentally induced lesion most closely mimics the disease process and that produce behavioral deficits most closely resembling patients' symptoms *(21)*. Clearly such models have the advantage of being able to address many different types of experimental question within the same paradigm, for example, whether the disease process can be prevented or delayed and whether a particular treatment effectively relieves all parkinsonian symptoms. An alternative argument is that it may not be necessary (or indeed possible) to mimic precisely all the symptoms of Parkinson's disease in a monkey in order to address specific experimental questions. Indeed, it would be impractical and immoral to induce severe parkinsonism in a monkey if that were not absolutely essential to achieve the experimental goals. Rather a balance must be struck between the needs of the experiment and the severity of the model system employed. For example, if the purpose of the experiment is to investigate whether a novel viral vector has the capacity to infect primate neurons in vivo, or whether a particular manipulation in primates increases neural graft survival, then it is not necessary to use a model that induces all the symptoms of Parkinson's disease. Similarly, if the aim of the study is to see whether a certain treatment can affect behavior then a model that provides well characterized behavioral deficits is required, but not necessarily one that induces all of the disease symptoms. However, if the goal of the study is to see whether the experimental treatment provides relief from the full range of parkinsonian symptoms, then clearly a model that provides the most complete symptomotolgy possible is required. What constitutes a "good" experimental model therefore depends in part on the particular purpose of individual experiments.

3.1. Neuroanatomical Considerations

The stereotaxic procedure for producing unilateral 6-OHDA lesions in marmosets is relatively straightforward and can be completed within a single surgical session lasting about 1 h. The toxin is injected into five sites distributed across the nigrostriatal bundle about 1 mm anterior of the substantia nigra *(4)*. Major effects of the lesion are almost complete loss of dopamine cells from the substantia nigra and reductions in dopamine levels of about 98% throughout the striatum, including the caudate nucleus, putamen, and nucleus accumbens. Noradrenaline and 5-hydroxytryptamine in the striatum are also affected, although to a lesser extent, as is dopamine in the frontal cortex. The lesion model is therefore primarily of dopaminergic dysfunction and does not mimic other neuroanatomical aspects of the disease such as the formation of Lewy bodies. Nor is the 6-OHDA lesion likely to be of use in investigations of the possible genetic bases of Parkinson's disease, including the role of synucleins *(22)*. Its main advantage is the ease with which massive dopamine depletions can reliably be produced. Because the dopaminergic lesion is so massive, the behavioral deficits persist for months and years. Nevertheless, the marmosets remain in excellent health and do not require hand feeding or any other extra care because the lesion is unilateral. The same 6-OHDA lesion performed bilaterally in marmosets, in a staged procedure, produces a marked parkinsonian condition that requires considerable postoperative care *(23)* and yet if the dopaminergic depletion were not so complete spontaneous recovery would be more problematic.

As a model of dopaminergic dysfunction in Parkinson's disease, the unilateral 6-OHDA lesion clearly does not mimic the disease process in two important respects. First, although many patients experience asymmetric symptoms, the imbalance in dopamine levels between the two sides of the diseased brain is unlikely ever to be as massive as that created by the unilateral 6-OHDA lesion. Second, the loss of dopaminergic function in patients typically occurs gradually over a number of years and therefore differs markedly from the experimental lesion created in a single surgical session. The progressive loss of dopamine neurons characteristic of the idiopathic disease has been recreated more accurately in monkeys by Hantraye and colleagues using a regimen in which low doses of MPTP were administered first daily and then weekly for

21 mo. Both the gradual appearance of parkinsonian symptoms and the pattern of dopamine cell loss from the substantia nigra closely resembled clinical features of the disease *(24,25)*. However, it would be unrealistic to use a primate model that takes nearly 2 yr to create to resolve every issue relevant to the development of novel treatments for Parkinson's disease.

Whether differences between the unilateral 6-OHDA lesion model and the idiopathic disease process are serious disadvantages depends on the specific goals of particular experiments, as discussed previously. For treatments that aim to increase dopamine levels in the striatum, the rate at which the dopaminergic lesion is created may not be crucial. However, if a treatment strategy aims to protect surviving intrinsic dopamine neurons from a chronic disease process, then a lesion model in which cell loss is more protracted may be preferred (although is not essential *[26]*). Recent studies of growth factors in rodents have favored a terminal lesion model in which the 6-OHDA is injected into the striatum rather than into the substantia nigra or nigrostriatal bundle *(27,28)*. Dopamine cell loss from the nigra then occurs gradually over a number of weeks *(29)*. We have recently made 6-OHDA lesions unilaterally in the putamen of marmosets to produce a primate version of this terminal partial lesion model *(30)*. Behavioral deficits lasting up to 6 mo after the lesion were recorded on some tests, although the deficits were milder and more variable than after our standard nigrostriatal bundle lesion. A clear relationship between behavior (measured here as an ipsilateral head bias) and dopamine cell loss from the nigra was revealed by these variable lesions (Fig. 1).

3.2. Functional Assessment

The functional assessment appropriate for a unilateral lesion model differs considerably from the assessment appropriate in the case of bilateral lesions. In marmosets, strong biases in behavior are generated by the unilateral 6-OHDA lesion so that actions with the ipsilateral side of the body into ipsilateral space are greatly favored above those made on the contralateral side. This in turn creates opportunities for assessment in which different aspects of the bias are quantified using a variety of behavioral tests, for example, measuring the bias in the position of the head, spontaneous and drug-induced rotation, reaching ability with the impaired compared with the unaffected hand, and contralateral neglect *(4,5)*. For bilateral lesions, rating scales in which an observer formally rates the severity of parkinsonian symptoms, often on a scale of 0–3, are the most important and commonly used method of assessment, together with measures of locomotor activity *(23,31–33)*. Behavioral tests that require the monkey to respond and interact with a test apparatus or stimulus in some way may not be possible if the bilateral lesion results in severe parkinsonism. Importantly, marmosets with unilateral 6-OHDA lesions are not overtly parkinsonian. We have therefore not routinely used observational rating scales to assess the behavioral deficits, although such scales may detect a subtle akinesia (*unpublished observations*). To the casual observer, the amount of locomotion about the home cage often appears normal (albeit directed in circles). However, reductions in activity levels of about 30% over 24-h periods were recorded in marmosets with unilateral 6-OHDA lesions using an infra-red movement detector attached to the front of the home cage *(34)*.

Tremor is a symptom of Parkinson's disease that neither marmosets with unilateral

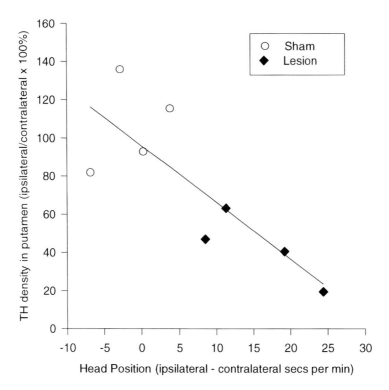

Fig. 1. The ipsilateral bias in head position induced by a 6-OHDA lesion of the putamen was significantly correlated with the density of TH immunoreactivity in the putamen ($r = 0.81$, $p < 0.01$) measured using a SeeScan© (Cambridge) image analyzer. Head position was recorded every second for 60 s for three tests on each of 4 d during the second week following the lesion.

nor bilateral 6-OHDA lesions exhibit *(4,23)*. Tremor has been described in monkeys with MPTP-induced parkinsonism, although it is not always a marked component of the syndrome and is most often described as an action tremor rather than the resting tremor typical of idiopathic Parkinson's disease *(15,35)*. Resting tremor was reported in the baboons that received MPTP chronically for 21 mo *(24)*. Rigidity is another symptom that we have not observed in marmosets with unilateral 6-OHDA lesions. Although marmosets with MPTP-induced bilateral lesions have been described as rigid *(15)*, this symptom does not feature as an item requiring (or perhaps permitting) quantitative assessment in most rating scales *(31)*.

Dyskinesias are a major long-term complication of L-Dopa therapy for many patients with Parkinson's disease and have been reported in several species of monkey made parkinsonian by MPTP *(35–38)*. Following a report by Pearce et al. *(32)* that dyskinesias, including chorea, choreoathetosis, and dystonias, were reliably provoked by L-Dopa in MPTP-lesioned marmosets, we were interested to see whether the same drug regimen would induce dyskinesias in marmosets with unilateral 6-OHDA lesions. Sinemet (12.5 mg/kg of L-Dopa 100 mg/Carbidopa 10 mg); Merck, Sharpe and Dohme, Hoddeson, U.K.) was administered orally in banana-flavored milk shake twice daily for 3 wk to four marmosets that had received their 6-OHDA lesions 12–18 mo

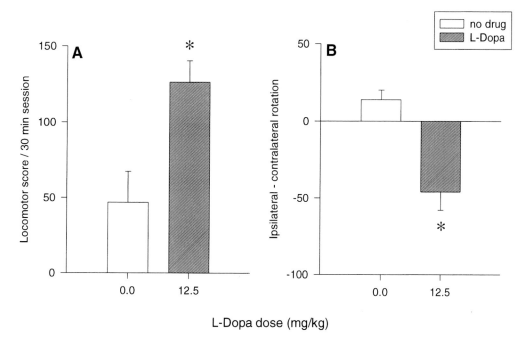

Fig. 2. Mean locomotor **(A)** and rotation **(B)** scores of four marmosets with unilateral 6-OHDA lesions of the nigrostriatal pathway treated either with no drug or with 12.5 mg/kg of L-Dopa (oral Sinemet). Locomotor and rotation scores were determined from video recordings of the marmosets' behavior during 30-min sessions in an observation cage starting 30 min after the administration of drug. No drug scores are means from eight sessions during the 2 wk preceding and following drug treatment, and L-Dopa scores from six sessions recorded during the 3-wk period of daily drug treatment. L-Dopa treatment produced a significant increase in locomotion ($t = 3.104$, $p < 0.05$) and contralateral rotation ($t = 4.032$, $p < 0.05$).

previously. The main effects of the drug were to increase locomotion and induce rotation away from the side of the lesion (i.e., contralaterally) (Fig. 2). There were no clear cases of drug-induced dyskinesias, although three of the four did exhibit mild stereotypies. In one case this comprised grooming of the contralateral foot, another repeated scent marking and chewing of the wooden perch, and in the third case a locomotor stereotypy in which the animal returned repeatedly to a particular corner of the cage. The fourth marmoset was simply more active. None of the stereotypies were obviously dyskinetic. Rather, normal sequences of behavior were repeated an abnormal number of times. Perhaps dyskinesias would have developed with repeated and/or intermittent dosing, but there was no opportunity to pursue this with these particular marmosets. We can conclude that dyskinesias are not readily induced by L-Dopa in marmosets with unilateral 6-OHDA lesions. However, the contralateral rotation and increased locomotion provide behavioral measures of responsiveness to the drug that can easily be quantified and used as part of a functional assessment program.

Akinesia is the parkinsonian symptom that is most dramatically reproduced by the unilateral 6-OHDA lesion. Because the lesion is unilateral the akinesia is restricted to the side of the body contralateral to the lesion. The spontaneous rotation so characteristic of monkeys with unilateral dopaminergic lesions, produced either by 6-OHDA or intracarotid MPTP, can perhaps best be understood as an inevitable consequence of the contralateral akinesia. Because movements are not initiated on the contralateral side, any movement is directed ipsilaterally and the animal turns in circles. We have commonly observed marmosets with unilateral 6-OHDA lesions "overshoot" and veer off ipsilaterally as they attempt to approach food or a piece of experimental apparatus positioned at the front of the home cage. The monkey is then unable to turn contralaterally to correct the direction of motion and instead completes an ipsilateral circuit to reach the target. The deficit appears to be primarily in the initiation of voluntary movements, as the contralateral limbs may be used normally in relatively automatic movements, for example, climbing up the front of the cage.

Rotation is a useful practical tool for following the progress of an experimental treatment in repeated tests over a period of months, for example, charting the development of neural grafts. The advantage of rotation is that individual lesioned animals generally show consistent rates of turning on test, although rates may vary considerably between animals. A strong rotater will continue to rotate predictably for a year or more after the lesion, so that any changes can more confidently be ascribed to the experimental treatment. Both amphetamine and apomorphine are useful for increasing ipsilateral and inducing contralateral rotation, respectively. A low dose of amphetamine (0.5 mg/kg) is most effective so as to avoid inducing stereotypies incompatible with rotation. Apomorphine (0.05 or 0.1 mg/kg) can reveal whether an animal has a good lesion even if that animal does not rotate well spontaneously or in response to amphetamine. However, strong conditioning effects are a disadvantage of using apomorphine interchangeably with other rotation tests. Thus, a drug-free marmoset that had previously experienced contralateral rotation induced by apomorphine will rotate contralaterally again when placed in the same test environment (*unpublished observations*). This will obviously complicate the interpretation of subsequent tests, especially spontaneous rotation and saline control tests. Another complication for rotation tests is the temporary switching of bias from ipsilateral to contralateral seen in about 10% of lesioned individuals, especially those with severe lesions and in situations of stress, which may be equivalent to the "paradoxical rotation" by rats noted by Ungerstedt *(4,38–40)*. Behavior during these episodes is characteristically different from normal, in particular head jerks to the contralateral side, so routine behavioral assessment is best avoided until the more usual ipsilateral bias returns.

A criticism of using rotation as the basis of functional assessment is that patients with Parkinson's disease do not rotate. One might predict that if the imbalance in dopamine levels between the two striata in patients were as great as in the lesioned marmosets, particularly in the dorsal caudate site critical for rotation, and if patients were given a low dose of amphetamine, they probably would rotate! Spontaneous biases turning toward one side have been recorded in patients by means of monitors attached to the belt *(41)*. Nevertheless, functional tests that assess motor abilities more obviously relevant to the difficulties experienced by patients are important. We have therefore focused on the marmosets' ability to reach out and touch target stimuli or retrieve small pieces of food. A number of tests have been developed for the purpose of quantifying hand and arm movements. A strong preference to use the hand ipsilateral to the lesion is revealed on most tests, for example, when retrieving small pieces of apple or bread from a conveyor belt or from inside tubes. The preference may be so strong that the monkey does not voluntarily attempt any actions at all with the impaired contralateral hand. The problem then is how to restrict use of the preferred hand to assess contralateral performance more directly. This has been done either by physical restraint or by designing the test apparatus so as to prevent or disadvantage use of the ipsilateral hand *(5,6)*. When required to use the contralateral hand to reach into a tube most lesioned monkeys succeed in taking food placed up to 3 cm inside by scrabbling with their fingers and hands, but fail to produce the extended arm reaches necessary to take food from 4 cm or further inside the tube. Other monkeys with slightly less severe deficits may eventually succeed taking food from the end of the tube (6 cm), but require many more attempts than they did before the lesion.

To gain more accurate measures of movement and response times we have used an automated test apparatus in which the marmosets are trained to hold down and then release a lever as they reach for a target stimulus that appears in front of them on a touch-sensitive screen. The lever is placed on the left and then on the right of the apparatus on alternate test sessions and, because the marmosets are required to lick at a central spout while holding the lever down, the body is positioned so that the correct hand must be used. Unilateral 6-OHDA lesions profoundly impair performance of this task so that the marmosets fail to complete trials, especially when required to use the contralateral hand. They appear unable to put together the sequence of movements necessary to complete a trial, that is, hold down the lever—lick at the spout—release the lever—reach to and touch the square, even though in other tests they may succeed in performing each of the separate component actions. Because so few trials are completed on the contralateral side, the test has not provided useful data on response and movement times following the lesion. Such a complete dopaminergic lesion produces akinesia rather than bradykinesia, that is, absence rather than slowing of movements. With slightly less complete lesions, in the range of 85–91% rather than the 98% dopamine depletions produced in the marmosets, Apicella and colleagues have succeeded in measuring slowed response and movement times by macaque monkeys performing a similar task with the hand contralateral to a unilateral 6-OHDA lesion *(42)*.

Another task that has the advantage, unlike the touch screen task, of requiring only a short period of training and yet provides detailed information on the accuracy of reaches, is analogous to the "staircase test" used to assess independent forelimb reaching in rats *(43)*. Marmosets are required to reach through vertical slots to the left and right of a Perspex box for pieces of bread placed on a series of steps ascending from each slot toward the midline. The design of the box is such that reaches with the right hand through the left slot and vice versa are prevented by an angled inner wall. Following a unilateral 6-OHDA lesion, few response attempts are made on the contralateral side. However, a striking effect in many lesioned monkeys is that those attempts that are made with the contralateral hand are often directed toward the medial before the lateral bread, even though paradoxically the former require a longer reach. Thus, reaches with the contralateral hand toward the midline are accomplished more readily than reaches into relatively contralateral space *(44)*. This observation illustrates the importance of neglect of targets in contralateral space as a component of the lesion-induced deficit. Neglect is also revealed to visual targets arriving on the conveyor belt from the contralateral side, and to somatosensory stimuli in the form of adhesive labels placed around the feet in a marmoset version of the test devised by Schallert et al. for rats *(45)*.

4. REPAIR STRATEGIES

4.1. Neural Grafting

The marmoset unilateral 6-OHDA lesion model was developed to assess the recovery that may be achieved by grafts of embryonic nigral tissue into the dopamine depleted striatum. Of the behavioral tests discussed earlier, amphetamine-induced rotation has proved particularly useful as the most sensitive and reliable indicator of graft viability in vivo, for example, in determining the importance of donor embryo age *(7)*. The test is sensitive as the effect of a functional graft is to reverse the direction of rotation from ipsilateral to contralateral *(5,6)*. Compared with other behavioral tests that may simply measure reductions in lesion-induced deficits, a rotation score that changes from positive to negative is easier to detect and can more confidently be attributed to the presence of a graft. Indeed, ampethamine-induced rotation is the only behavioral test from which the presence or absence of a graft can be predicted reliably from individual monkeys' scores.

If rotation were the only behavioral measure used, one might wrongly conclude that nigral grafts produce complete recovery of function. Considering the results from a wider selection of behavioral tests, graft-dependent recovery is revealed as partial rather than complete *(46)*. Thus, marmosets with 6-OHDA lesions and nigral grafts succeeded in using their previously impaired contralateral hand to retrieve food from inside a tube, but required more attempts than before the lesion and did not choose to use the contralateral hand when the ipsilateral was free *(5,6)*. Similarly, on the touch screen reaching task, trials completed with the contralateral hand improved significantly ($p < 0.01$) in a group of marmosets with nigral grafts compared with lesion alone controls, but the extent and rate of recovery were highly variable between subjects and performance did not return to prelesion levels *(47)*.

A fundamental question raised by these observations is whether the partial recovery represents the limits of what may be achieved with nigral grafts or whether more complete recovery might be possible with more extensive reinnervation of the striatum. The question has been considered in some detail by Björklund et al. *(48)* with reference to the incomplete recovery shown by rats with nigral grafts across a range of behavioral tests. Because the grafts are located ectopically in the striatum and do not reconstruct the dopaminergic pathway from the nigra to the striatum, recovery of any behaviors dependent on the flow of information in this pathway, for example, homeostatic information relayed from the hypothalamus, will inevitably be limited. The extent of reinnervation of the denervated striatum is likely to be another important factor determining the limits of recovery. The topographic organization of the striatum is such that grafts located at restricted striatal sites affect specific functional systems rather then producing general recovery *(6,48)*. Therefore, the restoration of more complex behaviors is likely to require more complete striatal reinnervation. Although the entire striatum has been targeted in marmosets, graft survival in this species is typically variable and reinnervation rarely complete *(5,7)*. The limits of functional recovery have therefore probably not been reached yet in the marmoset studies completed to date. An important goal for future marmoset work is to improve the extent and consistency of nigral graft survival, perhaps by employing the growth factors or antioxidant strategies that have proved successful in rodents *(9,10)*.

4.2. Dopamine Agonists

Several potential novel treatment strategies for Parkinson's disease depend upon diffuse release of dopamine in the striatum, for example, from encapsulated PC12 cells or dopamine-impregnated polymers *(3)*. These pharmacological strategies differ from neural transplantation in that transmitter release is nonregulated, whereas grafts offer the prospect of at least partial regulation of transmitter release via local neuronal circuits. It has been argued that more complete functional recovery may be achieved by regulated compared with nonregulated transmitter release *(49)*. However, there have been few behavioral studies from which the effects of neural grafts vs diffuse release of dopaminergic agents in primate striatum can be compared. Three cynomologous monkeys rendered hemiparkinsonian by intracarotid MPTP improved on a task in which the contralateral hand was used to remove food treats from a tray following the striatal implantation of polymer-encapsulated PC12 cells *(50)*. We decided to assess the effects of direct infusions of the dopamine agonist apomorphine into the striatum of marmosets with unilateral 6-OHDA lesions using the same behavioral tests previously used to assess the functional effects of neural grafts.

Following initial training and testing on the behavioral tasks, seven marmosets each underwent a unilateral 6-OHDA lesion of the nigrostriatal bundle followed during the same surgical session by implantation of cannulae directed into the caudate nucleus and into the putamen on the same side. After a 2-wk recovery period, a series of intrastriatal infusions of ascorbate saline vehicle, 1.5 and 3.0 μg of apomorphine were made into either the caudate nucleus, or the putamen, or into both structures simultaneously. The most striking behavioral effect of the infusions into the caudate nucleus was a dose-dependent reversal of rotation *(51)*. Thus, the spontaneous tendency to rotate ipsilaterally following the 6-OHDA lesion changed to contralateral rotation within 5 min of an apomorphine infusion into the caudate nucleus. The ipsilateral bias in the position of the head also reversed to become contralateral. In contrast, apomorphine infusions into the putamen affected neither rotation nor head position. The results confirm the observations made in the nigral graft experiments that dopaminergic activity in the caudate nucleus and not the putamen determines the direction of rotation *(6)*. For the reaching tasks, the clearest effects of the apomorphine infusions were observed when the marmosets were required to retrieve pieces of bread from inside tubes. Four monkeys required fewer attempts to retrieve the bread after the high dose of apomorphine was infused into the putamen than following infusions of either saline into the same site or drug or saline into the caudate nucleus. However, reaches by the three monkeys with the severest lesion-induced deficits remained impaired despite the infusions of apomorphine *(51)*. Thus, diffuse release of a dopamine agonist into the striatum can have similar effects to those of dopaminergic grafts on rotation and have some impact on reaching, although it may not be sufficient to support complete recovery in the severest cases.

4.3. Excitotoxic Lesions of the Subthalamic Nucleus

Recently there has been a resurgence of interest in the clinic in the use of pallidotomy and subthalamotomy as treatments for Parkinson's disease *(1,52)*. These treatment strategies are radically different from the dopamine replacement therapies considered previously in that they do not attempt to reconstruct basal ganglia circuitry but instead aim to restore the balance between interconnected pathways by destroying brain areas that become overactive as a result of the disease process. Beneficial effects have been reported in patients of posteroventral pallidotomy in the region that corresponds to the site of entry of the excitatory fibers projecting from the subthalamic nucleus (STN) *(53)*. There is also experimental evidence in MPTP-treated monkeys that sub-thalamotomy, pallidotomy, or infusions of glutamate antagonists into the pallidum reduce parkinsonian symptoms *(33,54,55)*. Basic issues related to the lesion approach include, to what extent can destruction of part of the basal ganglia output pathways restore normal function in that system, and are the benefits produced by the lesions comparable to those produced by nigral grafts? Again, we have used the marmoset unilateral 6-OHDA lesion model to examine the behavioral effects of STN lesions using the same tests previously used to assess the functional effects of neural grafts. Excitotoxic *N*-methyl-D-aspartate (NMDA) lesions of the STN made unilaterally on the same side as a prior 6-OHDA lesion dramatically reversed the side biases induced by the dopaminergic lesion: in several cases the ipsilateral rotational and head biases induced by the dopaminergic lesion became contralateral biases following the STN-lesion, whereas in sham STN-lesioned controls the ipsilateral biases remained *(34)*.

Overcompensation was also apparent in the "staircase" reaching task in which, instead of responding first on the ipsilateral side as they had done after the 6-OHDA lesion, the marmosets with additional STN lesions responded first on the contralateral side. However, deficits in skilled movements persisted in the STN lesion group in that they did not complete the "staircase" task any faster than the control group and remained impaired on the task in which they were required to reach into tubes.

Comparisons between the effects of STN lesions and nigral grafts in marmosets with unilateral 6-OHDA lesions illustrate the importance of using a range of behavioral tests to assess the consequences of different treatment strategies. Both treatments reduced, and in some cases reversed, amphetamine-induced rotation. Based on this single measure alone the nigral grafts and STN lesions would appear to be equally effective. However, when the wider range of behavioral measures are considered clear differences emerge (Table 1). For the nigral grafts, overcompensation was seen only in the amphetamine and spontaneous rotation tests. In the home cage, the marmosets with grafts continued to show ipsilateral biases in head position and in the tests of reaching and neglect, hence the conclusion that graft derived recovery is partial rather than complete *(5,6)*. In contrast, overcompensation toward contralateral biases characterized several aspects of the behavior of the marmosets with STN lesions, including head position and the order of responding at the staircase. This overcompensation did not, however, benefit successful completion of the skilled reaching tasks. In contrast, marmosets with nigral grafts have shown significant improvements in the use of the contralateral hand *(5,6,47)*. Overall then, there appear to be qualitative differences between the impact of nigral grafts compared with STN lesions on the behavioral deficits resulting from dopamine loss. Nigral grafts improve function on the lesioned side, but not back to the level of the intact side, whereas STN lesions produce a more active bias to initiate behaviors on the side affected by the prior dopaminergic lesion.

Table 1
Comparisons Between the Functional Effects of Nigral Grafts,
Apomorphine Infusions, and Subthalamic Lesions

	Nigral grafts	Apomorphine infusions	Subthalamic lesions
Head position	0	* cd	*
Rotation			
Spontaneous	*	* cd	*
Amphetamine	* cd		*
Apomorphine	+		0
Reaches into tubes	+ put	(+ put)	0
Staircase			
Neglect	0		*
Skilled reaching	(0?)		0

0, no effect; +, compensated; *, overcompensated; cd, caudate nucleus; put, putamen.

5. CONCLUSIONS

The hemiparkinsonian marmoset provides a useful model for assessing novel treatment strategies for Parkinson's disease in a primate species. Advantages of the model are the relative ease with which profound dopamine depletions can be achieved and the long-lasting and well characterized behavioral deficits produced by the 6-OHDA lesion. Because the model is primarily of dopaminergic dysfunction, the most appropriate uses for the model are to assess dopamine replacement strategies. The unilateral lesion generates strong biases in behavior that in turn create numerous opportunities for functional assessment, for example, quantifying skilled use of the impaired compared with the unaffected hand. However, marmosets with unilateral 6-OHDA lesions are not overtly parkinsonian and do not exhibit tremor, rigidity, and dyskinesias. Whether or not this is a disadvantage depends on the specific goals of individual experiments. We have used the marmoset model to compare the functional effects of neural grafts, dopamine agonist infusions, and excitotoxic lesions of the subthalamic nucleus. Qualitative differences between these treatment effects are revealed only when a range of behavioral measures are considered.

ACKNOWLEDGMENTS

We would like to thank our collaborators E. M. Torres, R. M. Ridley, and H. F. Baker. Our studies in this area are funded by The Wellcome Trust.

REFERENCES

1. Benabid A. L., Pollak, P., Gross, C., Hoffman, D., Benazzouz, A., Gao D. M., Laurent, A., Gentil, M., and Perret, J. (1994) Acute and long-term stimulation of subthalamic nucleus stimulation in Parkinson's disease. *Stereotact. Funct. Neurosurg.* **62,** 76–84.
2. Lindvall, O. (1997) Neural transplantation: does it work for Parkinson's disease? *NeuroReport* **8,** iii–x.
3. Dunnett, S. B., Annett, L. E., Kendall, A. L., Rosser, A. E., Watts, C., and Svendsen, C. D. (1997) Sources of cells for transplantation and gene therapy in Parkinson's disease,

in *Neurochemistry* (Teelken, A. W. and Korf, J., eds.) Plenum Press, New York, pp. 249–257.

4. Annett L. E., Rogers, D. C., Hernandez, T. D., and Dunnett, S. B. (1992) Behavioral analysis of unilateral monoamine depletion in the marmoset. *Brain* **115**, 825–856.

5. Annett, L. E., Martel, F. L., Rogers, D. C., Ridley, R. M., Baker, H. F., and Dunnett, S. B. (1994) Behavioral assessment of the effects of embryonic nigral grafts in marmosets with unilateral 6-OHDA lesions of the nigrostriatal pathway. *Exp. Neurol.* **125**, 228–246.

6. Annett, L. E., Torres, E. M., Ridley, R. M., Baker, H. F., and Dunnett, S. B. (1995) A comparison of the behavioral effects of embryonic nigral grafts in the caudate nucleus and in the putamen of marmosets with unilateral 6-OHDA lesions. *Exp. Brain Res.* **103**, 355–371.

7. Annett, L. E., Torres, E. M., Clarke, D. J., Ishida, Y., Barker, R. A., Ridley, R. M., Baker, H. F., and Dunnett, S. B. (1997) Survival of nigral grafts within the striatum of marmosets with 6-OHDA lesions depends critically on donor embryo age. *Cell Transplant.* **6**, 557–569.

8. Kopyov, O. V., Jacques, D. S., Lieberman, A., Duma, C. M., and Rogers, R. L. (1996) Clinical study of fetal mesencephalic intracerebral transplants for the treatment of Parkinson's disease. *Cell Transplant.* **5**, 327–337.

9. Nakao, N., Frodl, E. M., Duan, W-M., Widner, H., and Brundin, P. (1994) Lazaroids improve the survival of grafted rat embryonic dopamine neurons. *Proc. Natl. Acad. Sci. USA* **91**, 12,408–12,412.

10. Sinclair, S., Torres, E. M., and Dunnett, S. B. (1996) GDNF enhances dopaminergic cell survival and fibre outgrowth in embryonic nigral grafts. *NeuroReport* **7**, 2547–2552.

11. Hearn, J. P. (1987) Marmosets and tamarins, in *The UFAW Handbook on the Care and Management of Laboratory Animals*, 6th edit. (Poole, T., ed.), Churchill Livingstone, New York, pp. 569–581.

12. Stephan, H., Baron, G., and Schwerdtfeger, W. K. (1980) *The Brain of the Common Marmoset,* Springer-Verlag, Berlin.

13. Annett, L. E. and Ridley, R. M. (1992) Neural transplantation in primates, in *Neural Transplantation: A Practical Approach* (Dunnett, S. B. and Björklund, A., eds.), IRL Press, Oxford, pp. 123–138.

14. Windle, C. P., Baker, H. F., Ridley, R. M., Oerke, A-K., and Martin, R. D. (1998) Unreliable litters and prenatal reduction of litter size in the common marmosets (*Callithrix jacchus*). *J. Med. Primatol.,* in press.

15. Jenner, P. and Marsden, C. D. (1990) 1-Methyl-4-phenyl-1,2,3,6-tetrahydropyridine (MPTP): an update on its relevance to the cause and treatment of Parkinson's disease, in *Function and Dysfunction in the Basal Ganglia* (Franks, A. J., Ironside J. W., Mindham, R. H. S., Smith, R. J., Spokes, E. G. S., and Winlow, W., eds.), Manchester University Press, Manchester, pp. 140–160.

16. Close, S. P., Elliot, P., Hayes, A. G., and Marriott, A. S. (1990) Effects of classical and novel agents in a MPTP-induced reversible model of Parkinson's disease. *Psychopharmacology* **102**, 295–300.

17. Kendall, A. L., Rayment, F. D., Torres, E. M., Baker, H. F., Ridley, R. M., and Dunnett, S. B. (1998) Functional integration of striatal allografts in a primate model of Huntington's disease. *Nat. Med.* **4**, 727–729.

18. Marshall, J. W. B. and Ridley, R. M. (1996) Assessment of functional impairment following permanent middle cerebral artery occlusion in a non-human primate species. *Neurodegeneration* **5**, 275–286.

19. Ridley, R. M., Baker, H. F., Harder, J. A., and Pearson, C. (1996) Effects of lesions of different parts of the septo-hippocampal system in primates on learning and retention of information acquired before or after surgery. *Brain Res. Bull.* **40**, 21–32.

20. Roberts, A. C., DeSalvia, M. A., Wilkinson, L. S., Collins, P., Muir, J. L., Everitt, B. J.,

and Robbins, T. W. (1994) 6-hydroxydopamine lesions of the prefrontal cortex in monkeys enhance performance of an analogue of the Wisconsin card sort test: possible interactions with subcortical dopamine. *J. Neurosci.* **14**, 2531–2544.

21. Bezard, E. and Gross, C. E. (1998) Compensatory mechanisms in experimental and human Parkinsonism: towards a dynamic approach. *Prog. Neurobiol.* **55**, 93–116.

22. Clayton, D. F. and George, J. M. (1998) The synucleins: a family of proteins involved in synaptic function, plasticity, neurodegeneration and disease. *Trends Neurosci.* **21**, 249–254.

23. Mitchell, I. J., Hughes, N., Carroll, C. B., and Brotchie, J. M. (1995) Reversal of Parkinsonian symptoms by intrastriatal and systemic manipulations of excitatory amino acid and dopamine transmission in the bilateral 6-OHDA lesioned marmoset. *Behav. Pharmacol.* **6**, 492–507.

24. Hantraye, P., Varastet, M., Peschanski, M., Riche, D., Cesaro, P., Willer, J. C., and Maziere, M. (1993) Stable Parkinsonian syndrome and uneven loss of striatal dopamine fibres following chronic MPTP administration in baboons. *Neuroscience* **53**, 169–178.

25. Varastet, M., Riche, D., Maziere, M., and Hantraye, P. (1994) Chronic MPTP treatment reproduces in baboons the differential vulnerability of mesencephalic dopaminergic neurons observed in Parkinson's disease. *Neuroscience* **63**, 47–56.

26. Gash, D. M., Zhang, Z., Ovadia, A., Cass, W. A., Yi, A., Simmerman, L., Russell, D., Martin, D., Lapchak, P. A., Collins, F., Hoffer, B. J., and Gerhardt, G. A. (1996) Functional recovery in parkinsonian monkeys treated with GDNF. *Nature* **380**, 252–255.

27. Rosenblad, C., MartinezSerrano, A., and Björklund, A. (1998) Intrastriatal glial cell line-derived neurotrophic factor promotes sprouting of spared nigrostriatal dopaminergic afferents and induces recovery of function in a rat model of Parkinson's disease. *Neuroscience* **82**, 129–137.

28. BilangBleuel, A., Revah, F., Colin, P., Locquet, I., Robert, J. J., Mallet, J., and Horellou, P. (1997) Intrastriatal injection of an adenoviral vector expressing glial- cell-line-derived neurotrophic factor prevents dopaminergic neuron degeneration and behavioral impairment in a rat model of Parkinson disease. *PNAS* **94**, 8818–8823.

29. Sauer, H. and Oertel, W. H. (1994) Progressive degeneration of nigrostriatal dopamine neurons following intrastriatal terminal lesions with 6-hydroxydopamine: a combined retrograde and immunocytochemical study in the rat. *Neuroscience* **59**, 410–415.

30. Smyly, R. E., Annett, L. E., and Dunnett, S. B. (1997) Behavioral effects of 6-hydroxydopamine lesions of the putamen in the common marmoset. *Soc. Neurosci. Abstr.* **23**, 80.8, p. 191.

31. Smith, R. D., Zhang, Z., Kurlan, R., McDermott, M., and Gash, D. M. (1993) Developing a stable bilateral model of Parkinsonism in rhesus monkeys. *Neuroscience* **52**, 7–16.

32. Pearce, R. K. B., Jackson, M., Smith, L., Jenner, P., and Marsden, C. D. (1995) Chronic L-DOPA administration induces dyskinesias in the 1-methyl-4-phenyl-1,2,3,6-tetrahydropyridine-treated common marmoset (*Callithrix jacchus*). *Move. Disord.* **10**, 731–740.

33. Brotchie, J. M., Mitchell, I. J., Sambrook, M. A., and Crossman, A. R. (1991) Alleviation of Parkinsonism by antagonism of excitatory amino acid transmission in the medial segment of the globus pallidus in rat and primate. *Move. Disord.* **6**, 133–138.

34. Henderson, J. M., Annett, L. E., Torres, E. M., and Dunnett, S. B. (1998) Behavioural effects of subthalamic nucleus lesions in the hemiparkinsonian marmoset. *Eur. J. Neurosci.* **10**, 689–698.

35. Crossman, A. R. (1987) Primate models of dyskinesia: the experimental approach to the study of basal ganglia-related involuntary movement disorders. *Neuroscience* **21**, 1–40.

36. Falardeau, P., Bouchard, S., Bédard, P. J., Boucher, R., and Di Paolo, T. (1988) Behavioral and biochemical effect of chronic treatment with D-1 and/or D-2 dopamine agonists in MPTP monkeys. *Eur. J. of Pharmacol.* **150**, 59–66.

37. Kurlan, R., Kim, M. H., and Gash, D. M. (1991) Oral levodopa dose-response study in MPTP-induced hemiparkinsonian monkeys: assessment with a new rating scale for monkey parkinsonism. *Move. Disord.* **6**, 111–118.

38. Ungerstedt, U. (1976) 6-Hydroxydopamine-induced degeneration of the nigrostriatal dopamine pathway: the turning syndrome. *Pharmacol. Ther. B,* **2,** 37–40.
39. Boyce, S., Rupniak, N. M. J., Steventon, M. J., and Iversen, S. D. (1990) Characterisation of dyskinesias induced by L-dopa in MPTP-treated squirrel monkeys. *Psychopharmacology* **102,** 21–27.
40. Bankiewicz, K. S., Oldfield, E. H., Chiuej, C. C., Dopman, J. L., Jacobowitz, D. M., and Kopin, I. J. (1986) Hemiparkinsonism in monkeys after unilateral internal carotid artery infusion of 1-methyl-4-phenyl-1,2,3,6-tetrahydropyridine (MPTP). *Life Sci.* **39,** 7–16.
41. Bracha, H. S., Shults, C., Glick, S. D., and Kleinman, J. E. (1987) Spontaneous asymmetric circling behaviour in hemi-parkinsonism: a human equivalent of the lesioned-circling rodent behaviour. *Life Sci.* **40,** 1127–1130.
42. Apicella, P., Trouche, E., Nieoullon, A., Legallet, E., and Dusticier, N. (1990) Motor impairments and neurochemical changes after unilateral 6-hydroxydopamine lesion of the nigrostriatal dopaminergic system in monkeys. *Neuroscience* **38,** 655–666.
43. Montoya, C. P, Astell, S., and Dunnett, S. B. (1990) Effect of nigral and striatal grafts on skilled forelimb use in the rat. *Prog. Brain Res.* **82,** 459–466.
44. Annett, L. E., Torres, E. M., Clarke, D. J., and Dunnett, S. B. (1992) A spatial component in the neglect produced by unilateral 6-OHDA lesions in marmosets. *J. Psychopharmac.* BAP and EBPS Abstr., p 189.
45. Schallert, T., Upchurch, M., Lobaugh, N., Farrar, S. B., Spirduso, W. W., Gillam, P., Vaughn, D., and Wilcox, R. E. (1982) Tactile extinction: distinguishing between sensorimotor and motor asymmetries in rats with unilateral nigrostriatal damage. *Pharmacol. Biochem. Behav.* **16,** 455–462.
46. Annett, L. E. (1994) Functional studies of neural grafts in Parkinsonian primates, in *Functional Neural Transplantation* (Dunnett, S. B. and Björklund, A., eds.), Raven Press, New York, pp. 71–102.
47. Annett, L. E., Cummings, R. M., Torres, E. M., Kendall, A. L., Baker, H. F., Ridley, R. M., and Dunnett, S. B. (1998) Nigral grafts restore reaching to a touch sensitive screen in marmosets with unilateral 6-OHDA lesions, presented at the *American Society for Neural Transplantation, Fifth Annual Conference*, Clearwater, FL, April 1998.
48. Björklund, A., Dunnett, S. B., and Nikkhah, G. (1994) Nigral transplants in the rat Parkinson model, in *Functional Neural Transplantation* (Dunnett, S. B. and Björklund, A., eds.), Raven Press, New York, pp. 47–69.
49. Dunnett, S. B. and Björklund, A. (1994) Mechanisms of function of neural grafts in the injured brain, in *Functional Neural Transplantation* (Dunnett, S. B. and Björklund, A., eds.), Raven Press, New York, pp. 531–567.
50. Aebischer, P., Goddard, M., Signore, A. P., and Timpson, R. L. (1994) Functional recovery in hemiparkinsonian primates transplanted with polymer-encapsulated PC 12 cells. *Exp. Neurol.* **126,** 151–158.
51. Smyly, R. E., Annett, L. E., Torres, E. M., and Dunnett, S. B. (1998) Differential behavioral effects of apomorphine infusions into the caudate nucleus and putamen of marmosets with unilateral 6-OHDA lesions, in preparation.
52. Obeso, J. A., DeLong, M. R., Ohye, C., and Marsden, C. D. (1997) The basal ganglia and new surgical approaches for Parkinson's disease, in *Advances in Neurology,* Vol. 74, Lippincott-Raven, Philadelphia.
53. Laitinen, L. V., Bergenheim, A. T., and Hariz, M. I. (1992) Leksell's posteroventral pallidotomy in the treatment of Parkinson's disease. *J. Neurosurg.* **6,** 575–582.
54. Aziz, T. Z., Peggs, D., Agarwal, E., Sambrook, M. A., Crossman, A. R. (1992) Subthalamic nucleotomy alleviates Parkinsonism in the 1-methyl-4-phenyl-1,2,3,6,-tetrahydropyridine (MPTP)-expose primate. *Br. J. Neurosurg.* 6, 575–583.
55. Bergman, H., Wichmann, T., and DeLong, M. R. (1990) Reversal of experimental Parkinsonism by lesions of the subthalamic nucleus. *Science* **249,** 1436–1438.

11

Molecules for Neuroprotection and Regeneration in Animal Models of Parkinson's Disease

O. Isacson, L. C. Costantini, and W. R. Galpern

1. INTRODUCTION

Parkinson's disease (PD) is a progressive neurodegenerative disease characterized clinically by bradykinesia, rigidity, and resting tremor. The motor abnormalities are associated with a specific loss of dopamine (DA) neurons in the substantia nigra pars compacts (SNc) and the secondary depletion of striatal DA levels *(1)*. While the loss of striatal DA correlates with the severity of clinical disability, clinical manifestations of PD are not apparent until 80–85% of SNc neurons have degenerated and striatal DA levels are depleted by 60–80% *(2)*. Administration of L-Dopa, the precursor of DA, initially relieves Parkinsonian motor signs, but long-term use is associated with severe fluctuations in drug response. As pathologic changes precede the manifestation of clinical symptoms, it is reasonable to develop strategies to protect remaining DA neurons during the subclinical stage.

Clinical abnormalities and postmortem findings in PD patients have led to various explanations for the observed dopaminergic cell loss. Several lines of evidence suggest excitotoxicity coupled with a decline in mitochondrial energy metabolism as the cause of SNc DA cell loss *(3–5)*. Exposure to the mitochondrial toxin l-methyl-4-phenyl-1,2,5,6-tetrahydropyridine (MPTP) results in loss of DA cells in the SN and depletion of DA and its metabolites in the neostriatum *(6–8)*. The neurotoxicity of MPP is mediated by its active metabolite l-methyl-4-phenylpyridiniumion (MPTP$^+$), which inhibits mitochondrial oxidative phosphorylation *(9)* at complex-1 of the electron transport chain *(10)*. MPP$^+$ toxicity can be reduced by inhibition of glutamatergic input by glutamine antagonists or decortication, indicating that excitotoxicity is necessary, if not sufficient, to produce the death of DA neurons in this paradigm *(3,11,12)*. The finding that the neurochemical, anatomical, and behavioral abnormalities of MPTP-induced Parkinsonism closely resemble idiopathic PD suggests there may be a common final pathway of neuronal degeneration.

Postmortem analyses indicate increased lipid peroxidation in the SNc of PD brains, implying either an excess production of neurotoxic free radicals or a failure of the normal protective mechanisms to clear these radicals *(13,14)*. Several detoxifying sys-

From: Central Nervous System Diseases
Edited by: D. F. Emerich, R. L. Dean, III, and P. R. Sanberg © Humana Press Inc., Totowa, NJ

tems may be deficient in the PD brain, including decreased catalase and peroxidase activity *(15)* as well as reduced glutathione free radical scavenging *(16,17)*. In addition, postmortem and in vivo studies have demonstrated mitochondrial changes in PD patients. Similar to the inhibition of mitochondrial function produced by MPTP, defects have been found in complex I in SNc *(18)*, platelet *(19)*, and muscle *(20)* mitochondria preparations from PD subjects, further supporting a relationship between oxidative stress and neuronal degeneration.

Although there are multiple causes of neurodegenerative diseases including environmental, genetic, and age-associated factors, the treatments may be directed at similar underlying mechanisms via neuroprotective or reparative interventions. In a theoretical framework, one working model of neuronal damage and the prevention of cell death is the concept of "neuronal resilience" *(21)*. Depending on the status of the cell with respect to pretraumatic events and gene expression relevant to neuronal preservation, the neuron will exist far from or close to the threshold for irreversible neuronal damage. The neuron can thus be thought of as oscillating between protected and vulnerable conditions. This model of neuronal homeostasis suggests that a number of separate therapeutic measures, including delivery of neurotrophic factors (NTFs), may reduce the overall probability of degeneration in those neuronal populations that approach their specific threshold for degeneration *(21)*.

2. NEUROPROTECTION BY NEUROTROPHIC FACTOR INTERVENTIONS

2.1. Families of Neurotrophic Peptides

A range of NTFs have been shown to protect neurons against a spectrum of cellular insults and may thus have potential value in the treatment of neurodegenerative disorders *(22,23)*. Among the mechanisms by which NTFs protect are the maintenance of calcium homeostasis and the increase of antioxidant enzyme activities *(24,25)*. By decreasing cellular oxidative stress, or by interfering in the cascade to cell death, NTFs can reduce neuronal vulnerability and protect against ensuing neurodegenerative processes. Classically, NTFs are secreted during specific developmental stages by target tissues in limited quantities *(26)* and are important determinants of neuronal development and organization, affecting innervation of target tissue as well as survival of neurons *(27)*. Previously, it was thought that different neuronal populations were each responsive to only a single NTF. However, evidence indicates that there is overlap and redundancy, whereby a single NTF may affect more than one cell type, and a specific cell type may respond to several NTFs *(28)*. Moreover, actions of NTFs are associated not only with retrograde transport from the target tissue but also autocrine and paracrine mechanisms *(29,30)*. The site-specific NTF expression in the adult brain suggests various mechanisms of action in relation to the observed selective neuronal trophism. NTFs are important for neuronal maintenance in the adult brain, and insufficiency of such trophic support owing to decreased NTF supply or impaired target cell response may account for some of the cell death in neurodegenerative diseases *(31,32)*.

The neurotrophin family of NTFs includes the "classic" nerve growth factor (NGF) *(33)*, as well as the structurally related molecules brain-derived neurotrophic factor (BDNF) *(34,35)*, neurotrophin-3 (NT-3) *(36–40)*, and neurotrophin-4/5 (NT-4/5)

(41–43). The highest levels of NGF are found in the neocortex and hippocampus of adult brain, and the projection sites of basal forebrain cholinergic neurons, while lower levels are found within the striatum and other brain regions *(44,45).* BDNF mRNA is found throughout the adult brain, with the highest amounts localized to the hippocampus (exceeding that of NGF *[46]*) and cortex and lower amounts present in the striatum *(46,47)* and DA neurons of the SNc *(48,49).* NT-3 mRNA has been detected in the hippocampus, cerebellum, and cortex of adult brain *(36,38,39)* as well as DA neurons of the SNc *(48,49).* The maximum NT-4/5 expression is seen in the pons-medulla, hypothalamus, and cerebellum in adult brain, with lower expression found in the hippocampus, midbrain, striatum, and septum *(50).* The biological actions of the neurotrophins are mediated by a family of receptor tyrosine kinases, Trks. These high-affinity receptors contain an extracellular domain for ligand recognition, a transmembrane domain, and a cytoplasmic domain that possesses tyrosine kinase activity *(51).* These different receptors each show different and overlapping binding characteristics. The effects of NGF are mediated via the TrkA receptor *(52–54),* whereas BDNF *(55–57)* and NT-4/5 *(42,43,58)* activate the TrkB receptor. NT-4/5 also binds with low affinity to the TrkA receptor *(42,43).* NT-3 binds with high affinity to the TrkC receptor and with lower affinity to both TrkA and TrkB *(55–57,59,60).* Trk receptors are expressed widely throughout the adult brain, predominantly by neurons. TrkA receptors are the least abundant and are expressed by the cholinergic interneurons of the striatum, neurons of the basal forebrain, as well as in various brain stem regions *(61–64).* In situ hybridization studies indicate that there is similar regional expression of TrkB and TrkC mRNA *(61,65).* Transcripts for both receptors are found in numerous brain regions including the striatum, hippocampus, and cortex as well as SN and ventral tegmental area *(61,65).* In addition, the truncated form of TrkB is localized to the ependymal lining of ventricles, choroid plexus, and astrocytes *(65,66).*

Separate from the neurotrophin family is a family including ciliary neurotrophic factor (CNTF), a member of the α-helical cytokine superfamily *(67,68).* In the adult brain, CNTF mRNA is expressed at moderate levels in the cerebellum and brain stem and at low levels in other brain regions including the hippocampus, striatum, cortex, and septum *(69).* Transforming growth factor-β (TGF-β) is expressed in several isoforms, with TGF-β2 and 3 mRNA present in cortex, striatum, hippocampus, cerebellum, and brainstem *(70).* Glial cell line-derived neurotrophic factor (GDNF) is a member of the TGF-β superfamily. GDNF mRNA is widely distributed in the adult CNS and has been localized to the striatum, hippocampus, cortex, ventral mesencephalon, cerebellum, and spinal cord *(71–74).* Basic fibroblast growth factor (bFGF) is a mitogenic growth factor with trophic activity *(75),* and its mRNA is widely distributed throughout the brain in both astrocytes and neurons *(76).* Platelet-derived growth factor (PDGF) is also a mitogen and occurs as three isoforms, PDGF-AA, PDGF-AB, and PDGF-BB: transcripts for both PDGF chains A and B are found throughout the brain (77).

2.2. Immunophilin Ligands

More recently, nonpeptidergic molecules, called immunophilin ligands (based on their initial description in biological processes of the immune system), have shown protective and regenerative effects in neuronal systems. Trophic activity has also been observed with the immunosuppressant drug FK506 in a variety of in vitro and in vivo models. FK506 increases the ability of NGF to stimulate neurite outgrowth in the PC12 cell line and dorsal root ganglion explants *(78,79)* and stimulates nerve regrowth after sciatic nerve injury *(80)*. The neurotrophic effects of FK506 are due, in part, to its complexing with FK506-binding protein of 12 kDa (FKBP12), one of several immunophilins that is highly concentrated in the brain and localized almost exclusively to neurons *(81)*. The immunosuppressive properties of FK506 are a consequence of the inhibition of the phosphatase calcineurin by the FK506/FKBP12 complex. Inhibition of calcineurin augments phosphorylation of several substrates, leading to cytokine synthesis *(82)* and immunosuppression. These effects of FK506 render the drug inappropriate for chronic administration to neurodegenerative disease patients. However, the neurotrophic activities of FK506 can be separated from calcineurin inhibition and immunosuppression. Whereas FK506 can increase nerve regeneration after sciatic nerve crush, cyclosporin A, whose immunosuppressive properties are also attributed to the inhibition of calcineurin, cannot *(83)*. These data indicate that the trophic properties of FK506 involve calcineurin-independent mechanisms.

Initial findings that FK506 has trophic capacity in vitro and in vivo *(79,80,84,85)* sparked an interest in development of immunophilin ligands that possess neurotrophic activities, yet are not immunosuppressive. Based upon the FKBP/FK506 complex structure, several novel small-molecule immunophilin ligands have been designed that bind the immunophilin FKBP12, yet do not interact with calcineurin, and are thus devoid of immuno-suppressive activity. These nonimmunosuppressive immunophilin ligands demonstrate neurotrophic activity analogous to that obtained with FK506; they potentiate the effects of NGF on PC12 cells and sensory neurons in culture by promoting neurite extension *(78)*. We and others have now evaluated these compounds in models relevant to PD.

The effects of such nonimmunosuppressive immunophilin ligands may be mediated by their binding with FKBP12 and subsequent effects on Ca^{2+} homeostasis, consistent with the understanding that an optimal Ca^{2+} concentration is involved in neurotrophic effects *(86)*. FKBP12 complexes with several Ca^{2+} channels via the ryanodine receptor *(87)* and inositol 1,4,5-trisphosphate receptor *(88)*, and transient changes in concentration of neuronal intracellular Ca^{2+} can trigger various processes including structural modifications, neurotransmitter release, modulation of synaptic transmission, excitotoxic cell death, and gene expression *(89)*. Intracellular Ca^{2+} concentrations can be altered by flux from internal stores, and distinct Ca^{2+} channels in subcellular regions of the neuron generate highly compartmentalized Ca^{2+} signaling *(90)*, possibly contributing to the observed trophic effects.

3. NEUROPROTECTION AND REGENERATION IN PD MODELS

3.1. Effects of Peptidergic Neurotrophic Factors

The supplementation or replacement of a DA NTF may protect or slow the neuronal degeneration of PD. Several NTFs have shown trophic activity in the DA system, including BDNF, NT-3, NT-4/5, bFGF, TGF-β, and GDNF *(91–97)*. The effects of the neurotrophins BDNF *(91,92,98)*, NT-3 *(92)*, and NT-4/5 *(92,93)* as well as GDNF *(94)* on fetal DA neurons were first demonstrated in vitro. PDGF-BB *(99)*, TGF-β *(97)*, and bFGF also increase DA cell survival in vitro, yet the effect of bFGF is thought to be mediated by glia *(95,96,100)*. In addition to promoting survival of DA neurons in culture, the administration of these factors to the intact adult rat brain is associated with significant behavioral and neurochemical alterations. Supranigral delivery of BDNF enhances striatal DA turnover and decreases nigral DA turnover, as well as causes contralateral rotations and locomotor activity in amphetamine-treated rats *(101,102)*. The chronic administration of BDNF above the SN enhances the firing rate and number of electrically active DA neurons *(103)*. The localization of mRNA for BDNF and its receptor, TrkB, to the SN in adult brain suggests that BDNF may maintain SNc neuronal function in the intact brain, perhaps in an autocrine or paracrine manner *(48,49,61,65,92)*. Also, exogenous BDNF delivered to the striatum can act on SNc DA neurons via receptor-mediated retrograde transport *(104)*.

Similar to the in vivo effects of BDNF, supranigral infusion of NT-4/5 results in increased striatal DA turnover and release as well as contralateral rotation following the administration of amphetamine *(105)*, and NT-3 increases amphetamine-induced contralateral turning and decreases SN DA turnover *(102)*. Likewise, intranigral GDNF administration in intact adult rats increases spontaneous and amphetamine-induced locomotor behavior. These behavioral changes are associated with increased DA levels and turnover in the SN and increased DA turnover and decreased DA levels in the striatum *(106)*. In addition, sprouting of tyrosine hydroxylase (TH)-positive fibers near the injection site and increased striatal TH fiber staining were noted *(106)*. By injecting GDNF into the striatum, this factor is retrogradely transported to the SN DA neurons, suggesting that GDNF may act as a target-derived NTF *(107)*. These actions suggest that NTFs may be able to augment DA neuronal function in the adult brain. Of relevance to the neurodegenerative processes of PD, pretreatment of DA neurons with BDNF protects against the neurotoxic effects of MPP^+ and 6-hydroxydopamine (6-OHDA) in vitro, perhaps by increasing levels of the antioxidant enzyme glutathione reductase *(91,108,109)*. NT-4/5 *(93)*, bFGF *(110)*, GDNF *(111)*, and TGF-β *(112)* also protect against the toxic effects of MPP^+ in vitro.

Several DA NTFs can prevent cell death in models of neurodegeneration in vivo. BDNF has shown variable efficacy in protection in in vivo lesion paradigms. In rats with partial lesions of the nigrostriatal pathway induced by striatal infusion of 6-OHDA, concomitant supranigral administration of BDNF enhances striatal DA metabolism and reverses lesion-induced rotational asymmetry *(113)*. Yet, BDNF or NT-3 did not alter SN DA levels nor protect against the loss of striatal DA nerve terminals in this partial lesion paradigm. Initial studies using BDNF failed to show protection of DA neurons in the SNc following axotomy of the medial forebrain bundle in rat *(114,115)*. However, another study suggests that BDNF is neuroprotective in the axotomy paradigm *(116)*. BDNF-secreting fibroblasts implanted in the mesencephalon of adult rats attenuate the SNc DA cell loss caused by subsequent administration of the mitochondrial complex I inhibitor MPP^+ *(117)*, and increase DA levels in the SNc *(118)*. Similarly, intrastriatal grafts of BDNF-secreting fibroblasts prevent DA neuronal degeneration associated with intrastriatal administration of 6-OHDA *(119)*. GDNF is able to protect against SN neuronal degeneration in MPTP-treated mice *(120)* as well as in rats following axotomy of the medial forebrain bundle *(121)* or striatal 6-OHDA administration *(122,123)*. Interestingly, GDNF was shown not only to be neuro-protective, but also regenerative as administration following MPTP administration resulted in regeneration of TH fibers *(120)*. In addition, GDNF administration to the SN 4 wk after partial 6-OHDA lesioning of the MFB decreased apomorphine-induced rotational asymmetry, increased SN DA and DOPAC content, and spared 10% of the SN DA neurons *(124,125)*. Additional in vivo studies have demonstrated that NT-4/5 prevents 6-OHDA denervation-induced changes in striatal neurotransmitter gene expression *(126)*. Continuous infusion of CNTF near the SN prevents DA neuronal degeneration in the SNc following transection of the nigrostriatal pathway in adult rat *(127)*. However, in contrast to overall cellular protection effects of GDNF and BDNF, TH expression was only slightly preserved.

3.2. Immunophilin Ligand-Mediated Effects

We have observed trophic effects of an orally available, nonimmunosuppressive immunophilin ligand on the DA system in an MPTP model of PD in mice (Fig. 1) *(128)*. Our findings are consistent with and extend a study that demonstrated neu-rotrophic effects of a subcutaneously injected immunophilin ligand *(129)*. Utilizing an MPTP mouse model, our study demonstrated that 10-d oral administration of a nonimmunosuppressive immunophilin ligand (overlapping with MPTP for 5 d) pro-vides complete protection against MPTP-induced loss of striatal TH-positive fiber innervation. This effect can be compared to the approx 85% protection obtained by Steiner et al. *(129)* using subcutaneous injection of a nonimmunosuppressive immunophilin ligand. When the 10-d administration of immunophilin ligands was delayed until after the termination of MPTP treatment, the nonimmunosuppressive compound showed a trend toward recovery of striatal TH-positive fiber innervation, similar to effects seen by Steiner et al. *(129)*. However, when animals that were killed 5 d after the delayed immunophilin ligand treatment (rather than killed immediately) there was a significantly increased density of striatal TH-positive fibers relative to vehicle/MPTP ($p < 0.05$) (Costantini et al., *unpublished observations*).

By what mechanisms are these effects mediated? The observed protection or regeneration of striatal DA innervation in vivo may be obtained through mechanisms

---

Fig. 1. *(see facing page)* Oral administration of non-immunosuppressive immunophilin ligand protects against MPTP-induced loss of striatal DA innervation. **(A)** Time line of experi-mental paradigm. To produce DA degeneration, mice were treated with MPTP (20 mg/kg, i.p.) twice a day for 2 d (at 12-h intervals), then once per day for the following 3 d *(black bar)*. The nonimmunosuppressive immunophilin ligand (V-10,367) or FK506 was orally administered twice per day 30 min prior to MPTP for 5 d, and then given for 5 additional days *(gray bar)*. Animals were perfused on experimental d 15. **(B)** TH-positive fiber density of unlesioned, normal mouse striatum. **(C)** Loss of TH-positive fiber density after 5-d intraperitoneal MPTP treatment. **(D)** Complete protection of striatal TH-positive fiber density with oral administra-tion of 200 mg/kg/d of V-10,367. **(E)** Significant differences in striatal TH-positive fiber density among groups were apparent (ANOVA; $F(8,52) = 18.49$; $p < 0.0001$); V-10,367 prevented the loss of striatal TH-positive fiber density at all doses utilized (Tukey Kramer HSD, $*p < 0.05$: significantly higher than veh/MPTP; $\#p < 0.05$, significantly higher than all doses of FK506; $+p < 0.05$, significantly higher than V-10,367 400 mg/kg/d). *Error bars* represent SEM. (Data from Costantini et al. *[128]*.)

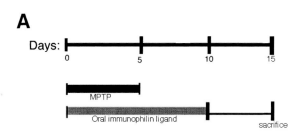

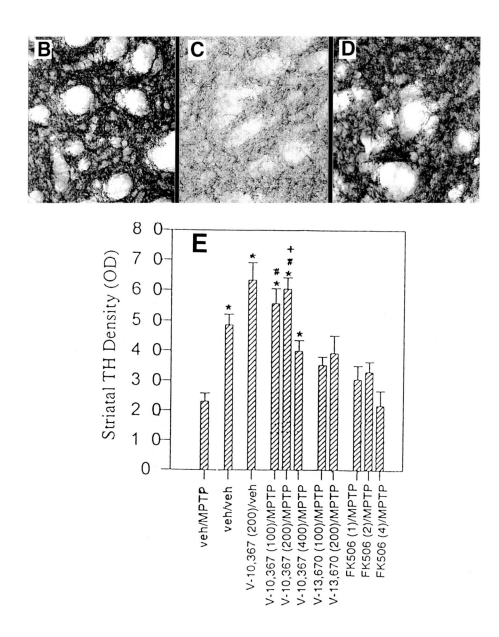

involving neuronal survival, metabolism, and/or neurite outgrowth. To investigate this question, we performed in vitro experiments demonstrating that both the nonimmuno-suppressive immunophilin ligand and FK506 have effects on neurite outgrowth. As discussed previously, earlier in vitro studies have shown that both FK506 and nonimmunosuppressive FKBP12 ligands enhance NGF-induced neurite outgrowth in the catecholaminergic PC12 cell line and the human neuroblastoma cell line SH-SY5Y *(78,129,130)*. Because DA neurons are directly affected in PD, the in vitro data obtained from primary cultures of E14 rat VM provided a more direct analysis of mechanisms involved in phenotypic DA survival, growth, and neurite branching than transformed cell lines. While we observed no effects on total or DA neuronal survival after immunophilin ligand treatments, differences were observed in branching vs elongation of TH-positive neurites. Even though both nonimmunosuppressive immunophilin ligand and FK506 affected neurite outgrowth of primary DA neurons at similar con-centrations, their effects were not identical: the nonimmunosuppressive immunophilin ligand caused increased branching of neurites whereas FK506 increased neurite elon-gation, indicating different mechanisms of action for these immunophilin ligands.

The demonstration that a nonimmunosuppressive immunophilin ligand increases branching of DA neurites in vitro suggests that the increased striatal TH-positive inner-vation after treatment in the MPTP-lesioned mouse may be due to stimulation of DA neurite branching in vivo. Another explanation for the preserved striatal TH-positive fibers in MPTP-lesioned mice after V-10,367 treatment could be normalization of TH levels. As there is no significant loss of TH-positive neurons in the SN using our MPTP paradigm, the decrease in TH-positive fiber density in the striatum may be a loss of phenotype rather than a loss of fibers. This reduction in TH protein is suggestive of a transient deficit of dopaminergic function. If this is the case, the normalization of TH levels after treatment with nonimmunosuppressive immunophilin ligand would poten-tially provide increased DA in the striatum. Because we have also found a spontaneous increase in striatal TH-positive fiber density at longer time points (30 d) after MPTP in this particular lesion paradigm (Costantini et al., *unpublished observations*), the major effects of nonimmunosuppressive immunophilin ligand may be to stimulate reexpression of TH. Nonetheless, because nonimmunosuppressive immunophilin ligand produced increased branching in vitro, terminal branching may have contrib-uted to the observed in vivo effects. To distinguish between an increase in TH levels caused by nonimmunosuppressive immunophilin ligand and branching/sprouting of TH-positive fibers, anterograde labeling of the nigrostriatal system posttreatment would be required.

The distinction between branching vs elongation of DA neurites has also been dem-onstrated when DA neurons are grown in the presence of growth factors *(131–133)* as well as their target striatal cells *(134)*. For instance, exposure of DA neurons to condi-tioned medium from VM induces growth of dendrite-like neurites (short, with a high number of branches), while striatal conditioned medium stimulates growth of axon-like neurites (long, with few branches) *(135)*. This suggests that these factors may differentially modify subsets of cytoskeletal proteins, cell adhesion molecules, or extracellular matrix components *(136)*. The actions of V-10,367 may also involve these modifications, and/or effects on growth cones which have been shown to serve a stabilizing function in outgrowth *(137)*. In contrast, neurite elongation, which was apparent only after FK506 treatment of DA neurons in our in vitro study, is consistent with previous studies showing enhanced neurite outgrowth in PC12 cells and sensory neurons after FK506 treatment *(78)*. Elongation appears to be generated in the neurite shaft *(137)*, in contrast to the branching mechanisms described previously. The neurite-lengthening effect of this immunosuppressive drug could occur in vivo; however such effects may not be sufficient to result in significant increases in TH-positive fiber density as was measured in our model. It has been previously reported that FK506 can increase DA levels in young (5-wk-old) MPTP-treated mice when given at doses of 10 mg/kg and above *(138)*. The doses used in the present study (4 mg/kg/d and lower) were designed to mimic a standard human immunosuppressive regimen. Thus, although it is possible that higher doses of FK506 might result in increased TH-positive fiber density, such doses may not be practical in a therapeutic setting.

In a wider perspective, the bell-shaped dose–response curves seen with both neurotrophins and immunophilin ligands on DA neuronal growth in vitro suggest similar mechanisms may be at play for these peptidergic and nonpeptidergic factors in the trophic effects observed. This could be achieved if the DA neurons are dependent on a basic neurotrophic "tone" and "activation profile," as demonstrated by neurotrophic peptide concentrations needed for trophic activation of specific set of neuronal receptors. These neurotrophin levels can be artificially increased by intracerebral infusion of peptides, but perhaps the same intracellular effects can be obtained by nonpeptidergic factors (such as the immunophilin ligands) at the second-messenger level by enhancing already existing neurotrophin "tone" in the dopaminergic nigrostriatal system.

4. IS THERE CLINICAL USE FOR NEUROTROPHIC FACTORS?

4.1. Delivery of Peptidergic Neurotrophic Factors to the Central Nervous System

The trophic support and neuroprotection afforded by NTFs indicate the possible therapeutic applicability of these factors to neurodegenerative disorders *(139–141)*. However, the clinical application of NTFs as a therapeutic modality for neurodegenerative disorders requires an effective means of delivery. Because NTFs cannot cross the blood–brain barrier (BBB) by themselves, additional strategies for delivery to the central nervous system must be considered. Various approaches to circumvent the BBB have been investigated, including direct intraparenchymal or intraventricular infusions. While these methods of delivery allow for dosage control and site specificity, there is often parenchymal damage at the cannula placement site. Furthermore, the delivery efficiency of factors may be limited by diffusion properties of the factor within brain parenchyma *(142)*. In addition, intraventricular administration of the TrkB ligand BDNF is ineffective owing to binding by truncated TrkB receptors that are present in the ependymal lining of the ventricles *(143)*. Similar difficulties may be expected for NT-4/5 as this factor also binds to the TrkB receptor. Continuous delivery could be achieved by a pump reservoir connected to a cannula. However, growth factor stability in the reservoir may be of concern. Moreover, approaches involving intracerebral cannulae introduce a high risk for infection. An additional approach to delivery would be to modify the NTF to enable BBB passage. It has been shown that NGF–transferrin receptor antibody conjugates are able to cross the BBB and protect neurons from excitotoxic lesions, suggesting that such an approach may provide a noninvasive route of delivery to the brain *(144,145)*.

As an alternative to direct infusion, methods for cell-mediated delivery of therapeutic proteins have been developed including ex vivo as well as in vivo gene transfer. Ex vivo gene transfer involves the genetic engineering of cells to express a transgene prior to transplantation (*see 146* for review). Intracerebral grafting of primary or immortalized fibroblasts genetically engineered to release trophic factors allows for site-specific biological delivery and has been shown to prevent cell loss in a variety of in vivo lesion paradigms (e.g., *147–149*). In contrast to the use of immortalized fibroblast lines, the use of primary fibroblasts decreases the likelihood of tumor formation following implantation and would allow for harvesting cells from the affected individual, thereby eliminating the possibility of immune rejection. Alternatively, cell division of immortalized fibroblasts could be arrested prior to implantation or cells could be encapsulated in a semipermeable membrane *(150)*. Difficulties with achieving stable, long-term gene expression by genetically modified cells remain an obstacle to this approach as gene expression is down-regulated over time after implantation *(151)*. Improved vector constructs with alternate promoters may resolve this issue. Moreover, the use of inducible promoters *(152)* may allow for regulated factor delivery by the genetically modified cells. Viral vectors have also been used to infect human neural progenitors prior to transplantation *(153,154)*, suggesting the potential utility of ex vivo genetic engineering of progenitor cells. Several groups have attempted to infect cells in vivo via herpes simplex virus *(155)*, adenovirus *(156,157)*, or adeno-associated virus *(158)* vectors. These approaches of genetic modification of cells in vivo hold promise for future therapeutic interventions, but current methodologies are severely limited by limited gene expression as well as potential cytopathogenicity, and the low rate of infectivity associated with virus vectors.

Optimal methods of delivery may vary for NTFs as some are soluble factors and others are heparin or extracellular matrix bound. Moreover, for each growth factor, effective dose ranges need to be determined as the dose–response relationship for many factors indicates higher doses may be less effective. Indeed such a dose–response effect has been reported in vitro for the trophic effects of NT-3 and NT-4/5 on DA neurons *(92)*, and studies of axotomized SN neurons as well as axotomized motor neuron rescue demonstrate a submaximal response with higher doses of BDNF *(116,159)*. As suggested by Vejsada et al. *(159)*, high levels of ligand may result in decreased affinity or down-regulation of the high-affinity TrkB receptor or an up-regulation of the inactive truncated TrkB receptor, thereby accounting for the decreased efficacy of higher doses of BDNF.

The multiple mechanisms for protection, coupled with the multiple sites for possible intervention, suggest that NTFs acting via different cellular mechanisms or at sequential sites in these neurodegenerative pathways may provide additive or synergistic protective effects. The precise mechanisms of NTF-mediated neuroprotection are not fully characterized, but the trophic and protective effects of NTFs indicate their promise for future therapeutic application. Yet, the clinical utility of NTFs is presently limited by delivery and dosage issues. Strategies to circumvent the BBB need to be refined as systemic delivery of NTFs may be associated with untoward effects via actions on additional cell populations. Furthermore, the overlap in NTF specificity indicates that the delivered factor is likely to affect more than the target neuronal population. The evaluation of dose–response relationships and long-term effects of NTF treatment will aid in the development of NTFs as a therapeutic approach for the treatment of neurodegenerative disorders.

4.2. Delivery of Nonpeptidergic Factors to the Central Nervous System

The trophic actions of the systemically administered nonimmunosuppressive immunophilin ligands are apparent at concentrations comparable to those obtained with classic neurotrophic peptides in similar models *(128,129,160–168)*. The necessity for intracerebral administration of neurotrophic peptides (such as osmotic pumps for GDNF) has hampered their clinical application, although systemic delivery methods, such as conjugation of the trophic factor to a transferrin receptor antibody *(169)*, have shown success in animal models. The neurite-branching and neurite-lengthening effects of immunophilin ligands observed in vitro and trophic effects observed after oral administration *(128)* encourage further investigation of the therapeutic potential of such compounds in PD.

REFERENCES

1. Agid, Y. (1991) Parkinson's disease: pathophysiology. *Lancet* **337,** 1321–1324.
2. Marsden, C. D. (1982) Basal ganglia disease. *Lancet* **2,** 1141–1147.
3. Beal, M. F., Hyman, B. T., and Koroshetz, W. (1993) Do defects in mitochondrial energy metabolism underlie the pathology of neurodegenerative diseases? *TINS* **16,** 125–131.
4. Difazio, M. C., Hollingsworth, Z., Young, A. B., and Penney, J., Jr. (1992) Glutamate receptors in the substantia nigra of Parkinson's disease brains. *Neurology* **42,** 402–406.
5. Young, A. B. (1993) Role of excitotoxins in heredito-degenerative neurologic diseases, in *Molecular and Cellular Approaches to the Treatment of Neurological Disease* (Waxman, S. G., ed.), Raven Press, New York, pp. 175–189.
6. Langston, J. W., Ballard, P., Tetrud, J. W., and Irwin, I. (1983) Chronic Parkinsonism in humans due to a product of meperidine-analog synthesis. *Science* **219,** 979–980.
7. Heikkila, R. E., Hess, A., and Duvoisin, R. C. (1984) Dopaminergic neurotoxicity of 1-methyl-4-phenyl-1,2,5,6-tetrahydropyridine in mice. *Science* **224,** 1451–1453.
8. Burns, R. S., Chiueh, C. C., Markey, S. P., Ebert, M. H., Jacobowitz, D. M., and Kopin, I. J. (1983) A primate model of Parkinsonism: selective destruction of dopaminergic neurons in the pars compacta of the substantia nigra by *N*-methyl-4-phenyl-1,2,3,6-tetrahydropyridine. *Proc. Natl. Acad. Sci. USA* **80,** 4546–4550.
9. Nicklas, W. J., Vyas, I., and Heikkila, R. E. (1985) Inhibition of NADH-linked oxidation in brain mitochondria by 1-methyl-4-phenyl-pyridine, a metabolite of the neurotoxin, 1-methyl-4-phenyl-1,2,5,6-tetrahydropyridine. *Life Sci.* **36,** 2503–2508.
10. Nicklas, W. J., Youngster, S. K., Kindt, M. V., and Heikkila, R. E. (1987) MTPT, MPP[+] and mitochondrial function. *Life Sci.* **40,** 721–729.
11. Srivastava, R., Brouillet, E., Beal, M. F., Storey, E., and Hyman, B. T. (1993) Blockade of 1-methyl-4-phenylpyridinium ion (MPP[+]) nigral toxicity in the rat by prior decortication or MK-801 treatment: a stereological estimate of neuronal loss. *Neurobiol. of Aging* **14,** 295–301.
12. Turski, L., Bressler, K., Rettig, K.-J., Loeschmann, P.-A., and Wachtel, H. (1991) Protection of substantia nigra from MPP[+] neurotoxicity by *N*-methyl-D-aspartate antagonists. *Nature* **349,** 414–419.

13. Dexter, D., Carter, C., Agid, F., Agid, Y., Lees, A. J., Jenner, P., and Marsden, C. D. (1986) Lipid peroxidation as cause of nigral cell death in Parkinson's disease. *Lancet* **ii,** 639–640.

14. Dexter, D. T., Carter, C. J., Wells, F. R., Javoy-Agid, F., Agid, Y., Lees, A., Jenner, P., and Marsden, C. D. (1989) Basal lipid peroxidation in substantia nigra is increased in Parkinson's disease. *J. Neurochem.* **52,** 381–389.

15. Ambani, L. M., Van Woert, M. H., and Murphy, S. (1975) Brain peroxidase and catalase in Parkinson disease. *Arch. Neurol.* **32,** 114–118.

16. Perry, T. L., Godin, D. V., and Hansen, S. (1982) Parkinson's disease: a disorder due to nigral glutathione deficiency? *Neurosci. Lett.* **33,** 305–310.

17. Kish, S. J., Morito, C., and Hornykiewicz, O. (1985) Glutathione peroxidase activity in Parkinson's disease brain. *Neurosci. Lett.* **58,** 343–346.

18. Schapira, A. H. V., Cooper, J. M., Dexter, D., Clark, J. B., Jenner, P., and Marsden, C. D. (1990) Mitochondrial complex 1 deficiency in Parkinson's disease. *J. Neurochem.* **54,** 823–827.

19. Parker, W. D., Jr., Boyson, S. J., and Parks, J. K. (1989) Abnormalities of the electron transport chain in idiopathic Parkinson's disease. *Ann. Neurol.* **26,** 719–723.

20. Shoffner, J. M., Watts, R. L., Juncos, J. L., Torroni, A., and Wallace, D. C. (1991) Mitochondrial oxidative phosphorylation defects in Parkinson's disease. *Ann. Neurol.* **30,** 332–339.

21. Isacson, O. (1993) On neuronal health. *Trends Neurosci.* **16,** 306–308.

22. Tatter, S. B., Galpern, W. R., and Isacson, O. (1995) Neurotrophic factor protection against excitotoxic neuronal death. *Neuroscientist* **1,** 286–297.

23. Mattson, M. P., Cheng, B., and Smith-Swintosky (1993) Mechanisms of neurotrophic factor protection against calcium- and free radical-mediated excitotoxic injury: implications for treating neurodegenerative disorders. *Exp. Neurol.* **124,** 89–95.

24. Cheng, B. and Mattson, M. P. (1991) NGF and bFGF protect rat hippocampal and human cortical neurons against hypoglycemic damage by stabilizing calcium homeostasis. *Neuron* **7,** 1031–1041.

25. Mattson, M. P., Lovell, M. A., Furukawa, K., and Markesbery, W. R. (1995) Neurotrophic factors attenuate glutamate-induced accumulation of peroxides, elevation of intracellular Ca^{2+} concentration, and neurotoxicity and increase antioxidant enzyme activities in hippocampal neurons. *J. Neurochem.* **65,** 1740–1751.

26. Barde, Y. A. (1988) What, if anything, is a neurotrophic factor? *Trends Neurosci.* **11,** 343–346.

27. Purves, D. (1986) The trophic theory of neural connections. *Trends Neurosci.* **9,** 486–489.

28. Korsching, S. (1993) The neurotrophic factor concept: a reexamination. *J. Neurosci.* **13,** 2739–2748.

29. Miranda, R. C., Sohrabji, F., and Toran-Allerand, C. D. (1993) Neuronal colocalization of mRNAs for neurotrophins and their receptors in the developing central nervous system suggests a potential for autocrine interactions. *Proc. Natl. Acad. Sci. USA* **90,** 6439–6443.

30. Kokaia, M., Bengzon, J., Metsis, M., Kokaia, M., Persson, H., and Lindvall, O. (1993) Coexpression of neurotrophins and their receptors in neurons of the central nervous system. *Proc. Natl. Acad. Sci. USA* **90,** 6711–6715.

31. Appel, S. H. (1981) A unifying hypothesis for the cause of amyotrophic lateral sclerosis, Parkinsonism, and Alzheimer disease. *Ann. Neurol.* **10,** 499–505.

32. Hefti, F. (1983) Is Alzheimer's disease caused by a lack of nerve growth factor? *Ann. Neurol.* **13,** 109–110.

33. Levi-Montalcini, R. (1987) The nerve growth factor 35 years later. *Science* **237,** 1154–1162.

34. Barde, Y.-A., Edgar, D., and Thoenen, H. (1982) Purification of a new neurotrophic factor from mammalian brain. *EMBO J.* **1,** 549–553.

35. Leibrock, J., Lottspeich, F., Hohn, A., Hofer, M., Hengerer, B., Masiakowski, P., Thoenen, H., and Barde, Y.-A. (1989) Molecular cloning and expression of brain-derived neurotrophic factor. *Nature* **341,** 149–152.

36. Hohn, A., Leibrock, J., Bailey, K., and Barde, Y.-A. (1990) Identification and characterization of a novel member of the nerve growth factor/brain-derived neurotrophic factor family. *Nature* **344,** 339–341.

37. Maisonpierre, P. C., Belluscio, L., Squinto, S., Ip, N. Y., Furth, M. E., Lindsay, R. M., and Yancopoulos, G. D. (1990) Neurotrophin-3: a neurotrophic factor related to NGF and BDNF. *Science* **247,** 1446–1451.

38. Ernfors, P., Ibanez, C. F., Ebendal, T., Olson, L., and Persson, H. (1990) Molecular cloning and neurotrophic activities of a protein with structural similarities to nerve growth factor: developmental and topographical expression in the brain. *Proc. Natl. Acad. Sci. USA* **87,** 5454–5458.

39. Jones, K. R. and Reichardt, L. F. (1990) Molecular cloning of a human gene that is a member of the nerve growth factor family. *Proc. Natl. Acad. Sci. USA* **87,** 8060–8064.

40. Rosenthal, A., Goeddel, D. V., Nguyen, T., Shih, A., Laramee, G. R., Nikolics, K., and Winslow, J. W. (1990) Primary structure and biological activity of a novel human neurotrophic factor. *Neuron* **4,** 767–773.

41. Hallbook, F., Ibanez, C. F., and Persson, H. (1991) Evolutionary studies of the nerve growth factor family reveal a novel member abundantly expressed in *Xenopus* ovary. *Neuron* **6,** 845–858.

42. Berkemeier, L. R., Winslow, J. W., Kaplan, D. R., Nikolics, K., Goeddel, D. V., and Rosenthal, A. (1991) Neurotrophin-5: a novel neurotrophic factor that activates trk and trkB. *Neuron* **7,** 857–866.

43. Ip, N. Y., Ibanez, C. F., Nye, S. H., McClain, J., Jones, P. F., Gies, D. R., Belluscio, L., LeBeau, M. M., Espinosa, R., Squinto, S. P., Persson, H., and Yancopoulos, G. D. (1992) Mammalian neurotrophin-4: structure, chromosomal localization, tissue distribution, and receptor specificity. *Proc. Natl. Acad. Sci. USA* **89,** 3060–3064.

44. Whittemore, S. R., Ebendal, T., Lärkfors, L., Olson, L., Seiger, A., Strömberg, I., and Persson, H. (1986) Developmental and regional expression of βnerve growth factor messenger RNA and protein in the rat central nervous system. *Proc. Natl. Acad. Sci. USA* **83,** 817–821.

45. Shelton, D. L. and Reichardt, L. F. (1986) Studies on the expression of the βnerve growth factor (NGF) gene in the central nervous system: level and regional distribution of NGF mRNA suggest that NGF functions as a trophic factor for several distinct populations of neurons. *Proc. Natl. Acad. Sci. USA* **83,** 2714–2718.

46. Hofer, M., Pagliusi, S. R., Hohn, A., Leibrock, J., and Barde, Y.-A. (1990) Regional distribution of brain-derived neurotrophic factor mRNA in the adult mouse brain. *EMBO J.* **9,** 2459–2464.

47. Wetmore, C., Ernfors, P., Persson, H., and Olson, L. (1990) Localization of brain-derived neurotrophic factor mRNA to neurons in the brain by in situ hybridization. *Exp. Neurol.* **109,** 141–152.

48. Gall, C. M., Gold, S. J., Isackson, P. J., and Seroogy, K. B. (1992) Brain-derived neurotrophic factor and neurotrophin-3 mRNAs are expressed in ventral midbrain regions containing dopaminergic neurons. *Mol. Cell. Neurosci.* **3,** 56–63.

49. Seroogy, K. B., Lundgren, K. H., Tran, T. M. D., Guthrie, K. M., Isackson, P. J., and Gall, C. M. (1994) Dopaminergic neurons in rat ventral midbrain express brain-derived neurotrophic factor and neurotrophin-3 mRNAs. *J. Comp. Neurol.* **342,** 321–334.

50. Timmusk, T., Belluardo, N., Metsis, M., and Persson, H. (1993) Widespread and developmentally regulated expression of neurotrophin-4 mRNA in rat brain and peripheral tissues. *Eur. J. Neurosci.* **5,** 605–613.

51. Meakin, S. O. and Shooter, E. M. (1992) The nerve growth factor family of receptors. *Trends Neurosci.* **15,** 323–330.

52. Kaplan, D. R., Martin-Zanca, D., and Parada, L. F. (1991) Tyrosine phosphorylation and tyrosine kinase activity of the trk proto-oncogene product induced by NGF. *Nature* **350,** 158–160.

53. Kaplan, D. R., Hempstead, B. L., Martin-Zanca, D., Chao, M. V., and Parada, L. F. (1991) The trk proto-ocncogene product: a signal transducing receptor for nerve growth factor. *Science* **252,** 554–558.

54. Klein, R., Jing, S., Nanduri, V., O'Rourke, E., and Barbacid, M. (1991) The trk proto-oncogene encodes a receptor for nerve growth factor. *Cell* **65,** 189–197.

55. Soppet, D., Escandon, E., Maragos, J., Middlemas, D. S., Reid, S. W., Blair, J., Burton, L. E., Stanton, B. R., Kaplan, D. R., Hunter, T., Nikolics, K., and Parada, L. F. (1991) The neurotrophic factors brain-derived neurotrophic factor and neurotrophin-3 are ligands for the trkB tyrosine kinase receptor. *Cell* **65,** 895–903.

56. Squinto, S. P., Stitt, T. N., Aldrich, T. H., Davis, S., Bianco, S. M., Radziejewski, C., Glass, D. J., Masiakowski, P., Furth, M. E., Valenzuela, D. M., DiStefano, P. S., and Yancopoulos, G. D. (1991) TrkB encodes a functional receptor for brain-derived neurotrophic factor and neurotrophin-3 but not nerve growth factor. *Cell* **65,** 885–893.

57. Klein, R., Nanduri, V., Jing, S., Lamballe, F., Tapley, P., Bryant, S., Cordon-Cardo, C., Jones, K. R., Reichardt, L. F., and Barbacid, M. (1991) The trkB tyrosine protein kinase is a receptor for brain-derived neurotrophic factor and neurotrophin-3. *Cell* **66,** 395–403.

58. Klein, R., Lamballe, F., Bryant, S., and Barbacid, M. (1992) The trkB tyrosine protein kinase is a receptor for neurotrophin-4. *Neuron* **8,** 947–956.

59. Lamballe, F., Klein, R., and Barbacid, M. (1991) TrkC, a new member of the trk family of tyrosine protein kinases, is a receptor for neurotrophin-3. *Cell* **66,** 967–979.

60. Cordon-Cardo, C., Tapley, P., Jing, S., Nanduri, V., O'Rourke, E., Lamballe, F., Kovary, K., Klein, R., Jones, K. R., Reichardt, L. F., and Barbacid, M. (1991) The trk tyrosine protein kinase mediates the mitogenic properties of nerve growth factor and neurotrophin-3. *Cell* **66,** 173–183.

61. Merlio, J.-P., Ernfors, P., Jaber, M., and Persson, H. (1992) Molecular cloning of rat trkC and distribution of cells expressing messenger RNAs for members of the trk family in the rat central nervous system. *Neuroscience* **51,** 513–532.

62. Richardson, P. M., Verge Issa, V. M. K., and Riopelle, R. J. (1986) Distribution of neuronal receptors for nerve growth factor in the rat. *J. Neurosci.* **6,** 2312–2321.

63. Steininger, T. L., Wainer, B. H., Klein, R., Barbacid, M., and Palfrey, H. C. (1993) High-affinity nerve growth factor receptor (Trk) immunoreactivity is localized in cholinergic neurons of the basal forebrain and striatum in the adult rat brain. *Brain Res.* **612,** 330–335.

64. Holtzman, D. M., Kilbridge, J., Li, Y., Cunningham, E. T., Lenn, N. J., Clary, D. O., Reichardt, L. F., and Mobley, W. C. (1995) TrkA expression in the CNS: evidence for the existence of several novel NGF-responsive neurons. *J. Neurosci.* **15,** 1567–1576.

65. Altar, C. A., Siuciak, J. A., Wright, P., Ip, N. Y., Lindsay, R. M., and Wiegand, S. J. (1994) In situ hybridization of trkB and trkC receptor mRNA in rat forebrain and association with high-affinity binding of [^{125}I]BDNF, [^{125}I]NT-4/5, and [^{125}I]NT-3. *Eur. J. Neurosci.* **6,** 1389–1405.

66. Klein, R., Conway, D., Parada, L., and Barbacid, M. (1990) The trkB tyrosine protein kinase gene codes for a second neurogenic receptor that lacks the catalytic kinase domain. *Cell* **61,** 647–656.

67. Lin, L.-F. H., Mismer, D., Lile, J. D., Armes, L. G., Butler III, E. T., Vannice, J. L., and Collins, F. (1989) Purification, cloning, and expression of ciliary neurotrophic factor (CNTF). *Science* **246,** 1023–1025.

68. Stockli, K. A., Lottspeich, F., Sendtner, M., Masiakowski, P., Carroll, P., Gotz, R., Lindholm, D., and Thoenen, H. (1989) Molecular cloning, expression and regional distribution of rat ciliary neurotrophic factor. *Nature* **342,** 920–923.

69. Stöckli, K. A., Lillien, L. E., Näher-Noé, M., Breitfeld, G., Hughes, R. A., Raff, M. C., and Thoenen, H. (1991) Regional distribution, developmental changes, and cellular localization of CNTF-mRNA and protein in the rat brain. *J. Cell Biol.* **115**, 447–459.

70. Unsicker, K., Flanders, K. C., Cissel, D. S., Lafyatis, R., and Sporn, M. B. (1991) Transforming growth factor beta isoforms in the adult rat central and peripheral nervous system. *Neuroscience* **44**, 613–625.

71. Schaar, D. G., Sieber, B.-A., Dreyfus, C. F., and Black, I. B. (1993) Regional and cell-specific expression of GDNF in rat brain. *Exp. Neurol.* **124**, 368–371.

72. Springer, J. E., Mu, X., Bergman, L. W., and Trojanowski, J. Q. (1994) Expression of GDNF mRNA in rat and human nervous tissue. *Exp. Neurol.* **127**, 167–170.

73. Strömberg, I., Bjorklund, L., Johansson, M., Tomac, A., Collins, F., Olson, L., Hoffer, B., and Humpel, C. (1993) Glial cell line-derived neurotrophic factor is expressed in the developing but not adult striatum and stimulates developing dopamine neurons *in vivo*. *Exp. Neurol.* **124**, 401–412.

74. Choi-Lundberg, D. L. and Bohn, M. C. (1995) Ontogeny and distribution of glial cell-line derived neurotrophic factor (GDNF) mRNA in rat. *Dev. Brain Res.* **85**, 80–88.

75. Baird, A. (1994) Fibroblast growth factors: activities and significance of non-neurotrophin neurotrophic growth factors. *Curr. Opin. Neurobiol.* **4**, 78–86.

76. Gonzalez, A. M., Berry, M., Maher, P. A., Logan, A., and Baird, A. (1995) A comprehensive analysis of the distribution of FGF-2 and FGFR1 in the adult brain. *Brain Res.* **701**, 201–226.

77. Sasahara, M., Fries, J. W., Raines, E. W., Gowan, A. M., Westrum, L. E., Frosch, M. P., Bonthron, D. T., Ross, R., and Collins, T. (1991) PDGF B-chain in neurons of the central nervous system, posterior pituitary, and in a transgenic model. *Cell* **64**, 217–227.

78. Steiner, J. P., Connelly, M. A., Valentine, H. L., Hamilton, G. S., Dawson, T. M., Hester, L., and Snyder, S. H. (1997) Neurotrophic actions of nonimmunosuppressive analogues of immunosuppressive drugs FK506, rapamycin, and cyclosporin A. *Nature Med.* **3**, 421–428.

79. Lyons, W. E., George, E. B., Dawson, T. M., Steiner, J. P., and Snyder, S. H. (1994) Immunosuppressant FK506 promotes neurites outgrowth in cultures of PC12 cells and sensory ganglia. *Proc. Natl. Acad. Sci. USA* **91**, 3191–3195.

80. Gold, B. G., Katoh, K., and Storm-Dickerson, T. (1995) The immunosuppressant FK506 increases the rate of axonal regeneration in rat sciatic nerve. *J. Neurosci.* **15**, 7509–7516.

81. Steiner, J. P., Dawson, T. M., Fotuhi, M., Glatt, C. E., Snowman, A. M., Cohen, N., and Snyder, S. H. (1992) High brain densities of the immunophilin FKBP colocalized with calcineurin. *Nature* **358**, 584–586.

82. Liu, J., Albers, M. W., Wandless, T. J., Alberg, D. G., Belshaw, P. J., Cohen, P., Mackintosh, C., Klee, C. B., and Schreiber, S. L. (1992) Inhibition of T cell signalling by immunophilin–ligand complexes correlates with loss of calcineurin phosphatase activity. *Biochemistry* **31**, 3896–3901.

83. Wang, M. S., Zeleny-Pooley, M., and Gold, B. G. (1997) Comparative dose-dependence study of FK506 and cyclosporin A on the rate of axonal regeneration in the rat sciatic nerve. *J. Pharmacol. Exp. Ther.* **282**, 1084–1093.

84. Dawson, T. M., Steiner, J. P., Dawson, V. L., Dinerman, J. L., Uhl, G. R., and Snyder, S. H. (1993) Immunosuppressant FK506 enhances phosphorylation of nitric oxide synthase and protects against glutamate toxicity. *Proc. Natl. Acad. Sci. USA* **90**, 9808–9812.

85. Sharkey, J. and Butcher, S. P. (1994) Immunophilins mediate the neuroprotective effects of FK506 in focal cerebral ischaemia. *Nature* **371**, 336–339.

86. Johnson, E. M., Koike, T., and Franklin, J. (1992) The "calcium set-point hypothesis" of neuronal dependence on neurotrophic factor. *Exp. Neurol.* **115**, 163–166.

87. Jayaraman, T., Brillantes, A., Timerman, A. P., Fleischer, S., Erdjument-Bromage, H., Tempst, P., and Marks, A. R. (1992) FK506 binding protein associated with the calcium release channel (ryanodine receptor). *J. Biol. Chem.* **267**, 9474–9477.

88. Cameron, A. M., Steiner, J. P., Roskams, A. J., Ali, S. M., Ronnett, G. V., and Snyder, S. H. (1995) Calcineurin associated with the inositol 1,4,5-trisphosphate receptor-SKBP12 complex modulates Ca^{2+} flux. *Cell* **83,** 463–472.

89. Mattson, M. P. (1996) Calcium and free radicals: mediators of neurotrophic factor and excitatory transmitter-regulated developmental plasticity and cell death. *Perspect. Dev. Neurobiol.* **3,** 79–91.

90. Subramanian, K. and Meyer, T. (1997) Calcium-induced restructuring of nuclear envelope and endoplasmic reticulum calcium stores. *Cell* **89,** 963–971.

91. Hyman, C., Hofer, M., Barde, Y.-A., Juhasz, M., Yancopoulos, G. D., Squinto, S. P., and Lindsay, R. M. (1991) BDNF is a neurotrophic factor for dopaminergic neurons of the substantia nigra. *Nature* **350,** 230–232.

92. Hyman, C., Juhasz, M., Jackson, C., Wright, P., Ip, N. Y., and Lindsay, R. M. (1994) Overlapping and distinct actions of the neurotrophins BDNF, NT-3, and NT-4/5 on cultured dopaminergic and GABAergic neurons of the ventral mesencephalon. *Neuroscience* **14,** 335–347.

93. Hynes, M. A., Poulsen, K., Armanini, M., Berkemeier, L., Phillips, H., and Rosenthal, A. (1994) Neurotrophin-4/5 is a survival factor for embryonic midbrain dopaminergic neurons in enriched cultures. *J. Neurosci. Res.* **37,** 144–154.

94. Lin, L.-F. H., Doherty, D. H., Lile, J. D., Bektesh, S., and Collins, F. (1993) GDNF: a glial cell-line derived neurotrophic factor for midbrain dopaminergic neurons. *Science* **260,** 1130–1132.

95. Knusel, B., Michel, P. P., Schwaber, J. S., and Hefti, F. (1990) Selective and nonselective stimulation of central cholinergic and dopaminergic development in vitro by nerve growth factor, basic fibroblast growth factor, epidermal growth factor, insulin and the insulin-like growth factors I and II. *J. Neurosci.* **10,** 558–570.

96. Mayer, E., Dunnett, S. B., Pellitteri, R., and Fawcett, J. W. (1993) Basic fibroblast growth factor promotes the survival of embryonic ventral mesencephalic dopaminergic neurons-I. Effects in vitro. *Neuroscience* **56,** 379–388.

97. Poulson, K. T., Armanini, M. P., Klein, R. D., Hynes, M. A., Phillips, H. S., and Rosenthal, A. (1994) TGFβ2 and TGFβ3 are potent survival factors for midbrain dopaminergic neurons. *Neuron* **13,** 1245–1252.

98. Knusel, B., Winslow, J. W., Rosenthal, A., Burton, L. E., Seid, D. P., Nikolics, K., and Hefti, F. (1991) Promotion of central cholinergic and dopaminergic neuron differentiation by brain-derived neurotrophic factor but not neurotrophin-3. *Proc. Natl. Acad. Sci. USA* **88,** 961–965.

99. Nikkhah, G., Odin, P., Smits, A., Tingstrom, A., Othberg, A., Brundin, P., Funa, K., and Lindvall, O. (1993) Platelet-derived growth factor promotes survival of rat and human mesencephalic dopaminergic neurons in culture. *Exp. Brain Res.* **92,** 516–523.

100. Engele, J. and Bohn, M. C. (1991) The neurotrophic effects of fibroblast growth factors on dopaminergic neurons in vitro are mediated by mesencephalic glia. *J. Neurosci.* **11,** 3070–3078.

101. Altar, C. A., Boylan, C. B., Jackson, C., Hershenson, S., Miller, J., Wiegand, S. J., Lindsay, R. M., and Hyman, C. (1992) Brain-derived neurotrophic factor augments rotational behavior and nigrostriatal dopamine turnover in vivo. *Proc. Natl. Acad. Sci. USA* **89,** 11,347–11,351.

102. Martin-Iverson, M. T., Todd, K. G., and Altar, C. A. (1994) Brain-derived neurotrophic factor and neurotrophin-3 activate striatal dopamine and serotonin metabolism and related behaviors: interactions with amphetamine. *J. Neurosci.* **14,** 1262–1270.

103. Shen, R.-Y., Altar, C. A., and Chiodo, L. A. (1994) Brain-derived neurotrophic factor increases the electrical activity of pars compacta dopamine neurons in vivo. *Proc. Natl. Acad. Sci. USA* **91,** 8920–8924.

104. Mufson, E. J., Kroin, J. S., Sobreviela, T., Burke, M. A., Kordower, J. H., Penn, R. D., and Miller, J. A. (1994) Intrastriatal infusions of brain-derived neurotrophic factor: retrograde transport and colocalization with dopamine containing substantia nigra neurons in rat. *Exp. Neurol.* **129,** 15–26.

105. Altar, C. A., Boylan, C. B., Fritsche, M., Jackson, C., Hyman, C., and Lindsay, R. M. (1994) The neurotrophins NT-4/5 and BDNF augment serotonin, dopamine, and GABAergic systems during behaviorally effective infusions to the substantia nigra. *Exp. Neurol.* **130,** 31–40.

106. Hudson, J., Granholm, A.-C., Gerhardt, G. A., Henry, M. A., Hoffman, A., Biddle, P., Leela, N. S., Mackerlova, L., Lile, J. D., Collins, F., and Hoffer, B. J. (1995) Glial cell-line derived neurotrophic factor augments midbrain dopaminergic circuits in vivo. *Brain Res. Bull.* **36,** 425–432.

107. Tomac, A., Widenfalk, J., Lin, L.-F. H., Kohno, T., Ebendal, T., Hoffer, B. J., and Olson, L. (1995) Retrograde axonal transport of glial cell-derived neurotrophic factor in the adult nigrostriatal system suggests a trophic role in the adult. *Proc. Natl. Acad. Sci. USA* **92,** 8274–8278.

108. Beck, K. D., Knusel, B., Winslow, J. W., Rosenthal, A., Burton, L. E., Nikolics, K., and Hefti, F. (1992) Pretreatment of dopaminergic neurons in culture with brain-derived neurotrophic factor attenuates toxicity of 1-methyl-4-phenylpyridinium. *Neurodegeneration* **1,** 27–36.

109. Spina, M. B., Squinto, S. P., Miller, J., Lindsay, R. M., and Hyman, C. (1992) Brain-derived neurotrophic factor protects dopamine neurons against 6-hydroxydopamine and *N*-methyl-4-phenylpyridinium ion toxicity: involvement of the glutathione system. *J. Neurochem.* **59,** 99–106.

110. Park, T. H. and Mytilineou, C. (1992) Protection from 1-methyl-4-phenylpyridinium (MPP$^+$) toxicity and stimulation of regrowth of MPP$^+$-damaged dopaminergic fibers by treatment of mesencephalic cultures with EGF and basic FGF. *Brain Res.* **599,** 83–97.

111. Hou, J.-G. G., Lin, L.-F. H., and Mytilineou, C. (1996) Glial cell line-derived neurotrophic factor exerts neurotrophic effects on dopaminergic neurons in vitro and promotes their survival and regrowth after damage by 1-methyl-4-phenylpyridinium. *J. Neurochem.* **66,** 74–82.

112. Krieglstein, K. and Unsicker, K. (1994) Transforming growth factor-β promotes survival or midbrain dopaminergic neurons and protects them against *N*-methyl-4-phenylpyridinium ion toxicity. *Neuroscience* **63,** 1189–1196.

113. Altar, C. A., Boylan, C. B., Fritsche, M., Jones, B. E., Jackson, C., Wiegand, S. J., Lindsay, R. M., and Hyman, C. (1994) Efficacy of brain-derived neurotrophic factor and neurotrophin-3 on neurochemical and behavioral deficits associated with partial nigrostriatal dopamine lesions. *J. Neurochem.* **63,** 1021–1032.

114. Knusel, B., Beck, K. D., Winslow, J. W., Rosenthal, A., Burton, L. E., Widmer, H. R., Nikolics, K., and Hefti, F. (1992) Brain-derived neurotrophic factor administration protects basal forebrain cholinergic but not nigral dopaminergic neurons from degenerative changes after axotomy in the adult rat brain. *J. Neurosci.* **12,** 4391–4402.

115. Lapchak, P. A., Beck, K. D., Araujo, D. M., Irwin, I., Langston, J. W., and Hefti, F. (1993) Chronic intranigral administration of brain-derived neurotrophic factor produces striatal dopaminergic hypofunction in unlesioned adult rats and fails to attenuate the decline of striatal dopaminergic function following medial forebrain bundle transection. *Neuroscience* **53,** 639–650.

116. Hagg, T. (1994) Neurotrophins (BDNF, NT-3, NT-4) prevent death of axotomized adult rat substantia nigra neurons. *Soc. Neurosci. Abstr.* **20,** 441.

117. Frim, D. M., Uhler, T. A., Galpern, W. R., Beal, M. F., Breakefield, X. O., and Isacson, O. (1994) Biologically delivered BDNF increases dopaminergic neuronal survival in a rat model of Parkinson's disease. *Proc. Natl. Acad. Sci. USA* **91,** 5104–5108.

118. Galpern, W. R., Frim, D. M., Tatter, S. B., Altar, C. A., Beal, M. F., and Isacson, O. (1996) Cell-mediated delivery of brain-derived neurotrophic factor enhances dopamine levels in an MPP$^+$ rat model of substantia nigra degeneration. *Cell Transplant.* **5,** 225–232.

119. Levivier, M., Przedborski, S., Bencsics, C., and Kang, U. J. (1995) Intrastriatal implantation of fibroblasts genetically engineered to produce brain-derived neurotrophic factor prevents degeneration of dopaminergic neurons in a rat model of Parkinson's disease. *J. Neurosci.* **15,** 7810–7820.

120. Tomac, A., Lindqvist, E., Lin, L.-F. H., Ögren, S. O., Young, D., Hoffer, B. J., and Olson, L. (1995) Protection and repair of the nigrostriatal dopaminergic system by GDNF in vivo. *Nature* **373,** 335–339.

121. Beck, K. D., Valverde, J., Alexi, T., Poulsen, K., Moffat, B., Vandlen, R. A., Rosenthal, A., and Hefti, F. (1995) Mesencephalic dopaminergic neurons protected by GDNF from axotomy-induced degeneration in the adult brain. *Nature* **373,** 339–341.

122. Kearns, C. M. and Gash, D. M. (1995) GDNF protects nigral dopamine neurons against 6-hydroxydopamine in vivo. *Brain Res.* **672,** 104–111.

123. Sauer, H., Rosenblad, C., and Björklund, A. (1995) Glial cell line-derived neurotrophic factor but not transforming growth factor ß3 prevents delayed degeneration of nigral dopaminergic neurons following striatal 6-hydroxydopamine lesion. *Proc. Natl. Acad. Sci. USA* **92,** 8935–8939.

124. Hoffer, B. J., Hoffman, A., Bowenkamp, K., Huettl, P., Hudson, J., Martin, D., Lin, L.-F. H., and Gerhardt, G. A. (1994) Glial cell-line derived neurotrophic factor reverses toxin-induced injury to midbrain dopaminergic neurons in vivo. *Neurosci. Lett.* **182,** 107–111.

125. Bowenkamp, K. E., Hoffman, A. F., Gerhardt, G. A., Henry, M. A., Biddle, P. T., Hoffer, B. J., and Granholm, A.-C. E. (1995) Glial cell-line derived neurotrophic factor supports survival of injured midbrain dopaminergic neurons. *J. Comp. Neurol.* **355,** 479–489.

126. Sauer, H., Wong, V., and Björklund, A. (1995) Brain-derived neurotrophic factor and neurotrophin-4/5 modify neurotransmitter-related gene expression in the 6-hydroxydopamine-lesioned rat striatum. *Neuroscience* **65,** 927–933.

127. Hagg, T. and Varon, S. (1993) Ciliary neurotrophic factor prevents degeneration of adult rat substantia nigra dopaminergic neurons in vivo. *Proc. Natl. Acad. Sci. USA* **90,** 6315–6319.

128. Costantini, L. C., Chaturvedi, P., Armistead, D. M., McCaffrey, P. G., Deacon, T. W., and Isacson, O. (1998) A novel immunophilin ligand: distinct branching effects on dopaminergic neurons in culture and neurotrophic actions after oral administration in an animal model of Parkinson's disease. *Neurobiol. Dis.* **5,** 97–106.

129. Steiner, J. P., Hamilton, G. S., Ross, D. T., Valentine, H. L., Guo, H., Connolly, M. A., Liang, S., Ramsey, C., Li, J. J., Huang, W., Howorth, P., Soni, R., Fuller, M., Sauer, H., Nowotnik, A. C., and Suzdak, P. D. (1997) Neurotrophic immunophilin ligands stimulate structural and functional recovery in neurodegenerative animal models. *Proc. Natl. Acad. Sci. USA* **94,** 2019–2024.

130. Gold, B. G., Zeleny-Pooley, M., Wang, M. S., Chaturvedi, P., and Armistead, D. M. (1997) A non-immunosuppressive FKBP-12 ligand increases nerve regeneration. *Exp. Neurol.* **147,** 269–278.

131. Aoyagi, A., Nishikawa, K., Saito, H., and Abe, K. (1994) CHaracterization of basic fibroblast growth factor-mediated acceleration of axonal branching in cultures rat hippocampal neurons. *Brain Res.* **661,** 117–126.

132. Beck, K. D., Knusel, B., and Hefti, F. (1993) The nature of the trophic action of brain-derived neurotrophic factor, sed(1-3)-insulin-like growth factor-1, and basic fibroblast growth factor on mesencephalic dopaminergic neurons developing in culture. *Neuroscience* **52,** 855–866.

133. Studer, L., Spenger, C., Seiler, R. W., ALtar, C. A., Lindsay, R. M., and Hyman, C. (1995) Comparison of the effects of the neurotrophins on the morphological structure of dopaminergic neurons in cultures of rat substantia nigra. *Eur. J. Neurosci.* **7,** 223–233.

134. Hemmendinger, L. M., Garber, B. B., Hoffmann, P. C., and Heller, A. (1981) Target neuron-specific process formation by embryonic mesencephalic dopamine neurons in vitro. *Proc. Natl. Acad. Sci. USA* **78,** 1264–1268.
135. Rousselet, A., Felter, L., Chamak, B., and Prochiantz, A. (1988) Rat mesencephalic neurons in culture exhibit different morphological traits in the presence of media conditioned on mesencephalic or striatal astroglia. *Dev. Biol.* **129,** 495–504.
136. Bixby, J. L. and Jhabvala, P. (1990) Extracellular matrix molecules and cell adhesion molecules induce neurites through different mechanisms. *J. Cell Biol.* **111,** 2725–2732.
137. Mattson, M. P. and Kater, S. B. (1987) Calcium regulation of neurite elongation and growth cone motility. *J. Neurosci.* **7,** 4034–4043.
138. Kitamura, Y., Itano, Y., Kubo, T., and Nomura, Y. (1994) Suppressive effect of FK506, a novel immunosuppressant, against MPTP-induced dopamine depletion in the striatum of young C57BL6 mice. *J. Neuroimmunol.* **50,** 221–224.
139. Lindsay, R. M., Altar, C. A., Cedarbaum, J. M., Hyman, C., and Wiegand, S. J. (1993) The therapeutic potential of neurotrophic factors in the treatment of Parkinson's disease. *Exp. Neurol.* **124,** 103–118.
140. Hefti, F. (1994) Neurotrophic factor therapy for nervous system degenerative diseases. *J. Neurobiol.* **25,** 1418–1435.
141. Jelsma, T. N. and Aguayo, A. J. (1994) Trophic factors. *Curr. Opin. Neurobiol.* **4,** 717–725.
142. Morse, J. K., Wiegand, S. J., Anderson, K., You, Y., Cai, N., Carnahan, J., Miller, J., DiStefano, P. S., Altar, C. A., Lindsay, R. M., and Alderson, R. F. (1993) Brain-derived neurotrophic factor (BDNF) prevents the degeneration of medial septal cholinergic neurons following fimbria transection. *J. Neurosci.* **13,** 4146–4156.
143. Kordower, J. H., Mufson, E. J., Granholm, A.-C., Hoffer, B., and Friden, P. M. (1994) Delivery of trophic factors to the primate brain. *Exp. Neurol.* **124,** 21–30.
144. Friden, P. M., Walus, L. R., Watson, P., Doctrow, S. R., Kozarich, J. W., Bäckman, C., Bergman, H., Hoffer, B., Bloom, F., and Granholm, A.-C. (1993) Blood–brain barrier penetration and in vivo activity of an NGF conjugate. *Science* **259,** 373–377.
145. Kordower, J. H., Charles, V., Bayer, R., Bartus, R. T., Putney, S., Walus, L. R., and Friden, P. M. (1994) Intravenous administration of a transferrin receptor antibody-nerve growth factor conjugate prevents the degeneration of cholinergic neurons in a model of Huntington's disease. *Proc. Natl. Acad. Sci. USA* **91,** 9077–9080.
146. Gage, F. H., Wolff, J. A., Rosenberg, M. B., Xu, L., Yee, J.-K., Shults, C., and Friedmann, T. (1987) Grafting genetically modified cells to the brain: possibilities for the future. *Neuroscience* **23,** 795–807.
147. Rosenberg, M. B., Friedmann, T., Robertson, R. C., Tuszynski, M., Wolff, J. A., Breakefield, X. O., and Gage, F. H. (1988) Grafting genetically modified cells to the damaged brain: restorative effects of NGF expression. *Science* **242,** 1575–1578.
148. Schumacher, J. M., Short, M. P., Hyman, B. T., Breakefield, X. O., and Isacson, O. (1991) Intracerebral implantation of nerve growth factor-producing fibroblasts protects striatum against neurotoxic levels of excitotoxic levels of excitatory amino acids. *J. Neurosci.* **45,** 561–570.
149. Kawaja, M. D., Rosenberg, M. B., Yoshida, K., and Gage, F. H. (1992) Somatic gene transfer of nerve growth factor promotes the survival of axotomized septal neurons and the regeneration of their axons in adult rats. *J. Neurosci.* **12,** 2849–2864.
150. Winn, S. R., Hammang, J. P., Emerich, D. F., Lee, A., Palmiter, R. D., and Baetge, E. E. (1994) Polymer-encapsulated cells genetically modified to secrete human nerve growth factor promote the survival of axotomized septal cholinergic neurons. *Proc. Natl. Acad. Sci. USA* **91,** 2324–2328.
151. Palmer, T. D., Rosman, G. J., Osborne, W. R. A., and Miller, A. D. (1991) Genetically modified skin fibroblasts persist long after transplantation but gradually inactivate introduced genes. *Proc. Natl. Acad. Sci. USA* **88,** 1330–1334.

152. Gossen, M. and Bujard, H. (1992) Tight control of gene expression in mammalian cells by tetracycline-responsive promoters. *Proc. Natl. Acad. Sci. USA* **89,** 5547–5551.

153. Sabaté, O., Horellou, P., Vigne, E., Colin, P., Perricaudet, M., Buc-Caron, M.-H., and Mallet, J. (1995) Transplantation to the rat brain of human neural progenitors that were genetically modified using adenovirus. *Nat. Genet* **9,** 256–260.

154. Martinez-Serrano, A., Lundberg, C., Horellou, P., Fischer, W., Bentlage, C., Campbell, K., McKay, R. D. G., Mallet, J., and Björklund, A. (1995) CNS-derived neural progenitor cells for gene transfer of nerve growth factor to the adult brain: complete rescue of axotomized cholinergic neurons after transplantation into the septum. *J. Neurosci.* **15,** 5668–5680.

155. Breakefield, X. O. and DeLuca, N. A. (1991) Herpes simplex for gene delivery to neurons. *New Biologist* **3,** 203–218.

156. Le Gal La Salle, G., Robert, J. J., Berrard, S., Ridoux, V., Stratford-Perricaudet, L. D., Perricaudet, M., and Mallet, J. (1993) An adenovirus vector for gene transfer into neurons and glia in the brain. *Science* **259,** 988–990.

157. Horellou, P., Vigne, E., Castel, M.-N., Barnéoud, P., Colin, P., Perricaudet, M., Delaère, P., and Mallet, J. (1994) Direct intracerebral gene transfer of an adenoviral vector expressing tyrosine hydroxylase in a rat model of Parkinson's disease. *NeuroReport* **6,** 49–53.

158. Kaplitt, M. J., Leone, P., Samulski, R. J., Xiao, X., Pfaff, D. W., O'Malley, K. L., and During, M. J. (1994) Long-term gene expression and phenotypic correction using adeno-associated virus vectors in the mammalian brain. *Nat. Genet.* **8,** 148–154.

159. Vejsada, R., Sagot, Y., and Kato, A. C. (1994) BDNF-mediated rescue of axotomized motor neurones decreases with increasing dose. *NeuroReport* **5,** 1889–1892.

160. Altar, C. A., Boylan, C. B., Fritsche, M., Jones, B. E., Jackson, C., Weigand, S. J., Lindsay, R. M., and Hyman, C. (1994) Efficacy of brain-derived neurotrophic factor and neurotrophin-3 on neurochemical and behavioral deficits associated with partial nigrostriatal dopamine lesions. *J. Neurochem.* **63,** 1021–1032.

161. Galpern, W. R., Frim, D. M., Tatter, S. B., Altar, C. A., Beal, M. F., and Isacson, O. (1996) Cell-mediated delivery of brain-derived neurotrophic factor enhances dopamine levels in an MPP$^+$ rat model of substantia nigra degeneration. *Cell Transplant.* **5,** 225–232.

162. Frim, D. M., Uhler, T. A., Galpern, W. R., Beal, M. F., Breakefield, X. O., and Isacson, O. (1994) Implanted fibroblasts genetically engineered to produce brain-derived neurotrophic factor prevent 1-methyl-4-phenylpyridinium toxicity to dopaminergic neurons in rat. *Proc. Natl. Acad. Sci. USA* **91,** 5104–5108.

163. Tomac, A., Lindquist, E., Lin, L. F., Ogren, S. O., Young, D., Hoffer, B. J., and Olson, L. (1995) Protection and repair of the nigrostriatal dopaminergic system by GDNF in vivo. *Nature* **373,** 335–339.

164. Kearns, C. M. and Gash, D. M. (1995) GDNF protects nigral dopamine neurons against 6-hydroxydopamine in vivo. *Brain Res.* **672,** 104–111.

165. Hoffer, B. J., Hoffman, A., Bowenkamp, K., Huettl, P., Hudson, J., Martin, D., Lin, L. F., and Gerhardt, G. A. (1994) Glial cell-line derived neurotrophic factor reverses toxin-induced injury to midbrain dopaminergic neurons in vivo. *Neurosci. Lett.* **182,** 107–111.

166. Hou, J. G., Lin, L. H., and Mytilineou, C. (1996) Glial cell-line derived neurotrophic factor exerts neurotrophic effects on dopaminergic neurons in vitro and promotes their survival and regrowth after damage by 1-methyl-4-phenylpyridinium. *J. Neurochem.* **66,** 74–82.

167. Sauer, H., Rosenblad, C., and Bjorklund, A. (1995) Glial cell-line derived neurotrophic factor but not transforming growth factor B3 prevents delayed degeneration of nigral dopaminergic neurons following striatal 6-hydroxydopamine lesion. *Proc. Natl. Acad. Sci. USA* **92,** 8935–8939.

168. Kreiglstein, K., Suter-Crazzolara, C., Hotten, G., Pohl, J., and Unsicker, K. (1995) Trophic and protective effects of growth/differentiation factor 5, a member of the transforming growth factor-beta superfamily, on midbrain dopaminergic neurons. *J. Neurosci. Res.* **42,** 724–732.
169. Backman, C., Rose, G. M., Hoffer, B. J., Henry, M. A., Bartus, R. T., Friden, P., and Granholm, A. C. (1996) Systemic administration of nerve growth factor conjugate reverses age-related cognitive dysfunction and prevents cholinergic neuron atrophy. *J. Neurosci.* **16,** 5437–5442.

12

Antisense Knockdown of Dopamine Receptors

Simranjit Kaur and Ian Creese

1. INTRODUCTION

Dopamine is one of the more extensively studied neurotransmitters in the central nervous system (CNS) and our knowledge of this neurotransmitter as well as its functions, both physiological and pathological, has continued to expand since the 1950s when it first came to the fore. Dopamine is known to mediate a myriad of functions in the CNS ranging from movement to emotion as well as being involved in a number of pathologies, including Parkinson's disease (PD), schizophrenia, Huntington's disease, and tardive dyskinesia (1–8).

Pathways in the CNS that utilize dopamine include the nigrostriatal pathway (A9), which originates in the substantia nigra pars compacta and projects to the dorsal striatum (9–12). It is primarily the degeneration of this pathway and the resulting large-scale depletion of dopamine that leads to the symptomatology of PD (1–2,13). PD is a progressive neurodegenerative disorder characterized mainly by bradykinesia, rigidity, and tremor (14). The dopamine precursor, L-Dopa, which is the mainstay treatment of PD, acts to increase dopamine levels in the remaining neurons of this pathway, and thus to alleviate the Parkinsonian symptoms (1).

The other major dopaminergic pathway is the mesocorticolimbic pathway (A10), which originates in the ventral tegmental area, and innervates limbic structures such as the ventral striatum; nucleus accumbens; the olfactory tubercle; and the medial prefrontal, cingulate, and entorhinal cortices. This pathway is thought to be involved in emotion and learning and memory. Alterations in the mesolimbic pathway may underlie the etiology of schizophrenia and may be where antipsychotic drugs act (3,5,15–16).

This laboratory has been investigating the use of antisense technology, a novel way to arrest gene expression, in determining the roles played by the individual dopamine receptors in behavior and in the responses produced by conventionally used dopaminergic agents (17–21). In this chapter, we discuss some of the recent dopamine antisense studies in the context of PD.

2. INTRODUCTION TO ANTISENSE KNOCKDOWN STRATEGY

Antisense technology has been coming into its own in the last few years as the inhibition of gene expression has increasingly been used in the investigation of the

From: Central Nervous System Diseases
Edited by: D. F. Emerich, R. L. Dean, III, and P. R. Sanberg © Humana Press Inc., Totowa, NJ

physiological and pathological functions of a number of neurotransmitters as well as in the therapy of conditions such as cancers *(22)* and as antiviral agents, including in AIDS *(23,24)*. The premise behind using antisense technology is that the concept behind it is simple and that, theoretically, it is a selective tool with which to target genes and thus specific functional proteins.

Antisense oligodeoxynucleotides are sequence-specific molecules of synthetic single-stranded DNA and are usually from 15 to 30 bases long. The oligodeoxynucleotides are designed to be complementary and selective to the chosen sequences within the targeted messenger RNA (mRNA) and "hybridize" to these sequences by nucleic acid basepairing. This, in turn, prevents the translation of the mRNA to the encoded protein molecule. The mechanisms by which antisense oligodeoxynucleotides inhibit gene expression include interfering with ribosome binding and processing of mRNA, interfering with mRNA conformation or mRNA splicing, and ribonuclease-H (RNase-H) activation of mRNA digestion (for review *see 25*) but the exact mechanisms involved are as yet unclear, especially in the CNS. In their unmodified form, these oligodeoxynucleotides can be easily synthesized and made in reasonable amounts and of good purity using automated methods *(25)*.

Antisense technology was initially used for the purposes of blocking viral replication or tumor cell activity (for review *see 26*) and has been used extensively in in vitro studies, less so in in vivo studies. This technology has also recently progressed to investigations into the CNS and has been used to arrest the synthesis of various receptors such as dopamine, neuropeptide, *N*-methyl-D-aspartate, and vasopressin *(17,21,27–32)*, G-proteins *(33)*, proteins such as SNAP-25 and neuropeptide Y *(34–38)*, enzymes such as tyrosine hydroxylase *(35,39,40)*, and early-onset genes such as c-*fos (41–43)*.

Another point in favor of the antisense knockdown as opposed to transgenic knockout strategy is that local infusion of the antisense oligodeoxynucleotides can be used selectively to target pre- or postynaptic receptors in a given brain region using local infusions. Also, a major difficulty in using transgenics is that developmental changes may occur to compensate for a reduction in the number of receptors. These changes may have a far-reaching effect on steps downstream of the receptor activation and possibly also on associated neurotransmitter systems, whereas antisense knockdown can be applied in the mature, normally developed CNS.

2.1. Practical Considerations in the Use of Antisense Technology

A number of factors should be considered to ensure optimal effectiveness of the antisense oligodeoxynucleotides. These include their stability at body temperature, uptake into cells, the specificity to the target mRNA, the use of unmodified vs modified oligodeoxynucleotides for added stability, the route of administration, and the length of the sequence so as to reduce the incidence of non-sequence-specific toxic effects and to ensure specificity.

2.1.1. The Design of Antisense Oligodeoxynucleotides

A number of chemical classes of oligodeoxynucleotides are available, and these include the unmodified phosphodiesters, negatively charged phosphorothioates, the nonionic methylphosphonates *(40)*, conjugated oligodeoxynucleotides, the α-oligomers, and the end-capped oligodeoxynucleotides with phosphorothioate groups *(26,44–45)*. There are a number of advantages and disadvantages inherent in using some of these types as opposed to others. For example, the natural phosphodiesters are sensitive to degradation by nucleases *(44,46)* but have less propensity to produce toxic effects whereas the chemically modified phosphorothioates have enhanced nuclease resistance but are liable to have non-sequence-specific effects *(39,47,48)*. Other factors that need to be taken into account are the affinity or the melting temperature of the oligodeoxynucleotides, which should be high, and the potential for the formation of secondary structures such as a hairpin within the antisense molecule itself, which would reduce its ability to hybridize with the target sequence, and this should be minimal *(49)*.

The unmodified as well as the chemically modified oligodeoxynucleotides enter the cell either by receptor-mediated endocytosis or nonselective pinocytosis after binding to a 80-kDa membrane-associated protein *(50–52)*. RNase-H, which degrades the RNA strand of the RNA–DNA duplex formed in the hybridization process, is thought to be involved in the inhibition of gene expression. Another possibility is that the RNA–DNA duplex may block the binding and translocation of ribosomes along the unhybridized mRNA and thus prevent the synthesis of the encoded protein. Antisense oligodeoxynucleotides that have greater affinity for the initiation codon of a mRNA strand often produce more effective knockdown than those that bind upstream or downstream of this codon.

Thus, the treatment regimen and paradigm is of great importance in choosing the type of oligodeoxynucleotides to be used. The phosphorothioates have been found to be the most stable of the different types of oligonucleotides in terms of exonuclease and endonuclease sensitivity *(53)*. Chemical modifications as well as groups linked at the end will alter the uptake and the intracellular distribution of oligodeoxynucleotides into intact cells. As an example, when the pharmacokinetics of phosphorothioates were compared with those of the phosphodiesters after infusion into the amygdala, the phosphorothioates were stable and had better cell penetrability whereas the phosphodiesters had better tissue retention *(47,54–55)*. Also hydrophobic groups and polycations, such as polylysine, are linked to antisense oligodeoxynucleotides to increase cellular uptake *(26,56)*. In this laboratory, we utilize mainly the "S"-modified phosphorothioate oligodeoxynucleotides, which have enhanced nuclease resistance and are taken up by neurons *(53,57)*.

2.1.2. The Limitations of Antisense Technology

Along with the obvious advantages of using antisense technology, such as specificity, there are a number of disadvantages as well, especially as it stands currently. Owing to poor blood–brain barrier penetrability and potential toxic side effects, including decreased blood clotting and cardiovascular problems, such as increased blood pressure and decreased heart rate *(58)*, it is unlikely that antisense technology will cross over successfully into the clinic as yet. Direct administration into the CNS is required at present. The nuclease-resistant phosphorothioate oligodeoxynucleotides have been shown to have non-sequence-specific detrimental effects, but progress is being made and it is likely that more specific, centrally acting antisense oligodeoxynucleotides may be developed that can cross the blood–brain barrier relatively easily in the not so distant future.

Another possible problem is the fact that the knockdown achieved by the antisense oligodeoxynucleotides is seldom complete, and owing to the pharmacological phenomenon of "spare receptors" it may be difficult to reconcile the knockdown obtained with any corresponding functional deficits. However, studies done previously with *N*-ethoxycarbonyl-2-ethoxy-1,2-dihydroquinoline (EEDQ), an irreversible dopamine receptor antagonist, showed that 70–90% receptor occupation may often be needed to produce a full behavioral response *(59–61)*.

To quantify the receptor knockdown, techniques such as homogenate radioligand binding assays or autoradiography are often utilized, especially if there are no obvious behavioral deficits, but a problem arises if there are no selective ligands that bind to the receptor in question.

3. CURRENTLY USED MODELS OF PD

The best model of PD to date, is the 1-methyl-4-phenyl-1,2,3,6-tetrahydropyridine (MPTP)-lesioned marmoset. The neurotoxicity produced by 1-methyl-4-phenylpyridinium ion (MPP^+), a metabolite of MPTP, is thought to mimic human PD. MPTP reduces the levels of dopamine and its metabolites in the striatum *(62,63)*. However, this model is far from ideal owing to the high cost of using primates (MPTP is ineffective in rats) and unlike human PD, which is progressive, the neurotoxic damage produced by MPTP is reversible.

Other models used to mimic PD are: (1) lesions produced by the selective neurotoxin 6-hydroxydopamine (6-OHDA) in rats; (2) akinesia induced by reserpine via dopamine depletion; (3) catalepsy induced by neuroleptics by blocking dopamine receptors; and (4) lesions of dopaminergic terminals produced by the systemic administration of very high doses of amphetamines. Each of these models, although serving a purpose in PD research, does not reproduce the human condition fully. For example, the 6-OHDA-lesioned rat model does not always look overtly Parkinsonian, and with unilateral lesions the only indication of cell damage is the presence of rotations after treatment with antiparkinsonian agents. The depletions produced by reserpine are not restricted to dopamine and there is no destruction of neurons, again unlike human PD. Also, the akinesia produced by reserpine is reversible *(64)*. In these models, it is difficult to ascertain the roles played by the individual dopamine receptors in the pathology as well as in the therapy of PD.

Thus, there is a need for a dopamine receptor subtype-specific method to address the question as to which dopamine receptors are involved in PD, and antisense technology appears a likely candidate to fill this need.

4. THE APPLICATION OF ANTISENSE TECHNOLOGY TO DOPAMINE RECEPTOR RESEARCH

In the past, dopamine receptors were classified into two subtypes based on their biochemistry, anatomical localization, and pharmacology: (1) dopamine D_1 receptors, located pre- and postsynaptically, which act by stimulating adenylyl cyclase via the G-protein, G_s *(65)* and (2) dopamine D_2 receptors, located pre- and postsynaptically, which inhibit adenylyl cyclase via the G-protein G_i *(66)*. However, recent molecular cloning studies have demonstrated the existence of at least five distinct dopamine receptors: the D_1 *(67)*, D_2 *(68)*, D_3 *(69)*, D_4 *(70)*, and D_5 *(71)* receptors.

These five receptors have been subdivided into two major families; the dopamine D_1-like receptor family comprising the D_1 and D_5 receptors and the dopamine D_2-like receptor family, which comprises the D_2, D_3, and D_4 receptors *(72,73)*. Within these families, there is some homology in the amino acid sequences, G-protein associations, and similarities in pharmacological affinities to justify this classification.

Owing to the lack of selective drugs that bind to the individual dopamine receptors, functions of these receptors cannot be ascertained conclusively, especially as the localization of the receptors overlaps to a certain extent. Thus, antisense technology is, in theory, a novel and ideal tool to target the individual receptors and thus investigate the functional roles of these receptors. Also, unlike pharmacological interventions that induce up-regulation of the receptors being investigated, antisense treatment is thought to be devoid of this problematic consequence *(74)*.

4.1. Treatment Protocols in Dopamine Antisense Experiments

The efficiency of the antisense dopamine receptor knockdown has been reported to be dependent on duration of administration and on the dose administered *(18,75)*. Appropriate controls must also be chosen so as to ensure that the knockdown seen is specific. Controls used commonly include mismatched oligodeoxynucleotides, in which a few bases are mismatched; "random" oligodeoxynucleotides, which contain the same bases as the antisense oligodeoxynucleotides but are in a random order; or the "sense" oligodeoxynucleotides, but these are problematic as they may hybridize to a specific sequence to which they are the antisense sequence *(76,77)*. In the studies done in this laboratory, random oligodeoxynucleotides have been utilized and have been found to be devoid of any receptor-specific effects. The amino acid sequences of the controls should be checked in GenBank to ensure that these sequences do not act as antisense oligodeoxynucleotides to other known sequences, but it should be kept in mind that until the complete genome is sequenced, "nonspecific" hybridization may occur to unknown sequences.

4.1.1. Optimal Dose of the Oligodeoxynucleotides

The dose of the oligodeoxynucleotides to be administered must be chosen with great care, as high doses of the antisense oligodeoxynucleotides may result in a higher incidence of hybridization to other related sequences, therefore curtailing specificity and increasing toxicity.

A study done in this laboratory investigated the effect of a range of concentrations (1–20 μg/μL administered at a rate of 1 μL/h) of the D_2 antisense oligodeoxynucleotide over a period of 3 d. The maximal D_2 receptor knockdown (58%) occurred at the highest dose administered (20 μg/μL) without any significant accompanying toxicity but this was not much higher than the knockdown observed at 10 μg/μL. Only a 14% reduction in receptor number was achieved with 1 μg/μL *(78)*.

4.1.2. The Optimal Route and Duration of Administration

Oligodeoxynucleotides are either administered as chronic injections or infusions through microosmotic pumps, focally into brain regions, or into the lateral cerebral ventricles *(17,79,80)*, as systemically administered oligodeoxynucleotides cannot cross the blood–brain barrier. Another reason why chronic administration is desirable is that the rate of protein turnover may be slow. Antisense oligodeoxynucleotides inhibit the synthesis of new receptors but have no effect on the receptors already present. For our purposes, chronic infusions via microosmotic pumps are ideal, as the half-life of dopamine D_2 receptors is usually 45–160 h depending on the conditions of the study *(81–83)*.

Although the oligodeoxynucleotides are inefficient at crossing the blood–brain barrier, focal administration into or close to the brain structures being investigated or continuous infusion into the cerebral ventricles results in adequate uptake into the neurons *(17,27–28,41,80)*.

The reductions in receptor number appear to increase with increasing duration of administration. Zhou et al. *(84)* reported that while dopamine D_2 receptors were significantly reduced after 1 d of repeated injection, the knockdown was almost complete after 6 d of treatment. Receptor numbers only recovered approx 2 d after the cessation of the antisense treatment. Our laboratory has shown increasing catalepsy with increasing duration of antisense infusion. D_2 antisense was infused into the cerebral ventricles or rats for 3 d and catalepsy was maximal on the third day of treatment *(17)*. Tepper et al. *(21)* administered D_2 or D_3 antisense oligodeoxynucleotides into the substantia nigra of rats over a duration of 3–6 d and found that the knockdown was maximal at 3 d but spontaneous rotations were observed to begin after 24 h of infusion.

4.2. Antisense Knockdown of Dopamine D_2 Receptors

4.2.1. Antisense Knockdown of Postsynaptic D_2 Receptors

The functional roles of the D_2 receptors have been investigated with antisense technology more so than those of the other dopamine receptors. D_2 receptors are abundant in the striatum and limbic regions, such as the nucleus accumbens and olfactory tubercle, and are thought to be involved in movement and emotion. D_2 mRNA is also expressed by enkephalin-containing neurons and by cholinergic neurons *(68,72,85,86)*. There are no existing fully selective agents for the D_2 receptors; as drugs acting at the D_2 receptor also have affinity for the D_3 and D_4 receptors, thus a novel way to determine the functional roles of these receptors is by the use of antisense technology. Successful D_2 receptor knockdown has been achieved in a number of studies carried out in the rodent brain by this laboratory and others *(17,75,79,84,87)*.

In vivo administration of D_2 antisense inhibited only the behaviors mediated by D_2 receptors and not those mediated by the other dopamine receptors or by muscarinic receptors *(17,75,84,87,88)*. D_2 antisense administered in vivo to unilaterally striatally 6-OHDA-lesioned mice reduced the response to the D_2-like receptor agonist quinpirole *(79,84)*. Zhou et al. *(84)* reported that the administration of D_2 antisense inhibited the rotations induced by quinpirole but had no effect on the rotations produced by a D_1-like agonist, SKF 38393, or a muscarinic receptor agonist, oxotremorine, in unilaterally 6-OHDA striatally lesioned mice.

An early study done by this laboratory infused a D_2 antisense 19-base S-oligodeoxynucleotide (10 μg/μL), corresponding to codons 2–8 of the D_2 mRNA (5′-AGGACAG GTTCAGTGGATC-3′) *(68)*, unilaterally into the lateral ventricle via subcutaneously implanted osmotic minipumps over a 3 d period. As the control, a random oligodeoxynucleotide was used, also 19-base S-modified at the same dose with the same base composition but in a randomized order (5′-AGAACGGCACTTAGTGGGT-3′). A 48% reduction in the number of D_2 receptors was demonstrated by saturation analysis after the antisense treatment with the largest decreases occurring in the nucleus accumbens (72% reduction) compared to the striatum (approx 50% reduction). To determine if this reduction in receptor number had functional consequences, catalepsy and locomotor activity were measured. A time-dependent increase in the cataleptic response and a reduction in spontaneous motor activity was observed over a 3 d administration period. Locomotor activity after treatment with quinpirole, a D_2-like agonist, was also decreased in the antisense-treated animals whereas responses to the D_1-like agonist, SKF 38393 were not affected. These data show that the knockdown was specific to D_2 receptors *(17)*. A more recent study done by this group has corroborated these findings *(75)*. In this study, the involvement of D_2 receptors was investigated in the response to amphetamine in rats, and it was observed that D_2 knockdown had no effect on the stereotypy response produced by high doses of amphetamine while reducing the locomotor response to quinpirole. This finding was ascribed to the mechanism of action of amphetamine, that is, it increases dopamine release, which also acts at D_1 receptors in addition to D_2 receptors.

The administration of D_2 antisense oligodeoxynucleotides into the lateral ventricle as chronic injections given every 12 h for up to 8 d produced a marked inhibition of D_2 receptor-mediated behaviors but only a small reduction in the number of dopamine D_2 receptors in the mouse striatum *(88)*. Administering the D_2 antisense after a irreversible D_2 receptor antagonist, fluphenazine-*N*-mustard (FNM), showed that the D_2 antisense reduced the rate of recovery of D_2 receptors in the mouse striatum as well as the restoration of normal motor activity after cessation of FNM treatment. In another study a D_2 receptor antisense oligonucleotide was administered to a primary culture of rat pituitary cells and was reported to result in a reduced number of D_2 receptors and also to prevent the inhibition of adenylyl cyclase by the D_2 agonist, bromocriptine, as well as reducing prolactin mRNA levels *(89)*. Again, these effects were specific only to the D_2 antisense, with the random oligodeoxynucleotide producing no such effects. D_2 antisense treatment, as in the previously mentioned studies, inhibited or reduced the synthesis of functional dopamine D_2 receptors.

The effects of site-specific D_2 antisense administration have also been investigated. D_2 antisense was either injected or infused unilaterally into the striata of rats and was found to induce ipsilateral rotations after apomorphine or quinpirole treatment *(78)*. These data strengthen the hypothesis that postsynaptic dopamine D_2 receptors are involved in motor activity. In a similar study, Rajakumar et al. *(90)* infused D_2 antisense bilaterally into the striatum and observed a reduction the stereotypic sniffing produced by high-dose apomorphine but had no effect on the behaviors seen after low-dose apomorphine administration. D_2 mRNA levels were found to be normal in the striatum and the substantia nigra but reduced D_2 receptor binding was seen, so the authors concluded that the response to high-dose apomorphine occurs via interactions at postsynaptic D_2 receptors whereas the behaviors induced by low-dose apomorphine are due to effects at the presynaptic D_2 receptors.

The observations made after D_2 knockdown correspond to earlier findings using D_2 receptor antagonists *(91,92)* or in D_2 receptor-deficient transgenic mice *(93)*. The D_2 knockout transgenic animals display akinesia and have significantly reduced spontaneous activity. Thus, it is conceivable that D_2 receptor knockdown may be used to model some of the motor deficits seen in Parkinsonian patients.

In PD, there is thought to be an initial up-regulation of dopamine receptors, including D_2 receptors, but as the disease progresses and further dopaminergic neurons are lost, the density of these receptors decreases in the caudate *(94–96)*. Thus, administering D_2 antisense continuously could prove to be useful in trying to ascribe functions to the dopamine D_2 receptors. Also, catalepsy and/or a reduction in spontaneous locomotor activity, which are taken to be Parkinsonian symptoms, are observed after D_2 antisense treatment.

4.2.2. Antisense Knockdown of Presynaptic D_2 Receptors

Dopamine neurons in the substantia nigra express D_2 receptor mRNA and show D_2 receptor binding and electrophysiological effects *(97–99)*. Dopamine agonists are reported to inhibit dopamine neuron firing *(100–102)*. Studies investigating the roles of dopamine D_2 autoreceptors have been done by Tepper et al. *(21)* using D_2 antisense infused directly into the substantia nigra over a period of 3–6 d. Using autoradiography, it was determined that there was a reduction in ipsilateral nigral D_2 receptors. The D_2 antisense oligodeoxynucleotide decreased the inhibitory effect of dopamine on nigral dopamine cell firing produced by apomorphine. Another finding obtained in this study was that D_2 autoreceptor knockdown increased somatodendritic and terminal excitability without affecting striatally induced inhibition of dopamine neurons, thus showing that D_2 autoreceptors are expressed by nigrostriatal neurons at somatodendritic and axon terminal areas.

To investigate the role of D_2 autoreceptors in the behavioral response to cocaine, Silvia et al. *(87)* administered D_2 antisense oligodeoxynucleotides unilaterally into the substantia nigra directly to target the autoreceptors for several days which resulted in contralateral rotations after cocaine administration. However, no alterations were seen in postsynaptic striatal D_2 receptors. The ability of sulpiride, a D_2 antagonist, to increase electrically stimulated dopamine release was also markedly reduced, without any effect on basal striatal dopamine release, which the authors reasoned is consistent with a decrease in the striatal D_2 autoreceptor number. Using autoradiography, it was ascertained that there is an approx 40% reduction in the nigral D_2 receptor population when compared to the untreated side. The authors suggest that nigrostriatal D_2 autoreceptors play a direct role in the reduction of the motor response to cocaine, and the absence of spontaneous rotation in antisense-treated animals suggests that autoreceptor effects may be compensated for during normal behavior.

Histochemistry studies have shown that tyrosine hydroxylase (TH), the enzyme involved in dopamine synthesis, is increased in the substantia nigra after D_2 antisense treatment *(78)*, indicating increased dopaminergic activity after D_2 receptor knockdown.

These findings again bear out the data obtained in other models of PD, that there is increased dopaminergic activity early in the disease, as compensatory mechanisms come into play. Therefore, there is a reduction in the activity of the D_2 autoreceptors and increased turnover of dopamine *(94)*.

4.3. Antisense Knockdown of Dopamine D_3 Receptors

The dopamine D_3 receptor is localized in limbic areas, such as the nucleus accumbens, the Islands of Calleja, the ventral tegmental area, and the hypothalamus as well as the substantia nigra *(69,103)*. It is difficult to discriminate pharmacologically between D_2 and D_3 receptors because most D_3 agonists and antagonists also have a significant affinity for D_2 receptors *(104–106)*.

Unilateral intrastriatal injection of D_3 antisense did not induce ipsilateral rotations after apomorphine or quinpirole administration *(78)*. It appears from this finding that D_3 receptors do not play a part in the regulation of motor activity by dopamine agonists. Nissbrandt et al. *(107)* reported that D_3 antisense, administered intracerebroventricularly, reduced D_3 binding, and increased dopamine synthesis in the nucleus accumbens, but had no effect in the striatum. D_3 antisense also did not counteract the effect of apomorphine on dopamine synthesis.

A behavioral study carried out using D_3 antisense showed that it did not induce catalepsy but increased spontaneous locomotion. Also, D_3 receptor knockdown markedly attenuated the stereotypic activity induced by amphetamine but had no effect on amphetamine-induced locomotion, which indicates that D_3 receptors may be involved in stereotypy *(75)*.

Accili et al. *(108)* reported increased locomotion in D_3 transgenic mice and pharmacological studies using antagonists with affinity for the D_3 receptor have also showed increased locomotion in rats *(109–111)*. The administration of D_3 agonists into the nucleus accumbens reduced spontaneous motor activity and negated the locomotion-enhancing effect of dopamine *(111,112)*.

The findings in these studies correspond to data obtained from previous pharmacological studies showing that 7-hydroxy-2-(di-*n*-propylamino)tetralin (7-OHDPAT), a partially selective D_3 agonist, and U99194A, a D_3 antagonist, increased and decreased spontaneous locomotor activity respectively *(109,113,114)*.

An investigation similar to that done with the D_2 autoreceptors was carried out by Tepper et al. *(21)* with D_3 antisense. As with D_2 antisense, D_3 antisense reduced the inhibitory effect of apomorphine on dopamine release as determined by rotational behavior, in addition to increasing somatodendritic as well as terminal excitability. Therefore, D_3 receptors were concluded to be expressed at the somatodendritic and terminal areas of nigrostriatal neurons. Interestingly, when D_2 and D_3 antisense were infused together, there was an additive effect on the electrophysiological and behavioral effects *(19,20)*. It thus appears that D_3 receptors may play an important role in the autoreceptor-mediated regulation of dopaminergic neurons.

The role of dopamine D_3 receptors in PD is unclear but it appears that one of the newer antiparkinsonian drugs, ropinirole, appears to have selectivity for the D_3 receptor *(115,116)*, although the mechanism for this is unclear, especially as postsynaptic D_3 receptors are reported to be unchanged in PD *(117)*. Here, antisense treatment may also be useful; by producing adequate knockdown of the dopamine D_3 receptor, and then administering these agonists, it can be determined if these drugs are producing their antiparkinsonian effect via actions at D_3 receptors, whether pre- or postsynaptic, and not postsynaptic D_2 receptors.

4.4. Antisense Knockdown of D_4 Receptors

There is a low density of dopamine D_4 receptors within the brain, with the highest numbers expressed in the frontal cortex, midbrain, amygdala, and medulla and low levels seen in the striatum and olfactory tubercle *(70,72)*.

The continuous infusion of D_4 S-modified antisense intracerebroventricularly over a period of 3 d resulted in decreased spontaneous locomotor activity and a reduced loco-motor response to quinpirole or *d*-amphetamine. The stereotypy response to high con-centrations of *d*-amphetamine was also decreased by the D_4 antisense. However, unlike D_2 knockdown, D_4 antisense did not induce catalepsy *(78)*.

The typical or classical neuroleptics induce motor side effects such as catalepsy, and this is thought to be due to their actions at dopamine receptors in the extrapyramidal system. The atypical neuroleptics, which act mainly in limbic areas and decrease loco-motor activity without inducing catalepsy *(118)*. Clozapine, categorized as an atypical neuroleptic, has been shown to have high affinity for the dopamine D_4 receptor *(70,119)*. This differential action of the typical vs atypical neuroleptics shows that the antipsychotic effects of the atypical neuroleptics may be mediated by the D_4 receptor. Dopamine D_4 receptors may not be involved in PD, especially as agents with high affinities at D_4 receptors do not produce motor deficits.

4.5. Antisense Knockdown of Dopamine D_1 Receptors

D_1 receptors are found in the striatum, nucleus accumbens, and olfactory tubercle *(120)* whereas dopamine D_5 receptors are localized in the hippocampus and the parafascicular nuclei of the thalamus *(121)*. D_1 receptors are thought to be involved in mediating grooming behavior and may need to be costimulated with D_2 receptors to produce stereotyped behavior. There have been only a few studies that have investi-gated the roles of dopamine D_1 receptors using antisense technology to date. Zhang et al. *(122)* reported that D_1 antisense administered into the cerebral ventricles reduced the grooming behavior induced by the D_1-like receptor agonist SKF 38393, but had no effect on D_2-like receptor-mediated quinpirole-induced stereotyped activity. The decrease in grooming was directly proportional to the duration of the antisense admin-istration. D_1 antisense also attenuated the rotations observed after SKF 38393 treat-ment in mice with 6-OHDA lesions of the striatum but not after quinpirole or oxotremorine administration. Cessation of the D_1 antisense administration resulted in the reinstatement of these behaviors. These observations all correlate with findings obtained using pharmacological agents acting primarily at the D_1 receptor *(91,123)*.

Dopamine D_1 receptors have recently been shown to be involved in PD, as agonists with high affinities for the D_1 receptor have become available and have been shown to be potent antiparkinsonian agents. Using antisense technology, it would be possible to differentiate the responses produced by the D_1 receptor itself and by the D_1 receptor acting in concord with the D_2 receptor.

As D_5 receptors are found in areas of the brain not directly involved in the patho-physiology of PD, they are not discussed here.

5. CONCLUSIONS

Antisense technology may provide a novel model for PD by allowing the knock-down of specific dopamine receptors in brain areas normally associated with PD and at the same time allow for the determination of the roles played by individual dopamine receptors in the pathology and/or therapy of PD. Previously, it was thought that most of the antiparkinsonian drugs *(124,125)*, exerted their beneficial effects via the D_2

receptor and that the D_1 receptor had no role in the antiparkinsonian effect. However, with the advent of drugs with better affinities for the D_1 receptor, such as CY 208-243 and A-77636 *(126,127)*, it has been discovered that these agents are also potent antiparkinsonian drugs, and thus do play a role in motor control.

The use of dopamine receptor antisense has to an extent clarified the possible functional roles played by these receptors in motor behaviors but it is clear that a lot more work is required to come to any conclusions. Cautious interpretation of data obtained from the antisense studies is also urged in light of some of the reported nonspecific effects that the oligodeoxynucleotides can exert.

REFERENCES

1. Hornykiewicz, O. (1966) Dopamine and brain function. *Pharmacol. Res.* **18,** 925–964.
2. Marsden, C. D. (1992) Dopamine and basal ganglia disorders in humans. *Semin. Neurosci.* **4,** 171–178.
3. Carlsson, A. (1988) The current status of the dopamine hypothesis of schizophrenia. *Neuropsychopharmacology* **1,** 179–186.
4. Nahmias, C., Garnett, E. S., Firnau, G., and Lang, A. (1985) Striatal dopamine distribution in Parkinsonian patients during life. *J. Neurol. Sci.* **69,** 223–230.
5. Davis, K. L., Kahn, R. S., Ko, G., and Davidson, M. (1991) Dopamine in schizophrenia: a review and reconceptualization. *Am. J. Psychiatry* **148,** 1474–1485.
6. Klawans, H. L. (1987) Chorea. *Can. J. Neurol. Sci.* **14 (Suppl. 3),** 536–540.
7. Tarsy, D. and Baldessarini, R. J. (1984) Tardive dyskinesia. *Annu. Rev. Med.* **35,** 605–623.
8. Fog, R. (1985) The effect of dopamine antagonists in spontaneous and tardive dyskinesia. *Psychopharmacology.* **2 (Suppl.),** 118–121.
9. Dahlstrom, A. and Fuxe, K. (1964) Evidence for the existence of monoamine-containing neurons in the central nervous system. I. Demonstration of monoamines in the cell bodies of brain stem neurons. *Acta Physiol. Scand.* **232 (Suppl.),** 1–55.
10. Ferger, B., Kropf, W., and Kuschinsky, K. (1994) Studies on electroencephalogram (EEG) in rats suggest that moderate doses of cocaine or *d*-amphetamine activate D_1 rather than D_2 receptors. *Psychopharmacology* **114,** 297–308.
11. Graybiel, A. M. and Ragadale, C. W. (1983) Biochemical anatomy of the striatum, in *Chemical Neuroanatomy* (Emson, P. C., ed.), Raven Press, New York, pp. 427–504.
12. Lindvall, O. and Bjorklund, A. (1983) Dopamine and norepinephrine-containing neuron systems: their anatomy in the rat brain, in *Chemical Neuroanatomy* (Emson, P. C., ed.), Raven Press, New York, pp. 229–255.
13. Poirier, L. J. and Sourkes, T. L. (1965) Influence of the substantia nigra on the catecholamine content of the striatum. *Brain* **88,** 181–192.
14. Marsden, C. D. (1984) The pathophysiology of movement disorders. *Neurol. Clin.* **2,** 435–459.
15. Creese, I., Burt, D. R., and Snyder, S. H. (1976) Dopamine receptor binding predicts clinical and pharmacological potencies of antischizophrenic drugs. *Science* **192,** 481–483.
16. Seeman, P., Lee, T., Chan-Wong, M., and Wong, K. (1976) Antipsychotic drug doses and neuroleptic/dopamine receptors. *Nature* **261,** 717–719.
17. Zhang, M. and Creese, I. (1993) Antisense oligodeoxynucleotide reduces brain dopamine D_2 receptors: behavioral correlates. *Neurosci. Lett.* **161,** 223–226.
18. Zhang, M., Ouagazzal, A., Sun, B.-C., and Creese, I. (1996) Regulation of motor behavior by dopamine receptor subtypes: an antisense knockout approach, in *The Dopamine Receptors* (Neve, K. and Neve, R., eds.), Humana Press, Totowa, NJ, pp. 425–455.

19. Martin, L. P., Kita, H., Sun, B.-C., Zhang, M., Creese, I., and Tepper, J. M. (1994) Electrophysiological consequences of D_2 receptor antisense knockouts in nigrostriatal neurons. *Soc. Neurosci. Abstr.* **20,** 908.

20. Sun, B.-C., Creese, I., and Tepper, J. M. (1995) Electrophysiology of antisense knockout of D_2 and D_3 dopamine receptors in nigrostriatal dopamine neurons. *Soc. Neurosci. Abstr.* **21,** 1661.

21. Tepper, J. M., Sun, B.-C., Martin, L. P., and Creese, I. (1997) Functional roles of dopamine D_2 and D_3 autoreceptors on nigrostriatal neurons analyzed by antisense knockdown in vivo. *J. Neurosci.* **17,** 2519–2530.

22. Bennett, C. F., Dean, N., Ecker, D. J,. and Monia, B. P. (1996) Pharmacology of antisense therapeutic agents, cancer and inflammation, in *Methods in Molecular Medicine: Antisense Therapeutics* (Agrawal, S., ed.), Humana Press, Totowa, NJ, pp. 13–46.

23. Leiter, J. M., Agrawal, S., Palese, P., and Zamecnik, P. C. (1990) Inhibition of influenza virus replication by phosphorothioate oligodeoxynucleotides. *Proc. Natl. Acad. Sci. USA* **87,** 3430–3434.

24. Lisziewicz, J., Sun, D., Metelev, V., Zamecnik, P., Gallo, R. C., and Agrawal, S. (1993) Long-term treatment of human immunodeficiency virus-infected cells with antisense oligonucleotide phosphorothioates. *Proc. Natl. Acad. Sci. USA* **90,** 3860–3864.

25. Cohen, J. (1992) Oligonucleotide therapeutics. *Trends Biotechnol.* **10,** 87–90.

26. Hélène, C. and Toulmé, J.-J. (1990) Specific regulation of gene expression by antisense, sense and antigene nucleic acids. *Biochem. Biophys. Acta* **1049,** 99–125.

27. Wahlestedt, C., Golanov, E., Yamamoto, S., Yee, F., Ericson, H., Yoo, H., et al. (1993a) Antisense oligodeoxynucleotides to NMDA-R1 receptor channel protect cortical neurons from excitotoxicity and reduce focal ischaemic infarctions. *Nature* **363,** 260–263.

28. Wahlestedt, C., Pich, E. M., Koob, G. F., Yee, F., and Heilig, M. (1993b) Modulation of anxiety and neuropeptide Y-Y1 receptors by antisense oligodeoxynucleotides. *Science* **259,** 528–531.

29. Landgraf, R., Gerstberger, R., Montkowski, A., Probst, J. C., Wotjak, C. T., Holsboer, F., and Engelman, M. (1995) V1 vasopressin receptor antisense oligodeoxynucleotide into septum reduces vasopressin binding, social discrimination abilities and anxiety-related behavior in rats. *J. Neurosci.* **15,** 4250–4258.

30. Weiss, B., Zhou, L.-W., and Zhang, S.-P. (1996) Dopamine antisense oligodeoxynucleotides as potential novel tools for studying drug abuse, in *Antisense Strategies for the Study of Receptor Mechanisms* (Raffa, R. B. and Poreca, F., eds), R. G. Landes, Austin, TX, pp. 71–91.

31. Standaert, D. G., Testa, C. M., Rudolf, G. D., and Holingsworth, Z. R. (1996) Inhibition of *N*-methyl-D-aspartate glutamate receptor subunit expression by antisense oligodeoxynucleotides reveals their role in striatal motor regulation. *J. Pharmacol. Exp. Ther.* **276,** 342–352.

32. Zhao, T.-J., Rosenberg, H. C., and Chiu, T. H. (1996) Treatment with an antisense oligodeoxynucleotide to the $GABA_A$ receptor $\gamma 2$ subunit increases convulsive threshold for β-CCM, a benzodiazepine 'inverse agonists', in rats. *Eur. J. Pharmacol.* **306,** 61–66.

33. Plata-Salamàn, C. R., Wilson, C. D., Sonti, G., Borkoski, J. P., and French-Mullen, J. M. H. (1995) Antisense oligodeoxynucleotides to G-protein α-subunit subclasses identity a transductional requirement for modulation of normal feeding dependent on $G\alpha_o$ subunit. *Mol. Brain. Res.* **33,** 72–78.

34. Akabayashi, A., Wahlestedt, C., Alexander, J. T., and Liebowitz, S. F. (1994) Specific inhibition of neuropeptide Y synthesis in arcuate nucleus by antisense oligonucleotides suppresses feeding behavior and insulin secretion. *Mol. Brain Res.* **21,** 55–61.

35. Skutella, T., Probst, J. C., Jirikowski, G. F., Holsboer, F., and Spanagel, R. (1994) Ventral tegmental area (VTA) injection of tyrosine hydroxylase phosphorothioates antisense oligonucleotide suppresses operant behavior in rats. *Neurosci. Lett.* **167,** 55–58.

36. Georgieva, J., Heilig, M., Nylander, I., Herrera-Marschitz, M., and Trerenius, L. (1995) *In vivo* antisense inhibition of prodynorphin expression in rat striatum: dose-dependence and sequence specificity. *Neurosci. Lett.* **92,** 69–71.

37. Meeker, R., Le Grand, G., Ramirez, J., Smith, T., and Shih, Y. H. (1995) Antisense vasopressin oligonucleotides: uptake, turnover, distribution, toxicity and behavioral effects. *J. Neuroendocrinol.* **7,** 419–428.

38. Catsicas, M., Osen-Sand, A., Staple, J. K., Jones, K. A., Ayala, G., Knowles, J., Grenningloh, G., Pich, E. M., and Catsicas, S. (1996) Antisense blockade of expression; SNAP-25 in vitro and in vivo, in *Methods in Molecular Medicine: Antisense Therapeutics* (Agrawal, S., ed.), Humana Press, Totowa, NJ, pp. 57–85.

39. Bergan, R., Connell, Y., Fahmy, B., Kyle, E., and Neckers, L. M. (1994) Aptameric inhibition of p210$^{bcr-abl}$ tyrosine kinase autophosphorylation by oligodeoxynucleotides of defined sequence and backbone structure. *Nucleic Acids Res.* **22,** 2150–2154.

40. McCarthy, M. M., Masters, D. B., Rimvall, K., Schwartz-Giblin, S., and Pfaff, D. W. (1994) Intracerebral administration of antisense oligodeoxynucleotides to GAD_{65} and GAD_{67} mRNA modulate reproductive behavior in the female rat. *Brain Res.* **636,** 209–220.

41. Heilig, M., Engel, J. A., and Soderpalm, B. (1993) c-fos antisense in the nucleus accumbens blocks the locomotor stimulant action of cocaine. *Eur. J. Pharmacol.* **236,** 339–340.

42. Chiasson, B. J., Hooper, M. L., Murphy, P. R., and Robertson, H. A. (1992) Antisense oligonucleotide eliminates in vivo expression of c-fos in mammalian brain. *Eur. J. Pharmacol.* **227,** 451–453.

43. Chiasson, B. J., Armstrong, J. N., Hooper, M. L., Murphy, P. R., and Robertson, H. A. (1994) The application of antisense oligonucleotide technology to the brain: some pitfalls. *Cell. Mol. Neurobiol.* **14,** 507–521.

44. Chiasson, B. J., Hong, M., Hooper, M. L., Armstrong, J. N., Murphy, P. R., and Robertson, H. A. (1996) Antisense therapeutics in the central nervous system. The induction of c-*fos*, in *Methods in Molecular Medicine: Antisense Therapeutics* (Agrawal, S., ed.), Humana Press, Totowa, NJ, pp. 225–245.

45. Widnell, K. L., Self, D. W., Lane, S. B., Russell, D. S., Vaidya, V. A., Miserendino, M. J. D., Rubin, C. S., Duman, R. S., and Nestler, E. J. (1996) Regulation of CREB expression: *in vivo* evidence for functional role in morphine action in nucleus accumbens. *J. Pharmacol. Exp. Ther.* **276,** 306–315.

46. Whitesell, L., Geselowitz, D., Chavany, C., Fahmy, B., Walbridge, S., Alger, J. R., and Neckers, L. M. (1993) Stability, clearance and disposition of intraventricularly administered oligodeoxynucleotides: implications for therapeutic application within the central nervous system. *Proc. Natl. Acad. Sci. USA* **90,** 4665–4669.

47. Chavany, C., Connell, Y., and Neckers, L. (1995) Contribution of sequence and phosphorothioate content to inhibition of cell growth and adhesion caused by c-*myc* antisense oligomers. *Mol. Pharmacol.* **48,** 738–746.

48. Matsukura, M., Shinozuka, K., Zon, G., Mitsuya, H., Reitz, M., Cohen, J. S., and Broder, S. (1987) Phosphorothioate analogs of oligodeoxynucleotides: novel inhibitors of replication and cytopathic effects of human immunodeficiency virus (HIV) *Proc. Natl. Acad. Sci. USA* **84,** 7706–7710.

49. Heilig, M. (1994) Antisense technology; prospects for treatment of neuropsychiatric disorders. *CNS Drugs* **1,** 405–409.

50. Loke, S. L., Stein, C. A., Zhang, X. H., Mori, K., Nakanishi, M., Subasinghe, C., Cohen, J. S., and Neckers, L. M. (1989) Characterization of oligonucleotide transport into living cells. *Proc. Natl. Acad. Sci. USA* **86,** 3474–3478.

51. Yabukov, L. A., Deeva, E. A., Zarytova, V. F., Ivanova, E. M., Ryte, S., Yurchenk, L. V., and Vlassov, V. V. (1989) Mechanism of oligonucleotide uptake by cells: Involvement of specific receptors? *Proc. Natl. Acad. Sci. USA* **86,** 6454–6458.

52. Neckers, L. M. (1993) Cellular internalization of oligodeoxynucleotides, in *Antisense Research and Applications* (Crooke, S. T. and Lebleu, B., eds.), CRC, Boca Raton, FL, pp. 451–456.

53. Agrawal, S., Temsamani, J., and Tang, J. Y. (1991) Pharmacokinetics, biodistribution and stability of oligodeoxynucleotide phosphorothioates in mice. *Proc. Natl. Acad. Sci. USA* **88,** 7595–7599.

54. Neckers, L. M. (1989) Antisense inhibitors of gene expression, in *Oligodeoxynucleotides* (Cohen, J. S., ed.) Macmillan, London, pp. 211–231.

55. Szklarczyk, A. and Kaczmarek, L. (1995) Antisense oligodeoxyribonucleotides: stability and distribution after intracerebral injection into rat brain. *J. Neurosci. Methods* **60,** 181–187.

56. Gewirtz, A. M., Stein, C. A., and Glazer, P. M. (1996) Facilitating oligonucleotide delivery: helping antisense deliver on its promise. *Proc. Natl. Acad. Sci. USA* **93,** 3161–3163.

57. Campbell, J. M., Bacon, T. A., and Wickstrom, E. (1990) Oligodeoxynucleotide phosphorothioate stability in subcellular extracts, culture media, sera and cerebrospinal fluid. *J. Biochem. Biophys. Methods* **20,** 259–269.

58. Gura, T. (1995) Antisense has growing pains. *Science* **270,** 575–577.

59. Hamblin, M. and Creese, I. (1983) Behavioral and radioligand binding evidence for irreversible dopamine receptor blockade by EEDQ. *Life Sci.* **32,** 2247–2255.

60. Meller, E., Bordi, F., and Bohmaker, K. (1989) Behavioral recovery after irreversible inactivation of D_1 and D_2 dopamine receptors. *Life Sci.* **44,** 1019–1026.

61. Saller, C. F., Kreamer, L. D., Adamovage, L. A., and Salama, A. I. (1989) Dopamine receptor occupancy in vivo: measurement using *N*-ethoxycarbonyl-2-ethoxy-1,2-dihydroquinoline (EEDQ). *Life Sci.* **45,** 917–929.

62. Close, S. P., Marriott, A. S., and Pays, S. (1985) Failure of SKF 38393-A to relieve Parkinsonism symptoms induced by 1-methyl-4-phenyl-1,2,3,6-tetrahydropyridine in the marmoset. *Br. J. Pharmacol.* **85,** 320–322.

63. Kinemuchi, H., Fowler, C. J., and Tipton, K. F. (1987) The neurotoxicity of 1-methyl-4-phenyl-1,2,3,6-tetrahydropyridine (MPTP) and its relevance to Parkinson's disease. *Neurochem. Int.* **11,** 359–373.

64. Ossowska, K. (1994) The role of excitatory amino acids in experimental models of Parkinson's disease. *J. Neural Transm.* **8,** 39–71.

65. Kebabian, J. W. and Greengard, P. (1971) Dopamine-sensitive adenylyl cyclase: possible role in synaptic transmission. *Science* **174,** 1346–1349.

66. Onali, P., Olianas, M. C., and Gessa, G. L. (1985) Characterization of dopamine receptors mediating inhibition of adenylate cyclase activity in rat striatum. *Mol. Pharmacol.* **28,** 138–145.

67. Sunahara, R. K., Niznik, H. B., Weiner, D. M., Stormann, T. M., Brann, M. R., Kennedy, J. L., Gelernter, J. E., Rozmahel, R., Yang, Y., Israel I., Seeman, P., and O'Dowd, B. F. (1990) Human dopamine D_1 receptor encoded by an intronless gene on chromosome 5. *Nature* **347,** 80–83.

68. Bunzow, J. R., Van Tol, H. H. M., Grandy, D. K., Albert, P., Salon, J., Chisre, M., Machida, C. A., Neve, K. A., and Civelli, O. (1988) Cloning and expression of a rat D_2 dopamine receptor cDNA. *Nature* **336,** 783–787.

69. Sokoloff, P., Giros, B., Martres, M.-P., Bouthenet, M.-L., and Schwartz, J..-C. (1990) Molecular cloning and characterization of a novel dopamine receptor (D_3) as a target for neuroleptics. *Nature* **347,** 146–151.

70. Van Tol, H. H. M., Bunzow, J. R., Guan, H.-C., Sunahara, R. K., Seeman, P., Niznik, H. B., and Civelli, O. (1991) Cloning of a human dopamine D_4 receptor gene with high affinity for the antipsychotic clozapine. *Nature* **350,** 614–616.

71. Sunahara, R. K., Guan, H.-C., O'Dowd, B. F., Seeman, P., Laurier, L. G., George, S. R.,

Torchia, J., Van Tol, H. H. M., and Niznik, H. B (1991) Cloning a human dopamine receptor gene (D_5) with higher affinity for dopamine than D_1. *Nature* **350,** 614–619.

72. Civelli, O., Bunzow, J. R., and Grandy, D. K. (1993) Molecular diversity of the dopamine receptors. *Annu. Rev. Pharmacol. Toxicol.* **32,** 281–307.

73. Civelli, O. (1995) Molecular biology of the dopamine receptor subtypes, in *Psychopharmacology: The Fourth Generation of Progress* (Bloom, F. E. and Kupfer, D. J., eds.), Raven Press, New York. pp. 155–161.

74. Weiss, B., Zhang, S.-P., and Zhou, L.-W. (1997) Antisense strategies in dopamine receptor pharmacology. *Life Sci.* **60,** 433–455.

75. Zhang, M., Ouagazzal, A.-M., and Creese, I. Antisense knockout of brain dopamine D_2 and D_3 receptors and its effects on motor behaviors, in press.

76. Wahlestedt, C. (1996) Antisense 'knockdown' strategies in neurotransmitter receptor research, in *Antisense Strategies for the Study of Receptor Mechanisms* (Raffa, R. and Porreca, F., eds.), R. G. Landes, Austin, TX, pp. 1–10.

77. Brysch, W. and Schlingensiepen, K.-H. (1994) Design and application of antisense oligonucleotides in cell culture, *in vivo* and as therapeutic agents. *Cell. Mol. Biol.* **14,** 557–568.

78. Sun, B.-C., Zhang, M., Ouagazzal, A.-M., Martin, L. P., Tepper, J. M., and Creese, I. (1996) Dopamine receptor function: An analysis utilizing antisense knockout *in vivo*, in *Pharmacological Regulation of Gene Expression in the CNS* (Merchant, K., ed.), CRC, Boca Raton, FL, pp. 51–78.

79. Weiss, B., Zhou, L.-W., Zhang, S.-P., and Qin, Z.-H. (1993) Antisense oligodeoxynucleotide inhibits D_2 dopamine receptor-mediated behavior and D_2 messenger RNA. *Neuroscience* **55,** 607–612.

80. Dragunow, M., Lawlor, P., Chiasson, B., and Robertson, H. (1993) c-fos antisense generates apomorphine and amphetamine-induced rotation. *NeuroReport* **5,** 305–306.

81. Hall, M. D., Jenner, P., and Marsden, C. D. (1983) Turnover of specific [^{3}H]spiperone and [^{3}H]*N,n*-propylnorapomorphine binding sites in rat striatum following phenoxybenzamine administration. *Biochem. Pharmacol.* **32,** 2973–2977.

82. Leff, S. E., Gariano, R., and Creese, I. (1984) Dopamine receptor turnover rates in rat striatum are age-dependent. *Proc. Natl. Acad. Sci. USA* **81,** 3910–3914.

83. Norman, A. B., Battaglia, G., and Creese, I. (1987) Differential recovery rates of rat D_2 dopamine receptors as a function of aging and chronic reserpine treatment following irreversible modification: a key to receptor regulatory mechanisms. *J. Neurosci.* **7,** 1484–1491.

84. Zhou, L.-W., Zhang, S.-P., Qin, Z.-H., and Weiss, B. (1994) *In vivo* administration of an oligodeoxynucleotide antisense to the D_2 dopamine receptor messenger RNA inhibits D_2 dopamine receptor-mediated behavior and the expression of D_2 dopamine receptors in mouse striatum. *J. Pharmacol. Exp. Ther.* **268,** 1015–1023.

85. Sibley, D. R. and Monsma, F. J. (1992) Molecular biology of dopamine receptors. *Trends Pharmacol.* **131,** 61–69.

86. Le Moine, C., Normand, E., Guitteny, A. F., Fouque, B., Teoule, R., and Bloch, B. (1990) Dopamine receptor gene expression by enkephalin neurons in the rat forebrain. *Proc. Natl. Acad. Sci. USA* **87,** 230–234.

87. Silvia, C. P., King, G. R., Lee, T. H., Xue, Z.-Y., Caron, M. G., and Ellinwood, E. H. (1994) Intranigral administration of D_2 dopamine receptor antisense oligodeoxynucleotides establishes a role for nigrostriatal D_2 autoreceptors in the motor actions of cocaine. *Mol. Pharmacol.* **46,** 51–57.

88. Qin, Z.-H., Zhou, L.-W., Zhang, S.-P, Wang, Y., and Weiss, B. (1995) D_2 dopamine receptor antisense oligodeoxynucleotide inhibits the synthesis of a functional pool of D_2 dopamine receptors. *Mol. Pharmacol.* **48,** 730–737.

89. Valerio, A., Belloni, M., Gorno, M. L., Tinti, C., Memo, M., and Spano, P. (1994) Dopam-

ine D_2, D_3 and D_4 receptor mRNA levels in rat brain and pituitary during aging. *Neurobiol. Aging* **15,** 713–719.

90. Rajakumar, N., Laurier, L., Niznik, H. B., and Stoessl, A. J. (1997) Effects of intrastriatal infusion of D_2 receptor antisense oligonucleotide on apomorphine-induced behaviors in the rat. *Synapse* **26,** 199–208.

91. Loschmann, P. A., Smith, L. A., Lange, K. W., Jahnig, P., Jenner, P., and Marsden, C. D. (1992) Motor activity following the administration of selective D_1 and D_2 dopaminergic drugs to MPTP-treated marmosets. *Psychopharmacology* **109,** 49–56.

92. Ouagazzal, A., Nieoullon, A., and Amalric, M. (1993) Effects of dopamine D_1 and D_2 receptor blockade on MK 801-induced hyperlocomotion in rats. *Psychopharmacology* **111,** 427–434.

93. Baik, J. H., Picetti, R., Saiardi, A., Thiriet, G., Dierich, A. Depaulis, A., Le Meur, M., and Borrelli, E. (1995) Parkinsonian-like locomotor impairment in mice lacking dopamine D_2 receptors. *Nature* **377,** 424–428.

94. Cooper, J. R., Bloom, F. E., and Roth, R. H. (1991) *The Biochemical Basis of Neuropharmacology*, Oxford University Press, Oxford, pp. 310–313 and 328–330.

95. Rinne, J. O., Laihinen, A., Lonnberg, P., Marjamaki, P., and Rinne, U. K. (1991) A post-mortem study on striatal dopamine receptors in Parkinson's disease. *Brain Res.* **556,** 117–122.

96. Stoof, J. C., Drukarch, B., De Boer, P., Westerink, B. H. C., and Groenewegen, H. J. (1992) Regulation of the activity of striatal cholinergic neurons by dopamine. *Neuroscience* **47,** 755–770.

97. Mengod, G., Villaro, M. T., Landwehrmeyer, G. B., Martinez-Mir, M. I., Niznik, H. B., Sunahara, R. K., Seeman, P., O'Dowd, B. F., Probst, A., and Palacios, J. M. (1992) Visualization of dopamine D_1, D_2 and D_3 receptor mRNAs in human and rat brain. *Neurochem. Int.* **20 (Suppl.),** 33S–43S.

98. Brouwer, N., Van Dijken, H., Ruiters, M. H., Van Willingen, J. D., and Ter Horst, G. J. (1992) Localization of dopamine D_2 receptor mRNA with non-radioactive in situ hybridization histochemistry. *Neurosci. Lett.* **142,** 223–227.

99. Tepper, J. M., Nakamura S., Young, S. J., and Groves, P. M. (1984) Autoreceptor-mediated changes in dopaminergic terminal excitability: effects of striatal drug infusions. *Brain Res.* **309,** 317–333.

100. Akaoka, H., Charlety, P., Saunier, C.-F., Buda, M., and Chouvnet, G. (1992) Inhibition of nigral dopaminergic neurons by systemic and local apomorphine: possible contribution of dendritic autoreceptors. *Neuroscience* **49,** 879–892.

101. Bunney, B. S. and Aghajanian, G. K. (1978) *d*-Amphetamine-induced depression of central dopamine neurons: evidence for mediation by both autoreceptors and a strio-nigral feedback pathway. *Naunyn Schmiedebergs Arch. Pharmacol.* **304,** 255–261.

102. Groves, P. M., Wilson, C. J., Young, S. J., and Rebec, G. V. (1975) Self-inhibition by dopaminergic neurons. *Science* **190,** 522–529.

103. Bouthenet, M.-L., Souil, E., Martres, M.-P., Sokoloff, P., Giros, B., and Schwartz, J.-C. (1991) Localization of dopamine D_3 receptor mRNA in the rat using in situ hybridization histochemistry: comparison with dopamine D_2 receptor mRNA. *Brain Res.* **564,** 203–219.

104. Gingrich, J. A. and Caron, M. G. (1993) Recent advances in the molecular biology of dopamine receptors. *Annu. Rev. Neurosci.* **16,** 299–321.

105. Burris, K. D., Pacheco, M. A., Filtz, T. M., Kung, M.-P., Kung, H. F., and Molinoff, P. B. (1995) Lack of discrimination of agonists for D_2 and D_3 dopamine receptors. *Neuropsychopharmacology* **12,** 335–345.

106. Seeman, P. and Van Tol, H. H. (1994) Dopamine receptor pharmacology. *Curr. Opin. Neurol. Neurosurg.* **6,** 602–608.

107. Nissbrandt, H., Ekman, A., Eriksson, E., and Heilig, M. (1995) Dopamine D_3 receptor antisense influences dopamine synthesis in the rat brain. *NeuroReport* **6,** 573–576.

108. Accili, D., Fishburn, C. S., Drago, J., Steiner, H., Lachowicz, J. E., Park, B. H., Gauda, E. B., Lee, E. J., Cool, M. H., Sibley, D. R., Gerfen, C. R., Westphal, H., and Fuchs, S. (1996) A targeted mutation of D_3 dopamine receptor is associated with hyperactivity in mice. *Proc. Natl. Acad. Sci. USA* **93,** 1945–1949.

109. Waters, N., Svensson, K., Haadsma-Svensson, S. R., Smith, M. W., and Carlsson, A. (1993) The dopamine D3-receptor: a postsynaptic receptor inhibitory on rat locomotor activity. *J. Neural Transm.* **94,** 11–19.

110. Sautel, F., Griffon, N., Sokoloff, S., Schwartz, J. H., Launay, C., Simon, P., Constentin, J., Schoenfelder, A., Garrido, F., Mann, A., and Wermuth, C. G. (1995) Nafadotride, a potent preferential dopamine D_3 receptor antagonist, activates locomotion in rodents. *J. Pharmacol. Exp. Ther.* **275,** 1239–1246.

111. Kling-Petersen, T., Ljung, E., and Svensson, K. (1995) Effect on locomotor activity after local application of D_3 preferring compounds in discrete areas of the rat brain. *J. Neural Transm.* **102,** 209–220.

112. Pugsley, T. A., Davis, M. D., Akunne, H. C., MacKenzie, G. R., Shih, Y. H., Damsma, G., Wikstrom, H., Whetzel, S. Z., Georgic, L. M., Cooke, L. W., Demattos, S. B., Corbin, A. E., Glase, S. A., Wise, L. D., Djikstra, D., and Heffner, T. G. (1995) Neurochemical and functional characterization of the preferentially selective dopamine D_3 agonist, PD 128907. *J. Pharmacol. Exp. Ther.* **275,** 1355–1366.

113. Svensson, K., Carlsson, A., Huff, R. M., Kling-Petersen, T., and Waters, N. (1994a) Behavioral and neurochemical data suggest functional differences between dopamine D_2 and D_3 receptors. *Eur. J. Pharmacol.* **263,** 235–243.

114. Svensson, K., Carlsson, A., and Waters, N. (1994b) Locomotor inhibition by the D_3 ligand R-(+)-7-OHDPAT is independent of changes in dopamine release. *J. Neural Transm.* **95,** 71–74.

115. Tulloch, I. F. (1997) Pharmacologic profile of ropinirole: a nonergoline dopamine agonist. *Neurology* **49 (Suppl. 1),** S58–62.

116. Hurley, M. J., Stubbs, C. M., Jenner, P., and Marsden, C. D. (1996) D_3 receptor expression within the basal ganglia is not affected by Parkinson's disease. *Neurosci. Lett.* **214,** 75–78.

117. Brooks, D. J., Torjanski, N., and Burn, D. J. (1995) Ropinirole in the symptomatic treatment of Parkinson's disease. *J. Neural Transm. Suppl.* **45,** 231–238.

118. Reynolds, G. P. (1992) Developments in the drug treatment of schizophrenia. *Trends Pharmacol. Sci.* **13,** 116–120.

119. Seeman, P., Guan, H.-C., and Van Tol, H. H. M. (1993) Dopamine D_4 receptors elevated in schizophrenia. *Nature* **365,** 441–445.

120. Clark, D. and White, F. J. (1987) D_1 dopamine receptor—the research for a function: a critical evaluation of the D_1/D_2 dopamine receptor classification and its functional implication. *Synapse* **1,** 347–388.

121. Meador-Woodruff, J. H., Mansour, A., Grandy, D. K., Damask, S. P., Civelli, O., and Watson, S. J. (1992) Distribution of D_5 dopamine receptor mRNA in rat brain. *Neurosci. Lett.* **145,** 209–212.

122. Zhang, S.-P., Zhou, L.-W., and Weiss, B. (1994) Oligodeoxynucleotide antisense to the D_1 dopamine receptor mRNA inhibits D_1 dopamine receptor-mediated behaviors in normal mice and in mice lesioned with 6-hydroxydopamine. *J. Pharmacol. Exp. Ther.* **271,** 1462–1470.

123. Blanchet, P. J., Grondin, R., Bedard, P. J., Shiosaki, K., and Britton, D. R. (1996) Dopamine D_1 receptor desensitization profile in MPTP-lesioned primates. *Eur. J. Pharmacol.* **309,** 13–20.

124. Beizer, J. L. (1995) Treatment options in Parkinson's disease. *Am. Pharmacol.* **NS35,** 21–30.

125. De Keyser, J., De Backer, J. P., Wilczak, N., and Herroelen, L. (1995) Dopamine agonists used in the treatment of Parkinson's disease and their selectivity for the D_1, D_2 and D_3 dopamine receptors in human striatum. *Prog. Neuropsychopharmacol. Biol. Psychiatr.* **19,** 1147–1154.

126. Temlett, J. A., Quinn, N. P., Jenner, P. G., Marsden, C. D., Pourcher, E., Bonnet, A.-M., Agid, Y., Markstein, R., and Lataste, X. (1989) Antiparkinsonian activity of CY 208-243, a partial D_1 agonist, in MPTP-treated marmosets and patients with Parkinson's disease. *Move. Disord.* **4,** 261–265.

127. Kebabian, J. W., Britton, D. R., DeNinno, M. P., Perner, R., Smith, L., Jenner, P., Schoenleber, R., and William, M. (1992) A-77636: a potent and selective D_1 receptor agonist with antiparkinsonian activity in marmosets. *Eur. J. Pharmacol.* **229,** 203–209.

13

Is Trophic Factor Gene Disruption a "Knockout" Model for Parkinson's Disease?

Ann-Charlotte Granholm and Barry Hoffer

1. INTRODUCTION

Parkinson's disease is a debilitating and progressive disease of the extrapyramidal motor system that occurs due to the degeneration of dopaminergic (DAergic) neurons in the ventral mesencephalic (VM) nucleus substantia nigra *(1)*. Despite numerous hypotheses and continued speculation, the triggering mechanisms for this degeneration of the DA neurons remain unknown. Recent studies have suggested that mitochondrial oxidative dysfunction may cause premature cell death and may be linked to accelerated apoptosis, excessive free and toxic radicals, deficient neurotrophic factors, or some combination thereof *(2–6)*. It has, for example, been shown that certain substances can give rise to significant apoptotic changes in these neurons, such as levodopa *(7)*, DA–melanin *(8)*, 1-methyl-4-phenyl-1,2,3,6-tetrahydropyridine (MPTP; *9*), hypo-ischemia *(10)*, as well as axonal transection *(11,12)*. The standard therapeutic intervention to date is the administration of the DA precursor levo-Dopa, or L-Dopa *(13,14)*. However, with the progression of the disease this drug is less effective, and continued administration also leads to significant negative side effects such as hallucinations and hyperkinesia *(1,13)*. Therefore, much research effort has been put forth during recent years to develop alternative treatment strategies.

Several different factors have been identified that are involved in differentiation, survival, and/or fiber outgrowth from midbrain DAergic neurons. Partial recovery of striatal DA levels were found in MPTP-lesioned mice following a stereotaxic injection of acidic fibroblast growth factor (aFGF; *15*), and similar trophic effects on DA neurons in the rodent have been reported for brain-derived neurotrophic factor (BDNF; *16–18*), insulin-like growth factor *(16,19)*, platelet-derived neurotrophic factor (PDGF; *20,21*), and ciliary neurotrophic factor (CNTF; *22*). Transforming growth factor-α (TGF-α *23*) has also been found to have in vivo effects on DA neurons. In 1993, a novel member of the TGF-β superfamily of growth factors was isolated and cloned by Lin and collaborators *(24)*. The factor was named glial cell line-derived neurotrophic factor (GDNF) and was found to be potent and highly selective for DA neurons, at least in tissue cultures of developing rodent mesencephalon. In the past 5 yr, we and others

From: Central Nervous System Diseases
Edited by: D. F. Emerich, R. L. Dean, III, and P. R. Sanberg © Humana Press Inc., Totowa, NJ

have worked on the characterization of the in vivo effects of GDNF, and it was found that this factor had significant effects on both intact and injured adult nigra DA neurons in rodents *(25–30)* and in nonhuman primates *(31)*, as well as on survival and fiber outgrowth from developing DA neurons in mesencephalic transplants to the 6-hydroxydopamine (6-OHDA)-lesioned striatum of rat *(32,33)*. To determine if absence of GDNF during development correspondingly leads to agenesis of DA neurons, we studied the development of this neural system in mice with a GDNF null mutation *(34,35)*. The results of these GDNF knockout studies are reviewed in detail below (pp. 231–232).

Most of the earlier work on trophic dependency during development was done on tissue culture preparations of selective neuronal populations. However, culture studies may not always accurately reflect the trophic relationships at a certain developmental time point in the whole animal in vivo. For example, trigeminal ganglion cells show a switching from BDNF/NT3 dependency to nerve growth factor NGF dependency in culture *(36,37)*. This change in trophic factor dependency may be caused by alterations in trophic factor receptors but may also be the result of altered biological processes in neurons that are expressed during in vitro conditions. It is known that neurons also switch trophic factor dependency several times during their development *in situ*, so that one factor may be crucial for the proliferation of a specific phenotype, another for migration, and yet a third factor for the axonal extension toward the target and establishment of synaptic contact. The signaling mechanisms within the neuron that determine the various trophic factor dependencies during the different phases of neural development have not been fully identified, but it is clear that the process of neural development is complicated and multifaceted, especially for the central nervous system (CNS). Because of the limitations of earlier methods, there was a need for whole animal models of trophic dependency during early development. Targeted gene disruption of trophic factors therefore revolutionized the field when first described a decade ago *(38–42; see 43* for technical details on the gene disruption animal model).

Even though gene targeting by null mutation of a certain gene (knockouts, K/O) has simplified studies of gene function in vivo, this model system has certain limitations. The early lethality of many trophic factor knockout mice prevents repeated assessment after birth, and prevents studies of maturation and age-related alterations (*see* Fig. 1). In addition, marked congenital defects can compromise other systems such that the primary effects of a null mutation are difficult to separate from secondary and tertiary effects *(44)*. The inserted gene cassette may also in itself exert unwanted side effects that may obscure direct effects of the knockout and further complicate interpretation of the data. Finally, the normal interactive relationships between different trophic factors may be obscured by a compensatory "takeover" by other, closely related, molecules. Such a compensation has been described, for example, by Erickson and collaborators *(45)*, who found that a double knockout of BDNF and NT-4 leads to a much greater reduction in DAergic neurons innervating the carotid body than with each trophic factor knockout by itself (Fig. 2). A prominent adaptability and cross-compensation for developing CNS neurons between different trophic factors has also been suspected for other knockouts, for example, the NGF–/– and the TrkA–/– animals, which appear to develop normal cholinergic neurons with the normal pattern of innervation, albeit with a reduced synthesis of the transmitter enzyme, choline acetyltransferase, and in

Knockout:	PNS abnormalities	CNS DA neurons	Other CNS MA neurons	DA levels in target	Behavioral changes
NURR-1	No	Agenesis	No effects	↓	-/- die at P2
Neurturin	Normal PNS	Normal CNS	No effects?	Normal ?	Normal life
GDNF	Kidney and enteric plexus agenesis or dysgenesis	No apparent effects	25% reduction in LC neurons	?	-/- die at birth
NT-4	Sensory neuron loss	Normal	Normal	Decreased DA in carotid body	Normal life span, no behavioral signs
BDNF	Sensory neuron loss	Normal	Normal	Decreased DA in carotid body Neuropeptide expression down	-/- die within several weeks of birth/balance deficiencies
D_{1A} rec.	No	Normal	Normal	Normal	Hyperactive/ lacking cocaine response
D_2 rec.	No	Normal	Normal	Normal	Cataleptic
D_3 rec.	No	Normal	Normal	Normal	Hyperactive/ lacking cocaine response
Dopamine	No	Normal dev. but no TH or DA	Normal dev. and normal NE levels	No DA in striatum	Aphagia, adipsia, and hypoactive
VMAT2	Impaired storage and release of DA	Other MA affected as well	↓ in striatum of +/- mice		-/- die a few days after birth

Fig. 1. Summary of the findings from the different knockout models discussed in the present chapter.

TrkA–/–, a reduction in axonal cholinesterase activity in the target areas *(42,46)*. It would be incorrect to deduct from these two different knockout experiments that NGF does not affect development, differentiation, and axonal growth of cholinergic neurons in the CNS, as a multitude of in vivo studies have indicated the opposite. Instead, it is more plausible to postulate that a compensatory change between different trophic factors within the same family occurs. The most puzzling finding from the trophic factor knockout studies is, however, that such compensation does NOT occur to the same extent in peripheral neurons. The NGF and TrkA knockout mice exhibit a significant decrease in both sensory and sympathetic neurons in the periphery even though these peripheral ganglion cells are known to respond to, and have receptors for, the other family members of the neurotrophin family *(44)*.

Recent developments have led to systems that can generate tissue-specific and/or temporally regulated knockouts, for example, the development of antisense technology (see, e.g., Kaur and Creese, *this volume*) and CreLoxP-mediated gene targeting *(47,48)*. Alternatively, classic knockouts can be rescued at a specified developmental time point *(44)*. Another way to determine if compensation by other, closely related, ligands occurs when the gene for one trophic factor is disrupted is to generate receptor knockouts instead of null mutations of the trophic factor itself *(41,42)*. These different novel model systems will allow for more detailed analysis of trophic factors and their role during

TARGET INNERVATION:

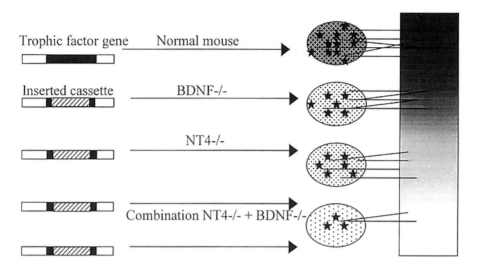

Fig. 2. Compensation by other growth factors following a selective knockout. A combination of gene disruption of the neurotrophins BDNF and NT4 leads to a more severe impairment in target innervation of DA fibers than either of the knockouts by themselves. (Adapted from Erickson et al., *ref. 45.*)

both development and in the adult brain, and hold great promise for future studies in this field.

The present chapter focuses on the data available to date from trophic factor knockout strains that may (or should) affect the nigrostriatal DA system. As can be seen from the previous description, these factors have potent effects when tested in both in vivo and tissue culture experiments, but they do not always have the same effects in knockout experiments. Different explanations for these confounds between in vivo results and gene disruption experiments are discussed.

2. NEUROTROPHIN KNOCKOUTS

The neurotrophins are one family of trophic molecules that have great homology with each other and that have been intensively studied for possible therapeutic use in neurodegenerative diseases *(49–52)*. The family consists of NGF, BDNF, neurotrophin 3 (NT-3), and neurotrophin 4/5 (NT-4/5), and the recently discovered neurotrophin 6 (NT-6, *53*). They are dependent on three high-affinity subtypes of tyrosine kinase receptors—TrkA, TrkB, and TrkC—for their biological activity *(54–56)*. Of the neurotrophins, both BDNF and NT-4 have been shown to exert biological activity on midbrain DA neurons *(17,18,57)*. Knockout mice for all of the neurotrophins, except for NT-6, as well as all of the Trk receptors have been described (for review *see 44,58*). It has been concluded from these gene disruption experiments that, at least in the peripheral nervous system (PNS), similar phenotypes result from the knockouts of the trophic molecules as from the receptor knockouts and, in addition, that these phenotypes are roughly in line with what the neurotrophic hypothesis would predict in terms of neuronal loss. In contrast with the in vivo results, however, it was surprisingly found that BDNF–/– mice did not develop any observable abnormalities of the substantia nigra, at least not in the early expression of this phenotype (the BDNF –/– animals die within weeks of birth; *see* Fig. 1). Similarly, adult NT-4 knockout mice had no obvious loss of midbrain DAergic neurons even though NT-4 mRNA is expressed in the mesencephalon *(59)* and DA neurons are responsive to NT-4 in tissue culture experiments *(57)*. The only alteration that was seen in DA neurons in either BDNF or NT-4 gene disrupted mice was a reduction in the peripheral DA innervation of the carotid body *(45)*. In addition, a regional alteration in neuropeptide and calcium-binding protein expression in cerebral cortex, hippocampus, and striatum was found in the BDNF-deficient mice *(45)*. Thus, it is possible that these peptidergic neurons are dependent on BDNF for their phenotypic expression, or, alternatively, are affected by secondary alterations in the target environment. As has been discussed previously, it was thought that the lack of alterations of DA neurons in the BDNF–/– mice might be due to compensation by other neurotrophins in these mice. The brain appears to be significantly more adaptive than peripheral nervous systems, as the BDNF–/– mice die shortly after birth due to peripheral neuronal agenesis.

All of the neurotrophin knockout mice demonstrated significant loss of peripheral neurons, including sympathetic, parasympathetic, and sensory neurons *(44)*. Double mutant mice that lacked both TrkA and TrkC, or TrkB and TrkC, showed an additive loss of dorsal root ganglion cells (DRG; 93% and 41%, respectively) compared to single knockouts. The authors concluded that this suggests that distinct TrkA, B, and C positive populations of neurons innervate different targets in the periphery. NT-4 and BDNF both use the TrkB receptor for their biological effects. Interestingly, the TrkB knockout mice exhibited deficits that were similar to those seen in a combined NT-4–BDNF knockout mouse with the exception of motor neurons *(44)*. Motor neuron numbers were found to be normal in NT-4–/–, in BDNF–/–, and in the combination knockout *(60)*, while TrkB knockouts showed a marked loss of facial and spinal cord motor neurons *(41)*. This strongly suggests that another ligand for TrkB exists that may compensate for the loss of both NT-4 and BDNF during development of motor neurons.

3. TGF-β KNOCKOUTS

The TGF-β family of trophic factors is a group of secreted proteins that inhibit proliferation of many cell types in the periphery, regulate the expression of extracellular matrix proteins, and are considered important regulators of cell proliferation and differentiation *(61)*. The genes for GDNF and neurturin, two subfamily members with actions on midbrain DA neurons in vitro and in vivo *(24–26,29–32,62,63)*, have been disrupted and the DA systems investigated in the CNS *(34,64–66)*.

In collaboration with Drs. Liya Shen and Heiner Westphal, we studied the early development of tyrosine hydroxylase (TH) immunoreactive neurons of the substantia nigra in GDNF null mutated mice and, surprisingly, found no evidence for a dysgenesis or agenesis of DA neurons at this developmental stage, despite the potent effects reported from in vivo experiments (Fig. 3). Both central and peripheral noradrenergic neurons appeared to be more affected by the GDNF gene disruption, as we found approximately a 25% decrease in the TH immunoreactivity in brain stem locus coeruleus neurons, as well as a marked reduction in sympathetic innervation of the head and neck *(34,35)*. A dependency on GDNF for the noradrenergic locus coeruleus neurons has also been reported in vivo in the adult rat, where GDNF-secreting grafted cells are able to rescue noradrenergic neurons from 6-OHDA-induced apoptosis *(67)*.

There are several possible explanations for the lack of effects of the GDNF null mutation on DA neurons of the substantia nigra. It is well known that the DA neurons undergo two major waves of apoptosis during development. The first wave occurs at 1–2 d postnatally, and the second wave occurs at 10–14 d of age *(68)*. If it is true that GDNF is effective in blocking apoptosis of DA neurons, as suggested by in vitro studies *(62,63,69)*, the GDNF knockout animals die before the DA neurons would have been subjected to their first wave of apoptosis, and would thus not need GDNF until later in life. The waves of apoptosis for DA neurons coincide with the time of synaptic contact with their striatal target. This notion is consistent with the hypothesis that GDNF is a target-derived trophic factor, and is thus not required until the neurons reach the target and establish contact. Another plausible explanation for the lack of effects of the GDNF null mutation on DA neurons is that another, closely related, trophic molecule is able to compensate and take over as a ligand for the GFR-α1–Ret receptor complex *(70,71)*. We know that at least one more member of this TGF-β subfamily, neurturin, shares receptors and signal transduction pathways with GDNF, which suggests that neurturin is a physiologically relevant ligand for the Ret receptor and may function in a compensatory fashion in the GDNF knockout animals *(72)*. Interestingly, neurturin knockout mice *(66)* do not share the same peripheral abnormalities that were found in GDNF knockout mice. They grow well and are fertile and also exhibit normal kidneys and enteric nerve plexa. They do not exhibit any alterations in DA morphology or function, as described in preliminary studies *(66)*. Another member of this TGF-β subfamily of growth factors, persephin, has recently been discovered *(73)*. Persephin has approx 40% homology with neurturin and GDNF, and has been found to protect adult DA neurons from 6-OHDA-induced lesions in the rat, as well as to act as a differentiation factor for DA neurons in tissue cultures of fetal midbrain tissue *(73)*. It would be interesting to study combined knockouts of these three TGF-β subfamily members, to see if the combined loss of Ret ligands might alter the fate of midbrain DAergic neurons. It would also be interesting to study aged neurturin knockout mice to determine if this factor may be more important for age-related plasticity and target plasticity than for early development, when many other factors are present in abundance. The GDNF-related trophic factors represent a novel group of neurotrophic factors, and their biological significance has not yet been fully elucidated.

3.1. GDNF Receptor Knockouts

As was discussed previously regarding the neurotrophin knockouts, additional information regarding biological effects of a trophic factor may be obtained if a receptor knockout is performed, rather than gene disruption of the trophic molecule itself. It has previously been shown that Ret receptor knockout animals display peripheral symptoms that are similar to those found in the GDNF knockout animals *(74)*. Recently, two different research groups *(75,76)* have reported the effects of deletion of the gene coding for the surface membrane binding portion of the receptor complex for GDNF, termed GFR-α1. Deletion of this protein using standard homologous recombination techniques produced mice with agenesis of kidneys and enteric nervous system, as well as reduced numbers of spinal motor and sensory neurons, analogous to the alterations

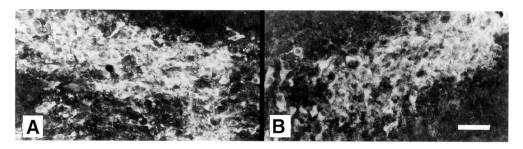

Fig. 3. The microphotographs show sections from the substantia nigra pars compacta from embryonic day 18 mouse embryos, incubated with antibodies directed against TH. The section in (**A**) is from a GDNF–/– fetus, while the section in (**B**) is from a wild-type fetus of the same litter. Note that there are no observable differences in the size, staining intensity, or packing density of TH-immunoreactive neurons in this CNS nucleus, at least not at this early stage. The GDNF–/– animals die at birth, which prevents studies of maturation and target innervation of the DA neurons. The scale bar in (B) represents 40 μm.

seen in GDNF knockout mice. CNS DAergic neurons, however, were not altered in terms of their morphological development in the GFR-α1 knockout mice. On the other hand, in vitro studies using fetal midbrain DA neurons or nodose ganglion tissues from GFR-α1 null mutations demonstrated a marked loss of GDNF sensitivity in these tissues *(75)*. However, tissue cultures of submandibular ganglion cells from GFR-α1 knockout fetuses retained their sensitivity to both GDNF and neurturin, despite the lack of the GPI-linked portion of the GDNF receptor complex *(75)*. Thus, GDNF and neurturin may mediate some of their activities through a second, unknown, receptor in some neuronal phenotypes. The combined results from the Ret and the GFR-α1 knockout studies indicate that both of these receptor components are necessary for normal development of kidneys and enteric nerve plexa, but that some adaptability still exists in the CNS DA neurons in terms of developmental dependency. The TGF-β subfamily knockouts and their receptor knockouts have given valuable information regarding peripheral development, and will continue to yield information regarding development of both the PNS and the CNS.

4. NURR 1 KNOCKOUTS

NURR1 is an immediate early gene product, and an orphan member of the steroid nuclear receptor superfamily *(77)*. These receptors function as transcription factors, regulating and programming developmental, physiological, and behavioral responses to a variety of chemical signals. NURR1 is expressed in many regions of the CNS, but its function and cognate ligand remain to be investigated *(77)*. NURR1 was shown to appear in midbrain neurons before expression of DA phenotype occurred in this brain region *(78)*. The NURR-1 knockout appears to be the best model so far for DA neuron agenesis, as NURR1 knockout mice were born with a complete agenesis of mesencephalic DA neurons. It was interesting to note from this study that the agenesis of DA neurons appeared to be restricted to the midbrain DA phenotype and unrelated to TH, as other TH-positive neurons in the brain stem were present and developed normally. The NURR-1–/– animals died 2 d postnatally of unrelated causes, most likely due to a lack of suckling ability. These animals are very interesting from several different perspectives, not only because of their complete lack of midbrain DAergic neuron development, but also because they display a completely normal development of peripheral sympathetic and sensory neurons, unlike the trophic factor knockouts described previously. It is not known what causes this selective loss of DA neurons, but the answer may be found in the expression patterns of NURR-1 during early development. NURR-1 appears in the central DA neurons during development, as early as embryonic day 10.5 *(78)*. In conclusion, the NURR-1 mouse represents a powerful tool for investigation of DA dependency for different behaviors. It was found that DA levels were significantly decreased in NURR-1 heterozygous mice but it is not known yet if motor behavior is impaired in these heterozygous mice. It would also be interesting to transplant ventral mesencephalon from embryonic day 11–12 of NURR-1 knockout fetuses to wild-type adult hosts, to determine if the presence of trophic factors may induce development of DA phenotype in such transplants.

5. DOPAMINE AND DA RECEPTOR KNOCKOUTS

5.1. DA Knockouts

A direct approach toward deleting DA in the brain was attempted in 1995, when Zhou and collaborators disrupted the TH gene in mice *(79)*. However, they soon realized that most TH–/– mice died *in utero*, and also that they had disrupted production not only of DA in the brain but also of norepinephrine (NE) in the PNS and CNS. They designed an extremely interesting model, in which they were able to disrupt the DA production in DAergic neurons, but maintain the production of NE in locus coeruleus neurons *(80)*. Their model is described schematically in the diagram below. Basically, they first produced TH–/– mice (that had a deficiency in the production of both NE and DA, and of which most died *in utero*). They then inserted the entire TH gene sequence, including a neocassette, into the dopamine-β-hydroxylase (DBH) gene, predicting that this would disrupt TH in all neurons that did not have the ability to produce NE through DBH. For simplistic reasons, they called the TH–/–:TH-DBH+/– combination offspring DA–/– mice, as this particular combination gave rise to lack of DA production, but maintained NE production in the locus coeruleus *(80)*.

1. Disruption of the TH gene → TH–/– mice → loss of both DA and NE
2. Insertion of the TH gene into DBH gene → TH-DBH–/– → loss of only NE, keep DA
3. Combination of TH+/– and TH-DBH+/–→ TH–/–+ TH-DBH+/–→ loss of only DA, keep NE

The DA–/– mouse is an excellent model for pharmacology and development of DA neurons, and probably one of the better approaches that have been described to date. The newborn DA–/– are grossly indistinguishable from wild-type littermates, but by postnatal d 10 they become hypoactive, adipsic and aphagic and will die if not treated with L-Dopa *(80)*. It is interesting to note that the DAergic neurons of the substantia nigra, as well as their projections to the striatum, appeared to develop normally in the DA–/– mice, despite a total lack of TH and DA in this pathway, as evidenced by immunocytochemistry using antibodies against L-aromatic amino acid decarboxylase (AADC), which is an enzyme that converts L-Dopa to DA. In addition, the authors found a quite normal development of many different parameters in the target area striatum in the DA–/– mice. For example, staining with DARPP32 (DA- and cyclic AMP-regulated phosphoprotein; a protein whose phosphorylation is regulated by D_1 receptor stimulation), expression of D_1 and D_2 receptor mRNA, as well as levels of GAD65 and GAD67 (γ-aminobutyric acid [GABA] synthesizing enzymes) all appeared normal in DA–/–, as compared to wild-type striata (80). They drew the conclusion that, despite earlier suggestions, DA is not necessary for development of the nigrostriatal system, and also not for expression of its receptors or for normal target development. These findings have shed light on the complex interrelationship between target and neuronal afferents during early neurogenesis. The DA–/– mice had significant alterations in their behavior, however, that demonstrated the dependence of DA for complex behavior in mammals. The hypoactivity observed in the DA–/– was easily counteracted by small doses of L-Dopa; the same dose of the drug had no effect on activity in wild-type mice. They also exhibited significant reductions in stereotypical behavior (rearing, sniffing, etc.) as compared to controls, a deficiency that was also rapidly reversed by the L-Dopa treatment. The DA–/– mice are interesting not only for studies of behavior and pharmacological assessments. The model can be used for development of novel treatment strategies in Parkinson's disease as well. A recent presentation by Szczypka et al. *(81)* demonstrated that the aphagia, hypoactivity, and adipsia found in the DA–/– could be counteracted by intrastriatal injections of adeno-associated virus-mediated L-Dopa, and that one injection resulted in sufficient amounts of L-Dopa for at least 9 wk in the mouse using this novel drug delivery system.

5.2. DA Receptor Knockouts

Mutant mice lacking the different subtypes of DA receptors have also been developed. Five distinct DA receptor subtypes have been recognized to date (for review, *see 82*). The DA receptor knockouts are particularly interesting in the perspective of understanding the interaction of DAergic interaction with other transmitter systems, such as the excitatory amino acid input to the striatum from cortex (see, e.g., *83,84*). Gene disruption of D_{1A}, D_2, and D_3 receptors has been generated and characterized physiologically (*85–90; see* Fig. 1). The DA receptor knockout mice generally appear to do well and survive until adulthood, but they manifest certain behavioral impairments (*see also* Fig. 1). The D_2 knockout mice are cataleptic *(88)*, whereas the D_{1A} and the D_3 receptor knockouts have been reported to be hyperactive and to display decreased rearing and also a lack of response to cocaine *(86,91)*. The differences in behavior of the different receptor knockouts clearly demonstrated the different functions of the DA receptor subtypes in a behaving animal, and just not on the cellular level. None of the DA receptor knockouts have been reported to have agenesis or dysgenesis of monoaminergic neurons, either in the CNS or PNS, and the neostriatal spiny neurons exhibited normal morphological features. It is possible that a combined knockout of all DA receptors would have more severe consequences for DAergic circuitry in the brain. Meanwhile, this model serves as an excellent tool for electrophysiological characterization of the different DA receptor subtypes, and crucial information has been gathered regarding DA transmission using this system.

6. MONOAMINE TRANSPORTER KNOCKOUTS

The actions of DA are mediated almost exclusively by G-protein-coupled membrane receptors, but the intra- and extracellular concentrations of monoamines are primarily controlled by two types of transporters. The sodium- and calcium-dependent plasma membrane transporters exhibit cell-specific gene expression between the different monoamines, and are responsible for transport of DA into the cell following release. The vesicular transporters that allow synaptic vesicles to engage in stimulated release by transporting monoamines from the cytoplasm to the vesicles, on the other hand, exhibit a lack of selectivity for different monoamines (*see* e.g., *92*). There are two subtypes of vesicular monoamine transporters, namely VMAT1 and VMAT2 *(93)*, of which VMAT2 is expressed primarily in the nervous system and in histaminergic cells of the gastrointestinal canal *(94)*. Mice that are homozygous for a gene disruption of VMAT2 die within a few days of birth, showing severely impaired monoamine storage and vesicular release *(95)*. In heterozygous adult mice, the extracellular striatal DA levels are decreased, and the potassium- and amphetamine-evoked DA release is significantly reduced. Because potassium-evoked DA release has been shown to be reduced also in the aged rat striatum *(96)*, the VMAT2+/– mice may serve as an excellent model system for age-related decreases in neuronal plasticity and function. These presynaptic alterations in DA storage and release are coupled with an increased sensitivity to applied neurotoxins such as MPTP, and also to an increased sensitivity to the locomotor effects of the DA agonist apomorphine, the psychostimulants cocaine and amphetamine, as well as ethanol *(95,97)*. They also do not develop sensitization to repeated cocaine administration (lack of adaptability in the system).

Recently, inactivation experiments of the gene encoding the plasma membrane DA transporter (DAT) have also been undertaken *(98,99)*. These studies clearly demonstrated the importance of DAT as the key element for maintaining DA levels in the brain, and also its role as an obligatory target for the behavioral and biochemical actions of DAergic drugs such as cocaine and amphetamine *(98)*. The studies on gene disruption of both the membrane and vesicular DA transporters provide concordant evidence for a significant role of the transporters in DAergic transmission in the brain, and these findings may have significance for the future treatment of vastly different diseases, such as drug addiction, schizophrenia, and Parkinson's disease.

7. TRANSPLANTATION EXPERIMENTS USING KNOCKOUT MICE AS DONOR OR RECIPIENT TISSUE

Transplantation of ventral mesencephalic neurons from fetuses is a novel treatment approach for Parkinson's disease, which has been developed during the last decade *(52,100–107)*. However, there is a low survival rate of DAergic neurons in the transplants and only approx 5–10% of the transplanted cells survive *(108–110)*. It is not known what causes this low survival rate, nor is it known which factors might be used to stimulate increased survival. Initial experiments have demonstrated that GDNF may enhance both cell survival and fiber outgrowth from ventral mesencephalic transplants into a unilaterally lesioned striatum in the rat *(32,33)*. In addition, in vitro experiments by Clarkson and collaborators have indicated that GDNF may also affect the rate of apoptosis of both human and rodent DA neurons in the culture dish *(62,63,69)*. These data suggest that treatment with trophic factors may enhance the survival of DA-rich transplants clinically. Several different lines of research have been recently developed using transplantation in combination with transgenic mice to determine factors that may enhance graft survival in the clinic. Two different examples of neural transplantation using fetal tissue from knockout animals are briefly described in the following

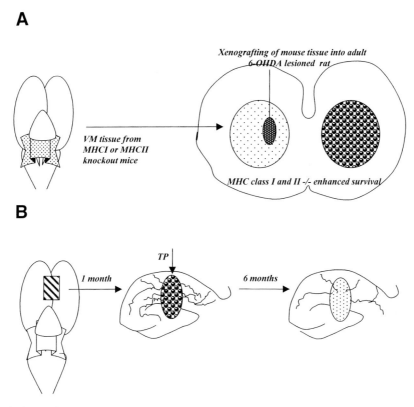

Fig. 4 Transplantation experiments using knockout animals. (**A**) Duan et al. (1998) found that transplants of ventral mesencephalon from MHC I and MHC II knockout mice survived significantly better than xenotransplants of mouse tissue from wild-type mice. (**B**) Trisomy 16 cross grafting experiments. Ts 16 fetal tissue survived well and grew in the hippocampal formation of adult wild-type mice. However, with time, there was a selective degeneration of cholinergic neurons in Ts 16, as compared to wild-type grafts.

paragraphs to demonstrate the potential of combining these two useful neuroscience models.

In the first set of experiments, Low and collaborators described experiments involving transplantation of tissue from major histocompatibility complex (MHC) class I and II knockout mice into rats, and found significant effects on survival when the genes for both of these cell surface molecules were disrupted (Fig. 4A; *111*). The authors showed that both MHC class I and MHC class II gene disruption significantly enhanced survival, and reduced rejection, of mouse tissue that was xenografted into an adult 6-OHDA-lesioned rat. At 42 d postgrafting, the wild-type, MHC class I, and MHC class II knockout mouse transplants exhibited 5 ± 5, 262 ± 76, and 216 ± 81 surviving DA neurons per section, respectively. This represented a significant increment in DA cell survival in both classes of MHC knockout transplants, compared to wild-type transplants. They argued that these two cell surface molecules affected survival of transplanted DA neurons. These findings are intriguing and could lead to a possible genetic manipulation strategy for transplantation of tissue between different human donors and recipients, as well.

Another interesting possibility for combining transgenic mouse studies with transplantation models has been demonstrated by Holtzman and collaborators, who have developed a grafting strategy for studies of trisomic mice with selective forebrain cholinergic deficits *(112)*. These authors transplanted tissue from a fetal mouse model of Down's syndrome, the trisomy 16 mouse, to wild-type mice to study the spontaneous neurodegeneration seen in the former. The trisomy 16 (Ts 16) mouse model served as a model for Down's syndrome, as the mouse chromosome 16 (MMU-16) is the chromosome most similar to human chromosome 21(HSA-21; *113)*. However, the mice with Ts 16 do not survive past birth, and age-related degeneration could therefore not be studied. Holtzman and collaborators *(112)* performed an experiment in which fetal forebrain tissue from Ts 16 or wild-type fetal brains was transplanted to the hippocampus of young adult mice. Transplanted neurons survived, and grew processes, in all grafts. However, there was a selective cholinergic degeneration with time in the grafts from the Ts 16 mice, compared to wild-type grafts (*see* Fig. 4B). Denervation of hippocampus produced a significant increase in the size of cholinergic neurons in Ts 16 grafts, consistent with trophic interaction from the denervated host hippocampus. This represents an interesting model for studies of trophic factor effects on neurodegeneration. It would be interesting to determine if spontaneous age-related degeneration also occurs in the DA system of trisomic mice, as the DA system also undergoes age-related impairment in the Down's syndrome patient.

There are some limitations to the Ts 16 model described previously, for example, the fact that the MMU-16 is larger than HSA-21, and contains genes that are located both on HSA-21 and on other human chromosomes. Therefore, another model for Down's syndrome was developed, that has a segmental trisomy only on the genetic material present on the HSA-21 responsible for the Down's syndrome genotype. One such model is the Ts65Dn mouse that has been described recently *(114)*. These mice survive until adulthood, experience a significant and spontaneous degeneration of forebrain cholinergic neurons at approx 6 mo of age, and exhibit specific memory impairments consistent with cholinergic hypofunction *(114,115)*. Grafting experiments between the Ts65Dn mice and wild-type adult hosts could determine if environmental and/or supplemental growth factor treatment in a wild-type host could prevent the spontaneous degeneration occurring in forebrain cholinergic neurons in Ts65Dn mice. Studies of cross-grafts between Ts65Dn fetuses and wild-type adult hosts are ongoing in our laboratory. We are also transplanting tissue from trophic factor knockout mice to wild-type hosts, to study if deprivation of a specific trophic factor in the target, for example, striatal tissue, causes alterations in DA innervation during development.

8. CONCLUDING REMARKS

Single gene disruption of trophic factor genes has already generated a wealth of information regarding mammalian development, and in particular the development of the PNS. It is likely that this approach is only in its infancy, and that many discoveries can be expected in the years to come both from the utilization of knockout mice that already exist and those that will be generated, in particular with temporally and/or regionally controlled gene disruptions. When studying single gene disruption and its effects on development of the brain, one must be cautious in the interpretation of data, as the brain appears to have an inherent adaptability to gene mutations that is unparalleled in the PNS. It is also clear from these initial studies that the actions of different growth factors are more complex, and in some instances also more interactive, than previously thought. Nevertheless, the gene disruption systems that have been reviewed here can give us valuable information regarding the signaling of specific trophic molecules during neuronal proliferation, migration, and differentiation as well as target innervation and phenotypic maintenance. By utilizing knockout animals in grafting experiments, we can take advantage of the single gene disruptions and study effects of one trophic factor in, for example, target innervation or synaptic development.

The different knockout models described in this chapter are clearly important for elucidation of factors that may regulate development of midbrain DAergic neurons. It is puzzling that none of the trophic factor knockouts lead to any impairments of these neurons, with the exception of the NURR1 knockout mice, which exhibit an exclusive and complete agenesis of midbrain DAergic neurons. Even the nigra neurons of DA–/– mice developed normal connections with the striatum, despite a complete absence of DA during the development and axonal extension into the target. This has yielded important information regarding the nigrostriatal system; apparently the projection is not sensitive to presence or absence of the transmitter itself, but will develop normal target connections in the absence of DA synthesis and release. The lack of effects of the other trophic factor knockouts demonstrates the marked adaptability that exists in the CNS and also suggests that conclusions should not be drawn from knockout experiments regarding trophic factor dependency of specific central neuronal phenotypes. However, the trophic factor knockout mice are valuable as model systems for transplantation experiments and, furthermore, combinatorial knockouts may render more specific information regarding the importance of several trophic ligands for developmental function.

ACKNOWLEDGMENTS

This work was supported by USPHS Grants AG12122, AG15239, and AG04418. The authors would like to thank Linda Sanders, Justin Mott, Stephanie Henry, and Nisha Srivastava for excellent assistance with the immunohistochemistry and microphotography. Dr. Granholm is supported by a career development award from the National Institutes on Aging (Grant no. AG00796).

REFERENCES

1. Schneider, J. S. (1998) GM1 ganglioside and neurotrophic factors in the treatment of Parkinson's disease. *Neuroscience News* **1,** 10–16.
2. Fahn, S. and Cohen, G. (1992) The oxidant stress hypothesis in Parkinson's disease: evidence supporting it. *Ann. Neurol.* **32,** 804–812.
3. Mochizuki, H., Goto, K., Mori, H., and Mizuno, Y. (1996) Histochemical detection of apoptosis in Parkinson's disease. *J. Neurol. Sci.* **137,** 120–123.
4. Tompkins, M. M., Basgall, E. J., Zamrini, E., and Hill, W. D. (1997) Apoptotic-like changes in Lewy-body associated disorders and normal aging in substantia nigra neurons. *Am. J. Pathol.* **150,** 119–131.
5. Ruberg, M., Brugg, B., Prigent, A., Hirsch, E., Brice, A., and Agid, Y. (1997) Is differential regulation of mitochondrial transcripts in Parkinson's disease related to apoptosis? *J. Neurochem.* **68,** 2098–2110.

6. Kosel, S., Egensperger, R., von Eitzen, U., Mehraein, P., and Graeber, M. B. (1997) On the question of apoptosis in the Parkinsonian substantia nigra. *Acta Neuropathol.* **93,** 105–108.

7. Ziv, I., Zilkha-Falb, R., Offen, D., Shirvan, A., Barzilai, A., and Melamed, E. (1997) Levodopa induces apoptosis in cultured neuronal cells—a possible accelerator of nigrostriatal degeneration in Parkinson's disease? *Move. Disord.* **12,** 17–23.

8. Offen, D., Ziv, I., Barzilai, A., Gorodin, S., Glater, E., Hochman, A., and Melamed, E. (1997) Dopamine-melanin induces apoptosis in PC12 cells: possible implications for the etiology of Parkinson's disease. *Neurochem. Int.* **31,** 207–216.

9. Tatton, N. A. and Kish, S. J. (1997) In situ detection of apoptotic nuclei in the substantia nigra compacta of MPTP-treated mice using terminal deoxynucelotidyl transferase labeling and acridine orange staining. *Neuroscience* **77,** 1037–1048.

10. Oo, T. F., Henchcliffe, C., and Burke, R. E. (1995) Apoptosis in substantia nigra following developmental hypoxic-ischemic injury. *Neuroscience* **69,** 893–901.

11. Venero, J., Revuelta, M., Cano, J., and Machado, A. (1997) Time course changes in the dopaminergic nigrostriatal system following transection of the medial forebrain bundle: detection of oxidatively modified proteins in substantia nigra. *J. Neurochem.* **68,** 2458–2468.

12. Marti, M. J., James, C. J., Oo, T. F., Kelly, W. J., and Burke, R. E. (1997) Early developmental destruction of terminals in the striatal target induces apoptosis in dopamine neurons of the substantia nigra. *J. Neuroscience* **17,** 2030–2039.

13. Koller, W. C. and Hubble, J. P. (1990) Levodopa therapy in Parkinson's disease. *Neurology* **40 (Suppl.),** 40–47.

14. Montgomery, E. B., Jr. (1992) Pharmacokinetics and pharmacodynamics of levodopa. *Neurology* **42,** 17–22.

15. Date, I., Notter, M., Felten, S., and Felten, D. (1990) MPTP-treated young mice but not aging mice show partial recovery of the nigrostriatal dopiminergic system by stereotaxic injection of acidic fibroblast growth factor (aFGF). *Brain Res.* **526,** 156–160.

16. Beck, K. D., Knusel, B., and Hefti, F. (1993) The nature of trophic action of brain-derived neurotrophic factor, des(1-3)-insulin-like growth factor-1, and basic fibroblast growth factor on mesencephalic DAergic neurons developing in culture. *Neuroscience* **52,** 855–866.

17. Hyman, C., Hofer, M., Barde, Y-A., Juhasz, M., Jancoupolous, G., Squinto, S., and Lindsay, R. (1991) BDNF is a neurotrophic factor for dopaminergic neurons of the substantia nigra. *Nature* **350,** 230–232.

18. Knusel, B., Winslow, J., Rosenthal, A., Burton, L., Seid, B., Nikolicks, K., and Hefti, F. (1991) Promotion of central cholinergic and dopaminergic neuron differentiation by brain-derived neurotrophic factor but not neurotrophin-3. *Proc. Natl. Acad. Sci. USA* **88,** 961–965.

19. Knusel, B., Michel, P. P., Schwaber, J. S., and Hefti, F. (1990) Selective and non-selective stimulation of central cholinergic and dopaminergic development in vitro by nerve growth factor, basic fibroblast growth factor, epidermal growth factor, insulin and the insulin-like growth factors I and II. *J. Neurosci.* **10,** 558–570.

20. Nikkah, G., Odin, P., Smits, A., Tingström, A., Ohtberg, A., Brundin, P., Funa, K., and Lindvall, O. Platelet-derived growth factor promotes survival of rat and human mesencephalic dopamine neurons in culture. *Exp. Brain Res.* **92,** 516–523.

21. Giacobini, M. M. J., Almstrom, S., Funa, K., and Olson, L. (1993) Differential effects of PDGF isoforms on dopamine neurons in vivo: AA enhances formation—BB supports cell survival. *Neuroscience* **57,** 923–929.

22. Magal, E., Burnham, P., Varon, S., and Louis, J-C. (1993) Convergent regulation by ciliary neurotrophic factor and dopamine of tyrosine hydroxylase expression in cultures of rat substantia nigra. *Neuroscience* **52,** 867–881.

23. Alexi, T. and Hefti, F. (1993) Trophic actions of transforming growth factor alpha on mesencephalic dopaminergic neurons in culture. *Neuroscience* **55,** 903–918.

24. Lin, L. F., Doherty, D. H., Lile, J. D., Bektesh, S., and Collins, F. (1993) GDNF: a glial cell line-derived neurotrophic factor for midbrain dopaminergic neurons. *Science* **260,** 1130–1132.

25. Hoffer, B. J., Hoffman, A. F., Bowenkamp, K., Huettl, P., Hudson, J., Martin, D., Lin, L.-F., and Gerhardt, G. A. (1994) Glial cell-line derived neurotrophic factor reverses toxin-induced injury to midbrain dopaminergic neurons in vivo. *Neurosci. Lett.* **182,** 107–111.

26. Hudson, J., Granholm, A-Ch., Gerhardt, G. A., Henry, M. A., Hoffman, A., Biddle, P., Leela, N. S., Mackerlova, L., Lile, J. D., Collins, F., and Hoffer, B. J. (1995) Glial cell line-derived neurotrophic factor augments dopaminergic circuits *in vivo. Brain Res. Bull.* **36,** 425–432.

27. Bowenkamp, K. E., Hoffman, A. F., Gerhardt, G. A., Henry, M. A., Biddle, P. T., Hoffer, B. J., and Granholm, A. C. H. (1995) Glial cell line-derived neurotrophic factor supports survival of injured midbrain dopaminergic neurons. *J. Comp. Neurol.* **355,** 479–489.

28. Tomac, A., Lindquist, E., Lin, L-F., Orgen, S. O., Young, D., Hoffer, B. J., and Olson, L. (1995) Protection and repair of the nigrostriatal DAergic system by GDNF. *Nature (Lond.)* **373,** 335–339.

29. Sauer, H., Rosenblad, C., and Björklund, A. (1995) Glial cell line-derived neurotrophic factor but not transforming growth factor beta 3 prevents delayed degeneration of nigral dopaminergic neurons following striatal 6-hydroxy dopamine lesion. *Proc. Natl. Acad. Sci. USA* **92,** 8935–8939.

30. Lindner, M. D., Winn, S. R., Baetge, E. E., Hammang, J. P., Gentile, F. T., Doherty, E., McDermot, P. E., Frydel, B., Ullman, M. D., and Schallert, T. (1995) Implantation of encapsulated catecholamine and GDNF-producing cells in rats with unilateral dopamine depletions and Parkinsonian symptoms. *Exp. Neurol.* **132,** 62–76.

31. Gash, D. M., Zhang, Z., Ovadia, A., Cass, W. A., Yi, A., Simmerman, L., Russell, D., Martin, D., Lapchak, P. A., Collins, F., Hoffer, B. J., and Gerhardt, G. A. Functional recovery in Parkinsonian monkeys treated with GDNF. *Nature* **380,** 252–255.

32. Granholm, A-Ch., Mott, J. L., Eken, S., van Horne, C., Bowenkamp, K., Hoffer, B. J., Henry, S., and Gerhardt, G. A. (1997b) Glial cell line-derived neurotrophic factor improves survival of ventral mesencephalic grafts to the 6-OHDA lesioned striatum. *Exp. Brain Res.* **116,** 29–38.

33. Sinclair, S. R., Clive, C. A., Svendsen, N., Torres, E. M., Fawcett, J. W., and Dunnett, S. B. (1996) GDNF enhances dopaminergic cell survival and fiber outgrowth in embryonic nigral grafts. *NeuroReport* **7,** 2547–2552.

34. Pichel, J. G., Shen, L., Sheng, H. Z., Drago, J., Grinberg, A., Lee, E. J., Ping Huang, S., Saarma, M., Hoffer, B. J., Sariola, H., and Westphal H. (1996) Defects in enteric innervation and kidney development in mice lacking GDNF. *Nature* **382,** 73–76.

35. Granholm, A.-Ch., Srivastava, N., Mott, J. L., Henry, S., Henry, M., Westphal, H., Pichel, J. P., Shen, L., and Hoffer, B. J. (1997) Morphological alterations in the peripheral and central nervous systems of mice lacking glial cell line-derived neurotrophic factor (GDNF): immunohistochemical studies. *J. Neurosci.* **17,** 1168–1178.

36. Buchman, V. L. and Davies, A. M. (1993) Different neurotrophins are expressed and act in a developmental sequence to promote the survival of embryonic sensory neurons. *Development* **118,** 989–1001.

37. Davies, A. M., Horton, A., Burton, L. E., Schmelzer, C., Vandlen, R., and Rosenthal, A. (1993) Neurotrophin 4/5 is a mammalian-specific survival factor for distinct populations of sensory neurons. *J. Neurosci.* **13,** 4961–4967.

38. Ernfors, P., Lee, K.-F., and Jaenisch, R. (1994) Mice lacking brain-derived neurotrophic factor develop with sensory deficits. *Nature* **368,** 147–150.

39. Farinas, I., Jones, K. R., Backus, C., Wang, X.-Y., and Reichardt, L. F. (1994) Severe sensory and sympathetic deficits in mice lacking neurotrophin-3. *Nature* **369,** 658–661.

40. Jones, K. R., Farinas, I., Backus, C., and Reichardt, L. F. (1994) Targeted disruption of the

BDNF gene perturbs brain and sensory neuron development but not motor neuron development. *Cell* **76,** 989–999.

41. Klein, R., Smeyne, R. J., Wurst, W., Long, L. K., Auerbach, B. A., Jouyner, A. L., and Barbacid, M. (1994) Targeted disruption of the trkB neurotrophin receptor gene results in nervous system lesions and neonatal death. *Cell* **75,** 113–122.

42. Smeyne, R. J., Klein, R., Schnapp, A., Long, L. K., Bryant, S., Lewin, A., Lira, S. A., and Barbacid, M. (1994) Severe sensory and sympathetic neuropathies in mice carrying a disrupted Trk/NGF receptor gene. *Nature* **368,** 246–249.

43. Ziljstra, M., Li, E., Sajjadi, F., Subramani, S., and Jaenisch, R. (1989) Germ-line transmission of a disrupted beta 2-microglobulin gene produced by homologous recombination in embryonic stem cells. *Nature* **342,** 435–438.

44. Conover, J. C. and Yancopoulos, G. D. (1997) Neurotrophin regulation of the developing nervous system: Analyses of knockout mice. *Rev. Neurosci.* **8,** 13–27.

45. Erickson, J. T., Conover, J. C., Borday, V., Champagnat, J., Barbacid, M., Yancopoulos, G. D., and Katz, D. M. (1996) Transgenic mice lacking brain-derived neurotrophic factor exhibit visceral sensory neuron losses distinct from mice lacking NT-4 and display a severe developmental deficit in control of breathing. *J. Neurosci.* **1617,** 5361–5371.

46. Crowley, C., Spencer, S. D., Nishimura, M. C., Chen, K. S., Pitts-Meek, S., Armanini, M. P., Ling, L. H., McMahon, S. B., Shelton, D. L., Levinson, A. D., and Phillips, H. S. (1994) Mice lacking nerve growth factor display perinatal loss of sensory and sympathetic neurons yet develop basal forebrain cholinergic neurons. *Cell* **76,** 1001–1011.

47. Gu, H., Zhou, Y. R., and Rajewksy, K. (1993) Independent control of immunological switch regions evidenced through CreloxP-mediated gene targeting. *Cell* **73,** 1155–1164.

48. Rossant, J. and Nagy, A. (1995) Genomic engineering: the new mouse genetics. *Nat. Med.* **16,** 592–594.

49. Davies, A. M. (1994) The role of neurotrophins in the developing nervous system. *J. Neurobiol.* **2511,** 1334–1348.

50. Ernfors, P., Wetmore, C., Olson, L., and Persson, H. (1990) Identification of cells in the rat brain and peripheral tissues expressing mRNA for members of the nerve growth factor family. *Neuron* **5,** 511–526.

51. Mufson, E. J., Kroin, J. S., Liu, Y., Sobreviela, T., Penn, R. D., Miller, J. A., and Kordower, J. H. (1996) Intrastriatal and intraventricular infusion of brain-derived neurotrophic factor in the cynomologous monkey: distribution, retrograde transport and co-localization with substantia nigra dopamine-containing neurons. *Neuroscience* **71,** 179–191.

52. Date, I. and Ohmoto, T. (1996) Neural transplantation and trophic factors in Parkinson's disease: special reference to chromaffin cell grafting, NGF support from pretransected peripheral nerve and encapsulated dopamine secreting cell grafting. *Exp. Neurol.* **137,** 333–344.

53. Götz, R., Koster, R., Winkler, C., Raulf, F., Lottspeich, F., Schartl, M., and Thoenen, H. (1994) Neurotrophin-6 is a new member of the nerve growth factor family. *Nature* **372,** 266–269.

54. Bothwell, M. (1991) Keeping track of the neurotrophin receptors. *Cell* **65,** 915–918.

55. Chao, M. V. (1992) Neurotrophin receptors: a window into neuronal differentiation. *Neuron* **9,** 583–593.

56. Barde, Y.-A. (1994) Neurotrophic factors: an evolutionary perspective. *J. Neurobiol.* **2511,** 1329–1333.

57. Hyman, C., Juhasz, M., Jackson, C., Wrigth, P., Ip, N. Y., and Lindsay, R. M. (1994) Overlapping and distinct actions of the neurotrophins BDNF, NT-3, and NT4/5 on cultured dopaminergic and GABAergic neurons of the ventral mesencephalon. *J. Neurosci.* **14,** 335–347.

58. Birling, M.-C. and Price, J. (1995) Influence of growth factors on neuronal differentiation. *Curr. Opin. Cell Biol.* **7,** 878–884.

59. Timmusk, T., Belluardo, N., Metsis, M., and Persson, H. (1993) Widespread and developmentally regulated expression of neurotrophin 4 mRNA in rat brain and peripheral tissues. *Eur. J. Neurosci.* **5,** 605–613.

60. Conover, J. C., Erickson, J. T., Katz, D. M., Bianchi, L. M., Poueymirou, W. T., McClain, J., Pan, L., Helgren, M., Ip, N. Y., Boland, P., and Friedman, B. (1995) Neuronal deficits not involving motor neurons in mice lacking BDNF and/or NT4. *Nature* **375,** 235–238.

61. Roberts, A. B. and Sporn, M. B. (1990) in *Peptide Growth Factors and Their Receptors,* (Sporn, M. B. and Roberts, A. B., eds., Springer Verlag, Heidelberg, pp. 419–472.

62. Clarkson, E. D., Zawada, W. M., and Freed, C. R. (1995) GDNF reduces apoptosis in dopaminergic neurons in vitro. *NeuroReport* **7,** 145–149.

63. Clarkson, E. D., Zawada, W. M., and Freed, C. R. (1997) GDNF improves survival and reduces apoptosis in human embryonic dopaminergic neurons in vitro. *Cell Tissue Res* **289,** 207–210.

64. Moore, M. W., Klein, R. D., Farinas, I., Sauer, H., Armanini, M., Phillips, H., Reichardt, L. F., Ryans, A. M., Carver-Moore, K., and Rosenthal, A. (1996) Renal and neuronal abnormalities in mice lacking GDNF. *Nature* **382,** 76–79.

65. Sanchez, M. P., Silos-Santiago, I., Frisen, J., He, B., Lira, S. A., and Barbacid, M. (1996) Renal agenesis and the absence of enteric neurons in mice lacking GDNF. *Nature* **382,** 70–73.

66. Heuckeroth, R. O., Tourtelotte, L., Gavrilina, G., Johnson, E., and Milbrandt, J. (1997) Characterization of the neurturin knockout mouse. *Abstr. Soc. Neurosci.* **668,** 1.

67. Arenas, E., Trupp, M., Akerud, P., and Ibanez, C. F. (1995) GDNF prevents degeneration and promotes the phenotype of brain noradrenergic neurons in vivo. *Neuron* **15,** 1465–1473.

68. Mahalik, T. J., Hanh, W. E., Clayton, G. H., and Owens, G. P. (1994) Programmed cell death in developing grafts of fetal substantia nigra. *Exp. Neurol.* **129,** 27–36.

69. Kaddis, F. G., Zawada, W. M., Schaack, J., and Freed, C. R. (1996) Conditioned medium from aged monkey fibroblasts stably expressing GDNF and BDNF improves survival of embryonic dopamine neurons in vitro. *Cell Tiss. Res.* **286,** 241–247.

70. Jing, S. Q., Wen, D. Z., Yu, Y. B., Holst, P. L., Luo. Y., Fang, M., Tamir, R., Antonio, L., Hu, Z., and Cupples, R. (1996) GDNF-induced activation of the Ret protein tyrosine kinase is mediated by GDNFRα, a novel receptor for GDNF. *Cell* **85,** 1113–1124.

71. Treanor, J., Goodman, L., de Sauvage, F., Stone, D. M., Poulsen, K. T., Beck, K. D., Gray, C., Armanini, M. P., Pollock, R. A., and Hefti, F. (1996) Characterization of a multicomponent receptor for GDNF. *Nature* **382,** 80–83.

72. Creedon, D. J., Tansey, M. G., Baloh, R. H., Osborne, P. A., Lampe, P. A., Fahrner, T. J., Heuckeroth, R. O., Milbrandt, J., and Johnson, E. M., Jr. (1997) Neurturin shares receptors and signal transduction pathways with glial cell line-derived neurotrophic factor in sympathetic neurons. *Proc. Natl. Acad. Sci. USA* **94,** 7018–7023.

73. Milbrandt, J., de Sauvage, F. J., Fahrner, T. J., Baloh, R. H., Leitner, M. L., Tansey, M. G., Lampe, P. A., Heuckeroth, R. O., Kotzbauer, P. T., Simburger, K. S., Golden, J. P., Davies, J. A., Vejsada, R., Kato, A. C., Hynes, M., Sherman, D., Nishimura, M., Wang, L. C., Vandlen, R., Moffat, B., Klein, R. D., Poulsen, K., Gray, C., Garces, A., and Johnson, E. M., Jr. (1998) Persephin, a novel neurotrophic factor related to GDNF and neurturin. *Neuron* **20,** 245–253.

74. Durbec, P. L., Larsson-Blomberg, L. B., Schuchardt, A., Costantini, F., and Pachnis, V. (1996) Common origin and developmental dependence of subsets of enteric and sympathetic neuroblasts. *Development* **122,** 349–358.

75. Cacalano, G., Farinas, I., Wang, L-C., Hagler, K., Forgie, A., Moore, M., Armanini, M., Phillips, H., Ryan, A. M., Reichardt, L. F., Hynes, M., Davies, A., and Rosenthal, A. (1998) GFRα1 is an essential receptor component for GDNF in the developing nervous system and kidney. *Neuron* **21,** 53–62.

76. Enomoto, H., Araki, T., Jackman, A., Heuckeroth, R. O., Snider, W. D., Johnson, E. M., Jr., and Milbrandt, J. (1998) GFRα-1 deficient mice have deficits in the enteric nervous system and kidneys. *Neuron* **21,** 317–324.

77. Saucedo-Cardenas, O., Kardon, R., Ediger, T. R., Lydon, J. P., and Conneely, O. M. (1997) Cloning and structural organization of the gene encoding the murine receptor transcription factor NURR1. *Gene* **187,** 135–139.

78. Zetterström, R. H., Solomin, L., Jansson, L., Hoffer, B. J., Olson, L., and Perlmann, T. (1997) Dopamine neuron agenesis in Nurr1-deficient mice. *Science* **276,** 248–250.

79. Zhou, Q-Y., Quaife, C. J., and Palmiter, R. D. (1995) Targeted disruption of the tyrosine hydroxylase gene reveals that catecholamines are essential for mouse fetal development. *Nature* **374,** 640–643.

80. Zhou, Q-Y. and Palmiter, R. D. (1995) DA-deficient mice are severely hypoactive, adipsic, and aphagic. *Cell* **83,** 1197–1209.

81. Szczypka, M. S., Mandel, R. J., Donahue, B. A., Snyder, R. O., Leff, S. E., and Palmiter, R. D. (1999) Viral gene delivery selectively restores feeding and prevents lethality of dopamine-deficient mice. *Neuron* **22,** 167–178.

82. Carter-Russell, H. R., Song, W-J., and Surmeier, D. J. (1995) Coordinated expression of dopamine receptors (D1-5) in single neostriatal neurons. *Soc. Neurosci. Abstr.* **21,** 1425.

83. Boyer, J. J., Park, D. H., Joh, T. H., and Pickel, V. M. (1984) Chemical and structural analysis of the relation between cortical inputs and tyrosine hydroxylase-containing terminals in rat neostriatum. *Brain Res.* **302,** 267–275.

84. Freund, T. F., Powell, J., Smith, A. D. (1984) Tyrosine hydroxylase immunoreactive boutons in synaptic contact with identified striatonigral neurons with particular reference to dendritic spines. *Neuroscience* **13,** 1189–1215.

85. Drago, J., Gerfen, C. R., Lachowicz, J. E., Steiner, H., Hollon, T. R., Love, P. E., Ooi, G. T., Grinberg, A., Lee, E. J., Huang, S. P., Bartlett, P. F., Jose, P. A., Sibley, D. R., and Westphal, H. (1994) Altered striatal function in a mutant mouse lacking D1A dopamine receptors. *Proc. Natl. Acad. Sci. USA* **91,** 12,564–12,568.

86. Xu, M., Moratalla, R., Gold, L. H., Hiroi, N., Koob, G. F., Graybiel, A. M., and Tonegawa, S. (1994) Dopamine D1 receptor mutant mice are deficient in striatal expression of dynorphin and in dopamine-mediated behavioral responses. *Cell* **79,** 729–742.

87. Xu, M., Caine, S. B., Cooper, D. C., Gold, A. M., Graybiel, A. M., Hu, X. T., Koeltzow, T. E., Koob, G., Moratalla, R., White, F. J., and Tonegawa, S. (1995) Analysis of D3 and D1 receptor mutant mice. *Soc. Neurosci. Abstr.* **21,** 363.

88. Balk, J. H., Picetti, R., Salardi, A., Thirlet, G., Dierich, A., Depaulls, A., LeMeur, M., and Borrelli, E. (1995) Parkinsonian-like locomotor impairments in mice lacking dopamine D2 receptors. *Nature* **377,** 424–428.

89. Accili, D., Fishbourne, C. S., Drago, J., Steiner, H., Lachowicz, J. E., Gerfen, C. R., Sibley, D. R., Westphal, H., Fuchs, S. (1996) A targeted mutation of the D3 dopamine receptor gene is associated with hyperactivity in mice. *Proc. Natl. Acad. Sci. USA* **93,** 1945–1949.

90. Levine, M. S., Altemus, K. L., Cepeda, C., Cromwell, H. C., Crawford, C., Ariano, M. A., Drago, J., Sibley, D. R., and Westphal, H. (1996) Modulatory actions of dopamine on NMDA receptor-mediated responses are reduced in D1A-deficient mutant mice. *J. Neurosci.* **16,** 5870–5882.

91. Miner, L. L., Drago, J., Chamberlain, P. M., Donovan, D., and Uhl, G. R. (1995) Retained cocaine conditioned place reference in D1 receptor deficient mice. *NeuroReport* **6,** 2314–2316.

92. Schuldiner, S., Shirvan, A., and Linial, M. (1995) Vesicular neurotransmitter transporters: from bacteria to humans. *Physiol. Rev.* **75,** 369–392.

93. Johnson, R. G., Jr. (1988) Accumulation of biological amines into chromaffin granules: a model for hormone and neurotransmitter transport. *Physiol. Rev.* **68,** 232–307.

94. Gonzalez, A. M., Walther, D., Pazos, A., and Uhl, J. (1994) Synaptic vesicular monoamine transporter expression: distribution and pharmacological profile. *Mol. Brain Res.* **22,** 219–226.

95. Wang, Y. M., Gainetdinov, R. R., Fumagalli, F., Xu, F., Jones, S. R., Bock, C. B., Miller, G. W., Wightman, R. M., and Caron, M. G. (1997) Knockout of the vesicular monoamine transporter 2 gene results in neonatal death and supersensitivity to cocaine and amphetamine. *Neuron* **19,** 1285–1296.

96. Hebert, M. A. and Gerhardt, G. A. (1998) Normal and drug-induced locomotor behavior in aging: comparison to evoked dopamine release and tissue content of F344 rats. *Brain Res.* **797,** 42–54.

97. Gainetdinov, R. R., Fumagalli, F., Wang, Y. M., Jones, S. R., Levey, A. I., Miller, G. W., and Caron, M. G. (1998) Increased MPTP neurotoxicity in vesicular monoamine transporter 2 heterozygote knockout mice. *J. Neurochem.* **70,** 1973–1978.

98. Jaber, M., Jones, S., Giros, B., and Caron, M. G. (1997) The dopamine transporter: a crucial component regulating dopamine transmission. *Move. Disord.* **12,** 629–633.

99. Uhl, G. R., Vandenbergh, D. J., and Miner, L. L. (1996) Knockout mice and dirty drugs. *Drug Addict. Curr. Biol.* **6,** 935–936.

100. Hoffer, B. J., Granholm, A-Ch., Stevens, J. O., and Olson, L. (1988) Catecholamine containing grafts in Parkinsonism: Past and present. *Clin. Res.* **36,** 189–195.

101. Lindvall, O., Rehncrona, S., Gustaavii, B., Brundin, P., Åstedt, B., Widner, H., Lindholm, T., Björklund, A., Leenders, K., and Rothwell, J. (1988) Fetal dopamine-rich mesencephalic grafts in Parkinson's disease. *Lancet* **2**(8626–8627), 1483–1484.

102. Granholm, A-Ch., Strömberg, I., Gerhardt, G. A., Seiger, Å., Olson, L., and Hoffer, B.J. (1989) Transplantation in Parkinson's Disease: experimental and Clinical Trials, in *Neural Regeneration and Transplantation* (Seil, F. J., ed.), Alan R. Liss, New York, pp. 227–237.

103. Freed, C. R., Breeze, R. E., Rosenberg, N. L., Schneck, S. A., Kriek, E., Qi, J-X., Lone, T., Zhang, Y-B., Snyder, J. A., Wells, T. H., Olson Ramig, L., Thompson, L., Maziotta, J. C., Huang, S. C., Grafton, S. T., Brooks, D., Sawle, G., Schroter, G., and Ansari, A. A. (1992) Survival of implanted dopamine cells and neurological improvement 12 to 46 months after transplantation for Parkinson's disease. *N. Engl. J. Med.* **327,** 1549–1555.

104. Hitchcock, E. (1995) Current trends in neural transplantation. *Neurol. Res.* **17,** 33–37.

105. Kordower, J. H., Freeman, T. B., Snow, B. J., Vingerhoets, F. J., Mufson, E. J., Sanberg, P. R., Hauser, R. A., Smith, D. A., Nauert, G. M., and Perl, D. P. (1995) Neuropathological evidence of graft survival and striatal reinnervation after transplantation of fetal mesencephalic tissue in a patient with Parkinson's disease. *N. Engl. J. Med.* **332,** 118–1124.

106. Lopez-Lozano, J. J., Bravo, G. G., Brera, B., Dargallo, J., Salmean, J., Uria, J., Insausti, J., and Millan, I. (1995) Long-term follow up in 10 Parkinson's disease patients subjected to fetal brain grafting into a cavity in the caudate nucleus: the Clinica Puerta de Hierro experience. CPH Neural Transplantation Group. *Transplant Proc.* **27,** 1395–1400.

107. Deacon, T., Schumacher, J., Dinsmore, J., Thomas, C., Palmer, P., Kott, S., Edge, A., Penney, D., Kassisieh, S., Dempsey, P., and Isacson, O. (1997) Histological evidence of fetal pig neural cell survival after transplantation into a patient with Parkinson's disease. *Nat. Med.* **3,** 350–353.

108. Rioux, L., Gaudin, D., Bui, L., Gregoire, L., DiPaolo, T., and Bedard P. (1991) Correlation of functional recovery after a 6-hydroxy dopamine lesion with survival of grafted fetal neurons and release of dopamine in the striatum of the rat. *Neuroscience* **40,** 123–131.

109. Freeman, T. B., Sanberg, P. R., and Nauert, G. M. (1995) The influence of donor age on the survival of solid and suspension intraparenchymal human embryonic nigral grafts. *Cell Transplant.* **4,** 141–154.

110. Rosenstein, J. M. (1995) Why do neural transplants survive? *Exp. Neurol.* **133,** 1–6.
111. Duan, W.-M., Westerman, M., Flores, T., and Low, W. C. (1998) Xenotransplantation of fetal dopamine neurons: From MHC I and MHC II knockout mice to adult rats: enhancement of cell survival. *Abstr. American Society for Neural Transplantation. Exp. Neurol.* **153,** 376.
112. Holtzman, D. M., Li, Y., DeArmnod, S. J., McKinley, M. P., Gage, F. H., Epstein, C. J., and Mobley, W. C. (1992) Mouse model of neurodegeneration: atrophy of basal forebrain cholinergic neurons in trisomy 16 transplants. *Proc. Natl. Acad. Sci. USA* **89,** 1383–1387.
113. Coyle, J. T., Oster-Granite, M. L., Reeves, R. H., and Gearhart, J. D. (1988) Down syndrome, Alzheimer syndrome, and the trisomy 16 mouse. *Trends Neurosci.* **11,** 390–394.
114. Holtzman, D. M., Santucci, D., Kilbridge, J., Chua-Couzens, J., Fontana, D. J., Daniels, S. E., Mohnson, R. M., Chen, K., Sun, Y., Carlson, E., Alleva, E., Epstein, C. J., and Mobley, W. C. (1996) Developmental abnormalities and age-related neurodegeneration in a mouse model of Downs syndrome. *Proc. Natl. Acad. Sci. USA* **93,** 13,333–13,338.
115. Fiedler, J. L., Epstein, C. J., Rapoport, S. I., Caviedes, R., and Caviedes, P. (1994) Regional alteration of cholinergic function in central neurons of trisomy 16 mouse fetuses, an animal model of human trisomy 21 (Down's syndrome). *Brain Res.* **658,** 27–32.

III

HUNTINGTON'S DISEASE

14

Operant Analysis of Striatal Dysfunction

Peter J. Brasted, Màté D. Döbrössy, Dawn M. Eagle, Falguni Nathwani, Trevor W. Robbins, and Stephen B. Dunnett

1. INTRODUCTION

Huntington's disease (HD) is a fatal inherited neurological disorder, the genetic basis of which has recently been identified as an expanded trinucleotide repeat in a previously unknown gene (the normal function of which remains mysterious) on chromosome 4 *(1)*. The disease is characterized pathologically by a primary progressive loss of medium spiny projection neurons within the neostriatum (caudate nucleus and putamen) in addition to fibrillary astrocytosis. In later stages of the disease the pathology spreads beyond the striatum itself to affect other cortical and subcortical systems, although predominantly those that are afferent and efferent to the striatum. HD is traditionally viewed as a movement disorder, and is typically characterized by dementia and psychiatric symptoms in addition to the more conspicuous choreic movements. This composition of impairments suggests a striatal influence in many response-related functions, mirroring the diversity of its neocortical afferents.

The anatomical, neurochemical, and morphological progression of the striatal atrophy in HD is well characterized. The disease process begins in the caudate nucleus and advances over time through the entire striatum in a medial to lateral and dorsal to ventral fashion *(2,3)*. Despite this widespread atrophy within the striatum, a preferential loss of specific neuronal types is consistently found. Specifically, the γ-aminobutyric acid-ergic (GABAergic) medium spiny projection neurons are destroyed, in contrast with the medium aspiny and large aspiny neurons which are spared *(4)*. Furthermore there is evidence to suggest that GABAergic neurons that coexpress enkephalin are lost prior to those that coexpress substance P *(5–7)*. The implication is that striatal output to the external segment of the globus pallidus (the "indirect" pathway) is affected earlier than projections to the internal segment of the globus pallidus and the substantia nigra pars reticulata (the "direct" pathway), which may be relevant to the development of symptoms such as chorea. The progression of this striatal pathology has also been subject to examination in the context of the striosome/matrix compartmentalization that characterizes the striatum. In a detailed analysis of HD brains at postmortem, Hedreen and Folstein *(7)* proposed two temporally discrete phases of cell loss. First, they reported that the dorsal to ventral progression of neuronal degeneration

From: Central Nervous System Diseases
Edited by: D. F. Emerich, R. L. Dean, III, and P. R. Sanberg © Humana Press Inc., Totowa, NJ

is specific only to those cells that are located in the matrix. Second, they demonstrated that the loss of striosomal neurons occurs early in the course of the disease, and is completed prior to the progressive loss of neurons in the matrix. The differential onset of behavioral symptoms has also been interpreted with respect to this pathological distinction, with the chorea cited potentially as the result of the initial striosomal loss *(7)*.

Despite the fact that the neural degeneration associated with HD centers almost exclusively on the striatum, there are reports of additional pathology, located both cortically and subcortically. The primary areas affected are the pyramidal projection cells in cortical areas that innervate the striatum *(8–10)*, and target areas of striatal outputs in the globus pallidus and substantia nigra *(11,12)*, leading to the suggestion that the extrastriatal changes may be direct anterograde and retrograde reactions to a primary striatal degeneration, rather than independent consequences of the disease process *(13)*, although direct evidence for and against this hypothesis remains sparse.

2. MOTOR, COGNITIVE, AND PSYCHIATRIC DEFICITS IN HD

The variety of impairments that are manifest in HD can be organized into the broad categories of motor, cognitive, and psychiatric symptoms, all of which worsen as the disease progresses. Although HD traditionally has been viewed as a movement disorder, the advent of genetic screening and the development of more sensitive clinical testing have indicated that it is the psychiatric and cognitive aspects that often present first *(14)*, and can certainly be the more debilitating for the patients and their families.

Although HD is regarded as a hyperkinetic disorder on account of the uncontrollable choreic movements, this often masks an underlying bradykinesia *(15)*. As the disease progresses this and other hypokinetic signs such as rigidity, dystonia, and oculomotor deficits become increasingly apparent *(16)*. Although elements of these symptoms, such as dystonia and chorea, are not well reproduced in rodent models of HD, other aspects of the hyperkinetic syndrome are well reproduced following striatal lesions in rats and other experimental animals *(17,18)*.

The neuropsychological deficits that are associated with HD are similar to those that are seen in patients with prefrontal cortical damage, as indeed is the case with other "subcortical dementias" *(19)*. Patients typically exhibit mnemonic impairments such as reduced digit spans *(20)* and reduced spatial spans *(21)* as well as deficits in paired associates learning *(20)* and procedural learning *(22)*. HD is also believed to impact upon "executive" functions such as planning and attention-shifting *(23–26)*. Although this cognitive profile is similar in many respects to that seen in "frontal" patients, these deficits are believed to be due to striatal cell loss rather than cortical atrophy, not least because the development of many deficits correlates with the severity of the hypokinetic symptoms that are closely and uniquely associated with striatal dysfunction *(27)*.

A large number of psychiatric symptoms have been reported in HD, perhaps reflecting a diversity in diagnoses as well as in pathology. Depression, aggression, and anxiety are commonly reported, often in the context of a "personality change," and patients also may exhibit inappropriate social and sexual behaviors *(16)*. However, any interpretation of these symptoms may be weakened by the uncertainty surrounding the knowledge that one is at risk of inheriting a fatal disease. Thus it is unclear the extent to which psychiatric symptoms represent a reaction to the diagnosis, although the observation that these may be some of the earliest appearing signs in otherwise symptom-free gene carriers suggests that the psychiatric/behavioral disturbance is an endogenous component of the disease.

3. EXCITOTOXIC LESIONS OF THE RAT STRIATUM AND THE NEUROPATHOLOGY OF HD

Early attempts to study striatal function involved the use of electrolytic lesions, but this technique has since been supplanted by the use of excitotoxins. This approach conveys two obvious advantages. First, excitotoxins selectively destroy intrinsic striatal neurons without damaging the fibers of passage, destroyed by electrolytic lesions, that project to and from the frontal cortices. Second, excitotoxins have provided the basis for animal models of HD. Although these techniques can never precisely model an ongoing pathology in the way that metabolic toxins perhaps can, it is a reliable and robust technique that has consistently produced behavioral sequelae that correlate well with the clinical symptoms of striatal damage.

Excitotoxins are glutamate analogues that can bind to different glutamate receptor subtypes. They are injected centrally, and induce cell death by increasing neuronal firing rate, which gives rise to a sustained period of cell depolarization. Kainic acid was used initially to model the neuropathology of HD *(28,29)*. Not only did the toxin produce selective striatal cell loss, but it also resulted in a reduction in the levels of choline acetyl transferase (ChAT) and glutamic acid dehydroxylase (GAD). These neurochemical markers for cholinergic and GABAergic neurons are also decreased in HD *(28)*. Since then, ibotenic acid and quinolinic acid have replaced kainic acid as the neurotoxins of choice in animal models of HD *(30,31)*. Although both toxins still cause loss of intrinsic striatal neurons, neither results in remote hippocampal damage or the seizures that can accompany kainate lesions. Furthermore, quinolinic acid, an endogenous (*N*-methyl-D-aspartate (NMDA) agonist, is reported to best model the human disease, on account of the preserved or increased levels of somatostatin and neuropeptide Y, in addition to the sparing of NADPH-diaphorase neurons, all of which resemble the postmortem findings reported in HD *(32,33)*. However, this difference between quinolinic and ibotenic acids is not found consistently, and in our hands the differences in apparent selectivity relate as much to the technical parameters of toxicity, concentration, and rate of delivery as to any differences in receptor selectivity or toxic profile of the two compounds. Moreover, to the extent that they have been compared on similar tasks (such as locomotor hyperactivity or deficits in skilled reaching) both toxins seem to produce rather similar behavioral effects for a similar sized lesion.

More recently, other models of HD have been developed that may more closely mimic the neuropathological process of the human disease. Striatal neurons appear to exhibit deficits in mitochondrial energy metabolism in HD striatum that can be mimicked by a variety of metabolic toxins such as malonate and 3-nitroproprionic acid (3-NPA) *(34,35)*. For example ,systemic administration of 3-NPA produces widespread disturbances in succinate dehydrogenase activity throughout the brain, but selective degeneration of the neostriatal neurons in both rats and monkeys *(36–38)*. Nevertheless, although lesions produced by the metabolic toxins may provide a good model to analyze the pathogenesis of the human disease, they are more variable than the excitotoxins, both in the doses required and in the extent of striatal damage that results. In our experience, the use of excitotoxins is preferable for behavioral studies in which stable and reproducible damage from animal to animal is critically important for tight experimental design.

In the wake of recent development in our understanding of the subdivisions in striatal circuitry, there has also arisen an interest in the possibilities of making selective lesions in the direct vs indirect pathways and in the striosome vs matrix compartments of the striatum *(39)*. Such selectivity may seem plausible from descriptions of the differential distribution of different subtypes of glutamate receptors in the striatum, and indeed there has been a claim that such selectivity can be achieved either by using supposedly selective excitotoxins *(40)* or by disrupting the dopamine innervation selectively from the striosomes by making 6-hydroxydopamine (6-OHDA) lesions at a brief window in early development *(41)*. However, neither of these strategies has yet proved sufficiently reliable and stable to provide the basis for systematic behavioral studies of striatal compartments, and the conventional excitotoxins remain the strategy of choice.

4. THE FUNCTIONAL ROLE OF THE STRIATUM IN EXPERIMENTAL ANIMALS

The striatum has always been recognized as a structure that is critical in mediating efficient motor control, a role that has been inferred both by the pathology of movement disorders and also its anatomical connectivity *(42)*. That excitotoxic lesions may model both the behavioral as well as the pathological aspects of HD was initially noted by Coyle and Schwarz *(28)*. Thus, bilateral striatal lesions in rats induce a marked hyperactivity *(17,18)* that is most prominent under circumstances where the rat would normally be active, such as during the night cycle or when food deprived *(43,44)*.

The tendency of rats to rotate to their contralateral side following unilateral striatal lesions is still widely used as a behavioral index of striatal damage and regeneration *(45–48)*. Why this motoric asymmetry should manifest itself in this way is not well understood. It is usually apparent only with the systemic administration of dopaminergic agonists such as amphetamine and apomorphine, and both the direction of rotation and response to different drugs can be critically dependent upon the locus of the striatal lesion *(47,48)*. Asymmetrical motor deficits are also apparent in un-drugged rats with striatal lesions, in tests varying from axial twisting of the body *(49)* to skilled paw use in manipulating, reaching, and grasping for food *(47,50,51)*. More recently, the interpretation of lesion data in terms of a general "motor" dysfunction has since been refined so that striatal damage is shown to produce "response-related" deficits (to be developed in **Subheading 6.**), reflecting the importance of cortico-striatal systems in producing stimulus-bound movement.

Although it has long been established that the striatum is integral in the mediation of effective motor output, it is only relatively recently that it has been assigned a significant role in cognition. Divac and colleagues showed that electrolytic lesions of discrete segments of the caudate in monkeys impaired performance in delayed response and spatial alternation tasks in much the same way as did lesions of areas of association cortex projecting to each striatal segment *(52)*. Excitotoxic lesions have since been shown to produce similar impairments in a variety of executive and other cognitive tasks in primates, cats, and rodents *(53–59)*.

Perhaps the most challenging aspect of examining HD through rodent models is to examine psychiatric or motivational components. Although difficult to examine more overt or florid aspects of psychiatric disturbance that are based on language and thought processes specific to humans, it is nevertheless possible to examine the role of the striatum in aspects of incentive motivation and the ability of rats to adapt their responses appropriately to reward and changes in reward. Classically, the ventral striatum has been considered primarily as the critical substrate for limbic and motivational aspects of striatal function *(60,61)*, but this reflects at least in part the fact that the dorsal or neostriatum has been relatively little investigated with regard to this aspect of complex function.

5. WHAT IS AN OPERANT TASK?

The term "operant" was originally defined by Skinner as a set of responses that "must *operate* upon nature to produce its reinforcement" *(62)*. Thus, in an "operant" task, a subject learns to respond in a particular way to gain reward (or to escape punishment) when a discrete cue, or stimulus, signals that this response will be effective.

The search for better methods to analyze operant behavior led Skinner to develop various types of automated test apparatus, that he called "operant chambers" but are now colloquially known as Skinner boxes. A conventional operant chamber typically has three main components:

1. The *operand*, something the animal can operate on or respond to. This is typically a lever to be pressed, but may be a chain to be pulled, a touch-sensitive panel to press, or indeed any other element to which a response can be measured.
2. *Discriminative stimuli* that can be switched on and off to signal the timing and location of the response. These are typically lights, but may be more complex stimuli such as visual patterns, or involve other modalities such as sounds (pure tone, buzzer, white noise, etc.) or indeed tactile or olfactory cues also.
3. *Reinforcers*, typically food or water, as positive stimuli, or an electric shock, as negative stimuli. A reinforcer is in fact anything that an animal will work for or learn a response to gain access to or to avoid.

The operant chamber is designed to present discriminative stimuli, to measure and record responses to the operand, and to deliver reinforcement (e.g., by delivering food pellets via a pellet dispenser, or delivering electric shock to the animal's feet via the grid floor), all under microcomputer control. The program that defines the contingencies under which reward is delivered defines the "schedule of reinforcement." One of the major triumphs of the operant approach to behavior by Skinner and his colleagues has been the detailed specification of the ways in which the behavior of animals comes under the control of the applied schedule of reinforcement so that the patterns of responding can be accurately predicted *(63)*.

Using operant principles, it is possible to teach animals to perform specific motor and cognitive tasks in a well-defined manner that permits a detailed and accurate behavioral analysis. Striatal lesions do not impair a rat's ability to learn a simple operant response. Rats readily learn to press a lever press for food reward following striatal damage *(56,64)*, and they can go on to acquire more complex schedules requiring conditional responses under the control of visual and spatial stimuli *(65,66)*. Nevertheless even on simple lever-pressing schedules they do exhibit impairments in the temporal organization of their responding, such as distributing their responses inefficiently or continuing to respond when that is no longer reinforced. Thus, they continue to respond at a higher rate and for longer than controls in "extinction" when a lever press is no longer reinforced *(64,66,67)*; they respond prematurely in a "differential reinforcement of low rates" (DRL) schedule in which a response is reinforced only if the rat has waited longer than a fixed time since the last response *(56)*; and they fail to show the normal scallop-shaped distribution of responding on a "fixed-interval" schedule during which a response is reinforced only if it occurs after a certain latency following the last reinforcement *(67)*. These disturbances in the temporal organization of responding are particularly apparent after lesions in the ventral parts of the striatum and nucleus accumbens *(56,66,67)*.

These measures of basic acquisition and maintenance of lever pressing under changing conditions of reward indicate that although a rat with a striatal lesion can learn the pressing response to acquire food, it may fail to select, initiate, or organize its responses over time in the most effective or efficient manner, and continue responding even at times when it is counterproductive to do so. Moreover, the nature of the striatal deficit is not homogeneous. Several studies have showed that the deficits in maintaining accurate performance in conditional visual and temporal discriminations are particularly associated with the lateral striatum *(56,66)*, and may be associated with difficulty in the acquisition of arbitrary stimulus-response habits *(66)*, a function of the striatum that transcends the motor, motivational, and cognitive domains.

The fact that behavior can be controlled in an operant chamber enables a far more precise control of the factors determining behavior than can be achieved by conventional observational and hand testing methods. Using different stimuli to define reinforced and unreinforced responses, we can determine not only an animal's speed and accuracy of learning, but also the nature of the sensory discriminations that it can make, and the cognitive problems it can solve. Varying the schedules of reinforcement enables us to determine an animal's level of motivation to work for particular rewards or to avoid punishment, as well as its sensitivity to changes in reward and punishment that underlies the emotional control of behavior in animals as well as in humans (e.g., frustration, anxiety).

The following sections elaborate the use of operant paradigms to study motor, cognitive, and motivational aspects of striatal function in the laboratory rat, as the basis for developing more sensitive and valid tests to evaluate brain damage and repair in striatal systems relevant to the development of new therapies.

6. OPERANT ANALYSES OF STRIATAL MOTOR DEFICITS

Although gross motor deficits such as hypokinesia, hyperkinesia, and drug-induced rotation can be assessed in an open field or photocell cage, the development of more sophisticated tasks has allowed more subtle lesion effects to be exposed. Perhaps the clearest example of this was the design of operant paradigms to clearly define the role of striatal dopamine in the rat. A unilateral striatal dopamine lesion (produced by central 6-OHDA injections) had previously been shown to produce a general contralateral impairment or "striatal neglect" *(68)*. Thus, animals with these lesions display an asymmetric posture and rotate in an ipsilateral manner *(69)*, they fail to respond to stimuli on the side contralateral to the lesion *(70,71)*, and they demonstrate an inefficient use of the contralateral limb for reaching and retrieving food *(51,70,72,73)*.

6.1. Sensorimotor Neglect

Operant tests were introduced to analyze this contralateral "neglect" in more detail. Turner introduced the fundamental design of using lateralized stimuli and responses to distinguish sensory, motor, and "sensorimotor" consequences of lesions within lateral hypothalamic circuits *(74)*. In this study, rats were suspended in a hammock and were trained to escape a lateralized footshock to the left or right side hindpaw by making a lateralized turn of the head to the left side. Two subgroups of trained rats then received unilateral lesion in either the lateral hypothalamus or the amygdala, made on the side ipsilateral to or contralateral to the trained response. Turner reported that a unilateral lesion would disrupt performance only if both the stimulus and response were to the same side, contralateral to the lesion. Animals lesioned on the right could still make the left response if the stimulus was applied to the right paw (i.e., they could still respond contralaterally to an ipsilateral stimulus, so there was no basic motor deficit), and animals lesioned on the left could still respond ipsilaterally to both ipsilateral and contralateral stimuli (so there was no basic sensory deficit). Rather the only deficits were seen in the animals lesioned on the right, which were unable to respond on the contralateral (left) side to escape a contralateral stimulus, leading to the concept of a "sensorimotor" deficit, an inability to make coordinated lateralized stimulus–response associations on the side contralateral to the lesion.

At the same time that Marshall and Turner used these observations to hypothesize a "sensorimotor" dysfunction associated with lateral hypothalamic lesions, Marshall was identifying damage of the nigrostriatal system as the primary substrate for the sensory neglect first described after lateral hypothalamic damage *(70,75)*. However, these authors did not undertake the next logical step and go on to evaluate nigrostriatal or striatal lesions explicitly on a lateralized conditioning task: that had to await the work of Robbins and Carli.

6.2. A Lateralized Choice Reaction Time Task

Influenced by the earlier Turner paradigm, Carli et al. *(76)* introduced an appetitive operant task using a novel "nine-hole box" apparatus (*see* Fig. 1A) to evaluate lateralized stimulus and response control in the basal ganglia, although this apparatus has since seen much wider application *(77)*. In the basic visual choice reaction time task only the central three holes are open; the others are all covered with a masking plate (Fig. 1B). Rats are trained to keep their noses held in the middle of the three holes for 100–1500 ms until the presentation of a light cue. This cue appears as a brief flash in one or other of the response holes on either side of the center hole. The actual timing and location of the cue is unpredictable. Rats are trained to respond to this cue in one of two ways to gain food reward. In the "Same" condition, they are required to make a nose-poke into the same response hole that had been previously lit. In the "Opposite" condition they must make a nose-poke in the response hole that had *not* previously been lit. In either condition, the response can be divided into two parts—a non-lateralized "Reaction Time" (RT) to withdraw the nose from the central hole, and a "Movement Time" (MT) to execute the lateralized response to the same or opposite

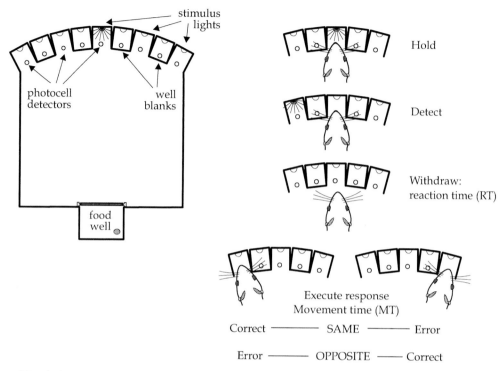

stimulus
lights

photocell
detectors

well
blanks

food
well

Hold

Detect

Withdraw:
reaction time (RT)

Execute response
Movement time (MT)

Correct ——— SAME ——— Error

Error ——— OPPOSITE ——— Correct

Fig. 1. Schematic illustration of the nine-hole box apparatus and the sequence of steps in a single trial of the lateralized choice reaction time task (the "Carli" task).

side (*see* Fig. 1B). A correct response for the different training conditions results in the delivery of a sucrose pellet to the food chamber at the rear of a chamber, whereas an incorrect choice is signaled by turning out all the lights for a brief period.

In the first report using this task, Carli et al. *(76)* trained rats on either the Same or the Opposite versions of the task prior to making a unilateral 6-OHDA lesion of the ascending dopamine inputs to the striatum. The lesion resulted in a marked slowing of the reaction time for correct contralateral responses but not for correct ipsilateral responses, even though the RT component of the response simply involves the latency to withdraw the nose from the central hole and is not as such lateralized. By contrast, the lateralized component of the response, the movement time to execute the nose-poke on one side or the other, was not itself affected by the lesions. Thus the unilateral 6-OHDA lesion deficit is in the output rather than input stage of the stimulus–response association; the deficit appears to involve an impairment in the selection and initiation of a response in contralateral space rather than a pure motor deficit *per se*.

In subsequent studies, a similar pattern of results has been shown in rats with excitotoxic lesions of intrinsic striatal neurons. Thus unilateral ibotenic acid lesions of the striatum induce a response bias toward the ipsilateral side, and those responses that are made toward the contralateral side take longer to initiate *(78,79)*. This accords with pharmacological studies in which bilateral injections of NMDA antagonists into the striatum increase the time taken to release a depressed lever when cued *(80)*. However, one important distinction between the excitotoxic lesion and the dopamine manipulation concerns the "temporal priming" of responding *(81)*. In control animals, reaction time invariably decreases as a function of the variable "delay" that precedes the imperative light stimulus. Thus, the greater the period of time that elapses between the start of the central nose-hold and this imperative cue, the quicker is the subsequent reaction time. This "delay-dependent" speeding of reaction time is abolished by dopamine depletion, so that the biggest increase in postoperative reaction times is seen at the longest delays. This contrasts with excitotoxic striatal lesions that increase reaction time uniformly at all delays, and suggests a specific role for striatal dopamine in "priming" an impending motor response as the onset of the action becomes more probable *(81)*.

6.3. A Lateralized Discrimination Task

The specific nature of the contralateral "neglect" has been quantified with the use of both dopaminergic and excitotoxic lesions *(82–84)*. In these studies, animals with unilateral lesions are required to choose between one of two responses as before, only now the response locations are either both located on the ipsilateral side, or both are located on the contralateral side. This methodology addresses the question of precisely what the deficit is "contralateral to," and thus can reveal how striatally mediated responses are coded in the unlesioned animal. So, in the most recent study, animals were trained to perform two discriminations, independently, on alternate days: to respond to the holes on the left on one day, and to the holes on the right on the next day *(84)*. Once the rats could perform both discriminations, they received unilateral striatal injections of quinolinic acid and postoperative performance was then assessed. The animals showed a marked disturbance of correct responding and an induced bias toward the "near" hole, that is, the hole closer to the center, in particular on trials on the contralateral side to the lesions *(see* Fig. 2).

This pattern of results allows a definitive comparison to be made between two specific hypotheses concerning the coding of striatally mediated response space. If responses were coded relative to an external referent ("allocentric" coding), such as the response holes themselves, then a bias might be expected toward the relatively ipsilateral hole. This would manifest itself as a bias toward the far hole when the task is performed to the ipsilateral side, and a bias toward the near hole when the task is performed to the contralateral side. Alternatively, if responses were coded with respect to the subject's body ("egocentric" coding), then one would predict responding to be disrupted only on the contralateral side. The data were consistent with the latter, egocentric, hypothesis: animals were relatively unbiased when responding to the ipsilateral holes, and certainly did not bias their responding toward the far (i.e., relatively contralateral) hole (Fig. 2). By contrast, when responding on the contralateral side, animals were completely unable to select responses to the far (i.e., relatively contralateral) hole (Fig. 2). This effect of intrinsic striatal lesions is again similar to the response profile seen following unilateral dopamine lesions *(83)*.

Together, these results indicate a large degree of independence between the two striata in mediating responding to each side of space. These studies illustrate well how the development of an operant paradigm can provide a systematic and well-controlled parametric framework within which to define the role of the rodent striatum in motor

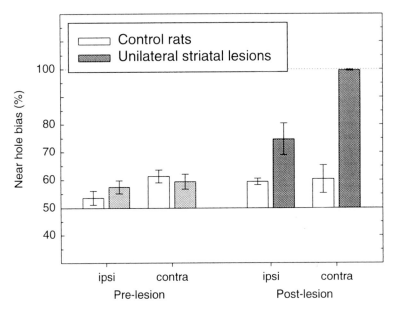

Fig. 2. Striatal lesions disrupt performance and increase near-hole bias on the contralateral side in the lateralized discrimination task. (Data from ref. *84.*)

behavior. In particular, this has involved the detailed analysis of the "striatal neglect" that occurs following unilateral lesions and has utilized the laterality of motor function that is evident within the striatum.

7. OPERANT ANALYSES OF STRIATAL COGNITIVE DEFICITS

7.1. Cognitive Deficits after Striatal Lesions

Although early case reports of HD noted a gradual "cognitive decline" in many patients, it was not until Rosvold's experimental work in the 1950s that the concept of the striatum as a structure involved purely in motor function was seriously challenged. In examining the functional relevance of the dense projection in primates from the prefrontal cortex to the caudate nucleus, it was shown that caudate lesions produced similar performance deficits as prefrontal cortex lesions on tasks such as delayed spatial alternation and delayed response *(85–87)*. This principle was extended empirically in primates, such that restricted striatal lesions produced impairments on the same tasks as those seen following lesions of the corresponding cortical areas *(52)*.

A similar functional organization in the rat was initially not found, largely because of ambiguity about the locus of prefrontal cortex (PFC) in rodents, or indeed whether they had prefrontal lobe at all *(88)*. However, once it was realized that PFC could be defined in terms of the projection areas of the mediodorsal nucleus of the thalamus, Leonard defined the PFC in the rat in terms of projections of the mediodorsal thalamus to a medial wall of cortex dorsal and anterior to the genu of the corpus callosum and to the orbital cortical strip along the dorsal bank of the rhinal sulcus *(89,90)*. Then, the identification of medial PFC projections to the anteromedial striatum and orbital PFC projections to the ventrolateral parts of the striatum *(89,91)* provided an anatomical foundation for subsequent behavioral experiments. A functional heterogeneity in the striatum that reflected distinct neocortical inputs was subsequently observed in rodents. Both aspirative lesions of the medial PFC *(88,92,93)* and excitotoxic lesions of the medial striatum *(55,57,94)* induced deficits in delayed alternation and spatial navigation tasks. Meanwhile, lesions of the orbital PFC and the ventrolateral striatum both produce "perseverative" errors in extinction and on a DRL schedule *(56,64,95,96)*.

The emphasis of studies investigating the "cognitive" functions of the striatum then turned to dissociating stimulus–response learning from other neuroanatomically distinct memory systems *(97,98)*. Meanwhile the idea of functional heterogeneity within the striatum based on corticostriatal connectivity became more complex in light of evolving concepts of neurochemical compartmentalization of the striatum *(99,100)* that did not correspond easily with disparate PFC innervations. However, the concept of functional heterogeneity based on the topographic pattern of cortico-striatal connections was resurrected with the proposal of cortico-striatal "loops" *(101,102)* as well as the emergence of disease modeling *(103,104)*. If impairments could be seen clinically that were believed to result from damage to the caudate, then the question arose as to whether similar deficits could be seen experimentally.

7.2. Delayed Matching Tasks

One such attempt was the introduction of delayed matching procedures designed to test short-term or working memory in experimental animals. An animal is shown a stimulus (the "sample") which is then removed; after a variable delay the animal is given a choice between two stimuli, one being the sample and the other novel; if the animal selects or responds to the sample it is rewarded with food, whereas a response to the other stimulus goes unrewarded or is mildly punished. In this way we can determine how well an animal can remember a single presentation of the stimulus, and on a series of trials we can develop a "forgetting function" of the amount remembered and hence the rate of forgetting over different time intervals. Delayed-matching-to-sample (DMTS) was first developed for use in primates with a large selection of novel junk objects as samples *(105)*. The visual modality is not particularly salient for rats, however, which are much better at learning about places in space, hence the development of delayed-matching-to-position (DMTP) for rats *(106)*. The design of a rat DMTP test in an operant box is illustrated in Fig. 3.

DMTS is sensitive to PFC damage in primates *(107–109)*. Consequently, we have developed DMTP to investigate the comparative contributions of the medial PFC and medial striatum in a working memory task in rats. Animals were trained to respond to one of two levers (left or right) in a sample trial (*see* Fig. 3). The lever was then retracted and a random variable delay period then commenced. The first panel press made after the delay was completed resulted in both levers being extended. A correct matching response on the same lever as that produced in the sample stage was rewarded with the delivery of a food pellet. Cortical lesions that are restricted to the anterior medial prefrontal cortex resulted in a progressive deficit such that accuracy was increasingly impaired at progressively longer delays. This delay-dependent pattern of impairment suggests a rather specific disturbance in short-term memory. However, a slightly larger lesion that extended into the anterior cingulate cortex produced a broader impairment at all delays, reflecting a more generalized deficit in the animal's ability to perform the matching rule *(110)*. Similarly, when we looked at impairments following striatal

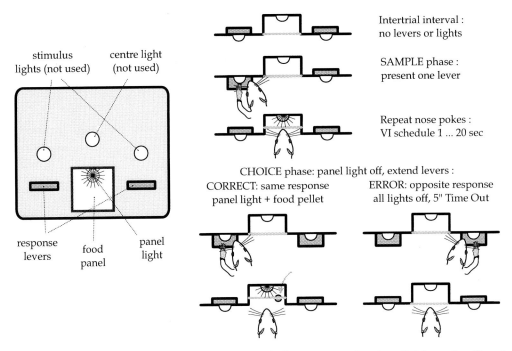

Fig. 3. Schematic illustration of the front panel of an operant chamber ("Skinner box") and the sequence of steps in a single trial of the delayed matching to position task.

lesions, lesions in the ventral striatum (which receives restricted prefrontal inputs) exhibited a clear delay-dependent deficit, whereas lesions in the dorsal striatum (which receives cingulate as well as frontal inputs) disrupted the animal's performance at all delays (Fig. 4) *(110)*.

More recently, we have undertaken a more detailed analysis of the nature of the deficit and recovery after striatal lesions. When animals are pretrained and then tested for retention after striatal lesions they show clear deficits in DMTP *(111,112)*, with the magnitude of the deficit attributable in part to the size of the lesion. Thus, large striatal lesions, as employed in our first study, produced random performance at all delays *(112)*, whereas smaller lesions produce a clear delay-dependent disruption, with the animals showing good performance at the shortest delays, but greater deficits at longer delays *(111)*. Nevertheless, animals will recover from this lesion if given repeated training (*see* Fig. 5) *(111)*. Similarly, if the animals are lesioned before training on the task, although they show small deficits early in the course of training, they can recovery and show little difference from control animals in overall ability to acquire the task *(65)*. The recovery seen after lesions is due to the ability of the animals to relearn the response contingency. This is the result of explicit training on the task as opposed to spontaneous recovery due to the simple passage of time, because if the animals are simply left in their home cage for a period of months before testing the initial deficit is as great as that seen in animals tested immediately post-lesion (*see* Fig. 5) *(111)*.

These studies illustrate, first, the way in which an operant task may be designed to tap discrete aspects of cognitive function—in this case short-term working memory, and second, that selective lesions within the fronto-striatal circuitry yield distinctive

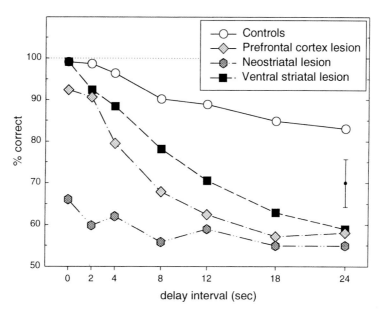

Fig. 4. Effects of lesions of the prefrontal cortex, the dorsal neostriatum, and the ventral striatum (nucleus accumbens) on the delayed matching to position task. (Data from ref. *110.*)

patterns of cognitive deficit associated with the particular cortical loop(s) that is (are) disturbed.

Awaiting more detailed analysis is the nature of the functional deficit introduced following damage at each level of the neural circuit. For example, we have been able to identify specific aspects of trial by trial interference by analyzing the effect of intertrial interval on intratrial forgetting *(113)*. However, this analysis has so far been undertaken only in intact rats *(113)*, and has been used to show that the increased rate of forgetting is not attributable to proactive interference in aged animals *(114)*. We might, for example, hypothesize that mPFC lesions would preferentially disrupt the retention of information in short-term memory in DMTP, whereas medial striatal lesion would disrupt the selection and initiation of an appropriate lateralized response in the same task, based on the plan of action organized in the neocortex. Indeed, our initial data suggest that whereas the prefrontal lesions may indeed disrupt short-term memory processes themselves, as revealed by clear-cut delay-dependent deficits in several variants of the task, the striatal lesions (although disrupting performance on a short-term memory task) do not involve a primary memory impairment *per se* but rather disrupt the animal's ability to acquire and maintain the appropriate stimulus–response (or "habit") contingencies involved in accurate task performance.

7.3. Delayed Alternation Tasks

The spatial delayed alternation (DA) task is another task that involves short-term working memory and requires animals to alternate their responding between two spatially distinct locations. Like the DMTP task, it is very sensitive to damage of the medial PFC. Indeed deficits in delayed alternation were one of the defining features of the prefrontal deficit described by Jacobsen in his classic primate studies in the 1930s, and replicated many times since *(115–117)*.

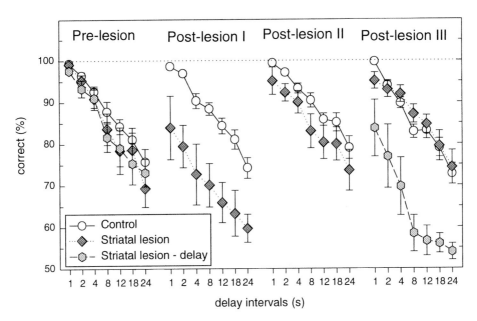

Fig. 5. Effects of acute striatal lesions on the delayed matching to position task, the recovery observed in successive blocks of training, and the stability of the deficit over time in animals not tested during the first two blocks (the "delay" subgroup). (Data from ref. *111*.)

Rats similarly exhibit clear deficits in delayed alternation after prefrontal lesions when tested in a T-maze *(92,112,118,119)*. There have been several attempts to adapt this classic task to the operant box *(120,121)*, although with varying degrees of success. In our adaptation, rats are trained to press the centrally located food panel until the end of a variable delay period (5–20 s). A panel press subsequent to the end of the delay period results in the extension of both the left and the right levers. On the first trial of the day, pressing either lever produces a food pellet reward. On all subsequent trials, the rat is required to press the lever that was *not* pressed on the previous trial. A correct press is rewarded with a food pellet whereas an incorrect press (repetition of the same lesion as on the previous trial) has no consequence. In either case, after pressing one lever, both levers are withdrawn and the timer for the next variable delay interval is started. The distinctive feature of our variant of operant delayed alternation, as in the DMTP task described previously, is that the animal is required to nose-poke at the central panel during the delay to trigger presentation of the two choice levers. This serves to keep the rat centralized between the two response locations and reduces the opportunity for it to adopt a simple mediating response strategy during the delay (i.e., simply waiting at the location where the correct lever will next appear). We vary the delay interval on each trial and thereby accumulate information about the animal's level of accuracy over different lengths of time that the last trial response must be held in memory.

In studies that are still ongoing, we find that lesions both of the medial PFC and of the medial striatum disrupt performance on this task. As might be predicted from the DMTP data, animals with medial PFC lesions are particularly impaired at the longer delays, whereas striatally lesioned animals are severely impaired (performing at chance

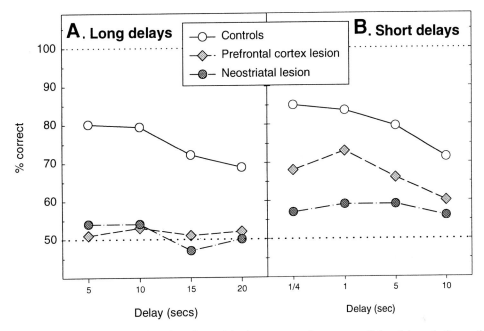

Fig. 6. Effects of striatal and prefrontal lesions on performance of the delayed alternation task. "Long" delays of 5–20 s were used between successive trials during the first post-lesion block of testing (**A**), which were reduced to "short" delays of 0.25–10 s in the second block (**B**). (Dunnett, S. B., Nathwani, F., and Brasted, P., *unpublished data*.)

levels) at all delays (Fig. 6). Detailed analyses of the different behavioral measures further demonstrate that the time taken to collect food reward is unaffected, suggesting neither a general motor deficit nor a motivational deficit as the basis for these impairments.

8. OPERANT ANALYSES OF STRIATAL MOTIVATIONAL DEFICITS

The third major class of deficits in the HD triad is psychiatric disturbance, which includes affective disorder, a variety of disinhibitory symptoms such as aggressive outbursts, and eating disorder including a marked weight loss. It is of course difficult to evaluate psychiatric disturbance *per se* in rats, but we can evaluate the effects of striatal lesions on changes in motivation and sensitivity to the stimuli of regulatory control. Thus, for example, one of the earliest deficits noted in animals with excitotoxic lesions of the striatum is a marked loss of body weight *(57,122,123)*, just like that seen in the patients, and an impairment in the animals sensitivity to the physiological stimuli signaling hunger and thirst *(123)*. Similarly, the hyperactivity is not general, but particularly apparent at times of arousal, such as associated with food deprivation *(18,43,44)*.

To investigate motivational factors in more detail, we turned to an operant progressive ratio schedule of lever pressing. In this schedule, the number of lever presses required for each reinforcer delivery increases step by step during a session. Thus the rat has to work progressively harder to obtain each successive reward. We can therefore analyze the extent to which an animal is prepared to work for different levels of reinforcement.

Optimal performance under a progressive ratio schedule requires precise regulation of instrumental behavior in relation to changing contingencies between response output and reward. This requires the coordination and integration of several distinct psychological processes, including motivational components, which can be monitored by performance on the schedule *(124,125)*. The primary measures of motivation under progressive ratio schedules are "breakpoint," which represents the amount of effort a rat is willing to expend to achieve a given level of reinforcement, and the "post-reinforcement pause," which represents the reluctance to resume lever pressing as reward becomes more costly. Decreased breakpoints are generally taken to reflect lower levels of motivation, and vice versa *(124,125)*.

We have used this task to characterize changes in motivation after striatal lesions *(126)*. Neither dorso-medial nor dorso-lateral striatal lesions had major effects on the primary motivational measures of the task. Thus, neither lesion induced any detectable change in either the breakpoint (*see* Fig. 7A) or the post-reinforcement pause, which was found to increase in all groups as a positively accelerated function of the number of lever presses required for reward *(127)*. However, striatal lesions did induce deficits in other aspects of task performance. First, the lateral striatal lesions produced a significant increase in the numbers of perseverative lever presses made by the animals between delivery and collection of the rewards (*see* Fig. 7B). Second, the lesions significantly lengthened the latencies to collect earned food pellets. Thus, neither medial nor lateral striatal lesions affected the animals' normal motivational ability to regulate their rates of responding, adapting efficiently to changes in the costs of earning each reward. Rather, the perseverative changes exhibited by the rats with striatal lesions suggest an "executive" difficulty in the efficient sequencing and switching of responses as signaled by discrete stimulus events in the environment, emphasizing again the close relationship between the striatum and the PFC.

9. ADVANTAGES OF OPERANT PARADIGMS

Operant paradigms for behavioral testing and analysis have several distinct advantages, both practical and theoretical, over traditional hand testing and observation.

The first and foremost practical advantage is that operant tasks permit a greater degree of experimental control. Tasks that require precise stimulus control and a variety of response options impose the requirement for careful specification of all the possible or potential outcomes of a trial and quantitative recording of each of the different responses an animal could make.

The tasks are run in a fully automated apparatus. This conveys the distinct advantage of reducing opportunities for experimental bias that accompanies the observational recording of behavior. Nevertheless, periodic observation of performance, via a video link or one-way mirror into the test chamber, is valuable to ensure that the animals are indeed engaging in the classes of response that are predicted, scheduled, and recorded.

The automated equipment and online computer control allow for considerable efficiency in data collection and numerical processing. With a bank of test chambers, multiple animals can be tested simultaneously. This allows the collection of data from many more trials than can be practically achieved with hand testing. This difference is

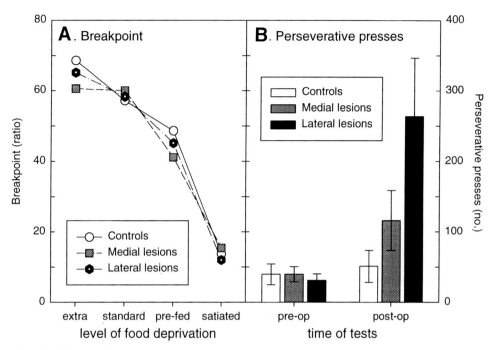

Fig. 7. Effects of striatal lesions on performance of a progressive ratio schedule. (**A**) Breakpoint under different levels of food deprivation was not affected by the lesions. (**B**) Perseverative lever presses after completion of each ration were markedly enhanced after striatal lesions. (Data from ref. *126.*)

well illustrated by contrasting DMTP or operant delayed alternation with classical T-maze alternation. Operant versions of the task can yield forgetting curves with tight variance estimates of choice accuracy (% correct) at seven or more delay intervals based on 100–200 trials per day over several weeks of testing. By contrast, classical T-maze alternation (or, e.g., spatial navigation learning in a Morris water maze) yield much more variable measures of learning based on only 4–10 trials per day. In parallel with allowing collection of many more trials, the automated paradigms allow many more parameters of performance to be recorded, including the incidence of different types of error and accurate measures of the speed and accuracy of correct responding.

In addition to the practical advantages, operant tests can provide a more refined analysis of complex behavior, as is clearly demonstrated in many of the studies outlined previously. Such a detailed analysis of specific aspects of performance has a rich history in the experimental analysis of behavior in normal animals, and there is a rich selection of procedures available for discriminating between specific aspects of motor, motivational, and cognitive function.

Whereas operant behavioural analyses have long been used for critical analysis of the psychopharmacology of drug action *(128)*, they have been less widely used in neurobiological studies of brain damage, regeneration, and repair. Perhaps the major experimental strength of operant paradigms is for behavioral assessment in experimental neurology where changes need to be monitored over long periods of time. In contrast to many other behavioral tests and measures, operant tasks can provide extremely stable levels of behavioral performance over many months. Such stable measures of functional capacity provide a powerful baseline for assessing changes attributable to an experimental treatment. Thus for our purposes, operant tasks provide effective and efficient functional assays of:

- drug effects in normal and brain damaged animals;
- deterioration of function associated with neurodegenerative events in the brain, whether due to explicit lesions, genetic mutations, or natural aging;
- stability of function associated with neuroprotective treatments, such as trophic molecules that prevent progression of a lesion;
- recovery of function associated with reparative treatments such as neural transplantation or growth factor induced plasticity.

In the present chapter, we have reviewed some applications of operant paradigms to the analysis of disturbed striatal function as the basis for developing improved animal models of HD, and which are being used to improve the validity of our evaluation of strategies for striatal repair using neural transplants *(129,130)*. It is clear that cortical and striatal lesions produce deficits on the same cognitive tasks, as predicted by their involvement in common fronto-striatal loops. Of greater interest, we have begun to differentiate the precise nature of the functional deficits associated with damage at different levels of the functional circuit, and this applies to the cognitive and motivational as well as motor domains of analysis.

Nevertheless, this level of analysis is still at an early stage. The goal of the present chapter is to outline the power of operant paradigms. The proof of that strategy will be in the extent to which it is successful in actually revealing over the next decade the principles of striatal organization and function that are at present still unknown.

ACKNOWLEDGMENTS

We gratefully acknowledge the financial support of the Medical Research Council and Wellcome Trust.

REFERENCES

1. Huntington's Disease Collaborative Research Group (1993) A novel gene containing a trinucleotide repeat that is expanded and unstable on Huntington's disease chromosomes. *Cell* **72,** 971–983.
2. Vonsattel, J.-P., Myers, R. H., and Stevens, T. J. (1985) Neuropathologic classification of Huntington's disease. *J. Neuropathol. Exp. Neurol.* **44,** 559–577.
3. Myers, R. H., Vonsattel, J.-P., Paskevich, P. A., Kiely, D. K., Stevens, T. J., Cupples, L. A., Richardson, E. P. and Bird, E. D. (1991) Decreased neuronal and increased oligodendroglial densities in Huntington's disease caudate nucleus. *J. Neuropathol. Exp. Neurol.* **50,** 742.
4. Graveland, G. A., Williams, R. S., and DiFiglia, M. (1985) Evidence for degenerative and regenerative changes in neostriatal spiny neurons in Huntington's disease. *Science* **227,** 770–773.
5. Albin, R. L., Qin, Y., Young, A. B., Penney, J. B., and Chesselet, M.-F. (1991) Preproenkephalin messenger RNA-containing neurons in striatum of patients with symptomatic and presymptomatic Huntington's disease: an in situ hybridisation study. *Ann. Neurol.* **30,** 542–549.

6. Albin, R. L., Reiner, A., Anderson, K. D., Dure, L. S., Handelin, B., Balfour, R., Whetsell, W. O., Penney, J. B., and Young, A. B. (1992) Preferential loss of striato-external pallidal projection neurons in presymptomatic Huntington's disease. *Ann. Neurol.* **31,** 425–430.

7. Hedreen, J. C. and Folstein, S. E. (1995) Early loss of neostriatal striosome neurons in Huntington's disease. *J. Neuropathol. Exp. Neurol.* **54,** 105–120.

8. Cudkowicz, M. and Kowall, N. W. (1990) Degeneration of pyramidal projection neurons in Huntington's disease cortex. *Ann.Neurol.* **27,** 200–204.

9. Sotrel, A., Paskevich, P. A., Kiely, D. K., Bird, E. D., Williams, R. S., and Myers, R. H. (1991) Morphometric analysis of the prefrontal cortex in Huntington's disease. *Neurology* **41,** 1117–1123.

10. Sotrel, A., Williams, R. S., Kaufmann, W. E., and Myers, R. H. (1993) Evidence for neuronal degeneration and dendritic plasticity in cortical pyramidal neurons of Huntington's disease—a quantitative Golgi study. *Neurology* **43,** 2088–2096.

11. Oyanagi, K., Takeda, S., Takahashi, H., Ohama, E., and Ikuta, F. (1989) A quantitative investigation of the substantia nigra in Huntington's disease. *Ann. Neurol.* **26,** 13–19.

12. Lange, H., Thorner, G., Hopf, A., and Schroder, K. F. (1976) Morphometric studies of the neuropathological changes in choreatic diseases. *J. Neurol. Sci.* **28,** 401–425.

13. Sanberg, P. R., Wictorin, K., and Isacson, O. (1994) Cell transplantation for Huntington's disease, R.G. Landes Company, Austin, TX.

14. Lawrence, A. D., Hodges, J. R., Rosser, A. E., Kershaw, A., ffrench-Constant, C., Rubinsztein, D. C., Robbins, T. W., and Sahakian, B. J. (1998) Evidence for specific cognitive deficits in preclinical Huntington's disease. *Brain* **121,** 1329–1341.

15. Van Vugt, J. P. P., Van Hilten, B. J., and Roos, R. A. C. (1996) Hypokinesia in Huntington's disease. *Move. Dis.* **11,** 384–388.

16. Harper, P. S. (1996) *Huntington's Disease*, W. B. Saunders, London.

17. Mason, S. T., Sanberg, P. R., and Fibiger, H. C. (1978) Kainic acid lesions of the striatum dissociate amphetamine and apomorphine stereotypy: similarities to Huntington's chorea. *Science* **201,** 352–355.

18. Sanberg, P. R. and Coyle, J. T. (1984) Scientific approaches to Huntington's disease. *CRC Crit. Rev. Clin. Neurobiol.* **1,** 1–44.

19. Brown, R. G. and Marsden, C. D. (1998) Subcortical dementia—the neuropsychological evidence. *Neuroscience* **25,** 363–387.

20. Butters, N., Sax, D., Montgomery, K., and Tarlow, S. (1978) Comparison of the neuropsychological deficits associated with early and advanced Huntington's disease. *Arch. Neurol.* **35,** 585–589.

21. Lange, K. W., Sahakian, B. J., Quinn, N. P., Marsden, C. D., and Robbins, T. W. (1995) Comparison of executive and visuospatial memory function in Huntington's disease and dementia of Alzheimer-type matched for degree of dementia. *J. Neurol. Neurosurg. Psychiatry* **58,** 598–606.

22. Knopman, D. and Nissen, M. J. (1991) Procedural learning is impaired in Huntington's disease—evidence from the serial reaction time task. *Neuropsychologia* **29,** 245–254.

23. Sprengelmeyer, R., Lange, H., and Hömberg, V. (1995) The pattern of attentional deficits in Huntington's disease. *Brain* **118,** 145–152.

24. Georgiou, N., Bradshaw, J. L., Phillips, J. G., and Chiu, E. (1996) The effect of Huntington's disease and Gilles de la Tourette's syndrome on the ability to hold and shift attention. *Neuropsychologia* **34,** 843–851.

25. Georgiou, N., Bradshaw, J. L., Phillips, J. G., and Chiu, E. (1997) Effect of directed attention in Huntington's disease. *J. Clin. Exp. Neuropsychol.* **19,** 367–377.

26. Lawrence, A. D., Sahakian, B. J., Hodges, J. R., Rosser, A. E., Lange, J. W., and Robbins, T. W. (1996) Executive and mnemonic functions in early Huntington's disease. *Brain* **119,** 1633–1645.

27. Brandt, J., Strauss, M. E., Larus, J., Jensen, B., Folstein, S. E., and Folstein, M. F. (1984) Clinical correlates of dementia and disability in Huntington's disease. *J. Cognit. Neuropsychol.* **6,** 401–412.

28. Coyle, J. T. and Schwarcz, R. (1976) Lesions of striatal neurones with kainic acid provides a model for Huntington's chorea. *Nature* **263,** 244–246.

29. McGeer, E. G. and McGeer, P. L. (1976) Duplication of the biochemical changes of Huntington's choreas by intrastriatal injection of glutamic and kainic acids. *Nature* **263,** 517–519.

30. Schwarcz, R., Whetsell, W. O., and Mangano, R. M. (1983) Quinolinic acid: an endogenous metabolite that produces axon-sparing lesions in rat brain. *Science* **219,** 316–318.

31. Schwarcz, R., Hökfelt, T., Fuxe, K., Jonsson, G., Goldstein, M., and Terenius, L. (1979) Ibotenic acid-induced neuronal degeneration: a morphological and neurochemical study. *Exp. Brain Res.* **37,** 199–216.

32. Beal, M. F., Ferrante, R. J., Swartz, K. J., and Kowall, N. W. (1991) Chronic quinolinic acid lesions in rats closely resemble Huntington's disease. *J. Neurosci.* **11,** 1649–1659.

33. Beal, M. F., Kowall, N. W., Swartz, K. J., Ferrante, R. J., and Martin, J. B. (1989) Differential sparing of somatostatin-neuropeptide Y and cholinergic neurons following striatal excitotoxic lesions. *Synapse* **3,** 38–47.

34. Beal, M. F., Hyman, B. T., and Koroshetz, W. (1993) Do defects in mitochondrial energy metabolism underlie the pathology of neurodegenerative diseases? *Trends Neurosci.* **16,** 125–131.

35. Borlongan, C. V., Koutouzis, T. K., and Sanberg, P. R. (1997) 3-Nitropropionic acid animal model and Huntington's disease. *Neurosci. Biobehav. Rev.* **21,** 289–293.

36. Beal, M. F., Brouillet, E. P., Jenkins, B. G., Ferrante, R. J., Kowall, N. W., Miller, J. M., Storey, E., Srivastava, R., Rosen, B. R., and Hyman, B. T. (1993) Neurochemical and histologic characterization of striatal excitotoxic lesions produced by the mitochondrial toxin 3-nitropropionic acid. *J. Neurosci.* **13,** 4181–4192.

37. Palfi, S. P., Ferrante, R. J., Brouillet, E., Beal, M. F., Dolan, R., Guyot, M. C., Peschanski, M., and Hantraye, P. (1996) Chronic 3-nitropropionic acid treatment in baboons replicates the cognitive and motor deficits of Huntington's disease. *J. Neurosci.* **16,** 3019–3025.

38. Page, K. J., Dunnett, S. B., and Everitt, B. J. (1998) 3-Nitroproprionic acid induced changes in the expression of metabolic and astrocyte mRNAs. *NeuroReport* **9,** 2881–2886.

39. Dunnett, S. B. and Everitt, B. J. (1998) Topographic factors affecting the functional viability of dopamine-rich grafts in the neostriatum, in *Fetal Transplantation in Neurological Disease* (Freeman, T. B. and Kordower, J. H., eds.), Humana Press, Totowa, NJ, pp. 135–169.

40. Figueredo-Cardenas, G., Anderson, K. D., Chen, Q., Veenman, C. L., and Reiner, A. (1994) Relative survival of striatal projection neurons and interneurons after intrastriatal injection of quinolinic acid in rats. *Exp. Neurol.* **129,** 37–56.

41. Gerfen, C. R., Baimbridge, K. G., and Thibault, J. (1987) The neostriatal mosaic: III. Biochemical and developmental dissociation of patch-matrix mesostriatal systems. *J. Neurosci.* **7,** 3935–3944.

42. Marsden, C. D. (1982) The mysterious motor function of the basal ganglia: the Robert Wartenberg lecture. *Neurology* **32,** 514–539.

43. Isacson, O., Dunnett, S. B., and Björklund, A. (1986) Graft-induced behavioral recovery in an animal model of Huntington disease. *Proc. Natl. Acad. Sci. USA* **83,** 2728–2732.

44. Mason, S. T. and Fibiger, H. C. (1979) Kainic acid lesions of the striatum mimic the spontaneous locomotor abnormalities of Huntington's disease. *Neuropharmacology* **18,** 403.

45. Schwarcz, R., Fuxe, K., Agnati, L. F., Hökfelt, T., and Coyle, J. T. (1979) Rotational behavior in rats with unilateral striatal kainic acid lesions: a behavioural model for studies on intact dopamine receptors. *Brain Res.* **170,** 485–495.

46. Dunnett, S. B., Isacson, O., Sirinathsinghji, D. J. S., Clarke, D. J., and Björklund, A. (1988) Striatal grafts in rats with unilateral neostriatal lesions. III. Recovery from dopamine-dependent motor asymmetry and deficits in skilled paw reaching. *Neuroscience* **24,** 813–820.

47. Fricker, R. A., Annett, L. E., Torres, E. M., and Dunnett, S. B. (1996) The locus of a striatal ibotenic acid lesion affects the direction of drug-induced rotation and skilled forelimb use. *Brain Res. Bull.* **41,** 409–416.

48. Norman, A. B., Norgren, R. B., Wyatt, L. M., Hildebrand, J. P., and Sanberg, P. R. (1992) The direction of apomorphine-induced rotation behavior is dependent on the location of excitotoxin lesions in the rat basal ganglia. *Brain Res.* **569,** 169–172.

49. Borlongan, C. V., Randall, T. S., Cahill, D. W., and Sanberg, P. R. (1995) Asymmetrical motor behavior in rats with unilateral striatal excitotoxic lesions as revealed by the elevated body swing test. *Brain Res.* **676,** 231–234.

50. Pisa, M. (1988) Motor functions of the striatum in the rat: critical role of the lateral region in tongue and forelimb reaching. *Neuroscience* **24,** 453–463.

51. Whishaw, I. Q., O'Connor, W. T., and Dunnett, S. B. (1986) The contributions of motor cortex, nigrostriatal dopamine and caudate-putamen to skilled forelimb use in the rat. *Brain* **109,** 805–843.

52. Divac, I., Rosvold, H. E., and Szwarcbart, M. K. (1967) Behavioral effects of selective ablation of the caudate nucleus. *J. Comp. Physiol. Psychol.* **63,** 184–190.

53. Divac, I. (1968) Effects of prefrontal and caudate lesions on delayed response in cats. *Acta Neurobiol. Exp.* **28,** 149–167.

54. Divac, I. (1971) Frontal lobe system and spatial reversal in the rat. *Neuropsychologia* **9,** 175–183.

55. Divac, I., Markowitsch, H. J., and Pritzel, M. (1978) Behavioural and anatomical consequences of small intrastriatal injections of kainic acid in the rat. *Brain Res.* **151,** 523–532.

56. Dunnett, S. B. and Iversen, S. D. (1982) Neurotoxic lesions of ventrolateral but not anteromedial neostriatum in rats impair differential reinforcement of low rates (DRL) performance. *Behav. Brain Res.* **6,** 213–226.

57. Sanberg, P. R., Lehmann, J., and Fibiger, H. C. (1978) Impaired learning and memory after kainic acid lesions of the striatum: a behavioral model of Huntington's disease. *Brain Res.* **149,** 1204–1208.

58. Pisa, M., Sanberg, P. R., and Fibiger, H. C. (1981) Striatal injections of kainic acid selectively impair serial memory performance in the rat. *Exp. Neurol.* **74,** 633–653.

59. Öberg, R. G. E. and Divac, I. (1979) Cognitive functions of the neostriatum, in *The Neostriatum* (Divac, I. and Öberg, R. G. E., eds.), Pergamon Press, Oxford.

60. Robbins, T. W., Cador, M., Taylor, J. R., and Everitt, B. J. (1989) Limbic-striatal interactions in reward-related processes. *Neurosci. Biobehav. Rev.* **13,** 155–162.

61. Robbins, T. W. and Everitt, B. J. (1996) Neurobehavioral mechanisms of reward and motivation. *Curr. Opin. Neurobiol.* **6,** 228–236.

62. Skinner, B. F. (1938) *The Behavior of Organisms,* Appleton-Century-Crofts, New York.

63. Ferster, C. B. and Skinner, B. F. (1957) *Schedules of Reinforcement,* Appleton-Century-Crofts, New York.

64. Sanberg, P. R., Pisa, M., and Fibiger, H. C. (1979) Avoidance, operant and locomotor behavior in rats with neostriatal injections of kainic acid. *Pharmacol. Biochem. Behav.* **10,** 137–144.

65. Döbrössy, M. D., Svendsen, C. N., and Dunnett, S. B. (1995) The effects of bilateral striatal lesions on the acquisition of an operant test of short-term memory. *NeuroReport* **6,** 2059–2053.

66. Reading, P. J., Dunnett, S. B., and Robbins, T. W. (1991) Dissociable roles of the ventral, medial and lateral striatum on the acquisition and performance of a complex visual stimulus—response habit. *Behav. Brain Res.* **45,** 147–161.

67. Reading, P. J., Torres, E. M., and Dunnett, S. B. (1995) Embryonic striatal grafts ameliorate the disinhibitory effects of ventral striatal lesions. *Exp. Brain Res.* **105,** 76–86.
68. Ljungberg, T. and Ungerstedt, U. (1976) Sensory inattention produced by 6-hydroxy-dopamine-induced degeneration of ascending dopamine neurons in the brain. *Exp. Neurol.* **53,** 585–600.
69. Ungerstedt, U. and Arbuthnott, G. W. (1970) Quantitative recording of rotational behaviour in rats after 6-hydroxydopamine lesions of the nigrostriatal dopamine system. *Brain Res.* **24,** 485–493.
70. Marshall, J. F., Richardson, J. S., and Teitelbaum, P. (1974) Nigrostriatal bundle damage and the lateral hypothalamic syndrome. *J. Comp. Physiol. Psychol.* **87,** 808–830.
71. Schallert, T., Upchurch, M., Wilcox, R. E., and Vaughn, D. M. (1983) Posture-independent sensorimotor analysis of inter-hemispheric receptor asymmetries in neostriatum. *Pharmacol. Biochem. Behav.* **18,** 753–759.
72. Evenden, J. L. and Robbins, T. W. (1984) Effects of unilateral 6-hydroxydopamine lesions of the caudate-putamen on skilled forepaw use in the rat. *Behav. Brain Res.* **14,** 61–68.
73. Abrous, D. N., Wareham, A. T., Torres, E. M., and Dunnett, S. B. (1992) Unilateral dopamine lesions in neonatal, weanling and adult rats: comparison of rotation and reaching deficits. *Behav. Brain Res.* **51,** 67–75.
74. Turner, B. H. (1973) Sensorimotor syndrome produced by lesions of the amygdala and lateral hypothalamus. *J. Comp. Physiol. Psychol.* **82,** 37–47.
75. Marshall, J. F., Turner, B. H., and Teitelbaum, P. (1971) Sensory neglect produced by lateral hypothalamic damage. *Science* **174,** 423–525.
76. Carli, M., Evenden, J. L., and Robbins, T. W. (1985) Depletion of unilateral striatal dopamine impairs initiation of contralateral actions and not sensory attention. *Nature* **313,** 679–682.
77. Robbins, T. W., Muir, J. L., Killcross, A. S., and Pretsell, D. (1993) Methods of assessing attention and stimulus control in the rat, in *Behavioural Neuroscience* Vol. I (Sahgal, A., ed.), IRL Press, Oxford, pp. 13–47.
78. Mittleman, G., Brown, V. J., and Robbins, T. W. (1988) Intentional neglect following unilateral ibotenic acid lesions of the striatum. *Neurosci. Res. Commun.* **2,** 1–8.
79. Mayer, E., Brown, V. J., Dunnett, S. B., and Robbins, T. W. (1992) Striatal graft-associated recovery of a lesion-induced performance deficit in the rat requires learning to use the transplant. *Eur. J. Neurosci.* **4,** 119–126.
80. Amalric, M., Baunez, C., and Nieoullon, A. (1995) Does the blockade of excitatory amino acid transmission in the basal ganglia simply reverse reaction time deficits induced by dopamine inactivation? *Behav. Pharmacol.* **6,** 508–519.
81. Brown, V. J. and Robbins, T. W. (1991) Simple and choice reaction time performance following unilateral striatal dopamine depletion in the rat. Impaired motor readiness but preserved response preparation. *Brain* **114,** 513–525.
82. Brown, V. J. and Robbins, T. W. (1989) Elementary processes of response selection mediated by distinct regions of the striatum. *J. Neurosci.* **9,** 3760–3765.
83. Brown, V. J. and Robbins, T. W. (1989) Deficits in response space following unilateral striatal dopamine depletion in the rat. *J. Neurosci.* **9,** 983–989.
84. Brasted, P., Humby, T., Dunnett, S. B., and Robbins, T. W. (1997) Response space deficits following unilateral excitotoxic lesions of the dorsal striatum in the rat. *J. Neurosci.* **17,** 8919–8926.
85. Rosvold, H. E. and Delgado, J. M. R. (1956) The effect on delayed alternation test performance of stimulating or destroying electrically structures within the frontal lobes of the monkey's brain. *J. Comp. Physiol. Psychol.* **49,** 365–372.
86. Rosvold, H. E. (1972) The frontal lobe system: cortical-subcortical interrelationships. *Acta Neurobiol. Exp.* **32,** 439–460.

87. Rosvold, H. E. and Szwarcbart, M. K. (1964) Neural structures involved in delayed response performance, in *The Frontal Granular Cortex and Behavior* (Warren, J. M. and Akert, K., eds.), McGraw-Hill, New York, pp. 1–15.

88. Divac, I., Wikmark, R. G. E., and Gade, A. (1975) Spontaneous alternation in rats with lesions in the frontal lobes: an extension of the frontal lobe syndrome. *Physiol. Psychol.* **3,** 39–42.

89. Leonard, C. M. (1969) The prefrontal cortex of the rat. I. Cortical projection of the mediodorsal nucleus. II. Efferent connections. *Brain Res.* **12,** 321–343.

90. Krettek, J. E. and Price, J. L. (1977) The cortical projections of the mediodorsal nucleus and adjacent thalamic nuclei in the rat. *J. Comp. Neurol.* **171,** 157–192.

91. Beckstead, R. M. (1979) An autoradiographic examination of corticocortical and subcortical projections of the mediodorsal-projection (prefrontal) cortex in the rat. *J. Comp. Neurol.* **184,** 43–62.

92. Larsen, J. K. and Divac, I. (1978) Selective ablations within the prefrontal cortex of the rat and performance of delayed alternation. *Physiol. Psychol.* **6,** 15–17.

93. Johnston, V. S., Hart, M., and Howell, W. (1974) The nature of the medial wall deficit in the rat. *Neuropsychologia* **12,** 503.

94. Dunnett, S. B. and Iversen, S. D. (1981) Learning impairments following selective kainic acid-induced lesions within the neostriatum of rats. *Behav. Brain Res.* **2,** 189–209.

95. Kolb, B., Sutherland, R. J., and Singh, R. K. (1975) Double dissociation of spatial impairments and perseveration following selective prefrontal lesions in rats. *J. Comp. Physiol. Psychol.* **88,** 806–815.

96. Neill, D. B. (1976) Frontal-striatal control of behavioral inhibition in the rat. *Brain Res.* **105,** 89–103.

97. Packard, M. G., Hirsh, R., and White, N. M. (1989) Differential effects of fornix and caudate nucleus lesions on two radial maze tasks: evidence for multiple memory systems. *J. Neurosci.* **9,** 1465–1472.

98. Adams, F. S., La Rosa, F. G., Kumar, S., Edwards-Prasad, J., Kentroti, S., Vernadakis, A., Freed, C. R., and Prasad, K. N. (1996) Characterization and transplantation of two neuronal cell lines with dopaminergic properties. *Neurochem. Res.* **21,** 619–627.

99. Gerfen, C. R. (1984) The neostriatal mosaic: compartmentalization of corticostriatal input and striatonigral output systems. *Nature* **311,** 461–464.

100. Graybiel, A. M. and Ragsdale, C. W. (1978) Histochemically distinct compartments in the striatum of human, monkey and cat demonstrated by acetylcholinesterase. *Proc. Natl. Acad. Sci. USA* **75,** 5723–5726.

101. Alexander, G. E., Crutcher, M. D., and DeLong, M. R. (1990) Basal ganglia-thalamocortical circuits: parallel substrates for motor, oculomotor, prefrontal and limbic functions. *Prog. Brain Res.* **85,** 119–146.

102. Alexander, G. E., DeLong, M. R., and Strick, P. L. (1986) Parallel organization of functionally segregated circuits linking basal ganglia and cortex. *Annu. Rev. Neurosci.* **9,** 357–381.

103. DeLong, M. R. (1990) Primate models of movement disorders of basal ganglia origin. *Trends Neurosci.* **13,** 281–285.

104. Albin, R. L., Young, A. B., and Penney, J. B. (1989) The functional anatomy of basal ganglia disorders. *Trends Neurosci.* **12,** 366–375.

105. D'Amato, M. R. (1973) Delayed matching and short-term memory in monkeys. *Psychol. Learn. Motiv.* **7,** 227–269.

106. Dunnett, S. B. (1985) Comparative effects of cholinergic drugs and lesions of nucleus basalis or fimbria-fornix on delayed matching in rats. *Psychopharmacology* **87,** 357–363.

107. Mishkin, M. and Manning, F. J. (1978) Non-spatial memory after selective prefrontal lesions in monkeys. *Brain Res.* **143,** 313–323.

108. Eacott, M. J., Gaffan, D., and Murray, E. A. (1994) Preserved recognition memory for small sets, and impaired stimulus identification for large sets, following rhinal cortex ablations in monkeys. *Eur. J. Neurosci.* **6,** 1466–1478.

109. Goldman-Rakic, P. S. (1989) Circuitry of primate prefrontal cortex and regulation of behavior by representational memory, in *Handbook of Physiology—The Nervous System V.* American Physiological Association, Baltimore, pp. 373–417.

110. Dunnett, S. B. (1990) Role of prefrontal cortex and striatal output systems in short-term memory deficits associated with ageing, basal forebrain lesions, and cholinergic-rich grafts. *Can. J. Psychol.* **44,** 210–232.

111. Döbrössy, M. D., Svendsen, C. N., and Dunnett, S. B. (1996) Bilateral striatal lesions impair retention of an operant test of short-term memory. *Brain Res. Bull.* **41,** 159–166.

112. Dunnett, S. B. (1990) Is it possible to repair the damaged prefrontal cortex by neural tissue transplantation? *Prog. Brain Res.* **85,** 285–297.

113. Dunnett, S. B. and Martel, F. L. (1990) Proactive interference effects on short-term memory in rats. 1. Basic parameters and drug effects. *Behav. Neurosci.* **104,** 655–665.

114. Dunnett, S. B., Martel, F. L., and Iversen, S. D. (1990) Proactive interference effects on short-term memory in rats. 2. Effects in young and aged rats. *Behav. Neurosci.* **104,** 666–670.

115. Rosenkilde, C. E. (1979) Functional heterogeneity of the prefrontal cortex in the monkey: a review. *Behav. Neur. Biol.* **25,** 301–345.

116. Jacobsen, C. F. (1936) Studies of cerebral function in primates. I. The functions of the frontal association areas in monkeys. *Comp. Psychol. Monogr.* **13,** 3–60.

117. Brutkowski, S., Mishkin, M., and Rosvold, H. E. (1963) Positive and inhibitory motor conditioned reflexes in monkeys after ablation of orbital or dorso-lateral surface of the frontal cortex, in *Central and Peripheral Mechanisms of Motor Functions* (Gutman, E. and Hnik, P., eds.), Czechoslovak Academy of Sciences, Prague, pp. 133–141.

118. Wikmark, R. G. E., Divac, I., and Weiss, R. (1973) Retention of spatial delayed alternation in rats with lesions in the frontal lobes. *Brain Behav. Evol.* **8,** 329–339.

119. Wilcott, R. C. (1986) Preoperative overtraining and effects of prefrontal lesions on delayed alternation in the rat. *Physiol. Psychol.* **14,** 87–89.

120. Mogensen, J., Iversen, I. H., and Divac, I. (1987) Neostriatal lesions impaired rats delayed alternation performance in a T-maze but not in a two-key operant chamber. *Acta Neurobiol. Exp.* **47,** 45–54.

121. van Haaren, F., van Zijderveld, G., van Hest, A., and de Bruin, J. P. C. (1988) Acquisition of conditional associations and operant delayed spatial response alternation—effects of lesions in the medial prefrontal cortex. *Behav. Neurosci.* **102,** 481–488.

122. Sanberg, P. R. and Fibiger, H. C. (1978) Body weight, feeding and drinking behaviors in rats with kainic acid lesions of striatal neurons: with a note on body weight symptomatology in Huntington's disease. *Exp. Neurol.* **66,** 444–466.

123. Dunnett, S. B. and Iversen, S. D. (1980) Regulatory impairments following selective kainic acid lesions of the neostriatum. *Behav. Brain Res.* **1,** 497–506.

124. Skjoldager, P., Pierre, P. J., and Mittleman, G. (1993) Reinforcer magnitude and progressive ratio responding in the rat: effects of increased effort, prefeeding and extinction. *Learn. Motiv.* **24,** 303–343.

125. Hodos, W. and Kalman, G. (1963) Effects of increment size and reinforcer volume on progressive ratio performance. *J. Exp. Anal. Behav.* **6,** 387–392.

126. Eagle, D. M., Humby, T., Dunnett, S. B., and Robbins, T. W. (1998) Effects of regional striatal lesions on motor, motivational and executive aspects of progressive ratio performance in rats. *Behav. Neurosci.,* in press.

127. Felton, M. and Lyon, D. O. (1966) The post-reinforcement pause. *J. Exp. Anal. Behav.* **9,** 131–134.

128. Iversen, S. D. and Iversen, L. L. (1981) *Behavioral Pharmacology*, Oxford University Press, New York and Oxford.
129. Björklund, A., Campbell, K., Sirinathsinghji, D. J. S., Fricker, R. A., and Dunnett, S. B. (1994) Functional capacity of striatal transplants in the rat Huntington model, in *Functional Neural Transplantation* (Dunnett, S. B. and Björklund, A., eds.), Raven Press, New York, pp. 157–195.
130. Dunnett, S. B. (1995) Functional repair of striatal systems by neural transplants: evidence for circuit reconstruction. *Behav. Brain Res.* **66,** 133–142.

15

Intrastriatal Injections of Quinolinic Acid as a Model for Developing Neuroprotective Strategies in Huntington's Disease

Dwaine F. Emerich

1. INTRODUCTION

Huntington's disease (HD) is an inherited neurodegenerative disease characterized by a relentlessly progressive movement disorder with psychiatric and cognitive deterioration. HD is found in all regions of the world, even in remote locations, with slight variations in prevalence rates. Overall, the prevalence of HD in the United States is between 5 and 10 per 100,000 *(1)*. On average patients suffer 17 yr of symptomatic illness. From the time of onset, an intractable course of mental deterioration and progressive motor abnormalities begins, invariably resulting in death. Clinically, HD is characterized be an involuntary choreiform (dancelike) movement disorder, psychiatric and behavioral changes, and dementia *(2)*. The age at onset is usually between the mid 30s and the late 50s, although juvenile (<20 yr of age) and late onset (>65 yr of age) HD occurs *(1,3)*. Although medications may reduce the severity of chorea or diminish behavioral symptoms, there are no effective treatments *(4)* as current approaches do not increase survival or substantially improve quality of life as it relates to cognitive state, gait disorder, or dysphagia *(5)*.

Pathogenetically, HD is a disorder characterized by a programmed premature death of cells, predominantly in the caudate nucleus and the putamen. Initially, the disease affects medium-sized spiny neurons that contain γ-aminobutyric acid (GABA). These cells receive afferent input from cortical glutamatergic and nigral dopaminergic neurons and provide efferent projections to the globus pallidus. Large aspiny interneurons and medium aspiny projection neurons are less affected and degenerate later in the disease process *(6–10)*. Other subcortical and cortical brain regions are involved, but the degree of degeneration varies and does not correlate with the severity of the disease *(2,9,10)* and is dwarfed by the striatal changes.

In 1993, it was discovered that an unstable expansion of a CAG trinucleotide repeat in the *IT15* gene located near the telomere of the short arm of chromosome 4 produces the disease *(11,12)*. Although it is believed that the HD mutation causes an increase in excitatory neurotransmission or produces a bioenergetic cell defect that confers

From: Central Nervous System Diseases
Edited by: D. F. Emerich, R. L. Dean, III, and P. R. Sanberg © Humana Press Inc., Totowa, NJ

increased sensitivity to ambient levels of excitation, the precise mechanism of neuronal degeneration remains unknown. Establishing a mechanism of cell death may offer hope for specific medical therapies to prevent or slow the degenerative process. This progress, although encouraging, does not obviate the need to develop novel interventions for patients in whom the neurodegenerative process is clinically manifest or in whom degenerative changes are inevitable.

An important approach for understanding and treating neurodegenerative diseases such as HD is the development of appropriate animal models that closely mimic the behavioral and neurobiological sequelae of the disorder. Such models aid in the further elucidation of the biological and behavioral expression of HD, and suggest unique therapeutic strategies for its treatment. This chapter evaluates the development of animal models of HD based on intrastriatal injections of the endogenous excitotoxic compound quinolinic acid (QA). First the behavioral pathology of HD is discussed. Next the neurochemical and cellular pathology of HD is reviewed. Against this backdrop, the QA model of HD is discussed with special reference to the behavioral, anatomical, and neurochemical parallels between the model and the human condition. Finally, the use of the QA model as system for developing new therapies is outlined using several recent examples detailing the use of neutrotrophic factors to prevent the cell death and behavioral deficits induced by QA.

2. BEHAVIORAL PATHOLOGY OF HUNTINGTON'S DISEASE

The behavioral changes characteristic of HD have been detailed elsewhere (13) and are briefly outlined here. The onset of the motor and mental features of HD is variable, with the symptoms manifesting themselves over a protracted time course. Although the onset of symptoms typically occurs at 35–40 yr of age, it may range from childhood to the eighth decade. In the absence of genetic testing, the variability of the clinical features of HD make a family history essential to avoid misdiagnosing the disease as cerebellar ataxia, Creutzfeldt–Jacob disease, Wilson's disease, tardive dyskinesia, or basal ganglia infarction.

The physical features of HD are constant irregular and involuntary choreiform movements of the entire body (2,13). These movements often appear unilaterally, but inevitably recruit all of the limbs (14). The face often appears quite grotesque because of the constant writhing contortions of the facial muscles. Dysthargia, dysphagia, and disturbances in ocular motility may also develop, and the individual's speech becomes unintelligible (14–18). Expression of the mental symptoms of HD are even more variable than the motor features and are generally considered to be similar to dementia (19–25). Initially, the individual may show signs of eccentric traits, a lack of grooming, irritability, impairments in memory, emotional instability, and delusions of paranoid grandeur. Personality disorders, such as outbursts of rage or violent temper, may occur. Conversely, some patients may be slow and apathetic. Depression is the most common early psychological symptom and may be associated with changes in the ability to manage daily work and household-related activities. Patients also show a lack of initiative as well as decreased attention and communicativeness. These early symptoms of HD can mimic those of schizophrenia and bipolar affectiveness, leading many patients to be misdiagnosed as having schizophrenia (26–28). It is generally agreed that the mental symptoms precede the motor changes, which probably also contributes to mis-diagnoses *(29,30)*.

Often, the HD patient does not appear demented because of a general preservation of intellect and language functions *(24,25,31,32)*. This preservation is likely the result of the sparing of cortically mediated cognitive functions *(33)*. In other degenerative disorders, such as Alzheimer's disease, the cognitive alterations occur against a back-drop of substantial cortical degeneration *(24,34)*. Accordingly, the mnemonic impair-ments observed in HD are referred to as a subcortical dementia and include a progressive loss of generalized cognitive powers, psychic apathy, and inertia, which progresses to an akinetic mute state *(35)*.

The often subtle differences between the dementia in various neurodegenerative dis-orders illustrate the notion that the cortical and subcortical dementia distinction is a relative one. This distinction is based primarily on anatomical grounds, but much like the behavioral expression of different diseases, the anatomical differences are not always clear cut. Some Parkinson's patients have changes in cortical morphology that resemble those seen in Alzheimer's patients. Moreover, Alzheimer's patients have well-documented changes in subcortical nuclei. Accordingly, the dementia associated with a given disease state is best considered to involve a relatively greater contribution of either cortical or subcortical pathology. Recent metabolic studies support this conten-tion *(36–39)*.

Most studies have made direct comparisons of the cognitive changes in Alzheimer's and HD patients *(40–44)*. Individuals afflicted with Alzheimer's disease have an inability to store or consolidate new material, show increased sensitivity to proactive interference, display poor recognition memory, and have rapid forgetting over time. In contrast, HD patients are impaired in their ability to initiate any systematic retrieval of already stored information, but exhibit superior retention of information and have nor-mal recognition memory. The subcortical dementia in HD patients also appears to be different from that observed in other subcortical disorders such as Parkinson's disease. Although many similarities exist in the dementia presented by these two patient popu-lations, HD patients have poorer free recall, impaired learning over trials, abnormal serial position effects, and increased perseveration *(45)*. These results indicate that disorders characterized by a predominant subcortical pathology are still differentiated by subtle differences in clinical symptoms.

Even in advanced cases, the pattern of dementia in HD is not diffuse and homoge-neous, but is characterized by a relative sparing of many higher cortical functions. In addition, the cognitive deficits do not develop uniformly; rather the memory disorders precede the more generalized intellectual deterioration. From the time of onset, the motor and mental symptoms of HD progress at an increasingly disabling rate. Initially, the individual may exhibit subtle shifts in personality or cognition that occur together with a mild chorea. As the neuronal degeneration of the disease progresses, the afflicted individual undergoes more intense and abrupt personality changes combined with sub-stantial cognitive impairments. As the mental features of HD change over time, the motor effects change from a choreic dyskinesia to a more disabling dystonic and Parkinsonian-like syndrome *(46)*. Finally, the severity of the cognitive and mood dis-turbances does not correlate well with the motor impairments. Table 1 lists the psycho-logical, psychiatric, and neuropsychological changes reported in HD.

Table 1
Psychological, Psychiatric, and Neuropsychological Fundings in HD

Psychiatric changes
- Patients are erratic, impulsive, and prone to emotional outbursts, rage, and violence and may be apathetic
- Lack of initiative, labile affect, decreased communicativeness
- Emotional instability, severe fluctuations in mood, irritability, and anxiety
- Eccentric traits, lack of grooming and delusions of grandeur and paranoia
- Depression (more common in females) and inability to manage work and household-related tasks
- Mental restlessness and general feelings of discomfort

Psychological changes
- Loss of memory
- Difficulty with organization, planning, and temporal ordering of information
- Easily distracted and disturbed by excessive sensory input
- Repetitive thought patterns
- Decreased spontaneity in the absence of strong motivation
- Mental apathy
- No deterioration of higher cortical functions
- Loss of psychomotor inhibition

Neuropsychological changes
- Impaired performance on neuropsychological test batteries
  - California Verbal Fluency Test
  - Wechsler Memory Scale
  - Wechsler–Bellevue Scale
  - Wechsler Adult Intelligence Scale
  - Halstead Neuropsychological Test Battery
  - Minnesota Multiphasic Personality Inventory
  - Bender Visual–Motor Gestalt Test
  - Shipley–Hartford Retreat Scale
  - Dementia Rating Scale
  - Boston Naming Test
- Impaired imagery recall
- Impaired trailmaking
- Increased anxiety
- Normal language performance. Reading, writing, picture naming, and tactual performance also normal
- Altered cognitive processes
  - Decreased procedural learning
  - Loss of immediate memory
  - Impaired delayed recall
  - Impaired semantic clustering
  - Loss of recognition memory
  - Increased susceptibility to priming effects
  - Poor improvement of learning over repeated trials
  - Abnormal serial position effects
  - Higher rates of perseveration
  - Normal rates of intrusion errors
  - Normal vulnerability to proactive and retroactive interference
  - Normal retention of information

3. ANATOMICAL AND NEUROCHEMICAL PATHOLOGY OF HD

Classically, HD is associated with a gross generalized atrophy of the cerebral cortex and basal ganglia affecting both the gray and white matter *(47)*. Histologically, this is accompanied by an extensive gliotic reaction and loss of small neurons in the striatum and in layers 3, 5, and 6 of the frontal and parietal cortices. The damage to the neostriatum results in compensatory, secondary hydrocephalus with a gross dilation of the ventricular system *(10,47–50)*.

The severe atrophy of the striatum is uniformly exhibited among HD patients with approx a 60–90% decrease in mass occurring in cases of juvenile onset. These decreases in striatal volume are related to a severe loss of the medium-sized spiny projection neurons, the major output neurons of the striatum *(51)*. Neurochemically, the loss of these neurons is associated with decreases in GABA, substance P, dynorphin, and enkephalin *(52)* (Table 2). On the other hand, local circuit aspiny neurons reactive for NADPH-diaphorase and somatostatin are relatively spared *(6–10,53)*. The large apsiny cholinergic neurons, although spared in the early stages of the disorder, may exhibit degenerative changes as the disease progresses (Table 3).

Other regions that have direct interconnections with the striatum also reveal marked degenerative changes. The globus pallidus exhibits a marked atrophy (typically about 50%) which results from the loss of strio-pallidal fibers as well as a substantial decrease in the number of pallidal neuronal perikarya *(54)*. The sulcal widening and the loss of neuronal weight in the cerebral cortex indicate that the cortex undergoes a substantial degeneration *(55)*. However, unlike the relatively ubiquitous cell loss observed in the striatum, the degenerative process is more restricted with cell loss occurring in the third, fifth, and sixth cortical layers. The small pyramidal cells of these deep cortical layers are the source of innervation to the striatum, suggesting that HD is associated with severe striatal atrophy concomitant with degeneration of both its major inputs and outputs *(56–59)*.

Although HD has generally been considered to be a neurodegenerative disorder characterized by the restricted loss of certain populations of neurons, it is clear that the neural degeneration is actually quite widespread. The pathological changes in neural regions removed from the basal ganglia are, however, typically not severe and do not correlate with the severity of the disease. Nonetheless, the pars reticulata of the substantia nigra, thalamic nuclei, subthalamic nucleus, cerebellum, hippocampus, hypothalamus, and a variety of brain stem regions, including the superior olive and red nucleus have all been shown to exhibit reactive gliosis and evidence of neuronal degeneration (Table 4).

4. EXCITOTOXIC MODELS OF HD

An important approach for understanding neurological diseases is to develop appropriate animal models. Such models help to elucidate the biological and behavioral manifestations of clinical disorders, and suggest novel therapeutic strategies for their prevention and/or treatment. Although animal models may not reproduce the complex etiologies, pathophysiology, or behavioral abnormalities associated with diseases of the nervous system, they do provide a practical way to explore questions concerning structure and function. Like other neurodegenerative diseases, therapy for HD is limited

Table 2
Changes in the Striatum in HD and Rats Intrastriatally Injected with QA

	HD	QA
Neurochemical indices		
GABA	Decreased	Decreased
Dopamine	Normal	Normal
Serotonin	Normal	Normal
Acetylcholine	Normal/decreased	Normal/decreased
Somatostatin	Normal/increased	Normal/decreased
Neuropeptide Y	Normal/increased	Normal/decreased
Substance P	Decreased	Decreased
Angiotenson	Decreased	Decreased
Enkephalin	Decreased	Decreased
Taurine	Normal	Decreased
Glutamate	Decreased	Decreased
Receptor binding		
GABA	Decreased	Decreased
Muscarinic	Decreased	Decreased
Serotonin	Decreased	Decreased
Dopamine	Decreased	Decreased
Adenosine	Decreased	Decreased
Anatomical indices		
Ventricular size	Increased	Increased
Striatal volume	Decreased	Decreased
Striatal GABAegic neurons	Decreased	Decreased
Striatal cholinergic neurons	Normal	Normal/decreased
Striatal NADPH-diaphorase neurons	Normal	Normal/decreased
Striatal somatostatin neurons	Normal	Normal/decreased
Striatal neuropeptide Y neurons	Normal	Normal/decreased
Striatal calretinin neurons	Normal/increased	Normal/decreased
Substantia nigra neurons	Decreased	Atrophied
Cortical neurons	Decreased	Atrophied

and does not favorably influence the progression of the disease. However, recent advances in our understanding of the genetic and pathogenetic events that cause neural degeneration provide optimism that novel experimental therapeutic strategies may be devised for the treatment of HD.

Various avenues have been explored to develop an animal model of HD *(for a review see 13)*. Although models focusing on hyperkinesias in animals resulting from genetic conditions and various types of brain lesions have been used, historically the majority of models have centered around motor abnormalities produced by short-term pharmacological stimulation of some of the neurochemical changes found in HD. Although useful, there are several inherent problems with pharmacological models of HD or chorea-like syndromes. Many of the compounds used in these studies have limited central nervous system (CNS) specificity, and affect more than one transmitter or

Table 3
Classes of Striatal Neurons Affected in HD

Morphology	Transmitter/peptide	Projection	Effect in HD
Medium spiny	GABA, substance P, dynorphin	Internal segment of the globus pallidus	Decreased
Medium spiny	GABA, substance P, dynorphin	Substantia nigra pars reticulata	Decreased
Medium spiny	GABA, enkephalin, dynorphin	External segment of the globus pallidus	Decreased
Medium spiny	GABA	Substantia nigra pars compacta	Decreased
Large aspiny	Acetylcholine	Striatal interneuron	Spared
Small aspiny	Somatostatin, neuropeptide Y	Striatal interneuron	Spared

receptor type. These compounds may also exhibit a high degree of behavioral toxicity or have a short duration of action and therefore cannot reproduce the chronicity of the pathology in HD. Finally, the dyskinesia produced in these models often bears little homology to the motor impairments observed in HD. This is not to say that the behavioral alterations observed are of no relationship to those in HD. In fact, it is unlikely that any rodent model of HD would be able to reproduce with precision the chorea found in HD. Considering that rats are quadrupeds and humans are bipeds, it is not surprising that chorea found in HD as such does not develop in these models. Although pharmacological manipulations of striatal function do not reproduce the chorea of HD, the resulting locomotor abnormalities may be generally considered to resemble those in the disease.

Given these considerations, it is difficult to establish causal or even suggestive relationships between drug-induced neural changes and specific behavioral alterations. Another approach to mimic the symptoms of HD is to mechanically or electrolytically lesion various brain areas, such as the striatum. However, this is an unsatisfactory approach in that lesion techniques almost invariably damage supportive structures as well as fibers that pass through and terminate in the damaged area. A better strategy for investigating the relationship between striatal damage and locomotor abnormalities might involve the use of selective cytotoxic compounds. Selective toxic compounds have been widely used in neurobiology to examine the functional and structural properties of the nervous system. Toxins have also been used to examine the function and molecular biology of ion channels, axoplasmic transport processes, neurotransmitter systems, and the principles of neurotransmission. In a related context, neurotoxins have been successfully used to examine the covariation between altered neurotransmitter dynamics and behavior, and to develop animal models of neurological disorders.

Glutamate is one of the major excitatory neurotransmitters found in the CNS. It can act, however, as a potent neurotoxin. Because the corticostriate projections utilize glutamate, a number of attempts have been made to develop animal models of HD

Table 4
Regional Pathology in HD Brain and in Animals
Following Intrastriatal Injections of QA

Region	Effect in HD brain	Effect following intrastriatal QA
Gross brain	Weight and size reduced	Relatively normal
Ventricular system	Dilation	Dilation of lateral ventricles
Caudate/putamen	Reduction in volume, neuronal loss, gliosis	Reduction in volume, neuronal loss, gliosis
Globus pallidus	Reduction in volume, neuronal loss, gliosis	Cell atrophy, gliosis
Nucleus accumbens	Slight neuronal loss, gliosis	Normal
Thalamus	Neuronal loss, cell atrophy, gliosis	Gliosis
Hypothalamus	Neuronal loss, cell atrophy, gliosis	Normal
Cortex	Neuronal loss, cell atrophy, gliosis	Cell atrophy, gliosis
Hippocampus	Atrophy, gliosis, neurofibriliary tangles	Normal
Cerebellum	Neuronal loss, cell atrophy, gliosis	Normal
Subthalamic nucleus	Neuronal loss, gliosis	Normal
Substantia nigra	Neuronal loss, cell atrophy, gliosis	Cell atrophy, gliosis
Brain stem nuclei	Neuronal loss, gliosis	Normal
Spinal cord	Gliosis	Normal

based on the relatively cytotoxic effects of excitotoxic compounds. These compounds include structural analogs of glutamate, such as kainic acid (KA), ibotenic acid (IA), and the endogenous tryptophan metabolite QA. When injected into the brains of rats, in extremely small doses, these compounds produce a marked and locally restricted toxic effect while sparing axons of passage and afferent nerve terminals. The behavioral, neurochemical, and anatomical consequences of excitoxicity resemble those observed in HD and have led to the speculation that the neuronal death occurring in HD is related to an underlying endogenous excitotoxic process *(60–64)*.

5. THE QUINOLINIC ACID MODEL

5.1. Neuropathology

QA, a metabolite of tryptophan has attracted a great deal of attention because of its powerful excitotoxic properties and wide distribution in both the rat and human brain *(65–68)*. High concentrations of its catabolic enzyme, quinolinic acid phosphoribosyltransferase, and immediate anabolic enzyme, 3-hydroxyanthranilic acid, have been detected within the caudate, suggesting that it normally serves a role in striatal functioning *(69–71)*. The striatum is also among the structures most vulnerable to the excitotoxic effects of QA with the cerebellum, amygdala, substantia nigra, septum, and hypothalamus being significantly less susceptible. Although these structures are affected in HD, they are less so than is the striatum, providing further support for a role of QA or some endogenous excitotoxin in the selective neuropathlogy in HD (Table 4).

QA has also been reported to exert a more selective degenerative effect in the striatum than either KA or IA, which more closely resemble the pathology of HD *(72–75)*. Like KA, QA injections cause depletions of GABAergic neurons while relatively sparing cholinergic neurons and axons of extrinsic origin. Beal and colleagues *(72–75)* have reported that unlike KA or IA, intrastriatal injections of QA spare somatostatin- and neuropeptide Y-containing neurons. Because this pattern of cell loss closely mimics that seen in HD, it has been suggested that this model most closely reproduces the neuropthology observed in the disease (Tables 2 and 4). However, controversy exists over whether these cell populations are actually spared following QA. Davies and Roberts *(76)* injected QA into the striatum of rats and found no evidence for a sparing of somatostatin-containing neurons. Likewise, Boegman *(77)* reported that QA did not spare neuropeptide Y-immunoreactive neurons following intrastriatal QA injection. Several other authors have similarly failed to find any obvious selective neurotoxic effects of QA, although the doses of QA used may be high enough to mask any neuronal sparing *(78–81)*. Supporting a possible dose-dependent cell sparing are observations that QA produces a more generalized toxicity within the primary lesion site, together with a more selective pattern of cell loss within the transition zone or peripheral aspect of the lesion zone *(74)*. Moreover, in vitro studies have shown that high doses of QA produced a generalized toxic effect in striatal cultures while lower doses were more selective and spared both NADPH-diaphorase and choline acetyltransferase (ChAT)-positive cells *(82)*. Similar results have been obtained in cultured cortical neurons *(83,84)*.

5.2. Behavioral Pathology

Although the anatomical and neurochemical experiments conducted to date generally support the use of QA as a model of HD, few systematic behavioral studies have been conducted in this model. Like the HD patient, the QA-lesioned rat exhibits a hyperkinetic syndrome. Intrastriatal injections of 150–300 nmol of QA produce significant increases in activity levels beginning 2 wk postsurgery *(85–87)*. These animals show transient losses in body weight that are qualitatively similar to that seen following KA, but smaller in overall magnitude. Together with the generalized increases in activity levels, the QA-lesioned rat shows deficits in tasks that require the use of fine motor control. For instance, injections of 225 nmol of QA produce significant deficits in an operant bar pressing task and result in a profound impairment in skilled paw use as measured by the ability to retrieve food pellets in a staircase test *(80)*. Relative to normal animals, QA-lesioned rats also exhibit differential responses when pharmacologically challenged. Calderon et al. *(88)* demonstrated that QA lesions abolish dopamine-mediated catalepsy. Bilateral QA (150–225 nmol) decreased the cataleptic response to the D1 receptor antagonist SCH23390 and the D2 receptor antagonist haloperidol. Exaggerated increases in locomotor activity are seen following bilateral QA lesions and amphetamine challenge, and asymmetric rotational behavior occurs following unilateral QA and apomorphine administration *(80)*.

Interestingly, there appears to be a sex-dependent regulation of the behavioral effects produced by QA administration. In one study, male, female, and ovariectomized female rats received bilateral intrastriatal injections of QA, and the extent of body weight loss was examined for 30 d *(89)*. Although all rats injected with QA exhibited an equivalent initial loss of body weight, the female rats rapidly regained weight and did not differ from controls after the first postsurgical day. In contrast, the male and female ovariectomized rats that received QA exhibited a significant loss of body weight that did not recover until 19 d following surgery. Similarly, QA produces locomotor hyperactivity in female but not in male or ovariectomized female rats *(79)*. These data suggest that sex and hormonal variables play an important role in the regulatory and locomotor changes following QA.

Because HD patients exhibit cognitive as well as motor changes, it is critical to examine any potential changes in learning and memory functions following QA. To date, few studies have systematically examined cognitive changes following QA, but the notable exceptions have suggested that QA-lesioned rats may have a spectrum of behavioral changes that extends beyond the motor realm alone. High doses of QA (225 nmol) produce significant impairments in spatial learning as demonstrated by their performance in the Morris water maze task *(86,87)*. Caution should be used in interpreting a strict cognitive effect in these studies. In each case, the doses of QA used produced significant motor deficits and at least one study has reported a marked thigmotaxis in the water maze following QA *(87)*. Although excitotoxic lesions of the striatum may indeed produce changes in cognition, detailed dose–response studies together with detailed batteries of behavioral tests are still needed to sort the nature of the behavioral deficits following excitotoxic striatal damage.

Although these data are similar to those obtained following administration of KA and other excitotoxins, and resemble the behavioral pathology observed in HD, they do not yet adequately permit the evaluation of the behavioral specificity of QA. The existence of no qualitative differences between QA and KA in these experiments may (1) support the lack of neural specificity observed by some investigators or (2) indicate that the behavioral analysis employed to date is insufficient, and that additional, more precise testing paradigms are required to validate the utility of the QA model. Regardless, the validity of QA as a model of HD must be based in part on the behavioral consequences of the compound and the resulting homology to HD itself (Table 5).

6. EXPERIMENTAL STRATEGIES FOR TREATING HD

Currently, therapy for HD is limited and does not favorably influence the progression of the disease. However, recent advances in our understanding of the genetic and pathogenetic events that cause neural degeneration provide optimism that novel experimental therapeutic strategies may be devised for the treatment of HD. The ability to devise novel therapeutic strategies in HD is tightly linked to the characterization of highly relevant animal models. The advent of appropriate animal models has permitted the evaluation of multiple therapeutic strategies. These strategies have generally fallen into one of two categories. The first encompasses replacement strategies including pharmacological manipulation of altered transmitter levels and replacement of degenerative neural systems via tissue/cell transplantation. The practical appeal of pharmacotherapy is offset by the anatomical and behavioral complexity of HD, together with

Table 5
**Behavioral and Elecrophysiological Changes Reported
in Rodents Following Intrastriatal Injections of QA**

Behavioral/functional measure	Effect
General	
Body weight lesions	Decreased, particularly following bilateral
Mortality	Increased following bilateral lesions
Motor tests	
Apomorphine-induced rotation	Increased
Skilled paw use (staircase test)	Impaired ability to retrieve food pellets
Bracing	Increased motor rigidity
Tactile adhesive removal paws	Increased latency to remove adhesive
Balance (rotarod)	Impaired balance
Locomotor activity	Hyperactivity following bilateral lesions
Haloperidol-induced catelepsy responsivity	Bilateral lesions abolish catalepsy
Amphetamine-induced locomotion increase	Bilateral lesions produce an exaggerated in activity
Cognitive Tests	
Morris water maze	Impaired spatial learning and memory
Delayed nonmatch to position operant task	Impaired
Electrophysiological changes	
Cortical EEG	Reduced voltage amplitude in frontal cortex
Evoked potentials	Reduced amplitude of evoked somatosensory potentials in parietal cortex, no changes in visually evoked potentials
Long-term potentiation	Normal hippocampal long-term potentiation

the difficulty of titrating effective treatment regimens in the face of the ongoing degeneration of multiple neural systems. The logic of replacing lost populations of neurons is equally, if not more, appealing. Animal studies have clearly demonstrated that grafted tissue can integrate within the excitotoxic lesioned brain and promote recovery of motor and cognitive function (90–95). Given the positive results acquired in the animal models to date, clinical trials evaluating fetal tissue grafting in HD patients have recently been initiated. Although replacement strategies hold hope for the treatment of HD, they do not affect the continued and insidious degeneration of the striatum and related structures. A treatment that stopped or slowed these changes might be more therapeutic than neurochemical or cell replacements. Toward this end, it has been suggested that trophic factors may slow the progression of cellular dysfunction and ultimately prevent the degeneration of populations of striatal neurons that are vulnerable in HD (96). Infusions of trophic factors or grafts of trophic factor-secreting cells prevent degeneration of striatal neurons destined to die from excitotoxic insult or mitochondrial dysfunction (80,81,87,96–105). The use of trophic factors in a neural protection strategy may be particularly relevant for the treatment of HD. Unlike for other neurodegenerative diseases, genetic screening can identify virtually all individuals at risk who will ultimately suffer from HD. This provides a unique opportunity to design treatment strategies that can intervene prior to the onset of striatal degeneration. Thus, instead of replacing neuronal systems that have already undergone extensive neuronal death, trophic factor strategies can be designed to support host systems destined to die at later time in the organism's life.

7. NEUROTROPHIC FACTOR THERAPY IN CNS DISEASES

7.1. Use of Genetically-Modified Cells for Treating CNS Disorders

If trophic factors prove to be a worthwhile therapeutic strategy, the method of delivery may be critical. To date, long-term administration of growth factors has been limited to intraventricular infusions using cannulae or pumps. These routes of administration require repeated injections or refilling of pump reservoirs to maintain specific drug levels and avoid the degradation of the therapeutic agent in solution. In addition, chronic low-dosage infusion of compounds is difficult to sustain using current pump technology. Current pump technology is also suitable for ventricular but not parenchymal delivery. An alternative method is the implantation of cells that have been genetically modified to produce a therapeutic molecule *(106–111)*. This avoids the problem of degradation and repeated refilling while allowing a localized distribution within the cerebrospinal fluid or parenchyma. The use of immortalized cell lines for delivery of trophic molecules avoids many of these concerns by providing a continuous *de novo* cellular source of the desired molecule, the dose of which theoretically can be adjusted with specific promoters.

One means of using genetically modified cells to deliver neurotrophic factors to the CNS is to encapsulate them within a semipermeable polymer membrane *(112,113)*. Single cells or small clusters of cells can be enclosed within a selective, semipermeable membrane barrier that admits oxygen and required nutrients and releases bioactive cell secretions, but restricts passage of larger cytotoxic agents from the host immune defense system. Use of a selective membrane both eliminates the need for chronic immunosuppression of the host and allows the implanted cells to be obtained from nonhuman sources, thus avoiding the cell-sourcing constraints that have limited the clinical application of generally successful investigative trials of unencapsulated cell transplantation. Polymer encapsulated cell implants have the advantage of being retrievable should the transplant produce undesired effects, or should the cells need to be replaced. Cross-species immunoisolated cell therapy has been validated in small and large animal models of Parkinson's disease *(112,114)*, HD *(80,81,87)*, amyotropic lateral sclerosis *(115)*, and Alzheimer's disease *(113,116)*.

7.2. Intrastriatal Transplants of Encapsulated NGF Secreting Cells into Intact Rats: Effects on Cholinergic and Noncholinergic Striatal Neurons

It appeared from initial studies that nerve growth factor (NGF) delivered from genetically modified cells could prevent the loss of both cholinergic and noncholinergic neurons in a rodent model of HD. To further characterize this neuroprotective effect, we examined the ability of polymer-encapsulated NGF-producing fibroblasts to influence normal, intact populations of striatal neurons *(103)*. Using an immunoiso-latory polymeric device, encapsulated baby hamster kidney cells (BHK) that were pre-viously transfected with a DHFR-based expression vector containing the human NGF gene (BHK-NGF) were transplanted unilaterally into the striatum of normal rats. Control rats received identical transplants of encapsulated BHK cells that were not transfected with the NGF construct (BHK-Control). Rats were killed 1, 2, or 4 wk post-implantation and cell size and optical density for immunohistochemical staining was performed. At each time point, ChAT-immunoreactive (ChAT-ir) neurons hypertro-phied in response to the BHK-NGF secreting grafts but not in response to control grafts. Furthermore, ChAT-ir striatal neurons displayed an enhanced optical density reflect-ing a putative increase in ChAT production in rats receiving BHK-NGF. In addition to changes in ChAT-ir, encapsulated BHK-NGF secreting grafts induced a significant hypertrophy of noncholinergic neuropeptide Y (NPY)-ir neurons. This hypertrophy was observed 1, 2, and 4 wk post-implantation. In contrast to ChAT-ir, the BHK-NGF grafts did not influence the optical density of NPY-ir neurons.

To determine whether the hypertrophy of cholinergic and noncholinergic striatal neurons induced by encapsulated BHK-NGF grafts was dependent upon the continued presence of the implant, rats received BHK-NGF grafts for 1 wk. The grafts were then retrieved and the animals allowed to survive for an additional 3 wk. The size of ChAT-ir and NPY-ir striatal neurons in these animals was compared to rats receiving BHK-NGF grafts for 4 wk. The hypertrophy of both ChAT-ir and NPY-ir induced by the transplant dissipated upon removal of the graft, indicating that chronic NGF is required for these effects to occur.

NGF delivered from encapsulated cells clearly influenced both cholinergic and noncholinergic neurons in a reversible manner. However, no data existed concerning the extent of diffusion of neurotrophic factors such as NGF from encapsulated cells. An understanding of the spread of NGF, or any therapeutic molecule, is an essential piece of information that would ultimately determine the feasibility of its delivery into discrete CNS sites. To assess the spread of NGF from the capsules, additional rats received grafts of either encapsulated BHK-NGF cells or BHK-Control cells and were killed 1 wk post-implantation using a specially designed fixation procedure and stain-ing protocol to visualize NGF. The spread of NGF secreted from the capsule was deter-mined to be 1–2 mm. Interestingly, we were able to visualize neurons within the striatum that bound, internalized, and transported NGF secreted from the BHK-NGF implant to the striatal neuron perikarya. In BHK-Control grafted animals, staining was limited to the host basal forebrain system and no staining was observed within the striatum. These data provide *de facto* evidence that NGF transplants can influence both cholinergic and noncholinergic neurons and that NGF diffuses a considerable distance in the striatum, and provide the basis for further evaluating this approach in animal models of HD.

7.2.1. Transplants of Encapsulated Cells Genetically Modified to Secrete NGF: Rescue of Cholinergic and Noncholinergic Neurons Following Striatal Lesions

Having determined that encapsulated NGF-producing fibroblasts could affect cho-linergic as well as noncholinergic striatal neurons, the next series of studies evaluated the neuroprotective effects of BHK-NGF cells in a rodent model of HD. Rats received unilateral NGF-producing or control BHK cell implants into the lateral ventricle *(98)*. Three days later, all animals received unilateral injections of QA (225 nmol) or saline vehicle into the ipsilateral striatum. Rats were killed 4 wk later. Nissl-stained sections revealed a comprehensive shrinkage of the size of the lesion within the striatum following NGF treatment with a concomitant diminution of glial fibrillary acidic protein (GFAP)-ir astrocytosis. Qualitatively, rats receiving BHK-NGF grafts displayed significantly more ChAT-ir and NADPH-positive neurons within the striatum ipsilateral to the lesion relative to BHK-Control grafted animals. As measured by enzyme-linked immunosorbent assay (ELISA), NGF was released by the encapsulated BHK-NGF cells prior to implantation and following removal. Morphology of retrieved capsules revealed numerous viable and mitotically active BHK cells. These results supported the initial studies by Frim et al. *(99–101)* and suggested that implantation of polymer-encapsulated NGF-secreting cells can be used to protect cholinergic and noncholinergic neurons from excitotoxic damage in a rodent model of HD.

To determine if the anatomical protection afforded by NGF treatment was manifested as a functional protection, QA-lesioned animals were tested on a battery of behavioral tests. Approximately 2 wk following surgery, animals were tested for apomorphine-induced rotation behavior. Animals receiving QA lesions together with BHK-Control cells showed a pronounced rotation response to apomorphine that increased over repeated test sessions. In contrast, animals that received BHK-NGF cells rotated significantly less than those animals receiving BHK-Control cells or QA alone. To characterize further the extent of behavioral recovery produced by NGF in QA-lesioned rats, a series of animals were bilaterally implanted with BHK-NGF or BHK-Control cells. One week later, these animals received bilateral intrastriatal injections of QA (150 nmol) and were tested for changes in locomotor activity and responsiveness to haloperidol in a catalepsy test. Animals receiving QA lesions without BHK cell implants or implants of BHK-Control cells showed a pronounced hyperactivity when tested in automated activity chambers as well as a diminished cataleptic response to haloperidol. In contrast, animals receiving BHK cells producing NGF showed a significant attenuation of the hyperactivity produced by QA. These same animals showed a normal cataleptic response to haloperidol, further indicating the anatomical protection afforded by NGF in this model was paralleled by a robust behavioral protection (Table 6).

7.3. Transplants of Encapsulated Cells Genetically Modified to Secrete CNTF: Rescue of Cholinergic and GABAergic Neurons Following Striatal Lesions

When delivered via genetically modified cells, NGF has a clear and potent effect upon both normal and damaged cholinergic and noncholinergic striatal neurons. It remains important, however, to determine whether these effects are limited to NGF and if not to compare the potency of NGF to other neurotrophic factors in these model systems. Ciliary neurotrophic factor (CNTF) is a member of the α-helical cytokine superfamily which has well documented functions in the peripheral nervous system *(117–125)*. Recently it has become clear that CNTF also influences a wide range of CNS neurons. CNTF administration prevents the loss of cholinergic, dopaminergic, and GABAergic neurons in different CNS lesion paradigms *(126–128)*. Importantly,

Table 6
Behavioral Protection Produced by NGF and CNTF in Rodent Models of HD

Behavioral/functional measure	Neurotrophic Factor	
	NGF	CNTF
General		
Body weight	+	+
Mortality	+	+
Motor tests		
Apomorphine-induced rotation	+	+
Skilled paw use (staircase test)	–	+/–
Bracing	NE	+
Tactile adhesive removal	NE	+
Balance (rotarod)	NE	+
Locomotor activity	+	+
Haloperidol-induced catelepsy	+	NE
Cognitive tests		
Morris water maze	NE	+
Delayed nonmatch to position operant task	NE	+

NE, not examined, +, positive outcome, –, negative outcome.

an initial study demonstrated that infusions of CNTF into the lateral ventricle prevented the loss of Nissl-positive striatal neurons following QA administration *(97)*. To confirm and expand upon these studies we conducted a series of studies examining the ability of CNTF to protect against QA lesions *(80)*. In these studies, animals received intraventricular implants of CNTF-producing BHK cells followed by QA lesions as described previously. Animals were again tested for rotation behavior, but were also examined for their ability to retrieve food pellets using a staircase apparatus. Rats receiving BHK-CNTF cells rotated significantly less than animals receiving BHK-Control cells. No behavioral effects of CNTF were observed on the staircase test. An analysis of Nissl-stained sections demonstrated that the size of the lesion was significantly reduced in those animals receiving BHK-CNTF cells (1.44 ± 0.34 mm^2) compared with those animals receiving control implants (2.81 ± 0.25 mm^2). Quantitative analysis of striatal neurons further demonstrated that both ChAT and glutamic acid decarboxylase (GAD) ir neurons were protected by BHK-CNTF implants. The loss of ChAT-ir neurons in those animals receiving CNTF implants was 12% compared to 81% in those animals receiving control cell implants. Similarly, the loss of GAD-ir neurons was attenuated in animals receiving CNTF-producing cells (20%) compared to those animals receiving control cell implants (72%). In contrast, a similar loss of NADPH-diaphorase-positive cells was observed in the striata of both implant groups (65–78%). Analysis of retrieved capsules revealed numerous viable and mitotically active BHK cells that continued to secrete CNTF.

A separate series of animals were tested on a battery of behavioral tests to determine the extent or behavioral protection produced by CNTF *(87)*. Bilateral infusions of high doses of QA produced a significant loss of body weight and mortality that was pre-

Table 7
Striatal ChAT and GAD Neurochemistry

	Group		
	Lesion + CNTF	Control	Lesion Only
Neurochemical measure			
ChAT	267.0 (11.2)	266.3 (13.3)	172.6 (38.7)[a]
GAD	26.4 (1.4)	29.5 (1.13)	15.8 (5.2)[a]

ChAT and GAD levels are expressed as means (± SEM) nmol ACh or GABA/mg of protein.
[a]$p < 0.05$ lesion + CNTF vs lesion only.

vented by prior implantation with CNTF-secreting cells. Moreover, QA produced a marked hyperactivity, an inability to use the forelimbs to retrieve food pellets in a staircase test, increased the latency of the rats to remove adhesive stimuli from their paws, and decreased the number of steps taken in a bracing test that assessed motor rigidity. Finally, the QA-infused animals were impaired in tests of cognitive function—the Morris water maze spatial learning task, and a delayed nonmatching to position operant test of working memory (Table 6). Prior implantation with CNTF-secreting cells prevented the onset of all the above deficits such that implanted animals were nondistinguishable from sham-lesioned controls. At the conclusion of behavioral testing, 19 d following QA, the animals were killed for neurochemical determination of striatal ChAT and GAD levels. This analysis revealed that QA decreased striatal ChAT levels by 35% and striatal GAD levels by 45%. In contrast, CNTF-treated animals did not exhibit any decrease in ChAT levels and only a 10% decrease in GAD levels (*see* Table 7).

These results extended previous findings and began to define the extent of both the quantitative and qualitative aspects of the behaviorally protective effects of CNTF. The observation that CNTF may prevent the occurrence of both motor and nonmotor deficits following QA has particular relevance for the treatment of HD which is characterized by a wide range of behavioral alterations. HD is a disorder most often associated with pronounced motor changes. However, neurological and psychiatric changes frequently occur as much as a decade prior to onset of the motor symptoms. Indeed, given the severity and persistence of the cognitive changes in HD, any viable neuroprotective strategy would necessarily need to exert beneficial effects on cognitive and psychiatric symptoms.

7.3.1. Transplants of Encapsulated Cells Genetically Modified to Secrete CNTF: Preservation of Basal Ganglia Circuitry in Nonhuman Primates

An essential prerequisite before the initiation of clinical trials for HD is the demonstration that trophic factors can provide neuroprotection in a nonhuman primate model of HD. Toward this end, three cynomolgus monkeys received unilateral intrastriatal implants of polymer-encapsulated BHK-CNTF cells *(81)*. The remaining three monkeys served as controls and received BHK-Control cells. One week later, all animals received injections of QA into the ipsilateral caudate and putamen. All monkeys were killed 3 wk later for histological analysis.

Within the host striatum, QA induced a characteristic lesion of intrinsic neurons

Table 8
Neuronal Cell Counts in QA-Lesioned Monkeys

Striatal region					
Caudate			Putamen		
Intact side	Lesioned/ implanted side	Percent loss	Intact side	Lesioned/ implanted side	Percent loss
GAD-positive neurons					
CNTF 18088 (437)	6516 (1233)	64.0[a]	17731 (604)	6374 (851)	64.1[a]
Control 19298 (1691)	2133 (784)	89.0	16674 (867)	978 (545)	94.1
ChAT-positive neurons					
CNTF 2259 (188)	1328 (147)	41.2[a]	2327 (209)	1069 (36)	54.1[a]
Control 2239 (164)	441 (189)	80.3	2933 (172)	354 (253)	87.9
NADPH-diaphorase-positive neurons					
CNTF 3339 (553)	1733 (489)	48.1[a]	3508 (824)	1647 (389)	53.1[a]
Control 3405 (428)	548 (318)	83.9	3393 (106)	458 (347)	86.5

[a]$p < 0.05$, CNTF vs Control.

together with a substantial atrophy of the striatum. In BHK-Control-implanted monkeys, Nissl-stained sections revealed extensive lesions in the caudate nucleus and putamen that were elliptical in shape. The lesion area in the three BHK-Control animals averaged 317.72 mm^3 ($\pm$26.01) in the caudate and 560.56 mm^3 ($\pm$83.58) in the putamen. Many of the neurons that could be identified were shrunken and displayed a dystrophic morphology. In contrast, the size of the lesion was significantly reduced in BHK-CNTF-implanted monkeys. The lesion area in the three BHK-CNTF monkeys averaged 83.72 mm^3 ($\pm$24.07) in the caudate and 182.54 mm^3 ($\pm$38.45) in the putamen. Numerous healthy-appearing Nissl-stained neurons were observed within the striatum of BHK-CNTF-implanted rats following the QA lesion even in regions proximal to the needle tract.

The neuroprotective effects of CNTF were examined in further detail by quantifying the loss of specific cell types within the striatum (Table 8). Lesioned monkeys receiving BHK-Control implants displayed a significant loss (caudate = 89.0%; putamen = 94.1%) of GAD-ir neurons within the striatum ipsilateral to the transplant. Many remaining GAD-ir striatal neurons appeared atrophic relative to neurons on the contralateral side. The excitotoxic degeneration of GAD-ir neurons in both the caudate nucleus and putamen was significantly attenuated in monkeys receiving implants of polymer-encapsulated BHK-CNTF cells as these monkeys displayed only a 64.0% and 64.1% reduction in GAD-ir neurons in the caudate and putamen respectively relative to the contralateral side. In addition to protecting the GABAergic cell bodies, CNTF implants sustained the striatal GABAergic efferent pathways. Enhanced DARPP-32 immunoreactivity (a marker for GABAergic terminals) was seen within the globus pallidus and pars reticulata of the substantia nigra. Optical density measurements revealed a significant reduction (49.0 $\pm$ 5.2%) in DARPP-32 immunoreactivity within the globus pallidus in BHK-Control animals ipsilateral to the lesion. This reduction was significantly attenuated (12.0 $\pm$ 4.3%) in BHK-CNTF-implanted animals. Likewise, the reduction in the optical density of DARPP-32-immunoreactivity in the pars reticulata of animals receiving BHK-Control implants (17 ± 1.8%) was significantly attenuated (4.3 ± 0.4%) by CNTF implants.

ChAT-ir and diaphorase-positve neurons were also protected by CNTF administration. Monkeys receiving BHK-Control implants displayed significant reductions in ChAT-postive neurons with the caudate (80.3%) and putamen (87.9%) compared to only a 41.2% (caudate) and 54.1% (putaman) in monkeys receiving BHK-CNTF implants (both $p < 0.05$). Similarly, the loss of NADPH-diaphorase-positive neurons in BHK-Control-implanted monkeys (caudate = 83.9%; putamen = 86.6%) was significantly attenuated (caudate = 48.1%; putamen = 53.1%) in monkeys receiving encapsulated CNTF implants.

These data clearly demonstrate that intrastriatal grafts of CNTF-producing fibroblasts protect GABAergic, cholinergic, and NADPH-diaphorase-containing neurons. However, the use of trophic factors, or any novel therapeutic strategy for HD, has been impeded by our inability to provide a rationale for how these approaches would influence critical nonstriatal regions such as the cerebral cortex that also degenerate in this disorder. Accordingly, we quantified the size of neurons from layer V of monkey motor cortex from a series of Nissl-stained sections in each of the control and CNTF-treated monkeys. QA produced a marked retrograde degeneration of cortical neurons in this region known to project to the striatum. While neuron number was unaffected, monkeys receiving BHK-Control implants displayed a significant atrophy (27%) of neurons in layer V of the motor cortex ipsilateral to the lesion. This atrophy was significantly attenuated (6%) in BHK-CNTF implanted animals. Further analysis demonstrated that the atrophy of cortical neurons was not due to general volumetric changes but rather occurred preferentially in the medium-sized (300–400 µm and 400–500 µm cross-sectional area) neurons of the motor cortex that project to the striatum.

Together these data provide the first demonstration that a therapeutic intervention can influence the degeneration of striatal neurons and disruption of basal ganglia circuitry in a primate model of HD. Not only are GABAergic neurons viable in CNTF-treated monkeys, but DARPP-32-ir reveals that the two critical GABAergic efferent projections from striatum to globus pallidus and the pars reticulata of the substantia nigra are sustained in these animals. Moreover, CNTF implants exerted a robust neuroprotective effect on the cortical neurons innervating the striatum. These data indicate that a major component of the basal ganglia loop circuitry, the cortical → striatal → globus pallidus/substantia nigra outflow circuitry is sustained by CNTF.

8. CONCLUSIONS

Ever since the initial discovery of a selective alteration in GABAergic parameters in HD, there has been a remarkable escalation in our understanding of the pathophysiology of the disease. Our concepts of the neurochemical and morphological sequelae of HD have grown from these early observations to the present-day complicated mosaic of neurobiological changes in the striatum and related structures. Although the cooperation of clinicians, basic scientists, and families of HD patients has contributed to this progress, an equally important contribution has been made from the development of animal models of HD. Investigators have taken advantage of new excitotoxic models that appear to mimic the neurobiological and behavioral characteristics of HD with remarkable homology. This research has substantially bolstered the hypothesis that a fundamental deficit in HD is a dysfunction of glutamatergic transmission or an increased susceptibility to excitotoxicity which results in the slow, progressive neural degeneration characteristic of HD.

The development of animal models holds great promise both for understanding the etiology of HD and for the development of therapeutic strategies. This chapter has highlighted one of the more exciting therapeutic possibilities which suggests that cellular delivery of neurotrophic factors may be one means of treating the neuropathological and behavioral consequences of excitotoxicity. Although the mechanisms by which these factors exert their beneficial effects remain unclear, their clearcut potency in both rodent and primate models of HD provides the hope that a means of preventing or slowing the relentlessly progressive motor and cognitive declines in HD may be forthcoming. Furthermore, excitotoxicity has been implicated in a variety of pathological conditions including ischemia, and neurodegenerative diseases such as Huntington's, Parkinson's, and Alzheimer's. Accordingly, biologically delivered neurotrophic factors may provide one means of preventing the cell loss and associated behavioral abnormalities of these and possibly other human disorders.

REFERENCES

1. Conneally, P. M. (1984) Huntington's disease: genetics and epidemiology. *Am. J. Hum. Genet.* **36,** 506–526.
2. Greenamyre, J. T. and Shoulson, I. (1994) Huntington's disease, in Neurodegenerative diseases (Calne, D., ed.) W. B. Saunders, Philadelphia, PA, pp. 685–704.
3. Farrer, L. A. and Conneally, P. M. (1985) A genetic model for age at onset in Huntington's disease. *Am. J. Hum. Genet.* **37,** 350–357.
4. Penney, J. B., Young, A. B., and Shoulson, I. (1990) Huntington's disease in Venezuela: 7 years of follow-up on symptomatic and asymptomatic individuals. *Move. Disord.* **5,** 93–99.
5. Shoulson, I. (1981) Huntington's disease. Functional capacities in patients treated with neuroleptic and antidepressant drugs. *Neurology* **31,** 1333–1335.
6. Ferrante, R. J., Kowall, N. W., and Beal, M. F. (1985) Selective sparing of a class of striatal neurons in Huntington's disease. *Science* **230,** 561–563.
7. Ferrante, R. J., Beal, M. F., Kowall, N. W., Richardson, E. P., and Martin, J. B. (1987) Sparing of acetylcholinesterase-containing striatal neurons in Huntington's disease. *Brain Res.* **415,** 178–182.
8. Graveland, G. A., Williams, R. S., and DiFiglia, M. (1985) Evidence of degenerative and regenerative changes in neostriatal spiny neurons in Huntington's disease. *Science* **227,** 770–773.
9. Kowall, N. W., Ferrante, R. J., and Martin, J. B. (1987) Pattern of cell loss in Huntington's disease. *Trends Neurosci.* **10,** 24–29.
10. Vonsattel, J. P., Ferrante, R. J., and Stevens, T. J. (1985) Neuropathologic classification of Huntington's disease. *J. Neuropathol. Exp. Neurol.* **44,** 559–577.
11. The Huntington's Disease Collaborative Research Group (1993) A novel gene containing a trinucleotide repeat that is expanded and unstable on Huntington's disease chromosomes. *Cell* **72,** 971–978.
12. Ashizawa, T., Wong, L-J. C., Richards, C. S., Caskey, C. T., and Jankovic, J. (1994) CAG repeat size and clinical presentation in Huntington's disease. *Neurology* **44,** 1137–1143.
13. Emerich, D. F. and Sanberg, P. R. (1992) Animal Models in Huntington's disease, in *Neuromethods,* Vol. 17: *Animal Models of Neurological Disease* (Boulton, A. A., Baker, G. B., Butterworth, R. F., eds.), Humana Press, Totowa, NJ, pp. 65–134.

14. Starr, A. (1967) A disorder of rapid eye movements in Huntington's chorea. *Brain* **90,** 545–564.
15. Dix, M. (1970) Clinical observations upon the vestibular responses in certain disorders of the central nervous system. *Adv. Otorhinolaryngol.* **17,** 118.
16. Davis, A. (1976) Emily—a victim of Huntington's chorea. *Nurs. Times* **72,** 449.
17. Young, A. B., Shoulson, I., Penny, J. B., Starosta-Rubenstein, S., Gomez, F., Travers, H., Ramos, M., Snodgras, S. R., Bonilla, A., Moreno, H., and Wexler, N. (1986) Huntington's disease in Venezuela: neurological features and functional decline. *Neurology* **36,** 244–249.
18. Podoll, K., Caspary, P., Lange, H. W., and Noth, J. (1988) Language functions in Huntington's disease. *Brain* **111,** 1475–1503.
19. Burns, A., Psych, M. R. C., Folstein, S., Brandt, J., and Folstein, M. (1990) Clinical assesment of irritability, aggression, and apathy in Huntington and Alzheimer disease. *J. Nerv. Ment. Dis.* **178,** 20–26.
20. Dewhurst, K., Oliver, J., Trick, K. L. K., and McKnight, A. L. (1979) Neuropsychiatric aspects of Huntington's disease. *Confinia. Neurol.* **31,** 258–268.
21. Knopman, D., and Nissen, M. J. (1991) Procedural learning is impaired in Huntington's disease: evidence from the serial reaction time task. *Neuropsychologia* **29,** 245–254.
22. Mann, J. J., Stanley, M., Gershon, S., and Rosser., M. (1980) Mental symptoms in Huntington's disease and a possible primary aminergic neuron lesion. *Science* **210,** 1369–1371.
23. Cummings, J. L. and Benson, D. F. (1988) Psychological dysfunction accompanying subcortical dimentias. *Annu. Rev. Med.* **39,** 53–61.
24. Brandt, J., Folstein, S. E., and Folstein, M. F. (1988) Differential cognitive impairment in Alzheimer's disease and Huntington's disease. *Ann. Neurol.* **23,** 555–561.
25. Heindel, W. C., Butters, N., and Salmon, D. P. (1988) Impaired learning of a motor skill in patients with Huntington's disease. *Behav. Neurosci.* **102,** 141–147.
26. Klawans, H. L., Goetz, C. G., and Westheimer, R. (1972) Pathophysiology of schizophrenia and the striatum. *Dis. Nerv. Syst.* **33,** 711–719.
27. Bowman, M. and Lewis, M. S. (1980) Site of subcortical damage in diseases which resemble schizophrenia. *Neuropsychology* **18,** 597–601.
28. Van Putten, T. and Menkes, J. H. (1973) Huntington's disease masquerading as chronic schizophrenia. *Dis. Nerv. Syst.* **34,** 54–56.
29. James, W. E., Meffered, R. B., and Kimbell, I. (1969) Early signs of Huntington's chorea. *Dis. Nerv. Syst.* **30,** 556–559.
30. Girotti, F., Marano, R., Soliver, E., Geminiani, G., and Scagliano, G. (1988) Relationship between motor and cognitive disorders in Huntington's disease. *J. Neurol.* **235,** 454–457.
31. Scholz, O. B. and Berlemann, C. (1987) Memory performances in Huntington's disease. *Int. J. Neurosci.* **35,** 155–162.
32. Saint-Cyr, J. A., Taylor, A. E., and Lang, A. E. (1988) Procedural learning and neostriatal dysfunction in man. *Brain* **111,** 941–959.
33. Reynolds, G. P., Pearson, S. J., and Heathfield, K. W. G. (1990) Dementia in Huntington's disease is associated with neurochemical deficits in the caudate nucleus, not the cerebral cortex. *Neurosci. Lett.* **113,** 95–100.
34. Coyle, J. T., Price, D. L., and Delong, M. R. (1983) Alzheimer's disease: a disorder of cholinergic innervation of cortex. *Science* **219,** 1184–1190.
35. McHugh, P. R. and Folstein, M. F. (1975) Psychiatric syndromes of Huntington's chorea: a clinical and phenomenologic study, in *Psychiatric Aspects of Neurologic Disease* (Blummer, F. B., ed.), Grune & Stratton, New York, pp. 267–281.
36. Buell, U., Costa, D. C., Kirsch, G., Moretti, J. L., Van Royen, E. A., and Schober, O. (1990) The investigation of dementia with single photon emission tomography. *Nucl. Med. Commun.* **11,** 823–841.

37. Jernigan, T. L., Salmon, D. P., Butters, N., and Hesselink, J. R. (1991) Cerebral structure on MRI. Part II: Specific changes in Alzheimer's and Huntington's diseases. *Biol. Psychiatry* **29,** 68–81.

38. Kuwert, T., Lange, G. W., Langen, K-J., Herzog, H., Aulich, A., and Feinendegen, L. E. (1990) Cortical and subcortical glucose consumption measured by PET in patients with Huntington's disease. *Brain* **113,** 1405–1423.

39. Leblhuber, F., Hoell, K., Reisecker, F., Gebetsberger, B., Puerhringer, W., Trenkler, E., and Deisenhammer, E. (1989) Single photon emission computed tomography in Huntington's chorea. *Psychiatry Res.* **29,** 337–339.

40. Hodges, J. R., Salmon, D. P., and Butters, N. (1991) The nature of the naming deficit in Alzheimer's and Huntington's disease. *Brain* **114,** 1547–1558.

41. Salmon, D. P., Kwo-on-Yeun, P. F., Heindel, W. C., Butters, N., and Thal, L. J. (1989) Differentiation of Alzheimer's disease and Huntington's disease with the dementia rating scale. *Arch. Neurol.* **46,** 1204–1208.

42. Pillon, B., Dubois, B., Ploska, A., and Agid, Y. (1991) Severity and specificity of cognitive impairment in Alzheimer's, Huntington's, and Parkinson's diseases and progressive supra-nuclear palsy. *Neurology* **41,** 634–643.

43. Randolph, C. (1991) Implicit, explicit, and semantic memory functions in Alzheimer's disease and Huntington's disease. *J. Clin. Exp. Neuropsychol.* **13,** 479–494.

44. Heindel, W. C., Salmon, D. P., and Butters, N. (1990) Pictoral priming and cued recall in Alzheimer's and Huntington's disease. *Brain Cognit.* **13,** 282–295.

45. Massman, P. J., Delis, D. C., and Butters, N. (1990) Are all subcortical dimentias alike? Verbal learning and memory in Parkinson's and Huntington's disease patients. *J. Clin. Exp. Neuropsychol.* **12,** 729–744.

46. Shoulson, I. (1986) Huntington's disease in *Disease of the Nervous System.* (Asbury, G. M. A. and McDonald, I., eds.), W. B. Saunders, Philadelphia, pp. 1258–1267.

47. Dulap, C. (1927) Pathological changes in Huntington's chorea with special reference to the corpus striatum. *Arch. Neurol. Psychiatry* **18,** 867–943.

48. Barr, A. N., Heinze, W. J., Dobben, G. D., Valvassori, G., and Sugar, O. (1978) Bicaudate index in computerized tomography of Huntington's disease and cerebral atrophy. *Neurology* **28,** 1196–1200.

49. Bruyn, G., Bots, G., and Dom, R. (1979) Huntington's chorea: current neuropathological status. *Adv. Neurol.* **23,** 83–93.

50. Lange, H. (1981) Quantitative changes of telencephalon, diencephalon, and mesencephalon in Huntington's chorea, postencephalitic and idiopathic Parkinson's disease. *Verh. Anat. Ges.* **75,** 923–925.

51. Reiner, A., Albin, D. L., Anderson, K. D., D'Amato, C. J., Penny, J. B., and Young, A. B. (1988) Differential loss of striatal projection neurons in Huntington's disease. *Proc. Natl. Acad. Sci. USA* **85,** 5733–5737.

52. Bird, E. (1980) Chemical pathology of Huntington's disease. *Annu. Rev. Pharmacol. Toxicol.* **20,** 533–551.

53. Beal, M. F., Ellison, D. W., Mazurek, M. F., Swarz, K. J., Malloy, J. R., Bird, E., and Martin, J. B. (1988) A detailed examination of substance P in pathologically graded cases of Huntington's disease. *J. Neurol. Sci.* **84,** 51–61.

54. Lange, H., Thorner, G., Hopf, A., and Schroeder, K. F. (1976) Morphometric studies of the neuropathological changes in choreatic diseases. *J. Neurol. Sci.* **28,** 401–425.

55. Forno, L. S. and Jose, C. (1973) Huntington's chorea: a pathological study. *Adv. Neurol.* **1,** 453–470.

56. Oka, H. (1980) Organization of the cortico-caudate projections. *Exp. Brain Res.* **40,** 203–208.

57. Cudkowicz, M. and Kowell, N. W. (1990) Degeneration of pyramidal projection neurons in Huntington's disease cortex. *Ann. Neurol.* **27,** 200–204.

58. Hedreen, J. C., Peyser, C. E., Folstein, S. E., and Ross, C. A. (1991) Neuronal loss in layers V and VI of cerebral cortex in Huntington's disease. *Neurosci. Lett.* **133,** 257–261.
59. Sortel, A., Paskevich, P. A., Kiely, D. K., Bird, E. D., Williams, R. S., and Myers, R. H. (1991) Morphometric analysis of the prefrontal cortex in Huntington's disease. *Neurology* **41,** 1117–1123.
60. Greenamyre, J. T. and O'Brien, C. F. (1986) The role of glutamate in neurotransmission and in neurologic disease. *Arch. Neurol.* **43,** 1058–1063.
61. Olney, J. W. (1989) Excitatory amino acids and neuropsychiatric disorders. *Biol. Psychiatry* **26,** 505–525.
62. Albin, R. L. and Greenamyre, J. T. (1992) Alternative excitotoxic hypothesis. *Neurology* **42,** 733–738.
63. Beal, M. F. (1992) Does impairment of energy metabolism result in excitotoxic neuronal death in neurodegenerative illnesses? *Ann. Neurol.* **31,** 119–130.
64. Beal, M. F. (1995) Aging, energy, and oxidative stress in neurodegenerative diseases. *Ann. Neurol.* **38,** 357–366.
65. Foster, A. C., Collins, J. F., and Schwarcz, R. (1983) On the excitotoxic properties of quinolinic acid, 2,3 piperidine dicarboxylic acids and structurally related compounds. *Neuropharmacology* **22,** 1331–1342.
66. Schwarcz, R., Whetsell, W. O., and Mangano, R. M. (1983) Quinolinic acid: an endogenous metabolite that produces axon-sparing lesions in rat brain. *Science* **219,** 316–318.
67. Schwarcz, R., Foster, A. C., French, E. D., Whetsell, W. O., and Kohler, C. (1984) Excitotoxic models for neurodegenerative disorders. *Life Sci.* **35,** 19–32.
68. Du, F., Okuno, E., Whetsell, W. O., Kohler, C., and Schwarcz, R. (1991) Immunohistochemical localization of quinolinic acid phosphoribosyltransferase in the human neostriatum. *Neuroscience* **42,** 397–406.
69. Foster, A. C., Whetsell, W. O., Bird, E. D., and Schwarcz, R. (1985) Quinolinic acid phosphoribosyltransferase in human and rat brain: Activity in Huntington's disease and in quinolate-lesioned rat striatum. *Brain Res.* **336,** 207–214.
70. Okuno, E., Kohler, C., and Schwarcz, R. (1987) Rat 3-hydroxyanthranilic acid oxygenase: purification from the liver and immunocytochemical localization in the brain. *J. Neurochem.* **49,** 771–780.
71. Reynolds, G. P., Pearson, S. J., Halket, J., and Sandler, M. (1988) Brain quinolinic acid in Huntington's disease. *J. Neurochem.* **50,** 1959–1960.
72. Ellison, D. W., Beal, M. F., Mazurek, M. F., Malloy, J. R., Bird, E. D., and Martin, J. B. (1987) Amino acid neurotransmiter abnormalities in Huntington's disease and the quinolinic model of Huntington's disease. *Brain* **110,** 1657–1673.
73. Beal, M. F., Kowall, N. W., Ellison, D. W., Mazurek, M. F., Swartz, K. J., and Martin, J. B. (1986) Replication of the neurochemical characteristics Huntington's disease by quinolinic acid. *Nature* **321,** 168–171.
74. Beal, M. F., Kowall, N. W., Swartz, K. J., Ferranti, R. J., and Martin, J. B. (1989) Differential sparing of somatostatin-neuropeptide Y and cholinergic neurons following striatal excitotoxin lesions. *Synapse* **3,** 38–47.
75. Beal, M. F., Ferrante, R. J., Swartz, K. J., and Kowall, N. W. (1991) Chronic quinolinic acid lesions in rats closely resemble Huntington's disease. *J. Neurosci.* **11,** 1649–1659.
76. Davies, S. W. and Roberts P. J. (1987) No evidence for preservation of somatostatin-containing neurons after intrastriatal injections of quinolinic acid. *Nature* **327,** 326–329.
77. Boegman, R. J., Smith, Y., and Parent, A. (1987) Quinolinic acid does not spare striatal neuropeptide Y-immunoreactive neurons. *Brain Res.* **415,** 178–182.
78. Sanberg, P. R., Zubrycki, E. M., Ragozinno, M. E., Lu, S. Y., Norman, A. B., and Shipley, M. T. (1990) NADPH-diaphorase-containing neurons and cytochrome oxidase activity following striatal quinolinic acid lesions and fetal striatal implants. *Prog. Brain Res.* **82,** 427–431.

79. Emerich, D. F., Zubricki, E. M., Shipley, M. T., Norman, A. B., and Sanberg, P. R. (1991) Female rats are more sensitive to the locomotor alterations following quinolinic acid-induced striatal lesions: effects of striatal implants. *Exp. Neurol.* **111,** 369–378.

80. Emerich, D. F., Winn, S. R., Lindner, M. D., Frydel, B. R., and Kordower, J. H. (1996) Implants of encapsulated human CNTF-producing fibroblasts prevent behavioral deficits and striatal degeneration in a rodent model of Huntington's disease. *J. Neurosci.* **16,** 5168–5181.

81. Emerich, D. F., Winn, S. R., Chen, E-Y., Chu, Y., McDermott, P., Baetge, E., and Kordower, J. H. (1997) Protection of basal ganglia circuitry by encapsulated CNTF-producing cells in a primate model of Huntington's disease. *Nature* **386,** 395–399.

82. Koh, J. Y. and Choi, D. W. (1988) Cultured striatal neurons containing NADPH-diaphorase or acetylcholinesterase are selectively resistant to injury by NMDA agonists. *Brain Res.* **446,** 374–378.

83. Koh, J. Y., Peters, S., and Choi, D. W. (1986) Neurons containing NADPH-diaphorase are selectively resistant to quinolinate toxicity. *Science* **234,** 73–76.

84. Whetsell, W. O. and Schwarcz, R. (1989) Prolonged exposure to micromolar concentration of quinolinic acid causes excitotoxic damage in organotypic cultures of rat corticostriatal system. *Neurosci. Lett.* **97,** 271–275.

85. Sanberg, P. R., Calderon, S. F., Giordano, M., Tew, J. M., and Norman, A. B. (1989) The quinolinic model of Huntington's disease: locomotor abnormalities. *Exp. Neurol* **105,** 45–53.

86. Block, F., Kunkel, M., and Schwarz, M. (1993) Quinolinic acid lesion of the striatum induces impairment in spatial learning and motor performance in rats. *Neurosci. Lett.* **149,** 126–128.

87. Emerich, D. F., Cain, C. K., Greco, C., Saydoff, J. A., Hu, Z-Y., Liu, H., and Lindner, M. D. Cellular delivery of human CNTF prevents motor and cognitive dysfunction in a rodent model of Huntington's disease. *Cell Transplant.,* in press.

88. Calderon, S., Sanberg, P. R., and Norman, A. B. (1988) Quinolinic acid lesions of rat striatum abolish D1- and D2-dopamine receptor-mediated catalepsy. *Brain Res.* **450,** 403–407.

89. Zubrycki, E. W., Emerich, D. F., and Sanberg, P. R. (1990) Sex differences in regulatory changes following quinolinic acid-induced striatal lesions. *Brain Res. Bull.* **25,** 633–637.

90. Bjorklund, A., Campbell, K., Sirinathsinghji, D. J., Fricker, R. A., and Dunnett, S. B. (1994) Functional capacity of striatal transplants in the rat Huntington model, in *Functional Neural Transplantation* (Dunnett, S. B. and Bjorklund, A., eds.), Raven Press, New York, pp. 157–195.

91. Isacson, O., Brundin, P., Kelly, P. A. T., Gage, F. H., and Bjorklund, A. (1984) Functional neuronal replacement by grafted striatal neurons in the ibotenic acid lesioned rat striatum. *Nature* **311,** 458–35.

92. Isacson, O., Dawbarn, D., Brundin, P., Gage, F. H., Emson, P. C., and Bjorklund, A. (1987) Neural grafting in a rat model of Huntington's disease: striosomal-like organization of striatal grafts as revealed by immunocytochemistry and receptor autoradiography. *Neuroscience* **22,** 401–497.

93. Isacson, O., Riche, D., Hantraye, P., Sofroniew, M. V., and Maziere, M. (1989) A primate model of Huntington's disease; cross-species implantation of striatal precursor cells to the excitotoxically lesioned baboon caudate-putamen. *Exp. Brain Res.* **75,** 213–220.

94. Wictorin, K. and Bjorklund, A. (1989) Connectivity of striatal grafts implanted into the ibotenic acid lesioned striatum: II. Cortical afferents. *Neuroscience* **30,** 297–311.

95. Wictorin, K., Simerly, R. B., Isacson, O., Wanson, L. W., and Bjorklund, A. (1989) Connectivity of striatal grafts implanted into the ibotemic acid lesioned striatum: III. Efferent projecting neurons and their relationships to host afferents within the grafts. *Neuroscience* **30,** 313–330.

96. Tatter, S. B., Galpern, W. R., and Isacson, O. (1995) Neurotrophic factor protection against excitotoxic neuronal death. *Neuroscientist* **5,** 286–297.

97. Anderson, K. D., Panayotatos, N., Cordoran, T,., Lindsay, R. M., and Wiegand, S. J. (1996) Ciliary neurotrophic factor protects striatal output neurons in an animal model of Huntington's disease. *Proc. Natl. Acad. Sci. USA* **93,** 7346–7351.

98. Emerich, D. F., Hammang, J. P., Baetge, E. E., and Winn, S. R. (1994) Implantation of polymer-encapsulated human nerve growth factor-secreting fibroblasts attenuates the behavioral and neuropathological consequences of quinolinic acid injections into rodent striatum. *Exp. Neurol.* **130,** 141–150.

99. Frim, D. M., Uhler, T. A., Short, M. P., Exxedine, Z. D., Klagsbrun, M., Breakefield, X. O., and Isacson O. (1993) Effects of biologically delivered NGF, BDNF, and bFGF on striatal excitotoxic lesions. *NeuroReport* **4,** 367–370.

100. Frim, D. M., Simpson, J., Uhler, T. A., Short, M. P., Bossi, S. R., Breakefield, X. O., and Isacson, O. (1993) Striatal degeneration induced by mitochondrial blockade is prevented by biologically delivered NGF. *J. Neurosci. Res.* **35,** 452–458.

101. Frim, D. M., Yee, W. M., and Isacson, O. (1993) NGF reduces striatal excitotoxic neuronal loss without affecting concurrent neuronal stress. *NeuroReport* **4,** 655–658.

102. Frim, D. M., Short, M. P., Rosenberg, W. S., Simpson, J., Breakefield, X. O., and Isacson, O. (1993) Local protective effects of nerve growth factor-secreting fibroblasts against excitotoxic lesions in the rat striatum. *J. Neurosurg.* **78,** 267–273.

103. Kordower, J. H., Chen, E-Y., Mufson, E. J., Winn, S. R., and Emerich, D. F. (1996) Intrastriatal implants of polymer-encapsulated cells genetically modified to secrete human NGF: trophic effects upon cholinergic and noncholinergic neurons. *Neuroscience* **72,** 63–77.

104. Martinez-Serrano, A. and Bjorklund, A. (1996) Protection of the neostriatum against excitotoxic damage by neurotrophin-producing genetically modified neural stem cells. *J. Neurosci.* **16,** 4604–4616.

105. Schumacher, J. M., Short, M. P., Hyman, B. T., Breakefield, X. O., and Isacson, O. (1991) Intracerebral implantation of nerve growth factor-producing fibroblasts protects striatum against neurotoxic levels of excitatory amino acids. *J. Neurosci.* **45,** 561–570.

106. Breakefield, X. O. (1989) Combining CNS transplantation and gene transfer. *Neurobiol. Aging* **10,** 647–648.

107. Gage, F. H., Wolf, J. A., Rosenberg, M. B., Xu, L., and Yee, J. K. (1987) Grafting genetically modified cells to the brain: possibilities for the future. *Neuroscience* **23,** 795–807.

108. Gage, F. H., Batchelor, P., Chen, K. S., Higgins, G. A., Koh, S., Deputy, S., Rosenberg, M. B., Fisher, W., and Bjorklund, A. (1989) NGF receptor reexpression and NGF-mediated cholinergic neuronal hypertrophy in the damaged adult neostriatum. *Neuron* **2,** 1177–1184.

109. Kawaja, M. D., Fagan, A. M., Firestein, B. L., and Gage, F. H. 1991. Intracerebral grafting of cultured autologous skin fibroblasts into the rat striatum:an assessment of graft size and ultrastructure. *J. Comp. Neurol.* **307,** 695–706.

110. Kawaja, M. D., Rosenberg, M. B., Yoshida, K., and Gage, F. H. (1992) Somatic gene transfer of nerve growth factor promotes the survival of axotomized septal neurons and the regeneration of their axons in adult rats. *J. Neurosci.* **12,** 2849–2864.

111. Levivier, M., Przedborski, S., Bencsics, C., and Kang, U-J. (1995) Intrastriatal implantation of fibroblasts genetically engineered to produce brain-derived neurotrophic factor prevents degeneration of dopaminergic neurons in a rat model of Parkinson's disease. *J. Neurosci.* **15,** 7810–7820.

112. Emerich, D. F., Winn, S. R., Christenson, L., Palmatier, M., Gentile, F. T., and Sanberg, P. R. (1992) A novel approach to neural transplantation in Parkinson's disease: use of polymer-encapsulated cell therapy. *Neurosci. Biobehav. Rev.* **16,** 437–447.

113. Emerich, D. F., Winn, S. R., Harper, J., Hammang, J. P., Baetge, E. E., and Kordower, J. H. (1994) Transplantation of polymer-encapsulated cells genetically modified to secrete human nerve growth factor prevents the loss of degenerating cholinergic neurons in non-human primates. *J. Comp. Neurol.* **349,** 148–164.

114. Kordower, J. H., Liu, Y-T., Winn, S. R., and Emerich, D. F. (1995) Encapsulated PC12 cell transplants into hemiparkinsonian monkeys: a behavioral, neuroanatomical and neurochemical analysis. *Cell Transplant.* **4,** 155–171.

115. Sagot, Y., Tan, S. A., Baetge, E., Schmalbruch, H., Kato, A. C., and Aebischer, P. (1995) Polymer encapsulated cell lines genetically engineered to release ciliary neurotrophic factor can slow down progressive motor neuronopathy in the mouse. *Eur. J. Neurosci.* **7,** 1313–1322.

116. Kordower, J. H., Winn, S. R., Liu, Y-T., Mufson, E. J., Sladek, J. R. Jr., Baetge, E. E., Hammang, J. P., and Emerich, D. F. (1994) The aged monkey basal forebrain: rescue and sprouting of axotomized basal forebrain neurons after grafts of encapsulated cells secreting human nerve growth factor. *Proc. Natl. Acad. Sci. USA* **91,** 10,898–10,902.

117. Apfel, S. C., Arezzo, J. C., Moran, M., and Kessler, J. A. (1993) Effects of administration of ciliary neurotrophic factor on normal motor and sensory peripheral nerves in vivo. *Brain Res.* **604,** 1–6.

118. Arakawa, Y., Sendtner, M., and Thoenen, H. (1990) Survival effect of ciliary neurotrophic factor (CNTF) on chick embryonic motoneurons in culture: comparison with other neurotrophic factors and cytokines. *J. Neurosci.* **10,** 3507–3515.

119. Forger, N. G., Roberts, S. L., Wong, V., and Breedlove, S. M. (1993) Ciliary neurotrophic factor maintains motoneurons and their target muscles in developing rats. *J. Neurosci.* **13,** 4720–4726.

120. Lin, L-F. H., Mismer, D., Lile, J. D., Armes, L. G., Butler, E. T., Vannice, J. L., and Collins, F. (1989) Purification, cloning, and expression of ciliary neurotrophic factor (CNTF). *Science* **246,** 1023–1025.

121. Masu, Y., Wolf, E., Holtmann, B., Sendtner, M., Brem, G., and Thoenen H. (1993) Disruption of the CNTF gene results in motor neuron degeneration. *Nature* **365,** 27–32.

122. Sendtner, M., Kreutzberg, G. W., and Thoenen, H. (1990) Ciliary neurotrophic factor prevents the degeneration of motor neurons after axotomy. *Nature* **345,** 440–441.

123. Sendtner, M., Schmalbruch, H., Stockli, K. A., Carroll, P., Kreutzberg, G. W., and Thoenen, H. (1992) Ciliary neurotrophic factor prevents degeneration of motor neurons in mouse mutant progressive motor neuronpathy. *Nature* **358,** 502–504.

124. Stockli, K. A., Lottspeich, F., Sendtner, M., Masiakowski, P., Carrol, P., Gotz, R., Lindholm, D., and Thoenen, H. (1989) Molecular cloning, expression, and regional distribution of rat ciliary neurotrophic factor. *Nature* **342,** 920–923.

125. Stockli, K. A., Lillien, L. E., Naher-Noe, M., Breitfeld, G., Hughes, R. A., Raff, M. C., Thoenen, H., and Sendtner, M. (1991) Regional distribution, developmental changes and cellular localization of CNTF-mRNA and protein in the rat brain. *J. Cell Biol.* **115,** 447–455.

126. Clatterbuck, R. E., Price, D. L., and Koliatsos, V. E. (1993) Ciliary neurotrophic factor prevents retrograde neuronal death in the adult central nervous system. *Proc. Natl. Acad. Sci. USA* **90,** 2222–2226.

127. Hagg, T. and Varon, S. (1993) Ciliary neurotrophic factor prevents degeneration of adult rat substantia nigra dopaminergic neurons in vivo. *Proc. Natl. Acad. Sci. USA* **90,** 6315–6319.

128. Hagg, T., Quon, D., Higaki, J., and Varon, S. (1993) Ciliary neurotrophic factor prevents neuronal degeneration and promotes low affinity NGF receptor expression in the adult rat CNS. *Neuron* **8,** 145–158.

16

Systemic Administration of 3-Nitropropionic Acid

A New Model of Huntington's Disease in Rat

Emmanuel Brouillet, Philippe Hantraye, and M. Flint Beal

1. INTRODUCTION

Huntington's disease (HD) is a neurodegenerative disorder characterized by dyskinetic abnormal movements and cognitive decline associated with progressive atrophy of the striatum (1). Generally onset of symptoms occurs in adults and the disease evolves over 10–15 yr toward a fatal outcome. The gene responsible for HD has been identified, and molecular studies of the corresponding encoded protein named huntingtin have made considerable progress (2). However, there are no appropriate phenotypic animal models of the disease based on transgenesis, and the mechanism underlying cell death in HD remains largely unknown.

A number of trigerring events leading to neuronal death have been identified by research in cellular biology, such as excitotoxicity (a term designating death resulting from excessive activation of glutamate receptors). The possibility that excitotoxicity may play a role in the etiology of HD was supported mainly by the finding that focal injection of glutamate receptor agonists into the striatum produced lesions with histological and neurochemical characteristics resembling those seen in HD (3). Recently this hypothesis was refined by suggesting that early energy impairment observed in HD patients may lead to the overactivation of NMDA receptors and relentless excitotoxic neuronal death (4–6). Experiments on neuronal cell cultures and laboratory animals in vivo have suggested that mild energy failure could indirectly produce activation of an N-methyl-D-aspartate (NMDA) receptor, triggering an excitotoxic cascade and neurodegeneration. Defects in energy metabolism in HD patients have been found in vivo using positron tomography and nuclear magnetic resonance spectroscopy. Biochemical analysis of postmortem tissue samples from HD patients showed a consistent decrease in activity of complex II–III (succinate dehydrogenase [SDH] and ubiquinone-cytochrome c oxidoreductase). The possibility that the defect in complex II–III activity seen in HD may have a causal role in the etiology of the disease is also suggested by the fact that well-characterized cases with biochemical defects in succinate dehydrogenase are associated with preferential striatal degeneration (7,8). In addition, poisoning with

From: Central Nervous System Diseases
Edited by: D. F. Emerich, R. L. Dean, III, and P. R. Sanberg © Humana Press Inc., Totowa, NJ

the succinate dehydrogenase irreversible inhibitor 3-nitropropionic acid (3-NPA) in humans results in striatal lesions *(9)*.

According to this hypothesis, the effect of systemic administration of 3-NPA is of particular interest. Since the pioneering studies of Gould and collaborators *(10–13)* reporting the neuropathological and neurological outcome of acute 3-NPA treatment in mice and rats, in particular the preferential vulnerability of the striatum, the effects of systemic administration of 3-NPA have been studied for the past 6 yr in the context of HD. Several groups have recently reevaluated the effects of acute or subacute (repeated injections) of 3-NPA administration either to further study the neurotoxic mechanisms of this toxin or use it for testing new neuroprotective therapies *(14–25)*. Alternatively, it was found that systemic administration of chronic low doses of 3-NPA produces motor deficits and selective striatal lesions highly reminiscent of HD *(14,15,17, 26–28)*. Here, we review the different aspects of 3-NPA neurotoxicity in rats and discuss how the model of progressive striatal degeneration using chronic administration of the toxin may lead to a better understanding of the mechanisms of cell death in vivo and testing of new therapeutic strategies for HD.

2. HUNTINGTON'S DISEASE

2.1. Clinical and Neuropathological Features

HD is an inherited dominant neurodegenerative disorder that is characterized by choreiform abnormal movements, cognitive deficits, and psychiatric manifestations associated with progressive striatal atrophy *(1,29)*. The onset, progression, and clinical expression of HD is variable even though it occurs in general during adulthood. The progression of the disease is correlated to the age at onset. In its common form (middle age onset), the disease evolves over 10–15 yr. Early symptoms consist typically of irritability, cognitive deficits, and choreiform movements. Later, bradykinesia and dystonia can be seen. Later, dystonia becomes more prominent and symptoms evolve toward rigidity. In patients with early onset (juvenile form), a rapid aggravation of symptoms occurs. In this case, choreiform movements are generally absent, whereas dystonia and bradykinesia evolving rapidly to rigidity are typical characteristics of juvenile and childhood HD. The most striking change in the HD brain is striatal atrophy, the severity of which correlates with the severity of psychiatric and motor symptoms *(30,31)*. Late in the course of the disease, other cerebral regions are affected such as the cerebral cortex, pallidum, subthalamic nucleus, and substantia nigra reticulata *(1)*.

The pathological process underlying HD does not affect uniformly all types of striatal cells *(32)*. There is a preferential degeneration of γ-aminobutyric acid-ergic (GABAergic) medium-sized spiny neurons as seen by marked depletions in GABA concentrations in the striatum of HD patients *(33,34)*. Concentrations of substance P and Met-enkephalin that colocalize with different subsets of GABA neurons in the striatum are also decreased *(35–40)*. Decreased numbers of neurons immunoreactive for calbindin D28k, a Ca^{2+}-binding protein present in a subset of striatal medium-sized spiny neurons, was also reported *(38,41–44)*. Interestingly, Golgi staining studies showed that many striatal GABAergic neurons show morphological abnormalities in moderate grades of HD *(41,45)*. Changes in immunoreactivity for calbindin or Golgi staining indicate the presence of proliferative dendritic changes early in the disease.

These morphological changes include increased size and density of dendritic spines, recurving of distal dendritic segments, and short-segment branching along dendrites. Another intriguing neuropathological characteristic of HD is the preferential degeneration of the GABAergic medium-sized spiny neurons as compared to the relative sparing of the medium-sized interneurons positive for NADPH-diaphorase and somatostatin *(46–48)*. The preservation of this subset of striatal interneurons is associated with an increase in somatostatin and neuropeptide Y concentrations *(49)*. The enzyme responsible for the labeling of the medium-sized aspiny neurons within the striatum using NADPH diaphorase activity turned to nitric oxide synthase (NOS) *(50,51)*. This striking sparing of NOS-positive interneurons and the possible neurotoxicity of NO led to the hypothesis that these interneurons may have a causal role in the pathogenesis of HD *(52,53)*. The large cholinergic interneurons are also spared in HD striatum, even though choline acetyltransferase activity and muscarinic receptors are significantly decreased, probably as a result of synapse loss due to the loss of neighboring neurons *(54)*. Concentrations of dopamine and its metabolites are not markedly decreased and tyrosine hydroxylase immunoreactivity is maintained in the HD striatum *(39,55–58)*. Another characteristic of HD neuropathology is the presence of intranuclear inclusion bodies and extracellular fibrillar deposition that consists of the ubiquitinated N-terminal part of huntingtin, the mutated protein in HD *(59)*.

The nature of neuronal death is HD is still debated. Magnetic resonance imaging (MRI) examination showed in certain cases, preferentially in patients with early onset, abnormalities in signal intensity indicating necrosis and a rapid process of degeneration *(60–62)*. In patients with the common form of HD, no indications of necrosis are observed. Interestingly, histological studies showed features reminiscent of apoptosis in HD cases *(63,64)*. These differences in disease expression indicate that necrosis and apoptosis may coexist in HD striatum.

2.2. Molecular Genetics

The gene responsible for HD has been localized and sequenced *(65,66)*. The characterization of the encoded protein huntingtin, including its localization and the discovery of other proteins interacting with it, has made considerable progress. However, its function remains unknown. The mutation of the HD gene consists in the expansion of a repetition of a triplet of nucleotides (CAG) and the resulting mutated huntingtin has an expanded polyglutamine stretch. One hypothesis for explaining the toxicity of the mutation is that the expanded polyglutamine tract may modify the normal interaction of huntingtin with other proteins or more probably generate a new protein–protein interaction triggering neuronal death *(67)*. Several candidate proteins have been identified. One of these candidates directly related to neuronal death is apopain (caspase 3), which plays a central role in apoptosis *(68)*. Another interesting protein interacting with huntingtin is GAPDH (glyceraldehyde-3-phosphate dehydrogenase) *(69)*, which in addition to its role in glycolysis has been recently implicated in apoptosis *(70,71)*. The exact link between the proteins potentially interacting with huntingtin and the pathological process occuring in HD remains to be clarified. Another hypothesis is a potential toxicity of the N-terminal part of the mutated human huntingtin. Nuclear and extracellular accumulation of this peptide fragment has been found in HD *(59)*. A similar accumulation has been found in transgenic mice overexpressing the N-terminal part of huntingtin with an expanded polyglutamine tract *(72)*. However, this rodent model lacks many characteristics of HD, in particular striatal neurodegeneration.

3. POSSIBLE INVOLVEMENT OF EXCITOTOXICITY IN HD

The mechanism leading to striatal atrophy in HD remains obscure. A chronic impairment in energy metabolism resulting in the activation of an excitotoxic cascade of cellular events has been suggested to play a key role in the etiology of many neurodegenerative diseases *(4–6,73,74)* and this hypothesis is particularly relevant for HD.

The initial reports showing that stereotaxic injection of kainate in the rat striatum produces axon-sparing lesions were the starting point of a wide literature on the neurotoxicity of glutamate analogues *(75,76)*. It was suggested that abnormal activation of glutamate receptors may produce the striatal atrophy that characterizes HD. However, kainate injection does not accurately reproduce the histological " signature " of HD *(77)*. Indeed, kainate kills both the medium-sized spiny neurons and NADPH interneurons. In contrast, striatal injection of quinolinate, an endogenous agonist of the NMDA receptor produces degeneration of GABAergic medium-sized spiny neurons with relative preservation of NADPH-diaphorase interneurons as seen with neurochemical mesurements of GABA and neuropeptides and histological evaluation *(77–80)*. An extensive histological and neurochemical characterization of quinolinate-induced striatal lesions in monkeys *(79)* further confirmed that the pattern of neurodegeneration induced by overactivation of NMDA receptors was very reminiscent of HD.

Striatal lesioning using direct infusion of excitotoxin could also replicate some behavioral aspects of HD. The long-term behavioral effects of striatal lesions produced by excitotoxins were first characterized in rats. For unilateral striatal lesions, the spontaneous motor symptomatology is barely detectable on visual inspection. The best way to show the presence of a functional motor deficit is to stimulate the nigrostriatal dopaminergic pathway by injecting dopamine agonists such as apomorphine or a dopamine releasing compound such as methamphetamine *(81,82)*. Then a typical "turning behavior" is observed. The increased locomotor activity (increased number of rotations) is considered to mimic to a certain extent the " hyperkinesia " seen in HD patients. At least, it reliably indicates to which extent the striatum is lesioned. Striatal lesions also lead to a spontaneous persistent nocturnal locomotor hyperactivity and cognitive impairment *(83–86)*. For bilateral striatal lesions produced by quinolinate focal injection, Sanberg's group showed by quantitative analysis (using the Digiscan system) that the nocturnal spontaneous locomotion of lesioned rats was significantly increased *(87)*. The most convincing results that excitotoxin-induced lesions of the striatum could lead to HD-like symptomatology were obtained in nonhuman primates. Indeed, in studies of monkeys with unilateral striatal lesions using either kainate *(88)*, ibotenic acid *(89,90)* or quinolinate *(91)*, injection of L-Dopa or apomorphine produces choreiform explosive involuntary movements, dyskinesia, and dystonia.

Thus, the striking histological and biochemical similarities between excitotoxic animal models and HD lend support to the hypothesis that glutamate receptors, and more specifically NMDA receptors, may play a central role in the etiology of HD *(3)*. In line with this, it was found that the density of NMDA receptors in the striatum of HD patients was decreased early in the time course of the disease, suggesting that striatal cells bearing these receptors may preferentially degenerate *(92)*.

4. REFINEMENT OF THE EXCITOTOXIC HYPOTHESIS: THE ROLE OF ENERGY METABOLISM

4.1. Evidence for Alteration in Energy Metabolism in HD

An interesting observation that may indicate an alteration in energy metabolism in HD is the presence of cachexia in advanced cases *(1)*. Despite adequate (or increased) diet and feeding, HD patients usually show weight loss. It is generally considered that this weight loss is not related to hyperkinesis which characterizes the disease because choreic, explosive movements tend to disappear in late-stage HD. It is tempting to speculate that cachexia may, at least in part, reflect abnormalities (possibly reduction) in consumption of energy substrates.

The initial studies by positron emission tomography (PET) carried out by Kuhl and collaborators *(93)* showed that cerebral glucose metabolism was affected in the striatum of HD patients. Marked decreases in glucose consumption could be attributable at least in part in symptomatic HD patients to striatal atrophy. However, substantial reductions in [^{18}F]-deoxyglucose incorporation in the striata of patients presenting early symptoms with no gross atrophy were seen, suggesting that severe metabolic impairment could precede bulk tissue loss. In 15 at-risk patients, 6 showed striatal glucose utilization that was more than 2 standard deviations lower than the normal mean value *(93)*. These initial observations were confirmed by a number of PET studies: alterations in striatal glucose metabolism were often seen in early or presymptomatic HD patients *(94–99)*.

Other evidence for alteration in energy metabolism in HD comes from the study of cerebral concentrations in lactate using proton nuclear magnetic resonance (NMR) spectroscopy in which substantial increases were seen in the striata but also in a brain region with no ongoing process of neurodegeneration, the occipital cortex *(100–102)*. This indicates an alteration of oxidative metabolism in HD, in line with the results of biochemical studies reviewed later.

A number of studies on postmortem samples from HD patients also demonstrated biochemical defects in the mitochondrial respiratory chain. A consistent defect in complex II–III was always found in the caudate nucleus of HD patients *(103–107)* as compared to that determined in the caudate of age-matched controls. This was characterized by a 39–59% decrease in succinate oxidation depending on the authors. Recently, Gu et al. *(106)* and Browne et al. *(104)* reported in well-controlled studies a 53% and 29% decrease in complex II–III activity, respectively. In the putamen, although an initial study on a small number of patients reported no substantial changes in complex II–III activity *(107)*, a more recent reevaluation on a large number of patients with normalization to citrate synthase activity showed a 69% decreased activity *(104)*. Complex IV (cytochrome oxidase) was also significantly affected in these regions although to a lesser extent as compared to the changes seen in complex II–III activity *(103,104,106)*. Interestingly, in none of these studies, complex I (NADH-dehydrogenase) was found altered in HD brain samples, suggesting that the mitochondrial defect in HD may involve selectively complex II–III and IV.

4.2. Mechanisms of Indirect Excitotoxicity In Vivo and In Vitro

A number of in vitro studies showed that an impairment of energy metabolism can produce secondary excitotoxicity without an elevation in glutamate concentration *(108–112)*. Oxygen/glucose deprivation or chemical hypoxia induced by mitochondrial toxins leads to activation of NMDA receptors. At low glutamate concentrations, suppression of energy supply triggers excitotoxicity which can be prevented by NMDA receptor antagonists (such as MK801 or D-z-amino-5-phosphonovalerate (APV)). An emerging idea from these studies is that the neurotoxic effects of glutamate and energy compromise are not additive but dramatically synergistic. Because it is known that intracellular calcium buffering after overactivation of NMDA receptors is dependent of mitochondria *(113,114)*, it is likely that a vicious cycle process develops once a certain threshold of deregulation of the NMDA receptor occurs. The mechanism underlying abnormal activation of NMDA receptors when energy metabolism impairment occurs may be due to the partial membrane depolarization produced by energy depletion. This depolarization would result in the relief of the voltage-dependent magnesium blockade of the NMDA receptor *(115)*. Elegant studies in vitro in cell culture support this view *(108–111)*.

In vivo, the possibility that a deficit in energy metabolism may lead to secondary excitotoxicity has been extensively studied (*see* for review *73,116*). Injection into the rat striatum of a number of mitochondrial toxins (aminooxyacetate, MPP$^+$, malonate, Mn^{2+}, 3-acetylpyridine) initially produces increased lactate production and ATP depletion, and later neuronal death that resembles that produced by the glutamate receptor agonists quinolinate and NMDA *(117–124)*. These lesions were characterized by a marked degeneration of GABAergic medium-sized spiny neurons, with sparing of cholinergic and NADPH-diaphorase-positive interneurons and could be prevented by NMDA antagonists or prior decortication which reduces glutamatergic afferents to the striatum. Thus, alteration of energy metabolism produced in vivo by mitochondrial toxins could indirectly activate NMDA receptors. Further supporting this view, subtoxic doses of the mitochondrial inhibitor malonate exacerbate the toxicity of NMDA *(125)*. This indirect activation of NMDA receptors after focal injection of mitochondrial toxins in vivo may possibly result from membrane depolarization triggered by energy compromise *(122)*. Thus, excitotoxicity now can be seen more generally as the lethal effects produced by the abnormal activation of ionotropic glutamate receptors even though glutamate concentrations may be in the physiological range.

As discussed in **Subheading 5.**, several in vitro and in vivo studies related to the neurotoxicity of the mitochondrial toxin 3-NPA, confirm the existence of this mechanism of indirect excitotoxicity.

5. SYSTEMIC ADMINISTRATION OF 3-NPA IN RATS: A NEW PHENOTYPIC MODEL OF HD

5.1. 3-NPA: A Fungal Neurotoxin in Human

The history of the determination of 3-NPA as a causal factor in the etiology of certain forms of basal ganglia degeneration in animals and in humans has been reviewed by Ludolph et al. *(126)*. In summary, the toxin 3-NPA, a metabolite of 3-nitropropanol, was initially identified as the active toxic compound of plants (such as *Indigofera* and *Astragalus*) responsible for livestock poisoning in the western United States. Intoxicated animals show motor abnormalities consisting of generalized weakness and incoordination of hindlimbs that evolves to paralysis.

Pioneering studies performed by Gould et al. *(10–13)* clearly illustrate that in laboratory animals, 3-NPA produces hypoxic-like cerebral lesions affecting preferentially the basal ganglia.

Numerous cases of poisoning with 3-NPA in humans have been reported in China for the past decade. Food poisoning were associated with ingestion of moldy (mildewed) sugarcane. It was found that moldy sugarcane was contaminated by the fungus *Arthrinium*, which produces large quantities of 3-NPA. Between 1972 and 1989 nearly 900 cases were reported, among whom 10% died and many left with irreversible disabilities. The first gastrointestinal signs of poisoning were often followed by coma that lasted several days. Although most of the patients recovered completely (based on neurological examination and evoked potential studies), a number of subjects showed irreversible neurological impairment. Symptoms included delayed onset dystonia, torsion spasms, facial grimacing, and jerk-like movements. CT scans revealed lesions of the basal ganglia, mainly the putamen, with the caudate less often affected. A recent report on 3-NPA poisoning in China confirmed these previous observations *(9)*.

5.2. 3-NPA: An Irreversible Inhibitor of Succinate Dehydrogenase In Vitro

Biochemical studies on submitochondrial preparations from heart beef or mitochondrial preparations from rat liver demonstrate that the toxin 3-NPA is a suicide inhibitor of the respiratory chain and Krebs cycle enzyme succinate dehydrogenase (SDH) (Fig.1). The exact mechanism leading to inactivation of the enzyme remains unclear. Alston et al. *(127)* initially suggested that the dianion form of 3-NPA is the inhibitory form and that the carbanion of the compound reacts with the flavin of SDH, to form a covalent adduct. Coles et al. *(128)* later challenged this hypothesis, by demonstrating that the carbanion of the dianion form of the toxin would preferentially bind to the substrate site, and would then be oxidized to 3-nitroacylic acid which may react with an essential thiol group of the substrate site. These in vitro studies showed that the active form of 3-NPA is the dianion form. Interestingly, the dianion form occurs at basic pH and reprotonation is slow. At physiological pH, the dianion form of 3-NPA would represent only 1% of total 3-NPA. Study of the kinetics of SDH inactivation by 3-NPA indicates that the toxin can rapidly (k_{obs} 1.2 min^{-1}) and stoichiometrically inactivate the enzyme. In the presence of succinate (or oxaloacetate), the rate of inactivation is dramatically slowed. This suggests that the rate of inhibition of SDH by 3-NPA in the living cell may be influenced by the activity of SDH itself, the redox state of its flavin, the concentration of endogenous substrate (succinate) or modulator (oxaloacetate), the temperature, and the pH.

5.3. Neurological and Neuropathological Effects of Acute 3-NPA Intoxication

Acute 3-NPA toxicity can be obtained by one or repeated daily intraperitoneal injections over a short period of time (2–10 d). The toxic dose varies depending on the age and strain of animals *(15,129)*. Recently published studies on systemic 3-NPA have

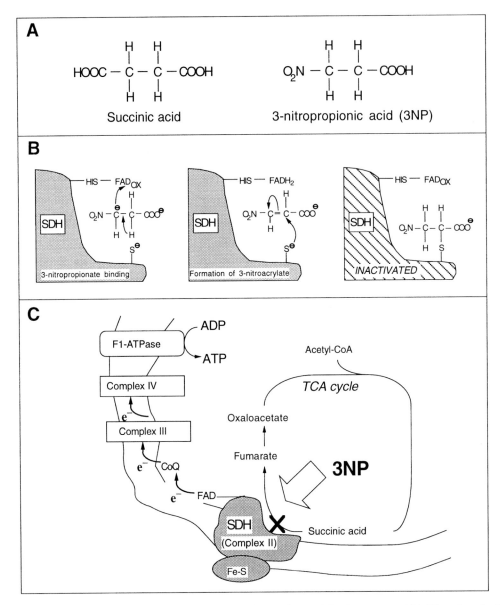

Fig. 1. 3-NPA is an irreversible inhibitor of succinate dehydrogenase. (**A**) The chemical structure of 3-NPA is close to that of succinic acid. (**B**) The carbanion of the dianion form of 3-NPA binds to the substrate site and is oxidized to 3-nitroacrylate which reacts with an essential thiol group. (From Coles et al., 1979.) (**C**) inactivation of SDH by 3-NPA blocks the tricarboxylic acid cycle and partially the respiratory chain.

been carried out with Sprague–Dawley rats *(14–18,25–27)*. Young animals (1 mo) are very resistant and up to 100 mg/kg is necessary to produce striatal lesions. In young adult animals (3–4 mo), a single injection of 25–30 mg/kg can be sufficient to produce cerebral lesions in almost all animals with a substantial rate of lethality in the following 12 h. A single injection of 20 mg/kg rarely produces lesions. However, 20 mg/kg/d for 2 or 3 d produces acute neurological signs of toxicity and cerebral lesions in more than

50% of the animals a few hours after the last injection. To optimize survival, injections of 10 mg/kg can be made twice a day (separated at least by 8–12 h) for 2–3 d. In old animals (6 mo or more), injected doses can be slightly decreased to reduce lethality. The age dependence of 3-NPA neurotoxicity is directly related to its capacity to produce energy failure at the level of the striatum, and does not result from an increased cerebral biodistribution with age *(15)*.

For acute intoxication (one injection), the neurological abnormalities develop rapidly, from general uncoordination, drowsiness, and general weakness (animals being unresponsive) to hindlimb paralysis without rigidity to the final stage of intoxication, recumbency, and death *(13)*. In some instances, death occurs within few hours without obvious severe neurological impairment. In a less severe paradigm of intoxication (daily intraperitoneal injection for 5–10 d), these symptoms develop slightly more progressively although from one day to the next, with fatal outcome in a substantial proportion of animals *(17)*. The main characteristic of acute and subacute 3-NPA toxicity is the consistent demonstration of the preferential vulnerability of the striatum toward the toxin. Neurochemical and histological evaluation of animals acutely treated with 3-NPA showed that the resulting striatal lesions were severe, affecting almost the entire striatum and generally extended caudally to the pallidum *(11–13,17)*. Neurochemical and histological evaluation showed that these striatal lesions were associated with marked decreases in GABA, substance P, somatostatin, and neuropeptide Y levels whereas a remarkable sparing of the dopaminergic afferents was found consistent with an excitotoxic mechanism of degeneration *(14)*. Confirming this possibility, striatal lesions produced by acute 3-NPA treatment can be significantly reduced by prior decortication *(14)*. When looking at the relationship between the degree of neurological impairment and the severity of striatal lesions, a significant correlation can be found, in line with the view that most of the symptoms result form striatal degeneration *(17)*. However, a number of observations clearly show that acute and subacute 3-NPA poisoning in rats is not an appropriate model of HD. The degree of SDH inhibition (70–80%) at onset of symptoms after 3-NPA injection is much higher than the partial reduction (30–50%) in activity of complex II–III seen in HD. Within the striatal lesions, there is no sparing of interneurons and glial invasion is massive after a week of survival. There is no transition between the core of the lesion (no neurons left) and apparently normal tissue *(17)*. In addition, subacute and acute 3-NPA toxicity is associated with extrastriatal cerebral lesions, the pallidum, hippocampus, thalamus, and substantia nigra reticulata being often affected *(13,14,17)*. These extrastriatal lesions are always associated with striatal lesions. It seems that in the case of acute 3-NPA poisoning, many factors are involved in the development of the striatal lesion. The disruption of the blood–brain barrier (BBB) after 3-NPA injection had been initially pointed out by Gould and collaborators *(12)*. The authors concluded that BBB dysfunction (as seen by histochemical detection of albumin extravasation) may participate in the development of large striatal lesions. In fresh tissue, signs of hemorrhage could be often seen when animals were killed a few hours after the first occurrence of symptoms or the following day. This was reevaluated more recently in a study showing the presence of immunological markers in the striatal lesions in animals intoxicated with 15 mg/kg/d of 3-NPA (i.p.) *(24)*. Acute 3-NPA injection produces in addition vasodilatation (with paradoxical hypertension) *(130)*, blood hyperoxygenation, and substantial increase in methemogobin levels as a result of nitrite production *(10,12)*.

Therefore, it appears that acute 3-NPA intoxication is not a suitable phenotypic model of HD even though this model may be particularly interesting for studying the mechanisms of neuronal death associated with acute mitochondrial dysfunction or for testing new neuroprotective strategies.

5.4. Chronic Administration of 3-NPA Reproduces Many Aspects of HD

In acute 3-NPA poisoning in rats, extrastriatal lesions are always associated with striatal lesions, suggesting a hierarchy of susceptibility, the striatum being the most vulnerable cerebral region. Consistent with this view, whatever the regimen of 3-NPA poisoning in rats, small striatal lesions are never associated with extrastriatal lesions. Thus there is the possibility that low-grade chronic impairment in energy metabolism using 3-NPA may lead to selective striatal lesions in rats, providing an adequate model of HD.

Chronic intoxication with 3-NPA was initially produced in adult rats with low doses (10–12 mg/kg/d) for 1 mo using osmotic minipumps *(14,15)*. This protocol of intoxication presents the advantage of steadily delivering the toxin, thus avoiding peaks of 3-NPA blood concentration which likely occurs after interperitoneal injection of the toxin. This steady delivery of 3-NPA closely reproduces the chronic impairment in energy metabolism seen in HD *(16)*. The main characteristics of this model are summarized in the following paragraph.

Chronic 3-NPA intoxication in rats leads to selective striatal lesions associated with motor abnormalities which are milder than those produced by acute intoxication. Although substantial interanimal variability is seen in the neurological and neuropathological outcomes, a certain proportion of 3-NPA-treated animals showed striking similarities to HD. Approximately 35–50% of rats chronically treated with 3-NPA show subtle bilateral and symmetrical striatal lesions. These lesions can be easily detected with conventional staining such as cresyl violet as palor in the striatum (Fig. 2). Within the lesion, cytochrome oxidase histochemistry is also profoundly reduced. In its rostral part, the lesion is localized in the dorsolateral aspect of the caudate-putamen, and in its caudal part it affects the most ventral part of the dorsolateral quadrant of the striatum. The histological nature of the lesion seen at high magnification differs radically from that obtained after acute intraperitoneal injection of the toxin. The lesion is not sharply bordered, but diffuse, and the loss of neuronal cells progressively increases from the untouched striatum to the center (core) of the lesion. Within the lesion, obvious neuronal loss can be seen (approx 30–35% neuronal loss) by Nissl stain with moderate gliosis. However, large cholinergic interneurons are preserved. NADPH-diaphorase histochemistry reveals that medium-sized aspiny interneurons are also spared within the lesion *(14,17)* (Fig. 3). Cell counts of NADPH-diaphorase-positive interneurons and Nissl-stained neurons confirmed a relative sparing of NADPH-diaphorase neurons within the striatal lesions, replicating the histological characteristics of HD. Interestingly, we found that during the time course of 3-NPA poisoning NADPH-diaphorase-positive interneurons showed morphological abnormalities at the onset of symptoms *(131)*. In symptomatic animals, dendritic varicosities were seen, and perikaryal swelling was evident. Similar morphological abnormalities had been reported in the HD

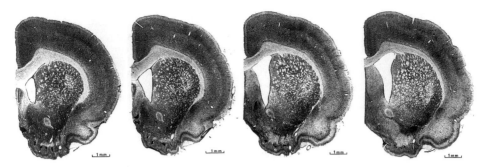

Fig. 2. Typical lesion of the dorsolateral striatum in a rat chronically treated with 3-NPA. Four Nissl-stained coronal sections of a rat treated chronically treated with 3-NPA (10 mg/kg/d for 1 mo). A palor can be seen in the lateral striatum corresponding to an area of neuronal loss and gliosis (*see* Fig. 3). No obvious extrastriatal lesions could be found in animals presenting such a striatal lesion.

striatum *(54,132)*. Immunohistochemistry for tyrosine hydroxylase showed that the dopamine terminals were essentially unaffected in the lesion *(15)*. *In situ* hybridization studies confirmed that chronic treatment with low doses (12 mg/kg/d) of 3-NPA using osmotic pumps leads to discriminant neuronal loss within the striatum: a marked decrease in met-enkephalin and substance P mRNA was found whereas somatostatin mRNA level was preserved *(28)*. Tyrosine hydroxylase mRNA content in the substantia nigra also remains unaffected. Alteration of glial cells has been less precisely studied. Increases in glial fibrillary acidic protein (GFAP) immunoreactivity were found in the vicinity of and within the striatal lesions *(14,17)*, whereas no change in GFAP mRNA was found *(28)*. Interestingly, chronic 3-NPA treatment produces certain morphological abnormalities of medium-sized spiny neurons highly reminiscent of HD *(14,15)*. These neurons, seen by Golgi staining, present a number of alterations of the dendrites, including increased dendritic spine density, recurved distal segments, and aberrant outgrowth of new branches. These signs of dendritic proliferation and outgrowth are corroborated by the increased expression of N-CAM immunoreactivity, a cell surface protein expressed during neuronal differentiation *(14,17)*. Whether these morphological abnormalities are simply the consequence of compensatory mechanisms taking place after degeneration of neighboring neurons or result from an intrinsic response of the neurons to chronic energy impairment is not known. Such abnormalities have been found after local excitotoxic lesions, favoring the former possibility *(3)*.

Behavioral studies indicate the existence of interesting similarities between HD symptoms and the motor abnormalities induced by 3-NPA in rats. Two types of studies have been conducted. One consists in studying the spontaneous behavior of the animals during their nocturnal phase. The other consists in detecting abnormalities in voluntary movement during a given motor task. The former has been performed in animals intoxicated with 3-NPA using repeated intraperitoneal injections (10 mg/kg every 4 d for 28 d), the latter in rats intoxicated by osmotic pumps implanted subcutaneously.

Quantitative analysis of spontaneous locomotor behavior of 3-NPA-treated animals was studied using the Digiscan system *(26,27,133)*. Results indicate that 3-NPA-treated animals have an early phase of hyperkinesia (first and second week of treatment) which

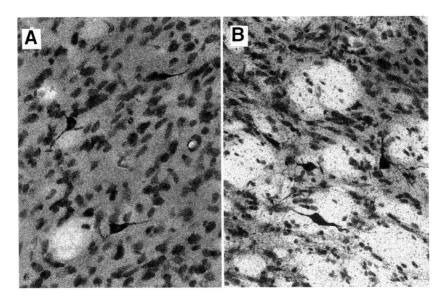

Fig. 3. Histological characteristics of the striatal lesions produced by chronic administration of 3-NPA. Each image represents a field of view at high magnification (objective ×20) in the dorsolateral striatum. Sections were reacted for NADPH-diaphorase histochemistry and counterstained with cresyl violet. (**A**) Section is from control animals. (**B**) Section is from 3-NPA-treated animals presenting dystonia and bradykinesia. Note the neuronal loss and gliosis in (**C**) and (**D**) and the relative sparing of NADPH-diaphorase-positive interneurons.

is followed by a phase of hypokinesia *(26)*. If 3-NPA intoxication is stopped during the hyperkinetic phase, symptoms persist. This interesting observation is reminiscent of the progression of HD symptomatology. Early motor symptoms are characterized by hyperkinetic behavior as a result of choreic movements, even though bradykinesia can be detected in certain voluntary movements. The disease progressing, bradykinesia remains, chorea decreases in severity, and incidence and dystonias are more prominent. Dystonia accompanied by rigidity is the major component of late-stage HD symptomatology *(1)*.

Neurological observation of rats chronically treated with 3-NPA shows a progression of symptom severity *(14,17)*. In animals responsive to 3-NPA treatment, the initial detectable symptoms consist of incoordination with a wobbling gate. These signs are usually followed by the appearance of dystonic postures of the hindlimbs. These initial motor deficits can be subtle and appear more readily detectable once the animal performs a voluntary motor task (such as staying on an inclined board or climbing the wall of its home cage). Initiation of movement can be erratic and the balance of the animal substantially affected (equilibrium). Then animals can evolve toward a more dramatic symptomatology in which hindlimb paralysis and often recumbency (lying on one side) are cardinal symptoms. At this stage further treatment can be fatal if the 3-NPA treatment is not stopped. After osmotic pump removal, the symptoms of recumbent animals evolve with a tendency to recovery. Animals regain responsiveness, moving themselves by using their forelimbs. Hindlimb paralysis with complete loss of muscle tone evolves toward a complete hindlimb extension with rigidity. A few weeks after chronic

3-NPA treatment is stopped, animals partially recover and very specific motor symptoms persist for more than 4 mo. In severely affected animals, these symptoms consist of marked bradykinesia and dystonia, with abnormal paw positioning easily detectable using ink footprints. In most animals, however, spontaneous symptoms are very subtle and cannot be scored objectively using a neurological scale. For this reason, we developed a quantitative approach to detect the extent of motor deficits in these animals *(17)*. Rats can be easily trained to cross an elevated board so that they perform a linear trajectory as fast as they can. Video recording of each run allows by image analysis (centroid tracking) the precise determination (every 40 ms) of the mean and peak tangential velocity, lateral velocity, step size, and acceleration of each animal (Fig. 4). Using this approach we found that animals chronically treated with 3-NPA for a month, and then remaining alive for 3 mo, showed decreased mean and peak velocities, indicating bradykinesia. The length of animal steps (inferred from acceleration peaks) shows a significant reduction. Plots of velocity showed "irregularities," suggestive of a (staccato-like) saccaded walk. Determination of lateral velocities confirmed the presence of gait abnormalities in 3-NPA-lesioned animals, consisting of a wobbling gait with an enlarged base as compared to age-matched control littermates. These abnormalities are reminiscent of HD *(1)*. Indeed, chorea results in general clumsiness in tasks requiring the use of the hands and arms while disturbances in the lower limbs appear as gait abnormalities. The walk of affected patients can be described as slow, stiff, and unsteady. These gait abnormalities have been quantitatively studied by an ultrasound transducer, showing that step size and walking speed were decreased *(134)*. Similarly, bradykinesia is a common feature of HD *(135)*. We also found that in the animals studied 3 mo after termination of 3-NPA intoxication, the severity of motor abnormalities significantly correlated with the degree of neuronal loss within the dorsolateral aspect of the striatum. In animals killed during the time course of 3-NPA intoxication at onset of symptoms the severity of motor impairment was also correlated to the degree of neuronal loss *(131)*.

5.5. Limitations and Current Drawback of the 3-NPA Rat Model of HD

The limitation of the 3-NPA rat model of HD results from two major characteristics. First, the anatomical organization of the striatum is markedly different in rodents and humans. In rats, the striatum is globally a homogeneous structure whereas in the primate the striatum is divided into the caudate nucleus and putamen. Second, motor behavior in rats obviously differs markedly from that of primates. Behavioral studies of nonhuman primates with ibotenate-, kainate-, or quinolinate-induced striatal lesions showed that after apomorphine injection, the repertoire of abnormal movement could be very close to that seen in HD patients *(40,88–90)*. Hyperkinetic abnormal movements resembling chorea and dyskinesia have never been observed in rats with excitotoxin lesions after apomorphine or amphetamine administration. Rats chronically treated with 3-NPA did not show clearly identifiable dyskinetic movements resembling chorea even though an hyperlocomotor activity has been reported early in the time course of intoxication *(26)*, and the presence of dystonia, bradykinesia, and gait abnormalities has been clearly identified *(17)*. Thus, the dyskinetic component of HD symptomatology may be part of a motor repertoire that can only be expressed in

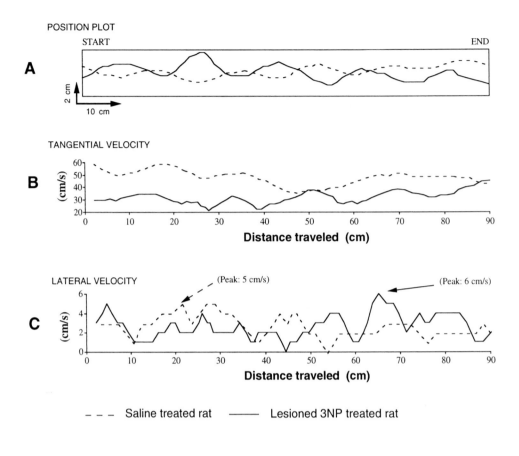

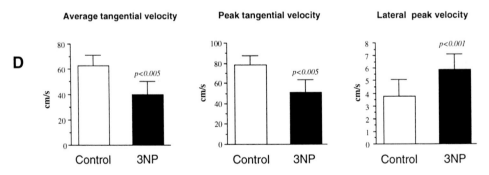

Fig. 4. Motor abnormalities in rats chronically intoxicated with 3-NPA. Animals were trained to cross an elevated board (120 cm long, 7 cm wide). After 5 d of training, test runs were recorded using a video camera fixed on the ceiling. Top view images from these runs were analyzed off-line using an automatic video-based motion tracking and analysis system. The animal center of mass was estimated every 40 ms with spatial resolution of 2 mm/pixel. **(A)** Raw data showing the position of a control animal (*dashed line*) and a 3-NPA-treated animal (*solid line*) while they cross the elevated board. Note the unsteady gait of the 3-NPA-treated rat. **(B)** Tangential velocity of a control rat and a 3-NPA-treated animal during their runs corresponding to position plots shown in (A). Note that the 3-NPA-treated animals has a lower velocity as compared to that of the control rat, indicating bradykinesia. **(C)** Lateral velocity of the control and 3-NPA-treated animals during their runs corresponding to position plots shown in (A). Note that the lateral velocity determined in the 3-NPA-treated animal is increased as compared to control, consistent with the presence of dystonia and wobbling gait. **(D)** Histograms showing the mean (± standard deviation) kinetic parameters determined in control animals and 3-NPA-treated animals with neuronal loss in the striatum.

primates. In line with this, chronic administration of 3-NPA in primates leads to apomorphine-inducible and spontaneous dystonia and choreiform movements *(136,137)*.

Another current limitation of the 3-NPA rat model of HD is the substantial interanimal variability toward 3-NPA toxicity. However, preliminary observations suggest that Lewis rats may be much more suitable than Sprague–Dawley rats *(129)*. Whereas chronic 3-NPA treatment produces HD-like striatal lesions in 30–35% of Sprague–Dawley rats, it leads to striatal lesions in nearly 100% of Lewis rats. Very recent data obtained in our laboratory suggested that the 3-NPA model of HD in Lewis rats is particularly valuable for studies testing new therapeutic strategies for which homogeneous groups of animals are required.

6. MECHANISM UNDERLYING THE NEUROTOXICITY OF CHRONIC 3-NPA

6.1. Does Chronic Administration of 3-NPA Reproduce the Partial Blockage of Complexe II–III Seen in HD?

In synaptosomal preparations, it was shown that the most rapid event produced by 3-NPA was a decrease in the phosphocreatine/creatine ratio and an increase in the lactate/pyruvate ratio. This was followed by a decrease in ATP/ADP and GDP/GTP ratios *(138)*. In living neurons in culture, 3-NPA neurotoxicity has been related to partial inhibition of SDH activity and significant decline in ATP cellular levels *(139–143)*.

In vivo, focal injection of 3-NPA into the striatum produces a marked increase in lactate as seen with NMR chemical shift imaging *(15,19)*. Freeze–clamp techniques also allowed detection of decreases in ATP level and increase in lactate concentrations after intrastriatal injection of 3-NPA *(14,15)*. An age-dependent increase in lactate concentration was detected in living rats by NMR chemical shift imaging in the striatum after systemic injection of 3-NPA *(14,19)*. Systemic injection of high 3-NPA doses produces an accumulation of succinate, GABA, and lactate as seen using NMR spectroscopy analysis of brain extracts from [1-^{13}C]glucose-injected mice *(144)*. Tsai et al. *(145)*, using NMR spectroscopy techniques on brain extracts from rats that received repeated injection of low doses (10 mg/kg/d for 1 wk) of 3-NPA, have shown that succinate, GABA, and lactate accumulate consistent with a blockade of the Krebs cycle. In addition, significant decreases in *N*-acetylaspartate (NAA) were found, supporting a general alteration of mitochondrial metabolism in neural cells. All these changes were found to be age dependent, markedly more pronounced in adult animals (4 and 8 mo old) than in young animals (1 mo old). No histological observation has been provided in this study, but this regimen of intoxication is likely to produce no obvious striatal damage in young animals, although mild neurological symptoms can be seen, such as incoordination. Thus, reduction in NAA concentrations may correspond to a state of neuronal dysfunction without obvious cell loss. In primates in vivo, NMR spectroscopy study of animals chronically treated with 3-NPA (8 wk of treatment) also showed early reduction in NAA and elevation of lactate in the striatum whereas no changes were found in the occipital cortex. Histological evaluation or NMR imaging of the animals confirmed that these changes preceded identifiable neuronal loss *(146)*.

The fact that a mitochondrial toxin produces neurochemical changes indicative of energy failure may seem hardly surprising. However, the main issue of these studies is that neurotoxicity can result from partial blockage of the energy metabolism machinery. They indicate that chronic and relentless blockade of the mitochondrial respiratory chain can produce depletion of energy stores associated progressively with symptom expression and neuronal suffering, and finally death. This situation is relevant for neurodegenerative diseases and particularly HD, for which the alteration of energy metabolism is only partial *(104,106)*. However, what is the precise relationship between the degree of energy impairment and the resulting neurotoxic effect? As a first attempt to answer this question, we assessed in vivo the relationship between the level of 3-NPA-induced SDH inhibition and the resulting neurological and neuropathological outcomes in rats *(16)*. The characterization of the effect of systemic injection of 3-NPA on the respiratory chain had been rarely evaluated. Only two studies performed by Gould et al. *(10,11)* reported that the activity of SDH in whole brain mitochondrial preparation was largely decreased (75–80%) after injection of toxic doses of 3-NPA in rodents. In our study, the nature, time course, and level of inhibition to produce striatal lesions was precisely reevaluated in the context of HD *(16)*. A quantitative histochemical method was developed to study the level of regional SDH inhibition resulting from intraperitoneal injection of 3-NPA or subcutaneous infusion using osmotic minipumps. This study showed that 3-NPA very rapidly entered the brain, and irreversibly inactivated SDH in the striatum to an extent similar to that in other brain regions. This work showed that the striatal lesions resulting from chronic intoxication with 3-NPA was associated with a 50–60% inhibition of SDH, mimicking the partial alteration of complex II–III observed in HD (Fig. 5). This further supports the hypothesis that the alteration of this complex may have a causal role in the etiology of the disease. This study also demonstrated that the preferential vulnerability of the striatum toward 3-NPA was not related to an increased inhibition of SDH in the striatum as compared to the cerebral cortex. This preferential vulnerability of the striatum seems rather to result from an intrinsic inability of the striatum to resist mild energy impairment. The exact mechanism underlying this propensity of the striatum to be " deregulated " by energy compromise is unknown.

6.2. 3-NPA-Induced Impairment in Energy Metabolism Activates the Excitotoxic Cascade

In vitro, activation of the glutamate receptors seems to play a crucial role in the neurotoxicity produced by 3-NPA (Fig. 6). Ludolph et al. *(139)* were the first to demonstrate that impairment in energy metabolism produced by 3-NPA was not solely responsible for neuronal degeneration but also involved glutamate toxicity. In frontal cortex explants, the toxicity of 3-NPA was found significantly reduced by the NMDA receptor antagonist MK801, and MK801 plus the non-NMDA receptor antagonist CNQX, although CNQX alone was not neuroprotective. In the presence of MK801, neurodegeneration produced by 3-NPA was decreased but ATP concentration remained depleted, showing that the sole blockade of NMDA receptors (and presumably the resulting excitotoxic cascade) was sufficent to prevent neuronal death *(141)*. A similar prevention or delay of 3-NPA neurotoxicity by glutamate receptor antagonists in dissociated neuronal culture was found by others *(140,142,143,147,148)*. However, when glutamate receptors are fully blocked by antagonists, 3-NPA exposure is still associated with a substantial delayed neuronal death (48 h) *(140)*. The morphological and molecular abnormalities of dying neurons are reminiscent of apoptosis *(140,149)*. Thus

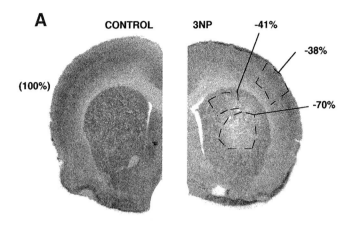

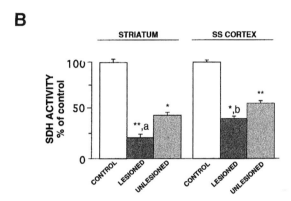

Fig. 5. Partial inhibition of SDH by 3-NPA is sufficient to produce striatal degeneration. SDH activity was revealed by incubating frozen sections with succinate (0.05 M) as a specific substrate and nitrobluetetrazolium (NBT) as electron acceptor, which eventually forms formazan. Regional percents of inhibition are determined by image analysis. Nonspecific staining is determined on adjacent sections incubated with NBT without succinate. **(A)** Digitized images of rat brain sections stained for SDH histochemistry showing partial decrease of SDH activity after 3-NPA administration (*right*) as compared to control animal (*left*). **(B)** Histograms showing the levels of SDH activity in striatum and cortex in animals chronically treated with 3-NPA (12 mg/kg/d for 7 d) as compared to control animals. Note that "lesioned" animals (showing striatal lesions) have higher SDH inhibition as compared to "unlesioned" 3-NPA treated animals. Results are means ± standard error of the mean. **$p < 0.001$; *$p < 0.01$ as compared to control levels. a, $p < 0.01$; b, $p < 0.05$ as compared to unlesioned 3-NPA-treated animals.

excitotoxicity and apoptosis may coexist during 3-NPA-induced degeneration as has been proposed for glutamate-induced excitotoxicity *(63,150–153)*.

Several observations strongly suggest that in vivo the striatal lesions produced by systemic injection of 3-NPA result from an indirect excitotoxic mechanism of cell death. First, the histological characteristics of lesions resulting from chronic poisoning with 3-NPA resemble that seen after focal injection of quinolinate *(14,15,17)*. Lesions

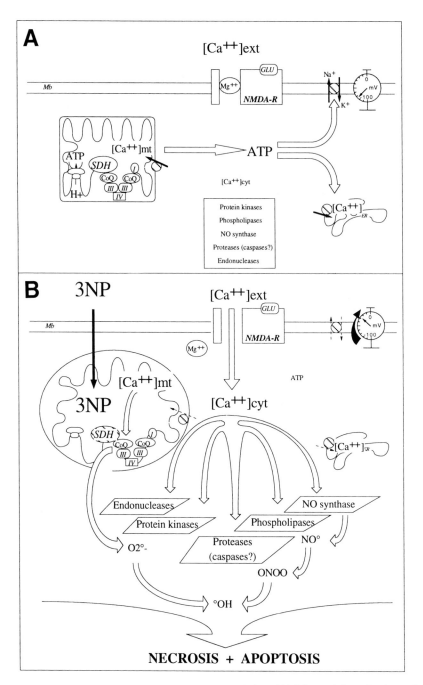

Fig. 6. Proposed mechanisms of 3-NPA neurotoxicity. **(A)** Normal functioning of the cell. Mitochondrial energy metabolism provides ATP which is used by ionic pumps to maintain membrane potential. Mitochondria can buffer large quantities of Ca^{2+} which enters through Ca^{2+} channels. **(B)** 3-NPA produces energy impairment which decreases ATP levels. Dysfunction of ATPase due to decreased ATP availability produces partial membrane depolarization, leading to the relief of the voltage-dependent Mg^{2+} block of the NMDA receptor channel. This produces massive entry of Ca^{2+} which activates Ca^{2+}-dependent enzymes such as NO synthase, phospholipases, proteases, and endonucleases and leads to increased production of free radicals. Cytoplasmic Ca^{2+} overload, NO production, and radical oxygen species in turn may disrupt mitochondrial function, leading to a vicious cycle. The final outcome of 3-NPA toxicity may be apoptosis or necrosis, depending on the cell considered, the severity of energy impairment, and the local environment (trophic factors, glutamate concentrations).

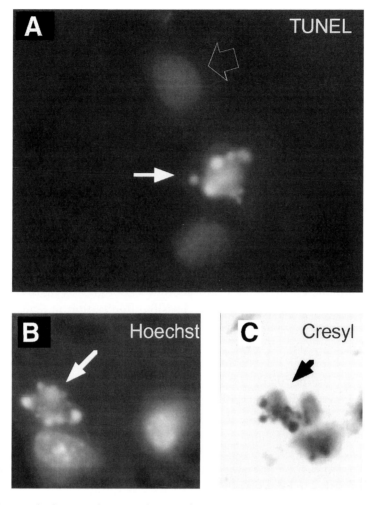

Fig. 7. Apoptotic features in the striatum of rats receiving chronic 3-NPA treatment. **(A)** Digitalized images of neuronal nuclei positive for TUNEL labeling with dUTP-fluorescein in Lewis rats chronically intoxicated with 3-NPA. Approximately 15–30% of striatal neurons were labeled. Two types of labeling were observed: diffuse labeling was the most frequently (75%) observed (A, *open arrow*). Intense labeling with apoptotic bodies was seen less frequently (25%) (A, *white arrow*). Nuclei with apoptotic bodies were also seen on sections stained with the DNA-binding dye Hoechst 33258 (**B**, *white arrow*) and cresyl violet (**C**, *black arrow*).

can be blocked by surgical removal of the corticostriatal glutamatergic afferents *(14)*. Riluzole, which decreases glutamate release, significantly reduces 3-NPA-induced striatal degeneration and associated neurological impairment *(18)*. Systemic injection of a subtoxic dose of 3-NPA potentiates the toxicity of quinolinate *(154)*. Finally, systemic injection of a toxic dose of 3-NPA increases the opening of the NMDA receptor as seen by increased [^{3}H]MK801 binding in the striatum *(28)*.

The possibility that apoptotic pathways may be activated in vivo by 3-NPA has been suggested recently. Young animals injected with high doses of 3-NPA showed positive TUNEL nuclei within the striatum with DNA laddering on agarose gel electrophoresis typical of apoptosis *(155)*. Because this may be inherent to young animals (physiological apoptotis occurs in the striatum in neonates), we reassessed this issue in adult Lewis rats during chronic 3-NPA intoxication (Fig. 7). Histological evaluation of the striata of symptomatic rats showed that approx 26% of striatal neurons had DNA strand breaks and 6% had nuclei with apoptotic bodies (Brouillet, Ménétrat, Altairac, and Hantraye, *unpublished observations*).

It is probable that abnormal activation of NMDA receptors resulting from 3-NPA-induced energy impairment leads in vivo to increased production of free radicals, in particular nitrogen radicals. Consistent with this, systemic administration of 3-NPA produces increased production of the peroxynitrite byproduct, 3-nitrotyrosine, of 8-hydroxy-2-deoxyguanosine (a marker of oxidative DNA damage), and of dihydroxy-benzoic acids (markers of hydroxyl radical production) in the striatum *(25)*. Blockade of the neuronal form of NOS by the antagonist 7-nitroindazol decreases the volume of the striatal lesions produced by systemic administration of 3-NPA. This was accompanied by a normalization of the 3-NPA-induced increases in concentrations of 3-nitrotyrosine and dihydroxybenzoic acid. It is likely that other components of the excitotoxic and apoptotic cascade may also play a major role such as phospholipases, proteases, or endonucleases. Demonstration of the importance of these components in 3-NPA neurotoxicity would suggest their possible implication in neurodegenerative diseases.

7. FUNDAMENTAL AND THERAPEUTIC IMPLICATIONS

7.1. A Model of Choice for Better Understanding Cell Death Mechanisms In Vivo

In vitro studies in cell cultures have permitted the identification of many components involved in excitotoxic and/or apoptotic cell death, but the relative importance of these components during a degenerative process evolving over weeks or months cannot be determined in vitro. The study of the mechanisms underlying the pathological neuronal death that may be involved in HD also requires appropriate in vivo models. The 3-NPA rat model of striatal degeneration reproduces not only various aspects of the HD, but also its dynamic aspect, including the delayed onset of symptoms and the aggravation of these symptoms from a hyperkinetic phase toward a hypokinetic phase, with bradykinesia, gait abnormalities, and dystonia. The severity of these symptoms is correlated with the severity of neuropathological changes in the striatum. The characteristics of the striatal degeneration are reminiscent of HD, in particular the sparing of dopaminergic afferents and interneurons and the presence of apoptotic features. This model also replicates many neurochemical and metabolic abnormalities of the disease, such as the decreased activity of complex II–III and brain lactate increases. Therefore, this model seems particularly relevant for the study of the mechanism of progressive neuronal death in vivo.

It is tempting to speculate from the number of behavioral and histological similarities that exist between HD and 3-NPA neurotoxicity that the 3-NPA rat model, in parallel with the studies related to the molecular biology of huntingtin, may help to uncover the possible mechanisms that lead to striatal atrophy in HD. For instance, two characteristics of HD are far from being elucidated: the age-dependent onset of the disease and the preferential vulnerability of the striatum. Although there is a clear inverse correlation between the number of repeats and the age at disease onset, it remains obscure why the disease generally is delayed to adulthood *(67)*. Similarly, the preferential vulnerability of the striatum cannot be explained simply by a preferential enrichment of the striatum in huntingtin *(156–158)*.

We showed that 3-NPA neurotoxicity has an age-dependent pattern of expression as for that of HD mutation and that this expression occurs first within the striatum. One possibility is that this delay in the onset of expression results from the normal decrease in efficiency in mitochondrial enzyme activity associated with aging *(6,159,160)*. A number of markers indicating abnormalities of oxidative metabolism increase with normal aging. Brain mitochondrial complexes I and IV showed significant decline with age, although complex II–III remains relatively unchanged *(161,162)*. Interestingly, the age-dependent increases in mitochondrial DNA deletions that may be indicative of mitochondrial dysfunction are more pronounced in the striatum as compared to other brain regions *(163,164)*. The results obtained on toxicity of systemically delivered 3-NPA strongly support the view that the striatum is indeed intrinsically highly vulnerable to energy impairment. Thus, the addition (or synergy) of the toxicity of the HD mutation (or 3-NPA administration), plus the age-related decline in oxidative metabolism, would reach a threshold that may generate sufficient perturbations for triggering neuronal death. This threshold of additive (or synergistic) toxicity may be achieved earlier in the striatum as compared to other brain regions.

That 3-NPA can lead to excitotoxic cell death and apoptosis may also provide clues concerning the cellular mechanisms of cell death in HD. Apoptosis as been proposed as a mechanism of cell death in HD *(63,64)*. Molecular biology studies of HD have already reinforced this possibility *(68)*. Huntingtin is a substrate of caspase 3, and polyglutamine expansion increased the cleavage of huntingtin by this cysteine protease involved in apoptosis. In addition, huntingtin interacts with GAPDH *(69)*, a key enzyme of glycolysis but also a protein with a role in neuronal apoptosis that has been demonstrated recently *(70,71)*. Increased cleavage of huntingtin with polyglutamine expansion by caspase 3 and its increased interaction with GAPDH may activate apoptotic pathways or " deinhibit " apoptotic pathways to an extent sufficient for clinical expression and neurodegeneration. It is not inconceivable that mitochondrial abnormalities associated with normal aging directly sensitize neurons toward apoptotic processes. The role of mitochondria in apoptosis is likely to be crucial *(165)*. Synergistically, apoptosis may at least in part be triggered or facilitated by the indirect activation of the excitotoxic cascade resulting from energy impairment. The 3-NPA rat model of progressive striatal degeneration may be a model of choice for the study of the probably complex interplay that may exist between mitochondrial dysfunction, apoptosis, and excitotoxicity.

7.2. A Model for Studying New Therapeutic Strategies

A number of potential therapeutic strategies could be examined using the 3-NPA rat model of HD. Any strategy aimed at blocking the cascade of events downstream of mitochondrial blockade and subsequent NMDA receptor activation can be tested with this model of striatal degeneration.

A straightforward approach to slowing neurodegeneration involving NMDA receptors consists of blocking the NMDA receptor using pharmacological agents (competitive or noncompetitive antagonists) or decreasing the concentrations of ambient glutamate. The latter strategy has been tested in the 3-NPA rat model of HD with riluzole, a compound that has been shown to decrease glutamate release. Significant neuroprotection was found associated with significant beneficial effects on the neurological impairment resulting from chronic 3-NPA treatment *(18)*. Another way of limiting the deleterious effects of NMDA receptor activation is to block the downstream enzymes activated by intracellular calcium increases. NOS is one of these enzymes that play a major role in excitotoxicity. The efficacy of a blockade of the neuronal form of NOS by pharmacological compounds have been tested in the 3-NPA rat model. The nNOS inhibitor 7-nitroindazole has showed significant neuroprotective effects against subacute 3-NPA intoxication in rats *(25)*.

Another possibility is to limit the extent of energy depletion produced by the mitochondrial toxin or increase the yield of electron transfer through the respiratory chain. The advantage of such a strategy is that it may not interfere with normal neural transmission processes in contrast to glutamate receptor antagonists. The toxicity of mitochondrial toxins given intrastriatally such as aminooxyacetic acid and malonate can be substantially decreased using this approach *(23,166–168)*. The most convincing experimental evidence that such a strategy may be promising is the recent finding that food supplementation with creatine is neuroprotective against mitochondrial toxins in rats *(22)*. Feeding of animals with creatine increases brain concentrations of phosphocreatine, the major pool of high-energy phosphorylated compounds. Animals supplemented with creatine are much less vulnerable to the toxicity of malonate injected into the striatum and 3-NPA given systemically.

Because apoptosis may be part of the complex scheme of mechanisms leading to 3-NPA-induced cell death, the 3-NPA rat model could be helpful in testing the neuroprotective effects of agents that could block in vivo the activation of certain enzymes involved in apoptosis such as caspases or endonucleases *(169)*.

Given that neurotrophic factor deprivation (withdrawal) in certain types of neurons leads to apoptosis, it is popularly assumed that supplementation of neuronal cells with neurotrophic factors may have neuroprotective effects. In this context, it has been shown that neurotrophic factors such as CNTF are neuroprotective against NMDA receptor mediated neurodegeneration *(170,171)*. It may be interesting to study the efficacy of these neurotrophins in the 3-NPA rat model of striatal degeneration. A pioneering study showed that fibroblasts transfected to express NGF and transplanted into the lateral ventricles in rats produced a significant protection of the striatum against acute 3-NPA toxicity *(20,21)*. Techniques to deliver these factors into the brain, such as viral transfection or encapsulation of genetically engineered cells, are in progress. The 3-NPA rat model may be suitable to optimize the delivery techniques and further help to elucidate the mechanisms underlying the neuroprotective effects of the factors tested in vivo during progressive degeneration.

8. CONCLUSION

The study of the mechanisms underlying pathological neuronal death requires appropriate in vivo models. This is particularly true for chronic, progressive processes of cell death underlying neurodegenerative diseases such as HD. In this context, the new rat model of HD using chronic systemic administration of 3-NPA is particularly interesting. It helped to determine to which extent an alteration of energy metabolism could play a role in the pathogenesis of HD. Although it may present limitations as compared to the 3-NPA primate model of HD in terms of anatomical and behavioral characteristics it may be particularly useful to examine the mechanisms of cell death resulting from chronic energy impairment and to test new therapeutic strategies.

ACKNOWLEDGMENTS

We thank our colleagues who were involved in the study of 3-NPA neurotoxicity in rats: Robert J. Ferrante, F. Condé, B. Jenkins, M. C. Guyot, S. Altairac, H. Ménétrat, S. Ouary, V. Mittoux, R. Henshaw, J. Schulz, and R. Matthews.

REFERENCES

1. Harper, P. S. (1991) *Huntington's Disease* (Harper, P. S., ed.), W. B. Saunders, London.
2. Sharp, A. H. and Ross, A. R. (1996) Neurobiology of Huntington's disease. *Neurobiol. Dis.* **3,** 3–15.
3. DiFiglia, M. (1990). Excitotoxic injury of the neostriatum is a model for Huntington's disease. *TINS* **13,** 286–289.
4. Albin, R. L. and Greenamyre, J. T. (1992) Alternative excitotoxic hypotheses. *Neurology* **42,** 733–738.
5. Beal, M. F. (1992) Does impairment of energy metabolism result in excitotoxic neuronal death in neurodegenerative illness? *Ann. Neurol.* **31,** 119–130.
6. Beal, M. F. (1995b) Aging, energy and oxidative stress in neurodegenerative diseases. *Ann. Neurol.* **38,** 357–366.
7. Martin, J. J., Van de Vyver, F. L., Scholte, H. R., Roodhooft, A. M., Ceuterick, C., Martin, L., and Luyt-Houwen, I. E. M. (1988) Defect in succinate oxidation by isolated muscle mitochondria in a patient with symmetrical lesions in the basal ganglia. *J. Neurol. Sci.* **84,** 189–200.
8. Bourgeron, T., Rustin, P., Chretien, D., Birch-Machin, M., Bourgeois, M., Viegas-Pequignot, E., Munnich, A., and Rotig, A. (1995) Mutation of a nuclear succinate dehydrogenase gene results in mitochondrial respiratory chain deficiency. *Nat. Genet.* **11,** 144–149.
9. He, F., Zhang, S., Qian, F., and Zhang, C. (1995) Delayed dystonia with striatal CT lucencies induced by a mycotoxin (3-nitropropionic acid). *Neurology* **45,** 2178–2183.
10. Gould, D. H. and Gustine, D. L. (1982) Basal ganglia degeneration, myelin alterations, enzyme inhibition induced in mice by the plant toxin 3-nitropropionic acid. *Neuropathol. Appl. Neurobiol.* **8,** 377–393.
11. Gould, H., Wilson, M. P., and Hamar, D. W. (1985) Brain enzyme and clinical alterations induced in rats and mice by nitroaliphatic toxicants. *Toxicol. Lett.* **27,** 83–89.
12. Hamilton, B. F. and Gould, D. H. (1987a) Correlation of morphological brain lesions with physiological alterations and blood–brain barrier impairment by 3-nitropropionic acid toxicity in rats. *Acta Neuropathol. (Berl.)* **74,** 67–74.
13. Hamilton, B. F. and Gould, D. H. (1987b) Nature and distribution of brain lesions in rats intoxicated with 3-nitropropionic acid: a type of hypoxic (energy deficient) brain damage. *Acta Neuropathol. (Berl.)* **72,** 286–297.
14. Beal, M. F., Brouillet, E., Jenkins, B., Ferrante, R., Kowall, N., Miller, J., Storey, E., Srivastava, R., Rosen, B., and Hyman, B. T. (1993a). Neurochemical and histologic characterization of the striatal lesions produced by the mitochondrial toxin 3-nitropropionic acid. *J. Neurosci.* **13,** 4181–4192.

15. Brouillet, E., Jenkins, B., Hyman, B., Ferrante, R. J., Kowall, N. W., Srivastava, R., Roy, D. S., Rosen, B., and Beal, M. F. (1993a) Age dependent vulnerability of the striatum to the mitochondrial toxin 3-nitropropionic acid. *J. Neurochem.* **60,** 356–359.

16. Brouillet, E., Guyot, M.-C., Mittoux, V., Altairac, S., Condé, F., Palfi, S., and Hantraye, P. (1998a) Partial inhibition of brain succinate dehydrogenase by 3-nitropropionic acid is sufficent to initiate striatal degeneration in rat. *J. Neurochem.* **70,** 794–805.

17. Guyot, M. C., Hantraye, P., Dolan, R., Palfi, S., Mazière, M., and Brouillet, E. (1997a) Quantifiable bradykinesia, gait abnormalities and Huntington's disease-like striatal lesions in rats chronically treated with 3-nitropropionic acid. *Neuroscience* **79,** 45–56.

18. Guyot, M. C., Palfi, S., Stutzmann, J. M., Mazière, M., Hantraye, P., and Brouillet, E. (1997b) Riluzole protects from motor deficits and striatal degeneration produced by systemic 3-nitropropionic acid intoxication in rats. *Neuroscience* **81,** 141–149.

19. Jenkins, B. G., Brouillet, E., Chen, Y. C., Storey, E., Schulz, J. B., Kirschner, P., Beal, M. F., and Rosen, B. R. (1996) Non-invasive neurochemical analysis of focal excitotoxic lesions in models of neurodegenerative illness using spectroscopic imaging. *J. Cereb. Blood Flow Metab.* **16,** 450–461.

20. Frim, D. M., Simpson, J., Uhler, T. A., Short, M. P., Bossi, S. R., Breakfield, X. O., and Isacson, O. (1993) Striatal degeneration induced by mitochondrial blockade is prevented by biologically delivered NGF. *J. Neurosci. Res.* **35,** 452–458.

21. Galpert, W. R., Matthews, R. T., Beal, M. F., and Isacson, O. (1996) NGF attenuates 3-nitrotyrosine formation in a 3NP model of Huntington's disease. *NeuroReport* **7,** 2639–2642.

22. Matthews, R. T., Yang, L., Jenkins, B. G., Ferrante, R. J., Rosen, B. R., Kaddurah-Daouk, R., and Beal, M. F. (1998a) Neuroprotective effects of creatine and cyclocreatine in animal models of Huntington's disease. *J. Neurosci.* **18,** 156–163.

23. Matthews, R. T., Yang, L., Browne, S., Baik, M., and Beal, M. F. (1998b) Coenzyme Q10 administration increases brain mitochondrial concentrations and exerts neuroprotective effects. *Proc. Natl. Acad. Sci. USA* **95,** 8892–8897.

24. Nishino, H., Shimano, Y., Kumazaki, M., and Sakurai, T. (1995) Chronically administered 3-nitropropionic acid induces striatal lesions attributed to dysfunction of the blood–brain barrier. *Neurosci. Lett.* **186,** 161–164.

25. Schulz, J..B., Matthews, R. T., Jenkins, B. G., Ferrante, R. J., Siwek, D., Henshaw, D. R., Cipolloni, P. B., Meccoci, P., Kowall, N. W., Rosen, B. R., and Beal, M. F. (1995b) Blockade of neuronal nitric oxide synthase protects against excitotoxicity in vivo. *J. Neurosci.* **15,** 8419–8429.

26. Borlongan, C. V., Koutousis, T. K., Freeman, T. B., Cahill, D. W., and Sanberg, P. R. (1995a) Behavioral pathology induced by repeated systemic injections of 3-nitropropionic acid mimics the motoric symptoms of Huntington's disease. *Brain Res.* **697,** 254–257.

27. Borlongan, C. V., Koutousis, T. K., Randall, T. S., Freeman, T. B., Cahill, D. W., and Sanberg, P. R. (1995b) Systemic 3-nitropropionic acid: behavioral deficits and striatal damage in adult rats. *Brain Res. Bull.* **36,** 549–556.

28. Wüllner, U., Young, A., Penney, J., and Beal, M. F. (1994) 3-Nitropropionic acid toxicity in the striatum. *J. Neurochem.* **63,** 1772–1781.

29. Kremer, B., Weber, B., and Hayden, M. R. (1992) New insights into the clinical features, pathogenesis and molecular genetics of Huntington's disease. *Brain Pathol.* **2,** 321–335.

30. Myers, R. H., Vonsattel, J. P., Stevens, T. J., Cupples, L. A., Richardson, E. P., Martin, J. B., and Bird, E. D. (1988) Clinical and neuropathologic assessment of severity in Huntington's disease. *Neurology* **38,** 341–347.

31. Vonsattel, J.-P., Myers, R. H., and Stevens, T. J. (1985) Neuropathological classification of Huntington's disease. *J. Neuropathol. Exp. Neurol.* **44,** 559–577.

32. Kowall, N., Ferrante, R. J., and Martin, J. B. (1987) Patterns of cell loss in Huntington's disease. *TINS* **10,** 24–29.

33. Beal, M. F., Ellison, D. W., and Martin, J. B. (1987) Inhibition in Huntington's disease. *J. Mind Behav.* **8,** 635–642.
34. Bird, E. D. and Iversen, L. L. (1977) Neurochemical findings in Huntington's chorea, in *Esaays in Neurochemistry and Neuropharmacology,* Vol. 1 (Youdim, M. B. H., Sharman, D. F., Lovenberg, W., and Lagnado, J. R., eds.), John Wiley & Sons, New York, pp. 177–195.
35. Buck, S. H., Burks, T. F., Brown, M. R., and Yamamura, H. I. (1981) Reduction in basal ganglia and sustantia nigra substance P levels in Huntington's disease. *Brain Res.* **209,** 464–469.
36. Aronin, N., Cooper, P. E., Lorenz, L. J., Bird, E. D., Sagar, S. M., Leeman, S. E., and Martin, J. B. (1983) Somatostatin is increased in the basal ganglia in Huntington's disease. *Ann. Neurol.* **13,** 519–526.
37. Beal, M. F., Ellison, D. W., Mazurek, M. F., Swartz, K. J., Malooy, J. R., Bird, E. D., and Martin, J. B. (1988a) A detailed examination of substance P in pathologically graded cases of Huntington's disease. *J. Neurol. Sci.* **84,** 51–61.
38. Richfield, E. K., Maguire-Zeiss, K. A., Vonkeman, H. E., and Voorn, P. (1995) Preferential loss of preproenkephalin versus preprotachykinin neurons from the striatum of Huntington's disease patients. *Ann. Neurol.* **38,** 852–861.
39. Spokes, E. G. S. (1980) Neurochemical alterations in Huntington's chorea: a study of postmortem brain tissue. *Brain* **103,** 179–210.
40. Storey, E. and Beal, M. F. (1993) Neurochemical substrates of rigidity and chorea in Huntington's disease. *Brain* **116,** 1201–1222.
41. Ferrante, R. J., Kowall, N. W., and Richardson, E. P., Jr. (1991) Proliferative and degenerative changes in striatal spiny-neurons in Huntington's disease: a combine study using the section-Golgi method and calbindin D28k immunochemistry. *J. Neurosci.* **11,** 3877–3887.
42. Goto, S., Hirano, A., and Rojas-Corona, R. R. (1989) An immunohistochemical investigation of the human neostriatum in Huntington's disease. *Ann. Neurol.* **25,** 298–304.
43. Kiyama, H., Seto-Ohshima, A., and Emson, P. C. (1990) Calbindin D28k as a marker for the degeneration of the striatonigral pathway in Huntington's disease. *Brain Res.* **525,** 209–214.
44. Seto-Oshima, A., Emson, P. C., Lawson, E., Mountjoy, C. Q., and Carrasco, L. H. (1988) Loss of matrix calcium-binding protein-containing neurons in Huntington's disease. *Lancet* **1,** 1252–1255.
45. Graveland, G. A., Williams, R. S., and DiFiglia, M. (1985) Evidence for degenerative and regenerative changes in neostriatal spiny neurons in Huntington's disease. *Science* **227,** 770–773.
46. Ferrante, R. J., Kowall, N. W., Beal, M. F., Richardson, E. P., and Martin, J. B. (1985) Selective sparing of a class of striatal neurons in Huntington's disease. *Science* **230,** 561–563.
47. Ferrante, R. J., Kowall, N. W., Beal, M. F., Martin, J. B., Bird, E. D., and Richardson, E. P. (1987c) Morphologic and histochemical characteristics of a spared subset of striatal neurons in Huntington's disease. *J. Neuropathol. Exp. Neurol.* **46,** 12–27.
48. Dawbarn, D., DeQuidt, M. E., and Emson, P. C. (1985) Survival of basal ganglia neuropeptide Y somatostatin neurones in Huntington's disease. *Brain Res.* **340,** 251–261.
49. Beal, M. F., Mazurek, M. F., Ellison, D. W., Swartz, K. J., MacGarvey, U., Bird, E. D., and Martin, J. B. (1988b) Somatostatin and neuropeptide Y concentrations in pathologically graded cases of Huntington's disease. *Ann. Neurol.* **23,** 562–569.
50. Hope, B. T., Michael, G. J., Knigge, K. M., and Vincent, S. R. (1991) Neuronal NADPH-diaphorase is a nitric oxide synthase. *Proc. Natl. Acad. Sci. USA* **88,** 2811–2814.
51. Dawson, T. M., Bredt, D. S., Fotuhi, M., Hwang, P. M., and Snyder, S. H. (1991) Nitric oxide synthase and neuronal NADPH-diaphorase are identical in brain and peripheral tissues. *Proc. Natl. Acad. Sci. USA* **88,** 7797–7801.

52. Bredt, D. S. and Snyder, S. H. (1992) Nitric oxide, a novel neuronal messenger. *Neuron* **8,** 3–11.

53. Dawson, V. L., Dawson, T. M., Bartley, D. A., Uhl, G. R., and Snyder, S.H. (1993) Mechanisms of nitric oxide mediated neurotoxicity in primary brain cultures. *J. Neurosci.* **13,** 2651–2661.

54. Ferrante, R. J., Beal, M. F., Kowall, N. W., Richardson, E. P., and Martin, J. B. (1987b) Sparing of acetylcholinesterase-containing striatal neurons in Huntington's disease. *Brain Res.* **411,** 162–166.

55. Beal, M. F., Matson, W. R., Swartz, K. J., Gamache, P. H., and Bird, E. D. (1990) Kynurenine pathway measurements in Huntington's disease striatum: evidence for reduced formation of kynurenic acid. *J. Neurochem.* **55,** 1327–1339.

56. Bird, E. D. and Iversen, L. L. (1974) Huntington's chorea. Post-mortem measurement of glutamic acid decarboxylase, choline acetyltransferase and dopamine in basal ganglia. *Brain* **97,** 452–472.

57. Ferrante, R. J. and Kowall, N. W. (1987a) Tyrosine hydroxylase-like immunoreactivity is distributed in the matrix compartment of normal human and Huntington's disease striatum. *Brain Res.* **416,** 141–146.

58. McGeer, P..L. and McGeer, E. G. (1976b) Enzymes associated with the metabolism of cathecholamines, acetylcholine and GABA in human controls and patients with Parkinson's disease and Huntington's chorea. *J. Neurochem.* **26,** 65–76.

59. DiFiglia, M., Sapp, E., Chase, K. O., Davies, S. W., Bates, G. P., Vonsattel, J. P., and Aronin, N. (1997) Aggregation of Huntingtin in neuronal intranuclear inclusions and dystrophic neurites in brain. *Science* **277,** 1990–1993.

60. Savoiardo, M., Strada, L., Oliva, D., Girotti, F., and D'Incerti, L. (1991) Abnormal MRI signal in the rigid form of Huntington's disease. *J. Neurol. Neurosurg. Psychiatry* **54,** 888–891.

61. Lenti, C. and Bianchini, E. (1993) Neuropsychological and neuroradiological study of a case of early-onset Huntington's chorea. *Dev. Med. Child Neurol.* **35,** 1007–1010.

62. Oliva, D., Carella, F., Savoiardo, M., Strada, L., Giovannini, P., Testa, D., Filippini, G., Caraceni, T., and Girotti, F. (1993) Clinical and magnetic resonance features of the classic and akinetic-rigid variants of Huntington's disease. *Arch. Neurol.* **50,** 17–19.

63. Portera-Caillau, C., Hedreen, J. C., Price, D. L., and Koliatsos, V. E. (1995) Evidence for apoptotic cell death in Huntington disease and excitotoxic animal models. *J. Neurosci.* **15,** 3775–3787.

64. Thomas, L. B., Gates, D. J., Richfield, E. K., O'Brien, T. F., Schweitzer, J. B., and Steindler, D. A. (1995) DNA end labeling (TUNEL) in Huntington's disease and other neuropathological conditions. *Exp. Neurol.* **133,** 265–272.

65. Gusella, J. F., Wexler, N. S., Conneally, P. M., Naylor, S. L., Anderson, M. A., Tanzi, R. E., Watkins, P. C., Ottina, K., Wallace, M. R., Sakagushi, A. Y., Young, A. B., Shouldson, I., Bonnila, E., and Martin, J. B. (1986) A polymorphic DNA marker genetically linked to Huntington's disease. *Nature* **306,** 234–238.

66. The Huntington's Disease Collaborative Group (1993) A novel gene containing a trinucleotide repeat that is expanded and unstable on Huntington's disease chromosomes. *Cell* **72,** 971–983.

67. Wellington, C. L., Brinkman, R. R., O'Kursky, J. R., and Hayden, M. R. (1997) To ard understanding the molecular pathology of Huntington's disease. *Brain Pathol.* **7,** 979–1002.

68. Goldberg, Y. P., Nicholson, D. W., Rasper, D. M., Kalchman, M. A., Koide, H. B., Graham, R. K., Bromm, M., Kazemi-esfarjani, P., Thornberry, N. A., Vaillancourt, J. P., and Hayden, M. R. (1996) Cleavage of huntingtin by apopain, a proapoptotic cysteine protease, is modulated by the polyglutamine tract. *Nat. Genet.* **13,** 442–449.

69. Burke, J. R., Enghild, J. J., Martin, M. E., Jou, Y.-S., Myers, R. M., Roses, A. D., Vance, J. M., Strittmatter, W. J. (1996) Huntingtin and DRPLA proteins selectively intereact with the enzyme GAPDH. *Nat. Med.* **2,** 347–350.
70. Ishitani, R., Sunaga, K., Tanaka, M., Aishita, H., and Chuang, D.-M. (1997) Over-expression of glyceraldehyde-3-phosphate dehydrogenase is involved in low K⁺-induced apoptosis but not necrosis of cultured cerebellar cells. *Mol. Pharmacol.* **51,** 542–550.
71. Ishitani, R. and Chuang, D.-M. (1996) Glyceraldehyde-3-phosphate dehydrogenase antisense oligodeoxynucleotides protect against cytosine arabinonucleoside-induced apoptosis in cultured cerebellar neurons. *Proc. Natl. Acad. Sci. USA* **93,** 9937–9941.
72. Davies, S. W., Turmaine, M., Cozens, B., DiFiglia, M., Sharp, A. H., Ross, C. A., Scherzinger, E., Wanker, E. E., Mangiarini, L., and Bates, G. P. (1997) Formation of neuronal intranuclear inclusions underlies the neurological dysfunction in mice transgenic for the HD mutation. *Cell* **90,** 537–548.
73. Beal, M. F. (1995a) *Mitochondrial Dysfunction and Oxidative Damage in Neurological Diseases.* Neuroscience Intelligence Unit, R.G. Landes Company, Austin, TX.
74. Henneberry, R. C. (1989) The role of neuronal energy in the neurotoxicity of excitatory amino acids. *Neurobiol. Aging* **10,** 611–613.
75. McGeer, E. G. and McGeer, P. L. (1976a) Duplication of biochemical changes of Huntington's chorea by intrastriatal injections of glutamic and kainic acid. *Nature* **263,** 517–519.
76. Coyle, J. T. and Schwarcz, R. (1976) Lesion of striatal neurons with kainic acid provides a model for Huntington's chorea. *Nature* **263,** 244–246.
77. Beal, M. F., Kowall, N. W., Ellison, D. W., Mazurek, M. F., Swartz, K. J., and Martin, J. B. (1986) Replication of the neurochemical characteristics of Huntington's disease by quinolinic acid. *Nature* **321,** 168–171.
78. Beal, M. F., Ferrante, R. J., Swartz, K. J., and Kowall, N. W. (1991a) Chronic quinolinic acid lesions in rats closely resemble Huntington's disease. *J. Neurosci.* **11,** 1649–1659.
79. Ferrante, R. J., Kowall, N. W., Cipolloni, P. B., Storey, E., and Beal, M. F. (1993) Excitotoxin lesions in primates as a model for Huntington's disease: histopathologic and neurochemical characterization. *Exp. Neurol.* **119,** 46–71.
80. Roberts, R., Ahn, A., Swartz, K. J., Beal, M. F., and DiFiglia, M. (1993) Intrastriatal injections of quinolinic acid or kainic acid: differential patterns of cell survival and effects of data analysis on outcome. *Exp. Neurol.* **124,** 274–282.
81. Schwarcz, R., Fuxe, K., Agnati, L. F., Hökfelt, T., and Coyle, J. T. (1979) Rotational behaviour in rats with unilateral striatal kainic acid lesions: a behavioural model for studies on intact dopamine receptors. *Brain Res.* **170,** 485–495.
82. Dunnett, S. B. and Iversen, S. D. (1982) Spontaneous and drug-induced rotation following localized 6-hydroxydopamine and kainic acid-induced lesions of the neostriatum. *Neuropharmacology* **21,** 899–908.
83. Isacson, O., Brundin, P., Kelly, P. A. T., Gage, F. H., and Borjklund, A. (1984) Functional neuronal replacement by grafted striatal neurones in the ibotenic acid lesioned rat striatum. *Nature* **311,** 458–460.
84. Isacson, O., Dunnett, S. B., and Bjorklund, A. (1986) Behavioural recovery in an animal model of Huntington's disease. *Proc. Natl. Acad. Sci. USA* **83,** 2728–2732.
85. Sanberg, P. R., Lehmann, J., and Fibiger, H. C. (1978) Impaired learning and memory after kainic acid lesions of the striatum: a behavioral model of Huntington's disease. *Brain Res.* **149,** 546–551.
86. Sanberg, P. R., Pisa, M., and Fibiger, H. C. (1979) Avoidance, operant and locomotor behavior in rats with neostriatal injections of kainic acid. *Pharmacol. Biochem. Behav.* **10,** 137–144.

87. Sanberg, B. R., Calderon, S. F., Giordano, M., Tew, J. M., and Norman, A. B. (1989) The quinolinic acid model of Huntington's disease: locomotor abnormalities. *Exp. Neurol.* **105,** 45–53.

88. Kanazawa, I., Kimura, M., Mutrata, M., Tanaka, Y., and Cho, F. (1986) Choreic movements induced by unilateral kainate lesions of the striatum and L-Dopa administration in monkey. *Neurosci. Lett.* **71,** 241–246.

89. Hantraye, P., Riche, D., Maziere, M., and Isacson, O. (1990) An experimental primate model of Huntington's disease: anatomical and behavioural studies of unilateral excitotoxic lesions of the caudate-putamen in the baboon. *Exp. Neurol.* **108,** 91–104.

90. Hantraye, P., Riche, D., Mazière, M., and Isacson O. (1992) Intrastriatal transplantation of cross-species fetal striatal cells reduces abnormal movements in a primate model of Huntington's disease. *Proc. Natl. Acad Sci. USA* **89,** 4187–4191.

91. Beal, M. F., Kowall, N. W., Ferrante, R. J., and Cipolloni, P. B. (1989) Quinolinic acid striatal lesions in primates as a model of Huntington's disease. *Ann. Neurol.* **26,** 137.

92. Young, A. B., Greenamyre, J. T., Hollingsworth, Z., Albin, R., D'Amato, C., Shoulson, I., and Penney, J. B. (1988) NMDA receptor losses in putamen from patients with Huntington's disease. *Science* **241,** 981–983.

93. Kuhl, D. E., Phelp, M. E., Markham, C. H., Metter, E. J., Riege, W. H., and Winter, J. (1982) Cerebral metabolism and atrophy in Huntington's disease determined by 18FDG and computated tomographic scan. *Ann. Neurol.* **12,** 425–434.

94. Garnett, E. S., Firnau, G., Nahmias, C., Carbotte, R., and Bartolucci, G. (1984) Reduced striatal glucose consumption and prolonged reaction time are early features in Huntington's disease. *J. Neurol. Sci.* **65,** 231–237.

95. Grafton, S. T., Mazziotta, J. C., Pahl, J. J., George-Hyslop, P. S., Haines, J. L., Gusella, J., Hoffman, J. M., Baxter, L. R., and Phelps, M. E. (1990) A comparison of neurological, metabolic, structural, and genetic evaluations in persons at risk for Huntington's disease. *Ann. Neurol.* **28,** 614–621.

96. Hayden, M. R., Martin, W. R. W., Stoessl, A. J., Clark, C., Hollenberg, S., Adam, M. J., Ammann, W., Harrop, R., Rogers, J., Ruth, T., Sayre, C., and Pate, B. D. (1986) Positron emission tomography in the early diagnosis of Huntington's disease. *Neurology* **36,** 888–894.

97. Kuwert, T., Lange, H. W., Langen, K.-J., Herzog, H., Aulich, A., and Feinendegen, L. E. (1990) Cortical and subcortical glucose consumption measured by PET in patients with Huntington's disease. *Brain* **113,** 1405–1423.

98. Kuwert, T., Lange, H. W., Boecker, H., Titz, H., Herzog, H., Aulich, A., Wang, B.-C., Nayak, U., and Feinendegen, L. E. (1993) Striatal glucose consumption in chorea-free subjects at risk of Huntington's disease. *J. Neurol.* **241,** 31–36.

99. Mazziotta, J. C., Phelps, M. E., Pahl, J. J., Huang, S. C., Baxter, L. R., Riege, W. H., Hoffmann, J. M., Kuhl, D. E., Lanto, A. B., Wapenski, J. A., and Markham, C. H. (1987) Reduced cerebral glucose metabolism in asymptomatic subjects at risk for Huntington's disease. *N. Engl. J. Med.* **316,** 356–362.

100. Jenkins, B. G., Koroshetz, W. J., Beal, M. F., and Rosen, R. (1993) Evidence for an energy metabolic defect in Huntington's disease using localized proton spectroscopy. *Neurology* **43,** 2689–2693.

101. Jenkins, B. G., Rosas, H. D., Chen, Y. C. I., Makabe, T., Myer, R., MacDonald, M., Rosen, B. R., Beal, M. F., and Koroshetz, W. J. (1998) 1H NMR spectroscopy studies of Huntington's disease; correlation with CAG repeat numbers. *Neurology* **50,** 1357–1365.

102. Koroshetz, W. J., Jenkins, B. G., Rosen, B. R., Beal, M. F. (1997) Energy metabolism defects in Huntington's disease and effects of coenzyme Q10. *Ann. Neurol.* **41,** 160–165.

103. Brennan, W. A., Bird, E. D., and Aprille, J. R. (1985) Regional mitochondrial respiratory activity in Huntington's disease brain. *J. Neurochem.* **44,** 1948–1950.

104. Browne, S. E., Bowling, A. C., MacGarvey, U., Baik, M. J., Berger, S. C., Muqit, M. K., Bird, E. D., and Beal, M. F. (1997) Oxidative damage and metabolic dysfunction in Huntington's disease: selective vulnerability of the basal ganglia. *Ann. Neurol.* **41,** 646–653.

105. Butterworth, J., Yates, C. M., and Reynolds, G. P. (1985) Distribution of phosphate-activated glutaminase, succinic dehydrogenase, pyruvate dehydrogenase and gamma-glutamyl transpeptidase in post-mortem brain from Huntington's disease and agonal cases. *J. Neurol. Sci.* **67,** 161–171.

106. Gu, M., Gash, M. T., Mann, V. M , Javoy-Agid, F., Cooper, J. M., and Shapira, A. H. (1996) Mitochondrial defect in Huntington's disease caudate nucleus. *Ann. Neurol.* **39,** 385–389.

107. Mann, V. M., Cooper, J. M., Javoy-Agid, F., Agid, Y., Jenner, P., and Schapira, A. H. V. (1990) Mitochondrial function and parental sex effect in Huntington's disease. *Lancet* **336,** 749.

108. Novelli, A., Reilly, J. A., Lysko, P. G., Hennebery, R. C. (1988) Glutamate becomes neurotoxic via the *N*-methyl-D-aspartate receptor when intracellular energy levels are reduced. *Brain Res.* **451,** 205–212.

109. Zeevalk, G. D. and Nicklas, W. J. (1990) Chemically induced hypoglycemia and anoxia: relationship to glutamate receptor-mediated toxicity in retina. *J. Pharmacol. Exp. Ther.* **253,** 1285–1292.

110. Zeevalk, G. D. and Nicklas, W. J. (1991) Mechanisms underlying initiation of excitotoxicity associated with metabolic inhibition. *J. Pharmacol. Exp. Ther.* **257,** 870–878.

111. Zeevalk, G. D. and Nicklas, W. J. (1992) Evidence that the loss of the voltage-dependent Mg^{++} block at the *N*-methyl-D-aspartate receptor underlies receptor activation during inhibition of neuronal metabolism. *J. Neurochem.* **59,** 1211–1220.

112. Marey-Semper, I., Gelman, M., and Levi-Strauss, M. (1995) A selective toxicity toward cultured mesencephalic dopaminergic neurons is induced by the synergistic effects of energy metabolism impairment and NMDA receptor activation. *J. Neurosci.* **15,** 5912–5918.

113. Schinder, A. F., Olson, E. C., Spitzer, N. C., and Montal, M. (1996) Mitochondrial dysfunction is a primary event in glutamate neurotoxicity. *J. Neurosci.* **116,** 6125–6133.

114. White, R. J. and Reynolds, I. J. (1995) Mitochondria and Na^+/Ca^{2+} exchange buffer glutamate-induced calcium loads in cultured cortical neurons. *J. Neurosci.* **15,** 1318–1328.

115. Nowak, L., Bregestouski, P., Ascher, P., Herbert, A., and Prochiantz, A. (1984) Magnesium gates glutamate-activated channels in mouse central neurons. *Nature* **307,** 462–465.

116. Turski, L. and Turski, W. A. (1993) Towards an understanding of the role of glutamate in neurodegenerative disorders: energy metabolism and neuropathology. *Experientia* **49,** 1064–1072.

117. Beal, M. F., Swartz, K. J., Hyman, B. T., Storey, E., Finn, S. F., and Koroshetz, W. (1991b) Aminooxyacetic acid results in excitotoxic lesions by a novel indirect mechanism. *J. Neurochem.* **57,** 1068–1073.

118. Beal, M. F., Brouillet, E., Jenkins, B., Henshaw, R., Rosen, B., and Hyman, B. T. (1993b) Age-dependent striatal excitotoxic lesions produced by the endogenous mitochondrial inhibitor malonate. *J. Neurochem.* **61,** 1147–1150.

119. Brouillet, E, Shinobu, L., McGarvey, U., and Beal, M. F. (1993b). Manganese injection into the rat striatum produces excitotoxic lesions by impairing energy metabolism. *Exp. Neurol.* **120,** 89–94.

120. Greene, J. G., Porter, R. H. P., Eller, R. V., and Greenamyre, J. T. (1993) Inhibition of succinate dehydrogenase by malonic acid produces an "excitotoxic" lesion in rat striatum. *J. Neurochem.* **61,** 1151–1154.

121. Greene, J. G. and Greenamyre, J. T. (1995) Characterization of the excitotoxic potential of the reversible succinate dehydrogenase inhibitor malonate. *J. Neurochem.* **64,** 430–436.

122. Greene, J. G. and Greenamyre, J. T. (1996) Manipulation of membrane potential modulates malonate-induced striatal excitotoxicity in vivo. *J. Neurochem.* **66,** 637–643.

123. Schultz, J. B., Henshaw, D. R., Jenkins, B. G., Ferrante, R. J., Kowall, N. W., Rosen, B. R., and Beal, M. F. (1995a) 3-Acetylpyridine produces age-dependent excitotoxic lesions in rat striatum. *J. Cereb. Blood Flow Metab.* **14,** 1024–1029.

124. Storey, E., Hyman, B., Jenkins, B., Brouillet, E., Miller, J., Rosen, B., and Beal, M. F. (1992) MPP$^+$ produces excitotoxic lesions in rats striatum due to impairment of oxidative metabolism. *J. Neurochem.* **58,** 1975–1978.

125. Maragos, W. F. and Silverstein, F. S. (1995) The mitochondrial inhibitor malonate enhances NMDA toxicity in the neonatal rat striatum. *Dev. Brain Res.* **88,** 117–121.

126. Ludolph, A. C., He, F., Spencer, P. S., Hammerstad, J., and Sabri, M. (1991) 3-Nitroproprionic acid—exogenous animal neurotoxin and possible human striatal toxin. *Can. J. Neurol. Sci* . **18,** 492–498.

127. Alston, T. A., Mela, L., and Bright, H. J. (1977) Nitropropionate, the toxic substance of indigofera, is a suicide inactivator of succinate dehydrogenase. *Proc. Natl. Acad. Sci. USA* **74,** 3767–3771.

128. Coles, C. J., Edmonson, D. E., and Singer, T.P. (1979) Inactivation of succinate dehydrogenase by 3-nitropropionate. *J Biol. Chem.* **255,** 4772–4780.

129. Brouillet, E., Ménétrat, H., Ouary, S., Altairac, S., Poyot, T., Mittoux, V., and Hantraye, P. (1998b) Major species difference in behavioral and neuropathological deficit observed in rats following chronic 3-nitropropionate treatment: implication for modeling Huntington's disease in rodents. *Soc. Neurosci. Abst.* **24,** 970.

130. Castillo, C., Valencia, I., Reyes, G., and Hong, E. (1993) 3-Nitropropionic acid, obtained from astragalus species, has vasodilatator and hypertensive properties. *Drug Dev. Res.* **28,** 183–188.

131. Guyot, M. C., Hantraye, P., Moya, K., Mazière, M., and Brouillet, E. (1995) Abnormal NADPH-diaphorase staining in striatal interneurons of rats chronically treated with 3-nitropropionic acid. *Soc. Neurosci. Abstr.* **21,** 490.

132. Marshall, P. E. and Landis, D. M. D. (1985) Huntington's disease is accompanied by changes in the distribution of somatostatin-containing neuronal process. *Brain Res.* **329,** 71–82.

133. Koutouzis, T. K., Borlongan, C. V., Scorcia, T., Creese, I., Cahill, D. W., Freeman, T. B. and Sanberg, P. R. (1994) Systemic 3-nitropropionic acid: long-term effects on locomotor behavior. *Brain Res.* **646,** 242–246.

134. Koller, W. C. and Trimble, J. (1985) The gait abnormality of Huntington's disease. *Neurology* **35,** 1450–1454.

135. Thomson, P. D., Berardelli, A., Rothwell, J. C., Day, B. L., Dick, S. P. R., Benecke, R., and Marsden, C. D. (1988) The coexistence of bradykinesia and chorea in Huntington's disease and its implications for theories of basal ganglia control of movement. *Brain* **111,** 223–244.

136. Brouillet, E., Hantraye, P., Ferrante, R. J., Dolan, R., Leroy-Willig, A., Kowall, N. W., and Beal, M. F. (1995) Chronic mitochondrial energy impairment produces selective striatal degeneration and abnormal choreiform movements in primates. *Proc. Natl. Acad. Sci. USA* **92,** 7101–7109.

137. Palfi, S., Ferrante, R. J., Brouillet, E., Beal, M. F., Dolan, R., Guyot, M. C., Peschanski, M., and Hantraye, P. (1996) Chronic 3-nitropropionic acid treatment in baboons replicates the cognitive and motor deficits of Huntington's disease. *J. Neurosci.* **16,** 3019–3025.

138. Erecinska, M. and Nelson, D. (1994) Effects of 3-nitropropionic acid on synaptosomal energy and transmitter metabolism: relevance to neurodegenerative brain diseases. *J. Neurochem.* **63,** 1033–1041.

139. Ludolph, A. C., Seeling, M., Ludolph, A., Novitt, P., Allen, C. N., and Sabri, M. I. (1992) 3-Nitropropionic acid decreases cellular energy levels and causes neurodegeneration in cortical explants. *Neurodegeneration* **1**, 155–161.

140. Pang, Z. and Geddes, J. W. (1997) Mechanisms of cell death induced by the mitochondrial toxin 3-nitropropionic acid: acute excitotoxic necrosis and delayed apoptosis. *J. Neurosci.* **17**, 3064–3073.

141. Riepe, M., Ludolph, A., Seeling, M., Spencer, P. S., and Ludolph, A. A. C. (1994) Increase of ATP levels by glutamate antagonists is unrelated to neuroprotection. *NeuroReport* **5**, 2130–2132.

142. Zeevalk, G. D., Derr-Yellin, E., and Nicklas, W. J. (1995a) NMDA receptor involvement in toxicity to dopamine neurons in vitro caused by the succinate dehydrogenase inhibitor 3-nitropropionic acid. *J. Neurochem.* **64**, 455–458.

143. Zeevalk, G. D., Derr-Yellin, E., Nicklas, W. J. (1995b) Relative vulnerability of dopamine and GABA neurons in mesencephalic culture to inhibition of succinate dehydrogenase by malonate and 3-nitropropionic acid and protection by NMDA receptor blockade. *J. Pharmacol. Exp. Ther.* **275**, 1124–1130.

144. Hassel, B. and Sonnewald, U. (1995) Selective inhibition of the tricarboxylic acid cycle of GABAergic neurons with 3-nitropropionic acid in vivo. *J. Neurochem.* **65**, 1184–1191.

145. Tsai, M. J., Goh, C. C., Wan, C., and Chang, C. (1997) Metabolic alterations produced by 3-nitropropionic acic in rat striata and cultured astrocytes: quantitative in vitro 1H nuclear magnetic resonance spectroscopy and biochemical characterization. *Neuroscience* **79**, 819–826.

146. Dautry, C., Condé, F., Brouillet, E., Mittoux, V., Beal, M. F., Bloch, G., and Hantraye, P. (1998) Early striatal metabolic impairment detected by localized NR spectroscopy in a chronic primate model of Huntington's disease. *Soc. Neurosci. Abstr.*, in press.

147. Fink, S. L., Ho, D. Y., and Sapolsky, R. M. (1996) Energy and glutamate dependency of 3-nitropropionic acid neurotoxicity in culture. *Exp. Neurol.* **138**, 298–304.

148. Weller, M. and Paul, S. M. (1993) 3-Nitropropionic acid is an indirect excitotoxin to cultured cerebellar granule neurons. *Eur. J. Pharmacol.* **248**, 223–228.

149. Behrens, M. I., Koh, J. Y., Muller, M. C., and Choi, D. (1996) NADPH diaphorase-containing striatal or cortical neurons are resistant to apoptosis. *Neurobiol. Dis.* **3**, 72–75.

150. Ankarcrona, M., Dypbukt, J. M., Bonfoco, E., Zhivotovsky, B., Orrenius, S., Lipton, S. A., and Nicotera, P. (1995) Glutamate-induced neuronal death: a succession of necrosis or apoptosis depending on mitochondrial function. *Neuron* **15**, 961–973.

151. Bonfoco, E., Krainc, D., Ankarcrona, M., Nicotera, P., and Lipton, S. A. (1995) Apoptosis and necrosis: two distinct events induced, respectively, by mild and intense insults with *N*-methyl-D-aspartate or nitric oxide/superoxide in cortical cell cultures. *Proc. Natl. Acad. Sci. USA* **92**, 7162–7166.

152. Charriaut-Marlangue, C., Aggoun-Zouaoui, D., Represa, A., and Ben-Ari, Y. (1996) Apoptotic features of selective neuronal death in ischemia, epilepsy and gp120 toxicity. *TINS* **19**, 109–114.

153. Didier, M., Bursztajn, S., Adamec, E., Passani, L., Nixon, R. A., Coyle, J. T., Wei, J. Y., and Berman, S.A. (1996) DNA strand breaks induced by sustained glutamate excitotoxicity in primary neuronal cultures. *J. Neurosci.* **16**, 2238–2250.

154. Simpson, J. R. and Isacson, O. (1993) Mitochondrial impairment reduces the threshold for in vivo NMDA-mediated neuronal death in the striatum. *Exp. Neurol.* **121**, 57–64.

155. Sato, S., Gobel, G. T., Honkaniemi, J., Li, Y., Kondo, T., Murakami, K., Sato, M., Copin, J.-C., and Chan, P. H. (1997) Apoptosis in the striatum of rats following intraperitoneal injection of 3-nitropropionic acid. *Brain Res.* **745**, 343–347.

156. DiFiglia, M., Sharp, E., Chase, K., Shwarz, C., Meloni, A., Young, C., Martin, E., Vonsattel, J-P., Carraway, R., Reeves, S. A., Boyce, F. M., and Aronin, N. (1995) Huntingtin is a cytoplasmic protein associated with vesicles in human and rat brain neurons. *Neuron* **14,** 1075–1081.

157. Li, S. H., Schiling, G., Young, W. S., Li, X. J., Margolis, R. L., Stine, O. C., Wagster, M. V., Abbott, M. H., Franz, M. L., Ranen, N. G., Folstein, S. E., Hedreen, J. C., and Ross, C. A. (1993) Huntington's disease gene (IT15) is widely expressed in human and rat tissues. *Neuron* **11,** 985–993.

158. Sharp, A. H., Loev, S. J., Schilling, G., Li, S. H., Li, X-J., Bao, J., Wagster, M. V., Kotzuk, J. A., Steiner, J. P., Lo, A., Hedreen, J., Sisodia, S., Snyder, S., Dawson, T. M., Ryugo, D. K., and Ross, C. A. (1995) Widespread expression of Huntington's disease gene (IT15) protein product. *Neuron* **14,** 1065–1074.

159. Wallace, D. C. (1992) Mitochondrial genetics: a paradigm for aging and degenerative diseases? *Science* **256,** 628–632.

160. Coyle, J. T. and Puttfarcken, P. (1993) Oxidative stress, glutamate and neurodegenerative disorders. *Science* **262,** 689–695.

161. Bowling, A., Mutisya, E. M., Walker, L. C., Price, D. L., Cork, L. C., and Beal, M. F. (1993) Age-dependent impairment of mitochondrial function in primate brain. *J. Neurochem.* **60,** 1964–1967.

162. Di Monte, D. A., Sandry, M. S., Jewell, S. A., Irwin, I., Delanney, L. E., and Langston, W. J. (1992) Age-related changes in mitochondrial energy metabolism in the squirrel monkey. *Soc. Neurosci. Abstr.* **18,** 1488.

163. Corral-Debrinski, M., Horton, T., Lott, M. T., Shoffner, J. M., Beal, M. F. and Wallace, D.C. (1992) Mitochondrial DNA deletions in human brain: regional variability and increase with advanced age. *Nat. Genet.* **2,** 324–329.

164. Soong, N. W., Hinton, D. R., Cortopassi, G., and Arnheim, N. (1992) Mosaicism for a specific somatic mitochondrial DNA mutation in adult human brain. *Nat. Genet.* **2,** 318–323.

165. Reed, J. C. (1997) Cytochrome c: can't live with it—can't live without it. *Cell* **91,** 559–562.

166. Brouillet, E., Henshaw, D., Schulz, J. B., and Beal, M. F. (1994) Aminooxyacetic acid striatal lesions attenuated by 1,3-butanediol and coenzyme Q10. *Neurosci. Lett.* **177,** 58–62.

167. Beal, M. F., Henshaw, D. R., Jenkins, B. G., Rosen, B. R., and Schulz, J. B. (1994) Coenzyme Q10 and nicotinamide block striatal lesions produced by the mitochondrial toxin malonate. *Ann. Neurol.* **36,** 882–888.

168. Greenamyre, J. T., Garcia-Osuna, M., and Greene, J. G. (1994) The endogenous cofactors, thioctic acid and dihydrolipoic acid are neuroprotective against NMDA and malonic acid lesions of striatum. *Neurosci. Lett.* **171,** 17–20.

169. Holtzman, D. M. and Deshmukh, M. (1997) Caspases: a treatment target for neurodegenerative disease. *Nat. Med.* **3,** 954–955.

170. Anderson, K. D., Panayotatos, N., Corcoran, T. L., Lindsay, R .M., and Wiegand, S. J. (1996) Ciliary neurotrophic factor protects striatal output neurons in an animal model of Huntington disease. *Proc. Natl. Acad. Sci. USA* **93,** 7346–7351.

171. Emerich, D. F., Winn, S. R., Hantraye, P. M., Peschanski, M., Chen, E. Y., Chu, Y., McDermott, P., Baetge, E. E., and Cordower, J. H. (1997) Protective effect of encapsulated cells producing neurotrophic factor CNTF in a monkey model of Huntington's disease. *Nature* **386,** 395–399.

17

Replicating Huntington Disease's Phenotype in Nonhuman Primates

Philippe Hantraye, Stéphane Palfi, Vincent Mittoux, Caroline Dautry, Françoise Condé, and Emmanuel Brouillet

1. INTRODUCTION

Huntington's disease (HD) is an inherited, autosomal dominant, neurodegenerative disorder characterized by involuntary choreiform movements, cognitive decline, and a progressive neuronal degeneration primarily affecting the striatum. At present there is no effective therapy, even palliative, against this disorder. The gene responsible for the disease has been localized on the short arm of chromosome 4 *(1)* and the molecular defect recently identified *(2)* as an abnormal repeat of CAG triplets in the 5′ coding region of a gene *(IT15)* encoding a protein (huntingtin) with unknown function. Despite the intense search for a cell pathology attached to this molecular defect, the mechanisms leading to neurodegeneration in HD still remain largely speculative *(3)*. Nevertheless, recent studies have suggested that abnormal interactions between the mutated huntingtin and other proteins could be involved in the pathogenesis of HD. Thus, huntingtin has been shown to interact with several proteins including a cytoplasmic protein that associates with microtubules, mitochondria, and synaptic vesicles (HAP-1, *4)*, glyceraldehyde phosphate dehydrogenase (GAPDH, *5)*, an unidentified calmodulin-associated protein *(6)*, a ubiquitin-associated protein (HIP-2, *7)*, and a protein homologous to the yeast cytoskeleton-associated protein sla2p (HIP-1, *8)*. These observations suggest that alterations in glycolysis, vesicle trafficking, or apoptosis could all be pathological mechanisms involved in HD. However, direct and indirect evidence for defects in mitochondrial energy metabolism (complex II–III deficiency) has been increasingly compelling over the past decade *(9–15)*. It may be then conceivable that a complex interplay and possibly a direct link exist between HD mutation, mitochondrial impairment, excitotoxicity/apoptosis, and striatal neurodegeneration. To test these pathophysiological hypotheses, we and others have developed animal models of HD using either excitotoxic or metabolically induced striatal lesions to produce a phenotypic model of this disorder. Since the discovery that there is a striking neuropathological similarity between excitotoxic striatal lesions in the rat and HD *(16,17)*, rats with such lesions have served as useful models of HD. Unfortunately, excitotoxic striatal lesions

From: Central Nervous System Diseases
Edited by: D. F. Emerich, R. L. Dean, III, and P. R. Sanberg © Humana Press Inc., Totowa, NJ

in rats do not mimic the dyskinesia and chorea seen in HD *(18–20)*. We have therefore developed different models in nonhuman primates, to determine the underlying mechanisms for HD, and later to use these models to determine the effects of new therapeutic approaches. The present chapter reviews the main characteristics of two primate models of HD, the excitotoxic lesion model and the chronic 3-nitropropionic acid (3-NPA) lesion model, and describe their use in experimental neurology.

2. HUNTINGTON'S DISEASE

It is beyond the scope of this chapter to give a full description of HD and the reader is referred to review articles on the topic for more detailed information *(21–23)*. However, as we intended to develop a phenotypic model of this pathology, a brief summary of the most relevant neuropathological and clinical features of HD is given below.

HD is an inherited dominant neurological disorder characterized by abnormal choreiform movements, cognitive deficits, and psychiatric manifestations associated with progressive striatal atrophy. In the common form of the disease, clinical symptoms develop very rapidly after onset and compose a three-part picture with motor symptoms, initially characterized by hyperkinesia evolving to bradykinesia; psychiatric disturbances, with aggressiveness and depression; and profound cognitive impairment. With the exception of patients bearing the largest triplet repeats (juvenile variant) who normally becomes clinically overt within a few years after birth, most HD patients start to express motor abnormalities at ages 30–40.

In the most common form of the disease, motor disabilities progress over a 10–15-yr period from a hyperkinetic to an akineto-rigid syndrome. Typically, the earliest motor signs are eye movement abnormalities, followed by the progressive appearance of orofacial dyskinesias; dyskinesias involving the head, neck, trunk, and arms; and finally chorea *(24)*. As the disease progresses, choreiform movements may disappear, the initial hyperkinetic syndrome being progressively replaced by a more hypokinetic syndrome in which bradykinesia, rigidity, and dystonia may predominate *(25)*.

Comparative neuropsychological testing of patients with HD, Parkinson's disease with dementia, or Alzheimer's disease points to HD as a model of subcortical dementia, even if this concept has been a matter of controversy (*see 26* for discussion). Nevertheless, the cognitive deficits observed in HD are very similar to those observed following lesions of the frontal cortex and perseverative behavior as well as severe impairment in set-shifting strategies ("cognitive flexibility") are key features of the frontal-type HD syndrome.

The most striking neuropathological manifestation of HD is the progressive degeneration of the striatum. The degree of striatal atrophy has been used by Vonsattel and collaborators *(21)* to categorize HD brains into five different grades (from grade 0, no striatal pathology, to grade 4, severe caudate–putamen and nucleus accumbens atrophy). At this latest stage of neurodegeneration, atrophy is also readily observed in the cerebral cortex (frontal and prefrontal areas), pallidum, subthalamic nucleus, various thalamic nuclei, and substantia nigra. Interestingly, all these areas have in common that they belong to the basal ganglia circuitry and, as such, are directly or indirectly connected to the striatum. Another interesting feature of HD striatal pathology is that not all striatal cells are equally affected by the degenerative process *(27)*. There is a preferential degeneration of γ-aminobutyric-ergic (GABAergic) medium-sized spiny neurons and a relative sparing of the other subpopulations of striatal cells, at least in the early course of the disease *(27,28)*. Even within the subpopulation of GABAergic spiny neurons, all cells are not similarly affected by HD. A double gradient of striatal degeneration has been described in the HD striatum, one progressing in a dorsoventral direction and another in a caudo-rostral direction *(21,29)*. As a consequence, the most vulnerable GABAergic neurons appear located mostly within the dorsal parts of both caudate and putamen nuclei. In the striatum, the calcium-binding protein calbindin-D28k expressed by the medium-sized spiny neurons is severely depleted *(30)*. Interestingly, both immunochemistry for calbindin and Golgi impregnation studies have shown that many of the striatal medium-sized spiny neurons that will degenerate in late-stage HD patients already display typical morphological abnormalities in the early course of the disease *(31,32)*. Finally, it can be noted that all major striatal afferences (in particular the dopaminergic, glutamatergic and serotoninergic afferents) are relatively unaffected by the degenerative process *(33–35)*.

3. EXCITOTOXIC STRIATAL LESION MODELS IN PRIMATES

The resemblance between the striatal degeneration seen in HD and that produced by intrastriatal injections of excitotoxins in the rat stimulated research on HD. Initial studies have shown that unilateral striatal excitotoxic lesions using kainic acid can produce neuropathological changes relatively similar to those of HD *(16,17)* and have supported the hypothesis of a direct glutamate receptor-mediated neurotoxicity in HD. Since then, rats with such lesions have consistently served as animal models of HD *(18–20,36,37)*. However, their behavioral manifestations include locomotor hyperactivity and deficits in memory and cognitive tests but not dyskinesias or chorea. The therapeutic predictions from the rodent model are therefore limited to some symptoms, which points to the necessity of developing a model in primates.

Similar to the rodent model, a unilateral lesion of the caudate–putamen complex has been used in different species of nonhuman primates, with either kainate *(38)* or ibotenate *(39,40)*, which are non-*N*-methyl-D-aspartate (NMDA) glutamate receptor agonists, or quinolinate *(41–43)*, an NMDA receptor agonist. In all cases, a severe depletion of neuronal markers has been observed, restricted to the striatal regions injected. More precisely, in the case of ibotenate, the lesioned areas were characterized by a drastic loss of projection neurons as well as interneurons, associated with a strong depletion in the number of processes and puncta in the neuropil, an intense astrocytic reaction, and a relative sparing of fibers *en passant* (Fig. 1). In contrast, in the case of quinolinate injections the neuronal loss seemed to be limited to projections neurons, sparing the cholinergic and NADPH-diaphorase-positive interneurons *(41)*. With ibotenate and quinolinate the loss of striatal projection neurons has been associated with a decrease of the striatal projections in the globus pallidus, according to the location and the extent of the excitotoxic striatal lesion *(39,44)*. Similarly, some atrophy might occur in the long term within the substantia nigra pars reticulata, and in cases of very extensive caudate–putamen lesions, a decrease in the size of the dopaminergic neurons was observed even in the substantia nigra pars compacta *(45)*. Together with the recent evidence for an atrophy of layer V neurons in cortical area 4 following severe

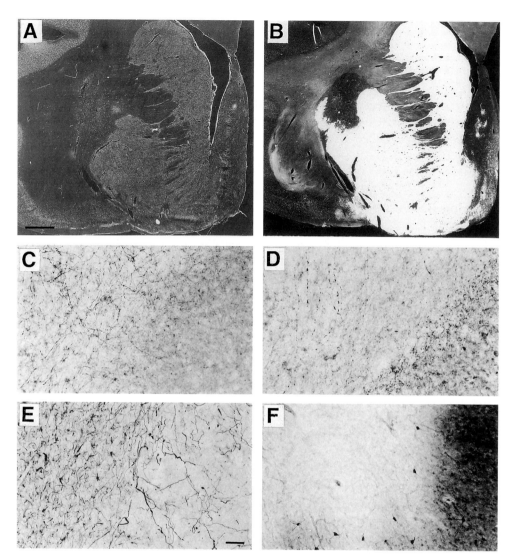

Fig. 1. Excitotoxic lesion induced by an injection of ibotenic acid (10 μL at a concentration of 35 μg/μL in phosphate buffer), into the right putamen of a baboon 2 yr prior to killing. (A, B) Direct printings through sections (**A**) immunostained for neuron-specific nuclear protein (Chemicon, diluted 1:1000) or (**B**) stained for acetylcholinesterase, showing the drastic loss in neurons as well as the depletion of acetylcholinesterase-positive processes in the neuropil. (C–F) Photomicrographs of the border between the lesioned area (*left*) and normal tissue (*right*) of sections immunostained for (**C**) tyrosine hydroxylase (Jacques Boy, France, diluted 1:3000), (**D**) calbindin–D28k (Swant, Switzerland, diluted 1/3000), (**E**) glial fibrillary acidic protein (Dako, Denmark, diluted 1:5000), and (**F**) NADPH-diaphorase. Note the relative sparing of tyrosine hydroxylase- and calbindin–D28k-positive fibers *en passant* (C, D), the strong glial reaction (E), and the loss of NADPH-positive neurons in the lesioned area (F). Nevertheless, in the transition zone between the lesioned and the intact striatum, the density of NADPH-positive neurons is normal (F). Calibration bar: (A, B): 300 μm; (C–F): 50 μm.

quinolinate-induced striatal degeneration *(44)*, these observations indicate that the excitotoxic striatal lesion may also lead with time to retrograde and/or transsynaptic neurodegeneration in regions interconnected to the caudate–putamen complex.

In behavioral studies, very few spontaneous abnormal movements are found whatever the excitotoxin used. These include some transient seizure-like involuntary movements in the contralateral limbs several hours after kainate injection or transient dystonic and/or choreatic-like movements after intrastriatal injections of ibotenate or quinolinate. All these symptoms generally subside by the second week following the excitotoxin injections *(38,39,46,47)*.

Although spontaneous dyskinesias disappear within the first 2 wk following the intrastriatal injections, reversible symptoms can always be elicited during a test session using dopamine agonists or dopamine releasing drugs such as apomorphine or L-Dopa *(39,40,46,47)*. These involuntary movements are characterized by a rather stereotyped, abrupt series of limb movements contralateral to the excitotoxin injection. Sometimes, series of involuntary movements are observed resembling ballistic movements, accompanied less frequently, by twisting movements of a forearm or a hindlimb. Superimposed to the choreic-like symptoms, a general hyperkinesia is also noted (Fig. 2). All these symptoms disappeared spontaneously afterwards, until another dopamine agonist injection is given, ensuring the survival and good health of the animals. The relevance of such studies to HD is based on the hypothesis that hyperkinesia and choreic symptoms in HD can be mimicked by increased dopaminergic neurotransmission and that dopaminomimetic drugs have been shown to precipitate choreic movements in persons at risk for HD *(48)*. As shown by Burns et al. *(49)*, the behavioral outcome following excitotoxic striatal lesions greatly depends on the location of the lesions. Thus, only selective unilateral or bilateral lesions located in the posterior putamen produce a long-lasting dyskinetic syndrome under dopamine agonist stimulation.

When discussing functional and behavioral features, it is apparent that the degeneration of the caudate–putamen may not be the only cause of both motor and cognitive manifestations of HD. Nevertheless, observations from the excitotoxic lesion models have indicated that lesions involving the posterior putamen were associated with motoric manifestations such as dyskinesias and dystonias, whereas lesions localized within the head of the caudate will probably lead to neuropsychiatric manifestations as it receives afferents from the dorsolateral prefrontal cortex *(50–52)*. It is therefore highly predictable that damaging the circuit at the striatal level will have similar effects as lesioning the frontal cortex itself *(53)*.

To sum up, the excitotoxic lesion model of HD has interesting features. First, dyskinesias in the model are evoked only by administration of dopamine agonists which may reduce the animal's discomfort despite a permanent striatal dysfunction. Second, location and extent of the striatal lesions are controlled by the experimentor which enables testing of the functional significance of regionally specific disruptions of qualitatively different motor and cognitive circuits. However this model does not provide a mean to understand HD pathology further and, for example, to elucidate the intimate mechanisms governing the typical dorsoventral gradient of striatal degeneration observed in HD patients.

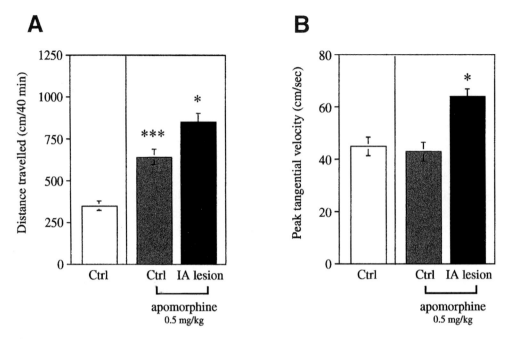

Fig. 2. VMA of apomorphine-induced abnormal movements in baboons following unilateral ibotenic acid striatal lesion (caudate and putamen). This kinematic study was performed more than 2 yr after the excitotoxin injections, according to a previously described procedure *(86,88)*. In **(A)**, as compared to the performances in untreated normal primates, administration of apomorphine (0.5 mg/kg) to control animals led to a significant increase in the distance traveled (locomotor hyperactivity) during a 40-min test session. This hyperactivity was found further increased following unilateral excitotoxic lesion of the caudate–putamen complex. In **(B)**, administration of apomorphine (0.5 mg/kg, i.m.) to IA-lesioned baboons resulted in a significant increase in the peak tangential velocity (hyperkinesia), which was not observed in the control (unlesioned) animals under the same experimental conditions. Brain sections corresponding to one of the IA-lesioned animals included in these VMA data are shown in Fig. 1.

4. REFINING THE EXCITOTOXIC HYPOTHESIS OF HUNTINGTON'S DISEASE: GENERAL ENERGY IMPAIRMENT AS A FACTOR OF SELECTIVE STRIATAL DEGENERATION

A number of in vitro and in vivo studies have shown that under various conditions, partial energy impairment can result in a secondary excitotoxic degeneration, not necessarily accompanied by detectable increases in extracellular glutamate concentrations *(54–70)*. Thus, oxygen or glucose deprivation and chemical hypoxia induced by mitochondrial toxins have all been shown to activate NMDA receptors and to induce excitotoxic insults, even in the presence of low (physiological) extracellular glutamate concentrations. Interestingly, in all cases, pretreatment with various NMDA receptor antagonists (such as MK801) prevented the cell losses, further confirming the excitotoxic nature of these lesions. Emerging from these data, one leading hypothesis has proposed that both glutamate and moderate energy impairment, such as the one produced by chronic treatment with a mitochondrial toxin, could act synergistically to produce cell death (for a review, *see* Brouillet et al., Chapter 16, and *ref. 71*). Various possible mechanisms were put forward. However, in the mechanism most frequently cited, the neurodegenerative process starts by a partial membrane depolarization resulting from energy depletion, which in turn induces the release of the voltage-dependent magnesium block on the calcium-gated NMDA receptor. This (indirect) activation of the calcium-gated NMDA receptor brings into play all the classical features of excitotoxicity including a sudden rise in $[Ca^{2+}]_i$ and the subsequent activation of numerous intracellular intermediates implicated in apoptotic and/or necrotic cell death pathways such as caspases, protein kinases, lipases, and endonucleases.

Interestingly, several in vivo and post-mortem observations have also confirmed the existence of a chronic, possibly early, metabolic impairment in HD. The pioneering studies using positron emission tomography (PET) and [^{18}F]fluorodeoxyglucose as a tracer showed that cerebral glucose metabolism can be severely reduced in the HD striatum, even in presymptomatic patients *(72–78)*. These initial observations have been recently substantiated with data obtained in vivo using localized proton nuclear magnetic resonance (NMR) spectroscopy and pointing to an early alteration in energy metabolism in HD *(12–14)*. Together with postmortem evidence for a 39–59% decrease in the mitochondrial complex II–III activity *(9–11,15)*, these findings have been interpreted as suggesting that the HD mutation could result in a general mitochondrial energy impairment, leading to respiratory chain failure, reorientation of the metabolism to the anaerobic pathway, and overproduction of lactate *(71)*.

5. CHRONIC 3-NPA TREATMENT IN PRIMATES AS A PHENOTYPIC MODEL OF HUNTINGTON'S DISEASE

In a search for an improved animal model of Huntington's disease, we looked for a mean to mimic, in the primate, the chronic succinate dehydrogenase/complex II deficiency hypothesized for HD. Initial studies in rodents, including the initial work in mice using large doses of the succinate dehydrogenase inhibitor 3-NPA *(79–81)* and recent studies using chronic intoxication regimen in the rat *(56,82)*, provided the necessary rationale (for discussion, *see* Brouillet et al., Chapter 16) to develop a primate model of progressive striatal degeneration, using chronic systemic 3-NPA injections.

Despite its interest as an animal replicate of HD, the chronic 3-NPA rat model still suffers from major limitations. Primarily, there is a very different repertoire and capacity of movements in the primate compared to the rat. As discussed earlier, rats chronically treated with 3-NPA did not show clearly identifiable dyskinetic movements resembling chorea, even though a hyperlocomotor activity has been reported early in the time course of intoxication *(83)*. Even in more chronic toxic treatments when the presence of dystonia, bradykinesia, and gait abnormalities could be evident, no choreiform movements could be identified *(82)*. It is possible therefore that the dyskinetic component of HD symptomatology is part of a motor repertoire that can be expressed only in primates. Then, the organization of the basal ganglia in the human is closer to that of the monkey than to that of the rat. For instance, in monkeys as in humans, and not in rats, the striatum is structurally divided into two parts, the caudate nucleus and the putamen. Therefore, although rodent studies are highly informative for initial exploratory studies, primate models are necessary to yield results with relevance and higher predictive validity for human therapy. Another advantage presented by primate models is that in vivo brain imaging techniques such as positron emission tomography, nuclear magnetic resonance imaging (MRI), or single-photon emission computed tomography can be used for determination of anatomo-pathological changes in relation with symptoms (84,85). Finally, primate models allow the final adjustment of therapeutic protocols prior to clinical trials.

The neurotoxicity of 3-NPA was studied in adult or adolescent macaques (*Macaca nemestrina* or *Macaca fascicularis*) and baboons (*Papio papio, Papio anubis*). Clinical examinations were performed before each injection of 3-NPA to detect any sign of acute intoxication or the presence of spontaneous abnormal movements. The 3-NPA solution was injected intramuscularly on a daily basis (86–88). During the 3-NPA intoxication, in vivo examinations with MRI and NMR spectroscopy were regularly performed, using a 0.5T MR magnet (General Electric). Behavioral deficits were studied using two methods: time-sampled neurological observations after intramuscular administration of 0.5 mg/kg of apomorphine and a video movement analysis (VMA) of the animal's displacements during the apomorphine test (36,86). Cognitive deficits were also assessed using the object retrieval detour task (ORDT), a test sensitive to frontal cortex or striatal dysfunction (87). Finally brains were processed for histological evaluation.

5.1. Acute 3-NPA Toxicity in Primates

Although we had already demonstrated in rats that chronic but not acute 3-NPA intoxication could produce behavioral and neuropathological changes rather reminiscent of HD (56,58), one adult macaque was acutely treated with 3-NPA for 5 d (12 mg/kg/d). Such a protocol produced a bilateral necrosis of the striatum as evaluated in vivo using MRI. The most severe changes (T2 hypersignals) were observed in the putamen; however, abnormal NMR signals were also recorded in the caudate nucleus and the hippocampus. The animal displayed spontaneous neurological symptoms consisting of severe loss of muscle tone, general bradykinesia, and dystonia. Histological evaluation confirmed the presence of massive neuronal loss within the putamen and caudate (86). This striatal pathology as well as these behavioral changes were more comparable to those observed in juvenile cases of HD than those developed in adult cases of HD.

5.2. The Chronic Primate Model

To develop a more progressive model of striatal degeneration, various regimens of intoxication were tested. A first series of aged animals were treated with a constant dose of 8 mg/kg/d of 3-NPA for 3–6 wk, in a search for the optimal dose capable of inducing a slow striatal degeneration (81). Under these conditions and during the entire period of the neurotoxic treatment, no acute sign of intoxication and no obvious changes in spontaneous behavior could be observed. The animals had normal food and drink intake and classical neurological examination could not disclose any abnormal movements. However, administration of apomorphine transiently induced the appearance of choreiform movements including jerk-like movements, dyskinesia of extremities, and dystonia. In two of these three animals, abnormal movements were of an explosive nature, very similar to those observed in baboons with unilateral ibotenic acid induced striatal lesions (39). In the third animal, the abnormal movements were less explosive and characterized mainly by twisting of the trunk and abnormal posturing of the lower limbs. Quantitative image analysis of the VMA session evidenced a marked 5–6-fold increase in the distance traveled as compared to age-matched control animals, demonstrating the presence of an increase in locomotor activity (hyperactivity) in these otherwise nonsymptomatic animals. Interestingly, neither in vivo nor in vitro examinations could disclose any obvious abnormalities in the striata of these animals, thus confirming the presymptomatic status of this striatal dysfunction. We concluded that, even in aged animals, the dose of 3-NPA used (8 mg/kg/d), was not sufficient to induce *spontaneous* abnormal movements close to those observed in HD patients.

The second series of animals was treated for 4 mo with 3-NPA at an initial dose of 10 mg/kg/d which was progressively increased (1 mg/kg increment at weekly intervals) until a final dose of 28 mg/kg/d was reached. During the first 6 wk of the protocol, no spontaneous or even apomorphine-induced abnormal movements could be observed in these animals, which remained essentially "nonsymptomatic." Between 8 and 10 wk of neurotoxic treatment, choreiform movements could be evoked in all 3-NPA-treated animals by apomorphine, indicating entry into the "presymptomatic" phase of the intoxication. This presymptomatic phase was characterized by the presence of frontal-type cognitive deficits and apomorphine-inducible abnormal movements without any spontaneous abnormal movements. The severity of the apomorphine-induced motor abnormalities tended to increase as the 3-NPA intoxication progressed. After 3 mo of intoxication, all animals began to show spontaneous foot dyskinesia and dystonia, therefore entering a "symptomatic" phase of the treatment. At this stage, postmortem evaluation confirmed the presence of bilateral striatal lesions, with no detectable extra-striatal lesions.

From all these results, a standard protocol was defined to induce selective and progressive striatal degeneration in primates. In this protocol, adult nonhuman primates (baboons or macaques) receive 3-NPA injections over a 25–30-wk time period to induce a progressive mitochondrial energy impairment without acute toxicity (starting dose 10–14 mg/kg/d, final dose 29–34 mg/kg/d, delivered in two daily doses and increments of 2 mg/kg/d [during the first 6 wk] and 0.5 mg/kg/d [for the remaining 19–24 wk]). According to the individual sensitivity and the age of the animals (older animals are more sensitive), the presymptomatic phase begin after 4–6 wk of intoxication, and the symptomatic phase following 10–12 wk of treatment (Fig. 3, intoxication protocol). Postmortem evaluations after this chronic protocol showed minor or no sign of striatal necrosis, as observed in vivo by MRI, but a severe loss of projecting neurons in the dorsolateral parts of the putamen and caudate nucleus *(88,89)*.

In summary, the time course and the type of striatal degenerescence (necrosis or selective loss of striatal GABAergic projection neurons) are determined by the initial dose of 3-NPA, the rate of increment of this dose, the age of the animal, and finally by a limited interindividual sensitivity toward the toxicity of 3-NPA.

5.2.1. Motor Deficits

Combining the use of the clinical scale for abnormal movements already validated in excitotoxically lesioned baboons and a video-based movement analysis system, we demonstrated that the chronic 3-NPA treatment mimics in primates various aspects of the motor syndrome typically associated with HD. Comparing motor performances of chronically treated animals at various stages (i.e., presymptomatic vs symptomatic

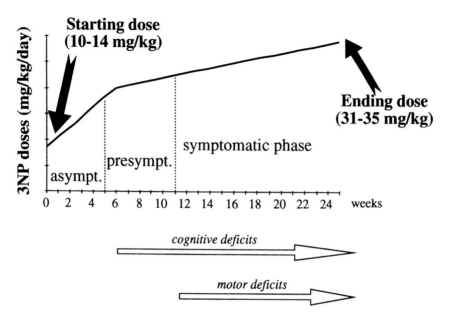

Fig. 3. Schematic representation of the experimental protocol of 3-NPA intoxication used in nonhuman primates (baboons and macaques). The 3-NPA treatment is given in two daily doses delivered 5 d a week which were augmented on a weekly basis by 2 mg/kg during the first 6 wk and by 0.5 mg/kg for the remaining treatment period. In our experience, we found that the starting dose greatly influences the striatal selectivity of the lesion. Accordingly, aged nonhuman primates (which are more sensitive to 3-NPA) should receive a starting dose of 10 mg/kg in two daily 5 mg/kg injections (9.00 a.m., 5.00 p.m.) whereas young adolescent baboons may receive up to 14 mg/kg as a starting dose. With such a neurotoxic regimen, frontal-type cognitive deficits have been detected in both baboons and macaques as soon as following 5–6 wk of 3-NPA treatment, whereas overt motor deficits (mostly leg dystonia) were observed only following 10–12 wk of chronic 3-NPA intoxication. MRI-detectable striatal lesions are usually observed after 10–12 wk of treatment, in conjunction with the appearance of spontaneous leg dystonia.

phases), it was apparent that although animals at a presymptomatic phase presented an hyperkinetic syndrome under apomorphine stimulation, the same animals studied during the symptomatic phase spontaneously displayed a hypokinetic syndrome under apomorphine resembling the motor deficits of late-stage HD patients (Fig. 4). As shown in Fig. 4, in the presymptomatic phase (Pre-sympt.), 3-NPA animals displayed an increased incidence of various abnormal movements (dyskinesia index, i.e., sum of incidences of all abnormal movements, Fig. 4A), including limb dyskinesia (Fig. 4B), orofacial dyskinesia (Fig. 4C), and to a lesser extent leg dystonia (Fig. 4D), as compared to their control (Ctrl) basal clinical score. In addition, quantitative kinematic parameters indicated that these animals were significantly hyperactive (increased distance traveled, Fig. 4E) and hyperkinetic (increased maximal speed, Fig. 4F), as compared to their performances before 3-NPA treatment. Continuation of the 3-NPA treatment for an an additional 14-wk period led to an aggravation of the motor symptoms and the appearance of spontaneous and persistent leg dystonia (symptomatic phase, Sympt.). Under apomorphine stimulation (0.5 mg/kg, i.m., as compared to the 1

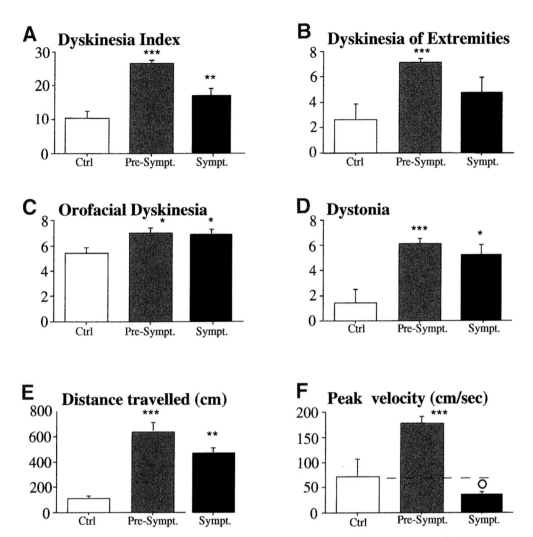

Fig. 4. Clinical and kinematic characterization of apomorphine-induced motor abnormalities observed in chronic 3-NPA-treated baboons at a presymptomatic (Pre-sympt) and symptomatic (Sympt) stage of the neurotoxic treatment. Following intramuscular apomorphine injection (1 mg/kg), presymptomatic animals displayed a significant increase in the dyskinesia index **(A)** which represents the sum (during a 40-min time period) of incidence of various categories of abnormal movements including dyskinesia **(B)**, orofacial dyskinesia **(C)**, and dystonia **(D)**. Kinematic VMA analysis demonstrates that at this stage of intoxication, animals are significantly hyperactive (increase locomotor activity) and hyperkinetic (increase in maximal speed) as compared to their prelesion (Ctrl) performances. This stage of 3-NPA intoxication resembles the early hyperkinetic state observed in the HD motor syndrome. Under similar experimental conditions, but with a lower dose of apomorphine (0.5 mg/kg), symptomatic animals display a significant increase in dyskinesia index (A) due to increased limb (B) and orofacial dyskinesias (C) and leg dystonia (D). In the kinematic study, although symptomatic animals still appear significantly hyperactive (increase in distance traveled), a significant decrease in their peak velocity (bradykinesia) could be demonstrated. As such, this stage of 3-NPA intoxication appears to better mimic the bradykinetic state of the HD motor syndrome, observed later in the course of this disorder.

mg/kg dose necessary to induce abnormal movements in the presymptomatic phase), the 3-NPA-treated animals displayed (Fig. 4) a further increase in the dyskinesia index (Fig. 4A) due to an increased incidence in dyskinesia of extremities and dystonia. Interestingly, kinematic parameters demonstrated an increase in locomotor activity (hyperactivity, Fig. 4E) and a decrease in maximal speed (bradykinesia, Fig. 4F) as compared to their control performances; these motor symptoms were more reminiscent of a late-stage HD.

5.2.2. Cognitive Deficits

As discussed previously, frontal-type cognitive deficits are typical features of HD. We examined the cognitive status of presymptomatic *(87)* and symptomatic animals *(88)* using the ORDT, a task especially designed to detect frontal-type deficits in human and nonhuman primates *(90)*. This test was selected because it requires complex sequential motor planning and is particularly sensitive for detecting frontal cortex or striatal dysfunction *(90–92)*. In brief, the ORDT assesses the ability of the animals to retrieve an object from inside a transparent box, open only on one side. The cognitive level and motor skills required for the subject to solve the task and retrieve the object can be modified by varying the location of the box relative to the subject, the location of the reward into the box, and finally the orientation of the open side of the box relative to the subject. The subject's responses are video recorded and measures of performance include number of "success" responses (retrieval of the reward on the first reach of the trial), number of "correct" responses (retrieval of the reward within the 60-s time period, whatever the strategy used by the baboon to get the reward), "barrier hits" responses (hitting the closed transparent side of the box instead of making a detour), and "motor problems" responses (reaching the correct [open] side of the box but failing to retrieve the reward). ORDT test sessions were performed every week for 2 wk during either the "presymptomatic" or the "symptomatic" phase of the 3-NPA treatment. Compared to controls, 3-NPA animals were not impaired in motor problems or correct responses so that they were globally as able as the controls to get the reward. However, 3-NPA-treated animals were significantly less sucessful in obtaining the reward on the first reach and were making more barrier hits as compared with controls, indicating that they were selectively impaired in their ability to respond using the right strategy. The deficit noted in 3-NPA-treated baboons suggests an impairment in organizational strategy that resembles the cognitive alterations typically associated with HD. Interestingly, further studies with ORDT in 3-NPA-treated baboons in the symptomatic phase indicated that despite the appearance of spontaneous abnormal movements and an overall aggravation of the motor deficits, the cognitive deficits remained very similar in intensity to the one already noted in the early phase of the neurotoxic treatment *(88)*.

5.2.3. Neuropathological Alterations

Brain examination of animals presenting spontaneous motor symptoms at the time they were killed invariably detected histological abnormalities. Essentially, two types of neuropathological patterns could be identified depending on the rate and duration of the 3-NPA-treatment. A first type, characterized by the presence of a necrotic lesion core, clearly identifiable at MRI (T2 hypersignal) with marked cell loss at histological evaluation (Fig. 5), and a second type characterized by no change at MRI and only a diffuse loss of projection striatal neurons (Fig. 6).

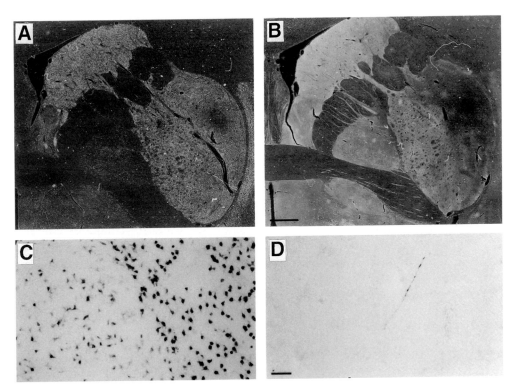

Fig. 5. Caudate–putamen complex of a baboon presenting a 3-NPA lesion, associated with a T2 hypersignal at MRI. (A, B) Direct printings through sections immunostained for **(A)** neuron-specific nuclear protein (Chemicon, diluted 1:1000) and **(B)** the calcium binding protein calbindin–D28k (Swant, Switzerland, diluted 1:3000). **(C)** Photomicrogaph of the border between the lesion area (*left*) and the normal tissue (*right*) from the section shown in (A). **(D)** Photomicrograph of immunoreactivity for calbindin–D28k in the core of the lesion. Note the severe neuronal loss and the total depletion of projection neurons in the core of the lesion. Moreover the border betwen the core of the lesion and the normal tissue is sharp in the case of the neuronal marker (A, C) whereas it is more diffuse and larger in the case of the marker of projection neurons (B), indicating the presence of a transition area in which projection neurons are more depleted than the other striatal neuronal types. Calibration bar: (A–D): 300 to µm, (C, D): 50 µm.

The first type of lesion (Fig. 5) was seen in animals with an initial dose of 10 mg/kg/d of 3-NPA, progressively increased to 30 mg/kg/d over a 20-wk time period (*86,87*). According to the degree of intoxication, histological evaluation showed an almost complete loss of neurons and neuronal processes, first in the dorsolateral part of the putamen (core of the lesion, Fig. 5A, C), then in the dorsomedial part of the caudate nucleus. Accordingly, the number of all striatal neuronal types strongly decreased, especially the number of projection neurons and processes immunoreactive for calbindin–D28k (Fig. 5B, D). In the transition area between the core of the lesion and the normal tissue, NADPH-diaphorase-positive interneurons and large cholinergic interneurons were found relatively preserved, as compared with projection neurons, and immunoreactivity for tyrosine hydroxylase was also spared, indicating that dopamine terminals are

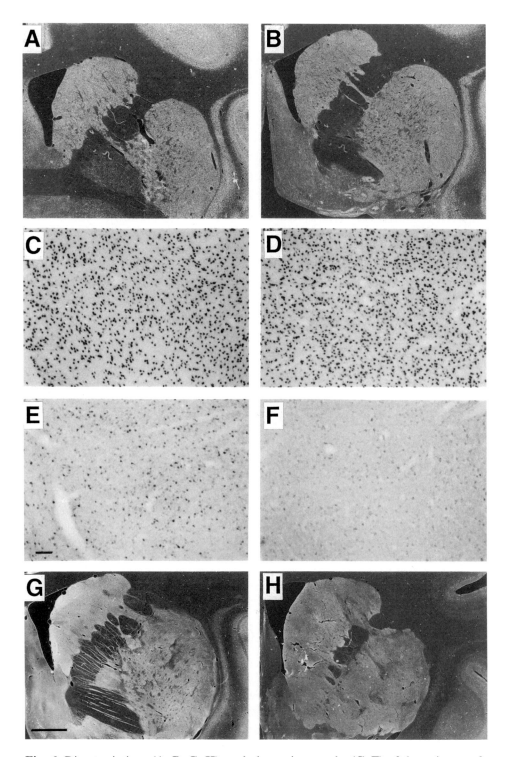

Fig. 6. Direct printings **(A, B, G, H)**, and photomicrographs **(C–F)** of the striatum of one control animal (A, C, E, and G) and one macaque intoxicated with 3-NPA (B, D, F, and H), but presenting motor and cognitive deficits without a T2 hypersignal on MRI. Distribution of neurons immunoreactive for (A–D) the neuron-specific protein (Chemicon, diluted 1:1000) and (E–H) the calcium binding protein calbindin–D28k (Swant, Switzerland, diluted 1:3000). Note the absence of obvious neuronal loss (B, D) but decrease of calbindin–D28k immunoreactivity (F, H), suggesting the presence of a neuronal dysfunction rather than of a cell loss within the dorsolateral parts of the caudate and putamen nuclei. Calibration bar: (A, B), (G, H): 300 µm; (C–F): 50 µm.

left largely unaffected by the toxin *(39)*. The astrocytic glial reaction is maximal in the transition area.

The second type of lesion was characterized by the absence of any detectable changes in MRI signal despite the presence of spontaneous persistent abnormal movements and frontal-type cognitive deficits. This type of lesion was observed in macaques *(88,89)* treated with the most progressive regimen of 3-NPA intoxication (25 wk of treatment, with an initial dose of 10 mg/kg/d and a final dose of 29 mg/kg/d). In these animals, no obvious neuronal depopulation could be detected within the striatum (Fig. 6). However, quantitative analysis of the distribution of calbindin–D28k-immunoreactive neurons revealed a clear dorsal to ventral loss of immunoreactive cells *(89)*. The other striatal neuronal populations were unchanged. This suggests that a selective dysfunction of the striatal projection neurons (as exemplified by the loss of calbindin–D28k immunoreactivity) may be sufficient to cause the appearance of clinically overt motor and cognitive deficits. These findings are of great relevance for the design of new therapeutic strategies for HD, especially for neuroprotective clinical trials in which the treatment should be introduced before a massive loss of striatal neurons has taken place.

6. CONCLUSION

As discussed in the foregoing sections, the striking similarities between HD and the chronic 3-NPA lesion model support the view that an early deficit in energy metabolism may be involved in the aetiology of HD. Indeed, the "indirect excitotoxicity" hypothesis is supported by the many behavioral and histological features reminiscent of HD pathology described in the symptomatic 3-NPA treated primates. Our results and those from other laboratories *(93,94)* suggest that the cell death pathways involved in the 3-NPA-induced toxicity share some common mechanisms with those brought into play by the HD mutation. For example, in both conditions, a strong activation of apoptotic pathways and a severe mitochondrial complex II deficiency have been reported *(93,94*; Brouillet et al., *unpublished data)*, suggesting that unraveling the mechanisms of 3-NPA toxicity may provide new insights into HD pathogenesis.

The primate 3-NPA lesion model of HD, in which the extent and the time of appearance of the lesion can be modulated by the rate of intoxication, is useful for designing and testing new therapeutic strategies, with different goals. In our model, the presymptomatic phase with only frontal-type cognitive deficits and apomorphine-inducible motor symptoms, but no obvious cell loss in the striatum, mimics the early stages of HD. It allows the assessment of neuroprotective strategies, as illustrated by our promising results obtained in 3-NPA-treated primates implanted during the presymptomatic phase with encapsulated fibroblasts genetically engineered to release *in situ* the ciliary neurotrophic factor *(89)*. Moreover, the possibility of also mimicking more advanced stages of HD with the symptomatic phase of our model allows testing of different therapeutic strategies such as the substitution of missing cells by fetal striatal allografts *(88)* or the evaluation of new pharmacological compounds potentially capable of alleviating HD symptoms *(95)*.

ACKNOWLEDGMENTS

The authors acknowledge with great appreciation the contributions and invaluable help in this project of Drs. M. Flint Beal, Robert F. Ferrante, Marc Peschanski, Danielle Riche, Ole Isacson and the continuing support of Dr. Mariannick Maziere and Prof.

André Syrota. C. Genty, C. Jouy, D. Mauchand, and F. Sergent are thanked immensely for their help and outstanding care of the primate colony. The present studies were supported by grants from CEA, INSERM, Association Française contre les Myopathies, and by fellowships from Association Huntington-France and Association France-Parkinson.

REFERENCES

1. Gusella, J. F., Wexler, N. S., Conneally, P. M., Naylor, S. L., Anderson, M. A., and Tanzi, R. E. (1983) A polymorphic DNA marker genetically linked to Huntington's disease. *Nature* **306**, 234–238.
2. The Huntington's Disease Collaborative Group (1993) A novel gene containing a trinucleotide repeat that is expanded and unstable on Huntington's disease chromosomes. *Cell* **72**, 971–983.
3. Sharp, A. H. and Ross, A. R. (1996) Neurobiology of Huntington's disease. *Neurobiol. Dis.* **3**, 3–15.
4. Gutekunst, C. A., Li, S. H., Yi, H., Ferrante, R .J., Li, X. J., and Hersch, S. M. (1998) The cellular and subcellular localization of huntingtin-associated protein 1 (HAP-1): comparison with huntingtin in rat and human. *J. Neurosci.* **18**, 7674–7686.
5. Burke, J. R., Enghild, J. J., Martin, M. E., Jou, Y.-S., Myers, R. M., Roses, A. D., Vance, J. M., and Strittmatter, W. J. (1996) Huntingtin and DRPLA proteins selectively intereact with the enzyme GAPDH. *Nature Med.* **2**, 347–350.
6. Bao, J., Sharp, A., Wagster, M., Becher, M., Schilling, G., Ross, C., Dawson,V., and Dawson, T. (1996) Expansion of polyglutamine repeat in huntingtin leads to abnormal protein interactions involving calmodulin. *Proc. Natl. Acad. Sci. USA* **93**, 5037–5042.
7. Kalchman, M. A., Graham, R. K., Xia, G., Koide, H. B., Hodgston, J. G., Graham, K. C., Goldberg, Y. P., Gietz, R. D., Pickart, C. M., and Hayden, M. R. (1996) Huntingtin is ubiquitized and interacts with a specific ubiquitin conjugated enzyme. *J. Biol. Chem.* **271**, 19,385–19,394.
8. Kalchman, M. A., Koide, H. B., McCutcheon, K., Graham, R. K., Nichol, K., Nishiyama, K., Kazemi-Esfarjani, P., Lynn, F. C., Wellington, C., Metzler, M., Goldberg, Y. P., Kanazawa, I., Gietz, R. D., and Hayden, M. R. (1997) HIP1, a human homolog of *S. cerevisiae* Sla2p, interacts with membrane-associated huntingtin in the brain. *Nat. Genet.* **16**, 44–53.
9. Brennan, W. A. J., Bird, E. D., and Aprille, J. R. (1985) Regional mitochondrial respiratory activity in Huntington's disease brain. *J. Neurochem.* **44**, 1948–1950.
10. Browne, S. E., Bowling, A. C., McGarvey, U., Baik, M. J., Berger, S. C., Muqit, M. K., Bird, E. D., and Beal, M. F. (1997) Oxidative damage and metabolic dysfunction in Huntington's disease: selective vulnerability of the basal ganglia. *Ann. Neurol.* **41**, 646–653.
11. Gu, M, Gash, M. T., Mann, V. M., Javoy-Agid, F., Cooper, J. M., and Schapira, A. H. (1996) Mitochondrial defect in Huntington's disease caudate nucleus. *Ann. Neurol.* **39**, 385–389.
12. Jenkins, B. G., Koroshetz, W. J., Beal, M. F., and Rosen, B. R. (1993) Evidence for impairment of energy metabolism in vivo in Huntington's disease using localized 1H NMR spectroscopy. *Neurology* **43**, 2689–2695.
13. Jenkins, B. G., Rosas, H. D., Chen, Y. C. I., Makabe, T., Myers, R., McDonald, M., Rosen, B. R., Beal, M. F., and Koroshetz, W. J. (1998) 1H-NMR spectroscopy studies of Huntington's disease, correlation with CAG repeat numbers. *Neurology* **50**, 1357–1365.
14. Koroshetz, W. J., Jenkins, B. G., Rosen, B. R., and Beal, M. F. (1997) Energy metabolism defects in Huntington's disease and effects of coenzyme Q10. *Ann. Neurol.* **41**, 160–165.

15. Mann, V. M., Cooper, J. M., Javoy-Agid, F., Agid, Y., Jenner, P., and Schapira, A. H. V. (1990) Mitochondrial function and parental sex effect in Huntington's disease. *Lancet* ii, **8717,** 749.

16. Coyle, J. T. and Schwarcz, R. (1976) Lesion of striatal neurons with kainic acid provides a model for Huntington's chorea. *Nature* **263,** 244–246.

17. McGeer, E. G. and McGeer, P. L. (1976) Duplication of biochemical changes of Huntington's chorea by intrastriatal injections of glutamic and kainic acid. *Nature* **263,** 517–519.

18. Deckel, A. W., Robinson, R. G., Coyle, J. T., and Sanberg, P. R. (1983) Reversal of long-term locomotor abnormalities in the kainic acid model of Huntington's disease by day 18 fetal striatal implants. *Eur. J. Pharmacol.* **93,** 287–288.

19. Dunnett, S. B. and Iversen, S. D. (1982) Spontaneous and drug-induced rotation following localized 6-hydroxydopamine and kainic acid-induced lesion of the neostriatum. *Neuropharmacology* **21,** 899–908.

20. Isacson, O., Dunnett, S. B., and Björklund, A. (1986) Behavioural recovery in an animal model of Huntington's disease. *Proc. Natl. Acad. Sci. USA* **83,** 2728–2732.

21. Vonsattel, J.-P., Myers, R. H., and Stevens, T. J. (1985) Neuropathological classification of Huntington's disease. *J. Neuropathol. Exp. Neurol.* **44,** 559–577.

22. Harper, P. S. (1991) *Huntington's Disease* (Harper, P. S., ed). London, W. B. Saunders.

23. Wellington, C. L., Brinkman, R. R., O'Kursky, J. R., and Hayden, M. R. (1997) Toward understanding the molecular pathology of Huntington's disease. *Brain Pathol.* **7,** 979–1002.

24. Kremer, B., Weber, B., and Hayden, M. R. (1992) New insights into the clinical features, pathogenesis and molecular genetics of Huntington's disease. *Brain Pathol.* **2,** 321–335.

25. Thomson, P. D., Berardelli, A., Rothwell, J. C., Day, B. L., Dick, S. P. R., Benecke, R., and Marsden, C. D. (1988) The coexistence of bradykinesia and chorea in Huntington's disease and its implications for theories of basal ganglia control of movement. *Brain* **111,** 223–244.

26. Podoll, K., Caspary, P., Lange, H. W., and Noth, J. (1988) Language functions in Huntington's disease. *Brain* **111,** 1475–1503.

27. Kowall, N. W., Ferrante, R. J., and Martin, J. B. (1987) Patterns of cell loss in Huntington's disease. *TINS* **10,** 24–29.

28. Ferrante, R. J., Kowall, N. W., Beal, M. F., Richardson, E. P., and Martin, J. B. (1985) Selective sparing of a class of striatal neurons in Huntington's disease. *Science* **230,** 561–563.

29. Hedreen, J. C. and Foldstein, S. E. (1995) Early loss of neostriatal neurons in Huntington's disease. *J. Neuropathol. Exp. Neurol.* **54,** 105–120.

30. Seto-Ohshima, A., Emson, P. C., Lawson, E., Mountjoy, C. Q., and Carrasco, L. H. (1988) Loss of matrix calcium-binding protein-containing neurons in Huntington's disease. *Lancet* **i,** 1252–1255.

31. Ferrante, R. J., Kowall, N. W., and Richardson, E. P., Jr. (1991) Proliferative and degenerative changes in striatal spiny-neurons in Huntington's disease: a combined study using the section-Golgi method and calbindin D28k immunochemistry. *J. Neurosci.* **11,** 3877–3887.

32. Graveland, G. A., Williams, R. S., and DiFiglia, M. (1985) Evidence for degenerative and regenerative changes in neostriatal spiny neurons in Huntington's disease. *Science* **227,** 770–773.

33. Bird, E. D. and Iversen, L. L. (1977) Neurochemical findings in Huntington's chorea, in *Essays in Neurochemistry and Neuropharmacology,* Vol. 1 (Youdim, M. B. H., Sharman, D. F., Lovenberg, W., and Lagnado, J. R., eds.), John Wiley & Sons, New York, pp. 177–195.

34. Kish, S. J., Shannack, K., and Hornykiewicz, O. (1987) Elevated serotonin and reduced dopamine on subregionally divided Huntington's disease striatum. *Ann. Neurol.* **22,** 386–389.

35. Beal, M. F., Matson, W. R., Swartz, K. J., Gamache, P. H., and Bird, E. D. (1990) Kynurenin pathway measurments in Huntington's diseased striatum: evidence for reduced kynurenic acid. *J. Neurochem.* **55,** 1327–1339.

36. Beal, M. F., Kowall, N. W., Ellison, D. W., Swartz, K. J., McGarvey, U., Bird, E. D., and Martin, J. B. (1986) Replication of the neurochemical characteristics of Huntington's disease by quinolinic acid. *Nature* **321,** 168–171.

37. Beal, M. F., Ferrante, R. J., Swartz, K. J., and Kowall, N. W. (1991) Chronic quinolinic acid lesions in rats closely resemble Huntington's disease. *J. Neurosci.* **11,** 1649–1659.

38. Kanazawa, I., Tanaka, Y., and Cho, F. (1986) Choreic movements induced by unilateral kainate lesions of the striatum and L-Dopa administration in monkey. *Neurosci. Lett.* **71,** 241–246.

39. Hantraye, P., Riche, D., Maziere, M., and Isacson, O. (1990) An experimental primate model of Huntington's disease: anatomical and behavioural studies of unilateral excitotoxic lesions of the caudate–putamen in the baboon. *Exp. Neurol.* **108,** 91–104.

40. Hantraye, P., Riche, D., Maziere, M., and Isacson, O. (1992) Intrastriatal transplantation of cross-species fetal striatal cells reduces abnormal movements in a primate model of Huntington's disease. *Proc. Natl. Acad. Sci. USA* **89,** 4187–4191.

41. Ferrante, R. J., Kowall, N. W., Cipolloni, P. B., Storey, E., and Beal, M. F. (1993) Excitotoxin lesions in primates as a model for Huntington's disease: histopathologic and neurochemical characterization. *Exp. Neurol.* **119,** 46–71.

42. Kendall, A. L., Rayment, F. D., Torres, E. M., Baker, H. F., Ridley, R. M., and Dunnett, S. B. (1998) Functional integration of striatal allografts in a primate model of Huntington's disease. *Nat. Med.* **4,** 727–729.

43. Schumacher, J. M., Hantraye, P., Brownell, A.-L., Riche, D., Madras, B. K., Davenport, P. D., Maziere, M., Elmaleh, D. R., Brownell, G. L., and Isacson, O. (1992) A primate model of Huntington's disease: functional neural transplantation and CT-guided stereotactic procedures. *Cell Transplant.* **1,** 313–322.

44. Emerich, D. F., Winn, S. R., Hantraye, P., Peschanski, M., Chen, E.-Y, Chu, Y., McDermott, P., Baetge, E. E., and Kordower, J. H. (1997) Encapsulated CNTF-producing cells protect monkeys in a model of Huntington's disease. *Nature* **386,** 395–399.

45. Lundberg, C., Wictorin, K., and Björklund, A. (1994) Retrograde degenerative changes in the substantia nigra pars compacta following an excitotoxic lesion of the striatum. *Brain Res.* **644,** 205–212.

46. Isacson, O., Hantraye, P., Riche, D., Schumacher, J. M., and Maziere, M. (1991) The relationship between symptoms and functional anatomy in the chronic neurodegenerative diseases: from pharmacological to biological replacement therapy in Hunington's disease, in *Intracerebral Transplantation in Movement Disorders* (Lindvall, O., Björklund, A., and Widner, H., eds.), Elsevier, Amsterdam, pp. 231–244.

47. Kanazawa, I., Kimura, M., Mutrata, M., Tanaka, Y., and Cho, F. (1990) Choreic movements in the macaque monkey induced by kainic acid lesions of the striatum combined with L-Dopa. *Brain* **113,** 509–535.

48. Paulson, G. W. (1976) Predictive tests in Huntington's disease, in *The Basal Ganglia* (Yahr, M. D., ed.) Raven Press, New York, pp. 317–329.

49. Burns, L. H., Pakzaban, P., Deacon, T. W., Brownell, A. L., Tatter, S. B., Jenkins, B., and Isacson, O. (1995) Selective putaminal excitotoxic lesions in nonhuman primates model the movement disorder of Huntington disease. *Neuroscience* **64,** 1007–1017.

50. Johnson, T. N., Rosvold, H. E., and Mishkin, M. (1968) Projections from behaviorally-defined sectors of the prefrontal cortex to the basal ganglia, septum, and diencephalon in the monkey. *Exp. Neurol.* **21,** 20–34.

51. Künzle, H. (1977) Projections from the primary somatosensory cortex to basal ganglia and thalamus in the monkey. *Exp. Brain Res.* **30,** 481–492.
52. Selemon, L. D. and Goldman-Rakic, P. S. (1985) Longitudinal topography and interdigitation of corticostriatal projections in the rhesus monkey. *J. Neurosci.* **5,** 776–794.
53. Battig, K., Rosvold, H. E., and Mishkin, M. (1960) Comparison of the effects of frontal and caudate lesions on delayed response and alternation in monkeys. *J. Comp. Physiol. Psychol.* **53,** 400–404.
54. Albin, R. L. and Greenamyre, J. T. (1992) Alternative excitotoxic hypotheses. *Neurology* **42,** 733–738.
55. Beal, M. F., Swartz, K. J., Hyman, B. T., Storey, E., Finn, S. F., and Koroshetz, W. (1991) Aminooxyacetic acid results in excitotoxic lesions by a novel indirect mechanism. *J. Neurochem.* **57,** 1068–1073.
56. Beal, M. F., Brouillet, E., Jenkins, B., Ferrante, R. J., Kowall, N. W., Miller, J. M., Storey, E., Srivastava, R., Rosen, B. R., and Hyman, B. T. (1993) Neurochemical and histological characterization of the striatal excitotoxic lesions produced by the mitochondrial toxins 3-nitropropionic acid. *J. Neurosci.* **13,** 1481–1492.
57. Beal, M. F., Brouillet, E., Jenkins, B., Henshaw, R., Rosen B., and Hyman, B. T. (1993) Age-dependent striatal excitotoxic lesions produced by the endogenous mitochondrial inhibitor malonate. *J. Neurochem.* **61,** 1147–1150.
58. Brouillet, E., Jenkins, B., Hyman, B., Ferrante, R. J., Kowall, N. W., Srivastava, R., Roy, D. S., Rosen, B., and Beal, M. F. (1993) Age-dependent vulnerability of the striatum to the mitochondrial toxin 3-nitropropionic acid. *J. Neurochem.* **60,** 356–359.
59. Brouillet, E., Shinobu, L., McGarvey, U., and Beal, M. F. (1993b) Manganese injection into the rat striatum produces excitotoxic lesions by impairing energy metabolism. *Exp. Neurol.* **120,** 89–94.
60. Brouillet, E., Hyman, B. T., Jenkins, B., Henshaw, R., Shulz, J. B., Sodhi, P., Rosen, B., and Beal, M. F. (1994) Systemic or local administration of azide produces striatal lesions by an energy impairment-induced excitotoxic mechanism. *Exp. Neurol.* **129,** 175–182.
61. Greene, J. G., Porter, R. H. P., Eller, R. V., and Greenamyre, J. T. (1993) Inhibition of succinate dehydrogenase by malonic acid produces an "excitotoxic" lesion in rat striatum. *J. Neurochem.* **61,** 1151–.
62. Greene, J. G., Sheu, S. S., Gross, R. A., and Greenamyre, J. T. (1998) 3-Nitropropionic acid exacerbates *N*-methyl-D-aspartate toxicity in striatal culture by multiple mechanisms. *Neuroscience* **84,** 503–510.
63. Greene, J. G. and Greenamyre, J. T. (1995) Characterization of the excitotoxic potential of the reversible succinate dehydrogenase inhibitor malonate. *J. Neurochem.* **64,** 430–436.
64. Greene, J. G. and Greenamyre, J. T. (1996) Manipulation of membrane potential modulates malonate-induced striatal excitotoxicity in vivo. *J. Neurochem.* **66,** 637–643.
65. Ludolph, A. C., He, F., Spencer, P. S., Hammerstad, J., and Sabri, M. (1991) 3-Nitroproprionic acid—exogenous animal neurotoxin and possible human striatal toxin. *Can. J. Neurol. Sci.* **18,** 492–498.
66. Novelli, A., Reilly, J. A., Lysko, P. G., and Hennebery, R. C. (1988) Glutamate becomes neurotoxic via the *N*-methyl-D-aspartate receptor when intracellular energy levels are reduced. *Brain Res.* **451,** 205–212.
67. Schultz, J. B., Henshaw, D. R., Jenkins, B. G., Ferrante, R. J., Kowall, N. W., Rosen, B. R., and Beal, M. F. (1995) 3-Acetylpyridine produces age-dependent excitotoxic lesions in rat striatum. *J. Cereb. Blood Flow Metab.* **14,** 1024–1029.
68. Zeevalk, G. D. and Nicklas, W. J. (1991) Mechanisms underlying initiation of excitotoxicity associated with metabolic inhibition. *J. Pharmacol. Exp. Ther.* **257,** 870–878.
69. Zeevalk, G. D. and Nicklas, W. J. (1992) Evidence that the loss of the voltage-dependent Mg^{++} block at the *N*-methyl-D-aspartate receptor underlies receptor activation during inhibition of neuronal metabolism. *J. Neurochem.* **59,** 1211–1220.

70. Zeevalk, G. D., Derr-Yellin, E., and Nicklas, W. J. (1995) NMDA receptor involvement in toxicity to dopamine neurons in vitro caused by the succinate dehydrogenase inhibitor 3-nitropropionic acid. *J. Neurochem.* **64,** 455–458.

71. Beal, M. F. (1994) Neurochemistry and toxin models in Huntington's disease. *Curr. Opin. Neurol.* **7,** 542–546.

72. Garnett, E. S., Firnau, G., Nahmias, C., Carbotte, R., and Bartolucci, G. (1984) Reduced striatal glucose consumption and prolonged reaction time are early features in Huntington's disease. *J. Neurol. Sci.* **65,** 231–237.

73. Grafton, S. T., Mazziotta, J. C., Pahl, J. J., George-Hyslop, P. S., Haines, J. l., Gusella, J., Hoffman, J. M., Baxter, L. R., and Phelps, M. E. (1990) A comparison of neurological, metabolic, structural, and genetic evaluations in persons at risk for Huntington's disease. *Ann. Neurol.* **28,** 614–621.

74. Hayden, M. R., Martin, W. R. W., Stoessl, A. J., Clark, C., Hollenberg, S., Adam, M. J., Ammann, W., Harrop, R., Rogers, J., Ruth, T., Sayre, C., and Pate, B. D. (1986) Positron emission tomography in the early diagnosis of Huntington's disease. *Neurology* **36,** 888–894.

75. Kuhl, D. E., Phelps, M. E., Markham, C. H., Metter, E. J., Riege, W. H., and Winter, J. (1982) Cerebral metabolism and atrophy in Huntington's disease determined by 18FDG and computed tomographic scan. *Ann. Neurol.* **12,** 425–434.

76. Kuwert, T., Lange, H. W., Langen, K. J., Herzog, H., Aulich, A., and Feinendegen, L. E. (1990) Cortical and subcortical glucose consumption measured by PET in patients with Huntington's disease. *Brain* **113,** 1405–1423.

77. Kuwert, T., Lange, H. W., Boecker, H., Titz, H., Herzog, H., Aulich, A., Wang, B. C., Nayak, U., and Feinendegen, L. E. (1993) Striatal glucose consumption in chorea-free subjects at risk of Huntington's disease. *J. Neurol.* **241,** 31–36.

78. Mazziotta, J. C., Phelps, M. E., Pahl, J. J., Huang, S. C., Baxter, L. R., Riege, W. H., Hoffman, J. M., Kuhl, D. E., Lanto, A. B., Wapenski, J. A., and Markham, C. H. (1987) Reduced cerebral glucose metabolism in asymptomatic subjects at risk for Huntington's disease. *N. Engl. J. Med.* **316,** 357–362.

79. Gould, D. H. and Gustine, D. L. (1982) Basal ganglia degeneration, myelin alterations, enzyme inhibition induced in ice by the plant toxin 3-nitropropionic acid. *Neuropathol. Appl. Neurobiol.* **8,** 377–393.

80. Gould, H., Wilson, M. P., and Hamar, D. W. (1985) Brain enzyme and clinical alterations induced in rats and mice by nitroaliphatic toxicants. *Tox. Lett.* **27,** 83–89.

81. Hamilton, B. F. and Gould, D. H. (1987) Correlation of morphological brain lesions with physiological alterations and blood–brain barrier impairment by 3-nitropropionic acid toxicity in rats. *Acta Neuropathol. (Berl.)* **74,** 67–74.

82. Guyot, M.-C., Hantraye, P., Dolan, R., Palfi, S., Mazière, M., and Brouillet, E. (1997) Quantifiable bradykinesia, gait abnormalities and Huntington's disease-like striatal lesions in rats chronically treated with 3-nitropropionic acid. *Neurosci.* **79,** 45–56.

83. Borlongan, C. V., Koutousis, T. K., Freeman, T. B., Cahill, D. W., and Sanberg, P. R. (1995) Behavioral pathology induced by repeated systemic injections of 3-nitropropionic acid mimics the motoric symptoms of Huntington's disease. *Brain Res.* **697,** 254–257.

84. Hantraye, P., Loc'h, C., Maziere, B., Khalili-Varasteh, M., Crouzel, C., Fournier, D., Yorke, J. C., Riche, D., Isacson, O., and Maziere, M. (1992) 6-[18F]Fluoro-L-Dopa uptake and [76Br]bromolisuride binding in the excitotoxically lesioned caudate–putamen of non-human primates studied using positron emission tomography. *Exp. Neurol.* **115,** 218–227.

85. Brownell, A.-L., Hantraye, P., Wüllner, U., Hamberg, L., Shoup, T., Elmaleh, D. R., Frim, D. M., Madras, B. K., Brownell, G. L., Rosen, B. R., and Isacson, O. (1994) PET- and MRI-based assessment of glucose utilization, dopamine receptor binding and hemodynamic changes after lesions to the caudate–putamen in primates. *Exp. Neurol.* **125,** 41–51.

86. Brouillet, E., Hantraye, P., Ferrante, R. J., Dolan, R., Leroy-Willig, A., Kowall, N. W., and Beal, M. F. (1995) Chronic mitochondrial energy impairment produces selective striatal degeneration and abnormal choreiform movements in primates. *Proc. Natl. Acad. Sci. USA* **92,** 7105–7109.

87. Palfi, S., Ferrante, R. J., Brouillet, E., Beal, M. F., Dolan, R., Guyot, M. C., Peschanski, M., and Hantraye, P. (1996) Chronic 3-nitropropionic acid treatment in baboons replicates the cognitive and motor deficits of Huntington's disease. *J. Neurosci.* **16,** 3019–3025.

88. Palfi, S., Condé, F., Riche, D., Brouillet, E., Dautry, C., Mittoux, V., Chibois, A., Peschanski, M., and Hantraye, P. (1998) Fetal striatal allografts reverse cognitive deficits in a primate model of Huntington's disease. *Nat. Med.* **4,** 963–966.

89. Mittoux, V., Joseph, J. M., Condé, F., Palfi, S., Zurn, A., Dautry, C., Poyot, T., Peschanski, M., Aebischer, P., and Hantraye, P. (1998) Encapsulated CNTF-producing fibroblasts reverse motor and cognitive deficits and protect striatal neurons in a chronic primate model of Huntington's disease. *Soc. Neurosci. Abstr.* **24,** 973.

90. Diamond, A. (1990) Developmental time course in human infants and infant monkeys, and the neural bases of inhibitory control in reaching. *Ann. NY Acad. Sci.* **608,** 637–669.

91. Schneider, J. S. (1992). Behavioral and neuropathological consequences of chronic exposure to low doses of the dopaminergic neurotoxin MPTP, in *The Vulnerable Brain and Environmental Risks* (Isaacson, R. L. and Jensen, K. F., eds.), New York, Plenum Press, pp. 293–308.

92. Taylor, J. R., Elsworth, J. D., Roth, R. H., Sladek, J. R., Jr., and Redmond, D. E., Jr. (1990) Cognitive and motor deficits in the acquisition of an object retrieval/detour task in MPTP-treated monkeys. *Brain* **113,** 617–637.

93. Sato, S., Gobel, G. T., Honkaniemi, J., Li, Y., Kondo, T., Murakami, K., Sato, M., Copin, J.-C., and Chan, P. H. (1997) Apoptosis in the striatum of rats following intraperitoneal injection of 3-nitropropionic acid. *Brain Res.* **745,** 343–347.

94. Alexi, T., Hughes, P. E., Knusel, B., and Tobin, A. J. (1998) Metabolic compromise with systemic 3-nitropropionic acid produces striatal apoptosis in Sprague–Dawley rats but not in BALB/c ByJ mice. *Exp. Neurol.* **153,** 74–93.

95. Palfi, S., Riche, D., Brouillet, E., Guyot, M.-C., Mary, V., Wahl, F., Peschanski, M., Stutzmann, J. M., and Hantraye, P. (1997) Riluzole reduces incidence of abnormal movements but not striatal cell death in a primate model of progressive striatal degeneration. *Exp. Neurol.* **146,** 135–141.

18

Transgenic Mouse Models of Huntington's Disease

Gillian P. Bates, Laura Mangiarini, and Stephen W. Davies

1. HUNTINGTON'S DISEASE

1.1. Symptoms and Disease Progression

Huntington's disease (HD) is an autosomal dominant progressive neurodegenerative disorder *(1)*. Onset is generally in mid-life but can range from early childhood to >70 yr and the duration of the illness is of the order of 15–20 yr. Anticipation is observed predominantly when the disease is inherited through the male line, with the result that 70% of juvenile patients inherit the disease from their fathers. The symptoms are complex and variable with recognized emotional, motor, and cognitive components. The motor disorder can differ markedly between the adult and juvenile forms of the disease. Adult-onset HD frequently presents with chorea that can vary from being barely perceptible to extremely severe. It involves all parts of the body, can have repetitive and stereotypic elements and may have a pseudo-purposive appearance *(1)*. Other motor abnormalities include dystonia and oculomotor dysfunction and the disease may progress to an akinetic state. Juvenile patients may never show chorea and are more likely to exhibit a "Parkinson-like" rigidity, myoclonus, tremor, and cerebellar dysfunction and to suffer from epileptic seizures *(1)*. Voluntary movement disorders include fine motor incoordination, dysarthria, and dysphagia. The emotional disorder is commonly depression and irritability and the cognitive component comprises a subcortical dementia. There is no effective therapy for this disease.

1.2. Neuropathology

Neuropathological analysis of HD postmortem brains shows the most striking atrophy to occur in the caudate nucleus which can be reduced to a rim of tissue in severe cases. The putamen and globus pallidus also undergo atrophy and there are subtle changes in the cerebral cortex *(2)*. Neuropathological changes in HD have been classified into five grades that progress from grade 0 in which no gross or microscopic abnormalities can be detected to grade 4 in which the most extreme atrophy is observed *(2)*. A 60% reduction in the cross-sectional area of the caudate, putamen, and globus pallidus increases with the higher grades of HD brains, indicating that these structures progressively degenerate with prolonged survival *(3)*. This specific progressive atro-

From: Central Nervous System Diseases
Edited by: D. F. Emerich, R. L. Dean, III, and P. R. Sanberg © Humana Press Inc., Totowa, NJ

phy appears absent from the small number of grade 0 brains that have been described (from patients with HD symptomology for between 2 and 13 yr) *(2,4,5)*. In addition, a 30% reduction in brain weight has been noted, associated with 20–30% areal reductions in cerebral cortex, white matter, hippocampus, amygdala, and thalamus *(3)*. This atrophy was similar for all grades of HD, did not appear to be associated with gliosis, and the neuronal density in these structures was assessed to be normal *(3)*.

1.3. The HD Gene: Structure and Expression

The HD gene is 180 kb in size, contains 67 exons, and was identified by positional cloning techniques *(6,7)*. This entire genomic region has been sequenced *(8)*. The CAG repeat that is expanded on HD chromosomes is located in exon 1 and encodes a tract of polyglutamine (polyQ) residues. Two ubiquitously expressed transcripts have been identified, of 10,366 bp (IT15) and 13,711 bp, arising by differential polyadenylation *(9)*. Huntingtin protein products translated from expanded alleles have been identified in protein extracts from HD patients *(10–13)*, indicating that the mutation does not block gene expression. Despite selective neuronal vulnerability, the HD transcript is widely expressed in brain and peripheral tissues *(14,15)*. Within the brain there is a widespread predominantly neuronal distribution *(14–16)* that could be detected at 20 wk gestation *(17)*. Immunocytochemistry, electron microscopy, and subcellular fractionations have shown that huntingtin is primarily a cytosolic protein, associated with vesicles and/or microtubules *(11,18–20)*. Within the striatum, the pattern of selective neuronal vulnerability is mirrored by huntingtin expression levels, the projection neurons staining more intensely on immunocytochemistry than the interneurons *(21,22)*.

1.4. The HD Mutation

The CAG repeat that lies within exon 1 of the HD gene is polymorphic in the normal population with an allele size range of $(CAG)_{6-39}$ repeat units *(23,24)*. Hence, the normal protein can tolerate a very wide variation in the length of the polyQ tract. The expanded size range extends from $(CAG)_{35}$ to $(CAG)_{180}$ repeats *(25,26)*, although repeats of greater than $(CAG)_{100}$ are very rare. The vast majority of adult-onset HD patients have expansions falling within the $(CAG)_{40-55}$ range. A negative correlation between repeat size and age at onset has the consequence that tracts of $(CAG)_{70}$ or more invariably cause the juvenile form of the disease *(27,28)*. Although repeat size contributes to age at onset, it is not the only component and other genetic factors clearly play a role.

The HD mutation shows both intergenerational and somatic instability. Expanded CAG repeats are unstable on approx 80% of all transmissions. Expansions and contractions are observed on both male and female transmission, but large repeat expansions are seen almost exclusively when inherited through the male line *(27,29,30)*. This underlies the association between anticipation and paternal inheritance; however, the molecular basis of repeat instability is not yet understood. A modest degree of instability is observed in certain brain regions and non-central nervous system (CNS) tissues that is unlikely to play a role in the pathogenesis of the disease *(29)*.

It appears that at a critical size threshold the polyQ expansions become pathogenic probably by allowing the huntingtin protein to form novel interactions, a mechanism supported by the identification of an antibody that specifically detects pathogenic polyQ tracts *(31)*.

2. A TRANSGENIC MOUSE MODEL OF HD

2.1. Generation of the R6 Transgenic Lines

The dominant nature of the HD mutation predicts that the introduction of a mutated copy of the gene into the mouse germ line might result in a model of HD, despite the presence of two copies of the endogenous mouse *Hdh* gene. Further support for this hypothesis came from the generation of *Hdh* knockouts in which the expression of the *Hdh* gene had been disrupted *(32–34)*. This showed that the HD phenotype is not induced by inactivation of either a single copy or both copies of *Hdh*, indicating that the HD mutation is unlikely to act by haploinsufficiency (reduction of the gene product to 50%) or a dominant negative mechanism (removal of a functional gene product). There are several approaches by which the HD mutation can be introduced into the mouse germ line. Standard transgenesis can be used to inject constructs into fertilized single cell embryos. These are generally cDNA constructs under the control of a suitable promoter or genomic fragments that include the endogenous sequences important for the expression and processing of the transgene. Genomic constructs are frequently more successful, as they ensure the appropriate transgene expression. Alternatively, a CAG expansion could be introduced into the *Hdh* gene by gene targeting.

To date, the only published transgenic mouse model of HD is one expressing a short truncated mutant version of the huntingtin protein. The R6 transgenic lines were produced by pronuclear injection of a genomic fragment encompassing the 5′ end of the HD gene *(35)*. This contained approx 1 kb of promoter sequences, all of exon 1, and 262 bp of intron 1. The CAG repeat expansions were of the order of 130 repeat units in an attempt to exploit the negative correlation between the length of the CAG repeat and age at onset of the disease and to accelerate the onset of a phenotype to within the lifetime of a mouse. A truncated protein corresponding to approx 3% of huntingtin is generated by this fragment. It contains the first 67 amino acids of huntingtin in addition to the length of the polyQ tract. The HDex lines carrying $(CAG)_{18}$ repeat tracts were also produced to control for any phenotypic effect due to the presence of the novel truncated exon 1 protein.

Four R6 lines were established: R6/1, R6/2, R6/5, and R6/0 *(35)*. All lines contain a single intact copy of the transgene with the exception of line R6/5, which has four intact copies. The CAG repeat expansions are of the order of R6/0, $(CAG)_{142}$; R6/2, $(CAG)_{145}$; R6/1, $(CAG)_{115}$; and R6/5, $(CAG)_{128-156}$. The CAG repeats are unstable on transmission and so a definitive CAG repeat size cannot be stated for each line *(36)*. The HDex6 and HDex27 lines had integrated approximately twenty and seven copies of the transgene respectively and each carried $(CAG)_{18}$ repeats. Reverse transcription-polymerase chain reaction (RT-PCR) and Northern and Western analysis showed that the transgene is expressed at the RNA and protein levels in all lines except R6/0 in every tissue and brain region studied *(35,37)*.

The CAG repeat was found to show both intergenerational and somatic repeat stability in lines R6/1, R6/2, and R6/5 *(36)*. The repeats were more likely to increase on paternal transmission and decrease when passed through the female line. The size of the expansions were found to increase with the age of the transmitting male in line R6/2. Somatic instability became apparent from approx 6 wk and increased with age in all three lines. Similarly, in all cases, certain brain regions (including cerebral cortex, striatum, and thalamus) and non-neuronal tissues (in particular liver and kidney) consistently showed the earliest and most pronounced instability. However, the spectrum of instability across all brain regions and tissues was found to differ between lines. Remarkably, line R6/0 did not show any evidence of somatic repeat instability. The mechanism by which instability is induced and the significance of the absence of instability in line R6/0 are not understood.

2.2. The Phenotype Exhibited by the R6 Lines

A progressive neurological phenotype occurs in lines R6/1, R6/2, and R6/5 but not in line R6/0 in which the transgene is not expressed, or in the HDex lines carrying $(CAG)_{18}$ repeat tracts (35). Line R6/2 has been studied most extensively. At birth and at weaning the transgenic mice are indistinguishable from their nontransgenic littermates. The onset of the phenotype is at approx 2 mo of age and in all transgenic mice includes an irregular gait; resting tremor; sudden, abrupt, irregularly timed shuddering movements; stereotypic grooming; and leg clasping when held by the tail. A proportion of the mice exhibit handling-induced epileptic seizures. Coincident with the onset of the motor disorder, the mice cease to gain weight and then their weight begins to fall. The phenotype progresses rapidly over the next month and by 12 wk the R6/2 mice weigh approx 60–70% of their normal littermates. Although sudden deaths can occur from 9 wk, they are rare before 12 or 13 wk. R6/2 mice are only occasionally kept beyond this age.

R6/2 females and approx 50% of R6/2 males are sterile. Therefore, it has not been possible to breed line R6/2 to homozygosity and only hemizygotes have been available for analysis. The same neurological phenotype has been observed in R6/1 hemizygotes (onset 4–5 mo) and R6/5 homozygotes (onset 9 mo). By 18 mo of age, a phenotype has not been observed in the R6/0 line, in which the transgene is not expressed, in the HDex lines carrying $(CAG)_{18}$ repeat tracts or in R6/5 hemizygotes. This gene dosage effect extends beyond line R6/5 in that the age at onset is reduced and the progression accelerated when lines are bred to homozygosity or combinations of R6/1, R6/2, and R6/5 transgenes are interbred. The absence of a phenotype in the HDex lines indicates that the disorder is caused by the polyQ expansions rather than arising as a result of the presence of the novel truncated huntingtin protein.

2.3. Neuronal Intranuclear Inclusions Are Present in the R6 Lines

An initial neuropathological analysis was carried out on thionin-stained serial sections throughout the entire brain and spinal cord from R6/2 mice of 12–13 wk of age. Because these mice express a novel truncated version of the huntingtin protein, care was taken not to bias the analysis to brain regions known to atrophy in HD. The results of this analysis were disappointing as no evidence of a specific neuronal degeneration could be detected (35). The only difference between the transgenic and littermate control brains appeared to be that at 12 wk of age the R6/2 brains weighed approx 20% less than those of their normal littermates. This reduction in size seemed to occur across all brain structures and yet the neuronal density appeared to be normal. A longitudinal study showed that this size reduction begins between 4 and 6 wk of age, prior to the loss in body weight *(37)*.

Immunocytochemistry with antibodies that detect the exon 1 huntingtin protein uncovered the first major significant difference between normal and transgenic mouse brains *(37)*. In normal mouse brain, antibodies to huntingtin detect neuronal staining in all brain regions with reaction product appearing in the cell bodies of neurons, dendrites, and axons, but never in the neuronal nucleus. This pattern of staining was observed in mice of all ages from line R6/0 (no expression of the transgene), and from the $(CAG)_{18}$ HDex lines. In contrast, exon 1 huntingtin antibodies identified an intense focus of staining in neuronal nuclei in symptomatic mouse brains from the R6/1 hemizygous, R6/2 hemizygous, and R6/5 homozygous lines. In R6/2 mice of 12 wk, this staining pattern was present in the cerebral cortex, striatum, cerebellar Purkinje cells, and spinal cord and far fewer were present in the globus pallidus, thalamus, hippocampus, and substantia nigra. To date the only other antibodies that have been found to detect this pattern of staining are those that recognize ubiquitin. Immunoelectron microscopy showed the reaction product to be localized to a neuronal intranuclear inclusion (NII) that could be identified by ultrastructural analysis alone. NII are devoid of a membrane, are slightly larger and distinct from the nucleolus, and only a single NII is present per nucleus. In 12-wk R6/2 mice NII have a pale granular and sometimes fibrillar morphology, whereas NII in older mice from the R6/5 line appeared to have a more fibrillar structure (Fig. 1). Other nuclear changes included invagination of the nuclear membrane and an apparent increase in the density of nuclear pores *(37)*.

One of the major advantages that a transgenic model affords is the ability to correlate neuropathological changes with the progression of the disease phenotype. The failure to detect neuronal cell death by 12–13 wk in line R6/2 suggests that the phenotype is a result of neuronal dysfunction rather that neurodegeneration. In this line, NII could first be detected by exon 1 huntingtin antibodies in the cortex at 3.5 wk followed by the striatum at 4.5 wk *(37)*. The appearance of NII is therefore considerably earlier than the onset of symptoms, suggesting that NII may play a causative role in the R6 phenotype.

2.4. Possible Structure of Neuronal Intranuclear Inclusions

Prompted by the observation that the huntingtin exon 1 protein containing expanded polyQ repeats was sufficient to generate a progressive neurological phenotype in transgenic mice, Scherzinger et al. generated exon 1 fusion proteins to study this fragment in vitro *(38)*. They prepared GST–exon 1 fusion proteins containing polyQ stretches of 20 and 30 in the normal range (GST–HD20 and –HD30) and 51, 83, and 122 in the pathogenic range (GST–HD51, –HD83, and –HD122). After in vitro purification, the GST–HD83 and GST–HD122 proteins were observed to form insoluble aggregates, detectable either by sodium dodecyl sulfate-polyacrylamide gel electrophotresis (SDS-PAGE), the aggregates remaining trapped in the stacking gel, or by a sensitive cellulose acetate filter assay developed by the authors *(38)*. They discovered that the GST tag aided solubility. Its removal with either Factor Xa or digestion of the exon 1 protein with trypsin resulted in a similar aggregation of –HD51; however, the –HD20 and –HD30 fusion proteins containing repeats in the nonpathogenic range remained soluble. It appears that the ability to self-aggregate might be the "gain of function" that pathogenic polyQ repeats impart to the huntingtin protein.

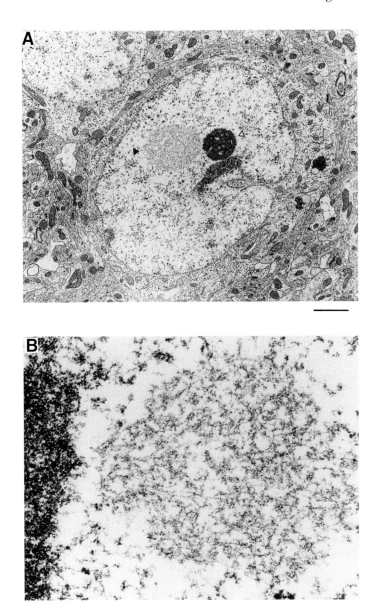

Fig. 1. Electron micrographs showing the morphology of neuronal intranuclear inclusions. **(A)** Striatal neuron from a 12-wk-old R6/2 mouse. NII (*filled arrow*) and nucleolus (*open arrow*) are indicated. Note the characteristic invagination to the nuclear membrane. Scale bar = 2 μm. **(B)** Electron micrograph of a striatal neuron from a 15-mo-old R6/5 homozygous mouse. The NII has a fibrillar morphology and the surrounding chromatin has a clumped appearance. Scale bar = 200 nm.

Electron microscopy of aggregates formed from the Factor Xa GST–HD51 cleavage product showed them to be composed of fibrils with a diameter of 10–12 nm. When stained with Congo red and viewed under cross-polarized light the aggregates appeared green and birefringent *(38)*. This property is characteristic of amyloid proteins *(39)*, suggesting that the fibrils are composed of a cross-β-structure *(40)*. Remarkably, it had already been predicted by Perutz that polyQ tracts could interact via a β-pleated sheet structure to form a polar zipper *(41)*. In addition, Scherzinger et al. were able to show that aggregates purified from the transgenic mouse brains had similar characteristics to those produced in vitro *(38)*. This work indicates that there is no requirement for the inclusions to contain additional proteins in vivo.

2.5. Neurotransmitter Receptor Analysis in the R6 Lines

Many changes in neurotransmitter receptors have been identified in HD striatum including decreases in glutamate, dopamine, γ-aminobutyric acid (GABA), and muscarinic cholinergic receptors *(42–45)*. To explore the relationship of glutamate and other receptors known to be affected in HD to the symptoms in R6/2 mice, receptors were examined in the brains of 4-, 8-, and 12-wk-old mice using receptor binding autoradiography, immunoblotting for receptor proteins, and *in situ* hybridization *(46)*. In 12-wk-old mice, analysis of ionotropic glutamate receptors N-methyl-D-aspartate [NMDA], α-amino-3-hydroxy-5-methyl-4-isooxazolepropionic acid [AMPA], and kainate receptors) showed varying alterations: NMDA receptor binding was unchanged in transgenic mice with respect to normal littermates whereas AMPA and kainate binding were decreased. Of the metabotropic glutamate receptors, group I binding was not statistically different whereas group II binding was decreased. Receptor binding was also decreased for the D_1 and D_2 families of dopamine receptors and the muscarinic acetylcholine receptors whereas no change was found for the $GABA_A$ and $GABA_B$ receptors. Decreases in D_1 and D_2 dopamine receptor binding was observed in the brains of mice at 8 wk and there was a trend toward a decrease in the brains of 4-wk-old mice *(46)*.

It appears that the observed changes in receptor binding may be a function of altered gene expression. At 12 wk, *in situ* hybridization showed marked decreases in mRNA expression for mGluR1, mGluR2, and mGluR3 metabotropic receptors and D_1 and D_2 dopamine receptors. This was unlikely to be a reflection of cell loss as there was no difference in mRNA levels for mGluR5, NMDA-RI, or β-actin. The decreases in mGluR1, D_1, and D_2 dopamine receptor mRNA signals in the striatum were statistically significant by 4 wk of age and the decrease in mGluR2 mRNA expression in the cortex was significant by 8 wk of age. Therefore, by 8 wk there are major alterations in the dopamine and glutamate neurotransmitter systems, both of which are important in striatal function. Changes are detectable as early as 4 wk, long prior to the onset of clinical symptoms and at approximately the same time that NII are first detected in certain brain regions. The changes are not consistent with loss of a specific neuronal cell type and it is possible that the mutant HD protein enters the nucleus and alters the expression of multiple neurotransmitter receptors through interaction with a common transcriptional control element *(46)*.

3. NEURONAL INTRANUCLEAR INCLUSIONS IN POLYGLUTAMINE DISEASE

3.1. NII in HD

The description of NII in the R6 lines led to the identification of these structures in HD postmortem brains *(47,48)*. Remarkably, an ultrastructural analysis of striatal and cortical biopsy material from HD patients had previously detected intranuclear structures identical to the NII seen in the transgenic mice *(49)*, but the significance of this report had not been realized. More recently, NII have been described in HD postmortem brains. NII are found at high frequency in juvenile-onset HD (38–52% of all cortical neurons); however, dystrophic neurites (axonal inclusions) were found to be more prominent in adult disease brains *(47)*. The distribution of NII appears more widespread than the pattern of HD-associated neurodegeneration, with NII most frequent in the neocortex and also present in the striatum, amygdala, hippocampus, red nucleus, and dentate nucleus of the cerebellum *(48)*. Both NII and dystrophic neurites are immunoreactive for N-terminal huntingtin and ubiquitin antibodies but could not be detected with antibodies raised against more C-terminal huntingtin epitopes *(47,48)*. One interpretation of this finding is that the aggregates contain only the N-terminus of huntingtin, which is supported by the detection of N-terminal 40-kDa fragments on Western blots from juvenile HD brains *(47)*. The identification of dystrophic neurites in cortical layer VI of a presymptomatic HD brain suggests that in patients, as in transgenic mice, these structures have formed prior to the onset of symptoms *(47)*.

3.2. NII in Other Polyglutamine Neurodegenerative Diseases

In addition to HD there are an increasing number of neurodegenerative diseases caused by a CAG/polyQ expansion which include dentatorubral pallidoluysian atrophy (DRPLA), spinal and bulbar muscular atrophy (SBMA), and the spinocerebellar ataxias SCA1, SCA2, SCA3 (also known as Machado Joseph disease), SCA6 (reviewed in *[50]*), and SCA7 *(51)*. Broad similarities in the inheritance patterns of these diseases and the comparability of the normal and pathogenic CAG repeat ranges would suggest that they share a common mutational mechanism. This is proving to be the case and, in addition to HD, NII have now been identified in SCA1 *(52)*, SCA3 *(53)*, and DRPLA *(48,54)* postmortem brains and in the cerebellar Purkinje cells of SCA1 transgenic mice *(52)*. In each case the inclusions are nuclear and are immunoreactive for antibodies to the polyQ-containing protein in question and to ubiquitin. As in HD, the distribution of inclusions in DRPLA is more widespread than the pattern of neuropathology, with inclusions being detected in the neocortex and the caudate nucleus in addition to the dentate nucleus of the cerebellum *(48)*.

4. CONCLUSION

These recent advances have led to a complete revision of the molecular mechanism that causes HD. It appears that huntingtin needs to be cleaved to form an N-terminal toxic fragment *(47,48,55)*. Cleavage may occur via specific proteases, for example, huntingtin has been shown to be cleaved by caspase 3 *(56)*, or may be the result of a more general proteolysis. It appears that only N-terminal fragments can move into the nucleus which may occur by passive diffusion or via an active transport mechanism *(55)*. These fragments can form highly insoluble amyloid-like aggregates via polyQ polar zippers *(38)*. In the transgenic lines and in juvenile HD, intranuclear inclusions are mostly observed; however, aggregates form in axons over a more protracted time course. If the seeding of the aggregates occurs via a random nucleation step, its rate is likely to be governed by the local fragment concentration and by its polyQ length. Transgenic mouse studies have clearly shown that a twofold increase in the concentration of the exon 1 protein accelerates the onset of the phenotype *(35)* associated with an increases the rate of NII formation. The work of Scherzinger et al. *(38)* demonstrates that further interactions are not a requirement of aggregate formation although it has yet to be ruled out that other factors do not aid nucleation in vivo. Inclusions have been found to be immunoreactive for huntingtin and ubiquitin antibodies and appear to become ubiquitinated once they have formed in vivo *(37)*.

The R6 transgenic lines have thus far led to profound insights into the molecular basis of HD. In line R6/2, NII have been identified in the brain from 3.5 wk *(37)*, alterations in specific neurotransmitter receptors have been identified by 4 wk *(46)*, the onset of the movement disorder is at approx 2 mo, and yet by 12 wk there is no evidence of a specific neurodegeneration in R6/2 brains. This suggests that the phenotype arises as a consequence of neuronal dysfunction rather than cell death. It remains to be established as to whether the formation of inclusions leads to the neuronal dysfunction, or whether this is brought about by earlier steps in the pathogenesis. This work has important implications with respect to the development of therapeutic strategies, the focus of which is likely to turn away from the prevention of cell death and the replacement of degenerate neurons.

REFERENCES

1. Harper, P. S., ed. (1991) *Huntington's Disease*, W. B. Saunders, London.
2. Vonsattel, J.-P., Myers, R. H., Stevens, T. J., Ferrante, R. J., Bird, E. D., and Richardson, E. P. (1985) Neuropathological classification of Huntington's disease. *J. Neuropathol. Exp. Neurol.* **44,** 559–577.
3. de la Monte, S. M., Vonsattel, J.-P., and Richardson, E. P. (1988) Morphometric demonstration of atrophic changes in the cerebral cortex, white matter and neostriatum in Huntington's disease. *J. Neuropathol. Exp. Neurol.* **47,** 516–525.
4. Myers, R. H., Vonsattel, J. P., Stevens, T. J., Cupples, L. A., Richardson, E. P., Martin, J. B., and Bird, E. D. (1988) Clinical and neuropathological assessment of severity in Huntington's disease. *Neurology* **38,** 341–347.
5. Hedreen, J. C. and Folstein, S. E. (1995) Early loss of early neostriatal neurons in Huntington's disease. *J. Neuropathol. Exp. Neurol.* **54,** 105–120.
6. Huntington's Disease Collaborative Research Group (HDCRG) (1993) A novel gene containing a trinucleotide repeat that is unstable on Huntington's disease chromosomes. *Cell* **72,** 971–983.
7. Ambrose, C. M., Duyao, M. P., Barnes, G., Bates, G. P., Lin, C. S., Srinidhi, J., Baxendale, S., Hummerich, H., Lehrach, H., Altherr, M., Wasmuth, J., Buckler, A., Church, D., Housman, D., Berks, M., Micklem, G., Durbin, R., Dodge, A., Read, A., Gusella, J., and MacDonald, M. E. (1993) Structure and expression of the Huntington's disease gene: evidence against simple inactivation due to an expanded CAG repeat. *Somat. Cell Mol. Genet.* **20,** 27–38.
8. Baxendale, S., Abdulla, S., Elgar, G., Bucks, D., Berks, M., Micklem, G., Durbin, R., Bates, G., Brenner, S., Beck, S., and Lehrach, H. (1995) Comparative sequence analysis of the human and puffer fish Huntington's disease gene. *Nat. Genet.* **10,** 67–75.
9. Lin, B., Rommens, J. M., Graham, R. K., Kalchman, M., MacDonald, H., Nasir, J., Delaney, A., Goldberg, Y. P., and Hayden, M. R. (1993) Differential 3' polyadenylation of the Huntington disease gene results in two mRNA species with variable tissue expression. *Hum. Mol. Genet.* **2,** 1541–1545.
10. Jou, Y.-S. and Myers, R. M. (1995) Evidence from antibody studies that the CAG repeat in the Huntington disease gene is expressed in the protein. *Hum. Mol. Genet.* **4,** 465–469.

11. Trottier, Y., Devys, D., Imbert, G., Sandou, F., An, I., Lutz, Y., Weber, C., Agid, Y., Hirsch, E. C., and Mandel, J.-L. (1995) Cellular localisation of the Huntington's disease protein and discrimination of the normal and mutated forms. *Nat. Genet.* **10,** 104–110.

12. Ide, K., Nobuyuki, N., Masuda, N., Goto, J., and Kanazawa, I. (1995) Abnormal gene product identified in Huntington's disease lymphocytes and brain. *Biochem. Biophys. Res. Commun.* **209,** 1119–1125.

13. Schilling, G., Sharp, A. H., Loev, S. J., Wagster, M. V., Li, S.-H., Stine, O. C., and Ross, C. A. (1995) Expression of the Huntington's disease (IT15) protein product in HD patients. *Hum. Mol. Genet.* **4,** 1365–1371.

14. Li, S. H., Schilling, G., Young, W. S., Li, X. J., Margolis, R. L., Stine, O. C., Wagster, M. V., Abbott, M. H., Franz, M. L., Ranen, N. G., Folstein, S. E., Hedreen, J. C., and Ross, C. A. (1993) Huntington's disease gene (it-15) is widely expressed in human and rat tissues. *Neuron* **11,** 985–993.

15. Strong, T. V., Tagle, D. A., Valdes, J. M., Elmer, L. W., Boehm, K., Swaroop, M., Kaatz, K. W., Collins, F. S., and Albin, R. L. (1993) Widespread expression of the human and rat Huntington's disease gene in brain and nonneuronal tissues. *Nat. Genet.* **5,** 259–263.

16. Landwehrmeyer, G. B., McNeil, S. M., Dure, L. S., Ge, P., Aizawa, H., Huang, Q., Ambrose, C. M., Duyao, M. P., Bird, E. D., Bonilla, E., de Young, M., Avila-Gonzales, A. J., Wexler, N. S., DiFiglia, M., Gusella, J. F., MacDonald, M. E., Penney, J. B., Young, A. B., and Vonsattel, J.-P. (1995) Huntington's disease gene: regional and cellular expression in brain of normal and affected individuals. *Ann. Neurol.* **37,** 218–230.

17. Dure, L. S., Landwehrmeyer, G. B., Golden, J., McNeil, S., Ge, P., Aizawa, H., Huang, Q., Ambrose, C. M., Duyao, M. P., Bird, E. D., DiFiglia, M., Gusella, J. F., MacDonald, M. E., Penney, J. B., Young, A. B., and Vonstattel, J.-P. (1994) IT15 gene expression in fetal human brain. *Brain Res.* **659,** 33–41.

18. Sharp, A. H., Loev, S. J., Schilling, G., Li, S.-H., Li, X.-J., Bao, J., Wagster, M. V., Kotzuk, J. A., Steiner, J. P., Lo, A., Hedreen, J., Sisodia, S., Snyder, S. H., Dawson, T. M., Ryugo, D. K., and Ross, C. A. (1995) Widespread expression of Huntington's disease gene (IT15) protein product. *Neuron* **14,** 1065–1074.

19. DiFiglia, M., Sapp, E., Chase, K., Schwarz, C., Meloni, A., Young, C., Martin, E., Vonstattel, J.-P., Carraway, R., Reeves, S. A., Boyce, F. M., and Aronin, N. (1995) Huntingtin is a cytoplasmic protein associated with vesicles in human and rat brain neurons. *Neuron* **14,** 1075–1081.

20. Gutekunst, C.-A., Levey, A. I., Heilman, C. J., Whaley, W. L., Yi, H., Nash, N. R., Rees, H. D., Madden, J. J., and Hersch, S. M. (1995) Identification and localisation of huntingtin in brain and human lymphoblastoid cell lines with anti-fusion protein antibodies. *Proc. Natl. Acad. Sci. USA* **92,** 8710–8714.

21. Kosinski, C. M., Cha, J. H., Young, A. B., Persichetti, F., MacDonald, M., Gusella, J. F., Penney, J. B., and Standaert, D. G. (1997) Huntingtin immunoreactivity in the rat neostriatum: differential accumulation in projection and interneurons. *Exp. Neurol.* **144,** 239–247.

22. Ferrante, R. J., Gutekunst, C.-A., Persichetti, F., McNeil, S. M., Kowall, N. W., Gusella, J. F., MacDonald, M. E., Beal, M. F., and Hersch, S. M. (1997) Heterogeneous topographic and cellular distribution of huntingtin expression in the normal human neostriatum. *J. Neurosci.* **17,** 3052–3063.

23. Stine, O. C., Pleasant, N., Franz, M. L., Abbott, M. H., Folstein, S. E., and Ross, C. A. (1993) Correlation between the onset age of Huntington's disease and length of the trinucleotide repeat in IT-15. *Hum. Mol. Genet.* **2,** 1547–1549.

24. Rubinsztein, D. C., Leggo, J., Coles, R., Almqvist, E., Biancalana, V., Cassiman, J.-J., Chotai, K., Connarty, M., Crauford, D., Curtis, A., Curtis, D., Davidson, M. J., Differ, A.-M., Dode, C., Dodge, A., Frontali, M., Ranen, N. G., Stine, O. C., Sherr, M., Abbott, M. H., Franz, M. L., Graham, C. A., Harper, P. S., Hedreen, J. C., Jackson, A., Kaplan, J.-C.,

Losekoot, M., MacMillan, J. C., Morrison, P., Trottier, Y., Novelletto, A., Simpson, S. A., Theilmann, J., Whittaker, J. L., Folstein, S. E., Ross, C. A., and Hayden, M. R. (1996) Phenotypic characterisation of individuals with 30–40 CAG repeats in the Huntington's disease (HD) gene reveals HD cases with 36 repeats and apparently normal elderly individuals with 36–39 repeats. *Am. J. Hum. Genet.* **59**, 16–22.

25. Barron, L. H., Warner, J. P., Porteous, M., Holloway, S., Simpson, S., Davidson, R., and Brock, D. J. H. (1993) A study of the Huntington's disease associated trinucleotide repeat in the Scottish population. *J. Med. Genet.* **30**, 1003–1007.

26. Sathasivam, K., Amaechi, I., Mangiarini, L., and Bates, G. P. (1997) Identification of an HD patient with a $(CAG)_{180}$ repeat expansion and the propagation of highly expanded CAG repeats in lambda phage. *Hum. Genet.* **99**, 692–695.

27. Duyao, M., Ambrose, C., Myers, R., Novelletto, A., Persichetti, F., Frontali, M., Folstein, S., Ross, C., Franz, M., Abbott, M., Gray, J., Conneally, M. P., Young, A., Penney, J., Hollingsworth, Z., Shoulson, I., Lazzarini, A., Falek, A., Koroshetz, W., Sax, D., Bird, E., Vonsattel, J., Bonilla, E., Alvir, J., Conde, J. B., Cha, J.-H., Dure, L., Gomez, F., Ramos, M., Sanchez-Ramos, J., Snodgrass, S., de Young, M., Wexler, N., Moscowitz, C., Penchaszadeh, G., MacFarlane, H., Anderson, M., Jenkins, B., Srinidhi, J., Barnes, G., Gusella, J., and MacDonald, M. (1993) Trinucleotide repeat length instability and age of onset in Huntington's disease. *Nat. Genet.* **4**, 387–392.

28. Telenius, H., Kremer, H. P. H., Theilmann, J., Andrew, S. E., Almquist, E., Anvret, M., Greenberg, C., Greenberg, J., Lucotte, G., Squitieri, F., Starr, E., Goldberg, Y. P., and Hayden, M. R. (1993) Molecular analysis of juvenile Huntington disease: the major influence on $(CAG)_n$ repeat length is the sex of the affected parent. *Hum. Mol. Genet.* **2**, 1535–1540.

29. Telenius, H., Kremer, B., Goldberg, Y. P., Theilmann, J., Andrew, S. E., Zeisler, J., Adam, S., Greenberg, C., Ives, E. J., Clarke, L. A., and Hayden, M. R. (1994) Somatic and gonadal mosaicism of the Huntington disease gene CAG repeat in brain and sperm. *Nat. Genet.* **6**, 409–413.

30. Ranen, N. G., Stine, O. C., Abbott, M. H., Sherr, M., Codori, A.-M., Franz, M. L., Chao, N. I., Chung, A. S., Pleasant, N., Callahan, C., Kasch, L. M., Ghaffari, M., Chase, G. A., Kazazian, H. H., Brandt, J., Folstein, S. E., and Ross, C. A. (1995) Anticipation and instability of IT15 (CAG)n repeats in parent-offspring pairs with Huntington's disease. *Am. J. Hum. Genet.* **57**, 593–602.

31. Trottier, Y., Lutz, Y., Stevanin, G., Imbert, G., Devys, D., Cancel, G., Sandou, F., Weber, C., David, G., Tora, L., Agid, Y., Brice, A., and Mandel, J.-L. (1995) Polyglutamine expansion as a pathological epitope in Huntington's disease and four dominant cerebellar ataxias. *Nature* **378**, 403–406.

32. Duyao, M. P., Auerbach, A. A., Ryan, A., Persichetti, F., Barnes, G. T., McNeil, S. M., Ge, P., Vonstattel, J.-P., Gusella, J. F., Joyner, A. L., and MacDonald, M. E. (1995) Inactivation of the mouse Huntington's disease gene homolog *Hdh. Sci.* **269**, 407–410.

33. Nasir, J., Floresco, S. B., O'Kusky, J. R., Diewert, V. M., Richman, J. M., Zeisler, J., Borowski, A., Marth, J. D., Phillips, A. G., and Hayden, M. R. (1995) Targeted disruption of the Huntington's disease gene results in embryonic lethality and behavioral and morphological changes in heterozygotes. *Cell* **81**, 811–823.

34. Zeitlin, S., Liu, J.-P., Chapman, D. L., Papaioannou, V. E., and Estradiatis, A. (1995) Increased apoptosis and early embryonic lethality in mice nullizygous for the Huntington's disease gene homologue. *Nat. Genet.* **11**, 155–163.

35. Mangiarini, L., Sathasivam, K., Seller, M., Cozens, B., Harper, A., Hetherington, C., Lawton, M., Trottier, Y., Lehrach, H., Davies, S. W., and Bates, G. P. (1996) Exon 1 of the Huntington's disease gene containing a highly expanded CAG repeat is sufficient to cause a progressive neurological phenotype in transgenic mice. *Cell* **87**, 493–506.

36. Mangiarini, L., Sathasivam, K., Mahal, A., Mott, R., Seller, M., and Bates, G. P. (1997) Instability of highly expanded CAG repeats in transgenic mice is related to expression of the transgene. *Nat. Genet.* **15,** 197–200.

37. Davies, S. W., Turmaine, M., Cozens, B. A., DiFiglia, M., Sharp, A. H., Ross, C. A., Scherzinger, E., Wanker, E. E., Mangiarini, L., and Bates, G. P. (1997) Formation of neuronal intranuclear inclusions (NII) underlies the neurological dysfunction in mice transgenic for the HD mutation. *Cell,* **90,** 537–548.

38. Scherzinger, E., Lurz, R., Turmaine, M., Mangiarini, L., Hollenbach, B., Hasenbank, R., Bates, G. P., Davies, S. W., Lehrach, H., and Wanker, E. E. (1997) Huntingtin encoded polyglutamine expansions form amyloid-like protein aggregates in vitro and in vivo. *Cell* **90,** 549–558.

39. Caputo, C. B., Fraser, P. E., Sobel, I. E., and Krischner, D. A. (1992) Amyloid-like properties of a synthetic peptide corresponding to the carboxy terminus of b-amyloid protein precursor. *Arch. Biochem. Biophys.* **292,** 199–205.

40. Sunde, M. and Blake, C. (1997) The structure of amyloid fibrils by electron microscopy and X-ray diffraction. *Adv. Prot. Chem.* **50,** 123–159.

41. Perutz, M. F., Johnson, T., Suzuki, M., and Finch, J. T. (1994) Glutamine repeats as polar zippers: their possible role in inherited neurodegenerative diseases. *Proc. Natl. Acad. Sci. USA* **91,** 5355–5358.

42. Dure, L. S., Young, A. B., and Penney, J. B. (1991) Excitatory amino acid binding sites in the caudate nucleus and frontal cortex of Huntington's disease. *Ann. Neurol.* **30,** 785–793.

43. London, E. D., Yamamura, H. I., Bird, E. D., and Coyle, J. T. (1981) Decreased receptor binding sites for kainic acid in brains of patients with Huntington's disease. *Biol. Psychiatry* **16,** 155–162.

44. Penney, J. B. and Young, A. B. (1982) Quantitative autoradiography of neurotransmitter receptors in Huntington's disease. *Neurology* **32,** 1391–1395.

45. Richfield, E. K., O'Brien, C. F., Eskin, T., and Shoulson, I. (1991) Heterogeneous dopamine receptor changes in early and late Huntington's disease. *Neurosci. Lett.* **132,** 121–126.

46. Cha, J.-H. J., Kosinski, C. M., Kerner, J. A., Alsdorf, S. A., Mangiarini, L., Davies, S. W., Penney, J. B., Bates, G. P., and Young, A. B. (1998) Altered brain neurotransmitter receptors in transgenic mice expressing a portion of an abnormal human Huntington's disease gene. *Proc. Natl. Acad. Sci. USA* **95,** 6480–6485.

47. DiFiglia, M., Sapp, E., Chase, K. O., Davies, S. W., Bates, G. P., Vonsattel, J.-P., and Aronin, N. (1997) Aggregation of huntingtin in neuronal intranuclear inclusions and dystrophic neurites in brain. *Science* **277,** 1990–1993.

48. Becher, M. W., Kotzuk, J. A., Sharp, A. H., Davies, S. W., Bates, G. P., Price, D. L., and Ross, C. A. (1998) Intranuclear neuronal inclusions in Huntington's disease and dentatorubral pallidoluysian atrophy: correlation between the density of inclusions and IT15 CAG repeat length. *Neurobiol. Dis.* **4,** 387–395.

49. Roizin, L., Stellar, S., and Liu, J. C. (1979) Neuronal nuclear-cytoplasmic changes in Huntington's chorea: electron microscope investigations, in *Advances in Neurology,* Vol. 23 (Chase, T. N., Wexler, N. S., and Barbeau, A., eds.), Raven Press, New York, pp. 95–122.

50. Reddy, P. S. and Housman, D. E. (1997) The complex pathology of trinucleotide repeats. *Curr. Opin. Cell Biol.* **9,** 364–372.

51. David, G., Abbas, N., Stevanin, G., Durr, A., Yvert, G., Cancel, G., Weber, C., Imbert, G., Saudou, F., Antoniou, E., Drabkin, H., Gemmill, R., Giunti, P., Benomar, A., Wood, N., Ruberg, M., Agid, Y., Mandel, J. L., and Brice, A. (1997) Cloning of the SCA7 gene reveals a highly unstable CAG repeat expansion. *Nat. Genet.* **17,** 65–70.

52. Skinner, P. J., Koshy, B. T., Cummings, C. J., Klement, I. A., Helin, K., Servadio, A., Zoghbi, H. Y., and Orr, H. T. (1997) Ataxin-1 with an expanded glutamine tract alters nuclear matrix-associated structures. *Nature* **389,** 971–974.

53. Paulson, H. L., Perez, M. K., Trottier, Y., Trojanowski, J. Q., Subramony, S. H., Das, S. S., Vig, P., Mandel, J.-L., Fischbeck, K. H., and Pittman, R. N. (1997) Intranuclear inclusions of expanded polyglutamine protein in spinocerebellar ataxia type 3. *Neuron* **19,** 1–20.

54. Igarashi, S., Koide, R., Shimohata, T., Yamada, M., Hayashi, Y., Takano, H., Date, H., Oyake, M., Sato, T., Sato, A., Egawa, S., Ikeuchi, T., Tanaka, H., Nakano, R., Tanaka, K., Hozumi, I., Inuzuka, T., Takahashi, H., and Tsuji, S. (1998) Suppression of aggregate formation and apoptosis by transglutaminase inhibitors in cells expressing truncated DRPLA protein with an expanded polyglutamine stretch. *Nat. Genet.* **18,** 111–117.

55. Martindale, D., Hackman, A., Wieczorek, A., Ellerby, L., Wellington, C., McCutcheon, K., Singaraja, R., Kazemi-Esfarjani, P., Devon, R., Kim, S. U., Bredesen, D. E., Tufaro, F., and Hayden, M. R. (1998) Length of huntingtin and its polyglutamine tract influences localisation and frequency of intracellular aggregates. *Nat. Genet.* **18,** 150–154.

56. Goldberg, Y. P., Nicholson, D. W., Rasper, D. M., Kalchman, M. A., Koide, H. B., Graham, R. K., Bromm, M., Kazemi-Esfarjani, P., Thornberry, N. A., Vaillancourt, J. P., and Hayden, M. R. (1996) Cleavage of huntingtin by apopain, a proapoptotic cysteine protease, is modulated by the polyglutamine tract. *Nat. Genet.* **13,** 442–449.

IV

TRAUMATIC BRAIN INJURY AND STROKE

19

Rigid Indentation Models
of Traumatic Brain Injury in the Rat

Richard L. Sutton

1. INTRODUCTION

The spectrum of pathobiological events occurring with traumatic brain injury (TBI) can be extremely diverse and heterogeneous in nature, reflecting not only the mechanism(s), location, and severity of the initial or primary damage but also the progression of reactive events that contribute to delayed, or secondary damage. Over the last five decades a variety of rodent models of TBI have been developed in attempts to mimic the clinical features of closed head injury. Given the diverse causes and biomechanical forces that contribute to the heterogeneity of human TBI, it is not realistic to expect that any single rodent model can fully replicate the spectrum of pathobiological events that contribute to morbidity and mortality in humans. This chapter reviews some of the objectives, applications, experimental findings, and limitations of some rat models of TBI currently in use. In this overview the models discussed include those that are generally categorized as rigid indentation or direct brain deformation (i.e., a percussion concussion injury induced after a craniectomy), most of which have been developed within the past decade. For overviews of TBI model developments in numerous animal species, including fluid percussion and impact acceleration (closed head injury) models of TBI in rodents, the readers are referred to previous reviews (1–8). Readers interested in penetrating or missile injury models of brain injury are referred to a recent review by Torbati (9). Although some duplication of materials included in other reviews of experimental head injury necessarily occurs in this chapter, this has been minimized as much as possible.

2. CONSEQUENCES, CAUSES, AND CHARACTERISTICS
OF HUMAN HEAD INJURY

The consequences of head injury depend on the magnitude and mechanisms of mechanical input forces and range from a relatively benign concussive syndrome characterized by a transient loss of consciousness with no detectable histopathology to prolonged coma and/or death which is usually associated with extensive damage to brain tissues and/or vasculature. Mild head injury in humans may produce no residual defi-

From: Central Nervous System Diseases
Edited by: D. F. Emerich, R. L. Dean, III, and P. R. Sanberg © Humana Press Inc., Totowa, NJ

cits in the majority of patients, but can also result in disability in as many of 10–18% of these individuals *(10)*. Morbidity in such cases may be related to axonal damage and/or to alterations in cerebral blood flow, cerebral metabolism, or neurochemistry *(11,12)*. The neuropathological sequelae of moderate-to-severe TBI in humans frequently include cortical contusion(s), epidural and/or subdural hematoma, intracranial hemorrhage, and/or diffuse axonal injury as well as secondary complications including ischemia, cerebral edema/elevated intracranial pressure, and multiple neurochemical abnormalities *(13–19)*. Survivors of these relatively severe injuries will have a higher incidence, duration, and extent of disability *(10,20)*. Motor impairments impacting upon mobility, self-care, and basic activities of daily living play an important role in the initial recovery phase and may persist for years after moderate-to-severe TBI, but are seldom a factor after mild TBI. Cognitive impairments are the most enduring sequelae after all severities of TBI, with the injuries producing impairments of attention, learning, and memory abilities *(15,20–23)*.

The principal mechanisms leading to primary brain damage after nonmissile head injury have traditionally been attributed to either contact damage or to acceleration/ deceleration forces *(3,6,7,19)*. Contact injuries, induced by a blow to the head (e.g., assault) or by the moving head contacting a stationary material (e.g., falls), are likely to be associated with focal injuries including epidural or subdural hematoma, contusion and/or laceration of the brain surface, and intracerebral or petechial hemorrhage/ hematoma. Acceleration/deceleration injury, induced by rapid movement of the brain mass within the cranial vault in the moments after head movement and impact (e.g., road traffic accidents), may be associated with focal damage such as acute subdural hematoma and contusion and/or with diffuse injury (diffuse axonal injury and diffuse brain swelling).

Surface contusions resulting from skull fracture or caused by movement and contact of the brain with bony protuberances within the skull upon head impact are characteristically focal injuries distributed on the frontal and temporal poles, frequently on the inferior surfaces of these lobes, or in tissue surrounding the Sylvian fissure. Such contusions typically produce damage to intraparenchymal vessels with resultant hemorrhage and edema formation within gray matter of the affected gyrus/gyri. Coup lesions refer to contusions occurring at the site of initial impact, with contrecoup lesions occurring at the point opposite to impact. The gliding contusion, thought to result from shear forces damaging vessels at the gray–white matter interface, is characterized as a hemorrhagic lesion generally localized in the parasagittal cortex and is frequently associated with deep (basal ganglia) hematoma and with diffuse axonal injury. Severe closed head injury in humans frequently (estimates range from 40% to 55%) results in a diffuse injury, characterized by diffuse brain swelling with either an occasional petechial hemorrhage or an otherwise normal CT scan. Diffuse axonal injury occurs frequently in white matter/fiber tracts of the cerebral hemispheres, the corpus callosum, and the brain stem, with occasional evidence of axonal injury seen in the cerebellum. This type of injury is thought to be the most common cause of persistent coma or vegetative state and severe disability, and histological evidence of diffuse axonal injury has been demonstrated in up to 28% of fatalities resulting from head injury *(16)*.

Some common complications arising in patients sustaining moderate-to-severe closed head injury that contribute to secondary injury, morbidity, and mortality include elevated intracranial pressure and ischemic brain damage. Intracranial pressure may increase due to an increase in cerebral blood volume, the presence of mass lesions due to hematoma, and/or brain swelling *(17)*. Elevated intracranial pressure, in concert with the secondary insults of hypoxia and hypotension occurring after TBI, contribute to the development of cerebral ischemia *(18,24)*. Clinical studies have shown that global or regional reductions in cerebral blood flow that correlate with functional outcomes are most likely to occur within the first few hours after TBI *(25–27)*.

3. DESIRED CHARACTERISTICS OF EXPERIMENTAL MODELS

Although the goals and objectives of individual investigators using experimental models of TBI may differ, it is apparent that some common characteristics and requirements apply to all such models. It is generally acknowledged that the model chosen needs to be reproducible and quantifiable with regard to producing a clinically relevant outcome (anatomical, physiological, neurochemical, or behavioral) and that the magnitude and type of mechanical input used to create the injury must be able to be graded, controlled, reproducible, and quantifiable. Ideally, the magnitude of mechanical input used to create the injury should be directly related to, and predictive of, the severity of the outcome variable(s) chosen for study *(4,5)*.

4. METHODS AND VARIATIONS USED IN RAT RIGID INDENTATION MODELS OF TBI

4.1. Weight-Drop Models

4.1.1. Lateral Weight-Drop onto a Dural Impounder

Various adaptations of the cortical compression model developed by Feeney and colleagues *(28,29)* have been frequently used by investigators employing weight-drop methods to induce brain trauma. In such models a craniotomy (usually circular) is generally placed over the lateral temporo-parietal cortex, leaving dura intact. Cortical compression is induced by dropping cylindrical objects of various mass from various distances down a guide tube and onto a cylindrical impounder resting upon the dura mater. The extent of tissue compression upon impact can be varied and controlled by altering the distance that the impounder recedes up into the guide tube *(see 28)*. The magnitude of injury with such weight-drop methods is traditionally expressed in g-cm (mass times distance dropped, e.g. a 20-g weight dropped from 20 cm is expressed as a 400 g-cm injury).

4.1.2. Lateral Weight-Drop Directly onto Dura

Some investigators *(30,32)* have utilized a modification of the weight-drop model wherein a lateral craniotomy is placed over the parietal cortex and injury is produced by dropping a weight down a guide tube to directly impact on the exposed dura. In published reports, the mass (e.g., 10 g) and distance (e.g., 5 cm) that the weight is dropped when using this technique are generally smaller than those reported when using a dural impounder, presumably to avoid creation of a penetrating injury. Control over the degree of tissue compression in this dural impact model is less that when weights are dropped onto a dural impounder.

4.2. Controlled Cortical Impact Models

The characteristic of controlled cortical impact (CCI) models of TBI is the use of stroke-constrained and pneumatically driven impactor to deliver impact forces to the exposed dura and underlying cortex. The level of injury can be altered by changing impact velocity (m/s; controlled by lowering or raising gas/air pressure to the pneumatic impactor) and/or by altering the depth of tissue compression *(5)*.

4.2.1. Central CCI

The original report using the CCI method of inducing TBI in the rat utilized a 10 mm diameter impactor with a spherically shaped tip that was attached to the lower end of the pneumatic piston to deliver graded impacts through a 10 mm diameter craniectomy centered over midline and between bregma and lambda *(32)*. Subsequent work *(103,111)* with this model has used this same method, albeit injury levels vary in these studies.

4.2.2. Lateral CCI

Subsequent researchers using the CCI method of experimental TBI have generally utilized smaller diameter impactor tips and placement of a lateral craniotomy over one tempero-parietal cortex. The initial report on this lateral CCI injury model utilized a 5 mm diameter impactor tip having a flattened surface with beveled edges *(33)*. As with central CCI, the lateral CCI method of TBI can be used to induce varying grades of injury severity by altering impact velocity (m/s) and/or by altering the depth of tissue compression.

4.2.3. Lateral CCI with Bilateral Craniotomy

In an effort to increase the extent of axonal damage, some investigators have altered the lateral CCI method by creating craniotomies bilaterally, and then inducing a unilateral CCI through one craniotomy site (e.g., *34*).

4.2.4. Bilateral Frontal CCI

Finally, some investigators have used the CCI method to induce a bilateral compression of the frontal cortices after creation of a midline craniotomy anterior to bregma (e.g., *35*).

4.3. Variations Potentially Influencing Injury Profiles

As has been partially outlined by other authors with regards to the rat CCI model *(2)*, and as suggested in the preceding outline of six general models, variations to the rat models of rigid indentation that may influence outcome variables studied have been numerous. Variations apparent in the rigid indentation models include alterations to the: (1) site of impact (e.g., central, bilateral frontal, or lateral [left or right hemisphere]); (2) attempts to replace the bone flap (craniotomy vs craniectomy) or seal the cranial defect; (3) size and shape (circular vs rectangular) of the craniotomy or craniectomy; (4) size and shape of the impactor tip; (5) distance or clearance between the edges of the impactor tip and the edges of the craniotomy/craniectomy; (6) the extent of tissue compression; (7) impact velocity or force; (8) angle of the impact velocity/

forces relative to brain surface; (9) use of secondary craniotomies; (10) anesthetics utilized, including presence or absence of ventilatory support; and (11) use of postoperative analgesics and/or antibiotics. Such variations, although not unexpected given the diverse interests of increasing numbers of investigators who have begun to utilize the rigid indentation models of TBI, can result in diverse injury profiles and has prevented standardization of the injury models. The influence of each of the above variables on the numerous pathobiological components of TBI or on the specific endpoints chosen for study by individual investigators will not be fully understood without extensive further research.

Standardization efforts would be greatly enhanced if all investigators itemized the specific methodology and any variations to a preceding model they utilize during any individual study. For example, some published reports have failed to give details on the shape or size of the impactor tip utilized, thus making comparisons between the results reported in different articles difficult. Investigators frequently utilize the terms "mild," "moderate," or "severe" in the description of injury levels used within individual articles. It should be recognized, however, that these terms are relative and there is no currently established standard in the field for these classifications of injury, particularly when trying to compare reported injury severities across different injury models.

In the material below (**Subheadings 5.–8.**), the generally shared characteristics of various rigid indentation models of rodent TBI are discussed as they relate to pathobiological events occurring after human TBI. It is hoped this chapter will illustrate the utility of model variants for producing outcomes or endpoints relevant to the human condition, as well as limitations and areas needing further work within each model. Illustration of these issues is aided by organization of material into a table, similar to that used to summarize several animal models of TBI *(36) (see* Table 1). References to publications using one of the six rigid indentation models under review, and reporting on one or more of several structural and functional changes associated with TBI, are provided within the body of Table 1.

5. STRUCTURAL CHANGES AND CELLULAR REACTIONS ASSOCIATED WITH RIGID INDENTATION

5.1. Contusion and Hemorrhage

In general, all rodent models of rigid indentation are capable of producing cerebral concussion and localized surface contusions *(28,31,33,35,37–39)*. Impact-dependent hemorrhagic cortical contusions gradually progress to cystic, glial-lined cavities at mild-to-moderate magnitudes of injury *(28,33,35,37,40)*, with almost total loss of contused cortical tissue and the underlying white matter occurring at late stages after more severe injuries *(28,33,41,42)*. Intraparenchymal hemorrhages and hemorrhagic lesions at the gray–white matter interface, similar to gliding contusions seen clinically, are reported in several of the rigid indentation models of TBI *(28,31–33,43,44)*. Subdural hematoma *(33,34)* and subarachnoid hemorrhage *(28,32,33,44,45)* have only occasionally been reported after lateral weight-drop or CCI and after central CCI. Contre coup lesions have generally been reported only in models utilizing bilateral craniotomies, with damage induced by extrusion of tissue through the contralateral opening at the time of impact *(34,38)*.

Table 1
References to Studies Reporting Various Structural and Functional Alterations in Rigid Indentation Models of Traumatic Brain injury in the Rat

	Surface contusion	Gliding contusion	Subdural hematoma	Subarachnoid hemorrhage	Intraparenchymal hemorrhage
Lateral weight drop					
Impounder	28,49, 106	28,43		28	28
Dural	31,41	31			31
Controlled cortical impact					
Central	32	32		32	32
Lateral	33,37,40, 42,116	33	33	33	33
Lateral with bilateral craniotomy	34,38,39	34,44	34	44,45	34,38
Bilateral–frontal	35,40				

	Axonal injury	Diffuse axonal injury	Neuronal cell loss	Glial reactions	Inflammatory reactions
Lateral weight drop					
Impounder			28,48, 43,49	28,29,49	
Dural			31,41,63		30,31,41, 61–65,68
Controlled cortical impact					
Central	32		32		
Lateral	33,46		33,37,40,42, 50–53,55,113, 115,116	37,50,58	60,66,67
Lateral with bilateral craniotomy	34,44,47		34,38,39,54	34,59	
Bilateral–frontal			35,40,56,57	35	

	Blood–brain barrier dysfunction	Measures of cerebral edema/ICP	Altered cerebral blood flow	Altered ionic flux	Altered neuro-chemistry
Lateral weight drop					
Impounder	43	43	83–85	43	83,88,95
Dural		30,31,63, 72–74	63,73,86		
Controlled cortical impact					
Central					
Lateral	69,70	33,75–80	75,76,78,87		42,90–94,98
Lateral with bilateral craniotomy					59,96,97
Bilateral–frontal	71	57,71,81,82			

	Altered electrical activity	Altered oxidative metabolism	Altered glucose utilization	Sensorimotor dysfunction	Cognitive dysfunction
Lateral weight drop					
Impounder	43,48, 89, 112,117	29,88,99	102	28,49,104–107	112
Dural					
Controlled cortical impact					
Central				32,103	103,111
Lateral		91,98,100	55,76	40,55,108, 109,115	40,53,108, 109,113,116
Lateral with bilateral craniotomy		101		38,39	39,96,97, 114
Bilateral–frontal				35,40	35,40,56,57

Subcortical hemorrhages are seldom reported in the rat models of rigid indentation. Moderate to high levels of TBI in the central CCI model resulted in bilateral hemorrhages in the hippocampi and intraparenchymal hemorrhages within the ponto-mesencephalic and cervicomedullary junctions *(32)*. A lateral weight-drop onto dura produces a hemorrhagic contusion extending through the ipsilateral parietal cortex, hippocampus, and thalamus as well as intraventricular hemorrhage *(31)*. Intraventricular hemorrhage has also been reported after more severe lateral CCI *(33)*.

5.2. Axonal Injury

A limited degree of axonal injury is produced by the various rigid indentation models. Axonal injury is presumably induced by shearing forces exerted upon fiber tracts and is more likely to be observed after more severe grades of injury induced by central CCI *(32)* or lateral CCI *(33,46)*. Lateral CCI with a bilateral craniotomy (with or without a dural opening) produces the largest extent of axonal and/or dendritic injury (illustrated via immunostaining of axon retraction balls, or decrease of neurofilament and/or cytoskeletal proteins 3 h to 2 wk post-injury) of all rigid indentation models *(34,44,47)*. It is important to note, however, that diffuse axonal injury of the type and extent seen clinically is not well duplicated by any of the rigid indentation models of TBI.

5.3. Neuronal Cell Loss

Neuronal cell loss, particularly in the cortex, has been frequently described in most models of rigid indentation (*see* Table 1). Ipsilateral loss of hippocampal neurons, most frequently within the CA3 and CA1 regions and in the hilar region of the dentate gyrus, has been reported to occur after lateral TBI induced via either weight drop onto a dural impounder *(48,49)* or via lateral CCI *(37,50–53)*. As with axonal injury, hippocampal cell loss may occur bilaterally if lateral CCI is induced after a bilateral craniotomy *(38,54)*. Evidence of damage to, or loss of, thalamic neurons has only rarely been reported after lateral weight-drop onto an impounder or lateral CCI *(49,51,55)*. Delayed, retrograde degeneration of neurons within the medial dorsal and ventrolateral nuclei of the thalamus has been reported after bilateral frontal CCI *(35,56,57)*. Lateral CCI with a bilateral craniotomy has been reported to induce bilateral neuronal injury (dystrophic neurons) within the thalamus, amygdala, and hypothalamus *(54)*.

5.4. Glial Reactions to TBI

Alterations in glial cell populations after TBI (i.e., reactive gliosis or astrogliosis), including increased synthesis of glial fibrillary acidic protein (GFAP), may indicate damage to both neurons and glia. Immunostaining has shown that lateral CCI leads to early (6 and 24 h) increases in numbers of GFAP-positive cells within the ipsilateral cortex, corpus callosum, hippocampus, striatum, and thalamus, with increased GFAP-positive cells also apparent in the contralateral cortex and hippocampus at 24 h *(50)*. Maximal astrogliosis in these regions was seen at 8 d post-injury, but persisted until 30 d after TBI *(50)*. The presence of macrophages in tissue bordering necrotic cortical cavitation *(28)* and gliosis within the ipsilateral hippocampus and thalamus *(49)* 1 mo after TBI induced via lateral weight-drop onto an impounder has also been reported. Other authors report that lateral CCI resulted in a robust increase in GFAP message (m)RNA in both ipsilateral cortex and hippocampus at 12 and 24 h after impact, with continued expression occurring within cortex only at 96 h post-injury whereas increased numbers of GFAP-immunoreactive astrocytes within both ipsilateral cortex and hippocampus were observed at 96 h after TBI *(58)*. Lateral CCI with bilateral craniotomy has been shown to induce reactive astrogliosis (GFAP immunoreactivity) in cortex surrounding the injury site by 3 d post-injury *(59)* and lateral CCI with a bilateral craniotomy and dural opening increases reactive astrocytes bilaterally in white matter tracts and the overlying cortex from 2 to 7 d post-injury *(34)*. Bilateral-frontal CCI results in a glial scar consisting of reactive astrocytes and ameboid microglia at the borders of the necrotic cortical cavity, as well as subcortical gliosis within the medial dorsal nucleus of the thalamus and the caudate-putamen, at 18 d post-injury *(35)*.

5.5. Inflammatory Reactions After TBI

In recent years there has been increasing awareness of a potential influence of the acute inflammatory response on secondary injury in clinical research as well as in numerous experimental brain injury models. The hallmark of such a response is the infiltration of circulating neutrophils into the brain parenchyma. Neutrophil accumulation within the contused hemisphere (cortex and hippocampus) has been shown to occur by 4–6 h, peak at 24–48 h, and be undetectable at 7–8 d after lateral CCI or TBI induced by a lateral weight-drop onto dura *(30,41,60–62)*. This neutrophil accumulation has been shown to be proportional to injury severity *(41,60)* and is associated with the up-regulation of cell adhesion molecules mediating neutrophil rolling, attachment, and migration across the endothelium and into brain parenchyma *(61,62)*. Neutrophil accumulation has been associated with formation of cerebral edema in some studies inducing TBI via a lateral weight-drop onto dura *(30)*, but not in others *(63)*. Delayed entry (4–7 d) of T lymphocytes and mononuclear phagocytes into pericontusional cortex has been demonstrated after TBI induced using a weight-drop onto dura *(64)*, and this inflammatory response has been shown to contribute to delayed edema formation *(65)*.

Proinflammatory cytokines are known to influence neutrophil trafficking via up-regulation of numerous cell adhesion molecules (on neutrophils and endothelium) and influencing synthesis and release of chemokines (chemoattractant cytokines) which mediate movement of neutrophils and lymphocytes through tissue. Increased levels of the proinflammatory cytokines have been found in cerebrospinal fluid and/or serum acutely after closed head injury in several human studies. Rat studies utilizing lateral CCI models of TBI have reported that mRNA for the proinflammatory cytokine interleukin-1β is increased within the pericontusional cortex at 6 h after injury *(66)* and the proinflammatory cytokine tumor necrosis factor-α is increased within the ipsilateral cortex and hippocampus at 6 and 24 h after TBI *(67)*. In contrast to these findings, mRNA and protein levels of interleukin-1β, tumor necrosis factor-α, and interleukin-6 were not detected within the pericontusional cortex until 4–6 d after lateral weight-drop onto dura *(68)*. Increased mRNA expression for the chemokines, macrophage inflammatory protein-1α, and macrophage inflammatory protein-2 within the ipsilateral cortex and hippocampus has also been found at 6 and 24 h after lateral CCI *(67)*.

6. FUNCTIONAL CHANGES ASSOCIATED
WITH RIGID INDENTATION

6.1. Blood–Brain Barrier Dysfunction and Cerebral Edema

In addition to the mechanical disruption of the cerebral vasculature and the infiltration of circulating leukocytes detailed in Subheadings 5.1. and 5.5., disruptions of the blood–brain barrier (increased permeability to intravascular dyes or albumin) after TBI induced via lateral weight-drop onto a dural impounder *(43)*, lateral CCI *(69,70)*, or bilateral-frontal CCI *(71)* have been demonstrated. This blood–brain barrier dysfunction likely contributes to the formation of cerebral edema (vasogenic component) and/or elevated intracranial pressure (ICP) reported in most models of TBI induced using rigid indentation.

After TBI induced via a lateral weight-drop onto dura a significant increase in brain water content (right–left hemisphere) occurs in the injured hemisphere at 8 and 24 h postinjury *(30,31,63)*, with the extent of edema formation being dependent upon the magnitude of TBI *(30)*. Superoxide anions have been shown to play an important role in edema formation in the pericontusional cortex (at 30 min, 6 h, and 3 d) after lateral weight-drop onto dura *(72)*. An age-dependent effect of TBI on acute edema onset has been demonstrated, with immature rats with lateral weight-drop onto dura showing significant edema at 2, 24, and 48 h whereas edema in mature rats reached significance only at 24 and 48 h *(73)*. Combined data for mature and immature rats indicated that ICP was only mildly increased at 24 h after this type of injury *(73)*. In a similar model, a significant increase in edema within cortex anterior to the impact site occurs at 2, 12, 24, and 48 h after TBI, is absent 3–5 d postinjury, and is present again at 6 and 7 d after injury *(74)*. The second peak in this "bimodal" edema formation, frequently observed clinically, was suggested to be related to a secondary inflammatory reaction *(74)*.

In isoflurane-anesthetized rats, it has been reported that intracranial hypertension (elevated ICP) and systemic hypotension combine to produce a reduction in cerebral perfusion pressure (mean arterial blood pressure minus ICP) and a decrease in cortical perfusion (laser Doppler flow) over the first 8 h after TBI, with the observed changes being proportional to the severity of lateral CCI *(75)*. The observed elevations of ICP in this study may have been due to an increase in cerebral blood volume, formation of cerebral edema, or both. Planimetry measures of hemispheric swelling after lateral CCI indicate significant edema is present in the ipsilateral cortex from 6 to 24 h postinjury, with peak tissue swelling occurring at 24 h after moderate-to-severe CCI *(33,76)*. Significant cortical edema at the site of impact, but not within pericontusional or contralateral cortex, occurs at 6 h after a moderate severity of lateral CCI *(77)*. Cerebral edema (brain water content within a 3 mm coronal section taken from the center of impact site) is significantly increased from 2 h through 7 d ipsilateral to a severe lateral CCI, with peak increases occurring at 1–2 d postinjury *(78)*. Magnetic resonance imaging techniques have indicated that the maximum edema extension after lateral CCI (assessed at 90 min, 24 and 72 h postinjury) occurs at 24 h after impact *(79)*. Other investigators, using a severe lateral CCI model *(80)*, have reported significant edema within ipsilateral pericontusional cortex and hippocampus at 6 and 24 h postinjury, with significant edema also occurring in the contralateral cortex and hippocampus by 24 h postinjury.

Bilateral–frontal CCI increases tissue water content in surrounding tissue at 24 h postinjury, with greater increases in male rats than in females and almost no increase in pseudopregnant females *(81)*. This cortical edema is reported to be present as early as 2 h and to endure for up to 7 d after bilateral–frontal CCI *(71,82)*, with significant increases enduring for only 1–3 d in other studies *(57)*.

6.2. Altered Cerebral Blood Flow

Measures of regional or local cerebral blood flow (lCBF) have been most thoroughly examined in lateral TBI models of rigid indentation. Although somewhat different magnitudes and duration of effects are reported by various investigators, a characteristic finding, in general agreement with the clinical literature, is that experimental TBI results in an early posttraumatic hypoperfusion.

Regional CBF measured using the hydrogen clearance technique indicated that CBF within the contused cortex underlying an impounder impacted via lateral weight-drop fell to approx 50% of normal from 3 to 90 min of impact, then gradually normalized over a 4-h period *(83)*. A regional CBF study using laser Doppler flowmetry after a lateral weight-drop onto an impounder found that ipsilateral flow in the cortex fell to 60% of pretrauma levels within 2 min and remained low over the ensuing 2-h period *(84)*. Flow within the contralateral cortex increased at 4 min after TBI, followed by mild hypoperfusion (78% of baseline) for 1 h. Two hours after lateral weight-drop onto an impounder there is widespread hypoperfusion in ipsilateral regions of the cortex, with less profound reductions in cortical CBF occurring contralateral to TBI *(85)*. By 24 h postinjury lCBF in cortical regions ipsilateral to injury, with the exception of the contusion core, recovered to normal levels whereas hyperemia was apparent in the contralateral cortex *(85)*.

At 2 h after TBI induced by lateral weight-drop onto dura mature rats exhibited a widespread, bilateral decrease in lCBF, whereas this hypoperfusion was greatly attenuated (only 4 of 14 regions analyzed) in immature rats *(73)*. Additional studies using the lateral weight-drop onto dura model indicate that this injury induces a heterogeneous pattern of lCBF, characterized by a low-flow in the contusion core with hyperemia within the peritrauma cortex and in distant structures (e.g., ipsilateral hippocampus and amygdala and contralateral parietal cortex, thalamus, and amygdala) at 24–48 h postinjury *(63,86)*. Further age-dependent effects were illustrated by the fact that the hyperemic response to TBI at 24 h postinjury was not observed in aged rats *(86)*.

Marked reductions in lCBF have been reported within the contusion core, with moderate-to-severe reductions in lCBF occurring in pericontusional tissue and the ipsilateral hippocampus within 30 min *(76,87)* or 2–6 h *(76,78)* of lateral CCI. Using laser Doppler, other investigators have reported both decreases and increases in CBF in the contralateral cortex during the first 6 h after lateral CCI *(75)*. More delayed measures of lCBF indicate that hypoperfusion tends to resolve more rapidly within the ipsilateral hippocampus (approx 24 h) than in the ipsilateral cortex (e.g., by 7–10 d) after lateral CCI *(76,78)*. Again, it has been noted that lCBF is heterogeneous after resolution of the initial hypoperfusion stages, with focal regions of low flow occurring immediately adjacent to regions of high flow (hyperemia) after lateral CCI *(78,87)*.

6.3. Alterations in Ionic Flux and Neurochemistry

In a model of TBI induced via weight-drop onto a dural impounder a transient increase in extracellular cortical levels of both excitatory (glutamate, aspartate) and inhibitory (γ-aminobutyric acid [GABA]) amino acid transmitters peaks within 10 min and then returns to basal levels within 20–30 min *(88)*. A similar time course for maximal release of both glutamate and aspartate in the cortex underlying the impounder impacted by a lateral weight-drop has been subsequently reported *(83)*. The amino acid taurine was also found to increase significantly in this model of TBI, and taurine showed increases proportional to the severity of injury *(88)*. Additional work using this TBI model has indicated that trauma produced a rapid and transient membrane depolarization with an immediate increase in the extracellular concentration of potassium and a concomitant decrease in levels of extracellular calcium (suggesting intracellular accumulation) within the contusion core and in cortex adjacent to impact *(43)*. Calcium levels in the pericontusional cortex did not return to basal levels until 50 min postinjury. These findings indicate that TBI induces rapid ionic shifts and negative electrical potential shifts resembling cortical spreading depression *(43)*. This form of spreading depression may explain the acutely elevated thresholds for microstimulation-evoked motor responses in cortical regions ipsilateral to contusion injury induced via weight-drop onto a dural impounder *(89)*.

Lateral CCI with bilateral craniotomy has also been shown to induce an acute increase in extracellular levels of aspartate and glutamate, with maximal increases in transmitter levels and time (30–60 min) to return to normal levels being proportional to the severity of injury *(59)*. These same authors have also found that lateral CCI results in maximal glutamate and aspartate release within 10 min of injury *(42)*. Subsequent investigation has indicated that interstitial glutamate levels ipsilateral to lateral CCI peak first in the contusion core and approx 10 min later in the pericontusional area, with extracellular levels returning toward baseline values by 20–30 min postinjury *(90)*. These latter authors presented evidence suggesting that glutamate release in the contusion core was due to physical disruption of the cell membranes or vasculature, whereas increased glutamate levels in the pericontusional cortex was due to a calcium-dependent release from depolarized nerve terminals.

Within the primary injury site, norepinephrine (NE) concentrations were found to be decreased from 29% to 38% at 30 min, 2.5 h and 24 h after lateral CCI delivered to the left cortex *(91)*. In the first 4 h after lateral CCI extracellular levels of NE were found to decrease bilaterally in cortex anterior to the contusion site, with reductions in ipsilateral cortex occurring more rapidly than in the contralateral cortex *(92)*. An initial increase in anterior cortex levels of NE ipsilateral to cortical impact (first 20 min postinjury) did not reach statistical significance. Other investigators have reported that NE turnover increases in the cortex ipsilateral to injury 30 min after lateral CCI, but marked differences in changes in NE turnover throughout the brain were site specific and dependent upon whether CCI was induced to the left or right cortex *(93)*. These same authors *(94)* report that NE turnover is decreased bilaterally throughout the brain at 6 and 24 h after lateral CCI to the right cortex. These findings are consistent with the report that extracellular NE levels within both cerebellar hemispheres were decreased (approx 50%) at 24–48 h after lateral weight-drop onto a dural impounder *(95)*.

A chronic and injury-dependent disturbance in cholinergic neurotransmission after lateral CCI with bilateral craniotomy is indicated by the fact that the cholinergic antagonist scopolamine can induce significant impairments of spatial memory in rats with a relatively mild grade of injury but no overt deficits at 2 wk postinjury and in rats with moderate grades of injury at 4, 6, and 8 wk postinjury *(96)*. Microdialysis studies have shown that scopolamine-evoked release of acetylcholine within the hippocampus ipsilateral to cortical impact is significantly reduced in rats 2 wk after a lateral CCI with bilateral craniotomy *(97)*.

6.4. Alterations in Cerebral Metabolism

6.4.1. Alterations in Energy-Related Metabolites and Oxidative Metabolism

Derangements of oxidative metabolism after TBI induced using a weight-drop onto a dural impounder are suggested by the marked decrease in activity for the oxidative enzyme α-glycerophospate dehydrogenase within the cortical mantle ipsilateral to injury for up to 4 d postinjury *(29)*. Cellular energy failure in cortical tissue surrounding TBI induced via weight-drop onto a dural impounder, indicated by increased extracellular levels of lactate and purines (adenosine, inosine, and hypoxanthine), was found to be proportional to injury severity *(88)*. Levels of the purines (adenosine and inosine) peaked within 10 min and returned to basal levels within 20–50 min *(88)*. Subsequent assessment of these same purines after lateral CCI revealed similar time course of TBI-induced increases over 40 min of postinjury sample collections. In this latter study no changes in dialysate concentrations of cyclic adenosine monophosphate were found, suggesting that breakdown of adenosine triphosphate contributed to the increase in adenosine and its metabolites *(98)*.

In the weight-drop onto a dural impounder injury model interstitial levels of lactate were elevated for up to 2 h after TBI *(88,99)*, and this increase in lactate levels was accompanied by a decrease (1 h) in extracellular glucose *(99)*. After lateral CCI tissue lactate levels within injured cortex, adjacent cortex and ipsilateral hippocampus were significantly increased from 30 min through 24 h postinjury. Phosphocreatinine levels were decreased within tissue from the contusion core only at 30 min after lateral CCI, indicative of energy depletion within the injury site acutely posttrauma. These authors suggested that the increase in lactate and decrease in high-energy phosphates after TBI could be caused by decreased CBF, mitochondrial dysfunction, ionic fluxes, and/or hyperglycolysis *(91)*. In vitro studies on isolated mitochondria after lateral CCI or lateral CCI with bilateral craniotomy have shown that there is a significant decrease in mitochondrial electron transfer and energy-transducing activity (state 3 respiratory rate, respiratory control, and P/O ratios) in the ipsilateral hemisphere from 1 h to 14 d postinjury, and these impairments are associated with increased calcium uptake *(100,101)*.

6.4.2. Alterations in Glucose Utilization

Measures of local cerebral metabolic rates of glucose (LCMRGlu) utilization indicate that a relatively mild-to-moderate level of lateral CCI in the rat results in hyperglycolysis in the pericontusional cortex and in the ipsilateral hippocamus immediately through 30 min after TBI. LCMRGlu in these regions was markedly decreased by 24 h postinjury, and by 10 d metabolic depression was more widespread in the ipsilateral cortex and also extended to the contralateral cortex and the ipsilateral thalamic nuclei *(76)*. A widespread depression of LCMRGlu in the first 2 d after TBI induced via lateral weight-drop onto a dural impounder has also been reported *(102)*. At 8–12 wk after lateral CCI to the left, but not right, parietal cortex a basal depression of LCMRGlu occurs within the cortex rostral and caudal to the site of injury *(55)*. Areas showing increased basal levels of LCMRGlu (defined as high activity areas) in the somatosensory cortex contralateral to left CCI were larger than corresponding areas in control animals, and vibrissal stimulation in these injured rats led to activation of caudal areas (normally processing visual and auditory stimuli) of the right cortex. This latter data suggest reorganization processes (sprouting, unmasking) have occurred within intact cortical regions remote from the primary injury *(55)*.

7. BEHAVIORAL ABNORMALITIES

7.1. Sensorimotor Dysfunctions

In general, increasing magnitude of central CCI to the rat brain results in more prolonged suppression of acute somatomotor reflexes (indicative of brief coma) and increased duration of vestibular and motor (beam balance and beam walking tasks) deficits *(32)*. Transient suppression of acute somatomotor reflexes/neurological responses has also been demonstrated after a moderate TBI induced using lateral CCI with bilateral craniotomy *(39)*. Beam-walking deficits of short (1 d) duration are reported after moderate levels of central CCI *(103)*.

TBI induced via lateral weight-drop onto a dural impounder has been shown to lead to deficits in maintaining posture on a balance beam for at least 2 wk *(28)* and deficits in beam-walking ability that persist for up to 1 mo postinjury *(104)*, with these sensorimotor deficits being dependent upon initial injury severity. The ability of this injury model to reliably produce beam-walking deficits of 2–4 wk duration has been shown repeatedly *(49,104–106)*. In addition to these deficits primarily assessing hindlimb function, rats with lateral weight-drop onto a dural impounder have injury-dependent contralateral forelimb deficits, revealed in tests of food pellet retrieval, that endure for at least 1 mo after injury *(107)*.

Lateral CCI has been shown to produce beam-walking deficits that endure for 1–2 wk postinjury *(108,109)*. In other studies examining lateral CCI to the right vs left parietal cortex, deficits in a beam-walking task employing placement of a wall/barrier adjacent to the beam to enable vibrissae stimulation have been observed for up to 30 d after left, but not right, cortical contusion *(55,110)*. These authors have also reported that left, but not right, lateral CCI produces significant sensory neglect as assessed by latency to remove tape stimuli applied to the vibrissae contralateral to injury *(55)*. Following lateral CCI significant deficits in forelimb placing occur during the first week postinjury, with deficits in a rotarod test of motor/ambulatory activity persisting for as long as 11 wk postinjury *(40)*. Beam balancing and beam-walking deficits of short (1–3 d) duration are reported after moderate levels of lateral CCI with bilateral craniotomy *(38,39)*.

Transient (1 wk duration for tongue protrusion, forelimb tactile placing) or no (pole balancing, overall spontaneous activity, rotarod test) sensorimotor deficits have been reported after bilateral–frontal CCI *(35,40)*. Rats with this type of injury, however, are significantly impaired on tests of sensorimotor neglect or inattention (forelimb tactile adhesive removal) at 8.5 mo postinjury *(40)*.

7.2. Cognitive Dysfunctions

Central CCI at a severity producing no overt hippocampal cell loss has been shown to result in spatial learning and memory deficits (assessed using a Morris water maze [MWM] task) at both 11–15 and at 30–34 d postinjury *(111)*. Similar spatial learning deficits were reported to endure 30–31 d after a slightly higher magnitude of central CCI *(103)*.

Following TBI induced via weight-drop onto a dural impounder, naive rats show significant deficits in spatial learning (anterograde amnesia) during MWM testing, whereas rats trained in the MWM prior to TBI exhibit mild and transient (1 d) retention (retrograde amnesia) deficits during reacquisition tests conducted in the week postinjury *(112)*. Spatial learning deficits in the MWM task during 3–4 wk postinjury have also been described after lateral CCI *(108,109)*. Deficits in MWM performance have been shown to be proportional to the initial injury severity for up to 2 wk after lateral CCI, during initial acquisition (learning) as well as in retention (memory) tests of spatial learning and memory ability *(53)*. When tested in the first 2 wk after lateral CCI, rats trained in eight-arm radial arm maze (RAM) tasks prior to surgery did not commit working memory errors (reentry into previously visited arms) but did exhibit significant reference memory deficits (entry into never baited arms) and required nearly twice as many trials to reach learning criterion as did noninjured control rats *(113)*. These latter authors present data analyses that indicate that rats with unilateral cortex contusion have an ipsiversive turning bias in the maze tasks and shifted from an allocentric to an egocentric strategy to relearn the RAM tasks. Other authors have reported that rats with lateral CCI exhibit significant working memory deficits on a RAM task when tested 6–7 mo postinjury and also exhibit significant acquisition deficits in a MWM task at 8 and 11 mo postinjury. Importantly, these animals were not impaired in tests of orientation, timing, and attention such as a delayed nonmatching to position task (1–11 wk postinjury) or during testing (11 mo postinjury) for ability to perform a task utilizing differential reinforcement of low rates of responding *(40)*.

Rats with moderate magnitudes of TBI induced via lateral CCI with bilateral craniotomy, but not those with milder injury, show significant impairments in spatial learning and memory (MWM testing) when tested between 10 and 20 d postinjury *(39,96,97,114)*.

Bilateral–frontal CCI is reported to produce significant spatial learning deficits (MWM test) when testing is conducted 7–18 d *(35,57)* or 10–20 d postinjury *(56)*. Rats with bilateral–frontal CCI were not impaired on tests of working memory in a RAM test at 6–7 mo postinjury, or on a MWM test of reference memory at 8 mo postinjury *(40)*. Animals with this form of injury were, however, significantly impaired in tests of orientation, timing, and attention such as a delayed nonmatching to position task (1–11 wk postinjury) or during testing (11 mo postinjury) for ability to perform a task utilizing differential reinforcement of low rates of responding *(40)*.

8. SECONDARY INSULTS AFTER RIGID INDENTATION

Acute secondary insults of a cerebral (e.g., intracranial hypertension, vasospasm) or systemic (e.g., hypoxia, hypotension) nature can contribute to ischemic injury subsequent to TBI in humans, increasing neuropathology and worsening functional outcomes *(18,24,25–27)*. The effects of purposefully imposed secondary insults on structural or functional outcome measures in rigid indentation models of TBI in the rat have been examined in lateral CCI models *(77,115,116)*. Imposition of hypoxia after lateral CCI did not increase cerebral edema formation in the ipsilateral cortex, assessed 6 h postinjury *(77)*. Hypoxemia was found to worsen beam-balance performance and delay recovery on this task after lateral CCI in a dose-dependent fashion *(115)*. These latter authors reported that this secondary insult increased CA3 hippocampal neuronal death (but not CA1 injury or the volume of cortical necrosis) at 7 d postinjury, and rats with hypoxia and lateral CCI exhibited evidence of necrotic and/or apoptotic cell death within the ipsilateral cortex and dentate gyrus (6–72 h postinjury) and the ipsilateral thalamus and CA3 region (24 and 72 h) of the hippocampus *(115)*. The addition of an ischemic injury (bilateral carotid occlusion) subsequent to lateral CCI has been shown to significantly increase hippocampal neuronal loss (CA1 and CA3 regions) and cortical contusion volume in a dose-dependent and time-dependent manner, with little effect seen if the ischemic insult was delayed until 24 h postinjury *(116)*.

9. SUMMARY AND FUTURE DIRECTIONS

Studies utilizing rigid indentation models of TBI in the rat indicate that these methods can consistently and reliably produce graded cerebral injuries with structural and functional outcomes that mimic many of the features of human TBI. These TBI models cannot replicate the full spectrum of human TBI, in that they do not produce prolonged coma, diffuse brain swelling, or diffuse axonal injury. Concern also exists regarding the frequency with which rigid indentation models of TBI may induce acute epileptiform, seizure-like activity *(33,48,117)*. Research with these and other rodent models of TBI has, however, contributed to our understanding of numerous pathobiological events and biochemical cascades set in motion by TBI. Further research utilizing variants of the six general models of rigid indentation considered in this chapter will no doubt facilitate greater understanding of the diverse and complex biochemical and molecular mechanisms contributing to neuropathology and functional outcomes after TBI.

Some relative strengths and weaknesses with regard to known structural and functional outcomes, and identification of areas needing additional research within the various models of rigid indentation, is illustrated by examination of Table 1. Of the six models of rigid indentation the lateral CCI model has been the most thoroughly characterized to date, but research on ionic fluxes and electrophysiological alterations induced by this type of injury is notably absent. As mentioned in the preceding discussion, work within some models of rigid indentation has indicated that profound differences in outcomes can occur dependent upon age (e.g., *73,86*), gender (e.g., *81*), and whether TBI was induced to the right or left hemisphere (e.g., *55*). Greater attention to, and research on, these important variables within each of the rigid indentation models of TBI will be needed in the future.

REFERENCES

1. Anderson, T. E. and Lighthall, J. W. (1996) The need for continued development of experimental models of brain injury, in *Neurotrauma* (Narayan, R. K., Wilberger, J. E., Jr., and Povlishock, J. T., eds.), McGraw-Hill, New York, pp. 1357–1365.

 2. Dixon, C. E. and Hayes, R. L. (1996) Fluid percussion and cortical impact models of traumatic brain injury, in *Neurotrauma* (Narayan, R. K., Wilberger, J. E., Jr., and Povlishock, J. T., eds.), McGraw-Hill, New York, pp. 1337–1346.

 3. Gennarelli, T. A. (1994) Animate models of human head injury. *J. Neurotrauma* **11**, 57–368.

 4. Lighthall, J. W. and Anderson, T. E. (1994) In vivo models of experimental brain and spinal cord trauma, in *The Neurobiology of Central Nervous System Trauma* (Salzman, S. K. and Faden, A. I., eds.), Oxford University Press, New York, pp. 3–11.

 5. Lighthall, J. W., Dixon, C. E., and Anderson, T. E. (1989) Experimental models of brain injury. *J. Neurotrauma* **6**, 83–98.

 6. McIntosh, T. K., Smith, D. H., Meaney, D. F., Kotapka, M. J., Gennarelli, T. A., and Graham, D. I. (1996) Neuropathological sequelae of traumatic brain injury: relationship to neurochemical and biomechanical mechanisms. *Lab. Invest.* **74**, 315–342.

 7. Povlishock, J. T. (1996) An overview of brain injury models, in *Neurotrauma* (Narayan, R. K., Wilberger, J. E., Jr., and Povlishock, J. T., eds.), McGraw-Hill, New York, pp. 1325–1336.

 8. Shohami, E., Cotev, S., and Shapira, Y. (1995) A closed head injury model, in *Central Nervous System Trauma: Research Techniques* (Ohnishi, S. T. and Ohnishi, T., eds.), CRC Press, Boca Raton, FL, pp. 235–245.

 9. Torbati, D. (1995) Gunshot wounds to the brain, in *Central Nervous System Trauma: Research Techniques* (Ohnishi, S. T. and Ohnishi, T., eds.), CRC Press, Boca Raton, FL, pp. 523–542.

10. Kraus, J. F. (1993) Epidemiology of head injury, in *Head Injury*, 3rd edit. (Cooper, P. R., ed.), Williams & Wilkins, Baltimore, pp. 1–25.

11. Levin, H. S., Eisenberg, H. M., and Benton, A. L. (1989) *Mild Head Injury*, Oxford University Press, New York.

12. Hovda, D. A., Becker, D. P., and Katayama, Y. (1992) Secondary injury and acidosis. *J. Neurotrauma* **9** (Suppl. 1), S47–S60.

13. Strich, S. J. (1970) Lesions in the cerebral hemispheres after blunt head injury. *J. Clin. Pathol.* (Suppl. Royal College of Pathologists) **4**, 166–171.

14. Willmore, L. J. (1990) Post-traumatic epilepsy: Cellular mechanisms and implications for treatment. *Epilepsia* **31 (Suppl. 3)**, S67–S73.

15. Auerbach, S. H. (1986) Neuroanatomical correlates of attention and memory disorders in traumatic brain injury: an application of neurobehavioral subtypes. *J. Head Trauma Rehabil.* **1,** 1–12.

16. Adams, J. H., Doyle, D., Ford, I., Gennarelli, T. A., Graham, D. I., and McClellan, D. R. (1989) Diffuse axonal injury in head injury: definition, diagnosis and grading. *Histopathology* **15**, 49–59.

17. Graham, D. I., Lawrence, A. E., and Adams, J. H. (1987) Brain damage in non-missile head injury secondary to high intracranial pressure. *Neuropathol. Appl. Neurobiol.* **13**, 209–217.

18. Graham, D. I., Ford, I., Adams, J. H., Doyle, D., Teasdale, G., Lawrence, A., and McClellan, D. R. (1989) Ischaemic brain damage is still common in fatal nonmissile head injury. *J. Neurol. Neurosurg. Psychiatry* **52**, 346–350.

19. Graham, D. I., Adams, J. H., Nicoll, J. A. R., Maxwell, W. L., and Gennarelli, T. A. (1995) The nature, distribution and causes of traumatic brain injury. *Brain Pathol.* **5**, 397–406.

20. Hall, K. M. and Johnston, M. V. (1994) Outcomes evaluation in traumatic brain injury rehabilitation. Part II: Measurement tools for a nationwide data system. *Arch. Phys. Med. Rehabil.* **75**, SC10–SC18.

21. Levin, H. S. (1992) Neurobehavioral recovery. *J. Neurotrauma* **9 (Suppl. 1)**, S359–S373.

22. Levin, H. S., Goldstein, F. C., High, W. M., Jr., and Eisenberg, H. M. (1988) Disproportionately severe memory deficit in relation to normal intellectual functioning after closed head injury. *J. Neurol. Neurosurg. Psychiatry* **51**, 1294–1301.

23. Levin, H. S., Grafman, J., and Eisenberg, H. M. (1987) *Neurobehavioral Recovery from Head Injury*, Oxford University Press, New York.

24. Miller, J. D. (1985) Head injury and brain ischaemia: implications for therapy. *Br. J. Anaesth.* **57,** 120–129.

25. Bouma, G. J., Muizelaar, J. P., Choi, S. C., Newton, P. G., and Young, H. F. (1991) Cerebral circulation and metabolism after severe traumatic head injury: the elusive role of ischemia. *J. Neurosurg.* **75,** 685–693.

26. Marion, D. W., Darby, J., and Yonas, H. (1991) Acute regional cerebral blood flow changes caused by severe head injuries. *J. Neurosurg.* **74,** 407–414.

27. Newton, M. R., Greenwood, R. J. Britton, K. E., Charlesworth, M., Nimmon, C. C., Carroll, M. J., and Doike, G. (1992) A study comparing SPECT with CT and MRI after closed head injury. *J. Neurol. Neurosurg. Psychiatry* **55,** 92.

28. Feeney, D. M., Boyeson, M. G., Linn, R. T., Murray, H. M., and Dail, W. G. (1981) Responses to cortical injury. I. Methodology and local effects of contusions in the rat. *Brain Res.* **211,** 67–77.

29. Dail, W. G., Feeney, D. M., Murray, H. M., Linn, R. T., and Boyeson, M. G. (1981) Responses to cortical injury. II. Widespread depression of the activity of an enzyme in cortex remote from a focal injury. *Brain Res.* **211,** 79–89.

30. Schoettle, R. J., Kochanek, P. M., Magargee, M. J., Uhl, M. W., and Nemoto, E. M. (1990) Early polymorphonuclear leukocyte accumulation correlates with the development of post-traumatic cerebral edema in rats. *J. Neurotrauma* **7,** 207–217.

31. Kochanek, P. M., Clark, R. S. B., Schiding, J. K., and Nemoto, E. M. (1995) Weight-drop model of traumatic brain injury: Assessment of the acute inflammatory response to cerebral trauma, in *Central Nervous System Trauma: Research Techniques* (Ohnishi, S. T. and Ohnishi, T., eds.), CRC Press, Boca Raton, FL, pp. 247–254.

32. Dixon, C. E., Clifton, G. L., Lighthall, J. W., Yaghmai, A. A., and Hayes, R. L. (1991) A controlled cortical impact model of traumatic brain injury in the rat. *J. Neurosci. Methods* **39,** 253–262.

33. Sutton, R. L., Lescaudron, L., and Stein, D. G. (1993) Unilateral cortical contusion injury in the rat: vascular disruption and temporal development of cortical necrosis. *J. Neurotrauma* **10,** 135–149.

34. Meaney, D. F., Ross, D. T., Winkelstein, B. A., Brasko, J., Goldstein, D., Bilston, L. B., Thibault, L. E., and Gennarelli, T. A. (1994) Modification of the cortical impact model to produce axonal injury in the rat cerebral cortex. *J. Neurotrauma* **11,** 599–612.

35. Hoffman, S. W., Fulop, Z., and Stein, D. G. (1994) Bilateral frontal cortical contusion in rats: behavioral and anatomic consequences. *J. Neurotrauma* **11,** 417–431.

36. Povlishock, J. T., Hayes, R. L., Michel, M. E., and McIntosh, T. K. (1994) Workshop on animal models of traumatic brain injury. *J. Neurotrauma* **11,** 723–732.

37. Goodman, J. C., Cherian, L., Bryan R. M., Jr., and Robertson, C. S. (1994) Lateral cortical impact injury in rats: pathologic effects of varying cortical compression and impact velocity. *J. Neurotrauma* **11,** 587–597.

38. Dixon, C. E., Markgraf, C. G., Angileri, F., Pike, B. R., Wolfson, B., Newcomb, J. K., Bismar, M. M., Blanco, A. J., Clifton, G. L., and Hayes, R. L. (1998) Protective effects of moderate hypothermia on behavioral deficits but not necrotic cavitation following cortical impact injury in the rat. *J. Neurotrauma* **15,** 95–103.

39. Long, D. A., Ghosh, K., Moore, A. N., Dixon, C. E., and Dash, P. K. (1996). Deferoxamine improves spatial memory performance following experimental brain injury in rats. *Brain Res.* **717,** 109–117.

40. Lindner, M. D., Plone, M. A., Cain, C. K., Frydel, B., Francis, J. M., Emerich, D. F., and Sutton, R. L. (1998) Dissociable long-term cognitive deficits after frontal versus sensorimotor cortical contusions. *J. Neurotrauma* **15,** 199–216.

41. Clark, R. S. B., Schiding, J. K., Kaczorowski, S. L., Marion, D. W., and Kochanek, P. M.

(1994) Neutrophil accumulation after traumatic brain injury in rats: comparison of weight drop and controlled cortical impact models. *J. Neurotrauma* **11,** 499–506.

42. Palmer, A. M., Marion, D. W., Botscheller, M. L., and Redd, E. E. (1993) Therapeutic hypothermia is cytoprotective without attenuating the traumatic brain injury-induced elevations in interstitial concentrations of aspartate and glutamate. *J. Neurotrauma* **10,** 363–372.

43. Nilsson, P., Hillered, L., Olsson, Y., Sheardown, M., and Hansen, A. J. (1993) Regional changes in interstitial K⁺ and Ca²⁺ levels following cortical compression contusion trauma in rats. *J. Cereb. Blood Flow Metab.* **13,** 183–192.

44. Posmantur, R. M., Kampfl, A., Taft, W. C., Bhattacharjee, M., Dixon, C. E., Bao, J., and Hayes, R. L. (1996) Diminished microtubule-associated protein 2 (MAP2) immunoreactivity following cortical impact brain injury. *J. Neurotrauma* **13,** 125–137.

45. Newcomb, J. K., Kampfl, A., Posmantur, R. M., Zhao, X., Pike, B. R., Liu, S-J., Clifton, G. L., and Hayes, R. L. (1997) Immunohistochemical study of calpain-mediated breakdown products to α-spectrin following controlled cortical impact injury in the rat. *J. Neurotrauma* **14,** 369–383.

46. Marion, D. W. and White, M. J. (1996) Treatment of experimental brain injury with moderate hypothermia and 21-aminosteroids. *J. Neurotrauma* **13,** 139–147.

47. Posmantur, R. M., Hayes, R. L., Dixon, C. E., and Taft, W. C. (1994) Neurofilament 68 and neurofilament 200 protein levels decrease after traumatic brain injury. *J. Neurotrauma* **11,** 533–545.

48. Krobert, K. A., Salazar, R. A., Sutton, R. L., and Feeney, D. M. (1992) Temporal evolution of histopathology and unit activity in rat hippocampal CA3 region after focal cortical contusion. *J. Neurotrauma* **9,** 64.

49. Weisend, M. P. and Feeney, D. M. (1994) Traumatic brain injury induces changes in brain temperature that are related to behavioral and anatomical outcome. *J. Neurosurg.* **80,** 120–132.

50. Sutton, R. L. (1995) Histopathology after traumatic brain injury, in *Central Nervous System Trauma: Research Techniques* (Ohnishi, S. T. and Ohnishi, T., eds.), CRC Press, Boca Raton, FL, pp. 497–508.

51. Sutton, R. L., Lescaudron, L., and Stein, D. G. (1999) Unilateral cortical contusion injury in the rat: Temporal development of histopathology in hippocampus and subcortical regions. *J. Neurotrauma*, submitted.

52. Baldwin, S. A., Gibson, T., Callihan, C. D., Sullivan, P. G., Palmer, E., and Scheff, S. W. (1997) Neuronal cell loss in the CA3 subfield of the hippocampus following cortical contusion utilizing the optical dissector method for cell counting. *J. Neurotrauma* **14,** 385–398.

53. Scheff, W. W., Baldwin, S. A., Brown, R. W., and Kraemer, P. J. (1997) Morris water maze deficits in rats following traumatic brain injury: lateral controlled cortical impact. *J. Neurotrauma* **14,** 615–627.

54. Colicos, M. A., Dixon, C. E., and Dash, P. K. (1996) Delayed, selective neuronal death following experimental cortical impact injury in rats: possible role in memory deficits. *Brain Res.* **739,** 111–119.

55. Dunn-Meynell, A. and Levin B. E. (1995) Lateralized effect of unilateral somatosensory cortex contusion on behavior and cortical reorganization. *Brain Res.* **675,** 143–156.

56. Roof, R. L., Braswell, L., Duvdevani, R., and Stein, D. G. (1994) Progesterone facilitates cognitive recovery and reduces secondary neuronal loss caused by cortical contusion injury in male rats. *Exp. Neurol.* **129,** 64–69.

57. Janis, L. S., Hoane, M. R., Conde, D., Fulop, Z., and Stein, D. G. (1998) Acute ethanol administration reduces the cognitive deficits associated with traumatic brain injury in rats. *J. Neurotrauma* **15,** 105–115.

58. Hinkle, D. A., Baldwin, S. A., Scheff, S. W., and Wise, P. M. (1997) GFAP and S100β expression in the cortex and hippocampus in response to mild cortical contusion. *J. Neurotrauma* **14,** 729–738.

59. Palmer, A. M., Marion, D. W., Botscheller, M. L., Swedlow, P. E., Stryen, S. D., and DeKosky, S. T. (1993) Traumatic brain injury-induced excitotoxicity assessed in a controlled cortical impact model. *J. Neurochem.* **61,** 2015–2024.

60. Sutton, R. L., Grachek, R. A., and Vargo, J. M. (1996) Acute neutrophil accumulation in hippocampus ipsilateral to cortical contusion. *J. Neurotrauma* **13,** 616.

61. Carlos, T. M., Clark, R. S. B., Franicola-Higgins, D., Schiding, J. K., and Kochanek, P. M. (1997) Expression of endothelial adhesion molecules and recruitment of neutrophils after traumatic brain injury in rats. *J. Leukocyte Biol.* **61,** 279–285.

62. Whalen, M. J., Carlos, T. M., Clark, R. S. B., Marion, D. W., DeKosky, S. T., Heineman, S., Schiding, J. K., Memarzadeh, F., and Kochanek, P. M. (1997) The effect of brain temperature on acute inflammation after traumatic brain injury in rats. *J. Neurotrauma* **14,** 561–572.

63. Uhl, M. W., Biagas, K. V., Grundl, P. D., Barmada, M. A., Schiding, J. K., Nemoto, E. M., and Kochanek, P. M. (1994) Effects of neutropenia on edema, histology, and cerebral blood flow after traumatic brain injury in rats. *J. Neurotrauma* **11,** 303–315.

64. Holmin, S., Mathiesen, T., Shetye, J., and Biberfeld, P. (1995) Intracerebral inflammatory response to experimental brain contusion. *Acta Neurochir.* **132,** 110–119.

65. Holmin, S. and Mathiesen, T. (1995) Dexamethasone and colchicine reduce inflammation and delayed edema following experimental brain contusion. *Acta Neurochir.* **138,** 418–424.

66. Goss, J. R., Styren, S. D., Miller, P. D., Kochanek, P. M., Palmer, A. M., Marion, D. W., and DeKosky, S. T. (1995) Hypothermia attenuates the normal increase in interleukin 1β RNA and nerve growth factor following traumatic brain injury in the rat. *J. Neurotrauma* **12,** 159–167.

67. Sutton, R. L., Sheng, W. S., and Chao, C. C. (1997) Expression of tumor necrosis factor and chemokines after traumatic brain injury in rat. *J. Neurotrauma* **14,** 789.

68. Holmin, S., Schalling, M., Hojeberg, B., Nordqvist, A.-C. S., Skeftruna, A.-K., and Mathiesen, T. (1997) Delayed cytokine expression in rat brain following experimental contusion. *J. Neurosurg.* **86,** 493–504.

69. Dhillon, H. S., Donaldson, D., Dempsey, R. J., and Prasad, M. R. (1994) Regional levels of free fatty acids and evans blue extravasation after experimental brain injury. *J. Neurotrauma* **11,** 405–415.

70. Smith, S. L., Andrus, P. K., Zhang, J-R., and Hall, E. D. (1994) Direct measurement of hydroxyl radicals, lipid peroxidation, and blood–brain barrier disruption following unilateral cortical impact head injury in the rat. *J. Neurotrauma* **11,** 393–404.

71. Duvdevani, R., Roof, R. L., Fulop, Z., Hoffman, S. W., and Stein, D. G. (1995) Blood–brain barrier breakdown and edema formation following frontal cortical contusion: Does hormonal status play a role? *J. Neurotrauma* **12,** 65–75.

72. Yunoki, M., Kawauchi, M., Ukita, N., Noguchi, Y., Nishio, S., Ono, Y., Asari, S., Ohmoto, T., Asanuma, M., and Ogawa, N. (1997) Effects of lecithinized superoxide dismutase on traumatic brain injury in rats. *J. Neurotrauma* **14,** 739–746.

73. Grundl, P. D., Biagas, K. V., Kochanek, P. M., Schiding, J. K., Barmada, M. A., and Nemoto, E. M. (1994) Early cerebrovascular response to head injury in immature and mature rats. *J. Neurotrauma* **11,** 135–148.

74. Holmin, S. and Mathiesen, T. (1995) Biphasic edema development after experimental brain contusion in rat. *Neurosci. Lett.* **194,** 97–100.

75. Cherian, L., Robertson, C. S., Contant C. F., Jr., and Bryan R. M., Jr., (1994) Lateral cortical impact injury in rats: cerebrovascular effects of varying depth of cortical deformation and impact velocity. *J. Neurotrauma* **11,** 573–585.

76. Sutton, R. L., Hovda, D. A., Adelson, P. D., Benzel, E. C., and Becker, D. P. (1994) Metabolic changes following cortical contusion: Relationships to edema and morphological changes. *Acta Neurochir.* **60 (Suppl.),** 446–448.

77. Nida, T. Y., Biros, M. H., Pheley, A. M., Bergman, T. A., and Rockswold, G. L. (1995) Effect of hypoxia or hyperbaric oxygen on cerebral edema following moderate fluid percussion or cortical impact injury in rats. *J. Neurotrauma* **12,** 77–85.

78. Kochanek, P. M., Marion, D. W., Zhang, W., Schiding, J. K., White, M., Palmer, A. M., Clark, R. S. B., O'Malley, M. E., Styren, S. D., Ho, C., and DeKosky, S. T. (1995) Severe controlled cortical impact in rats: assessment of cerebral edema, blood flow, and contusion volume. *J. Neurotrauma* **12,** 1015–1025.

79. Unterberg, A. W., Stroop, R., Thomale, E.-W., Kiening, K. L., Pauser, S., and Vollmann, W. (1996) Characterization of brain edema following controlled cortical impact injury in rats. *Acta Neurochir.* **70 (Suppl.),** 106–108.

80. Baskaya, M. K., Rao, A. M., Puckett, L., Prasad, M. R., and Dempsey, R. J. (1996) Effect of difluoromethylornithine treatment on regional ornithine decarboxylase activity and edema formation after experimental brain injury. *J. Neurotrauma* **13,** 85–92.

81. Roof, R. L., Duvdevani, R., and Stein, D. G. (1993) Gender influences outcome of brain injury: progesterone plays a protective role. *Brain Res.* **607,** 333–336.

82. Roof, R. L., Duvdevani, R., Heyburn, J. W., and Stein, D. G. (1996) Progesterone rapidly decreases brain edema: treatment delayed up to 24 hours is still effective. *Exp. Neurol.* **138,** 246–251.

83. Koizumi, H., Fujisawa, H., Ito, H., Maekawa, T., Di, X., and Bullock, R. (1997) Effects of mild hypothermia on cerebral blood flow-independent changes in cortical extracellular levels of amino acids following contusion trauma in the rat. *Brain Res.* **747,** 304–312.

84. Nilsson, P., Gazelius, B., Carlson, H., and Hillered, L. (1996) Continuous measurement of changes in regional cerebral blood flow following cortical compression contusion trauma in the rat. *J. Neurotrauma* **13,** 201–207.

85. Scremin, O. U., Li, M. G., and Jenden, D. J. (1997) Cholinergic modulation of cerebral cortical blood flow changes induced by trauma. *J. Neurotrauma* **14,** 573–586.

86. Biagas, K. V., Grundl, P. D., Kochanek, P. M., Schiding, J. K., and Nemoto, E. M. (1996) Posttraumatic hyperemia in immature, mature, and aged rats: autoradiographic determination of cerebral blood flow. *J. Neurotrauma* **13,** 189–200.

87. Bryan, R. M., Cherian, L., and Robertson, C. (1995) Regional cerebral blood flow after controlled cortical impact injury in rats. *Neurosurg. Anesth.* **80,** 687–695.

88. Nilsson, P., Hillered, L., Ponten, U., and Ungerstedt, U. (1990) Changes in cortical extracellular levels of energy-related metabolites and amino acids following concussive brain injury in rats. *J. Cereb. Blood Flow Metab.* **10,** 631–637.

89. Boyeson, M. G., Feeney, D. M., and Dail, W. G. (1991) Cortical microstimulation thresholds adjacent to sensorimotor cortex injury. *J. Neurotrauma* **8,** 205–217.

90. Maeda, T., Katayama, Y., Kawamata, T., and Yamamoto, T. (1998) Mechanisms of excitatory amino acid release in contused brain tissue: effects of hypothermia and in situ administration of Co^{2+} on extracellular levels of glutamate. *J. Neurotrauma* **15,** 655–664.

91. Prasad, M. R., Ramaiah, C., McIntosh, T. K., Dempsey, R. J., Hipkens, S., and Yurek, D. (1994) Regional levels of lactate and norepinephrine after experimental brain injury. *J. Neurochem.* **63,** 1086–1094.

92. Sutton, R. L. and Krobert, K. A. (1993) Acute changes in cortical noradrenaline levels of the anesthetized rat following cortical contusion in the rat: a microdialysis study. *J. Neurotrauma* **10** (Suppl. 1), S181.

93. Levin, B. E., Brown, K. L., Pawar, G., and Dunn-Meynell, A. (1995) Widespread and lateralized effects of acute traumatic brain injury on norepinephrine turnover in the rat brain. *Brain Res.* **674,** 307–313.

94. Dunn-Meynell, A., Pan, S., and Levin, B. E. (1994) Focal traumatic brain injury causes widespread reductions in rat brain norepinephrine turnover from 6 to 24 h. *Brain Res.* **660,** 88–95.

95. Krobert, K. A., Sutton, R. L., and Feeney, D. M. (1994) Spontaneous and amphetamine-evoked release of cerebellar noradrenaline after sensorimotor cortex contusion: an in vivo microdialysis study in the awake rat. *J. Neurochem.* **62,** 2233–2240.

96. Dixon, C. E., Liu, S. J., Jenkins, L. W., Bhattachargee, M., Whitson, J. S., Yang, K. Y., and Hayes, R. L. (1995) Time course of increased vulnerability of cholinergic neurotransmission following traumatic brain injury in the rat. *Behav. Brain Res.* **70,** 125–131.

97. Dixon, C. E., Bao, J., Long, R. L., and Hayes, R. L. (1996) Reduced evoked release of acetylcholine in the rodent hippocampus following traumatic brain injury. *Pharmacol. Biochem. Behav.* **53,** 679–686.

98. Bell, M. J., Kochanek, P. M., Carcillo, J. A., Mi, Z., Schiding, J. K., Wisniewski, S. R., Clark, R. S. B., Dixon, C. E., Marion, D. W., and Jackson, E. (1998) Interstitial adenosine, inosine, and hypoxanthine are increased after experimental traumatic brain injury in the rat. *J. Neurotrauma* **15,** 163–170.

99. Lewen, A. and Hillered, L. (1998) Involvement of reactive oxygen species in membrane phospholipid breakdown and energy perturbation after traumatic brain injury in the rat. *J. Neurotrauma* **15,** 521–530.

100. Xiong, Y., Peterson, P. L., Muizelaar, J. P., and Lee, C. P. (1997) Amelioration of mitochondrial dysfunction by a novel antioxidant U-101033E following traumatic brain injury in rats. *J. Neurotrauma* **14,** 907–917.

101. Xiong, Y., Gu, Q., Peterson, P. L., Muizelaar, J. P., and Lee, C. P. (1997) Mitochondrial dysfunction and calcium pertubation induced by traumatic brain injury. *J. Neurotrauma* **14,** 23–34.

102. Queen, S., Chen, M., and Feeney, D. M. (1997) Amphetamine alleviates the reduced local cerebral glucose utilization produced by traumatic brain injury. *Brain Res.* **777,** 42–50.

103. Dixon, C. E., Hamm, R. J., Taft, W. C., and Hayes, R. L. (1994) Increased anticholinergic sensitivity following closed skull impact and controlled cortical impact traumatic brain injury in the rat. *J. Neurotrauma* **11,** 275–287.

104. Sutton, R. L., Weaver, M. S., and Feeney, D. M. (1987) Drug-induced modifications of behavioral recovery following cortical trauma. *J. Head Trauma Rehab.* **2,** 50–58.

105. Feeney, D. M., Bailey, B. Y., Boyeson, M. G., Hovda, D. A., and Sutton, R. L. (1987) The effect of seizures on recovery of function following cortical contusion in the rat. *Brain Injury* **1,** 27–32.

106. Feeney, D. M. and Westerberg, V. S. (1990) Norepinephrine and brain damage: alpha noradrenergic pharmacology alters functional recovery after cortical trauma. *Can. J. Psychol.* **44,** 233–252.

107. Feeney, D. M. and Sutton, R. L. (1988) Catecholamines and recovery of function after brain damage, in *Pharmacological Approaches to the Treatment of Central Nervous System Injury* (Stein, D. G. and Sabel, B. A., eds.), Plenum Press, New York, pp. 121–142.

108. Sutton, R. L. (1994) U-74389G treatments after cortical contusion in the rat. *Soc. Neurosci. Abstr.* **20,** 192.

109. Sutton, R. L., Coffey, C., Cole, T., and Rockswold, G. L. (1995) Amphetamine treatments after severe cortical contusion in the rat. *J. Neurotrauma* **12,** 965.

110. Dunn-Meynell, A. A., Yarlagadda, Y., and Levin, B. E. (1997) α1-Adrenoceptor blockade increases behavioral deficits in traumatic brain injury. *J. Neurotrauma* **14,** 43–52.

111. Hamm, R. J., Dixon, C. E., Gbadebo, D. M., Singha, A. K., Jenkins, L. W., Lyeth, B. G., and Hayes, R. L. (1992) Cognitive deficits following traumatic brain injury produced by controlled cortical impact. *J. Neurotrauma* **9,** 11–20.

112. Sutherland, R. J., Sutton, R. L., and Feeney, D. M. (1993) Traumatic brain injury in the rat produces anterograde but not retrograde amnesia and impairment of hippocampal LTP. *J. Neurotrauma* **10 (Suppl.1),** S162.

113. Soblosky, J. S., Tabor, S. L., Matthews, M. A., Davidson, J. F., Chorney, D. A., and Carey, M. E. (1996) Reference memory and allocentric spatial localization deficits after unilateral cortical brain injury in the rat. *Behav. Brain Res.* **80,** 185–194.

114. Dixon, C. E., Ma, X., and Marion, D. W. (1997) Effects of CDP-choline treatment on neurobehavioral deficits after TBI and on hippocampal and neocortical acetylcholine release. *J. Neurotrauma* **14,** 161–169.

115. Clark, R. S. B., Kochanek, P. M., Dixon, C. E., Chen, M., Marion, D. W., Heineman, S., DeKosky, S. T., and Graham, S. H. (1997) Early neuropathologic effects of mild or moderate hypoxemia after controlled cortical impact injury in rats. *J. Neurotrauma* **14,** 179–189.

116. Cherian, L., Robertson, C. S., and Goodman, J. C. (1996) Secondary insults increase injury after controlled cortical impact in rats. *J. Neurotrauma* **13,** 371–383.

117. Nilsson, P., Ronne-Engstrom, E., Flink, R., Ungerstedt, U., Carlson, H., and Hillered, L. (1994) Epileptic seizure activity in the acute phase following cortical impact trauma in rat brain. *Brain Res.* **637,** 227–232.

20

Rodent Ischemia Models of Embolism and Ligation of the Middle Cerebral Artery

Clinical Relevance to Treatment Strategies of Stroke

Cesario V. Borlongan, Hitoo Nishino, Yun Wang, and Paul R. Sanberg

1. INTRODUCTION

This chapter focuses on two models of cerebral ischemia: embolism and ligation of the middle cerebral artery (MCA). Advantages and disadvantages of MCA embolism and ligation are discussed in relation to establishing an appropriate model of stroke. In addition, the clinical relevance of each technique to the development of experimental treatment strategies is outlined, highlighting recent novel therapeutic modalities, including neural transplantation and intracerebral infusion of neurotrophic factors. Prior to discussing these two treatment strategies, an overview of the current status of pharmacologic intervention for stroke is provided, and we present critical problems (i.e., therapeutic window) inherent in drug therapy that limit its efficacy in the clinic. It is believed that the experimental evidence presented here will encourage further utilization of embolism and ligation models of cerebral ischemia in rodents as suitable animal models of stroke, and more importantly, should caution researchers and clinicians alike about critical scientific issues (i.e., early stage development of treatment strategies for stroke) that warrant validation in the laboratory setting prior to proceeding with clinical interventions.

2. STROKE

One of the leading causes of death in the United States is stroke, killing two in every five Americans and costing annually more than 100 billion dollars *(1)*. To date, pharmacologic drugs are the treatment of choice for stroke; two major groups of drugs, namely thrombolytic agents and *N*-methyl-D-aspartate (NMDA) antagonists *(2–6)*, have been introduced to the clinic, but inconclusive results have been reported on whether such drugs actually alleviate central nervous system (CNS) dysfunctions associated with stroke. Unfortunately, most of these drugs have not been fully validated in animal models of stroke.

The use of thrombololytic agents (e.g., anticoagulants such as heparin, aspirin, and ticlopidine, and platelet inhibitors) is based on their action in combating the ischemia-

From: Central Nervous System Diseases
Edited by: D. F. Emerich, R. L. Dean, III, and P. R. Sanberg © Humana Press Inc., Totowa, NJ

induced blood clots, thereby ensuring normal cerebral blood flow. While the CNS accounts for only about 2% of total body weight, it utilizes 15% of cardiac output, and therefore it is not surprising that interruption to cerebral blood flow is the most common etiology of cerebral ischemia. Thrombolytic therapy directly reverses this hypoperfusion caused by ischemia, but it has been found also to trigger increased bleeding in treated stroke patients *(7)*. The timing of thrombolytic treatment is a decisive factor because a 1-h difference in its administration may prove detrimental rather than beneficial to the stroke patient. For example, injection of the thrombolytic drug streptokinase within 3 h of ischemic stroke could prevent the ischemic cascades, but at 4 hr, it could exacerbate CNS damage and even induce fatalities *(8,9)*. Thus, there is a very limited effective therapeutic window for current thrombolytic drugs following a stroke episode.

Critical examination of the benefit/risk ratio also persists in the use of NMDA receptor antagonists (i.e., MK-801), calcium channel blockers, caspase inhibitors, and protein kinase inhibitors. Accumulating evidence proposes that ischemic injury is accompanied by aberrant metabolic consequences, such as oxidative stress and free radical accumulation, which are harmful to cellular homeostasis *(4,10)*. In animal models of ischemic stroke, injection of NMDA antagonists, calcium channel blockers, caspase inhibitors, and protein kinase inhibitors (hereafter referred to as cell maintenance drugs) have been shown to exert a neuroprotective effect for maintenance of CNS vitality *(4,11)*. Similar to thrombolytic agents, the timing of administration of cell maintenance drugs determines their efficacious outcome. For example, MK-801 protects against ischemia-related apoptosis only when injected prior to but not after 3 or 6 h of occlusion *(4)*. Caspase inhibitors seem to extend the therapeutic window to up to within 6 h of occlusion, but only when the CNS insult involves brief ischemia *(4)*. Whereas the cascade of biochemical events leading to cell death, as well as neurologic dysfunctions, are prevented in animal models of stroke *(6,12)* in clinical trials, calcium channel blockers only partially reduce morbidity of posttraumatic subarachnoid hemorrhage, and only in younger patients *(5)*. Protein kinase inhibitors also have been demonstrated to block cell death events associated with experimental stroke *(3)*, but their utility in the clinic warrants additional investigation *(13)*. Drugs such as melatonin and thioredoxin *(10,14)* targeting abnormal accumulation of free radicals or reactive oxygen species, have been shown recently to ameliorate ischemia-induced CNS damage. However, maintenance or reversal of abnormal motor and cognitive functions has yet to be demonstrated in these free radical scavenger drugs.

Based on the knowledge from pharmacologic therapy for stroke-related CNS dysfunctions, two critical issues need to be addressed in evaluating the efficacy of future treatment strategies for stroke: first the timing of neuroprotective intervention (either via thrombolytic therapy or cell maintenance therapy) during a stroke episode is very important and, second, the therapeutic window is very limited, that is, only acute stage stroke can be potentially treated. Here, we discuss neural transplantation therapy as a possible treatment for chronic stage stroke, and neurotrophic factor therapy for acute stage stroke or as a protective strategy for patients who are at risk of developing the disease. Because of the debilitation following a stroke episode, there is a demand for a rapid translation of experimental therapeutic modalities from the laboratory to the clinic. Nevertheless, the potential benefits of a therapy should be assessed against the risk of developing side effects or of succumbing to unknown outcomes including fatality. Of note, 3 of 15 malpractice claims in a stroke setting are related to medications *(15)*. Clearly, critical evaluation of pharmacologic therapy and other new treatment strategies needs to be undertaken prior to their large-scale clinical usage.

3. EMBOLISM AND LIGATION

Stroke-induced neuronal damage arises from a two-stage insult involving ischemia and reperfusion injury. In adult rats, these two processes can be mimicked by either MCA embolism or ligation followed by reperfusion. In the following sections, the terms embolism and ligation of MCA should be interpreted to mean the surgical procedure involving transient ischemia followed by reperfusion, and we concur with the view that the reperfusion process actually creates more widespread CNS damage than that produced by the ischemia itself. In addition to characterizing neuronal and behavioral alterations, the models of embolism and the ligation of the MCA have been utilized to examine the potential benefits of neural transplantation and neurotrophic factor therapies in ameliorating stroke-related dysfunction. Although most studies presented here describe work done in our laboratories, we cite additional reports on similar topics by other researchers.

3.1. Embolism

Embolism *(16–18)* is one of the many standardized procedures for creating a stroke model in rodents. This model is one of the more noninvasive surgical procedures used to induce a stroke model. We describe here the rat model of embolic stroke; a mouse model has been recently developed *(19)*, and a photothrombotic or laser-guided occlusion technique has been introduced and appears even less invasive than the conventional embolic procedure *(20,21)*. Variations of the embolic procedure have made possible localization of the ischemia-induced infarcted area in the brain compared with that of chronic global ischemia which produces large areas of necrotic tissues or disappearance of a chunk of tissues altogether. It appears that necrotic cell death dominates global ischemia, whereas apoptotic cell death accompanies transient, focal ischemia. Alternatively, the duration of occlusion of the MCA may determine which mechanism of cell death promotes CNS damage. Of note, the apoptotic cell death caused by embolic ischemia can be replicated using an acute (10 min) global ischemia *(22)*. In general, the apoptotic cell death produces small CNS damage, and infarctions are visible only after a long delay post-ischemia *(23)*. Thus, the embolism technique that creates localized CNS infarction allows a more in-depth examination of the progression of the disease—from death of a small population of cells at early periods post-ischemia to apoptosis at later time points. Furthermore, this procedure allows a more extended survival of the ischemic animals that can then be tested behaviorally for deterioration of motor and cognitive functions or for characterization of effects of treatment interventions at different stages of the disease.

The embolic technique involves the use of a neurofilament that passes through the carotid artery and reaches the junction of the MCA, thus blocking the blood flow from the common carotid artery as well as from the circle of Willis. Based on our pilot studies, approx 15–17 mm of the embolic filament needs to be inserted from the junction of the external and internal carotid arteries to block the MCA. The preferred embolus size is 4–0, made of nonabsorbable material (e.g., nylon), with the diameter of the embolic tip tapered to 24–26-gauge size using rubber cement. These lengths and sizes of the tip of the embolus are recommended for animals weighing between 250 and 350 g, and additional adjustments are needed for larger animals. Nevertheless, even with animals weighing 250–350 g there exists some variation in the arterial structure (e.g., longer internal carotid artery or larger diameter of the artery) that would require changing the insertion length as well as the size of the tip of the embolus. To circumvent this problem, the availability of sensitive behavioral parameters is used to verify whether or not successful MCA occlusion has been obtained immediately following insertion of the embolus. For example, just after recovery from anesthesia, ischemic animals demonstrate biased elevated swing activity, spontaneous rotation, and a characteristic body posture (*see* Subheading 5.) compared to animals with incomplete blockage of the MCA. Thus, animals that do not show these behavioral hallmarks could be reanesthetized and successful ischemia could be accomplished by inserting the embolus deeper into the internal carotid artery or by using an embolus with larger diameter tip. In this case, we recommend the use of anesthetic gases (e.g., halothane or isofluorane) that would allow efficient behavioral testing, as well as adjustment of the embolus when necessary.

3.2. Ligation

The method of ligation described by Chen et al. *(24)* involves surgically opening the cranium (craniotomy) to isolate and gain access to the MCA. The ligation technique consists of ligating the MCA for 90 min and then opening the artery for reperfusion for 24 h. MCA ligation can induce motor behaviors in rats resembling stroke symptoms *(25)*. The severity of stroke can be graded, which is helpful to evaluate the preventive effect of therapeutic strategies, such as trophic factors or their regulatory agents (*see* **Subheading 7.**). The ligation of the right MCA is accompanied by occlusion of the bilateral common carotids (CCAs) *(24)*. Following anesthesia, bilateral CCAs are identified and isolated through a ventral midline cervical incision. A craniotomy of about 2×2 mm^2 is then made in the right squamosal bone. The right MCA is ligated with a 10–O suture while the CCAs are simultaneously ligated with nontraumatic microsurgical clips. These ligations are removed after 90 min, which has been shown to induce maximal infarctions in rats *(23)*. The craniotomy is then covered with Gelfoam and the wounds sutured. Reperfusion is usually allowed for 24 h and animals are then killed for histological examination.

Because of the invasive surgery inherent in the ligation model, the trauma associated with this procedure can promote additional problems for the animal. For example, deformity of the facial musculature, especially of the eye ipsilateral to the ischemia, is clearly visible post-surgery. The craniotomy ligation model, however, is much more consistent in producing CNS infarction compared with the embolic model, which is highly dependent on the size of, and some developmental variations in, the structure of the MCA or common carotid artery. Most studies utilizing the ligation model report a short survival time for the ischemic animals, and therefore suggest that this method is more suitable for investigations of pathophysiological alterations associated with stroke rather than analysis of long-term behavioral symptoms (but *see* the test introduced by

Bederson and colleagues, *[25]* described in Subheading 5). The locus of infarction associated with the ligation model is the frontal cerebral cortex and a large area of necrosis is usually observed in the ischemic region. Because the basal ganglia remains intact in MCA-ligated animals, behavioral alterations arising from the ischemia are assumed to be secondary to cortical dysfunction. In the embolic model, the striatum, particularly the lateral aspect, in addition to fronto-parietal cortex, is damaged, and therefore ischemia-induced behavioral deficits cannot be definitively associated with one CNS region. Thus, the choice of target CNS areas for therapeutic intervention should consider the ischemic regions produced in each model.

4. BRAIN HISTOCHEMICAL/PATHOLOGICAL ALTERATIONS IN ISCHEMIC ANIMALS

Ischemic animals develop CNS alterations that are reminiscent of stroke. The differences in localization of infarcted areas following embolism and ligation can offer some clues as to which CNS areas or cell populations and types are responsible for specific behavioral dysfunctions. In MCA-embolic ischemic animals, specific types of neurons degenerate, including γ-aminobutyric acid-ergic (GABAergic), cholinergic, and dopaminergic cells *(17,18,26–29)*. Neurons secreting substance P are found over-expressed in ischemic rat brains *(30)*. Because the striatum is at the core of infarction, these specific types of striatal neurons that are altered suggest their high level of vulnerability to stroke insult and further indicate that therapeutic strategy should focus on rescuing these specific neurons *(1,31)*. In contrast, most of the damaged neurons in the MCA-ligated ischemic animals are located in the frontal cortical region. Types of neurons shown to be vulnerable to ligation of MCA include GABA, cholinergic, and pyramidal neurons *(26,32,33)*. There is a massive infarction of the dorsal cortex, which is maximal at 24 h following the ligation *(34)*. The pyramidal neurons are generally undetectable within the remaining tissue of the infarcted area. Using triphenyl tetrazolium chloride (TTC) staining, which is a marker for mitochondrial activity, the regions of infarction in both embolic and MCA-ligated models can be localized (Figs. 1 and 2).

Examination of brains from embolic ischemic animals revealed histological disturbances at both acute (9 d post-ischemia–reperfusion) or chronic (28 d) stages characterized by loss of glial fibrillary acidic protein (GFAP) and microtubule associated protein (MAP-2)-positive cells, and infiltration of immunoglobulin G (IgG) and leukocytes in ischemic striatum and cortex. In the penumbra of the infarcted cortex, there is a disappearance of GFAP-positive glia and IgG was extravasated, but pyramidal neurons can still be detected. Furthermore, a high degree of reactive gliosis is detected along the outside lining of the penumbra, forming a wall of GFAP-positive astrocytes. The soma of these reactive astroglia are positively labeled by a marker (antibody against aquaporin-4) for the end-feet of astroglia in intact brain. These histopathological alterations in ischemic rats resemble some of the features observed in clinical cerebral ischemia. The minimal tissue loss in embolic ischemia may parallel the early stage of human stroke. In contrast, the massive infarction (i.e., tissue disappearance) and largely necrotic tissue of the penumbra of the core of infarcted cortex in ligation ischemia seem to mirror that seen in end-stage stroke.

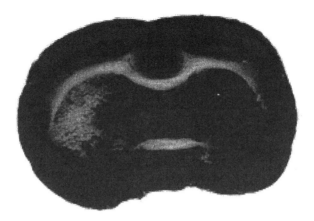

Fig. 1. Photomicrograph of a brain section from an MCA-occluded animal. Triphenyl-tetrazolium chloride staining, a marker of mitochondrial activity (*black shade*), reveals that the lateral aspect of the striatum is damaged at 72 h post-ischemia–reperfusion. Minor cortical damage (*more anterior section, not shown*) underlying the striatum is also detected at the same time period.

5. BEHAVIORAL ASSAYS FOR ISCHEMIC ANIMALS

As stated earlier, embolic ischemic animals have been reported to survive longer than ligated ischemic animals, and thus more exhaustive behavioral tests can be administered to the former group of animals. Behavioral parameters that have characterized dysfunctions in the embolic ischemic animals include the elevated body swing test (EBST), drug-induced rotational test, catalepsy test, digiscan locomotor activity test, forelimb akinesia test and the body posture examination. For the ligated ischemic animals, the postural reflex test by Bederson and colleagues *(25)* has been used. In this postural reflex test, rats are examined for the degree of abnormal posture when suspended by 1 m above the floor. Normal rats extend both forelimbs straight. Depending on the degree of infarction (i.e., caused by right MCA ligation), rats with minimal infarction may appose the right forelimb to the chest muscle and extend the left forelimb straight; rats with moderate infarction may show decreased resistance to lateral push in addition to the stereotypic posture displayed by minimally infarcted rats; and maximally infarcted rats may twist the upper half of their bodies in addition to the aforementioned behaviors. In addition, the passive avoidance test *(35,36)* has been shown to be a good measure of memory performance in embolic ischemic animals.

Among these arrays of behavioral tests, we note that the EBST and passive avoidance test appear to be more sensitive than the other tests, and observed dysfunctions in these two tests correlate highly with the degree of infarction caused by either embolism or ligation of the MCA. The EBST *(37)* involves lifting the animal by the tail and counting the number of body swings ipsilateral or contralateral to the lesion. Ischemic animals display a contralateral biased swing activity (>75%) compared to normal animals. On the other hand, the passive avoidance test *(35,36)* consists of the step-down apparatus, with electric shock (2 mA) given during the training period via the metal grid floors. The learning and memory performance of the animals is measured by the

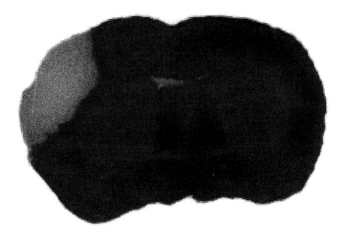

Fig. 2. Photomicrograph of a brain section from an MCA-ligated animal. Triphenylte-trazolium chloride shows that the dorsolateral aspect of the frontal cortex is damaged at 24 h post-ischemia–reperfusion, while the basal ganglia remains intact.

number of step-downs and the length of time required to learn the task, and the reten-tion of avoiding the electric shock (i.e., staying on the platform) after 24 h following the acquisition. Ischemic animals take a significantly longer time to learn the task, and could not stay on the platform for the required 3 min during the retention test *(1,35,36)*.

As noted earlier, animals can be tested during the 1-h embolism or immediately after recovery from anesthesia in both groups of ischemic animals. The absence of behav-ioral deficits in animals exposed to MCA ischemia–reperfusion can be used as a crite-rion for excluding animals that would subsequently be utilized to test treatment interventions. The importance of the behavioral tests to reliably exclude animals in which ischemia–reperfusion neuronal damage was not successfully induced is that they differentiate direct treatment effects from some residual effects in the host animal. Incomplete ischemia–reperfusion arising from technical problems in embolus or liga-tion can create only partial or no CNS damage at all, and would otherwise confound the interpretation of the results if not detected by appropriate behavioral assays. These different tests are also used to measure recovery of function in ischemic animals fol-lowing experimental treatment modalities. For example, neural transplantation therapy has been evaluated in ischemic animals using the EBST and the passive avoidance test *(1)*. Accordingly, the stability of motor/memory performance dysfunctions over time in ischemic animals needs to be demonstrated in the behavioral tests; fluctuations in behavioral deficits should be solely and reliably attributed to treatment effects and not to spontaneous recovery that can accompany repeated tests (i.e., training effects) or some other transient behavioral abnormalities. We again recommend the use of EBST and the passive avoidance test in characterizing treatment effects, based on our obser-vation that the performance of embolic ischemic animals in these tests is significantly altered over a long period of time (at least 6 mo post-ischemia), and therefore subse-quent normalization of ischemia-induced dysfunctions in these tests can be reliably ascribed to the experimental treatment.

6. NEURAL TRANSPLANTATION

Clinical trials of intracerebral transplantation of fetal cells have been initiated in a limited number of Parkinson's disease patients *(38,39)*. Encouraging results have been noted, including long-term survival of transplanted cells as well as significant behavioral improvement in transplanted patients. Because of logistical and ethical concerns regarding use of fetal cells, genetically engineered, porcine, Sertoli, adrenal medulla, and other nonfetal cells have been introduced as alternative graft sources for transplantation therapy *(40–43)*.

We have recently examined the utility of transplanting human neuroteratocarcinoma cell line-derived (hNT) neurons into the brains of embolic ischemic animals *(1)*. The main rationale for undertaking such transplantation in ischemic animals is the localized damage (i.e., striatal infarction) found in this model of stroke which resembles that observed in Huntington's disease, a neurodegenerative disease characterized primarily by striatal lesions. Because laboratory studies *(41,42,44,45)* and recently clinical data *(46)* from several Huntington's disease patients have demonstrated amelioration of motor dysfunction, as well as partial reconstruction of the striatal circuitry following transplantation of fetal striatal cells, neural transplantation therapy is potentially beneficial for animals with ischemic striatal damage. Of note, previous studies have shown that transplanted fetal striatal cells in ischemic animals resulted in normalization of neurotransmitter (i.e., striatal GABA) release as well as motor dysfunctions *(17,18, 29,47)*. Because of inherent problems with the use of fetal cells as noted earlier, we examined the efficacy of transplanting hNT cells in ischemic animals. The hNT cells attain neuron-like features following retinoic acid treatment, and they have been shown to survive in and integrate with normal striata of rats *(48)*. We find that transplantation of hNT neurons into the striata of ischemic animals promotes functional recovery *(1)*. Of note, there was robust recovery in animals transplanted with hNT neurons that is significantly better than that observed in animals transplanted with fetal striatal cells. The EBST and passive avoidance test both revealed that as early as 1 mo posttransplantation, and continuing for the next 6 mo, animals transplanted with hNT neurons displayed near normal motor function and memory performance compared to control ischemic animals. Those animals that recovered had surviving hNT neurons when their brains were analyzed histologically using antibody to human neural cell adhesion molecule. Furthermore, when serial sections were stained for antibody to low molecular weight neurofilament, the grafted hNT neurons demonstrated positively stained fibers with features similar to those of the host tissue. Ongoing studies are now focusing on possible secretion of neurotrophic factors and neurotransmitters by the hNT neurons that may underlie the observed functional recovery in ischemic animals.

The positive effects noted in transplantation of fetal striatal cells or human cloned hNT neurons in ischemic animals suggest that recovery of motor/cognitive functions noted in chronic late-stage stroke may be ameliorated by reconstructing CNS morphology or circuitry as well as normalization of neurotransmitter release. This opens possible treatment options for end-stage stroke patients who have been nonresponsive to pharmacologic therapy.

7. NEUROTROPHIC FACTOR TREATMENT

We and others recently found that neurotrophic factors prevent the loss of dopaminergic function in Parkinsonian rats and decrease stroke-induced cortical infarction and nitric oxide (NO) release *(34,49)*. These data suggest that neurotrophic factors may play important roles in therapy for these neurodegenerative disorders.

As mentioned earlier, grafting of fetal nigral tissues into the striata of hemiparkinsonian rats can diminish drug-induced rotational behavior. However, we found that the restoration of dopamine (DA) function is limited to the graft site *(50,51)*. Neurotrophic factors such as glial cell line-derived neurotrophic factors (GDNF) have been shown to enhance survival as well as neurite outgrowth of DAergic tissue in vitro and in vivo, and may play an important role during neuronal damage. For example, GDNF or neurotrophin-4 (NT-4) has been demonstrated to increase the survival of corticospinal neurons in vitro *(52)*. In addition, GDNF protects against 6-hydroxydopamine-induced lesions in the nigrostriatal DA neurons *(53)* and also promotes the fiber outgrowth of DAergic transplants in the anterior chamber of the eye *(54)* or in the substantia nigra *(55)*. GDNF receptor-α mRNA and GDNF mRNA are up-regulated in the sciatic nerve after injury in adult mice *(56)*. The expression of transforming growth factor (TGF)-β1 mRNA is increased in regenerating renal tubules following acute ischemic injury *(56)*. Similarly, TGF-β1 transcript expression is enhanced in the hippocampus following transient forebrain ischemia *(57)*. These data suggest that GDNF and other TGF-β superfamily molecules have protective effects against ischemia, and indeed we have shown that pretreatment with GDNF in ligated ischemic animals significantly reduced infarction. This protective action by GDNF appears to be mediated by production of NO reduction and release which occurs during the ischemia–reperfusion process and has been implicated in the neurodegeneration process.

Recent studies have indicated that 1,25-(OH)(2)-D3 is a potent inducer of GDNF expression in C6 glioma cells *(58)*. Retinoic acid can further increase the effect of 1,25-(OH)(2)-D3 *(59)*. It is still not clear if 1,25-(OH)(2)-D3 and retinoic acid can increase GDNF levels in vivo. Since GDNF does not easily cross the blood–brain barrier (BBB), it would be advantageous to use an agent that penetrates the BBB to increase the GDNF levels in the brain. Preliminary studies in our laboratory have shown that application of 1,25-(OH)(2)-D3 enhances the GDNF-mediated protection against ischemic damage in rats *(60)*.

One major obstacle in proceeding with clinical trials of neurotrophic factor therapy is the stroke's limited therapeutic window, which has been shown earlier to be an inherent problem with pharmacologic therapy. To be of clinical use, neurotrophic factors must be protective even if delivered after the onset of ischemia. The choice of route and duration of treatment should take into account the time span of excitotoxic insult associated with stroke, which lasts for 4 h in rats and about 48 h in humans *(61)*. A combination therapy may be the next logical step toward achieving a protective/reparative strategy that could be efficiently delivered during the extremely limited therapeutic window and could reverse the progressive metabolic consequences (e.g., apoptosis and free radical accumulation) of ischemia. One such combination therapy, the intracerebral transplantation of neuronal cells and GDNF treatment, has been initi-ated already in animal models of Parkinson's disease and other neurodegenerative dis-orders *(55)*. Further studies will be required to determine if such combination therapy induces anatomical restoration, coupled with functional recovery, following an experi-mental ischemic stroke.

8. SUMMARY

Animal models of stroke offer a platform to investigate neuronal dysfunctions, as well as motor/cognitive alterations, associated with this disease. Laboratory evidence directed toward understanding the mechanisms of cerebral ischemia has been gener-ated using both the embolism and ligation MCA models. These models appropriately parallel many of the neurohistopathological and behavioral hallmarks of stroke. Some differences exist between these two models and should be taken into consideration when investigating specific regions of stroke. For example, the CNS areas of infarction are primarily localized to the striatum and cortex in the embolism and ligation model, respectively. Accordingly, the target CNS areas for treatment intervention may differ in each model. We have demonstrated that neural transplantation therapy may have potential benefits for the embolism model, while treatment with the neurotrophic fac-tor GDNF can attenuate ligation-induced infarction. A combined therapy using neural transplantation and GDNF treatment has been explored in animal models of human disease, and may prove beneficial for treatment of stroke. Understanding the basic mechanisms of stroke and development of new, or refinement of existing, treatment strategies for the disease may be exploited using such animal models.

ACKNOWLEDGMENT

The authors appreciate critical comments provided by Barry J. Hoffer.

REFERENCES

1. Borlongan, C. V., Tajima, Y., Trojanowski, J. Q., Lee, V. M., and Sanberg, P. R. (1998a) Transplantation of cryopreserved human embryonal carcinoma-derived neurons (NT2N cells) promotes functional recovery in ischemic rats. *Exp. Neurol.* **149,** 310–321.
2. Lipton, S. A. (1996) Similarity of neuronal cell injury and death in AIDS dementia and focal cerebral ischemia: potential treatment with NMDA open-channel blockers and nitric oxide-related species. *Brain Pathol.* **6,** 507–517.
3. Satoh, S., Ikegaki, I., Suzuki, Y., Asano, T., Shibuya, M., and Hidaka, H. (1996) Neuroprotective properties of a protein kinase inhibitor against ischemia-induced neuronal damage in rats and gerbils. *Br. J. Pharmacol.* **118,** 1592–1596.
4. Endres, M., Namura, S., Shimizu Sasamata, M., Waeber, C., Zhang, L., Gomex Isla, T., Hyman, B. T., and Moskowitz, M. A. (1998) Attenuation of delayed neuronal death after mild focal ischemia in mice by inhibition of the caspase family. *J. Cereb. Blood Flow Metab.* **18,** 238–247.
5. Zornow, M. H. and Prough, D. S. (1996) Neuroprotective properties of calcium-channel blockers. *New Horiz.* **4,** 107–114.
6. Wood, N. I., Barone, F. C., Benham, C. D., Brown, T. H., Campbell, C. A., Cooper, D. G., et al. (1997) The effects of SB 206284A, a novel neuronal calcium-channel antagonist, in models of cerebral ischemia. *J. Cereb. Blood Flow Metab.* **17,** 421–429.

7. Schellinger, P. D., Orberk, E., and Hacke, W. (1997) Antithrombotic therapy after cerebral ishcemia. *Fortschr. Neurol. Psychiatr.* **65,** 425–434.

8. Donnan, G. A., Davis, S. M., Chambers, B. R., Gates, P. C., Hankey, G. J., McNeil, J. J., Rosen, D., Stewart, Wynne, E. G., and Tuck, R. R. (1996) Streptokinase for acute ischemic stroke with relationship to time of administration: Australian Streptokinase (ASK) Trial Study Group. *JAMA* **276,** 961–966.

9. Mattle, H. P. and Ringelstein, E. B. (1996) Therapy and prevention of cerebral ischemia. *Ther. Umsch.* **53,** 573–584.

10. Takagi, Y., Tokime, T., Nozaki, K., Gon, Y., Kikuchi, H., and Yodoi, J. (1998) Redox control of neuronal damage during brain ischemia after middle cerebral artery occlusion in the rat: immunohistochemical and hybridization studies of thioredoxin. *J. Cereb. Blood Flow Metab.* **18,** 206–214.

11. Onal, M. Z. and Fisher, M. (1997) Acute ischemic stroke therapy. A clinical overview. *Eur. Neurol.* **38,** 141–154.

12. Campbell, C. A., Mackay, K. B., Patel, S., King, P. D., Stretton, J. L., Hadingham, S. J., and Hamilton, T. C. (1997) Effects of isradipine, an L-type calcium channel blocker, on permamnent and transient focal cerebral ischemia in spontaneously hypertensive rats. *Exp. Neurol.* **148,** 45–50.

13. Shiota, T., Bernanke, D. H., Parent, A. D., and Hasui, K. (1996) Protein kinase C has two major roles in lattice compaction enhanced by cerebrospinal fluid from patients with subarachnoid hemorrhage. *Stroke* **27,** 1889–1895.

14. Nishino, H., Kumazaki, M., Sakurai, T., Sanberg, P. R., and Borlongan, C. V. (1998) Melatonin blocks ischemia-reperfusion induced neuronal damage. *Soc. Neurosci.* **24,** 213.

15. Fink, S. and Chaudhuri, T. K. (1997) Stroke and malpractice claims. *South Med. J.* **90,** 901–902.

16. Nagasawa, H. and Kogure, K. (1989) Correlation between cerebral blood flow and histologic changes in a new rat model of middle cerebral artery occlusion. *Stroke* **20,** 1037–1043.

17. Nishino, H., Koide, K., Aihara, N., Kumazaki, M., Sakurai, T., and Nagai, H. (1993a) Striatal grafts in the ischemic striatum improve pallidal GABA release and passive avoidance. *Brain Res. Bull.* **32,** 517–520.

18. Nishino, H., Aihara, N., Czurko, A., Hashitani, T., Isobe, Y., Ichikawa, O., and Watari, H. (1993b) Reconstruction of GABAergic transmission and behavior by striatal cell grafts in rats with ischemic infarcts in the middle cerebral artery. *J. Neural. Transplant. Plast.* **4,** 147–155.

19. Hata, R., Mies, G., Wiessner, C., Fritze, K., Hesselbarth, D., Brinker, G., and Hossmann, K.-A. (1998) A reproducible model of cerebral artery occlusion in mice: hemodynamic, biochemical, and magnetic resonance imaging. *J. Cereb. Blood Flow Metab.* **18,** 367–375.

20. Zhao, W., Ginsberg, M. D., Prado, R., and Belayev, L. (1996) Depiction of infarct frequency distribution by computer-assisted image mapping in rat brains with middle cerebral artery occlusion. Comparison of photothrombotic and intraluminal suture models. *Stroke* **27,** 1112–1117.

21. Wood, N. I., Sopesen, B. V., Roberts, J. C., Pambakian, P., Rothaul, A. L., Hunter, A. J., and Hamilton, T. C. (1996) Motor dysfunctions in a photothrombotic focal ischemia model. *Behav. Brain Res.* **78,** 113–120.

22. Schmidt-Kastner, R., Fliss, H., and Hakin, A. M. (1997) Subtle neuronal death after short forebrain ischemia in rats detected by in situ end-labeling for DNA damage. *Stroke* **28,** 163–169.

23. Du, C., Hu, R., Csemansky, C. A., Hsu, C. Y., and Choi, D. W. (1996) Very delayed infarction after mild focal cerebral ischemia: a role for apoptosis? *J. Cereb. Blood Flow Metab.* **16,** 195–201.

24. Chen, S. T., Hsu, C. Y., Hogan, E. L., Maricq, H., and Balentine, J. D. (1986) A model

of focal ischemic stroke in the rat: reproducible extensive cortical infarction. *Stroke* **17,** 738–743.

25. Bederson, J. B., Pitts, L. H., Tsuji, M., Nishimu, M. C., Davis, R. L., and Barowski, H. (1986) Rat middle cerebral artery occlusion: evalution of the model and development of a neurologic examination. *Stroke* **17,** 472–476.

26. Witte, O. W. and Stoll, G. (1997) Delayed and remote effects of focal cortical infarctions: secondary damage and reactive plasticity. *Adv. Neurol.* **73,** 207–227.

27. Yamada, K., Goto, S., Yoshikawa, M., and Ushio, Y. (1996) Gabaergic transmission and tyrosine hydroxylase expression in the nigral dopaminergic neurons: an in vivo study using a reversible ischemia model of rats. *Neuroscience* **73,** 783–789.

28. Ogawa, N., Asanuma, M., Tanaka, K., Hirata, H., Kondo, Y., Goto, M., Kawauchi, M., and Ogura, T. (1996) Long-term time course of regional changes in cholinergic indices following transient ischemia in the spontaneuosly hypertensive rat brain. *Brain Res.* **712,** 60–68.

29. Aihara, N., Mizukawa, K., Koide, K., Mabe, H., and Nishino, H. (1994) Striatal grafts in infarct striatopallidum increase GABA release, reorganize GABAA receptor and improve water-maze learning in the rat. *Brain Res. Bull* **33,** 483–488.

30. Yu, Z., Cheng, G., Huang, X., Li, K., and Cao, X. (1997) neurokinin-1 receptor antagonist SR 140333: a novel type of drug to treat cerebral ischemia. *NeuroReport* **8,** 2117–2119.

31. Borlongan, C. V., Cameron, D. F., Saporta, S., and Sanberg, P. R. (1997a) Intracerebral transplantation of testis-derived sertoli cells promotes functional recovery in female rats with 6-hydroxydopamine-induced hemiparkinsonism. *Exp. Neurol.* **148,** 388–392.

32. Ouchi, Y., Tsukada, H., Kakiuchi, T., Nishiyama, S., and Futatsubashi, M. (1998) Changes in cerebral blood flow and postsynaptic muscarinic cholinergic activity in rats with bilateral carotid artery ligation. *J. Nucl. Med.* **39,** 198–202.

33. Marinov, M., Schmarov, A., Natschev, S., Stamenov, B., and Wassman, H. (1996) Reversible focal ischemia model int he rat: correlation of somatosensory evoked potentials and cortical neuronal injury. *Neurol. Res.* **18,** 73–82.

34. Wang, Y., Lin, S. Z., Chiou, A. L., Williams, L. R., and Hoffer, B. J. (1997) GDNF protects aganist ischemia-induced injury in the cerebral cortex. *J. Neurosci.* **17,** 4341–4348.

35. Borlongan, C. V., Cahill, D. W., and Sanberg, P. R. (1995a) Locomotor and passive avoidance deficits following occlusion of the middle cerebral artery. *Physiol. Behav.* **58,** 909–917.

36. Borlongan, C. V., Martinez, R., Shytle, R. D., Freeman, T. B., Cahill, D. W., and Sanberg, P. R. (1995b) Striatal dopamine-mediated motor behavior is altered following occlusion of the middle cerebral artery. *Pharmacol. Biochem. Behav.* **52,** 225–229.

37. Borlongan, C. V. and Sanberg, P. R. (1995) Elevated body swing test: a new behavioral parameter for rats with 6-hydroxydopamine-induced hemiparkinsonism. *J. Neurosci.* **15,** 5372–5378.

38. Kordower, J. H., Freeman, T. B., Snow, B. J., Vingerhoets, F. J., Mufson, E. J., Sanberg, P. R., et al. (1995) Neuropathological evidence of graft survival and striatal reinnervation after the transplantation of fetal mesencephalic tissue in a patient with Parkinson's disease. *N. Engl. J. Med.* **332,** 1118–1124.

39. Freeman, T. B., Olanow, C. W., Hauser, R. A., Nauert, G. M., Smith, D. A., Borlongan, C. V., et al. (1995) Bilateral fetal nigral transplantation into the postcommissural putamen in Parkinson's disease. *Ann. Neurol.* **38,** 379–88.

40. Sanberg, P. R., Borlongan, C. V., Othberg, A. I., Saporta, S., Freeman, T. B., and Cameron, D. F. (1997) Testis-derived Sertoli cells have a trophic effect on dopamine neurons and alleviate hemiparkinsonism in rats. *Nat. Med.* **3,** 1129–1132.

41. Borlongan, C. V., Koutouzis, T. K., Poulos, S. G., Saporta, S., and Sanberg, P. R. (1998b) Bilateral fetal striatal grafts into the 3-nitropropionic acid-induced hypoactive model of Huntignton's disease. *Cell Transplant* **7,** 131–136.

42. Borlongan, C. V., Poulos, S. G., Cahill, D. W., and Sanberg, P. R. (1998c) Effects of fetal striatal transplants on motor asymmetry in ibotenic acid model of Huntington's disease. *Psychobiology* **26,** 49–52.
43. Borlongan, C. V., Koutouzis, T. K., Jorden, J. R., Martinez, R., Rodriguez, A. I., Poulos, S. G., Freeman, T. B., McKeown, P., Cahill, D. W., Nishino, H., and Sanberg, P. R. (1997b) Neural transplantation as an experimental treatment modality for cerebral ischemia. *Neurosci. Biobehav. Rev.* **21,** 79–90.
44. Isacson, O., Deacon, T. W., Pakzaban, P., Galpern, W. R., Dinsmore, J., and Burns, L. H. (1995) Transplanted xenogeneic neural cells in neurodegenerative disease models exhibit remarkable axonal target specificity and distinct growth patterns of glial and axonal fibres. *Nat. Med.* **1,** 1189–1194.
45. Nakao, N., Grasbon-Frodl, E. M., Widner, H., and Brundin, P. (1996) DARPP-32-rich zones in grafts of lateral ganglionic eminence govern the extent of functional recovery in skilled paw reaching in an animal model of Huntington's disease. *Neuroscience* **74,** 959–970.
46. Philpott, L. M., Kopyov, O. V., Lee, A. J., Jacques, S., Duma, C.M., Caine, S., Yang. M., and Eagle, K. S. (1997) Neuropsychological functioning following fetal striatal transplantation in Huntington's chorea: three case presentations. *Cell Transplant.* **6,** 203–212.
47. Onizuka, K., Fukuda, A., Kunimatsu, M., Kumazaki, M., Sasaki, M., Takaku, A., and Nishino, H. (1996) Early cytopathic features in rat ischemia model and reconstruction by neural graft. *Exp. Neurol.* **137,** 324–32.
48. Trojanowski, J. Q., Kleppner, S. R., Hartley, R. S., Miyazono, M., Fraser, N. W., Kesari, S., and Lee, V. M. (1997) Transfectable and transplantable postmitotic human neurons: a potential "platform" for gene therapy of nervous system diseases. *Exp. Neurol.* **144,** 92–97.
49. Snyder, S. H. (1996) No NO prevents parkinsonism. *Nat. Med.* **2,** 965–966.
50. Wang, Y., Perng, S. L., Lin, J. C., and Tsao, W. L. (1994a) Cholecystokinin facilitates methamphetamine-induced dopamine overflow in rat striatum and fetal ventral mesencephalic grafts. *Exp. Neurol.* **130,** 279–287.
51. Wang, Y., Wang, S. D., Lin, S. Z., and Liu, J. C. (1994b) Restoration of dopamine overflow and clearance from the 6-hydroxydopamine lesioned rat striatum reinnervated by fetal mesencephalic grafts. *J. Pharmacol. Exp. Ther.* **270,** 814–821.
52. Junger, H. and Varon, S. (1997) Neurotrophin-4 (NT-4) and glial cell line-derived neurotrophic factor (GDNF) promote the survival of corticospinal motor neurons of neonatal rats in vitro. *Brain Res.* **762,** 56–60.
53. Gash, D. M., Zhang, Z., Ovadia, A., Cass, W. A., Yi, A., Simmerman, L., et al. (1996) Functional recovery in parkinsonian monkeys treated with GDNF. *Nature* **380,** 252–255.
54. Stromberg, I., Bjorklund, L., and Forander, P. (1997) The age of striatum determines the pattern and extent of dopaminergic innervation: a nigrostriatal double graft study. *Cell Transplant.* **6,** 287–296.
55. Wang, Y., Tien, L. T., Lapchak, P. A., and Hoffer, B. J. (1996) GDNF triggers fiber outgrowth of fetal ventral mesencephalic grafts from nigra to striatum in 6-OHDA-lesioned rats. *Cell Tissue Res.* **286,** 225–233.
56. Basile, D. P., Rovak, J. M., Martin, D. R., and Hammerman, M. R. (1996) Increased transforming growth factor-beta 1 expression in regenerating rat renal tubules following ischemic injury. *Am. J. Physiol. Renal Fluid Elect.* **39,** F500–F509.
56. Naveilhan, P., ElShamy, V. M., and Emfors, P. (1997) Differential regulation of mRNAs for GDNF and its receptors Ret and GDNFR alpha after sciatic nerve lesion in the mouse. *Eur. J. Neurosci.* **9,** 1450–1460.
57. Knuckey, N. W., Finch, P., Palm, D. E., PRimiano, M. J., Johanson, C. E., Flanders, K. C., and Thompson, N. L. (1996) Differential neuronal and astrocytic expression of transform-

ing growth factor beta isoforms in rat hippocampus following transient forebrain ischemia. *Mol. Brain Res.* **40,** 1–14.

58. Neil Verity, A., Wyatt, T. L., Hajos, B., Eglen, R. M., Baecker, P. A., and Johnson, R. M. (1998) Regulation of glial cell line-derived neurotrophic factor release from rat C6 glioblastoma cells. *J. Neurochem.* **70,** 531–539.

59. Naveilhan, P., Neveu, I., Wion, D., and Brachet, P. (1996) 1,25-Dihydroxyvitamin D-3, an inducer of glial cell line-derived neurotrophic factor. *NeuroReport* **7,** 2171–2175.

60. Wang, Y., Lin, S. Z., Chiang, Y. H., and Hoffer, B. J. (1998) Co-administration of 1,25-dihydroxyvitamin D3 and retinoic acid attenuates cortical infarction induced by middle cerebral arterial ligation. *Soc. Neurosci. Abstr.* **24,** 1950.

61. Dyker, A. G. and Lees, K. R. (1998) Duration of neuroprotective treatment for ischemic stroke. *Stroke* **29,** 535–542.

21

A Primate Model of Hypertensive Cerebrovascular Disease

Mark B. Moss

1. INTRODUCTION

Cerebrovascular disease (CVD) in humans has been shown to produce a variety of cognitive impairments ranging from selective deficits to wide-range dementia. Among the risk factors for CVD (e.g., age, diabetes mellitus, serum lipids, obesity, cardiac disease), arterial hypertension has been identified as key *(1)* affecting more than 25% of the adult population of the United States *(2)*. Gross effects of extreme hypertension are well known and include a four times greater risk for CVD than normotensive individuals *(3)*. For the most part, hypertension is an asymptomatic disorder, but recently it has been the focus of attention on its possible detrimental effects on cognitive function. Indeed, over the past two decades, evidence has accumulated to suggest that hypertension in humans produces, in many cases, a significant impairment in several domains of cognitive function. But because of the inherent limitations of human research, even with recent advances in magnetic resonance and positron emission imaging technology, our understanding of the neurobiological basis for hypertensive related cognitive impairment is unknown. It is also unknown to what extent the changes in cognition that are associated with hypertension represent the first stage in the development of CVD and vascular dementia.

Animal studies using rodents, rabbits, and, to a lesser extent, nonhuman primates have contributed significantly to our understanding of the underlying mechanisms and pathological changes associated with hypertension and CVD, but we have yet to address the question of the neural basis for the neuropsychological consequences of hypertension. This chapter describes the development of a primate model of hypertensive CVD incorporating a multidisciplinary study that includes the assessment of cognitive function.

2. DEVELOPMENT OF THE MODEL

As clearly demonstrated by the contributions contained in this volume, the use of animal models provides a major approach to the study of human disease. The assessment of disease processes is difficult and rather limited with human subjects owing to

From: *Central Nervous System Diseases*
Edited by: D. F. Emerich, R. L. Dean, III, and P. R. Sanberg © Humana Press Inc., Totowa, NJ

lack of control over extraneous variables such unknown health histories, use of medications, control of diet, and unknown time of onset. It is also difficult to obtain behavioral, physiological, and morphological data from the same individuals and to thoroughly exclude individuals with insidious disease processes such as the early stages of Alzheimer's disease and other forms of dementia. Therefore, the development of animal models to study disease states provides clear advantages over research in humans. In cases of human disease that affect the higher cortical function, incorporating into animal models the assessment of cognition using behavioral tests that have been adapted for use from established neuropsychological tests in humans presents a added advantage.

Although this approach typically provides more reliable and more precise data, the relevance to the human condition is a function of the adequacy with which the experimental animal exhibits the human traits under study. In general, the use of animal models is indicated when the experimental methods are inconvenient or impossible to apply to human subjects, for example, specialized preparation and treatment of tissue. Our general goal was to determine whether the rhesus monkey is a suitable animal model of human hypertensive CVD and vascular dementia. Much had been learned about various aspects of these disorders through series of clinically relevant human and animal model investigations. However, relatively few investigations have used the monkey as a model of hypertensive CVD, particularly one in which cognitive function is carefully profiled. Thus, one major objective of this model was to establish the role of hypertension in cognitive impairment and decline and the relationship of this decline to specific alterations in the brain. A second and related goal was to determine the mechanisms that underlie the development of neuropathology as a consequence of hypertension.

2.1. Rationale for the Monkey

In addition to the advantage of developing animal models of human disease, we thought the particular strength of a nonhuman primate model lies in the close relationship of central nervous system (CNS) structure and connectivity between humans and nonhuman primates, and in the use of behavioral tasks in monkeys that have been adapted for use for the clinical assessment and differential diagnosis of patients with CVD and vascular dementia as well as other age-related disorders. Studies from the literature have shown that several species, including the monkey *(2,4,5)*, are suitable for the study of neurobiological consequences of hypertension and atherosclerosis *(see 6 for a review)*. However, for the purposes of our model, the monkey is ideally suited because many of the cognitive tasks used to assess patients with hypertensive CVD and vascular dementia can be adapted for use in this species. There are no counterparts of these tests that can be used in the rat or rabbit. The use of this species was also based on the long history of the demonstrated utility of monkeys as a model for conditions affecting humans in the various fields of medical science. There exists an extensive body of knowledge about the normal and abnormal biology of this species, an important factor from the standpoint of establishing correlations with observed CVD changes. Finally, whereas human life and health histories are often incomplete or nonexistent, extensive medical and social histories are available on the monkeys used for our model.

One issue often raised concerning the rhesus monkey is the age-equivalency and life span of these animals. In summary, the typical adult life-span of the rhesus monkey may be considered to extend from 5 up to 30 years, and it has been estimated that the life-span ratio of human to monkey is approx 3:1. On this basis, the monkeys that are describe below that were part of our 12-mo and 39-mo experimental protocol might be considered in human terms as being hypertensive for approx 3 yr and 10 yr, respectively.

3. COGNITION AND HYPERTENSION

The study of the effects of hypertension on intellectual function in humans was initiated more than 50 yr ago. Over this period of time, evidence has accumulated to show that hypertension produces impairment in cognition, but to a greater extent in some domains than in others. The earlier studies were conducted at a time when antihypertensive medications were not available and variables such as age and education were typically not considered *(7–9)*. The degree of impairment described in many of these studies was likely related to the fact that subjects often had severe, uncontrolled hypertension with frank neurologic signs. Therefore, it was not surprising that many investigators concluded that hypertension not only produced marked impairments in intellectual function, but also produced marked grossly visible damage to the brain.

More recent studies that have controlled for many of these factors have still shown that patients are impaired on a variety of cognitive tasks including those of general intelligence *(10,11)*, and memory function *(12)*, but without evidence of frank damage to the brain. A well controlled study by Schmidt et al *(13)* assessed a group of hypertensive individuals and compared performance to normotensive control subjects. They found that the hypertensive group was more impaired on tasks of verbal memory and total learning, but no difference in performance was observed on tasks of visual memory, attention, vigilance, and reaction time. Systolic blood pressure has been shown to be a sensitive measure of cognitive status as indicated by negative correlations with WAIS performance in a group of hypertensive male subjects *(14)*. Deficits have also been found when hypertensive subjects were compared to normotensive subjects in WAIS performance subtest scores *(15)* and in verbal scores when studied longitudinally over 5–6-yr intervals *(16)*. Findings also show that, with longer exposure, blood pressure measures accurately predict cognitive decline *(17)*.

In a study conducted in Sweden on 1736 community-based human subjects, both systolic and diastolic blood pressures were significantly related to baseline performance on the Mini Mental Status Exam (MMSE) and baseline systolic pressure was significantly related to follow-up performance *(18)*. Of direct relevance to the present animal model, untreated hypertension in humans has recently been found to be inversely related to both a composite index of cognitive performance and individual scores on tests of attention and memory in the Framingham Heart Study *(19–21)*. In a longitudinal study of 3735 Japanese-American men living in Hawaii, systolic blood pressure proved to be a significant predictor of reduced cognitive function 16 yr later *(22)*. Finally, cognitive tests have shown a consistent trend towards poorer performance in hypertensive subjects with significant deficits in verbal learning in a study of inner city subjects in England *(23)*.

Attentional measures such as symbol/digit substitution, continuous attention, reaction time, paired word association, and inspection time threshold have all been found to be significantly impaired in a group of untreated mild to moderate hypertensive subjects when compared to the performance of age-matched controls *(24)*. Similarly, regression analyses revealed that part of the impairment attributed to age in a study of visual selective attention was in fact due to blood pressure level in a study of unmedicated mild hypertensive subjects ranging from 18 to 78 yr of age *(25)*.

Taken together, the weight of evidence strongly suggested that hypertension produces impairment in the domains of attention, memory, and executive function (abstraction, set shifting) but to a much lesser extent in those of visuospatial skills, psychomotor speed, and verbal skills. Accordingly, we decided to assess the effects on cognition, with particular regard to attention, memory, and executive function, in our primate model of hypertensive CVD.

3.1. Hypertension and Age

Because the prevalence and incidence of hypertension increases with age, many studies have focussed on the contribution of hypertension to age-related cognitive change *(26–28)*. However, the effects of hypertension on cognition may not be uniform across the age range. In fact, and perhaps somewhat surprisingly, the effects of high blood pressure on cognition may be greater in younger than in older subjects *(29)*, and this effect appears to be independent of demographic, psychosocial, and education-related factors *(30)*. Hypertension was found to be negatively associated with WAIS verbal scores in younger (21–39 yr) but not older (45–65 yr) subjects while its effect on performance scores was greater for younger than older subjects *(31)*. Similarly, hypertension in middle-aged adults was associated with a disproportionate decline in performance on tests of psychomotor speed and an increase in error rate on a test of visual selective attention *(25)*. Certainly, additional work is needed to further understand the way in which these two variables interact.

4. COGNITION AND CVD

There is little disagreement among workers in the field that CVD can produce significant cognitive impairment, and if widespread, a dementia state. However, the precise nature and type of dementia states with cerebrovascular etiology are not clear. For example, the terms multiinfarct dementia, Binswanger leukoencephalopathy, small vessel disease, lacunar state, or cerebrovascular dementia have all been used to describe a cognitive decline associated with CVD. In addition, with the advent of more sensitive magnetic resonance imaging (MRI) technology, white matter abnormalities have been more readily detected and have given rise to a condition that is also associated with cognitive impairment. This condition, without known pathogenesis, has been coined by Hachinski et al. *(32)* as leukoaraiosis. Several investigators have identified a relationship between changes in white matter detected on MRI and decrement in cognitive performance *(13,33)*. For example, the study by Schmidt revealed that performance on a battery of cognitive tests, including memory and executive function, was significantly worse in a group of hypertensive elderly subjects with lesions of the white matter on MRI as compared with a matched group of patients with normal white matter or small focal lesions.

Indeed, the determination of the prevalence of cerebrovascular dementia has been a difficult challenge to investigators in the field. One of the earliest studies to assess the prevalence of hypertensive CVD was Akesson *(34)*, who, using hospital and nursing home records, found a prevalence of 0.21%. Later studies using community samples have yielded figures ranging from 1.86% *(35)* to 2.7% *(36)*. In the same study, age-specific rates revealed a range of 1.9% in the 65–74 yr range, to 4.3% in the 75–84-yr range. Incidence findings suggest that cerebrovascular dementia accounts for more than 20% of all dementia and likely contributes to another 15–20% of cases.

With regard to our primate model, we have found that the animals in our 39-mo studies evidence more severe and widespread cognitive impairment than those in the 12-mo studies. By virtue of the extent and range of impairment seen in the monkeys with 39 mo of hypertension, that is, an impairment in more than one cognitive domain that includes memory function, one might consider this as, at least by clinical definition, an early dementia state.

5. PRODUCTION OF HYPERTENSION

The basis of this model is the production of hypertension in the monkey which is achieved by surgical coarctation the thoracic aorta *(37,38)*. Prior to surgery, monkeys are trained to perform in a Wisconsin General Test Apparatus on two behavioral tasks. They are then assigned to one of the experimental groups or the control group in a predetermined fashion based on entry into the study (we have found no significant difference [$p > 0.5$] between the groups in preoperative performance). All monkeys are housed in the AAALAC approved Laboratory of Animal Science Center at Boston University Medical Center and are maintained on Purina monkey chow for 6 mo before entry into the study. At surgery, animals are initially sedated with Ketalar (ketamine hydrochloride) and blood pressure is measured using an Arteriosonde and an ECG is recorded. The animals are anesthetized using sodium pentobarbital and are then intubated orotracheally and connected to a respirator. The monkey is placed in a lateral position with its left side up. A left anterolateral thoractomy is done along the fifth intercostal space. The lung is retracted medially exposing the thoracic aorta at the posterior mediastinum. A segment of the thoracic aorta just below the level of the hilum of the left lung is mobilized and dissected without injuring the mediastinal and intercostal branches. The external diameter of the same segment is measured with a caliper. A 1 cm segment is then narrowed to luminal diameter of 2.0–2.5 mm using surgical callipers and a Casteneda partial occlusion vascular clamp (Pilling Instruments, Fort Washington, PA, USA (Fig. 1). A supporting band of umbilical tape is then drawn around the coarcted segment and sutured without further constriction of the vessel. The coarctation of the aorta results in a decrease in luminal area of about 75–80% (Fig. 2), as indicated by autopsy findings. During the immediate 3-wk postoperative period, the monkeys are given angiotensin I converting enzyme inhibitors, diuretics, and digoxin to prevent heart failure. During the baseline period, and at 2–3 mo intervals throughout the experimental period, measurements are made of body weight and blood pressure.

The blood pressure in the brachial artery is monitored indirectly by the ultrasonic cuff method weekly in the postoperative period and then at intervals of 2 mo with the use of the Arteriosonde *(39)*. Direct measurements of the intraarterial pressure in the brachial and femoral arteries also are performed on the day of the surgical coarctation and at 3, 6, and 12 mo after the surgery. After exposing and cannulating the brachial and femoral arteries, the arterial pressure in these arteries is simultaneously measured with strain gauge transducers attached to a Beckman dynograph recorder. Direct

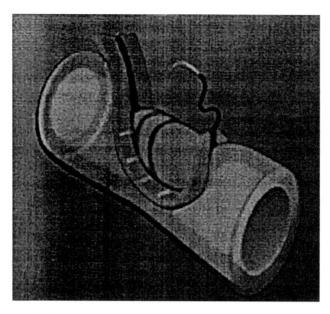

Fig. 1. Illustration depicting the partial clamping of the thoracic aorta during the coarctation surgery.

measurements of brachial arterial pressure tended to be higher than the indirect measurements *(40)* (Fig. 3).

6. BLOOD PRESSURE FINDINGS

Blood pressure data for 25 hypertensive monkeys who have completed our series of studies in either the 12-mo or 39-mo protocols, using direct intraarterial measurement, revealed an average systolic blood pressure of 185 ± 5.8 mm Hg and an average diastolic blood pressure of 113 ± 3.0 mm Hg. This compares with average systolic and diastolic blood pressures for 20 operated control monkeys of 124 ± 2.7 mm Hg and an average diastolic blood pressure of 81 ± 1.6 mm Hg, respectively.

7. BEHAVIORAL ASSESSMENT

Monkeys were trained both preoperatively and then postoperatively at 6 mo, 12 mo, 24 mo, and 39 mo. This approach allowed both postoperative retention of preoperatively learned tasks and postoperative acquisition of new tasks. Data are presented in this chapter on the 6-mo performance on attentional and memory tasks, on 12-mc performance on the memory tasks, and on 39-mo performance on the executive system function tasks. The details of these behavioral studies have recently been submitted for publication *(41,42)*.

7.1. Studies on Attention

Only a few studies have assessed the effects of hypertension on attentional skills. In each case the digit span subtest of the WAIS-R was used as an index of attention. Among these four studies, one showed a strong trend toward significance *(13)* and three showed a significant impairment *(11,15,30)* in groups of hypertensive subjects.

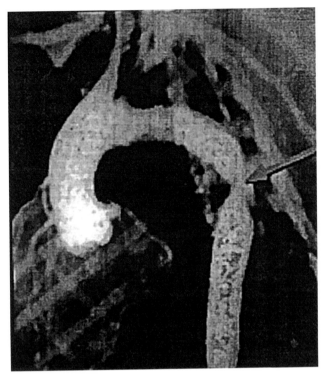

Fig. 2. Arteriogram showing the coarctation of the thoracic aorta (arrow) in the monkey.

Taken together, the weight of evidence from studies in humans suggests that attentional function, at least measured by digit span, may be impaired as a consequence of hypertension. But the domain of attention encompasses at least several interrelated abilities such as orienting, alerting, selection, and vigilance *(43–45)*, and the effect of hypertension on these elements of attention has not been systematically assessed in human or animal studies.

As part of our program, the effects of hypertension on attention was studied in 25 young adult rhesus monkeys after 6 mo of hypertension. Their performance was compared to that of 11 operated controls that underwent every stage of the surgical procedures up to but not including the actual narrowing of the aorta. Testing took place in a computer-controlled darkened testing chamber that contained a reward dispensing cup, a set of speakers, and a 19-in. color computer monitor covered with a resistive touch screen.

The monkeys were acclimated to the testing chamber and then completed a "pretraining" task that consisted of 50 consecutive responses to a stimulus that appeared in any one of 12 spatial locations on the screen in one session. Each response was rewarded. Formal testing began the day after the monkey had completed pre-training.

7.1.1. Simple Attention

For the test of simple attention, the monkeys were required to touch the same target stimulus on the touch screen that they had seen during the pretraining phase. Mixed intertrial intervals of 5, 10, 20, 40, and 60 s were used in a pseudo-random fashion to

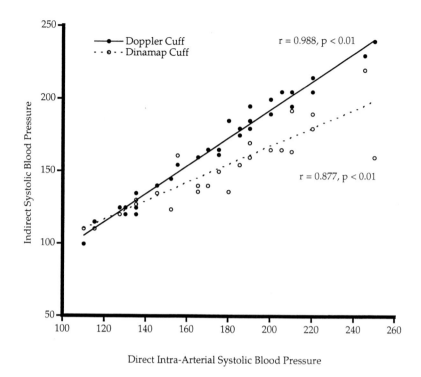

Fig. 3. The Intraarterial, Doppler and Dinamap methods of measuring systolic blood pressure all produce values that are highly significantly correlated with one another. However, the value that each technique identifies as the pressure differs significantly from one another (i.e., Intra-arterial vs Doppler, $p < 0.0001$, Intra-arterial vs Dinamap, $p < 0.0001$, Doppler vs Dinamap, $p < 0.0001$). On average, the Intra-arterial produces the highest pressure, while the Doppler produces a value that is 6.35 mm Hg below the Intra-arterial pressure. Similarly, the Dinamap produces a value that is 25.03 mm Hg below the Intra-arterial pressure. For the Doppler, this difference remains relatively constant. However, as can be seen on the graph, the difference between intra-arterial and the Dinamap pressures is not constant for as the pressure rises, the difference between the methods increases.

prohibit the monkey from anticipating the appearance of the stimulus. As during the pretraining phase the target stimulus appears pseudo-randomly in one of 12 spatial locations on the screen. When the stimulus was touched, the latency to touch was recorded, the touch screen became black, food reward was delivered, and the next intertrial interval began. If the monkey did not touch the stimulus on the screen within 60 s, a nonresponse was recorded, no reward was delivered, the touch screen became black, and the next intertrial interval began. Testing continued in this fashion for 50 trials per day for 2 consecutive days.

With this initial measure of performance, we found significant correlations ($p < 0.01$) with systolic and mean blood pressure and the latency to respond measure at the 10-s intertrial interval. Initial inspection of the data suggested an analysis based on the stratification of blood pressure. Indeed, when we grouped the hypertensive animals into borderline (135–150 mm Hg) and significantly hypertensive groups (above 150 mm Hg), we found that the monkeys in the hypertensive group were significantly impaired when compared to the monkeys in the normotensive group and marginally more impaired when compared to the borderline hypertensive group ($p = 0.07$). No significant difference was found between the performance of the monkeys in the normotensive group and the borderline group.

7.1.2. Cued Attention

Testing of cued attention was conducted in a fashion similar to that for simple attention with two exceptions. First, only 10- and 20-s intertrial intervals were used. Second, 2 s before the appearance of the target stimulus, a bright yellow cue, the same size as the target stimulus, appeared on the touch screen for 1 s before disappearing. The screen was then blank for 1 s before the appearance of the target stimulus. The cue was either valid (appearing in the same spatial location as the target stimulus) or invalid (appearing in a different spatial location than the target stimulus). The cues appeared on every trial in an intermixed fashion for an overall total of 75 positive cues and 25 negative cues (44 positive and 6 negative during the first day and 31 positive and 19 negative during the second day). Testing took place on the cued phase for 50 trials per day for 2 consecutive days.

On the assessment of this component of attention, as with simple attention, we found a significant correlation between the systolic, diastolic, and mean blood pressure measures and the latency to respond measures ($p < 0.01$) for the 10-s intertrial interval with a valid cue and for both types of cue at the 20-s intertrial interval. Group comparisons revealed that the monkeys in the hypertensive group ($p < 0.04$) were significantly impaired when compared to the monkeys in the borderline and normotensive groups. No significant difference ($p > 0.89$) was found between the performance of the borderline and normotensive groups.

7.1.3. Vigilance

A measure of vigilance was extracted from the data collected during the simple and cued phases of testing. The score for the vigilance test was the number of trials in which the monkey failed to respond to the target stimulus within 60 s of its appearance across the 4 d of testing (200 trials) simple and cued attention. The latency to respond measures from these trials was removed from the data set to avoid bias for between-subjects comparisons.

The results of this analysis showed no significant correlation between the number of nonresponses and blood pressure. Nor did it reveal any significant difference along this variable among the groups, a finding suggesting in part that motivation did not play a role in the attention findings.

Thus, on our studies of attention, monkeys with hypertension were impaired on a task that required orienting to, and then responding by touching, a randomly presented visual stimulus. Unlike normotensive animals, hypertensive monkeys did not benefit from the presentation of a cue that preceded the target stimulus. The effect did not appear related to motivational state as there was no difference in the number of missed trials. Rather, the findings suggest a reduction in the speed of processing in the stimulus–response chain.

7.2. Studies on Memory

As mentioned in the Introduction to this chapter, the weight of evidence suggests strongly that in humans, memory function appears quite vulnerable to the effects of hypertension. We assessed the effect of hypertension using two tasks of memory function *(42)*, the delayed nonmatching to sample task (DNMS) and the delayed recognition span task (DRST), the latter of which has been used extensively with human subjects *(46)*.

The same 36 animals that were used for the attention tasks described above participated in the memory studies we administered preoperatively, and at 6 and 12 mo postoperatively. Preoperatively, all monkeys were trained initially in a Wisconsin General Testing Apparatus (WGTA). They were then administered the basic condition of the Delayed non-matching-to-sample (DNMS) described below.

7.2.1. DNMS

The DNMS task assesses the subject's ability to identify a novel from a familiar stimulus over varying delay intervals. Various forms of this task have been used to assess memory function in monkeys following either transection of the fornix *(47–49)*, or limited removal of selected temporal lobe structures (e.g., hippocampus or amygdala *[48–51]*). In addition, this task has been used to evaluate and quantify some aspects of recognition memory in patients with Alzheimer's disease and age-matched controls *(52)*. Preoperatively, animals were administered the basic task.

For the basic task, the trial begins with a sample object presented over the central baited food well. The animal was permitted to displace the object and obtain the reward. Ten seconds later the recognition trial was begun with the sample object presented over an unbaited lateral well and a new, unfamiliar object presented over a baited lateral well. To now obtain the reward, the animal must recognize the original sample object and choose the unfamiliar, novel object. Twenty seconds later, a different sample object was presented over the baited central well followed 10 s later by another recognition trial. The position of the two objects varied, on successive recognition trials, from left to right lateral wells in a predetermined order and a noncorrection procedure was used. Thirty trials a day were given until the animals reach a learning criterion of 90 correct responses in 90 consecutive trials or to a maximum of 1000 trials. Objects were drawn from a pool of 1500 "junk" objects, and in each daily session, 60 of the objects were used. The 1500 objects were randomly recombined to produce new sets of pairs so that the pairings presented were new and unique on each trial.

Following administration of the basic task, the 10-s delay between the presentation of the sample object and the recognition trial was increased, in stages, first to 2 min and then to 10 min. Ten trials a day were given with the monkey remaining in the testing apparatus during the delay interval. A total of 50 trials were given with each of the two delays.

Six months postoperatively, all monkeys were readministered the DNMS basic task. Following this, they were trained in an automated test apparatus described on p. 417 and were administered the delayed recognition span test.

The delayed recognition span task is a short-term memory test that was designed to investigate recognition memory in monkeys following bilateral removal of the hippocampus *(53)*. It requires the subject to identify, trial-by-trial, a new stimulus among an increasing array of serially presented, familiar stimuli. The task is administered using two different classes of stimulus material, spatial location or pattern shape. In this way we will be able to characterize any recognition memory deficits that may occur as a general impairment or one that is material specific.

Testing on the delayed recognition span test took place in a computer-controlled testing chamber with a darkened interior. The chamber contained a reward dispensing cup, a set of speakers, and a 19-in. color computer monitor covered with a resistive touch screen. White noise was played in the background to mask extraneous sounds. Between test trials, the computer monitor blackened and the touch screen was deactivated. Monkeys were tested 5 d per week and M&Ms® or Skittles® candies were used as rewards.

The monkeys were acclimated to the testing chamber and completed a "pretraining" task consisting of touching a stimulus that appeared for 50 trials per day in a pseudo-random fashion in 12 different spatial locations on the screen for food reward.

For the spatial condition of the delayed recognition span task, the computer touch screen was programmed to display 12 nonoverlapping positions, arranged in a 3×4 matrix. Yellow circles were used as stimuli with the background color of the screen being black. On the first trial of the first chain of trials, a circle appeared in one of the 12 positions that was rewarded. The animal was allowed to touch the circle and obtain the reward. The screen was blanked and reactivated 10 s later with a second positively rewarded circle (identical to the first) on the screen with the first circle reappearing in its original location. The animal was required to touch the new circle to obtain the reward. Similarly, each successive correct response was followed by the addition of a new circle until the animal made an error (i.e., chooses one of the previously chosen circles). Ten such chains of trials were presented each day, 5 d per week for a total of 5 d (50 trials). The exact position of the circles in each chain was determined randomly and was unique within and across sessions.

The pattern form of the delayed recognition span task was administered in much the same way as the spatial form. However, for this condition of the task, on each trial the spatial location of the previously correct stimulus was changed in a predetermined random fashion so that the animal was able to identify the new stimulus based only on visual, rather than spatial, cues. The stimuli for the pattern condition were drawn from a pool of 600 "clip-art" images. The images were drawn from the pool in a predetermined fashion to ensure unique combinations on each trial.

The first findings on memory assessment revealed no significant difference among the groups on the DNMS measures at 6 mo postoperatively. Monkeys with moderate/severe or borderline hypertension relearned the DNMS task as efficiently as operated controls. In contrast, however, on both the spatial and pattern conditions of the DRST, the performance of the moderate/severe hypertensive monkeys was significantly impaired with respect to the performance of the control monkeys, suggesting that, in addition to attentional function, hypertension diminishes the memory "load" capacity by 6 mo.

At 12 mo postoperatively, the findings revealed that the monkeys in the moderate/severe group, unlike those in the borderline group or the operated controls, evidenced a significant decline in memory function. The moderate/severe monkeys now showed impaired performance on the 2-min delay condition of the DNMS. In addition, they continued to show impairment on the spatial and pattern conditions of the DRST relative to the borderline and operated control groups. Unlike either of these groups, they also showed a marked decline relative to themselves at 6 mo, on the spatial condition of the DRST.

We generated a global measure of memory function from the performance of the monkeys at 6 and 12 mo to have a single measure of cognitive impairment that could be related to the blood pressure measures. This measure was based on performance of the delayed recognition span test and the delay conditions of the DNMS tests and is referred to as the cognitive impairment index. To combine the scores from the two tests we used the mean and standard deviation from the population of control animals to standardize the scores by converting them to z-scores. We then averaged z-scores from the two tests to generate a measure of cognitive impairment. As can be seen in Fig. 4, this measure correlates significantly with both the systolic and diastolic blood pressures.

7.3. Studies on Executive Function

Although much is known about how memory is affected by various disease processes, little is know about the changes in executive system functions. Executive system functions encompass many cognitive skills necessary to perform high levels of cognitive abilities and includes skills such as cognitive flexibility, cognitive tracking, divided attention, ability to establish and maintain set, monitoring and modification of response pattern, and abstraction. A variety of tests of executive system have been developed and in studies of humans with hypertension, performance on these tests has generally been found to be impaired (11,54).

One well established human test of executive system function is the Wisconsin Card Sorting Task (55,56). It was developed to assess cognitive flexibility, cognitive tracking, the ability to identify abstract categories and to maintain and shift cognitive set according to changing contingencies (57–59). The task requires the patient to sort a series of cards based on three stimulus dimensions—color, form, and number—utilizing feedback information from the administrator. The WCST has been used to assess deficits that are associated with a variety of disease processes and injuries. Studies with the WCST have demonstrated impaired performance by individuals due to frontal lobe dysfunction marked by characteristic disturbances including perseverative responses, an inability to shift set once established, and an inability to utilize information from the environment to modify response (60,61). Milner (61) reported the ability to shift from one mode of solution to another on the WCST is more impaired by frontal than posterior cerebral lesions.

As part of an ongoing study of CVD in the rhesus monkey, we utilized the principles of the WCST to develop a test of executive system functions for use with nonhuman primates. Our test, the Conceptual Set Shifting Task (CSST), requires the monkey to establish a cognitive set based on a reward contingency, maintain that set for a period of time, and then shift the set as the reward contingency changes. In this study, the CSST was used to assess executive system functioning and frontal lobe integrity of monkeys with sustained hypertension to further our understanding of the relationship between hypertension and cognition.

For this study we used a subset of 10 of the monkeys used in the previous tasks on attention and memory which consisted of five hypertensive animals and five control animals. All monkeys were tested in the same automated General Testing Apparatus in which they had performed the attention and delayed recognition span tasks, for 80 trials a day, 5 d per week. During testing a non-correctional approach was used and M&M's and Skittles were given as rewards.

Cognitive Impairment Index

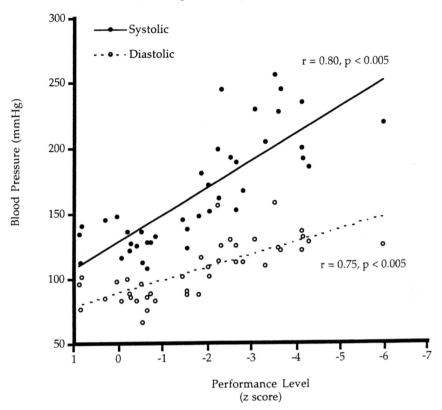

Fig. 4. Cognitive impairment index generated from the performance scores on the delayed nonmatching to sample and delayed recognition span test at the 6 and 12 mo testing intervals. The level of impairment on this index was significantly and linearly related to the level of both systolic and disastolic blood pressure in the monkeys in this study.

The monkeys were required to complete a pretraining task that required them to touch a stimulus on the computer screen 20 consecutive trials per day. The day after the monkeys completed this pretraining task, they began a simple three-choice discrimination task. The task presented the monkey with a pink square, orange cross, and a brown 12-point star. The three appeared in pseudo-random order in nine different spatial locations on the screen. The pink square was the positive stimuli for all trials and the monkey was rewarded with a food treat when he chose this stimuli. To make criterion, the monkey had to choose the pink square for ten consecutive trials.

The testing day after completing the discrimination task, the monkey began the CSST. On each trial of the CSST three stimuli appeared in a pseudo-random pattern on the computer screen (Fig. 5). The stimuli differed in color (red, green, or blue) and shape (triangle, star, and circle). All possible combinations of stimuli appeared on the screen over a 4-d cycle and if more than 4-d were needed to reach criterion the 4-d cycle was repeated until criterion was reached.

Testing consisted of an acquisition category (red) and then three concept shift categories (triangle, blue, and star). During acquisition the monkey was required to choose

Red

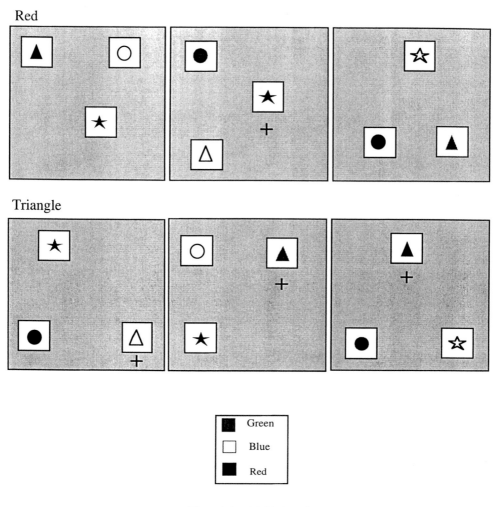

Triangle

15 sec inter-trial interval

Fig. 5. On each trial of the CSST the monkey is presented with three stimuli that vary in shape and color. During the first concept condition, the monkey must choose the red stimulus regardless of its shape as illustrated in the top three screens of this figure. Once the monkey chooses the correct stimulus on 10 consecutive trials, the computer switches the rewarded stimuli on the same testing day, without alerting the monkey. In the second concept condition the monkey must choose the triangle shaped stimulus, regardless of the color as illustrated in the bottom three screens of this figure. Again, when the monkey chooses the correct stimuli for 10 consecutive trials, the computer switches the rewarded stimuli on the same testing day, without alerting the monkey. Testing continues in this fashion for the blue and star concept conditions.

the red stimulus regardless of its shape to obtain a food reward. Once the monkey chose this stimulus on ten consecutive trials the program switched the rewarded contingency during the same testing session, without alerting the monkey. Now, the monkey had to choose the stimulus shaped like a triangle regardless of its color to obtain a food reward. Again, when the monkey reached a criterion of 10 consecutive responses, the computer switched the rewarded contingency within the same testing session, without alerting the monkey. The blue stimulus then had to be chosen, regardless of its shape, in order to obtain a food reward. Finally, the last category, star, was rewarded, when criterion was reached on the blue category.

The principal findings of this study include: (1) no significant difference between the hypertensive and normotensive monkeys on the simple three choice discrimination task; (2) hypertensive monkeys were impaired at abstracting the initial concept of color on the CSST and subsequently were impaired when shifted to the concept of shape, when shifted back to the concept of color and again when shifted back to the concept of shape; (3) hypertensive monkeys as a group demonstrated a greater tendency to perseverative in their response pattern when shifting categories than normotensive monkeys.

The simple discrimination task required the monkey to determine the positive stimuli out of an array of three stimuli based on a reward contingency. Both the hypertensive and normotensive groups of monkeys were able to complete this task successfully. This suggests that the two groups of monkeys were able to learn a stimulus reinforcement contingency at the same rate and that the impairment seen on the CSST is most like one of abstraction and cognitive flexibility.

8. NEUROPATHOLOGY

The neurobiological basis of the cognitive deficits associated with hypertension remain unclear. We have begun to explore the possible neuroanatomical bases for the impairment noted in attention. Yet, examinations of the brains of these animals using MRI even at 1 yr following surgery showed no evidence of stroke or infarct. Investigators *(62)* have suggested that hypertension produces subclinical pathological changes in the brain prior to the appearance of acute events such as strokes. These findings are in line with those from the human literature demonstrating that hypertension produces subclinical pathological changes in the brain *(28)*.

To date, we have identified at least four types of neuropathologies associated with hypertension in our primate model. The first, and most conspicuous, finding, using standard nissl and myelin stains was minute areas of infarction in both gray and white matter *(63)*.

The microinfarcts were irregularly shaped and of relatively uniform size with an average maximum diameter of slightly less than 1/2 mm. In the gray matter these lesions were characterized by a total loss of neurons and in the white matter by marked loss of myelinated fibers. The microinfarcts appeared with greater frequency in the white matter of the forebrain, particularly in the capsular system and the hemispheric white matter of the corona radiata and centrum semiovale. The area with the next highest density was in the cerebral cortex, with the remainder scattered throughout the forebrain, brain stem, and cerebellum.

The relationship of the microinfarcts observed in the cases in this primate model to the classically described focal lesions in chronic hypertension in the human brain is uncertain. The lesions are smaller than lacunar infarcts, which according to Fisher *(64)* measure 0.5–15 mm in diameter. They also differ from lacunar infarction in that they are not associated with obvious segmental degenerative changes in penetrating arteries or vascular occlusions, conspicuous changes in the brains with lacunar infarction *(64,65)*.

In contrast, the size of the microinfarcts is similar to that reported in hypertensive encephalopathy. However, the appearance of the lesions, the associated vascular changes, and the distribution pattern of the lesions is different. In human hypertensive encephalopathy, Chester et al. *(66)* reported that arteries in the brain and retina show segmental fibrinoid necrosis with thrombosis of arterioles and capillaries. In these brains, the vascular changes are associated with petechial hemorrhages and microscopic or miliary infarction.

The distribution pattern and morphology of the microinfarcts in the hypertensive monkeys also differs from that found in rat models of acute, severe hypertension and in rat models of severe established hypertension. In these models the lesions occur predominately in the cerebral cortex in the location of the arterial borderzones and to a lesser extent in the thalamus *(67–69)*. The characteristic lesions are microinfarcts with rarefaction of the neuropil and preserved neurons, cyst formation, and occasional hemorrhage *(67,70,71)*. Ogata et al. *(70)* felt that these lesions were due to brain edema and, in a serial section study of five stroke-prone, spontaneously hypertensive rats, Ogata et al. *(71)* noted that they were related to single or multiple arterial occlusions. These arteries, like those noted in man by Chester et al. *(66)*, showed fibrinoid degeneration. Thus, not only is the distribution of the infarcts different than that found in humans and in animal models of severe hypertension, there are differences in the details of the pathology and associated vascular changes.

The microinfarcts observed in the hypertensive monkeys also do not appear to be related to leukoaraiosis. This term, originally coined by Hachinski *(32)*, denotes a hypodensity in the white matter beneath the lateral ventricles in CT scans or intense white matter signal in this location with MRI. In a postmortem MRI study, van Swieten et al. *(33)* noted this change in a moderate to severe degree in 10% of the brains from individuals between 60 and 69 yr of age and in 50% of brains from individual between 80 and 89 yr of age. The MRI appearance of periventricular change correlated well with histological evidence of loss of myelinated fibers, gliosis, and abnormally thick arterioles measuring up to 150 μm in diameter. In some cases there were small white matter infarcts. The monkeys lack both the periventricular loss of myelin and the vascular changes noted by van Swieten et al. *(33)*.

A further consideration is the relationship of these lesions to Binswanger's subcortical leukoencephalopathy, a disease originally described by Binswanger in 1894 *(84)* and later called Binswanger's disease by Alzheimer *(72)*. It has received extensive attention in the literature, with several excellent reviews *(73–76)*. The hallmark of Binswanger's disease is a dementia that occurs in association with the loss of subcortical myelin that is greater in the subventricular zone, sparing of the subcortical myelinated "U" fibers, and severe disease of the medullary arteries that supply the affected region. The vast majority of individuals, but not all, have systemic hypertensive *(74,75,77)*. Fisher *(76)*, in his review, describes scattered foci of white matter destruction ". . . from typical lacunar infarcts 3–6 mm in diameter at one end of the spectrum, down to merely spongy looseness of the tissue without frank necrosis." In two cases personally observed by Fisher *(76)*, serial sections failed to reveal evidence of vascular occlusions. Lacunar infarction occurs in 87% of the cases *(76)*, indicating a close relationship of this change to Binswanger's disease. This relationship prompted Román

(75) to propose the term lacunar dementia to encompass both. The lesions in the hypertensive monkeys are smaller than those noted in Binswanger's encephalopathy and there is no evidence of diffuse white matter loss, subventricular accentuation of the lesions, or severe disease of penetrating arteries. Further, unlike in Binswanger's disease, the lesions are scattered throughout both white and gray matter rather.

The microinfarcts observed in these hypertensive monkeys resemble those reported by Garcia et al. *(78)*. These authors studied five hypertensive monkeys with coarctations of the aorta of 2 mo to 2 yr duration. They noted grossly visible "spongy" lesions and selected these for detailed electron microscopic study. These measured 0.5 mm in diameter, were present in all five monkeys in a random pattern, and were most abundant at 2 yr after the coarctation. In these areas, the capillaries were dilated with flattened endothelium, interrupted endothelial linings, increased thickness and deformity of the basement membrane, and deposition of collagen and osmophilic material.

In summary, we have found that the occurrence of multiple microinfarcts in the brains of hypertensive monkeys that occur in both the white and gray matter. Their relatively uniform size suggests that they are due to ischemia in the territory of a particular caliber of blood vessel. They occurred in relationship to hypertrophic arteries that show no evidence of more advanced hypertensive cerebrovascular disease such as occlusion or segmental pathology, suggesting that they may be an early change in natural history of hypertensive neuropathology. In consort with behavioral testing that shows evidence of a progressive decline in cognitive function *(79,80)*, the microinfarcts appear to be developing up to the time the animal is killed.

Operated control animals had either brief clamping of the aorta or were operated on up to the point of clamping the aorta. None of these animal showed focal lesions in the Nissl-stained sections, suggesting that the lesions seen in the hypertensive monkeys can not be directly attributed to the surgical procedures.

9. CONCLUSION

To date, this primate model of hypertensive cerebrovascular disease has yielded several important observations. First, the monkey appears as a suitable model to assess hypertensive related cognitive impairment and its relationship to alterations in the brain. We have been able to demonstrate the feasibility of inducing and maintaining moderate levels of chronic hypertension in monkeys with very low mortality and little to no health complications. Second, we have demonstrated impairments in domains of cognitive function as a consequence of hypertension that parallel those seen in humans and, in several instances using nearly identical behavioral tasks. Third, although not reported here, the model has permitted the application of a full range of MRI and PET imaging studies. Finally, the use of different perfusion protocols has allowed the demonstration of previously undescribed neuropathologies that can now be quantified and evaluation with respect to variables such as severity of blood pressure and degree of cognitive impairment.

As with any suitable animal model, once firmly established, within certain limitations, other neurobiologic variables can then be explored. With the establishment of this model, other goals can be pursued including determining the effects of hypertension on specific sensory systems, the use of computerized telemetry to monitor diurnal rhythms and spiking of blood pressure, to assess the effect of acute fluctuations in blood pressure on the integrity of the blood–brain barrier in hypertension, and to track the course of changes on MRI in hypertensive monkeys.

The use of animal models has provided important information toward the understanding of hypertension and its relationship to CVD, but at the same time, has been criticized for lack of relevance to the human clinical condition. For example, it has been noted by several authors *(81–83)* that studies in animal models have not yet translated into effective drug treatment for humans. This is related, in part, to the difficulty in assessing the clinical and cognitive function of the animal. With this in mind, we have worked very hard to identify and optimize conceptual and operational similarities between our animal model and the human condition.

Thus, by the careful selection of species and relevant cognitive, as well as physiological outcome measures, we believe animal models can continue to provide us with critical information about the neurobiological basis of hypertensive related CVD as well as other human disease states.

10. ACKNOWLEDGMENTS

The work described in this chapter represents the efforts of several individuals. The coarct model and surgical procedures were developed and carried out by Dr. William Hollander and Dr. Somnath Prusty. The behavioral studies were conducted with significant contributions from Dr. Ron Killiany, Tara Moore, and Beverly Duryea. The neuropathological studies were conducted by Drs. Tom Kemper, Gene Blatt, and Alan Peters. Without the contributions of these talented individuals, this program would never come to fruition. The studies described were funded solely by program project grant PO1 NS31649 from the NINDS.

REFERENCES

1. Forette, F. and Boller, F. (1991) Hypertension and the risk of dementia in the elderly. *Am. J. Med.* **90,** 14S–19S.
2. Kaplan, N. M. (1983) Hypertension: prevalence, risks, and effect of therapy. *Ann. Intern. Med.* **98,** 705–708.
3. Sandok, B. A. and Whisnant, J. P. (1983) Hypertension and the brain: clinical aspects, in *Hypertension* (Genest, J., Kuchel, O., Hamet, P., and Cantin, M. eds.), McGraw-Hill, New York, pp. 777–791.
4. Findlay, J. M., Weir, B. K., Kanamaru, K., Grance, M., and Baughman, R., (1990) The effect of timing of intrathecal fibrinlytic therapy on cerebral vasospasm in a primate model of subarachnoid hemorrhage. *Neurosurgery* **26,** 201–206.
5. Macdonald, R. L., Weir, B. K., Young, J. D., and Grace, M. G. (1992) Cytoskeletal and extracellular matrix proteins in cerebral arteries following subarachnoid hemorrhage in monkeys. *J. Neurosurg.* **76,** 81–90.
6. Jokinen, M. P., Clarkson, T. B., and Prichard, R. W. (1985) Recent advances in molecular pathology: Animal models in atherosclerosis research. *Exp. Molec. Path.* **42,** 1–28.
7. Simonson, E. and Enzer, N. (1941) State of motor centers in circulatory insufficiency. *Arch. Intern. Med.* **68,** 498–512.
8. Enzer, N., Simonson, E., and Blanksein, S. S. (1942) Fatigue of patients with circulatory insufficiency, investigated by means of fusion frequency of flicker. *Ann. Intern. Med.* **16,** 702–707.

9. Apter., N. S., Halstead, W. C., and Heimburger, R. F. (1951) Impaired cerebral functions in essential hypertension. *Am. J. Psychiatry.* **107,** 808–813.

10. Wilkie, F. and Eisdorfer, C. (1971) Intelligence and blood pressure in the aged. *Science* **172,** 959–962.

11. Boller, F., Vrtunski, P. B., Mack, J. L., and Kim, Y. (1977) Neuropsychological correlates of hypertension. *Arch. Neurol. 34, 701–705.*

12. Franceschi, M., Tancredi, O., Smirne, S., Mercinelli, A., and Canal, N. (1982) Cognitive processes in hypertension. *Hypertension* **4,** 226–229.

13. Schmidt, R., Fazekas, F., Offenbacher, H., Lytwyn, H., Blematl, B., Niederkorn, K., Horner, S., Payer, F., and Friedl, W. (1991) Magnetic resonance imaging white matter lesions and cognitive impairment in hypertensive individuals. *Arch. Neurol.* **48,** 417–420.

14. Vanderploeg, R. D., Goldman, H., and Kleinman, K. M. (1987) Relationship between systolic and diastolic blood pressure and cognitive functioning in hypertensive subjects: An extension of previous findings. *Arch. Clin. Neuropsychiatry* **2,** 101–109.

15. Mazzucchi, A., Mutti, A., Poletti, A., Ravanetti, C., Novarini, A., and Parma, M. (1986) Neuropsychological deficits in arterial hypertension. *Acta Neurol. Scand.* **73,** 619–627.

16. Schultz, N. R., Elias, M. R., Robbins, M. A. Streeten, D.H., et al. (1989) A longitudinal study of the performance of hypertensive and normotensive subjects on the Wechsler Adult Intelligence Scale. *Psychol. Aging* **4,** 496–499.

17. Sands, L. P. and Meredith, W. (1992) Blood pressure and intellectual functioning in late midlife. *J. Gerontol.* **47,** P81–84.

18. Guo, Z., Fratiglioni, L., Winblad, B., and Viitanen, M. (1997) Blood pressure and performance on the Mini-Mental State Examination in the very old. Cross-sectional and longitudinal data from the Kungsholmen Project. *Am. J. Epidemiol.* **145,** 106–1113.

19. Elias, M. F., Wolf, P. A., D'Agostino, R. B., Cobb, J., and White, L. R. (1993) Untreated blood pressure level is inversely related to cognitive functioning: the Framingham Study. *Am. J. Epidemiol.* **138,** 353–364.

20. Elias, M. F., D'Agostino, R. B., Elias, P. K., and Wolf, P. A. (1995a) Neuropsychological test performance, cognitive functioning, blood pressure and age: The Framingham Heart Study. *Exp. Aging Res.* **21,** 369–391.

21. Elias, M. F., D'Agostino, R. B., Elias, P. K., and Wolf, P. A. (1995b) Blood pressure, hypertension and age as risk factors for poor cognitive performance. *Exp. Aging Res.* **21,** 393–417.

22. Launer, L. J., Masaki, K., Petrovitch, H., Foley, D., Havlik, and R. J. (1995) The association between midlife blood pressure levels and later life cognitive function. *JAMA* **274,** 1846–1851.

23. Battersby, C., Hartley, K., Fletcher, A. E., et al. (1993) Cognitive function in hypertension: a community based study. *J. Hum. Hypertens.* **7,** 117–123.

24. Kalra, L., Jackson, L. H., and Swift, C. G. (1993) Psychomotor performance in elderly hypertensive patients. *J. Hum. Hypertens.* **7,** 279–284.

25. Madden, D. J. and Blumenthal, J. A. (1998) Interaction of hypertension and age in visual selective attention performance. *Health Psychol.* **17,** 76–83.

26. Wallace, R. B., Lemke, J. H., Morris, M. C., Goodenberger, M., Kohout, F., and Hinrichs, J. V. (1985) Relationship of free-recall memory to hypertension in the elderly: the Iowa 65+ rural health study. *J. Chronic Dis.* **38,** 475–481.

27. Farmer, M. E., Kittner, S. J., Abbott, R. D., Wolz, M. M., Wolf, P. A., and White, L. R. (1990) Longitudinally measured blood pressure, antihypertensive medication use, and cognitive performance: The Framingham Heart Study. *J. Clin. Epidemiol.* **43,** 475–480.

28. Waldstein, S. R., Ryan, C. M., Manuck, S. B., Parkinson, D. K., and Bromet, E. J. (1991) Learning and memory function in man with untreated blood pressure elevation. *J. Consult. Clin. Psychol.* **59,** 513–517.

29. Elias, M. F., Robbins, M. A., Schultz, N., and Pierce, T. W. (1990) Is blood pressure an important variable in research on aging and neuropsychological test performance? *J. Geront. Psychol. Sci.* **45,** 128–135.

30. Waldstein, S. R., Jennings, J. R., Ryan, C. M., Muldoon, M. F., Shapiro, A. P., Polefrone, J. M., Fazzari, T. V., Manuck, S. B., (1996) Hypertension and neuropsychological performance in men: Interactive effects of age. *Health Psychol.* **15,** 102–109.

31. Schultz, N. R., Dineen, J. T., Elias, M. F., Pentz, C. A., and Wood, W. G. (1979) WAIS performance for different age groups of hypertensive and control subjects during the administration of a diuretic. *J. Gerontol.* **31,** 246–253.

32. Hachinski, V. C. (1987) Leuko-araiosis. *Arch. Neurol.* **44,** 21–23.

33. van Swieten, J. C., van den Hout, H. W., van Ketek, B. A., Hijdra, A., Wokke, J. H. J., and van Gijn, J. (1991) Periventricular lesions in the white matter on magnetic resonance imaging in the elderly. *Brain* **114,** 761–774.

34. Akesson, H. D. (1969) A population of senile and arteriosclerotic psychosis. *Hum. Hered.* **19,** 546–566.

35. Broe, G. A., Akhtar, A. J., Andrews, G. R., Caird, F. I., Gilmore, A. J., and McLennan, W. M., (1976) Neurological disorders in the elderly at home. *J. Neurol. Neurosurg. Psychiatry* **39,** 362–366.

36. Sulkava, R., Wikstrom, J., Aromaa, A., Raitasola, R., Lehtinen, V., Lehtela, K., and Palo, K. (1985) Prevalence of severe dementia in Finland. *Neurology* **35,** 1025–1029.

37. Hollander, W., Madoff, I., Paddock, J., and Kirkpatrick, B. (1976) Aggravation of atherosclerosis by hypertension in a subhuman primate model with coarctation of the aorta. *Circ. Res.* **38 (Suppl. II),** 63–72.

38. Prusty, S., Kemper, T. L., Moss, M. B., and Hollander, W. (1988) Occurrence of stroke in a non-human primate model of cerebrovascular disease. *Stroke* **19,** 84–90.

39. Hollander, W., Prusty, S., Kirkpatrick, B., Paddock, J., and Nagrai, S. (1977) Role of hypertension in ischemic heart disease and cerebral vascular disease in the cynomolgus monkey with coarctation of the aorta. *Circ. Res.* **40 (Suppl. I),** I–70-I–83.

40. Hollander, W., Prusty, S., Killiany, R. J., and Moss, M. B. (1999) Comparison of direct and indirect methods, including telemetry for measurement of blood pressure in the hypertensive monkey, submitted.

41. Killiany, R. J., Duryea, B., Rosene, D. L., and Moss, M. B. (1999) Impaired attention in a primate model of hypertensive cerebrovascular disease, submitted.

42. Moss, M. B., Killiany, R. K., and Rosene, D. L. (1999) Impaired memory function as a consequence of hypertension in a primate model of cerebrovascular disease, submitted.

43. Posner, M. I. and Peterson, S. E. (1990) The attention system of the human brain, in: *Annual Review of Neuroscience,* Vol. 13, pp 25–42, Annual Reviews Inc., Palo Alto, CA.

44. Parasuraman, R. and Davies, R., (1984) *Varieties of Attention.* Academic Press, New York.

45. Hasher, L. and Zacks, R. (1979) Automatic and effortful processes in memory. *J. Exp. Psychol.* **108,** 356–388.

46. Albert, M. S. and Moss, M.B. (1998) Neuropsychological profiles of normal aging, in *Cerebral Cortex, Neurodegenerative and Age-Related Changes in Structure and Function of Cerebral Cortex.* (Peters, A. and Morrison, J., eds), Plenum Press, N.Y., In press.

47. Gaffan, D. (1974) Recognition impaired and association intact in the memory of monkeys after transection of the fornix. *J. Comp. Physiol. Psych.* **86,** 1100–1109.

48. Mahut, M., Zola-Morgan, S., and Moss, M. B. (1982) Hippocampal resections impair associative learning and recognition memory in the monkey. *J. Neurosci.* **2,** 1214–1229.

49. Saunders, R. C., Murray, E. A., and Mishkin, M. (1984) Further evidence that amygdala and hippocampus contribute equally to recognition memory. *Neuropsychologia* **22,** 785–796.

50. Mishkin, M. (1978) Memory in monkeys severely impaired by combined but not separate removal of amygdala and hippocampus. *Nature* **273,** 297–298.

51. Murray, E. A. (1992) Medial temporal lobe structures contributing to recognition memory: The amygdaloid complex versus the rhinal cortex, in *The Amygdala: Neurobiological Aspects of Emotion, Memory and Mental Dysfunction.* (Aggleton, J.P. e d.), Wiley-Liss, New York, pp. 453–470.

52. Albert. M. and Moss, M. B., (1984) The assessment of memory disorders in patients with Alzheimer's Disease, in *Neuropsychology of Memory* (Squire, L. and Butters, N. eds.), Guilford Press, New York.

53. Rehbein, L. and Mahut, H. (1983) Long-term deficits in associative and spatial recognition memory after early hippocampal damage in monkeys. *Soc. Neurosci. Abstr.* **9,** 639.

54. Blumenthal, J. A., Madden, D. J, Pierce, T. W. et al. (1993) Hypertension affects neurobehavioral functioning. *Psychosom. Med.* **55,** 44–50.

55. Berg, E. A. (1948). A simple objective test for measuring flexibility in thinking. *J. Gen. Psychol.* **39,** 15–22.

56. Grant, D. A. and Berg, E. A. (1948) A behavioral analysis of degree of reinforcement and ease of shifting to new responses in a Weigl-type card sorting problem. *J. Exp. Psychol.* **34,** 404–411.

57. Nagahama, Y., Fukuyama, H., Yamauchi, H., Matsuzaki, S., Konishi, B., Shibasaki, H., and Kimurs, J. (1996) Cerebral Actiation During Performance of a Card Sorting Task. *Brain* **119,** 1667–1675.

58. Damasio, A. R. and Anderson, S. W., (1993) The Frontal Lobes. In *Clinical Neuropsychology*, Kenneth Heilman and Edward Valenstein, Eds. Oxford University Press, Oxford.

59. Fristoe, N. M., Salthouse, T. A., and Woodward, J. L. (1997) Examination of Age-Related Deficits on the Wisconsin Card Sorting Task. *Neuropsychology* **11,** 428–436.

60. Fuster, J. M. (1997) *The Prefrontal Cortex; Anatomy, Physiology, and Neuropsychology of the Frontal Lobes.* Lippincott-Raven, NY.

61. Milner, B. (1963) Effects of different brain lesions on card sorting. *Arch. Neurol.* **9,** 90–100.

62. Shapiro, A. P., Miller, R. E., King, H. E., Ginchereau, E. H., and Fitzgibbon, K. (1982) Behavioral consequences of mild hypertension. *Hypertension* **4,** 355–360.

63. Kemper, T. L., Moss, M. B., Hollander, W., and Prusty, S. (1999) Microinfarction as a result of hypertension in a primate model of cerebrovascular disease. *Acta Neuropathol.,* in press.

64. Fisher, C. M. (1969) The arterial lesion underlying lacunes. *Acta Neuropathol.* **12,** 1–15.

65. De Reuck, J. and vander Eecken, H. (1976) The arterial angioarchitecture in lacunar state. *Acta Neurol. Belg.* **76,** 142–149.

66. Chester, E. M., Agamanolis, D. P., Banker, B. Q., and Victor, M. (1978) Hypertensive encephalopathy: A clinicopathological study of 20 cases. *Neurology* **28,** 928–939.

67. Robertson, D. M., Dinsdale, H. B., Hayashi, T., and Tu, J. (1970) Cerebral lesions in adrenal regeneration hypertension. *Am. J. Pathol.* **59,** 115–131.

68. Dinsdale, H. B., Robertson, D. M., and Haas, R. A. (1973) Acute hypertension and prolonged focal cerebral ischemia. *Trans. Am. Neurol. Assoc.* **98,** 162–165.

69. Nag, S. (1984) Cerebral changes in chronic hypertension: combined permeability and immunohistochemical studies. *Acta Neuropathol.* **62,** 178–184.

70. Ogata, J., Fujishima, M., Tamaki, K., Nakatomi, Y., Ishitsuka, T., and Omae, T. (1980) Stroke-prone spontaneously hypertensive rats as an experimental model of malignant hypertension. I. Light- and electron-microscopic study of the brain. *Acta Neuropathol.* **51,** 179–184.

71. Ogata, J., Fujishima, M., Tamaki, K., Nakatomi, Y., Ishitsuka, T., and Omae, T. (1981) Vascular changes underlying cerebral lesions in stroke-prone spontaneously hypertensive rats. *Acta Neuropathol.* **54,** 183–188.

72. Alzheimer, A. (1902) Die Seelenstorungen auf arteriosklerotischer Grundlage. *Allg. Z. Psychiatr. Psychischerichtl. Med.* **59,** 695–711.

73. Olszewski, J. (1962) Subcortical arteriosclerotic encephalopathy. *World Neurol.* **3,** 359–375.

74. Babikian, V. and Roper, A. H. (1987) Binswanger's disease: a review. *Stroke* **18,** 2–12.
75. Román, G. C. (1987) Senile dementia of the Binswanger's type: A vascular form of dementia in the elderly. *JAMA* **258,** 1782–1788.
76. Fisher, C. M. (1989) Binswanger's encephalopathy: a review. *J. Neurol.* **236,** 65–79.
77. Loizou, L. A., Jefferson, J. M., and Smith, W. T. (1982) Subcortical arteriosclerotic encephalopathy (Binswanger's type) and cortical infarcts in a young normotensive patient. *J. Neurol. Neurosurg. Psychiatr.* **45,** 409–417.
78. Garcia, J. H., Ben-David, E., Conger, K. A., Geer, J. C., and Hollander, W. (1981) Arterial hypertension injuries brain capillaries. Definition of the lesions. Possible pathogenesis. *Stroke* **12,** 410–413.
79. Moss, M. B., Kemper, T., Rosene, D. L., Prusty, S., and Hollander, W. (1989) Neuropsychological and pathological consequences of hypertension and high cholesterol diet: a primate model of cerebrovascular disease. *J. Clin. Exp. Neuropsychiatry* **12,** 68.
80. Moss, M. B., Rosene, D. L., Kemper, T. L., Prusty, S., and Hollander, W. (1993) Hypertension-induced memory impairment and microinfarction in monkeys. *Soc. Neurosci. Abstr.* **19,** 196.
81. Pulsinelli, W. A. and Buchan, A., (1989) The utility of animal ischemia models in predicting pharmocotherapeutic response in the clinical setting. in *Cerebrovascular Disease* (Ginsberg, M. D. and Dietrich, W. D., eds.), New York: Raven Press, pp 87–91.
82. Wiebers, D. O., Adams, H. P., and Whisnant, M. D., (1990) Animal models of stroke: Are they relevant to human disease? *Stroke* **21,** 1–3.
83. Molinari, G. F. (1988) Why model strokes? *Stroke* **19,** 1195–1197.
84. Binswanger, O. (1894) Die Abgrenzung der allgemeinen progressiven Paralyse. *Berl. Klin. Wochenschrift* **31,** 1103–1105; 1137–1139; 1180–1186.

V

INNOVATIONS LEADING TO CLINICAL THERAPY

22

Nicotinic Therapeutics for Tourette Syndrome and Other Neuropsychiatric Disorders

From Laboratory to Clinic

R. Doug Shytle, Archie A. Silver, Mary B. Newman, and Paul R. Sanberg

1. NICOTINIC THERAPEUTICS

Despite the evidence dissociating the harmful effects of cigarette smoking from the pharmacological effects of nicotine *(1)*, the potential therapeutic properties of nicotine have often been ignored *(2)*. On the other hand, a large body of empirical evidence from both animal and human studies suggests that research aimed at understanding nicotine pharmacology may be an important area for future drug development *(2–5)*. Nicotine modulates a family of ligand-gated ion channels know as acetylcholinergic nicotinic receptors (nAChRs). Those that bind to nicotine with high affinity and comprise 90% of nAChRs in the brain have been characterized as $\alpha4\beta2$ nAChRs, while those with lower affinity to nicotine but higher affinity to the snake toxin, α-bungarotoxin, have been characterized as $\alpha7$ nAChRs *(3)*. Several pharmaceutical companies are currently developing compounds that have more selective actions at various central nicotinic receptors with better safety profiles than nicotine. Some of these are discussed later in the chapter, but first we will discuss an area of research that may well have been the first systematic attempt to bring nicotine therapeutics from the animal laboratory to the clinic.

The story of the therapeutic use of nicotine in the treatment of Tourette syndrome has its roots in the animal laboratory. Using neuroleptic-induced catalepsy as an animal model of therapeutic efficacy, it was determined that nicotine potentiated the cataleptic effects of neuroleptics. Human clinical studies were subsequently initiated and confirmed that the combination of nicotine and neuroleptics, such as haloperidol, provided better treatment than neuroleptics given alone.

2. TOURETTE SYNDROME

Tourette syndrome (TS) is a childhood onset neuropsychiatric disorder characterized largely by the expression of sudden, rapid and brief, recurrent, usually nonrhythmic, stereotyped motor movements (motor tics) or sounds (vocal tics) that are experienced as irresistible but can be suppressed for varying lengths of time *(6,7)*. The

From: Central Nervous System Diseases
Edited by: D. F. Emerich, R. L. Dean, III, and P. R. Sanberg © Humana Press Inc., Totowa, NJ

symptoms usually begin in childhood and range from relatively mild to very severe over the course of a patient's lifetime *(8)*. Behavioral and emotional problems are also often present in these patients *(9)*. TS is frequently treated with the neuroleptic haloperidol, which is effective in about 70% of cases *(10)*. Although neuroleptics are effective in controlling the motor and vocal tics of TS, they have side effects that often lead to poor compliance. A therapy that could magnify the therapeutic effects of neuroleptics while at the same time reduce the incidence of neuroleptic-induced side effects would be a great advance in the treatment of TS.

2.1. Neuroleptic-Induced Catalepsy as an Animal Model

Although a simple behavioral task, neuroleptic-induced catalepsy has proven to be an effective behavioral tool for assessing the extrapyramidal effects of neuroleptics. Considerable research has been conducted to determine the behavioral and neural mechanisms underlying the neuroleptic cataleptic response in animals. For example, using selective neuronal lesion techniques with excitotoxic glutamate analogs such as kainic acid and quinolinic acid, Caldern et al. *(11)* demonstrated that the cataleptic response induced by haloperidol was mediated via the striatum, as lesions of the striatum attenuated the cataleptic effects of haloperidol. Neuropharmacologically, this motoric response is directly correlated with a neuroleptic's ability to antagonize dopamine receptors in the striatum.

Neuroleptic-induced catalepsy in animals has traditionally been used to model the extrapyramidal side effects of neuroleptics. However, the same behavioral effect can be viewed as a therapeutic model when used to test potential drugs for treating hyperkinetic movement disorders, such as TS.

Although the pathogenesis of TS is still not known, excessive striatal dopamine has been proposed *(12,13)*. This hypothesis is based largely on the therapeutic effectiveness of neuroleptics that block D2 dopamine receptors *(10)* and recent SPECT image findings showing differences in D2 dopamine receptor binding in the striatum of monozygotic twins who were discordant for TS *(14)*. Thus, it has been suggested that striatal dopamine excess or receptor hypersensitivity may play a role in the etiology of this disorder. In this senerio, the striatal cholinergic interneurons and striatopallidal γ-aminobutyric acid (GABA) output neurons are abnormally shut down owing to hyperinnervation of dopaminergic neurons acting via D2 receptors. As a consequence, the globus pallidus is disinhibited and unable to gate neuronal activity of thalamocortical projections to motor areas of the cortex, thus resulting in the expression of motor tics *(15)*. Haloperidol and other neuroleptics given to TS patients would then act therapeutically to reinstate inhibitory control over thalamocortical fibers by disinhibiting striatopallidal GABA output neurons. Thus, the neural circuitry underlying the cataleptic response to neuroleptics is the same by which neuropeltics are thought to help reduce the abnormal expression of tics in persons with TS.

2.2. Effect of Nicotine on the Cataleptic Response to Haloperidol

The use of nicotine in the treatment of TS stemmed from the finding of D. E. Moss and colleagues that cannabinoids strongly potentiated (up to 100-fold) neuroleptic-induced hypokinesis in rats *(7)*. Because cannabinoids produce their effects on the extrapyramidal motor system through a nicotinic cholinergic mechanism, nicotine was used in subsequent animal studies *(16–18)*. Nicotine administered intraperitoneally, in a dose of 0.1 mg/kg, potentiated the catalepsy, as measured by the bar test, induced in the animals by resperine, fluphenazine, and haloperidol. These effects could be blocked by mecamylamine, a nicotinic receptor antagonist. Further animal studies with four experimental conditions (haloperidol and nicotine, nicotine alone, haloperidol alone, control) using haloperidol in a dose of 0.3 mg/kg and nicotine 0.1 mg/kg confirmed that low doses of nicotine did indeed potentiate haloperidol-induced catalepsy in rats *(19)*. Nicotine alone, however, was ineffective. Further studies suggested that nicotine in doses of 0.1, 0.2, and 0.3 mg/kg had no effect with the lowest dose of haloperidol (0.1 mg) but had marked effect on catalepsy induced by 0.2 and 0.4 mg/kg doses of haloperidol *(20, 21)*. These effects were independent of the dose of nicotine used.

2.3. Effect of Nicotine on the Locomotor Response to Haloperidol

Since lower doses of haloperidol more closely approximate the doses of haloperidol used clinically, a more sensitive measure than catalepsy was needed to evaluate the effects of a low dose of haloperidol in conjunction with nicotine. Therefore, separate studies were conducted using Digiscan Animal Activity Monitors for assessing locomotion. The experiments evaluated the ability of nicotine to potentiate the locomotor inhibiting effects of haloperidol. Each animal was placed individually into one of eight open field boxes in an automated Digiscan-16 Animal Activity Monitor System (Omnitech Electronics, Columbus, OH, USA) for a 1-h habituation period immediately prior to the 6-h nocturnal testing period *(20,21)*. Haloperidol produced a dose-related decrease in locomotion. Whereas the lowest dose of haloperidol (0.1 mg/kg) decreased only total distance and the number of stereotypic movements, 0.4 mg/kg of haloperidol significantly decreased all variables with the exception of rest time and average speed. *Post hoc* analyses demonstrated that nicotine significantly potentiated the hypoactivity produced by haloperidol. Following 0.1 mg/kg of haloperidol, nicotine potentiated the haloperidol-induced decrease in total distance, number of movements, average distance per move, all vertical activity measures, stereotypy time, and clockwise rotations. Following 0.4 mg/kg of haloperidol, nicotine significantly potentiated the decreases in horizontal activity, number of movements, and vertical movements.

These findings indicated that nicotine is able to potentiate the locomotor effects of a dose of haloperidol that was subcataleptic. Moreover, this study also revealed that nicotine in conjunction with a low dose of haloperidol (0.1 mg/kg) produced decreases in locomotion equivalent to those produced by a high (0.4 mg/kg) dose of haloperidol alone. This suggested that in patients with TS nicotine adjunctive therapy may allow lower doses of haloperidol to be used without decreasing clinical effectiveness *(21)*.

2.4. Effect of Nicotine on D1and D2 Dopamine Receptor-Mediated Catalepsy

Because haloperidol blocks both D1 and D2 dopamine receptors in the striatum, it was important to know which dopamine receptor was involved with nicotine's potentiating effect. In another study, it was demonstrated that whereas nicotine potentiates haloperidol-induced catalepsy, it failed to potentiate the cataleptic response to the selective D1 antagonist, SCH23390 *(20)*. Thus, the evidence supports the involvement of D2 but not D1 receptor blockade in mediating the nicotinic potentiation of neuroleptic-induced catalepsy. The fact that dopamine D2 receptors inhibit intrinsic cholinergic activity within the striatum more so than D1 receptors suggests that the cholinergic synapse is the site of action for nicotine in producing its potentiating effect of neuroleptic-induced catalepsy.

2.5. Significance of Preclinical Findings

Although neuroleptic-induced catalepsy does not reflect a direct animal model for TS, it does provide a model for understanding the therapeutic actions of neuroleptics and nicotine in TS patients. This model's validity was supported by the observation that dopamine D2 receptors were implicated in both TS *(14)* and the potentiating action of the nicotine/neuroleptic combination *(20)*. The findings from these preclinical experiments suggested that nicotine may have some potential benefit in the treatment of TS.

2.6. From the Laboratory to Clinic

The potential usefulness of combining nicotine with neuroleptic treatment for TS has received support from human studies *(19,22–29)*. The first open trial consisted of chewing nicotine gum three times daily in 10 TS patients concurrently treated with haloperidol. A decrease in tic frequency and severity was noted as well as a subjective improvement in concentration and attention in 8 of the 10 subjects *(19)*. A second study of 10 additional patients receiving haloperidol revealed a quantitative reduction in tic frequency following nicotine gum chewing. This reduction occurred both during the 30-min period of gum chewing and at 1 h postadministration *(25)*. A subsequent controlled trial also yielded similar results, with nicotine gum plus haloperidol reducing both tic severity and frequency, whereas nicotine gum alone reduced only tic frequency and placebo gum alone had no effect *(26)*.

Because of gastrointestinal side effects of nicotine gum and the short duration of action, Silver et al. *(30,31)* examined the effects of transdermal nicotine patchesl (7 mg/24 h) in 11 TS patients who were not responding to current neuroleptic treatment. Using a video camera to record tics at baseline and at 3 h post-application, they noted a 47% reduction in tic frequency and a 34% reduction in tic severity following transdermal nicotine patch (TNP) application. Surprisingly, in two patients, the effect of a single nicotine patch persisted for a variable length of time after patch removal. Similar benefits of nicotine in TS patients were subsequently reported by an independent group *(27–29)*. These long-term therapeutic responses to the TNP were again found in another study comprising 20 TS patients who were followed for various lengths of time following the application of the patch *(24,32)*.

2.7. A New Hypothesis

Although the therapeutic interaction between nicotine and haloperidol in TS patients would be consistent with facilitation of striatal ACh release underlying the nicotine potentiation of haloperidol-induced catalepsy by nicotine in rats, nicotine also has the ability to increase striatal concentrations of several other neurotransmitters, one of which is dopamine. This is paradoxical because the potentiation of haloperidol would require a relative decrease in dopamine release. Another possibility was that nicotine was causing an inactivation of the nicotinic receptors involved with the presynaptic release of dopamine. Although a mixed agonist/antagonist effect of nicotine may be involved with the total therapeutic response to transdermal nicotine found in TS, the available evidence from animal studies suggests that a prolonged inactivation of nAChRs following exposure to transdermal nicotine may be responsible for the long-term therapeutic response seen in some TS patients *(32)*. Thus, the difference in the short duration of action of nicotine gum and the longer action of transdermal nicotine may be attributed to a difference in length of nAChR inactivation.

2.8. From Nicotine to Mecamylamine

A nicotine-induced "receptor inactivation" hypothesis would be consistent with reports that mecamylamine, a noncompetitive nAChR antagonist, also potentiates neuroleptic-induced catalepsy *(17)* possibly by blocking nAChRs involved with the tonic release of striatal dopamine *(33,34)*. Mecamylamine (3-methylaminoisocamphane HCl) was developed and characterized by Merck & Co. as a ganglionic blocker with hypotensive actions *(35)*. Although mecamylamine was marketed for many years as an antihypertensive agent (Inversine®), it has been replaced by more effective antihypertensive medications. The doses used to control blood pressure in adults range from 10 to 90 mg/d, but are on average 25 mg/d (package insert).

In addition to its peripheral ganglionic blocking actions, mecamylamine crosses the blood–brain barrier and functions as a selective nicotinic receptor antagonist at doses that do not have a significant effect on parasympathetic function *(36,37)*. As a result, mecamylamine blocks most of the physiological, behavioral, and reinforcing effects of tobacco and nicotine *(38)*.

Recently, mecamylamine, given in oral doses ranging from 2.5 to 7.5 mg per day to several TS patients who either had not or currently were not responding to traditional pharmacological treatment *(39)*, resulted in a wide range of therapeutic benefits. Although tic severity improved in some of the patients, a more consistent observation was that many of the behavioral and emotional symptoms of TS improved even in those patients in whom severity of tics was unchanged. Most patients have reported feeling less tense, less irritable, with fewer mood swings. Patients with rage attacks have also reported improvement. While placebo effects and spontaneous remission of symptoms are obvious possibilities for the reported improvement, many patients continued to report improvement in symptoms while taking mecamylamine on a daily basis for more than 6 mo.

3. OTHER ANIMAL MODELS IN NICOTINE THERAPEUTICS

3.1. Cognitive Enhancing Effects of Nicotine and Its Therapeutic Sigificance

It has long been recognized that nicotine improves performance on a variety of cognitive tasks in nicotine-naive animals *(40,41)*. In fact, studies investigating the therapeutic potential of nicotine for Alzheimer's disease *(42)* and attention deficit hyperactivity disorder (ADHD) *(43,44)* are consistent with cognitive enhancing effects in aged rats *(45)* and in rats with lesions of the basal forebrain *(46)*.

Because nicotine improves cognitive function in animals, several pharmaceutical companies have been characterizing novel drugs that have more selective effects and have a better safety profile than nicotine *(47)*. For example, scientists at Abbot Laboratories were the first to study the actions of the novel nAChRs activator, ABT-418. This drug, which has high affinity to the $\alpha4\beta2$ nAChR, was found to improve cognitive function under various experimental conditions in both rats and primates *(48–50)*. Moreover, this drug was virtually devoid of the side effects produced by nicotine. Whereas phase II clinical trials with ABT-418 in Alzheimer's disease patients was discontinued owing to short-lived improvements, ABT-418 was recently found to perform superior to placebo in a human clinical study designed to investigate the drug's effect in cognitively impaired adults with ADHD *(51)*.

A very good behavioral model for understanding attentional processes is the well known prepulse inhibition (PPI) of a stimulus-evoked behavioral response *(52)*. The use of PPI as a model for attentional processes is based on the observation that PPI is deficient in schizophrenics *(53)* and may therefore be related to cognitive-attentional deficits found in this population. Interestingly, nicotine has recently been found to normalize PPI deficits found in schizophrenics *(54)* and in animal models of sensory gating *(55)*. These findings, together with the fact that between 70% and 90% of all schizophrenic individuals smoke tobacco *(56)*, suggest that schizophrenic individuals may smoke tobacco to self-medicate because of the cognitive activating effects of nicotine.

Another nicotine receptor ligand, GTS-21, which was developed by scientists interested in improving cognitive function in elderly patients with Alzheimer's disease, has been found to have neuropharmacological actions that differ markedly from both nicotine and most other synthetically generated nAChR ligands *(57)*. This drug, while also improving water maze performance in aged rats *(40,58)*, acts as a partial agonist at the $\alpha7$ nAChR and an antagonist at the $\alpha4\beta2$ nAChR. These findings may have particular relevance to persons with schizophrenia, as significantly fewer $\alpha7$ nAChRs are found in the hippocampus of schizophrenic individuals when compared to normals. Because GTS-21 produces prolonged improvements in PPI over that obtained by nicotine, the possibility exists that this drug may have marked therapeutic effects in schizophrenia. Thus, animals models of sensory gating would be a logical choice for nicotine scientists who are interested in designing potentially novel antipsychotics.

3.2. Analgesic Effects of Nicotine and Its Therapeutic Significance

Another animal model of potential therapeutic interest is nicotine's ability to increase latencies to respond to noxious thermal stimulation in mice *(59)*. Because this effect was completely blocked by mecamylamine, but not by the opioid receptor antagonist naloxone, scientists at Abbot recognized the possibility that potent nonaddictive analgesics could be developed with selective actions at nAChRs. By developing analogs of the highly potent nicotinic analgesic *(60)* epibatidine, a toxin isolated from frogs, Abbot scientists screened nearly 500 compounds for their ability to increase the latency to respond to noxious thermal stimulation in mice *(61)*. One of these compounds, ABT-594 was found the produce naloxone-insensitive analgesic effects at a potency equivalent to that of morphine *(62)*. Phase I clinical trials with this drug in humans with chronic pain are currently underway.

4. SUMMARY AND CONCLUSIONS

In this chapter, we reviewed the evidence, consistent with findings initially observed in animals, that administration of nicotine (either 2 mg nicotine gum or 7 mg transdermal nicotine patch) potentiates the therapeutic properties of neuroleptics in treating TS patients and that a single patch may be effective for a variable number of days. This research clearly demonstrates that a relatively simple behavioral effect of a drug, such as neuroleptic-induced catalepsy in animals, can be used quite successfully to model the therapeutic effects of the same drug in humans.

We also briefly reviewed how other animal models are currently being used in the field of nicotine research to develop potentially useful therapeutic agents. Advances in the understanding of the molecular biology and neuropharmacology of nAChRs may provide targets for the development of novel and selective modulators of nAChRs in the brain. This contention is supported by the dissimilar behavioral effects observed following systemic administration of currently available nicotinic ligands. Novel animal models such as PPI, coupled with a thorough understanding of the neuropharmacology and functional characteristics of more of the putative human nAChR subtypes, will facilitate the discovery of more efficacious and less toxic nicotine-like drugs that may provide potential novel therapeutic agents for a variety of central nervous system conditions.

ACKNOWLEDGMENT

The authors are grateful to Layton Bioscience, Inc., and its CEO Gary Snable for contributions made in this project and for their continued support.

REFERENCES

1. Shytle, R. D., Silver, A. A., and Sanberg, P. R. (1996) Nicotine, tobacco and addiction [letter]. *Nature* **384,** 18–19.
2. Jarvik, M. E. (1991) Beneficial effects of nicotine. *Br. J. Addict.* **86,** 571–575.
3. Decker, M. W., Brioni, J. D., Bannon, A. W., and Arneric, S. P. (1995) Minireview: Diversity of neuronal nicotinic acetylcholine receptors: lessons from behavior and implications for CNS therapeutics. *Life Sci.* **56,** 545–570.
4. Brioni, J. D., Decker, M. W., Sullivan, J. P., and Arneric, S. P. (1997) The pharmacology of (–)-nicotine and novel cholinergic channel modulators. *Adv. Pharmacol.* **37,** 153–214.
5. Sanberg, P. R., Silver, A. A., Shytle, R. D., Philipp, M. K., Cahill, D. W., Fogelson, H. M., et al. (1997) Nicotine for the treatment of Tourette's syndrome. *Pharmacol. Ther.* **74,** 21–25.
6. Group, T.S.C.S. (1993) Definitions and Classification of Tic Disorders. *Arch. Neurol.* **50,** 1013–1016.
7. Montgomery, S. P., Moss, D. E., and Manderscheid, P. Z. (1985) Tetrahydrocannabinol and levonatradol effects on extrapyramidal motor behaviors: neuroanatomical location and hypothesis of mechanism, in *Marijuana 84*, (Harvey, D. J. and Paton, W. D. M., eds.), IRL Press, Oxford, pp. 295–302.
8. Robertson, M. M. (1989) The Gilles de la Tourette syndrome: the current status. *Br. J. Psychiatry* **154,** 147–169.
9. Coffey, B. J. and Park, K. S. (1997) Behavioral and emotional aspects of Tourette syndrome. *Neurol. Clin.* **15,** 277–289.

10. Shapiro, E. S., Shapiro, A. K., Fulop, G., Hubbard, M., Mandeli, J., Nordie, J., et al. (1989) Controlled study of haloperidol, pimozide, and placebo for the treatment of Gilles de la Tourette's syndrome. *Arch. Gen. Psychiatry* **46**, 722–730.

11. Calderon, S. F., Sanberg, P. R., and Norman, A. B. (1988) Quinolinic acid lesions of rat striatum abolish D1- and D2-dopamine-receptor-mediated catalepsy. *Brain Res.* **450**, 403–407.

12. Singer, H. S., Butler, I. J., Tune, L. E., Seifert, W. E. Jr, and Coyle, J. T. (1982) Dopaminergic dsyfunction in Tourette syndrome. *Ann. Neurol.* **12**, 361–366.

13. Singer, H. S., Hahn, I. H., and Moran, T. H. (1991) Abnormal dopamine uptake sites in postmortem striatum from patients with Tourette's syndrome. *Ann. Neurol.* **30**, 558–562.

14. Wolf, S. S., Jones, D. W., Knable, M. B., Gorey, J. G., Lee, K. S., Hyde, T. M., et al. (1996) Tourette syndrome: prediction of phenotypic variation in monozygotic twins by caudate nucleus D2 receptor binding. *Science* **273**, 1225–1227.

15. Leckman, J. F., Knorr, A. M., Rasmusson, A. M., and Cohen, D. J. (1991) Basal ganglia research and Tourette's syndrome [letter]. *Trends Neurosci.* **14**, 94.

16. Emerich, D., Sanberg, P., Manderscheid, P., McConville, B., Rich, J., El-Etri, M., et al. (1990) Differential effect of nicotine on D1 vs D2 antagonist-induced catalepsy. *Soc. Neurosci. Abstr.* **16**, 247.

17. Moss, D. E., Manderscheid, P. Z., Kobayashi, H., and Montgomery, S. P. (1988) Evidence for the nicotinic cholinergic hypothesis of cannabinoid action within the central nervous system: exrapyramidal motor behaviors, in *Marijuana: An International Research Report*, Chesher, G., Consroe, P., and Musty, R. (eds.), Australian Government Printing Service, Canberra, Australia, pp. 359–364.

18. Moss, D. E., Manderscheid, P. Z., Montgomery, S. P., Norman, A. B., and Sanberg, P. R. (1989) Nicotine and cannabinoids as adjuncts to neuroleptics in the treatment of Tourette syndrome and other motor disorders. *Life Sci.* 44, 1521–1525.

19. Sanberg, P. R., McConville, B. J., Fogelson, H. M., Manderscheid, P. Z., Parker, K. W., Blythe, M. M., et al. (1989) Nicotine potentiates the effects of haloperidol in animals and in patients with Tourette syndrome. *Biomed. Pharmacother.* **43**, 19–23.

20. Emerich, D. F., Norman, A. B., and Sanberg, P. R. (1991) Nicotine potentiates the behavioral effects of haloperidol. *Psychopharmacol. Bull.* **27**, 385–390.

21. Emerich, D. F., Zanol, M. D., Norman, A. B., McConville, B. J., and Sanberg, P. R. (1991) Nicotine potentiates haloperidol-induced catalepsy and locomotor hypoactivity. *Pharmacol. Biochem. Behav.* **38**, 875–880.

22. Sanberg, P. R., Fogelson, H. M., Manderscheid, P. Z., Parker, K. W., Norman, A. B., and McConville, B. J. (1988) Nicotine gum and haloperidol in Tourette's syndrome [letter]. *Lancet* **1**, 592.

23. Silver, A. A. and Sanberg, P. R. (1993) Transdermal nicotine patch and potentiation of haloperidol in Tourette's syndrome [letter]. *Lancet* **342**, 182.

24. Silver, A. A., Shytle, R. D., Philipp, M. K., and Sanberg, P. R. (1996) Case study: long-term potentiation of neuroleptics with transdermal nicotine in Tourette's syndrome. *J. Am. Acad. Child. Adolesc. Psychiatry* **35**, 1631–1636.

25. McConville, B. J., Fogelson, M. H., Norman, A. B., Klykylo, W. M., Manderscheid, P. Z., Parker, K. W., et al. (1991) Nicotine potentiation of haloperidol in reducing tic frequency in Tourette's disorder (published erratum appears in *Am. J. Psychiatry* [1991] **148**, 1282) (see comments). *Am. J. Psychiatry* **148**, 793–4.

26. McConville, B. J., Sanberg, P. R., Fogelson, M. H., King, J., Cirino, P., Parker, K. W., et al. (1992) The effects of nicotine plus haloperidol compared to nicotine only and placebo nicotine only in reducing tic severity and frequency in Tourette's disorder. *Biol. Psychiatry* **31**, 832–840.

27. Dursun, S. M., Reveley, M. A., Bird, R., and Stirton, F. (1994) Longlasting improvement of Tourette's syndrome with transdermal nicotine [letter]. *Lancet* **344**, 1577.

28. Dursun, S. M., Bird, R., and Reveley, M. A. (1995) Differential effects of transdermal nicotine patch on the symptoms of Tourette's syndrome. *Br. J. Clin. Pharmacol.* **39,** 100P–101P.

29. Dursun, S. M. and Reveley, M. A. (1997) Differential effects of transdermal nicotine on microstructured analyses of tics in Tourette's syndrome: an open study. *Psychol. Med.* **27,** 483–487.

30. Silver, A. A. and Sanberg, P. R. (1993) Transdermal nicotine patch and potentiation of haloperidol in Tourette's Syndrome. *Lancet* **342,** 182.

31. Sanberg, P. R. (1980) Haloperidol-induced catalepsy is mediated by post-synaptic dopamine receptors. *Nature* **284,** 472–473.

32. Shytle, R. D. Silver, A. A., Philipp, M. K., McConville, B. J., and Sanberg, P. R. (1996) Transdermal Nicotine for Tourette's Syndrome. *Drug Dev. Res.* **38,** 290–298.

33. Ahtee, L. and Kaakkola, S. (1978) Effect of mecamylamine on the fate of dopamine in striatal and mesolimbic areas of rat brain: interactin with morphine and haloperidol. *Br. J. Pharmacol.* **62,** 213–218.

34. Clarke, P. B. S., Hommer, D. W., Pert, A., and Skirboll, L. R. (1985) Electrophysiological actions of nicotine on substantia nigra single units. *Br. J. Pharmacol.* **85,** 827–835.

35. Stone, C. A., Torchiana, M. L., Navarro, A., and Beyer, K. H. (1956) Ganglionic blocking properties of 3-methylaminoisocamphane hydrochloride (mecamylamine): a secondary amine. *J. Pharmacol.* **117,** 169–183.

36. Martin, B. R., Martin, T. J., Fan, F., and Damaj, M. I. (1993) Central actions of nicotine antagonists. *Medical Chemistry Research* **2,** 564–577.

37. Banerjee, S., Punzi, J. S., Kreilick, K., and Abood, L. G. (1990) [3H] Mecamylamine Binding to Rat Brain Membranes: Studies with mecamylamine and nicotine analogues. *Biochem. Pharmaco.* **40,** 2105–2110.

38. Martin, B. R., Onaivi, E. S., and Martin, T. J. (1989) What is the nature of mecamylamine's antagonism of the central effects of nicotine? *Biochem. Pharmacol.* **38,** 3391–3397.

39. Sanberg, P. R., Silver, A. A., and Shytle, R. D. (1998) Treatment of Tourette's Syndrome with Mecamylamine. *Lancet* **352,** 705–706.

40. Arendash, G. W., Sengstock, G. J., Sanberg, P. R., and Kem, W. R. (1995) Improved learning and memory in aged rats with chronic administration of the nicotinic receptor agonist GTS-21. *Brain Res.* **674,** 252–259.

41. Levin, E. D., (1992) Nicotinic systems and cognitive function. *Psychopharmacology (Berl.)* **108,** 417–431.

42. Newhouse, P. A., Potter, A., and Levin, E. D. (1997) Nicotinic system involvement in Alzheimer's and Parkinson's diseases. Implications for therapeutics. *Drugs Aging* **11,** 206–228.

43. Levin, E. D., Conners, C. K., Sparrow, E., Hinton, S. C., Erhardt, D., Meck, W. H., et al. (1996) Nicotine effects on adults with attention-deficit/hyperactivity disorder. *Psychopharmacology (Berl.)* **123,** 55–63.

44. Conners, C. K., Levin, E. D., Sparrow, E., Hinton, S. C., Erhardt, D., Meck, W. H., et al. (1996) Nicotine and attention in adult attention deficit hyperactivity disorder (ADHD). *Psychopharmacol. Bull.* **32,** 67–73.

45. Arendash, G. W., Sanberg, P. R., and Sengstock, G. J. (1995) Nicotine enhances the learning and memory of aged rats. *Pharmacol. Biochem. Behav.* **52,** 517–523.

46. Levin, E. D., Christopher, N. C., Briggs, S. J., and Rose, J. E. (1993) Chronic nicotine reverses working memory deficits caused by lesions of the fimbria or medial basalocortical projection. *Brain Res. Cogn. Brain Res.* **1,** 137–143.

47. Arneric, S. P., Sullivan, J. P., Decker, M. W., Brioni, J. D., Bannon, A. W., Briggs, C. A., et al. (1995) Potential treatment of Alzheimer disease using cholinergic channel activators (ChCAs) with cognitive enhancement, anxiolytic-like, and cytoprotective properties. *Alzheimer Dis. Assoc. Disord.* **9 (Suppl 2),** p. 50–61.

48. Arneric, S. P., Sullivan, J. P., Briggs, C. A., Donnelly-Roberts, D., Anderson, D. J.,

Raszkiewicz, J. L., et al. (1994) (S)-3-methyl-5-(1-methyl-2-pyrrolidinyl) isoxazole (ABT 418): a novel cholinergic ligand with cognition-enhancing and anxiolytic activities: I. In vitro characterization. *J. Pharmacol. Exp. Ther.* **270**, 310–318.

49. Decker, M. W., Curzon, P., Brioni, J. D., and Arneric, S. P. (1994) Effects of ABT-418, a novel cholinergic channel ligand, on place learning in septal-lesioned rats. *Eur. J. Pharmacol.* **261**, 217–222.

50. Buccafusco, J. J., Jackson, W. J., Terry., A. V. Jr, Marsh, K. C., Decker, M. W., and Arneric, S. P. (1995) Improvement in performance of a delayed matching- to-sample task by monkeys following ABT-418: a novel cholinergic channel activator for memory enhancement. *Psychopharmacology (Berl.)* **120**, 256–266.

51. Wilens, T., Biederman, J., Spencer, T., Bostic, J., Prince, J., Monuteaux, B. S., et al. (1998) A controlled trial of ABT-418 for attention deficit hyperactivity disorder in adults. *NCDEU Abstr.* **38**, 137.

52. Swerdlow, N., Braff, D., Geyer, M., and Koob, G. F. (1986) Central dopamine hyperactivity in rats mimics abnormal acoustic startle response in schizophrenics. *Biol. Psychiatry* **21**, 23–33.

53. Braff, D .L., Grillon, C., and Geyer, M. A. (1992) Gating and habituation of the startle reflex in schizophrenic patients. *Arch. Gen. Psychiatry* **48**, 1069–1074.

54. Adler, L. E., Hoffer, L. D., Wiser, A., and Freedman, R. (1993) Normalization of auditory physiology by cigarette smoking in schizoprenic patients. *Amer. J. Psychiatry* **150**, 1856–1861.

55. Acri, J. B., Morse, D. E., Popke, E. J., and Grunberg, N. E. (1994) Nicotine increases sensory gating measured as inhibition of the acoustic startle reflex in rats. *Psychopharmacology* **114**, 369–374.

56. Hughes, J. R., Hatsukami, D. K., Mitchell, J. E., and Dahlgren, L. A. (1986) Prevalence of smoking among psychiatric outpatients. *Am. J. Psychiatry* **143**, 993–997.

57. Briggs, C. A., Anderson, D. J., Brioni, J. D., Buccafusco, J. J., Buckley, M. J., Campbell, J. E., et al. (1997) Functional characterization of the novel neuronal nicotinic acetylcholine receptor ligand GTS-21 in vitro and in vivo. *Pharmacol. Biochem. Behav.* **57**, 231–241.

58. Woodruff Pak, D. S., Li, Y. T., and Kem, W. R. (1994) A nicotinic agonist (GTS-21), eyeblink classical conditioning, and nicotinic receptor binding in rabbit brain. *Brain Res.* **645**, 309–317.

59. Damaj, M. I., Welch, S. P., and Martin, B. R. (1994) Nicotine-induced antinociception in mice: role of G-proteins and adenylate cyclase. *Pharmacol. Biochem. Behav.* **48**, 37–42.

60. Puttfarcken, P. S., Manelli, A. M., Arneric, S. P., and Donnelly-Roberts, D. L. (1997) Evidence for nicotinic receptors potentially modulating nociceptive transmission at the level of the primary sensory neuron: studies with F11 cells. *J. Neurochem.* **69**, 930–938.

61. Donnelly-Roberts, D. L., Puttfarcken, P. S., Kuntzweiler, T. A., Briggs, C. A., Anderson, D. J., Campbell, J. E., et al. (1998) ABT-594 [(R)-5-(2-Azetidinylmethoxy)-2-chloropyridine]: a novel, orally effective analgesic acting via neuronal nicotinic acetylcholine receptors: I. In vitro characterization. *J Pharmacol. Exp. Ther.* **285**, 777–786.

62. Bannon, A. W., Decker, M. W., Holladay, M. W., Curzon, P., Donnelly-Roberts, D., Puttfarcken, P. S., et al. (1998) Broad-spectrum, non-opioid analgesic activity by selective modulation of neuronal nicotinic acetylcholine receptors [see comments]. *Science* **279**, 77–81.

23

Neural Grafting for Parkinson's and Huntington's Disease

Ben Roitberg, Peter Shin, Joseph Sramek, and Jeffrey H. Kordower

1. INTRODUCTION

Parkinson's disease (PD) was first described by James Parkinson in 1817 and consists of a combination of the following cardinal features: (1) bradykinesia–akinesia, (2) resting tremor, (3) rigidity, and (4) postural instability. The clinical syndrome results from degeneration of dopaminergic neurons in the substantia nigra pars compacta secondary to marked decreases of dopamine (DA) at axon terminals in the striatum (i.e., caudate and putamen). The exact pathogenesis of PD is unknown although several theories exist. The discovery that the chemical agent 1-methyl-4-phenyl-1,2,3,6-tetrahydropyridine (MPTP) can cause Parkinsonism raised the possibility that PD can be caused by an environmental toxin. Other theories center around the possibility that PD is caused by oxidative stresses from defective elimination of oxyradicals in the nigrostriatal system. The mainstay of treatment for PD has been pharmacological (levodopa, dopamine receptor agonists, monoamine oxidase B inhibitors, anticholinergics, etc.). L-Dopa, which is the immediate metabolic precursor of DA, is the most frequently used therapeutic agent. Long-term use is associated with the development of several severe side effects including dyskinesias and fluctuations in response (the on–off phenomenon) which eventually limit its use. Surgical techniques have also been used to alleviate the symptoms of PD. These include ablative (thalamotomy, pallidotomy) or disruptive (deep brain stimulation) procedures aimed at augmenting thalamocortical innervation as well as reconstructive approaches aimed at replacing the degenerative nigrostriatal pathway, such as neural transplantation.

Another therapeutic approach is the replacement of lost neurons using neural transplantation strategies. Neural transplantation research for the treatment of PD has evolved considerably over the past two decades. Preclinically, Stenevi et al. *(1)* were the pioneers in transplantation research, first grafting monoaminergic neurons to central nervous system (CNS) sites. In 1976, they attempted to transplant cervical ganglion from newborn and adult rats, and ventromedial mesencephalon, dorsolateral pons, and median pontine raphe region from embryonic, newborn, and adult rats into three

From: Central Nervous System Diseases
Edited by: D. F. Emerich, R. L. Dean, III, and P. R. Sanberg © Humana Press Inc., Totowa, NJ

different CNS locations: the caudal diencephalon and caudate nucleus, onto the dorsal surface of the caudate nucleus and onto the pial covering in the choroidal fissure. The most consistent survival of central monoaminergic neurons occurred when embryonic tissue was transplanted in the choroidal fissure. These experiments established for the first time that (1) the age of the donor tissue, (2) the site of transplantation, and (3) the technique of transplantation are all important in predicting neural transplant survival.

To study the behavioral and histopathological results of neural transplantation for Parkinsonism, a good, predictable animal model was required. Fortunately, Ungerstedt *(2)* initially developed the unilateral 6-hydroxydopamine (6-OHDA) model in rats in 1968. His efforts were based upon earlier studies that showed 6-OHDA caused degeneration of sympathetic nerve terminals in the peripheral nervous system *(2a)*. Injections were made into several regions of the brain known to contain monoaminergic neurons including the caudate and putamen, the zona compacta of the substantia nigra (SN), and the area dorsolateral to the nucleus interpeduncularis. Fluorescence microscopy revealed depletion of dopamine at dopamine nerve terminals and cell bodies. Furthermore, injection of 6-OHDA into the SN caused depletion not only of the DA cell bodies of the SN, but also of the DA terminals in the caudate nucleus and putamen. Ungerstedt also noted motor asymmetry in the rats. From this initial study, Ungerstedt developed quantitative rotational models in rats by administering amphetamine, which causes presynaptic release of DA and hence rotation toward the lesioned side *(3)* and apomorphine which acts on postsynaptic DA receptors and hence rotation away from the lesioned side owing to DA receptor supersensitivity *(4)*. These models allow a behavioral confirmation of unilateral striatal lesions as the number of rotations strongly correlates with the degree of the striatal lesion.

Using this model, Bjorklund and Steveni *(5)* and Perlow et al. *(6)* independently were able to show graft survival and behavioral changes after transplantation of embryonic rat ventral mesencephalon into unilateral 6-OHDA lesioned rats. Bjorklund's *(5)* technique was to create a cavity in the parietal cortex and place the solid graft on the dorsal surface of the head of the caudate–putamen. Histologically, 19 of 23 transplants survived with some graft viability lasting for at least 7 mo. The number of DA neurons surviving varied from very few to as many as 4000. In 16 of the 19 surviving specimens, there was DA fiber outgrowth across the border of the transplant and into the neostriatum. Using Ungerstedt's rotational model, Bjorklund and co-workers were able to eliminate amphetamine-induced turning in rats with large transplants and extensive fiber ingrowth proving that the grafts were functional as well. Perlow et al. *(6)* placed grafts consisting of embryonic ventral mesencephalon into the lateral ventricle ipsilateral to the 6-OHDA-lesioned striatum. A reduction of turning behavior in response to apomorphine in rats receiving ventral mesencephalic transplants was observed. Histochemical studies showed good DA graft survival. Ingrowth from the grafts was limited to the part of the caudate adjacent to the graft. In 1980, Dunnett et al. *(7)* demonstrated that the embryonic nigral transplants could improve spontaneous rotation and choice behavior in addition to rotation. However, there was no difference between transplanted rats and controls in a battery of sensorimotor testing. As a follow-up to his original study, Bjorklund et al. *(8)* performed staged bilateral 6-OHDA lesioning in rats followed by unilateral ventral mesencephalic transplants placed on the dorsal surface of the right striatum. After the second lesion, rats had severe akinesia, aphagia, and adipsia. Following grafting, there was recovery in certain aspects of motor behavior, as well as sensory attention and sensorimotor orientation on the side contralateral to the transplant. One possible explanation for these findings is that the ingrowth of DA fibers was confined to the dorsal parts of the head of the caudate–putamen, whereas the ventral and lateral parts remain denervated in this model *(8)*. To test this hypothesis, Dunnett et al. *(9)* placed similar solid fetal ventral mesencephalic grafts in a lateral cortical cavity, thus innervating the ventral and lateral parts of the neostriatum, and was able to ameliorate deficits in sensorimotor attention. Conversely, grafts in this location did not affect motor asymmetry in 6-OHDA-lesioned rats. This may be due to the limitations of graft-derived innervation when tissue is placed within the lateral ventricle or other fluid-filled spaces.

To circumvent this problem, Bjorklund et al. *(10)* developed the cell suspension neural grafting technique in which trypsin is used to dissociate a solid graft into a suspension consisting of single cells or small cell aggregates. This technique allowed stereotactic placement of cells within the striatum as opposed to on the surface of the striatum with solid grafts or in ventricular sites, with minimal damage to surrounding tissue. This technique also resulted in behavioral benefit with respect to motor asymmetry testing. Brundin et al. *(11)* characterized factors that affect cell viability in suspensions of embryonic CNS tissue. The four factors that affected in vitro viability were: (1) the handling of the suspension after trypsinization, (2) the storage time of the suspension after dissociation, (3) the gestational age of the donor fetuses, (4) the brain region from which the cells were obtained. These in vitro criteria allowed for prediction of graft survival and functionality in vivo with the 6-OHDA rat model. This group also established that a minimum of about 120 surviving neurons was required to produce a functional effect in rotation testing in the rat model. This number is equivalent to approx 1–2% of the normal number of dopamine neurons in the nigrostriatal system. Bjorklund et al. *(12)* using the cell suspension neural grafting technique, was able to show that the grafts survived in both DA target areas (i.e., striatum) and nontarget areas (i.e., parietal cortex, substantia nigra, lateral hypothalamus) but that significant fiber outgrowth occurred only in the DA target areas. The fiber outgrowth radiated on average 1–2 mm from the implant site, necessitating multiple implants to achieve extensive coverage of the striatum. These observations provided further evidence that fiber outgrowth depends on specific interactions with the surrounding host tissue. Biochemical analysis of striatal DA levels demonstrated a restoration to 13–18% of normal with recovery of motor asymmetry on rotation testing occurring with as little as 3% of the normal striatal DA levels *(13)*. Behavioral recovery was also observed when the grafts were placed in the dorsal neostriatum *(14)* with compensation of amphetamine-induced rotation occurring more rapidly than in prior studies using the solid grafting technique. Using this technique, placement of suspension grafts in the ventral–lateral neostriatum was found to ameliorate sensorimotor deficits while placement into nontarget areas was without effect. The specificity of graft–host interactions was further elucidated by Mahalik et al. *(15)* who, using tyrosine hydroxylase (TH) immunoreactive staining and electron microscopy, was able to show that labeled axons made contact with unlabeled dendritic processes and labeled dendrites made contact with unlabeled axons in a normal morphological fashion. These findings suggest specific graft–host interactions and possibly functional synaptic transmission of DA from graft to host.

The success of these early transplant efforts led to more widespread interest in this area and to neurotransplantation in nonhuman primates. This was heralded by the discovery of a new animal model of Parkinsonism. In 1979, a single case of Parkinsonism was reported after the intravenous injection of an illicit narcotic, a reverse ester of meperidine. This mixture also contained the product MPTP. Shortly thereafter, several more cases were reported *(16)*. These patients gradually developed a syndrome characterized by akinesia, rigidity, flexed posture, and resting tremor. The clinical symptoms were reversed by L-Dopa or DA agonists. One of these patients died of a drug overdose 2 yr after the onset of MPTP-induced Parkinsonism and a postmortem was performed. Neuropathologically, there was neuron degeneration in the substantia nigra, abundant extraneuronal melanin pigment, and an astrocytic response with glial scarring *(16)*. In 1983, Burns et al. *(17)* applied MPTP intravenously to primates and was able to reproduce a syndrome similar to Parkinsonism. The rhesus monkeys exhibited akinesia, rigidity, a flexed posture, and a postural tremor all of which were reversible with L-Dopa. Decreased levels of homovanillic acid (HVA), a product of DA metabolism, occurred in the primate cerebrospinal fluid (CSF) along with a 90% reduction of substantia nigra neurons and a marked decrease of striatal DA. Bankiewicz et al. *(18)* administered MPTP via the right carotid artery to nonhuman primates and was able to produce a hemiparkinsonian primate model. In this model, there is very little effect to the contralateral nigrostriatal system secondary to the rapid conversion of lipophilic MPTP which readily crosses the blood–brain barrier into nonlipophilic MPP^+. Clinically, the hemiparkinsonism was manifest on the left side by the development of bradykinesia, rigidity, and dystonic posturing over a 3-wk period. The administration of L-Dopa reversed these motor effects. Neurochemically, DA and HVA levels were reduced by >95% when compared with the uninfused side in the same monkeys. Pathologically, the number of TH-immunoreactive (TH-ir) cell bodies in the lesioned substantia nigra compacta (A9 area) was markedly reduced when compared with the untreated side. This model has been useful for several reasons: first, it provides a unique internal control; second, the monkeys are still able to care for themselves without the need for L-Dopa or DA agonists.

In 1986, one of the first successful attempts at neurotransplantation in a nonhuman primate was published by Redmond et al. *(19)* and Sladek et al. *(20)*. The monkeys were lesioned with systemic MPTP initially, then fetal monkey nigral allografts (CRL 17 cm) were placed within the striatum bilaterally into two monkeys and in cortical regions in a control monkey. Prior to transplantation, all monkeys showed classic Parkinsonian signs as well as markedly reduced levels of CSF HVA. Posttransplant, a reduction in Parkinsonian signs was accompanied by a gradual increase in CSF HVA in the two monkeys with striatal implants but did not occur in the cortically transplanted monkey or in nontransplanted controls. The monkeys were killed at 69 d posttransplant and immunohistochemistry revealed numerous TH-positive neurons in the striatum with extensive neuritic arborization at the graft site. In this study, fetal cortex was transplanted as a control. These grafts did appear to survive but there was no evidence of neuritic arborization or neuronal migration outside of the transplant boundary. Graft rejection did not appear to be a problem in any of the transplants. This was the first study to show survival of fetal neurons in a nonhuman primate and suggested specificity of connections from transplanted fetal neural tissue that paved the way for further studies. A more extended reversal of MPTP-induced Parkinsonian symptoms in nonhuman primates was shown by Sladek et al. *(21)* in 1988 and Taylor et al. *(22)* in 1990. Solid grafts of fetal ventral mesencephalic tissue were placed bilaterally without precavitation in either the caudate nucleus or cingulate cortex or cerebellar grafts were placed in the caudate nucleus. Significant behavioral improvement as determined quantitatively by Parkinsonian summary scores and healthy behavior scores *(22)* was observed only following transplantation of ventral mesencephalic tissue into the caudate nucleus. This improvement was observed up until 7.5 mo, at which time the animals were killed for histochemical analysis. The grafts appeared to have surviving TH-positive neurons with TH-positive fiber outgrowth past the graft–host interface and no signs of tissue rejection were present. Bankiewicz et al. *(23)* performed transplants of primate fetal mesencephalic tissue into a hemiparkinsonian primate model *(18)*. The technique differed from that of Redmond and Sladek in that 2–5 wk prior to transplantation, cavities were surgically created in the medial caudate region to provide a well-vascularized site for the graft. Bankiewicz was also able to show improvement in Parkinsonian symptoms that persisted for 7 mo until the monkeys were killed. On histochemical analysis, there were TH-positive cells and fibers within the graft, but in contrast to results from other investigators *(19–21,24)*, there was no evidence that the TH-positive fibers cross into the host striatum. There were, however, TH-positive fibers from the ventral tegmental area oriented toward the graft and it was concluded by Bankiewicz et al. *(23)* that the motor recovery observed was due to sprouting of remaining host DAergic fibers and that the graft may act as a stimulus for this sprouting. Using the same animal model with nonDAergic fetal transplants (i.e., cerebellar tissue, spinal cord tissue), Bankiewicz et. al. *(25)* were able to show behavioral improvement and sprouting of DAergic fibers from areas of the host brain not affected by MPTP (ventral tegmental region). This further supported their hypothesis that behavioral recovery was secondary to some trophic influence of either the graft or the hosts response to injury rather than from direct effects of the graft alone. Sladek et al. *(24)* in addressing these differences, concluded that these findings were based on less than optimal graft placement and survival and that the use of "cavity" formation predisposed to glial scar formation which may act to inhibit sprouting across the graft–host margin.

Sladek et al. *(24)* also looked at graft survival as a function of donor gestational age. Prior studies suggest that neurogenesis of substantia nigra neurons occurs between embryonic d 36 and 43 with rapid axon elongation occurring soon thereafter *(26)*. In MPTP-treated monkeys, Sladek et al. grafted a 44-embryonic-d-old donor in which neuritic outgrowth would be expected to be minimal and a 49-d estimated gestational age (EGA) donor from which more extensive outgrowth would be expected. The DA cell survival and outgrowth of neurites into host tissue exceeded those in previous reports of Sladek et al. *(20,21)* with an average of 1500–3500 TH-positive neurons/graft site from the embryonic 44-d-old donor and an average of 150–400 TH-positive neurons/graft site from the embryonic 49-d-old donor. DA levels were also examined from micropunches in proximity to the graft sites and were found to be elevated from 4- to 14-fold when compared with the untransplanted side. Even though there was viability of neurons from donors of different ages, the number of surviving neurons from the younger graft was an order of magnitude higher than that from the older one.

The practical problems associated with transplantation of fetal tissue for neuro-degenerative disorders include availability, long-term storage, and transportation of donor tissue. Collier et al. *(27)* looked at cryopreservation of fetal primate DA neurons and found that there was survival of tissue (CRL 8.0–19.5 cm) that was cryopreserved for up to 28 d. Although there was viability of graft tissue in culture and 50 d after transplantation into primate striatum, the freeze stored tissue appeared to yield fewer TH-positive neurons than fresh tissue grafts. Behavioral improvement was not assessed in this study. Redmond et al. *(28)* were able to show viability of cryopreserved human fetal neural tissue (9–12 wk gestation) in both cell culture and after transplantation into primate neostriatum. The primates were placed on cyclosporin A for immunosuppres-sion and there was no evidence of rejection of the xenografts 70 d after transplantation. These findings open up the possibilities of banking fetal tissue and allow for safety testing and viability testing of the tissue prior to transplantation.

Although the results of fetal tissue grafts were promising in various animal models of Parkinsonism, ethical and practical considerations prompted investigators to search for alternative sources of donor tissue in a parallel fashion. Freed et al. *(29)*, in 1981, chose adrenal medullary tissue as an alternative to fetal grafts for several reasons. First, the adrenal medulla produces DA as an intermediary in the synthesis of epinephrine; second, adrenal chromaffin cells assume a more neuronal morphology when grown as grafts in the anterior eye chamber or in culture in the absence of corticosteroids; and finally, intraocular adrenal medulla grafts have been shown to innervate intraocular grafts of cerebral cortex. Using the 6-OHDA rat model, Freed et al. *(29)* were able to show that intraventricular adrenal medullary grafts survived at 2 mo posttransplant and that there was a significant reduction in apomorphine-induced rotation in the adrenal medullary transplanted rats vs the sciatic nerve graft controls. On histochemical analy-sis, however, there was no evidence of reinervation of the caudate by the grafts. The initial hypothesis for the reduction in rotational behavior postulated the diffusion of DA from the graft into the striatum which decreased postsynaptic supersensitivity of DA receptors.

2. CLINICAL ASPECTS OF FETAL NEURAL TRANSPLANTATION

Since the first report of transplanting large pieces of occipital cortex from one ani-mal to another within and across species in 1890 *(30)*, a wealth of information has been gathered that demonstrates transplanted neural tissues and genetically engineered cells survive and function in the nervous systems of animals and humans. This abundance of data, mostly from animal models of neurodegenerative disease, have led to the concept that neural transplantation may be therapeutic alternative for various neuropathologi-cal entities *(31–40)*. Parkinson's disease (PD) was chosen for initial clinical neural transplantation trials for a number of reasons. First, PD pathogenesis is believed prima-rily to affect one population of neurons, the DAergic cells of substantia nigra pars compacta (SNc). This specificity provides opportunity for an investigator to replace only one population of chemospecific neurons, thus reducing complications associated with trying to obtain and graft multiple populations of donor cells. Second, systemic administration of DA is an effective treatment in the early stages of PD. This implies that postsynaptic cells requiring nigrostriatal DA require nonspecific tonic stimulation for its function rather than phasic stimulation requiring much more complex interaction between the pre- and postsynaptic cells. Lastly, the target area for implantation, the caudate and putamen, although large, is relatively circumscribed.

The first published report of neural transplantation to treat a clinical condition in humans was in patients with PD. Bjorklund et al. in 1982 *(41)* grafted adrenal medullary tissue into the striatum of a patient affected by PD. The treatment was unsuccessful, but deemed safe and this study propelled other clinical investigations into studying neural transplantation as a possible therapy for PD in humans. In 1987 Madrazo et al. *(42)* reported clinical benefit following adrenal cell transplantation into the right caudate of two patients. Dramatic clinical improvement was reported in these two patients for 10 and 3 mo posttransplantation. This report was met with great enthusiasm and started a flurry of investigations into employing adrenal medullary tissue as a source of DAergic cells for transplantation. Multiple clinical investigations could not confirm the findings of Madrazo et al. At best, follow-up study showed mild to moderate improvement lasting almost 18 mo postoperatively, after which the patients' clinical function came back to their baseline level *(43–45)*. Furthermore, of the several transplanted patients who came to autopsy, only one patient showed survival of just a few grafted TH-positive cells *(46)*. Sporadic clinical investigations are ongoing using adrenal chromaffin cells along with peripheral nerve cografts as a source of nerve growth factor for the transplanted cells *(47)*. However, the major focus of neural transplantation has turned to use of fetal DAergic cells as the donor source for the graft.

Clinical testing using fetal DAergic cells for neural transplantation to treat PD began in 1985 but was not reported until 1987 by Jiang et al. *(48)*. A 20-wk-old fetal ventral mesencephalon (FVM) was grafted into a cavity created in the caudate nucleus of a patient suffering from PD that was made 20 d prior. Transient improvement was noted and there was no complication associated with the procedure. Similar trials were then undertaken by Swedish *(49)*, British *(50,51)*, and Mexican *(52)* groups. Lack of major complications and moderate clinical benefit seen from these studies sparked an interest in pursuing neural transplantation as a therapeutic alternative for PD. Clinical trials began worldwide *(53–59)*. To date, more than 150 fetal transplantation procedures for PD have been reported all over the world. A multicenter double-blinded randomized trial is currently under way to determine therapeutic efficacy of fetal neural transplantation for PD.

2.1. Lund Program

Lindvall and co-workers pioneered clinical fetal grafting for PD. A total of 16 patients have been transplanted with embryonic neural tissues starting 1984–1985 to present. These studies are ongoing to determine efficacy of neural transplantation in treatment of patients with Parkinsonian syndrome. Up to date 16 patients have been and are currently being followed by this group. These patients were diagnosed as having idiopathic PD for 5–15 yr except three patients who were suffering from MPTP-induced Parkinsonian syndrome. They were implanted with FVM tissues from 6 to 9 wk gestational age. FVM tissues from three to seven different fetuses were used to prepare a cell suspension for implantation into one side (caudate and putamen) of a patient. From two to five passes of the stereotactic needle were required to deposit into putamen and one to two passes into caudate for implanting FVM tissue suspension. Clinical and neurophysiological status were followed *(49,60–64)*. Pre- and serial post-operative 6-L[^{18}F]fluorodopa (F-Dopa) positron emission tomographic (PET) scans were done *(65)* to noninvasively assess graft viability. PET scan studies using F-Dopa have shown decreased uptake in the striatum, particularly in the posterior putamen of PD patients compared to normal volunteer subjects *(66)*. The patients were immuno-suppressed with cyclosporin A, azathioprine, and prednisone.

The first two patients were unilaterally implanted in the caudate and the putamen. Immediately after transplantation there was worsening of the Parkinsonian symptoms in these patients. These two patients showed only minor improvement in Parkinsonian symptoms after 2 yr of observation. The side contralateral to the implanted site was seen to be most improved. Minor increases in F-Dopa uptake were measured in grafted striatum compared to the nongrafted site 12 mo postoperatively in both patients. No change in F-Dopa uptake was evident in the 6 mo postoperative PET scan. More than 9 yr of follow-up showed no persistent improvement in these symptoms.

For the second set of patients in the series, Patients 3 and 4, approx 30% reductions in "off" phase and the number of daily "off" periods were consistently noted. Prolon-gation and pattern of effect of single-dose L-Dopa were also observed. L-Dopa was withdrawn from patient 4 by 30 mo postoperative because of improvement in symp-toms. Moreover, during the "off" phase there was a decrease in Parkinsonian symptoms contralateral to the grafted side. In one patient, dexterity and arm movement motor scores were significantly increased but not for the other patient. These clinical improvements started appearing approx 8–16 wk postoperative. The fetal neural grafts did not alleviate tremor in any of the patients from the two series. Serial postoperative F-Dopa PET scans showed gradual increase in F-Dopa uptake at the grafted site in both patients of the second series as compared to the contralateral side, which showed decline in the uptake. In fact, there was normalization of F-Dopa uptake in the grafted striatum of Patient 4 by 72 mo posttransplantation. Lindvall et al. attributed the differ-ences in the result of the two series to some technical changes from that employed in the first series. Use of a thinner stereotactic needle was thought to be the most impor-tant factor. Use of a buffered solution instead of saline tissue medium, differences in the cannula loading of the fetal tissue, a shorter time of tissue storage before transplan-tation, and implanting the same amount of tissue only in the putamen instead of in both caudate and putamen were other changes.

A third study in the series was carried out by Lindvall's group with the knowledge that there was a marked decline of F-Dopa uptake in the nongrafted regions of the basal ganglia of transplanted patients *(62,64)*. Staged bilateral implants were implemented 10–23 mo apart in four new patients, designated patients 7, 8, 9, and 10. The same technique as described earlier was used for transplantation. Immunosuppression was also started similar to the previous description but discontinued 12 mo after the last transplantation. Serial F-Dopa PET scans were also performed.

All cases showed F-Dopa PET scan evidence of graft survival but clinical responses were variable between individual patients. Patient 8 did not improve significantly after transplantation. This patient at a later evaluation was diagnosed to have grossly atypi-cal multisystem atrophy or grossly atypical PD. Patient 9 developed dementia after the second transplant and the authors were not able to evaluate the patient's clinical status.

On the other hand, Patients 7 and 10, who were relatively young onset PD patients with L-Dopa-induced dyskinesias, reportedly showed promising response. During the first 6 mo postoperatively the L-Dopa-induced dyskinesia was worse but then reportedly improved. Other fine motor activities were also reported improved. Group data of the four patients after first surgery showed a decrease in L-Dopa dose of 10% and 20% at the end of the first and second postoperative years and 34% and 40% decrease in the "off" phase in the similar time period. The L-Dopa induced "on" phase increased 45% and 58% serially during the first two consecutive years. United Parkinson's Disease Rating Scale (UPDRS) motor score was shown to be decrased 26% by the second year posttransplantation. After grafting, Patients 7 and 10 were able to reduce medications and return to their professional lives.

Five patients were also reportedly transplanted with FVM tissues treated with lazaroid *(64)*. A preliminary report by Widner stated that his group used only three donor tissues per side instead of four to seven donors. F-Dopa PET study showed clear graft survival at 9 mo posttransplantation with similar progression of clinical benefits reported in previous series. In fact, within this group, one patient was reportedly taken off all antiparkinsonian medication and returned to his full-time employment 10 mo after surgery.

At about the same time the trials were being done on patients 3 and 4, a very interesting study was concurrently going on *(67)*. Widner et al. implanted FVM tissues into the striatum bilaterally of two patients, 43 and 30 yr old, with Parkinsonism secondary to injection of MPTP. MPTP selectively destroys DAergic cells of SNc of humans *(68)* and monkeys *(69,70)*, inducing a condition that resembles idiopathic PD *(71)*. Preoperatively, two MPTP patients were rated at Hoehn and Yahr stage IV and in one patient stage V during practically defined "off" periods. Neither patient was able to carry on independent activities of daily living. For the first time, a matched clinical comparison could be made for the therapeutic efficacy of neural transplantation between human and laboratory models. In nonhuman primates, FVM cell transplantation has been shown to improve Parkinsonian symptoms induced by MPTP *(72–78)*. The same transplantation technique was used as described earlier.

These patients have been followed for more than 5 yr postoperatively. They displayed a marked decrease in rigidity and bradykinesia. Hoehn and Yahr scores were decreased approx 1.5 stages in both patients. Gradual improvement in activities of daily living parameters started approx 3–9 mo postoperatively and continued to improve for up to 24 mo in both patients. Improvement in motor performance appeared approx 12 mo postoperatively with sustained improvement noted at the 24 mo follow-up visit. A significant reduction in dyskinesia was noted for the two patients. One of the patients who had spent most of her preoperative "on" period with dyskinesia was able to spend > 50% of the "on" time without dyskinesia. F-Dopa PET scans showed no increase in striatal F-Dopa uptake 6 mo postoperatively. However, a significant bilateral increase in striatal F-Dopa uptake was noted at 12 mo posttransplantation. This was a significant finding because increase in striatal F-Dopa uptake was associated with the temporal pattern of clinical improvement. It is important to note that although immunosuppressants were discontinued at 12–18 mo postoperatively, the patients continued to improve clinically at 24 mo.

An additional MPTP-exposed patient was transplanted with the same surgical technique and evaluation methods *(64)*. The patient had very severe Parkinsonian symptoms and was intolerant of L-Dopa because of its side effects, including hallucinations and dyskinesias. All L-Dopa and agonists were discontinued because of their side effects prior to surgery. The patient was unable to walk, talk, or move her arm voluntarily for more than 2 yr. One year after transplantation, her FluoroDOPA-PET reportedly showed clear improvement from the preoperative scan. Reduction in rigidity and improvement in initiation ability was noted. The latest follow-up of 2.5 yr after implantation reported that the patient was able to manage most of her chores of daily living without any help and able to ambulate on her own for short periods. Her speech was also reportedly intelligible. The patient was not taking any antiparkinsonian medication.

Widner and collaborators hypothesized that the extent of Parkinsonian symptomatic amelioration noted in these two patients is attributed to using larger volumes of tissues to implant into the striatum bilaterally, and differences in pathophysiology of MPTP-induced and idiopathic Parkinsonism. Use of seven to eight fetuses for implanting FVM tissues into one patient is the most complete attempt at grafting of bilateral striatum in any clinical series reported up to that time. By providing a greater number of DAergic cells, perhaps a greater distribution of striatum was innervated exhibiting enhanced clinical amelioration of Parkinsonian symptoms. Another explanation, not exclusive of the former, is that idiopathic Parkinsonism is a progressive degenerative disease whereas MPTP-induced Parkinsonism is a one-time destruction of DAergic cells of the SNc. Possibly, in idiopathic PD the continual degenerative nature of the disease process offsets the improvement that occurs after transplantation. These two factors should be taken into account when comparing the transplantation studies from the MPTP-induced Parkinsonian model of nonhuman primates *(76,79)* with the clinical studies on neural transplantation in human idiopathic PD patients.

2.2. United Kingdom Trials

In 1988 Hitchcock et al. *(50,51)* first reported results following stereotactic implants of electively aborted FVM tissue into the right caudate nucleus of two patients suffering from PD. An additional 10 patients were entered into the study. At 6 mo and 1 yr posttransplantation follow-up of 12 patients were reported *(80,81)*. All patients entered into the study had progressive worsening Parkinsonian symptoms and had been suffering from PD for 6–26 yr. FVM tissues of 11–18 wk gestational age were used. These FVM tissues were older than those tissues utilized by other groups *(37,49,55,56,60–63,82)* who were experiencing success with neural transplantation. In fact, preclinical studies have demonstrated that donor cells of this age fail to survive *(83)*. No immunosuppression was utilized. Three patients were dropped from the study because they failed to comply with further follow-up and changed their drug regimen. These investigators reported that three patients had sustained clinical improvement for longer than 12 mo with significant reduction in levodopa (L-Dopa) dosage. Three patients had mild improvement with increase time spent in the "on" phase. L-Dopa dosage decrease also accompanied reduction in dyskinesia along with improvement of motor scores. Clinical improvement in motor state occurred within the first 3 postoperative months in both "on" and "off" states. These changes were seen to plateau at approx 6 mo postoperatively. Three patients had significant worsening or no improvement in Parkinsonian symptoms at 1 yr postoperatively.

2.3. American Trials

American trials began with Freed et al. *(55,56,82)*, who reported on transplantation of FVM tissues into the striata of seven patients suffering from PD with a disease duration of 7–20 yr. All patients were Hoehn and Yahr stages 3–4 prior to surgery. FVM tissues of 6–7 wk gestational age from one embryo were used for each patient. Using the growth pattern observed after implanting human fetal cells into immunosuppressed rats as a guide to surgical technique *(84–87)*; FVM tissues were injected stereotactically into the patient's striatum with 4 mm spacing. For the first two patients, the tissue was implanted unilaterally into caudate and putamen contralateral to the side with the worse symptoms. The last five patients received the tissues bilaterally, stereotactically placed in the striatum. Cyclosporin and prednisone were used in every other patient to assess the need for immunosuppression in these patients. In addition, to minimize the risk of graft rejection, all patients and the fetal tissue were matched for ABO blood group. The 1-yr postoperative clinical course of the first patient who received the implants unilaterally was reported in 1990 *(55,56)*, then in 1992 reports on the remaining patients were published *(82)*.

After the FVM transplantation, Freed and co-workers reported that Hoehn and Yahr scores were improved by 1.2 points ($p < 0.01$). Statistically significant improvement in activities of daily living scores appeared 3 mo postoperatively during "on" periods and at 12 mo postoperatively during "off" periods. In those two of the seven patients receiving unilateral implants, one patient was not improved during the 33 mo of follow-up. More interestingly, in one patient who received a unilateral implant, a significant improvement in motor function of hands and feet was noted with increase in activities of daily living. The side contralateral to the implant site showed a greater increase than the ipsilateral side although bilateral improvement was observed. Dyskinesia contralateral to the implanted side appeared 11 mo postoperatively, necessitating reduction in antiparkinsonian medication. In the five patients who received implants into the putamen bilaterally, statistically significant improvement in facial expression, postural control, gait performance, and body bradykinesia was noted. One interesting observation of note was that autonomic signs of PD such as urinary urgency and constipation improved in two patients, one patient with a unilateral graft and one with bilateral implants. The L-Dopa requirement decreased by 39% ($p < 0.01$). Lastly, no benefit of immunosuppression was noted. In fact, Freed et al. concluded immunosuppression may have interfered with the graft development.

Cryopreserved FVM tissues were used to implant into the right caudate nucleus of four patients with severe PD by Spencer and co-workers *(88–90)*. These fetal tissues were 7–11 wk gestational age and were cryopreserved for up to 10 mo in liquid nitrogen before being utilized. FVM from only one fetus was used per patient. Two to four targets were selected for implantation into the caudate at 4 mm apart. Cyclosporin was started 2 d prior to surgery and continued for 6 mo postoperatively in all patients except for a patient (Patient 2) who for medical reasons had to stop the immunosuppression 2 mo postoperatively. F-Dopa PET scans were performed on only one patient pre- and postoperatively. Three nonoperated control patients, matched with the grafted group for age and disability, were followed at the same time to compare the therapeutic effectiveness of the fetal nigra graft. One patient, Patient 2, was not included in the group statistics because he died 4 mo postoperatively from events unrelated to the surgery. Prior to his death, he showed minor improvement in autonomic function, but no improvement in his motor or other neurologic functions. The clinical courses of three operated and control patients were followed for up to 18 and 12 mo, respectively. There was no statistically significant difference in objective testing scores between grafted and control patients. This result may not be surprising because preclinical studies showed fetal nigral cells > 9 wk gestational age fail to survive transplantation *(83)*. Furthermore, cryopreservation has been shown to be associated with decreased fetal graft viability *(91)*.

An interesting addition to the study by Spencer et al. came with the patient who died 4 mo postoperatively. Histologic studies of this patient were reported by Redmond et al. *(57)*. On autopsy it was found that the patient suffered from striato-nigral degeneration rather than from PD. Viable graft sites were seen in the caudate. The graft–host margin was readily visible accompanied by capillary infiltration across this margin. Viable neurons were observed in the graft along with synaptic network seen with electron microscope but there was no mention of axons crossing the graft–host margin or synapse formation between grafted neuron and the host neuron. Grafted TH-immunoreactive cells were not found in the graft, although they were seen in the host SNc. However, they were able to identify neuromelanin granules in the neurons of grafted cells. It is noteworthy that those granules are typical of adult-type SNc DA cells and are not typically seen in normal prenatal cells *(92)*. This expression of neuromelanin may be pathologic. No evidence of immune rejection was observed although the immunosuppressive agent was discontinued 10 wk prior to patient death.

Markham et al. *(93)* in 1994 published another clinical series on the first 6 of 13 PD patients implanted unilaterally with FVM tissue derived from one or two fetuses 7–9 wk gestational age. The implantation was performed on the side of the striatum contralateral to the side exhibiting the most severe symptoms. A single pass was made into the caudate and three to four passes were made into the putamen. Cyclosporin was started perioperatively and continued indefinitely. Antiparkinsonian medications were also reduced postoperatively. Four of the six patients, followed postoperatively for at least 1 yr were reported to have modest to moderate improvement in their clinical symptoms. F-Dopa PET scan showed distinct quantitative improvement in three of the four modestly to moderately improved patients. One of the four patient's PET result was still pending at the time of publication. They concluded that given the results of the literature and their own result, neural transplantation should not be considered as a therapeutic alternative for PD.

2.4. French Trials

Peschanski and Defer reported 24 mo outcome of five unilaterally transplanted PD patients with FVM tissues between 6 and 9 wk gestational age *(94,95)*. In an accompanying article, Remy et al. *(96)* reported a series of F-Dopa PET scans performed on these five patients before and up to 24 mo after transplantation. Patients ranged in age from 48 to 67 yr old and had been suffering from PD from 10 to 32 yr. They were determined to be Hoehn and Yahr stages IV–V during "off" periods. These patients were unilaterally implanted with FVM tissues derived from one to three fetuses. The tissues were placed into the caudate and putamen of the side contralateral to the most severe symptoms, except for the first patient who received the grafts only in the putamen. Three tracks were used to stereotactically implant the FVM tissues into the putamen and one track into the head of the caudate. Cyclosporin, azathioprine, and prednisone were started 2 d preoperatively and cyclosporin was continued for 7 mo postoperatively, at which time it was discontinued: azathioprine and prednisone were continued indefinitely.

Nigral transplants were reported to improve skilled movements such as finger dexterity, hand–arm simultaneous movements and rapid pronation–supination. Also a moderate decrease in the amount of time spent in the "off" state, "on-off" fluctuations, and freezing episodes was noted; but other symptoms were not affected. These clinical improvements such as finger dexterity and increase in the "on" state were statistically correlated with the F-Dopa PET scan uptake increase. In fact, correlation coefficients of 0.544 and 0.514 were shown between F-Dopa uptake coefficient and finger dexterity for grafted and nongrafted putamen, respectively. Moreover, a correlation coefficient of 0.673 was determined between F-Dopa uptake coefficient and percent of time spent in the "on" period. A greater improvement was observed during the "off" period than the "on" period postoperatively. Bilateral improvement was observed in terms of movement speed, with the amplitude and precision of movement in the limb contralateral to the grafted striatum showing the most improvement. These clinical improvements appeared approx 3–6 mo postoperatively. These changes were stable for up to 36 and 28 mo postoperatively in the first and second patient of the series respectively. Gait, walking, and speech disturbances did not show any improvement. Results from this group indicated the possibility that increasing the volume of host striatum receiving the reinnervation may enhance the clinical condition.

An interesting observation in this study was the delayed appearance of lateralized dyskinesia in the limbs contralateral to the grafted side. This dyskinesia was seen appearing 8–26 mo postoperatively in three patients. All patients in this series were reported to have symmetrical bilateral dyskinesia preoperatively that was greatly improved initially by the graft. Then lateralized dyskinesia reportedly emerged. The side that continued to display greater motor improvement in terms of amplitude and precision was the side with worse dyskinesia, suggesting enhanced but unregulated DAergic tone. Freed et al. have reported a similar finding on one of their unilaterally implanted patients who developed similar dyskinesia 11 mo postoperatively on the side contralateral to the implanted striatum despite improvement in skilled movements.

2.5. Spanish and Mexican Trials

Most of the human neural transplantation studies in the recent past have utilized stereotaxic surgery for implantation of FVM cells into the striata of PD patients. Preferentially, these cells were implanted into the putamen, because motor control in the basal ganglia (97–99) and loss of DAergic cell projections from SNc (100–104) in PD have been found to preferentially lie in the posterior putamen. Still, Madrazo and co-workers (42,52,105,106) published a series of reports on transplanting a block of FVM tissue into the caudate of four patients: one woman and three men ages 45–52 yr who had been suffering from PD for 9–16 yr duration. These patients were Hoehn and Yahr stage III–IV. The tissues acquired from spontaneously aborted fetuses were 12–14 wk gestational age. Right frontal craniotomy was used to implant 2 × 2 × 3mm blocks of FVM tissue into the right caudate of a patient within 4 h of fetal tissue acquisition. Only one fetus per patient was used. Immunosuppression with cyclosporin A and prednisone was started perioperatively and continued indefinitely except for prednisone, which was tapered down to "off" in 6 mo. Improvement in bradykinesia, rigidity, gait disturbance, postural imbalance, and facial expression was noted both by the patient and by observers along with increased sensitivity to L-Dopa medication from 19 to 50 mo postoperative follow-up period.

Similarly, Lopez-Lozano et al. implanted 10 PD patients with FVM tissue blocks into the caudate nucleus using open craniotomy as done by Madrazo's group *(107,108)*. All patients in the study were in Hoehn and Yahr stages IV–V. One block of FVM tissue from a single fetus of 6–8 wk gestational age was utilized per patient. Pre- and postoperative Hoehn and Yahr, UPDRS, and Northwestern University Disability Scores (NUDS) were utilized. Immunosuppressants were started 1 d preoperative to indefinitely postoperative, except for four patients in whom the immunosuppressants were discontinued because of medical complications. Although clinical decline was observed for the first postoperative month, statistically significant improvement was reported in 8 of the 10 patients at 6–7 mo posttransplantation. Clinical improvement persisted in 7 of the 10 patients, but plateaued at 36 mo postoperatively. Thereafter there was a trend toward decline in clinical status. Clinical improvements in both "on" and "off" phases were noted. A statistically significant increase in time spent in "on" was seen that started approx 7 mo postoperatively and persisted until the end of the study 60 mo posttransplantation. Moreover, a significant improvement in dyskinesia and reduction in administration of antiparkinsonian medication was noted that appeared in a stepwise fashion over the 60 mo of follow-up. They concluded that improvement observed was the result of reinnervation of the caudate nucleus by fetal cells, although no physiolgic or anatomic data were available to support this conclusion.

2.6. Anatomic and Physiologic Evidence of Graft Survival in Humans

Although there has been an abundance of clinical reports on patients who were implanted with FVM tissues for treatment of PD, anatomic data have been scarce. Until recently it has been unclear whether fetal nigral grafts survive and innervate the surrounding structures and whether clinical improvement was associated with graft viability and growth. This question was answered in 1995 by Freeman et al. *(109)*, when they published a clinical report on four PD patients implanted bilaterally with FVM tissues 6 1/2–9 wk old in the postcommissural putamen. Three to four embryos were used per side and the deposits were situated approx 5 mm apart. Cyclosporin A was given for 6 mo postoperatively and subsequently discontinued. F-Dopa PET scan was done pre- and at 6 mo postoperatively. A report of 6 mo postoperative follow-up showed improvement in UPDRS "off" scores, and significant reduction of "off" time along with a decrease in percent of time spent in "on" with dyskinesia. Moreover, an increase in F-Dopa uptake in a PET scan was noted in all four patients.

Two patients died 18 and 19 mo postoperatively from events unrelated to the surgery. The first patient died from massive pulmonary embolus after an ankle fusion

18 mo after the FVM transplantation. The brain was acquired within 4 h of patient death and placed in Zamboni fixative. We published a series of reports on neuro-anatomic and metabolic studies of the patient's brain *(110–114)*. Implants were found to be larger than the volume that was originally transplanted. An abundance of TH-ir neurons were identified within the graft. These TH-ir cells displayed morphological features similar to those seen in a normal substantia nigra. TH-ir cells sent out fibers that crossed the graft–host boundary and innervated up to 5–7 mm from the graft sites. Electron microscopic studies indicated synapses between transplanted TH-ir cells and host cells in the putamen. No host DAergic cell sprouting was observed. *In situ* hybridization of the graft showed robust TH mRNA gene expression. To assess functionality of the graft, cytochrome oxidase histochemistry was performed. Densitometric measurements showed two- to threefold higher cytochrome oxidase activity within the grafted postcommissural putamen than in the nongrafted caudate or anterior putamen. Specifically, adjacent section staining of TH-ir and cytochrome oxidase showed significant codistribution of the staining pattern, indicating DAergic cells had high metabolic activity. Lastly, although there was decreased density of blood vessels within the graft as compared to the host putamen, these vessels were nonfenestrated and of the CNS type. The second patient also showed similar findings *(113)*.

2.7. Technical Issues

Olanow and co-workers *(115,116)* elaborated upon some important problems that need to be solved in order to offer neural transplantation as a therapeutic option. Factors considered to be most important are donor age, method of tissue storage, type of graft, site of implantation, volume of grafted tissue, tissue distribution, necessity of immunosuppression, and patient selection. Another important consideration is technique of tissue implantation. The answers to many of these issues are still unclear and have yet to be settled. In the following we briefly discuss current body of knowledge regarding these issues.

Donor age is a critical variable for transplantation. The most favorable donor age is thought to be when DAergic cells first appear in the subventricular zone to the time just prior to neuritic process extension. It has been determined that TH-ir cells appear in the human mesencephalic subventricular zone around 5 1/2–6 1/2 wk gestation with first identifiable neuritic processes identified at 8 wk gestation. Neuritic process are seen to reach the striatum at 9 wk gestation *(117)*. Implantation of human FVM tissues into immunosuppressed rats showed maximal survival of the suspension graft if the FVM tissues were 5 1/2–8 wk gestational age and 6 1/2–9 wk gestational age for solid grafts *(83)*. In fact, in human trials, those groups that used FVM tissues greater than 9 wk gestational age showed no significant clinical recovery or meaningful anatomic TH-ir cell survival *(48,57,81)*. However, those groups that used the cells within the optimal range showed meaningful clinical recovery and indirect evidence of anatomic recovery by F-Dopa PET scan *(61,94,109)*. Moreover, postmortem studies showed TH-ir cell survival in the transplanted patients *(110,111)*.

Development of an optimal storage technique is essential if fetal neural transplantation is to continue. The tissue needs to be stored to screen for various infectious agents and possible donor–recipient antigen matching. It seems at this time, more than one donor fetuses will be required to obtain necessary DAergic cell number survival in the host striatum. Groups have tried different ways to store fetal tissues prior to grafting. Fetal tissues, cryopreserved for up to 10 mo, have been utilized by Spencer et al. *(88)*, which showed a trend toward clinical improvement. Tissues cryopreserved for greater than 1 yr were also shown to be associated with a significant decline in cell survivability after transplantation in rat model *(91)*. Human amniotic fluid *(82)* and chemically defined "hibernation medium" *(109)* have also been utilized with clinical benefit. Most other clinical series that showed clinical benefit to date have used the tissues within 24–48 h of tissue acquisition.

Most clinical series have utilized cell suspension grafts. Although more homogeneous tissue distribution is achieved with suspension graft, the cells are more likely to be injured during the preparation of the tissues. In fact, the optimal window of donor age for the suspension graft is approx 1 wk earlier than for the solid graft *(83)*. Lopez-Lozano et al. *(107,108)* used solid human graft implantation using open craniotomy and reported significant clinical benefit in patients suffering from PD. Human FVM xenografts into rat models have shown equivalent survival, after transplantation as a solid piece or suspension *(83)*. The advantage of a solid graft is that tissue preparation is easier and cytoarchitectural relationships are better preserved.

Approximately 60,000 DAergic neurons have been shown to project to the human putamen *(118)*. Although this number provides a goal for cell number replacement for FVM cell transplantation, the precise number of cell replacements necessary to provide adequate functional recovery is as yet unknown. At this time, investigators can only guess as to the number of cells necessary to obtain full functional recovery in PD patients from existing basic sciences data. Approximately 5–10% of transplanted FVM tissue consist of DAergic cells and of these it has been shown that only 5–10% survive the grafting *(119)*. Preclinical data showed that approx 20,000–40,000 DA-containing neurons from single human fetuses survive transplantation in immunosuppressed rats *(87,120)*. However, at this time we do not know what percentage of surviving DA-containing neurons make functional connections with host striatal cells nor do we know the distribution of connection. The only thing that we know at this time is that clinically replacement of greater number of cells results in better clinical recovery in humans *(52–54,57,109,115,120)*.

Even if the correct number of cells is replaced, if the site of cell replacement is inappropriate, functional recovery is unlikely. Basic studies using animal models have shown that the posterior two thirds (postcommissural putamen) differs from the caudate and anterior putamen in its embryological development *(121)* and anatomic connectivity. The postcommissural putamen has been shown to have reciprocal anatomic and functional connections with precentral motor fields whereas caudate and anterior putamen have connections with prefrontal association cortex and frontal eye fields *(97,99)*. Pathological *(100–102)* and functional F-Dopa PET scan studies *(103)* show greater loss of DAergic innervation to the posterior putamen then caudate or anterior putamen in patients suffering from idiopathic PD. More interestingly, functional recovery in rodents *(122,123)* and primate *(77)* PD models show site-specific recovery of particular function, thus theoretically making postcommissural putamen the better site for graft implantation to improve motor deficits.

Distribution of fetal cell replacement within the striatum is another important consideration. Based on DA diffusion studies in the brain parenchyma *(124,125)* and laboratory observations, it has been hypothesized that transplanted human fetal nigral cells can innervate up to a 2.5 mm radius. Thus to achieve confluent innervation of the host striatum, various investigators have spaced the cell deposits at 5 mm apart. Postmortem anatomic studies in humans have shown that transplanted TH-ir cells extend neuritic processes up to 5–7 mm from the grafted site in host striatum *(110,111)*. However, even with this type of innervation pattern seen with histologic study, complete reversal of clinical symptoms has not been achieved.

The need for immunosuppression in fetal neural transplantation is a very important question because of its potential for morbidity and significant costs. Some grafting studies using immunosuppressive agents had to be discontinued because of complications associated with their use *(108)*. Although most groups proceeding with fetal neural transplantation have used immunosuppressive agents such as cyclosporin A, azathioprine, and steroids, the necessity of immunosuppression in clinical settings has not been firmly established. The existing data using animal models and human clinical trials are variable. In nonimmunosuppressed rats and nonhuman primate models of PD, fetal allografts have been shown to survive *(126,127)* and clinical improvement has been suggested in patients who received fetal grafts without immunosuppression *(51,80,82)*.

In light of these variable findings, some groups have examined the effect of immunosuppression on clinical outcome and histologic findings. Freed et al. *(82)*, who implanted ABO antigen matched FVM cells into half of the transplanted patients, showed greater clinical improvement in patients who were not immunosuppressed as compared to the immunosuppressed group. Similarly, Ansari et al. *(128)* noted a statistically significant reduction in L-Dopa requirement comparing pre- and postoperative patients in the nonimmunosuppressed group, whereas no differences in L-Dopa requirements were seen in the immunosuppressed group. Various investigators have been able to show continued clinical improvement even after the discontinuation of immunosuppressive agents *(64,67,94)*, suggesting that long-term immunosuppression may not be necessary. However, studies of immune marker expressions in the two patients who had fetal nigral transplant *(129,130)* and died 18 and 19 mo postoperatively demonstrated the presence of immune cells including microglia, macrophages, T cells, and B cells. These markers were seen within the fetal grafts in both patients who had received only a 6-mo course of cyclosporin A. Even with the presence of the immune cells, however, numerous TH-ir cells were noted and no other evidence of graft rejection was seen. A possible explanation raised from the study was that immune cells may secrete trophic factors that may ultimately enhance graft viability and/or neurite outgrowth.

On the other hand, Lopez-Lozano et al. *(131)* reported three of four patients in whom cyclosporin A had to be discontinued, 27–42 mo postoperatively because of complications associated with the medication. These patients experienced significant clinical deterioration after discontinuation of immunosuppressants. They attributed this deterioration to possible ABO antigen mismatch and open craniotomy instead of stereotactic needle injection for the surgical procedure.

In summary, a review of the existing literature from around the world showed significant clinical benefits in using fetal neural transplantation for treating idiopathic PD patients and some iatrogenically induced Parkinsonism. The greatest clinical benefit was seen in the "off" state. Moreover, reduction in the dose of L-Dopa was possible. In fact, Lindvall et al. reported on two patients suffering from idiopathic PD who were followed for more than 5 yr and were able to return to work after neural transplantation *(64)*. Other groups such as Freeman et al. and Pechanski et al. were able to show long-term clinical improvements in neural transplant patients. However, the range of observed clinical benefits was variable and it is extremely difficult to assess or predict what particular function neural transplantation will improve.

In realization of the fact that neural transplantation is still in its infancy as a therapeutic modality for treating CNS diseases, we are heartened by the fact that some questions have been answered because of the efforts of various investigators in the field. Namely, it can be stated that sustained clinical improvement can seen after neural transplantation. It is also reasonable to surmise that implanting a greater number of cells yields a better clinical outcome *(67,94,105)*. Although unilateral implantation of fetal grafts leads to bilateral improvement, greater improvement is seen contralateral to the implanted site; thus most beneficial clinical effects can be achieved by bilateral transplantation *(94)*. Most importantly, transplanted fetal TH-ir cells do survive and actually make functional connections with the host striatal cells. There are still many more questions to be answered for neurosurgeons to offer neural transplantation as a therapeutic option for a patient suffering from end-stage idiopathic PD at this time.

2.8. Huntington's Disease

Huntington's disease (HD) is a severe autosomal dominant neurologic disorder inherited with almost 100% penetrance. It is distinguished by a constellation of psychiatric, cognitive, and motor symptoms. There are about 25,000 cases in the United States, many of them descendants of just six individuals who had immigrated from one village in England *(132,133)*. The usual age at onset is in the fourth or fifth decade of life, but the disease can begin at any age, even in childhood *(132,133)*. The psychiatric and cognitive decline may precede the chorea, and frequently manifests itself with alterations in character, depression, and suicidal tendencies. The deterioration is inexorable, with complete disability and death inevitable, usually occurring within 17 yr of symptom onset *(132)*.

The primary neuropathological changes in HD are atrophy of the basal ganglia, with severe neuronal loss in the caudate nucleus and putamen *(132,134)*. The striatal projection neurons are preferentially affected *(135,134)*. Early in the disease, the striatal projection to the external globus pallidus is preferentially affected. The loss of these medium-sized spiny neurons is extensive, up to 90% *(133,136)*. The main neurotransmitter of these cells is γ-aminobutyric acid (GABA), and they also contain enkephalin and substance P. The GABAergic projection neurons receive a glutamatergic input from the entire cortex in an innervation pattern that is topographically organized. Striatal interneurons are relatively spared in HD. Striatal degeneration may underlie in part the dementia seen in HD *(137)* in addition to the motor deficits. However, in advanced cases, cortical atrophy is seen as well. Functional decline precedes actual atrophy as demonstrated by PET and SPECT studies of patient and of persons at risk *(132,138)*.

The gene for HD was found in 1993 *(139)*. The defect associated with the disease is an expanded CAG repeat on chromosome 4. The gene codes for a protein (Huntingtin), which is widely expressed, but the function of which remains unknown *(139)*.

There is no effective treatment for HD. The severity of this disease increases the urgency to find a treatment, and risks associated with experimental techniques become more acceptable. The initial successes of neural transplantation in various models, notably animal models of Parkinson's disease, and later in limited clinical trials *(140–142)*, prompted investigation into the potential of cell transplantation to treat HD *(143,144)*. The rationale for using transplantation is quite solid—there is a potential to reconstitute lost structures, at least partially, with cells that do not have the genetic defect that caused the disease. Nevertheless, many questions had to be answered before human trials could be attempted. First, a reliable animal model of the disease had to be established, both to help understand the etiology and pathophysiology of the human disease, and to test treatments. Then, parameters pertaining to the graft itself, the donor, the transplantation technique, and the host response had to be elucidated.

2.8.1. Excitotoxic Lesions of the Striatum in Rodents Is a Useful Model of Huntington's Disease

In 1976, the first animal model was developed *(145,146)*. The researchers noted that systemic administration of monosodium glutamate in rats caused widespread neurotoxic effects. This toxicity was limited to dendrites and soma, while the axons were unresponsive. Whereas systemic administration of kainic acid, a rigid analogue of glutamate and a potent neurotoxin, caused high incidence of seizures and death, local injection into the striatum produced focal striatal degeneration. The axons passing through that area, or terminating in it, were spared. The damage increased with increasing dose of kainic acid (KA). At the 2.5 µg dose, there was a marked loss of intrinsic neurons, while the DAergic innervation to the striatum was not affected. Moreover, the neurochemical and histological changes were paralleled by a marked spontaneous abnormal rotatory behavior by the rats.

Further studies confirmed the findings, and found additional similarities between the KA animal model and HD *(147,148)*. Characteristic stereotypy and impaired memory were found in the lesioned rats, making the model increasingly attractive. The presence of cognitive decline in a model where the lesion is restricted to the striatum also provided strong support to the concept that the striatum is implicated in complex psychological functions *(147)*. Other excitotoxins, such as ibotenic acid, a synthetic analogue of glutamate, were found to have similar properties to KA and have also been used in preclinical HD studies *(149)*.

The excitotoxicity model was further refined when quinolinic acid (QA) was found *(149)*. QA is an endogenous tryptophan metabolite, and is a potent neurotoxin. It produced lesions similar to those caused by KA or IA. Moreover, QA appeared to be more selective, damaging the medium spiny GABAergic projection neurons and selectively sparing cholinergic interneurons or neurons expressing nicotinamide adenine diphosphate hydrogenase (NADPH). In contrast to KA lesions rats treated with larger QA doses did not display neuronal degeneration in distal brain areas *(149,150)*. Also, the projection neurons that contained substance P were affected more than somatostatin/ neuropeptide Y containing neurons. These neurochemical findings closely paralleled those in HD *(150)*. The activity of quinolinic acid phosphoribosyl transferase, the degradative enzyme of QA, was increased both in QA-lesioned rats and in postmortem samples from HD patient brains *(151,152)*. Concentrations of GABA as well as glutamate were decreased in brains from both patients with HD and QA-lesioned rats

(153). Follow-up of the lesioned rats at 6 mo and 1 yr revealed striatal atrophy and persistence of neurochemical changes, such as reduced GABA, substance P, and choline acetyltransferase (ChAT) activity; increases in somatostatin and neuropeptide Y activity; elevated concentrations of 5-hydroxytryptophan and sparing of NADPH-diaphorase neurons. The findings resembled the neurochemical features of HD, more than the KA model.

Despite some controversy *(154,155)*, the QA excitotoxicity model of HD for many years was the standard animal model of this disease *(156)*. QA excitotoxicity duplicated the histological and neurochemical changes of HD, and suggested that glutamate-induced excitotoxicity is the basic mechanism of HD *(156)*. This notion was supported by the finding that skin fibroblasts of HD patients were more susceptible to the toxic effects of glutamate in culture medium. Interestingly, glutamate treatment produced a decrease in the gluthatione levels in the affected cells, and glutamate-induced cell death could be inhibited by cotreatment with antioxidants.

Most glutamate-induced toxicity is induced by activation of the NMDA receptor rather than the K or Q receptors. QA is a relatively selective NMDA agonist *(136)*. The NMDA receptor activates a Ca^{2+} channel, and the intracellular accumulation of this ion may lead to cell death. Increasing the extracellular calcium levels increased the sensitivity of neurons to glutamate toxicity. This effect was abolished by MK-801 and *N*-methyl-D-aspartate (NMDA) receptor antagonists *(156a)*. Calbindin–D28k, a calcium binding protein, is increased in the cytoplasm of the medium-sized spiny neurons in vitro as well as in vivo. The pattern of this increase in the rat QA model was similar to that in HD patients *(157)*.

The excitotoxicity model has imperfections, which are addressed later in this chapter. Nevertheless, the understanding of HD was remarkably advanced by the advent of a reliable animal model, which duplicated many of the anatomical and neurochemical findings of HD.

Maybe the most important benefit of the excitotoxic lesion model in rodents was in development of the fetal striatal cell graft paradigm, to reverse the deficits caused by this injury, as a first step toward application of the new understanding of the disease to the treatment of human patients.

3. ANIMAL MODELS

3.1. The Metabolic Theory of HD and a New Animal Model

The QA excitotoxicity model is very useful. QA is an endogenous molecule, but its production tended to be lower in HD patients, which does not support QA overproduction as the primary factor in neurodegeneration in HD *(158)*. Moreover, the neuron death in the human neurodegenerative diseases tends to be gradually progressive, unlike the sudden cell death from QA injection or from glutamate release associated with ischemia. Although excitotoxicity may play an important role in HD, other intrinsic vulnerabilities of the degenerating neurons are likely.

Metabolic defects have been implicated in a variety of neurological disorders *(159,160)*. Reduced oxidative metabolism would leave cells energy deficient and more vulnerable to excitatory amino acids. The delayed onset and gradual nature of neurodegenerative diseases may be related to progressive impairment of mitochondrial function, such as occurs normally with aging *(160)*. This progressive impairment could be accelerated in patients who have a genetic defect of mitochondrial metabolism.Indeed, mitochondrial toxins, most notably citric acid cycle inhibitor 3-nitropropionic acid (3-NPA), has been shown to produce a pattern of striatal degeneration closely resembling HD. This toxiticity could be blocked by NMDA antagonists, further linking metabolic impairment and excitotoxicity *(160,161)*. Direct injection of 3-NPA into the rat striatum produced a severe localized lesion, with extensive necrosis of neurons of all types *(162)*. Remarkably, low-dose systemic 3-NPA administration every 4 d for 28 d produces marked spontaneous changes in rat locomotor behavior *(161,162)*. Chronic systemic 3-NPA treatment led to lesions selective for dorsolateral striatum and to spontaneous bradykinesia and gait abnormalities in rats *(163)*. Manipulating the time course of the injections, researchers could produce persistent hyperactivity, or hypoactivity, correlating to early and advanced HD, respectively *(164)*. Generally, smaller total doses of 3-NPA led to hyperactivity, and higher doses or more injections resulted in hypoactivity, thus replicating the progression of HD *(165)*. Bordelon et al. *(166)* recently provided another link between QA toxicity and metabolic dysfunction. QA injection caused a 30–60% decrease in oxygen consumption, ATP, NAD, aspartate, and glutamate 12 h following administration. These declines presumably represented rapidly progressive mitochondrial dysfunction.

Mitochondrial dysfunction is increasingly associated with human neurodegenerative diseases. Mitochondrial toxin MPTP caused Parkinsonism in humans. Complex I deficiency may play a role in PD *(167)*, and Alzheimer's disease was recently associated with accumulation of genetic defects in mitochondria *(168)*. Complex II/III deficiency may contribute to HD *(167)*. The potential of the 3-NPA model to provide further insight into the pathogenesis and treatment of HD gave the impetus to expand this model to primates. Chronic administration of 3-NPA to baboons caused first apomorphine induced, and as the treatment continued, spontaneous dystonia and dyskinesia. Lesions in the caudate and putamen were seen on MRI, and the histological evaluation showed selective depletion of calbindin neurons, sparing of NADPH-diaphorase neurons, astrogliosis, and other changes typical of HD *(169)*. Some of the more devastating symptoms of HD are not motor, but cognitive and psychiatric. Palfi et al. *(170)* demonstrated the triad of spontaneous abnormal movements, cognitive impairment, and progressive striatal degeneration in the baboon.

Both the QA excitotoxicity model and the 3-NPA model are useful in cell transplantation research. Mitochondrial metabolic defects and excitotoxicity are probably intertwined in the pathogenesis of HD, and both are energy-depleting processes. Although the 3-NPA model demonstrates a biphasic deficit, it is also more time consuming, and may be less practical for studies of neuron replacement therapy. Replacement should be done only after the 3-NPA treatment is discontinued; otherwise the transplanted neurons will be exposed to its toxic effect. The questions of stability of the systemic 3-NPA effect over time after stopping the injections and direct comparison of the effects of 3-NPA with those of QA await further research.

3.2. Primate Models

Although the rat model helped establish the basic model of HD and was used to test therapeuric approaches, such as fetal cell transplantation, it is not solely adequate to determine a clinically relevant therapy. There are obvious differences between the rat and primate brain. The monkey brain is 20–30 times larger, primates have a very different movement repertoire, and anatomically, the striatum in primate brain is divided by the internal capsule into caudate and putamen. Primate studies were the next step on the road to clinical application of neural transplantation for HD.

In 1989, Hantraye and co-workers injected the NMDA agonist ibothenic acid (IA) into the caudate–putamen of baboons. The unilateral injection produced rare spontaneous dyskinesias. However, when apomorphine was administered to those animals following the lesion, it produced a variety of dyskinesias and choreiform and dystonic movements. Other studies were done with KA and QA, and L-Dopa was also used as a stimulant to induce motion abnormalities closely resembling HD *(171)*. The primate excitotoxicity model demonstrated a selective vulnerability of striatal neurons similar to that seen in HD *(172)*.

Using his model, xenografts (rat derived donors) of fetal striatal transplants were shown to partially reverse the behavioral effects of the lesion. When the cyclosporin treatment was stopped, the grafts were rejected, and the symptoms reappeared. Animals who received cyclosporin only did not show an improvement in apomorphine-induced abnormal movements *(173)*.

Others used different excitotoxins such as the NMDA agonist QA *(174)*. Using this model, a number of technical issues relating to transplantation, histopathology, neurochemistry, and MR imaging were elucidated *(173,175,176)*.

Ferrante et al. *(176)* characterized the QA lesion in detail. It showed remarkable similarities to HD. Medium-sized spiny neurons were selectively lost, and GABA immunoreactivity was depleted without significant loss of somatostatin immunoreactivity. The matrix part of the patch–matrix striatal organization was selectively atrophied. Most remarkably, pretreatment of three of the monkeys with MK-801, a noncompetitive NMDA antagonist, prevented QA toxicity. The possibility was raised that excitotoxicity mediated by NMDA receptors contributes to the pathogenesis of HD. Beyond its use in the basic understanding of HD and the basal ganglia circuitry *(176,177)*, the nonhuman primate model is relevant to the development of therapeutic strategies for humans. The success of fetal grafts in the nonhuman primate model led the way to the first clinical experiments.

3.3. The Rodent Model of Striatal Transplantation

The first report of successful fetal striatal grafts in the excitotoxic lesion model was 1981 by Wally Deckel *(178)*. This landmark study was rapidly followed by others. Deckel *(179)* and Isacson *(180,181)* reported reversal of some behavioral deficits caused by intrastriatal injections of excitotoxins. In these studies, rats received injections of fetal striatal cell suspensions bilaterally, into previously lesioned striata. The grafts survived well, in a patch–matrix organization. Nocturnal hyperactivity, cognitive impairments on T-maze test, and abnormal responses to amphetamine and apomorphine were partially corrected by fetal striatal implants *(181–183)*. These effects were paralleled by partial normalization in neurotransmitters, such as GABA and acetylcholine. The utilization of glucose followed the same pattern of partial recovery.

As opposed to the ectopic striatal placement of nigral grafts, the striatal grafts grew best when placed homotopically within the striatum. Evidence for graft integration was also found. Dense DAergic innervation was found within graft-derived patches as well as other evidence of graft functional integration *(184)*. Interestingly, the striatal grafts showed good survival in these studies, and better integration into the host tissue relative to nigral grafts which typically have a 5–10% survival rate *(185)*.

In early striatal grafting studies, the parameters mediating successful grafts were delineated. Most researchers used fetal rat or mouse tissue from d E14–E17, when the striatal neurons differentiate, but have not yet extended long axons and exhibit the maximal potential for growth *(186,187)*. The fetal grafts were shown to undergo growth and maturation at a rate similar to that of the normal striatum *(188)*. The microenvironment of the grafts proved crucial to their survival. Grafts survived best when implanted 1 wk after a lesion in the target area, worse when implanted into the intact brain, and worst when the transplant was done immediately after the lesion *(189)*.

Clearly, some changes occurred that created a better environment for the delayed transplants. Further in vivo and tissue culture studies *(190–192)* demonstrated the role of astrocytes and trophic factors in transplantation. Usually, astrocytes from both the graft and the host form a border between them, which represents a barrier to the growing axons *(188)*. However, under certain conditions astrocytes provide essential support for the transplant. Early in the development of the astroglial response to the transplant, the astrocytes provide important trophic support and produce laminins and other factors that guide the elongating graft axons into the host brain *(189,192)*, but subsequent migration of astrocytes to the area of the graft may eventually limit its integration. We will return to the later developments and the use of trophic factors in transplantation and brain regeneration.

Another critical factor to consider was the precise source of the transplanted fetal striatal tissue. Early experiments used both medial and lateral ganglionic eminences as a source of "fetal striatal primordia" *(188,193)*. The grafts tended to produce a distinct pattern of acetylcholinesterase (AChE) rich patches within areas of AChE-poor tissue. Initially, these were thought to represent a process of compartmentalization resembling the normal striatum *(184,194)*. In 1990, Graybiel et al. proposed a different interpretation: the patches were the sorting out of the striatal tissue from the mixed striatal and nonstriatal cells in the primordia. This reaggregation occurred even when a cell suspension was transplanted. The markers of striatal cells were found on the patch cells, and the nonpatch areas contained types of neurons not normally found in the striatum. Grafting tissue derived from the lateral ganglionic eminence enriched the AChE-rich fraction, and improved graft integration into the host *(195)*. These findings were correlated with improved functional recovery after grafts of lateral ganglionic eminence *(196)*.

Both whole pieces of tissue and cell suspensions have been used for transplantation. The suspensions, prepared by trypsination and mechanical separation, were successfully used in stereotactic neuronal grafting into deep areas of the brain *(185)*. Transplantation of whole pieces of neural tissue avoids the trauma of separation, but the central portions of the implants tend to grow less well and are vascularized less rapidly *(197)*. Both methods are currently used. The solid graft blood vessels anastomose with the host vessels, and the cell suspension grafts have the potential to become rapidly vascularized by the host *(197)*.

3.4. Early Xenotransplantation

The transplantation of CNS cells across species presents additional challenges, mainly related to immunological rejection, but also to optimal timing and size of the graft. Nevertheless, xenotransplantation was thoroughly investigated for two reasons: first, to explore the potential for future clinical use of xenotransplantation, and second, as a research tool.

Transplants of lateral ganglionic eminence cells from various species into the lesioned rat striatum resulted in anatomical integration and functional reconstitution of the rat neural circuits *(198)*. Human to rat transplants were instrumental in showing the ability of human fetal tissue to grow and reinnervate the host brain, as a step toward clinical application of neural grafting *(198–200)*. The most important single determinant of xenograft success was the use of immunosuppression with cyclosporin A. The risk of rejection increases with the phylogenetic distance between the donor and the recipient. Mouse to rat transplants had a 30% survival rate even without immunosuppression, but virtually all the pig to rat or human to rat transplants were rejected if not treated with cyclosporin A.

Mouse to rat transplants proved valuable in conclusively determining the origin of grafted axona sprouting into the host following transplantation. Lund et al. *(201)* took advantage of a mouse-specific monoclonal antibody to show immature graft axons growing into the host brain. Similar techniques were later used in the primate model and in the human to rat model *(198)*. Although research in neurotransplantation has come a long way since the first experiments in the rodent model, it still provides a necessary first step in the investigation of new approaches and new treatments aimed at neuroprotection or improvement of transplant outcome *(202)*.

4. NEUROPROTECTION AND CELL TRANSPLANTATION FOR NEUROPROTECTION IN HD

Most cell transplant experiments involve replacement strategies for neurons lost to a degenerative disease. However, it is arguably better to prevent the loss in the first place. HD is singularly amenable to neuroprotection strategies. This is the only neurodegenerative disease in which we know the precise genetic defect, and can predict who will come down with the disease.

Since the original studies of Levy-Montalcini, great progress has been made in our knowledge of trophic factors in the brain and their function. In 1987, the basis was laid for the practical use of nerve growth factor (NGF) for neuroprotection in the CNS. Three groups initially showed a dramatic increase in survival of axotomized medial septum cholinergic neurons after intraventricular infusion of NGF *(203,204)*. Following this initial demonstration of in vivo effects of NGF, a plethora of studies have revealed the potent effects of NGF in a variety of animal models.

With regard to HD, a number of trophic factors were tested to determine which could best protect the striatum in the animal models of HD. Glial cell line derived neurotrophic factor (GDNF) and ciliary neurotrophic factor (CNTF) were found to be effective *(205–209)*. Intracerebral administration of CNTF protected the striatal output neurons, whereas brain-derived neurotrophic factor (BDNF) and neurotrophin-3 did not prevent the death of those neurons following intrastriatal injection of QA *(207)*.

CNTF receptor immunoreactivity was identified in globus pallidus, subthalamic nucleus, red nucleus, and substantia nigra pars compacta in *Cebus apella* monkeys, indicating that CNTF supports many areas in the basal ganglia *(210)*. Therefore, CNTF was chosen for extensive further study, with significant neuroprotective effects against excitotoxic injury in both the rodent and the primate models *(208,211)*.

Brain protection by neurotrophic factors appeared to be a highly attractive strategy, but the central question remained that of their delivery. Systemic administration was not effective, and several clinical studies failed to show a significant benefit. To be effective, the factors had to be delivered directly to the neurons that needed them. Several strategies crystallized. First, direct injection into the brain was tried. Although simple, it has the theoretical disadvantage that it does not provide continuous protection, unless a constant infusion pump is used. Infusion of fluid directly into the brain may produce local swelling and mass effect. Nevertheless, ongoing human trials in amyotrophic lateral sclerosis (ALS) utilize constant intrathecal infusion of CNTF with an implanted pump. Another strategy, direct biological delivery by transplantation of cells engineered to produce trophic factors, was made possible by independent advances in genetic therapy *(212)*.

Specific cell replacement therapy for neurodegenerative diseases works best when the original cell types and their connections are reconstituted. However, if cells are used only as delivery systems, no such constraints exist. Therefore, nonneuronal cells and xenografts can be good vehicles, provided the issues of graft rejection are solved. Emerich et al. *(208)* transfected baby hamster kidney cells with an expression vector for human CNTF (hCNTF), and implanted them into the rat striatum using polymeric capsules. The pores in the polymer are large enough to allow CNTF out and nutrients in, but too small to admit immune system cells. Cognitive dysfunction, motor deficits, and striatal degeneration were effectively prevented. Another advantage was the ability to retrieve the capsules for analysis at any time. The encapsulated grafts showed excellent viability and no rejection. Despite concern about loss of transgene expression over time *(213)*, the grafts continued to secrete CNTF thoughout the time of the study *(208,211)*. Remarkable prevention of the striatal lesion by the encapsulated CNTF-producing cells was also shown in the primate model, then shown also in the primate *(211)*. CNTF also prevented retrograde atrophy of layer V neurons in the motor cortex.

Biological delivery of trophic factors provided an unexpected advantage. We have seen that intrastriatal NGF selectively protected cholinergic neurons when infused directly, but did not protect other neurons *(214)*. Indeed, only the cholinergic neurons express the high-affinity trkA NGF throughout life. Fibroblasts transfected with the gene for NGF were implanted into the rat corpus callosum, and decreased the size of subsequent excitotoxic lesions by 80% *(215)*. A similar effect was shown when striatal degeneration was induced by the mitochondrial toxin 3-NPA *(216)*. Even in the absence of any lesion, NGF delivered by encapsulated cells induced hypertrophy of both cholinergic and noncholinergic neuropeptide Y-immunoreactive neurons *(214)*. The mechanism responsible for this phenomenon is unknown, but may involve secretion of a secondary factor, affecting neurons that do not have a trkA NGF receptor. Possibly, the different routes of delivery affect different radical scavenger mechanisms, or differentially affect astrocytes, which then may change the permeability of the blood–brain barrier. Nonneuronal cell transplants for biological delivery of trophic factors are excellent vehicles, very safe if encapsulated, and easily obtainable. They appear to have advantages over direct infusion of trophic factors, and are close to immediate human application.

Nevertheless, the nonneural cells do not naturally secrete neurotrophic factors, and they may lose transgene expression over too long a time course for human clinical usefulness *(213)*. Moreover, the secretion of the factors is tonic, unregulated by host feedback systems, and potentially detrimental. Therefore, the future cell transplants for trophic factor delivery may focus on neurons, astrocytes, or neural stem cells in an attempt to best approximate physiologic conditions. Indeed, Kordower et al. *(217)* transplanted mouse stem cells from transgenic mice in which the glial fibrillary acid protein (GFAP) promoter directs the expression of human NGF. The cells were epidermal growth factor (EGF) responsive. The polymer encapsulated graft of these cells into the rat striatum markedly reduced the size of subsequent QA lesions.

5. HOW DO GRAFTS WORK?

Grafts can have several effects, from detrimental to full integration. They can damage the host brain by producing a pressure effect and by the initial trauma associated with the transplantation procedure. Grafts can diffusely secrete growth factors or neurotransmitters, they can form synapses with host neurons and provide tonic reinnervation of the host brain, and finally, the graft may be fully integrated into the recipient circuitry, with reciprocal innervation. Clearly, there are many beneficial effects of cell grafting that do not require synaptic connection with the host. Even transplants containing only glial cells improved a learning deficit in rats after frontal cortex ablation *(189)*. Intraparenchymal fetal striatal transplants resulted in functional recovery long before any anatomical connections were established *(218)*. Rat allografts tend to show only scant striatonigral efferents, but are a strong stimulant for sprouting or regeneration of adult nerve fibers *(219)*. But, if the effects of the grafts were largely neurohumoral, why transplant cells and go to the trouble of specimen collection and the risk of rejection or infection? Local delivery of medication can be accomplished in other ways. However, reciprocal innervation can provide the essential and precise modulation of graft function, and a more physiologic outcome. Years of intensive research showed a high degree of functional and anatomical integration of the grafts, and accumulated evidence of graft effects in the animal HD model that go beyond tonic secretion of substances. Long growth of axons into the host brain was demonstrated in the mouse to rat transplants *(201)*. Xu et al. *(220)* demonstrated restoration of the corticostriatal projection similar to the normal connections expected in the rat neostriatum *(221)*. Extensive DAergic afferents from the host substantia nigra were also seen *(222)*. Many neuronal types characteristic of the normal striatum were found in the fetal grafts, with the appropriate relation to both afferent cortical and thalamic fibers and efferent neurons projecting to the globus pallidus *(223)*. Wictorin et al. *(224)* demonstrated growth of the grafted fetal neurons along the internal capsule tracts, to reinstate the synaptic input into the host globus pallidus. Evidently, fetal neurons were able to overcome the inhibitory effect of the oligodendroglia on adult axon elongation. Moreover, human fetal neurons transplanted into the rat were shown to grow well and reach substantia nigra even as far as the spinal cord *(199)*. The longer period of growth and the greater length of the human axons in their original environment may have contributed to their success. Ultrastructural evidence was consistent with normal maturation of the transplanted neostriatum, at light and electron microscopic levels, although the density of the synapses was lower in the transplants than in the host tissue *(225,226)*. These anatomical data, together with the functional recovery and biochemical data reviewed earlier, began to suggest that the grafts can at least partially reconstitute the neural circuits in the host brain.Striatal tissue was more effective than peripheral nerve or adrenal medulla grafts in protecting the host striatum against QA toxicity *(187)*. Xu et al. *(227)* found that cortical and thalamic stimulation resulted in synaptic responses within the rat fetal grafts, measured by in vivo intracellular recording. Ultimately, Campbell, from the University of Lund team *(228)*, measured GABA overflow from transplants with intracerebral microdialysis, and demonstrated inhibition of graft GABA release by the host DAergic projection and its stimulation by the host glutamatergic corticostriatal projection.The issue of reconstituting lost circuits is more important in HD than in PD models, because nigral cells are mainly needed to provide tonic secretion of DA, and striatal cells homotopically transplanted into the striatum are expected to reach their normal targets for optimal effect. It appears that the fetal cells can, albeit partially, restore the host striatal function and connectivity.

6. HUMAN TRIALS AND FUTURE DIRECTIONS

Animal experiments in rodents and primates established models of HD; and proved survival, marked anatomical and functional integration of homotopic fetal allografts and xenografts, and functional improvement of HD signs in the animals. Moreover, transplants of cells genetically modified to secrete trophic factors protected striatal neurons from degeneration both in the excitotoxicity and the mitochondrial dysfunction models. Although many questions remained, such as the optimal quantity of cells and number of injections, ways to increase graft survival, and duration of immunosuppression, they could not be answered but with clinical trials. The first transplant strategy for HD was homotopic fetal striatal allografts.

After the reports of success following neural grafts in rodents, and initial successes in the human PD trials, and even before the results of the primate model experiments were in, the first attempts to treat humans with fetal cell allografts were reported from Czechoslovakia and Mexico *(229,230)*. These initial reports claimed some functional improvement in the patients, but no standardized evaluation was done, and they were not followed with long-term studies or larger series. In the West, encouraged by the successes of neural transplantation in PD, researchers were discussing the rationale for intracerebral transplantation in HD. Several assumptions had to be made: (1) The results of animal experiments can be extrapolated to clinical trials. (2) Transplantation of striatal neurons is adequate. (3) The graft will not be susceptible to the effects of the ongoing degenerative process, or will even slow it *(231,232)*. Consensus emerged that recognized the excellent correlation between success of transplantation in animal models of PD and the human clinical trials. The anatomical appearance of the grafts in humans was identical to that in the monkey brain *(233)*, and functional recovery was seen. Therefore, given the disabling nature of HD, and the successes of grafting in animal models, the logical next step was human trials *(231,232)*.

In 1995, enough data accumulated to allow determination of the optimal gestational age and anatomical location of the graft. Lateral ventricular eminence tissue taken before postovulatory wk 22 was likely to give good results *(141)*. At the same time, Quinn et al. *(234)* presented an assessment protocol based on the Core Assessment Program for Intracerebral Transplantation (CAPIT) which is used for PD transplantation studies. They called the protocol CAPIT-HD. The way was paved for systematic clinical trials. The human transplantation experiments are still at an early stage. Nevertheless, initial results are encouraging. Philpot et al. *(235)* report improvement of some cognitive symptoms in three patients 4–6 mo following transplantation.

Current advances in fetal intracerebral allografts notwithstanding, human fetuses are a limited, and at least in the United States, controversial source of donor tissue *(236)*. Therefore, the use of xenografts has tremendous appeal. Potentially, animals such as pigs can be a continual source of appropriate fetal tissue for transplantation. We have seen earlier in this chapter that xenografts can become integrated in the host brain circuitry, and restore lost function. Two major barriers remain to widespread clinical use of xenotransplants—graft rejection and transmission of known and unknown pathogens *(237)*.

The most common method of immunosuppression is the systemic administration of cyclosporin A, but the drug has many side effects. Despite the relative immunological privilege of the brain, xenografts are usually rejected, unless immunosupprssion is used *(238,239)*. Many methods were tried, with some success, to provide immunoprotection without immunosuppression. They ranged from anti-interleukin II *(240)*, through masking of donor MHC class I antigens *(241)*, anti-T-cell antibodies *(242,243)*, to cotransplantation with Sertoli cells *(244)*. Polymer encapsulation eliminates the risk of graft rejection, but is not useful for neural cell transplants where the goal is to provide functional integration *(244,245)*. Graft rejection appears to be a controllable problem. Furthermore, patients may be willing to undergo a course of immunosuppressive therapy, if the graft provides them with a functional improvement. The problem of transmissible pathogens, including retroviruses, remains unsolved at this time *(237)*.

In the future, neural transplantation is most likely to have long-term success and widespread application for human neurodegenerative diseases, when the optimal combination of the crucial variables is found, that is, a safe and plentiful cell source, with good function and integration, absence of graft rejection, adequate trophic support by factor expression or cotransplantation, and protection against oxidative stress.

REFERENCES

1. Stenevi, U., Bjorklund, A., and Svendgaard, N. (1972) Transplantation of central and peripheral monoamine neurons to the adult rat brain: techniques and conditions for survival. *Brain Res.* **114,** 1–20.
2. Ungerstedt, U. (1968) 6-Hydroxy-dopamine induced degeneration of central monoamine neurons. *Eur. J. Pharmacol.* **5,** 107–110.
2a. Tranzer, J. P. and Thoener, H. (1973) Selective destruction of adrenergic nerve terminals by chemical analogues of G-hydroxydopamine. *Experientia* **29,** 314–315.
3. Ungerstedt, U. (1971) Striatal dopamine release after amphetamine or nerve degeneration revealed by rotational behavior. *Acta Physiol. Scand. Suppl.* **367,** 49–68.
4. Ungerstedt, U. (1971) Postsynaptic supersensitivity after 6-hydroxy-dopamine induced

degeneration of the nigro-striatal dopamine system. *Acta Physiol. Scand. Suppl.* **367,** 69–93.

5. Bjorklund, A. and Stenevi, U. (1979) Reconstruction of the nigrostriatal dopamine pathways by intracerebral nigral transplants. *Brain Res.* **177,** 555–560.

6. Perlow, M. J., Freed, W. J., Hoffer, B. J., Seiger, A., Olson, L., and Wyatt, R. J. (1979) Brain grafts reduce motor abnormalities produced by destruction of nigrostriatal dopamine system. *Science* **204,** 643–647.

7. Dunnett, S., Bjorklund, A., Stenevi, U., and Iversen, S. (1981) Behavioral recovery following transplantation of substantia nigra in rats subjected to 6-OHDA lesions of the nigrostriatal pathway. I. Unilateral lesions. *Brain Res.* **215,** 147–161.

8. Bjorklund, A., Stenevi, U., Dunnett, S. B., and Iversen, S. D. (1981) Functional reactivation of the deafferented neostriatum by nigral transplants. *Nature* **289,** 497–499.

9. Dunnett, S. B., Bjorklund, A., Stenevi, U., and Iversen, S. D. (1981) Grafts of embryonic substantia nigra reinnervating the ventrolateral striatum ameliorate sensorimotor impairments and akinesia in rats with 6-OHDA lesions of the nigrostriatal pathway. *Brain Res.* **229,** 209–217.

10. Bjorklund, A., Schmidt, R. H., and Stenevi, U. (1980) Functional reinnervation of the neostriatum in the adult rat by use of intraparenchymal grafting of dissociated cell suspensions from the substantia nigra. *Cell Tissue Res.* **212,** 39–45.

11. Brundin, P., Isacson, O, and Bjorklund, A. (1985) Monitoring of cell viability in suspensions of embryonic CNS tissue and its use as a criterion for intracerebral graft survival. *Brain Res.* **331,** 251–259.

12. Bjorklund, A., Stenevi, U., Schmidt, R. H., Dunnett, S. B., and Gage, F. H. (1983) Intracerebral grafting of neuronal cell suspensions and survival and growth of nigral cell suspensions implanted in different brain sites. *Acta Physiol. Scand. Suppl.* **522,** 9–18.

13. Schmidt, R. H., Bjorklund, A., Stenevi, U., Dunnett, S. B., and Gage, F. H. (1983) Intracerebral grafting of neuronal cell suspensions and activity of intrastriatal nigral suspension implants as assessed by measurements of dopamine synthesis and metabolism. *Acta Physiol. Scand. Suppl.* **522,** 19–28.

14. Dunnett, S. B., Bjorklund, A., Schmidt, R. H., Stenevi, U., and Iversen, S. D. (1983) Intracerebral grafting of neuronal cell suspensions and behavioural recovery in rats with unilateral 6-OHDA lesions following implantation of nigral cell suspensions in different forebrain sites. *Acta Physiol. Scand. Suppl.* **522,** 29–37.

15. Mahalik, T. J., Finger, T. E., Stromberg, I., and Olson, L. (1985) Substantia nigra transplants into denervated striatum of the rat: ultrastructure of graft and host interconnections. *J. Comp. Neurol.* **240,** 60–70.

16. Langston, J. W., Ballard, P., Tetrud, J. W., and Irwin, I. (1983) Chronic parkinsonism in humans due to a product of meperidine-analog synthesis. *Science* **219,** 979–980.

17. Burns, R. S., Chiueh, C. C., Markey, S. P., Ebert, M. H., Jacobowitz, D. M., and Kopin, I. J. (1983) A primate model of parkinsonism: selective destruction of dopaminergic neurons in the pars compacta of the substantia nigra by N-methyl-4-phenyl-1,2,3,6-tetrahydropyridine. *Proc. Natl. Acad. Sci. USA* **80,** 4546–4550.

18. Bankiewicz, K. S., Oldfield, E. H., Chiueh, C. C., Doppman, J. L., Jacobowitz, D. M., and Kopin, I. J. (1986) Hemiparkinsonism in monkeys after unilateral internal carotid artery infusion of 1-methyl-4-phenyl-1,2,3,6-tetrahydropyridine (MPTP). *Life Sci.* **39,** 7–16.

19. Redmond, D. E., Sladek, J. R Jr, Roth, R. H., Collier, T. J., Elsworth, J. D., Deutch, A. Y., and Haber, S. (1986) Fetal neuronal grafts in monkeys given methylphenyltetrahydropyridine. *Lancet* **1,** 1125–1127.

20. Sladek, J. R., Redmond, D. E. Jr, Collier, T. J., Haber, S. N., Elsworth, J. D., Deutch, A. Y., and Roth, R. H. (1987) Transplantation of fetal dopamine neurons in primate brain reverses MPTP induced parkinsonism. *Prog. Brain Res.* **71,** 309–323.

21. Sladek, J. R., Redmond, D. E. Jr, Collier, T. J., Blount, J. P., Elsworth, J. D., Taylor, J. R., and Roth, R. H. (1988) Fetal dopamine neural grafts: extended reversal of methylphenyl-tetrahydropyridine induced parkinsonism in monkeys. *Prog. Brain Res.* **78,** 497–506.

22. Taylor, J. R., Elsworth, J. D., Roth, R. H., Sladek, J. R. Jr, Collier, T. J., and Redmond, D. E. Jr. (1991) Grafting of fetal substantia nigra to striatum reverses behavioral deficits induced by MPTP in primates: a comparison with other types of grafts as controls. *Exp. Brain Res.* **85,** 335–348.

23. Bankiewicz, K. S., Plunkett, R. J., Jacobowitz, D. M., Porrino, L., di Porzio, U., London, W. T., Kopin, I. J., and Oldfield, E. H. (1990) The effect of fetal mesencephalon implants on primate MPTP-induced Parkinsonism. *J. Neurosurg.* **72,** 231–244.

24. Sladek, J. R., Elsworth, J. D., Roth, R. H., Evans, L. E., Collier, T. J., Cooper, S. J., Taylor, J. R., and Redmond, D. E. Jr. (1993) Fetal dopamine cell survival after transplantation is dramatically improved at a critical donor gestational age in nonhuman primates. *Exp. Neurol.* **122,** 16–27.

25. Bankiewicz, K. S., Plunkett, R. J., Jacobowitz, D. M., Kopin, I. J., and Oldfield, E. H. (1991) Fetal nondopaminergic neural implants in parkinsonian primates. *J. Neurosurg.* **74,** 97–104.

26. Levitt, P. and Rakic, P. (1982) The time of genesis, embryonic origin and differentiation of the brain stem monoamine neurons in the rhesus monkey. *Dev. Brain Res.* **4,** 35–57.

27. Collier, T. J., Redmond, D. E. Jr, Sladek, C. D., Gallagher, M. J., Roth, R. H., and Sladek, J. R. Jr. (1987) Intracerebral grafting and culture of cryopreserved primate dopamine neurons. *Brain Res.* **436,** 363–366.

28. Redmond, D. E. Jr, Naftolin, F., Collier, T. J., Leranth, C., Robbins, R. J., Sladek, C. D., Roth, R. H., and Sladek, J. R. Jr. (1988) Cryopreservation, culture, and transplantation of human fetal mesencephalic tissue into monkeys. *Science* **242,** 768–771.

29. Freed, W. J., Morihisa, J. M., Spoor, E., Hoffer, B. J., Olson, L., Seiger, A., and Wyatt, R. J. (1981) Transplanted adrenal chromaffin cells in rat brain reduce lesion-induced rotational behavior. *Nature* **292,** 351–352.

30. Thompson, W. (19XX) Successful brain grafting. *NY Med. J.* **51,** 701–702.

31. Borges, L. F. (1988) Historical development of neural transplantation. *Appl. Neurophysiol.* **51,** 265–277.

32. Lindvall, O. (1991) Prospects of transplantation in human neurodegenerative disease. *Trends Neurosci.* **14,** 376–388.

33. Madrazo, I., Franco-Bourland, R., Aguilera, M., Ostrosky-Solis, F., Cuevas, C., Castrejon, H., Magallon, E., and Madrazo, M. (1991) Development of human neural transplantation. *Neurosurgery* **29,** 165–177.

34. Boyer, K. L. and Bakay, R. A. E. (1995) The history, theory, and present status of brain transplantation. *Neurosurg. Clin. North Am.* **6,** 113–125.

35. Jones, D. G. and Harris, R.J.(1996) Neural transplantation. *N. Zeal. Med. J.* **109,** 369–371.

36. Junn, F. C. (1998) Transplantation to improve function in the diseased central nervous system, in *Text Book of Stereotactic and Functional Neurosurgery* (Gildenberg, P. L., Tasker, R. R., and Franklin, P. O., eds.), McGraw-Hill, New York, pp. 1115–1122.

37. Gage, F. H. and Fisher, L. J. (1991) Intracerebral grafting: a tool for the neurobiologist. *Neuron* **6,** 1–12.

38. Emerich, D. F., Borlongan, C. V., Freeman, T. B., Cahill, D. W., and Sanberg, P. R. (1994) Cell transplantation for central nervous system disorders. *Crit. Rev. Neurobiol.* **8,** 125–162.

39. Sanberg, P. R., Borlongan, C. V., Wictorin, K., and Isacson, O. (1998) Fetal-tissue transplantation for Huntington's disease: preclinical studies, in *Cell Transplantation for Neurological Disorders: Toward Reconstruction of the Human Central Nervous System* (Freeman, T. B. and Widner, H., eds.), Humana Press, Totowa, NJ, pp. 77–94.

40. Kopyov, O. V., Jacques, S., Kurth, M., Philpott, L., Lee, A., Patterson, M., Duma, C., Lieberman, A., and Eagle, K. S. (1998) Fetal transplantation for Huntington's disease: clinical studies, in *Cell Transplantation for Neurological Disorders: Toward Reconstruction of the Human Central Nervous System* (Freeman, T. B. and Widner, H., eds.), Humana Press, Totowa, NJ, pp. 95–134.

41. Backlund, E. O., Granberg, P. O., Hamberger, B., Knutsson, E., Martensson, A., Sedvall, G., Seiger, A., and Olson, L. (1985) Transplantation of adrenal medullary tissue to striatum in Parkinsonism. *J. Neurosurg.* **62**, 169–173.

42. Madrazo, I., Drucker-Colin, R., Diaz, V., Martinez-Mata, J., Torres, C., and Becerril, J. J. (1987) Open microsurgical autograft of adrenal medulla to the right caudate nucleus in two patients with intractable Parkinson's disease. *N. Engl. J. Med.* **316**, 831–834.

43. Bakay, R. A. (1993) Neurotransplantation: a clinical update. *Acta Neurochir. Suppl.* **58**, 8–16.

44. Bakay, R. A., Allen, G. S., Apuzzo, M., Borges, L. F., Bullard, D. E., Ojemann, G. A., Oldfield, E. H., Penn, R., Purvis, J. T., and Tindall, G. T. (1990) Preliminary report on adrenal medullary grafting from the American Association of Neurological Surgeons Graft Project. *Prog. Brain Res.* **82**, 603–610.

45. Goetz, C. G., Stebbins, G. T. 3d, Klawans, H. L., Koller, W. C., Grossman, R. G., Bakay, R. A., and Penn, R. D. (1991) United Parkinson Foundation Neurotransplantation Registry on Adrenal Medullary Transplants: presurgical, and 1- and 2-year follow-up. *Neurology* **41**, 1719–1722.

46. Kordower, J. H., Cochran, E., Penn, R. D., and Goetz, C. G. (1991) Putative chromaffin cell survival and enhanced host-derived TH-fiber innervation following a functional adrenal medulla autograft for Parkinson's disease. *Ann. Neurol.* **29**, 405–412.

47. Watts, R. L., Subramanian, T., Freeman, A., Goetz, C. G., Penn, R. D., Stebbins, G. T., Kordower, J. H., and Bakay, R. A. (1997) Effect of stereotaxic intrastriatal cografts of autologous adrenal medulla and peripheral nerve in Parkinson's disease: two-year follow-up study. *Exp. Neurol.* **147**, 510–517.

48. Jiang, N., Jiang, C., Tang, Z., Zhang, F., Li, S., and Jiang, D. (1987) Human foetal brain transplant trials in the treatment of Parkinsonism. *Acta Acad. Med. Shanghai* **14**, 1.

49. Lindvall, O., Rehncrona, S., Brundin, P., Gustavii, B., Astedt, B., Widner, H., Lindholm, T., Bjorklund, A., Leenders, K. L., Rothwell, J. C., et al. (1989) Human fetal dopamine neurons grafted into the striatum in two patients with severe Parkinson's disease: a detailed account of methodology and a 6-months follow-up. *Arch. Neurol.* **46**, 615–631.

50. Hitchcock, E. R., Clough. C., Hughes, R., and Kenny, B. (1988) Embryos and Parkinson's disease. *Lancet* **1**, 1274.

51. Hitchcock, E. R., Clough, C. G., Hughes, R. C., and Kenny, B. G. (1989) Transplantation in Parkinson's disease: stereotactic implantation of adrenal medulla and foetal mesencephalon. *Acta Neurochir. Suppl.*(Wien), **46**, 48–50.

52. Madrazo, I., Leon, V., Torres, C., Aguilera, M. C., Varela, G., Alvarez, F., Fraga, A., Drucker-Colin, R., Ostrosky, F., Skurovich, M., et al. (1988) Transplantation of fetal substantia nigra and adrenal medulla to the caudate nucleus in two patients with Parkinson's disease. *N. Engl. J. Med.* **318**, 51.

53. Molina, H., Quinones, R., Alvarez, L., Galarraga, J., Piedra, J., Suarez, C., Rachid, M., Garcia, J. C., Perry, T. L., Santana, A., Carmenate, H., Macias, R., Torres, O., Rojas, M. J. Cordova, F., and Munoz, J. L. (1991) Transplantation of human fetal mesencephalic tissue in caudate nucleus as treatment for Parkinson's disease: the Cuban experience, in *Intracerebral Transplantation in Movement Disorders* (Lindvall, O. Bjorklund, A., and Widner, H., eds.), Elsevier, Amsterdam, pp. 99–110.

54. Lopez-Lozano, J. J., Bravo, G., Brera, B., Uria, J., Dargallo, J., Salmean, J., Insausti, J., Cerrolaza, J., and CPH Neural Transplantation Group. Can an analogy be drawn between the clinical evolution of Parkinson's patients who undergo autoimplantation of adrenal

medulla and those of fetal ventral mesencephalon transplant recipients? in *Intracerebral Transplantation in Movement Disorders* (Lindvall, O., Bjorklund, A., and Widner, H., eds.), Elsevier, Amsterdam, pp. 87–98.

55. Freed, C. R., Breeze, R. E., Rosenberg, N. L., Schneck, S. A., Wells, T. H., Barrett, J. N., Grafton, S. T., Huang, S. C., Eidelberg, D., and Rottenberg, D. A. (1990) Transplantation of human fetal dopamine cells for Parkinson's disease: results at 1 year. *Arch. Neurol.* **47,** 505–512.

56. Freed, C. R., Breeze, R. E., Rosenberg, N. L., Schneck, S. A., Wells, T. H., Barrett, J. N., Grafton, S. T., Mazziotta, J. C., Eidelberg, D., and Rottenberg, D. A. (1990) Therapeutic effects of human fetal dopamine cells transplanted in a patient with Parkinson's disease. *Prog. Brain Res.* **82,** 715–721.

57. Redmond, D. E. Jr, Leranth, C., Spencer, D. D., Robbins, R., Vollmer, T., Kim, J. H., Roth, R. H., Dwork, A. J., and Naftolin, F. (1990) Fetal neural graft survival. *Lancet* **336,** 820–822.

58. Subrt, O., Tichy, M., and Vlayaka, V. (1990) Grafting of fetal dopamine neurons in Parkinson's disease, in *Eric K. Fernstrom Symposium: Intracerebral Transplantation in Movement Disorders: Experimental Basis and Clincial Experiences*, Lund, Sweden, June 20–22.

59. Dymecki, J., et al. (1990) Human fetal dopamine cell transplantation in Parkinson's disease, in *Eric K. Fernstrom Symposium: Intracerebral Transplantation in Movement Disorders: Experimental Basis and Clinical Experiences*, Lund, Sweden, June 20–22.

60. Lindvall, O., Rehncrona, S., Brundin, P., Gustavii, B., Astedt, B., Widner, H., Lindholm, T., Bjorklund, A., Leenders, K. L., Rothwell, J. C., et al. (1990) Neural transplantation in Parkinson's disease: The Swedish experience. *Prog. Brain Res.* **82,** 729–734.

61. Lindvall, O., Widner, H., Rehncrona, S., Brundin, P., Odin, P., Gustavii, B., Frackowiak, R., Leenders, K. L., Sawle, G., Rothwell, J. C., et al. (1992) Transplantation of fetal dopamine neurons in Parkinson's disease: one-year clinical and neurophysiological observations in two patients with putaminal implants. *Ann. Neurol.* **31,** 155–165.

62. Wenning, G. K., Odin, P., Morrish, P., Rehncrona, S., Widner, H., Brundin, P., Rothwell, J. C., Brown, R., Gustavii, B., Hagell, P., Jahanshahi, M., Sawle, G., Bjorklund, A., Brooks, D. J., Marsden, C. D., Quinn, N. P., and Lindvall, O. (1997) Short- and long-term survival and function of unilateral intrastriatal dopaminergic grafts in Parkinson's disease. *Ann. Neurol.* **42,** 95–107.

63. Lindvall, O., Sawle, G., Widner, H., Rothwell, J. C., Bjorklund, A., Brooks, D., Brundin, P., Frackowiak, R., Marsden, C. D., Odin, P., et al. (1994) Evidence for long-term survival and function of dopaminergic grafts in progressive Parkinson's disease. *Ann. Neurol.* **35,** 172–180.

64. Widner, H. (1998) The Lund Transplant Program for Parkinson's Disease and patients with MPTP-induced Parkinsonism, in *Cell Transplantation for Neurological Disorders: Toward Reconstruction of the Human Central Nervous System* (Widner, F., ed.), Humana Press, Totowa, NJ, pp. 1–17.

65. Sawle, G. V., Bloomfield, P. M., Bjorklund, A., Brooks, D. J., Brundin, P., Leenders, K. L., Lindvall, O., Marsden, C. D., Rehncrona, S., Widner, H., et al. (1992) Transplantation of fetal dopamine neurons in Parkinson's disease: PET[^{18}F]-6-L-fluorodopa studies in two patients with putaminal implants. *Ann. Neurol.* **31,** 166–173.

66. Leenders, K. L., Salmon, E. P., Tyrrell, P., Perani, D., Brooks, D. J., Sager, H., Jones, T., Marsden, C. D., and Frackowiak, R. S. (1990) The nigrostriatal dopaminergic system assessed in vivo by positron emission tomography in healthy volunteer subjects and patients with Parkinson's disease. *Arch. Neurol.* **47,** 1290–1298.

67. Widner, H., Tetrud, J., Rehncrona, S., Snow, B., Brundin, P., Gustavii, B., Bjorklund, A., Lindvall, O., and Langston, J. W. (1992) Bilateral fetal mesencephalic grafting in two

patients with Parkinsonism induced by 1-methyl-4-phenyl-1,2,3,6-tetrahydropyridine (MPTP). *N. Engl. J. Med.* **327,** 1556–1563.

68. Langston, J. W., Ballard, P., Tetrud, J. W., and Irwin, I. (1983) Chronic Parkinsonism in humans due to a product of meperidine-analog synthesis. *Science* **219,** 979–980.

69. Burns, R. S., Chiueh, C. C., Markey, S. P., Ebert, M. H., Jacobowitz, D. M., and Kopin, I. J. (1983) A primate model of Parkinsonism: selective destruction of dopaminergic neurons in pars compacta of the substantia nigra by *N*-methyl-4-phenyl-1,2,3,6-tetrahydropyridine. *Proc. Natl. Acad. Sci. USA* **80,** 4546–4550.

70. Langston, J. W., Forno, L. S., Rebert, C. S., and Irwin, I. (1984) Selective nigral toxicity after systemic administration of 1-methyl-4-phenyl-1,2,3,6-tetrahydropyridine (MPTP): seven cases. *Brain Res.* **292,** 390–394.

71. Ballard, P. A., Tetrud, J. W., and Langston, J. W. (1985) Permanent human Parkinsonism due to 1-methyl-4-phenyl-1,2,3,6-tetrahydropyridine (MPTP): seven cases. *Neurology* **35,** 949–956.

72. Bakay, R. A. E., Barrow, D. L., Fiandaca, M. S., Iuvone, P. M., Schiff, A., and Collins, D. C. (1987) Biochemical and behavioral correction of MPTP Parkinson-like syndrome by fetal cell transplantation. *Ann. NY Acad. Sci.*, **495,** 623–640.

73. Freed, C. R., Richards, J. B., Sabol, K. E., and Reite, M. L. (1988) Fetal substantia nigra tranplants lead to dopamine cell replacement and behavioral improvement in bonnet monkeys with MPTP induced Parkinsonism, in *Pharmacology and Functional Regulation of Dopaminergic Neurons* (Beart, P. M., Woodruff, G. N., and Jackson, D. M., eds.), Macmillan Press, London, pp. 353–360.

74. Fine, A., Hunt, S. P., Oertel, W. H., Nomoto, M., Chong, P. N., Bond, A., Waters, C., Temlett, J. A., Annett, L., Dunnett, S., et al. (1988) Transplantation of embryonic marmoset dopaminergic neurons to the corpus striatum of marmosets rendered Parkinsonian by 1-methyl-4-phenyl-1,2,3,6-tetrahydropyridine. *Prog. Brain Res.* **78,** 479–489.

75. Elsworth, J. D., Redmond, D. E. Jr, Sladek, J. R. Jr., Deutch, A. Y., Collier, T. J., and Roth, R. H. (1989) Reversal of MPTP-induced Parkinsonism in primates by fetal dopamine cell transplants, in *Function and Dysfunction in the Basal Ganglia* (Franks, A. J., et al., eds.), Manchester University Press, Manchester, pp. 161–180.

76. Bankiewicz, K. S., Plunkett, R. J., Jacobowitz, D. M., Porrino, L., di Porzio, U., London, W. T., Kopin, I. J., and Oldfield, E. H. (1990) The effect of fetal mesencephalon implants on primate MPTP-induced Parkinsonism: histochemical and behavioral studies. *J. Neurosurg.* **72,** 231–244.

77. Dunnett, S. B. and Annett, L. E. (1991) Nigral transplants in primate models of Parkinsonism, in *Intracerebral Transplantation in Movement Disorders* (Lindvall, O., Bjorklund, A., and Widner, H., eds.), Elsevier, Amsterdam, pp. 27–51.

78. Taylor, J. R., Elsworth, J. D., Roth, R. H., Sladek, J. R. Jr, Collier, T. J., and Redmond, D. E. Jr. (1991) Grafting of fetal substantia nigra to striatum reverses behavioral deficits induced by MPTP in primates: a comparison with other types of grafts as controls. *Exp. Brain Res.*, **85,** 335–348.

79. Redmond, D. E. and Sladek, J. R. J. (1986) Fetal neuronal grafts in monkeys given methylphenyltetrahydropyridine. *Lancet* **1,** 1125–1127.

80. Hitchcock, E. R., Kenny, B. G., Clough, C. G., Hughes, R. C., Henderson, B. T., and Detta, A. (1990) Stereotactic implantation of foetal mesencephalon (STIM): the UK experience. *Prog. Brain Res.* **82,** 723–728.

81. Henderson, B. T. H., Clough, C. G., Hughes, R. C., Hitchcock, E. R., and Kenny, B. G. (1991) Implantation of human fetal ventral mesencephalon to the right caudate nucleus in advanced Parkinson's disease. *Arch. Neurol.* **48,** 822–827.

82. Freed, C. R., Breeze, R. E., Rosenberg, N. L., Schneck, S. A., Kriek, E., Qi, J. X., Lone, T., Zhang, Y. B., Snyder, J. A., Wells, T. H., et al. (1992) Survival of implanted fetal dopam-

ine cells and neurological improvement 12 to 46 months after transplantation of Parkinson's disease. *N. Engl. J. Med.* **327,** 1549–1555.

83. Freeman, T. B., Sanberg, P. R., Nauert, G. M., Boss, B. D., Spector, D., Olanow, C. W., and Kordower, J. H. (1995) The influence of donor age on the survival of solid and suspension intraparenchymal human embryonic nigral grafts. *Cell Transplant.* **4,** 141–154.

84. Bjorklund, A., Schmidt, R. H., and Stenevi, U. (1980) Functional reinnervation of the neostriatum in the adult rat by use of intraparenchymal grafting of dissociated cell suspensions from the substantia nigra. *Cell Tissue Res.* **212,** 39–45.

85. Brundin, P., Nilsson, O. G., Strecker, R. E., Lindvall, O., Astedt, B., and Bjorklund, A. (1986) Behavioural effects of human fetal dopamine neurons grafted in a rat model of Parkinson's disease. *Exp. Brain Res.* **65,** 235–240.

86. Stromberg, I., Bygdeman, M., Goldstein, M., Seiger, A., and Olson, L. (1986) Human fetal substantia nigra grafted to the dopamine-derived striatum of immunosuppressed rats: evidence for functional reinnervation. *Neurosci. Lett.* **71,** 271–276.

87. Brundin, P., Strecker, R. E., Widner, H., Clarke, D. J., Nilsson, O. G., Astedt, B., Lindvall, O., and Bjorklund, A. (1988) Human fetal dopamine neurons grafted in a rat model of Parkinson's disease: immunological aspects, spontaneous and drug-induced behaviour, and dopamine release. *Exp. Brain Res.* **70,** 192–208.

88. Spencer, D. D., Robbins, R. J., Naftolin, F., Marek, K. L., Vollmer, T., Leranth, C., Roth, R. H., Price, L. H., Gjedde, A., Bunney, B. S., et al. (1992) Unilateral transplantation of human fetal mesencephalic tissue into the caudate nucleus of patients with Parkinson's disease. *N. Engl. J. Med.* **327,** 1541–1548.

89. Redmond, D. E. Jr, Robbins, R. J., Naftolin, F., Marek, K. L., Vollmer, T. L., Leranth, C., Roth, R. H., Price, L. H., Gjedde, A., Bunney, B. S., Sass, K. J., Elsworth, J. D., Kier, E. L., Matuch, R., Hoffer, P. B., Gulanski, B. I., Serrano, C., and Spencer, D. D. (1992) Cellular replacement of dopamine deficit in Parkinson's disease using human fetal mesencephalic tissue: preliminary results in four patients, in *Molecular and Cellular Approaches to the Treatment of Brain Disease* (Waxman, S. G., ed.), Raven Press, New York, pp. 325–359.

90. Redmond, D. E. Jr, Roth, R. H., Spencer, D. D., Naftolin, F., Leranth, C., Robbins, R. J., Marek, K. L., Elsworth, J. D., Taylor, J. R., Sass, K. J., et al. (1993) Neural transplantation for neurodegenerative diseases: past, present, and future. *Ann. NY Acad. Sci.* **695,** 258–266.

91. Collier, T. J., Gallagher, M. J., and Sladek, C. D. (1993) Cryopreservation and storage of embryonic rat mesencephalic dopamine neurons for one year: comparison to fresh tissue in culture and neural grafts. *Brain Res.* **623,** 249–256.

92. Braak, H. and Braak, E. (1986) Nuclear configuration and neuronal types of the nucleus niger in the brain of the human adult. *Hum. Neurobiol.* **5,** 71–82.

93. Markham, C. M., Rand, R. W., Jacques, D. B., Diamond, S. G., Kopyov, O. V., and Snow, B. (1994) Transplantation of fetal mesencephalic tissue in Parkinson's patients. *Stereotact. Funct. Neurosurg.* **62,** 134–240.

94. Pechanski, M., Defer, G., N'Guyen, J. P., Ricolfi, F., Monfort, J. C., Remy, P., Geny, C., Samson, Y., Hantraye, P., Jeny, R., et al. (1994) Bilateral motor improvement and alteration of L-Dopa effect in two patients with Parkinson's disease following intrastriatal transplantation of foetal ventral mesencephalon. *Brain* **117,** 487–499.

95. Defer, G., Geny, C., Ricolfi, F., Fenelon, G., Monfort, J. C., Remy, P., Villafane, G., Jeny, R., Samson, Y., Keravel, Y., Gaston, A., Degos, J. D., Peschanski, M., Cesaro, P., and Nguyen, J. P. (1996) Long-term outcome of unilaterally transplanted Parkinsonian patients: I. Clinical approach. *Brain* **119,** 41–50.

96. Remy, P., Samson, Y., Hantraye, P., Fontaine, A., Defer, G., Mangin, J. F., Fenelon, G., Geny, C., Ricolfi, F., Frouin, V., et al. (1995) Clinical correlates of [^{18}F]-fluorodopa uptake in five grafted Parkinsonian patients. *Ann. Neurol.* **38,** 580–588.

97. Kunzle, H. (1975) Bilateral projections from precentral motor cortex to the putamen and other parts of the basal ganglia: an autoradiographic study in *Macaca fascicularis*. *Brain Behav. Evol.* **88,** 195–209.

98. Alexander, G. E. and Delong, M. R. (1985) Microstimulation of the primate neostriatum. II. Somatotopic organization of striatal microexcitable zones and their relation to neuronal response properties. *J. Neurophysiol.* **53,** 1417–1430.

99. Alexander, G. E., Crutcher, M. D., and DeLong, M. R. (1990) Basal ganglia-thalamocortical circuits: parallel substrates for motor, oculomotor, 'prefrontal' and 'limbic' functions. *Prog. Brain Res.* **85,** 119–146.

100. Bernheimer, H., Birkmayer, W., Hornykiewicz, O., Jellinger, K., and Seitelberger, F. (1973) Brain dopamine and the syndromes of Parkinson and Huntington: clinical, morphological and neurochemical correlations. *J. Neurosci.* **20,** 415–455.

101. Nyberg, P., Nordberg, A., and Webster, P. (1983) Dopaminergic deficiency is more pronounced in putamen than in nucleus caudatus in Parkinson's disease. *Neurochem. Pathol.* **1,** 193–202.

102. Kish, S. J., Shannak, K., and Hornykiewicz, O. (1988) Uneven pattern of dopamine loss in the striatum of patients with idiopathic Parkinson's disease: pathophysiologic and clinical implications. *N. Engl. J. Med.* **318,** 876–880.

103. Leenders, K. L., Salmon, E. P., Tyrrell, P., Perani, D., Brooks, D. J., Sager, H., Jones, T., Marsden, C. D., and Frackowiak, R. S. (1990) The nigrostriatal dopaminergic system assessed in vivo by positron emission tomography in healthy volunteer subjects and patients with Parkinson's disease. *Arch. Neurol.* **47,** 1290–1297.

104. Szabo, J. (1980) Organization of the ascending striatal afferents in monkeys. *J. Comp. Neurol.* **189,** 307-321.

105. Madrazo, I., Franco-Bourland, R., Ostrosky-Solis, F., Aguilera, M., Cuevas, C., Zamorano, C., Morelos, A., Magallon, E., and Guizar-Sahagun, G. (1990) Fetal homotransplants (ventral mesencephalon and adrenal tissue) to the striatum of Parkinsonian subjects. *Arch. Neurol.* **47,** 1281–1285.

106. Madrazo, I., Franco-Bourland, R., Aguilera, M., Madrazo, M., and Dector, T. (1991) Open microsurgery for human brain transplantation. Experience in the treatment of Parkinson's disease, in *Intracerebral Transplantation in Movement Disorders* (Lindvall, O., Bjorklund, A., and Widner, H., eds.), Elseviers, Amsterdam, pp. 153–162.

107. Lopez-Lozano, J. J., Bravo, G., Brera, B., Dargallo, J., Salmean, J., Uria, J., Insausti, J., and Millan, I. (1995) Long-term follow-up in 10 Parkinson's disease patients subjected to fetal brain grafting into a cavity in the caudate nucleus: the Clinica Puerta de Hierro experience. *Transplant. Proc.* **27,** 1395–1400.

108. Lopez-Lozano, J. J., Bravo, G., Brera, B., Millan, I., Dargallo, J., Salmean, J., Uria, J., and Insausti, J. (1997) Long-term improvement in patients with severe Parkinson's disease after implantation of fetal ventral mesencephalic tissue in a cavity of the caudate nucleus: 5-year follow up in 10 patients. *J. Neurosurg.* **86,** 931–942.

109. Freeman, T. B., Olanow, C. W., Hauser, R. A., Nauert, G. M., Smith, D. A., Borlongan, C. V., Sanberg, P. R., Holt, D. A., Kordower, J. H., Vingerhoets, F. J., et al. (1995) Bilateral fetal nigral transplantation into the postcommissural putamen in Parkinson's disease. *Ann. Neurol.* **38,** 379–388.

110. Kordower, J. H., Freeman, T. B., Snow, B. J., Vingerhoets, F. J., Mufson, E. J., Sanberg, P. R., Hauser, R. A., Smith, D. A., Nauert, G. M., Perl, D. P., et al (1995) Neuropathological evidence of graft survival and striatal reinnervation after the transplantation of fetal mesencephalic tissue in a patient with Parkinson's disease. *N. Engl. J. Med.* **332,** 1118–1124.

111. Kordower, J. H., Rosenstein, J. M., Collier, T. J., Burke, M. A., Chen, E. Y., Li, J. M., Martel, L., Levey, A. E., Mufson, E. J., Freeman, T. B., and Olanow, C. W. (1996) Func-

tional fetal nigral grafts in a patient with Parkinson's disease: chemoanatomic, ultrastructural and metabolic studies. *J. Comp. Neurol.* **370,** 203–230.

112. Staley, J. K., Mash, D. C., Basile, M., Weiner, W., Levey, A. L., and Kordower, J. H. (1996) Functional fetal nigral grafts in a patient with Parkinson's disease: effects on dopamine synaptic markers. *Soc. Neurosci. Abstr.* **22,** 318.

113. Kordower, J. H., Rosenstein, J. M., Collier, T. J., Chen, E.-Y., Li, J. M., Mufson, E. J., Sanberg, P., Freeman, T. B., and Olanow, C. W. (1996) Functional fetal nigral grafts in a second patient with Parkinson's disease: a postmortem analysis. *Soc. Neurosci. Abstr.* **22,** 318.

114. Kordower, J. H., Hanbury, R., and Bankiewicz, K. S. (1998) Neuropathology of dopaminergic transplants in patients with Parkinson's disease, in *Cell Transplantation for Neurological Disorders: Toward Reconstruction of the Human Central Nervous System* (Freeman, T. B. and Widner, H, eds.), Humana Press, Totowa, NJ, pp. 51–75.

115. Olanow, C. W., Kordower, J. H., and Freeman, T. B. (1996) Fetal nigral transplantation as a therapy for Parkinson's disease. *Trends Neurosci.* **19,** 102–109.

116. Olanow, C. W., Freeman, T. B., and Kordower, J. H. (1997) Transplantation strategies for Parkinson's disease, in *Movement Disorders: Neurologic Principles and Practice* (Watts, R. L. and Koller, W. C., eds.), McGraw-Hill, New York, pp. 221–236.

117. Freeman, T. B., Spence, M. S., Boss, B. D., Spector, D. H., Strecker, R. E., Olanow, C. W., and Kordower, J. H. (1991) Development of dopaminergic neurons in the human substantia nigra. *Exp. Neurol.* **113,** 344–353.

118. Pakkenberg, B., Moller, A., Gundersen, H. J., Mouritzen, D. A., and Pakkenberg, H. (1991) The absolute number of nerve cells in substantia nigra normal subjects and in patients with Parkinson's disease estimated with an unbiased stereological method. *J. Neurol. Neurosurg. Psychiatry* **54,** 30–33.

119. Brundin, P., Isacson, O., and Bjorklund, A. (1985) Monitoring of cell viability in suspensions of embryonic CNS tissue and its use as a criterion for intracerebral graft survival. *Brain Res.* **331,** 251–259.

120. Frodl, E. M., Duan, W. M., Sauer, H., Kupsch, A., and Brundin, P. (1991) Human embryonic dopamine neurons xenografted to rat: effect of cryopreservation and varying regional source of donor cells on transplant survival and morphology. *Brain Res.* **647,** 286–298.

121. Bayer, S. A. (1984) Neurogenesis in the rat neostriatum. *Int. J. Dev. Neurosci.* **2,** 163–175.

122. Dunnett, S. B., Bjorklund, A., Stenevi, U., and Iversen, S. D. (1981) Behavioral recovery following transplantation of substantia nigra in rats subjected to 6-OHDA lesions of the nigrostriatal pathway. *Brain Res.* **215,** 147–161.

123. Dunnett, S. B., Bjorklund, A., Schmidt, R. H., Stenevi, U., and Iversen, S. D. (1983) Intracerebral grafting of neuronal cell suspensions IV. Behavioral recovery in rats with unilateral 6-OHDA lesions following implantation of nigral cell suspensions in different forebrain sites. *Acta Physiol. Scand.* **522,** 29–37.

124. Horellou, P., Brundin, P., Kalen, P., Mallet, J., and Bjorklund, A. (1990) In vivo release of DOPA and dopamine from genetically engineered cells grafted to the denervated rat striatum. *Neuron* **5,** 393–402.

125. Sendeldeck, S. L. and Urquhart, J. (1985) Spatial distribution of dopamine, methotrexate, and antipyrine during continuous intracerebral microperfusion. *Brain Res.* **328,** 251–258.

126. Sladek, J. R. Jr, Redmond, D. E. Jr, Collier, T. J., Blount, J. P., Elsworth, J. D., Taylor, J. R., and Roth, R. H. (1988) Fetal dopamine neural grafts: extended reversal of methylphenyltetrahydropyridine-induced Parkinsonism in primates: transplantation into the mammalian CNS. *Prog. Brain Res.* **78,** 497–506.

127. Fiandaca, M. S., Kordower, J. H., Hansen, J. T., Jiao, S. S., and Gash, D. M. (1988) Adrenal medullary autografts into the basal ganglia of cebus monkeys: injury induced regeneration. *Exp. Neurol.* **102,** 76–91.
128. Ansari, A. A., Mayne, A., Freed, C. R., Breeze, R. E., Schneck, S. A., O'Brien, C. F., Kriek, E. H., Zhang, Y. B., Mazziotta, J. C., Hutchinson, M., et al. (1995) Lack of a detectable systemic humoral/cellular allogeneic response in human and nonhuman primate recipients of embryonic mesencephalic allografts for the therapy of Parkinson's disease. *Transplant. Proc.* **27,** 1401–1405.
129. Hanbury, R., Styren, S. D., Sanberg, P. R., Freeman, T. B., Olanow, C. W., and Kordower, J. H. (1996) Functional fetal nigral grafts in two Patients with Parkinson's disease: immunological studies. *Soc. Neurosci. Abstr.* **22,** 318.
130. Kordower, J. H., Styren, S., Clarke, M., DeKosky, S. T., Olanow, C. W., and Freeman, T. B. (1997) Fetal grafting for Parkinson's disease: expression of immune markers in two patients with functional fetal nigral implants. *Cell Transplant.* **6,** 213–219.
131. Lopez-Lozano, J. J., Bravo, G., Brera, B., Dargallo, J., Salmean, J., Uria, J., Insausti, J., Martinez, R., Sanchez, P., de la Torre, C., and Moreno, R. (1997) Regression of Parkinsonian fetal ventral mesencephalon grafts upon withdrawal of cyclosporine A immunosuppression. *Transplant. Proc.* **29,** 977–980.
132. Adams, R. and Victor, M. (1989) *Principles of Neurology,* McGraw-Hill Inc., New York, NY, pp. 61–64 and 932–935.
133. Cote, L. and Crutcher, M. D. (1991) The basal ganglia, in *Principles of Neural Science,* 3rd edit. (Kandel, E. R., Schwartz, J. H., and Jessel, T. M., eds.), Appleton & Lang, New York, pp. 647–659.
134. Reiner, A., Albin, R. L., Anderson, K. D., D'Amato, C. J., Penney, J. B., and Young, A. B. (1988) Differential loss of striatal projection neurons in Huntington's disease. *Proc. Natl. Acad. Sci. USA* **85,** 5733–5737.
135. Albin, R. L., Reiner, A., Anderson, K. D., Dure, L. S. 4th, Handelin, B., Balfour, R., Whetsell, W. O. Jr, Penney, J. B., and Young, A. B., (1992) Preferential loss of striatoexternal pallidal projection neurons in presymptomatic Huntington's disease. *Ann. Neurol.* **31,** 425–430.
136. Beal, M. F., Ferrante, R. J., Swartz, K. J., and Kowall, N. W. (1991) Chronic quinolinic acid lesions in rats closely resemble Huntington's disease. *J. Neurosci.* **11,** 1649–1659.
137. Reynolds, G. P., Pearson, S. J., and Heathfield, K. W. G. (1990) Dementia in Huntington's disease is associated with neurochemical deficits in the caudate nucleus, not the cerebral cortex. *Neurosci. Lett.* **113,** 95–100.
138. Leblhuber, F., Hoell, K., Reisecker, F., Gebetsberger, B., Puehringer, W., Trenkler, E., and Deisenhammer, E. (1989) Single photon emission computed tomography in Huntington's chorea. *Psychiatr. Res.* **29,** 337–339.
139. The Huntington's Disease Collaborative Research Group (1993) A novel gene containing a trinucleotide repeat that is expanded and unstable on Huntington's disease chromosomes. *Cell* **72,** 971–983.
140. Freed, C. R., Breeze, R. E., Rosenberg, N. L., Breeze, R. E., Rosenberg, N. L., Schneck, S. A., Kriek, E., Qi, J. X., Lone, T., Zhang, Y. B., Snyder, J. A., Wells, T. H., et al. (1992) Survival of implanted fetal dopamine cells and neurologic improvement 12 to 46 months after transplantation for Parkinson's disease. *N. Engl. J. Med.* **327,** 1549–1555.
141. Freeman, T. B., Sanberg, P. R., and Isacson, O. (1995) Development of the human striatum: implications for fetal striatal implantation in the treatment of Huntington's disease. *Cell Transplant.* **4,** 539–545.
142. Lindvall, O. (1991) Prospects of transplantation in human neurodegenerative diseases. *Trends Neurosci.* **14,** 376–384.

143. Dunnett, S. B. and Bjorklund, A. (1987) Mechanisms of function of neural grafts in the adult mammalian brain. *J. Exp. Biol.* **132,** 265–289.

144. Bjorklund, A., Lindvall, O., Isacson, O., Brundin, P., Wictorin, K., Strecker, R. E., Clarke, D. J., and Dunnett, S. B. (1987) Mechanisms of action of intracerebral neural implants: studies on nigral and striatal grafts to the lesioned striatum. *TINS* **10,** 509–515.

145. Coyle, J. T. and Schwarcz, R. (1976) Lesion of striatal neurones with kainic acid provides a model for Huntington's chorea. *Nature* **263,** 244–246.

146. McGeer, E. G. and McGeer, P. L. (1976) Duplication of biochemical changes of Huntington's chorea by intrastriatal injections of glutamic and kainic acids. *Nature* **263,** 517–518.

147. Sanberg, P. R., Lehmann, J., and Fibiger, H. C. (1978) Impaired learning and memory after kainic acid lesions of the striatum: a behavioral model for Huntington's disease. *Brain Res.* **149,** 546–551.

148. Mason, S. T., Sanberg, P. R., and Fibiger, H. C. (1978) Kainic acid lesions of the striatum dissociate amphetamine and apomorphine stereotypy: similarities to Huntington's chorea. *Science* **201,** 352–355.

149. Schwarcz, R., Whetsell, W. O., and Mangano, R. M. (1983) Quinolinic acid: an endogenous metabolite that produces axon sparing lesions in rat brain. *Science* **219,** 316–318.

150. Beal, M. F., Kowall, N. W., Ellison, D. W., Mazurek, M. F., Swartz, K. J., and Martin, J. B. (1986) Replication of the neurochemical characteristics of Huntington's disease by quinolinic acid. *Nature* **321,** 168–171.

151. Foster, A. C., Whetsell, W. O. Jr, Bird, E. D., and Schwarcz, R. (1985) Quinolinic acid phosphoribosyltransferase in human and rat brain: activity in Huntington's disease and in quinolinate-lesioned rat striatum. *Brain Res.* **336,** 207–214.

152. Du, F., Okuno, E., Whetsell, W. O. Jr, Kohler, C., and Schwarcz, R. (1991) Immunohistochemical localization of quinolinic acid phosphoribosyltransferase in the human neostriatum. *Neuroscience* **42,** 397–406.

153. Ellison, D. W., Beal, M. F., Mazurek, M. F., Malloy, J. R., Bird, E. D., and Martin, J. B. (1987) Amino acid neurotransmitter abnormalities in Huntington's disease and the quinolinic acid animal model of Huntington's disease. *Brain* **110,** 1657–1663.

154. Davies, S. W. and Roberts, P. J. (1987) No evidence for preservation of somatostatin-containing neurons after intrastriatal injections of quinolinic acid. *Nature* **327,** 326–329.

155. Davies, S. W. and Roberts, P. J. (1988) Model of Huntington's disease. Letter to *Science* **241,** 474–475.

156. Sanberg, P. R. and Johnston, G. A. (1981) Glutamate and Huntington's disease. *Med. J. Aust.* **2,** 460–465.

156a. Hahn, J. S., Aizenman, E., and Lipton, S. A. (1988) Central mammalian neurons normally resistant to glutamate toxicity are made sensitive by elevated extracellular Ca^{2+}: toxicity is blocked by the *N*-methyl-D-aspartate antagonist MK801. *Proc. Natl. Acad. Sci. USA* **85,** 6556–6560.

157. Huang, Q., Zhou, D., Sapp, E., Aizawa, H., Ge, P., Bird, E. D., Vonsattel, J. P., and DiFiglia, M. (1995) Quinolinic acid-induced increases in calbindin D28k immunoreactivity in rat striatal neurons in vivo and in vitro mimic the pattern seen in Huntington's disease. *Neuroscience* **65,** 397–407.

158. Heyes, M. P., Swartz, K. J., Markey, S. P., and Beal M. F. (1991) Regional brain and cerebrospinal fluid quinolinic acid concentrations in Huntington's disease. *Neurosci. Lett.* **122,** 265–269.

159. Lohr, J. B. (1991) Oxygen radicals and neuropsychiatric illness. Some speculations. *Arch. Gen. Psychiatry* **48,** 1097–1106.

160. Beal, M. F. (1992) Does impairment of energy metabolism result in excitotoxic neuronal death in neurodegenerative illnesses? *Ann. Neurol.* **31,** 119–130.

161. Koutouzis, T. K., Borlongan, C. V., Scorcia, T., Creese, I., Cahill, D.W., Freeman, T. B.,

and Sanberg, P. R. (1994) Systemic 3-nitropropionic acid: long term effects on locomotor behavior. *Brain Res.* **646,** 242–246.

162. Borlongan, C. V., Koutouzis, T. K., Freeman, T. B., Hauser, R. A., Cahill, D. W., and Sanberg, P. R. (1997) Hyperactivity and hypoactivity in a rat model of Huntington's disease: the systemic 3-nitropropionic acid model. *Brain Res. Protoc.* **1,** 253–257.

163. Guyot, M. C., Hantraye, P., Dolan, R., Palfi. S., Maziere, M., and Brouillet, E. (1997) Quantifiable bradykinesia, gait abnormalities and Huntington's disease-like striatal lesions in rats chronically treated with 3-nitropropionic acid. *Neuroscience* **79,** 45–56.

164. Borlongan, C. V., Koutouzis, T. K., Freeman, T. B., Cahill, D. W., and Sanberg, P. R. (1995) Behavioral pathology induced by repeated systemic injections of 3-nitropropionic acid mimics the motoric symptoms of Huntington's disease. *Brain Res.* **697,** 254–257.

165. Borlongan, C. V., Nishino, H., and Sanberg, P. R. (1997) Systemic, but not intraparenchymal, administration of 3-nitropropionic acid mimics the neuropathology of Huntington's disease: a speculative explanation. *Neurosci. Res.* **28,** 185–189.

166. Bordelon, Y. M., Chesselet, M. F., Nelson, D., Welsh, F., and Erecinska, M. (1997) Energetic dysfunction in quinolinic acid-lesioned rat striatum. *J. Neurochem.* **69,** 1629–1639.

167. Cooper, J. M. and Schapira, A. H. (1997) Mitochondrial dysfunction in neurodegeneration. *J. Bioenerg. Biomembr.* **29,** 175–183.

168. Davis, R. E., Miller, S., Herrnstadt, C., Ghosh. S. S., Fahy, E., Shinobu, L. A., Galasko, D., Thal, L. J., Beal, M. F., Howell, N., and Parker, W. D., Jr. (1997) Mutations in mitochondrial cytochrome c oxidase genes segregate with late-onset Alzheimer disease. *Proc. Natl. Acad. Sci. USA* **94,** 4526–4531.

169. Brouillet, E., Hantraye, P., Ferrante, R. J., Dolan, R., Leroy-Willig, A., Kowall, N. W., and Beal, M. F. (1995) Chronic mitochondrial energy impairment produces selective striatal degeneration and abnormal choreiform movements in primates. *Proc. Natl. Acad. Sci. USA* **92,** 7105–7109.

170. Palfi, S., Ferrante, R. J., Brouillet, E., Beal, M. F., Dolan, R., Guyot, M. C., Peschanski, M., and Hantraye, P. (1996) Chronic 3-nitropropionic acid treatment in baboons replicates the cognitive and motor deficits of Huntington's disease. *J. Neurosci.* **16,** 3019–3025.

171. Kanazawa, I., Kimura, M., Murata, M., Tanaka, Y., and Cho, F. (1990) Choreic movements in the macaque monkey induced by kainic acid lesions of the striatum combined with L-DOPA. *Brain* **113,** 509–535.

172. Isacson, O., Hantraye, P., Maziere, M., Sofroniew, M. V., and Riche, D. (1990) Apomorphine-induced dyskinesias after excitotoxic caudate–putamen lesions and the effects of neural transplantation in non-human primates. *Prog. Brain Res.* **82,** 523–533.

173. Hantraye, P., Riche, D., Maziere, M., and Isacson, O. (1992) Intrastriatal transplantation of cross-species fetal striatal cells reduces abnormal movements in a primate model of Huntington disease. *Proc. Natl. Acad. Sci. USA* **89,** 4187–4191.

174. Storey, E., Cipolloni, P. B., Ferrante, R. J., Kowall, N. W., and Beal, M. F. (1994) Movement disorder following excitotoxin lesions in primates. *NeuroReport* **5,** 1259–1261.

175. Schumacher, J. M., Hantraye, P., Brownell, A. L., Riche, D., Madras, B. K., Davenport, P. D., Maziere, M., Elmaleh, D. R., Brownell, G. L., and Isacson, O. (1992) A primate model of Huntington's disease: functional neural transplantation and CT guided stereotactic procedures. *Cell Transplant.* **1,** 313–322.

176. Ferrante, R. J., Kowall, N. W., Cipolloni, P. B., Storey, E., and Beal, M. F. (1993) Excitotoxin lesions in primates as a model for Huntington's disease: histopathologic and neurochemical characterization. *Exp. Neurol.* **119,** 46–71.

177. DeLong, M. (1990) Primate models of movement disorders of basal ganglia origin. *TINS* **13,** 281–285.

178. Deckel, A. W., Robinson, R. G., Coyle, J. T., and Sanberg, P. R. (1983) Reversal of long-term locomotor abnormalities in the kainic acid model of Huntington's disease by day 18 fetal striatal implants. *Eur. J. Pharmacol.* **93,** 287–288.

179. Deckel, A. W., Moran, T. H., Coyle, J. T., Sanberg, P. R., and Robinson, R. G. (1986) Anatomical predictors of behavioral recovery following fetal striatal transplants. *Brain Res.* **365,** 249–258.

180. Deckel, A. W. and Robinson, R. G. (1987) Receptor characteristics and behavioral consequences of kainic acid lesions and fetal transplants of the striatum. *Ann. NY Acad. Sci.* **495,** 556–580.

181. Isacson, O., Dunnett, S. B., and Bjorklund, A. (1986) Graft-induced behavioral recovery in an animal model of Huntington disease. *Proc. Natl. Acad. Sci. USA* **83,** 2728–2732.

182. Norman, A. B., Calderon, S. F., Giordano, M., and Sanberg, P. R. (1988) Striatal tissue transplants attenuate apomorphine-induced rotational behavior in rats with unilateral kainic acid lesions. *Neuropharmacology* **27,** 333–336.

183. Sanberg, P. R., Calderon, S. F., Garver, D. L., and Norman, A. B. (1987) Brain tissue transplants in an animal model of Huntington's disease. *Psychopharmacol. Bull.* **23,** 476–478.

184. Isacson, O., Pritzel, M., Dawbarn, D., Brundin, P., Kelly, P. A., Wiklund, L., Emson, P. C., Gage, F. H., Dunnett, S. B., and Bjorklund, A. (1987) Striatal neural transplants in the ibotenic acid-lesioned rat neostriatum. Cellular and functional aspects. *Ann. NY Acad. Sci.* **495,** 537–554.

185. Brundin, P., Isacson, O., and Bjorklund, A. (1985) Monitoring of cell viability in suspensions of embryonic CNS tissue and its use as a criterion for intracerebral graft survival. *Brain Res.* **331,** 251–259.

186. Wictorin, K., Brundin, P., Sauer, H., Lindvall, O., and Bjorklund, A. (1992) Long distance directed axonal growth from human dopaminergic mesencephalic neuroblasts implanted along the nigrostriatal pathways in 6-hydroxydopamine lesioned rats. *J. Comp. Neurol.* **323,** 475–494.

187. Pearlman, S. H., Levivier, M., Collier, T. J., Sladek, J. R. Jr, and Gash, D. M. (1991) Striatal implants protect the host striatum against quinolinic acid toxicity. *Brain Res.* **84,** 303–310.

188. Wictorin, K. (1992) Anatomy and connectivity of intrastriatal striatal transplants. *Prog. Neurobiol.* **38,** 611–639.

189. Nieto-Sampedro, M., Kesslak, J. P., Gibbs, R., and Cotman, C. W. (1987) Effects of conditioning lesions on transplant survival, connectivity, and function. *Ann. NY Acad. Sci.* **495,** 108–118.

190. Lindsay, R. M., Emmett, C., Raisman, G. R., and Seeley, P. J. (1987) Application of tissue culture and cell-marking techniques to the study of neural transplants. *Ann. NY Acad. Sci.* **495,** 35–52.

191. Kromer, L. F. and Cornbrooks, C. J. (1987) Identification of trophic factors and transplanted cellular environments that promote axonal regeneration. *Ann. NY Acad. Sci.* **495,** 207–225.

192. Whitaker-Azmitia, P. M., Ramirez, A., Noreika, L., Gannon, P. J., and Azmitia, E. C. (1987) Onset and duration of astrocytic response to cells transplanted into the adult mammalian brain. *Ann. NY Acad. Sci.* **495,** 10–23.

193. Isacson, O., Brundin, P., Kelly, P. A., Gage, F. H., and Bjorklund, A. (1984) Functional neuronal replacement by grafted striatal neurones in the ibotenic acid-lesioned rat striatum. *Nature* **311,** 458–460.

194. Gerfen, C. R. (1984) The neostriatal mosaic: compartmentalization of corticostriatal input and striatonigral output systems. *Nature* **311,** 461–462.

195. Pakzaban, P., Deacon, T. W., Burns, L. H., and Isacson, O. (1993) Increased proportion of acetylcholinesterase-rich zones and improved integration in host striatum of fetal grafts derived from the lateral but not the medial ganglionic eminence. *Exp. Brain Res.* **97,** 13–22.

196. Nakao, N., Grasbon-Frodl, E. M., Widner, H., and Brundin, P. (1996) DARPP-32-rich zones in grafts of lateral ganglionic eminence govern the extent of functional recovery in skilled paw reaching in an animal model of Huntington's disease. *Neuroscience* **74,** 959–970.

197. Broadwell, R. D., Charlton, H. M., Ebert, P., Hickey, W. F., Villegas, J. C., and Wolf, A. L. (1990) Angiogenesis and the blood–brain barrier in solid and dissociated cell grafts within the CNS. *Prog. Brain Res.* **82,** 95–101.

198. Pakzaban, P. and Isacson, O. (1994) Neural xenotransplantation: reconstruction of neuronal circuitry across species barriers. *Neuroscience* **62,** 989–1001.

199. Wictorin, K., Brundin, P., Gustavii, B., Lindvall, O., and Bjorklund, A. (1990) Reformation of long axon pathways in adult rat central nervous system by human forebrain neuroblasts. *Nature* **347,** 556–558.

200. Pundt, L. L., Kondoh, T., Konrad, J. A., and Low, W. C. (1996) Transplantation of the human fetal striatum into a rodent model of Huntington's disease ameliorates locomotor deficits. *Neurosci. Res.* **24,** 415–420.

201. Lund, R. D., Chang, F.-L. F., Hankin, M. H., and Lagenaur, C. F. (1985) Use of a species-specific antibody for demonstrating mouse neurons transplanted to rat brains. *Neurosci. Lett.* **61,** 221–226.

202. Goto, S., Yamada, K., Yoshikawa, M., Okamura, A., and Ushio, Y. (1997) GABA receptor agonist promotes reformation of the striatonigral pathway by transplant derived from fetal striatal primordia in the lesioned striatum. *Exp. Neurol.* **147,** 503–509.

203. Hefti, F. (1994) Neurotrophic factor therapy for nervous system degenerative diseases. *J. Neurobiol.* **25,** 1418–1435.

204. Kromer, L. F. (1987) Nerve growth factor treatment after brain injury prevents neuronal death. *Science* **235,** 214–216.

205. Araujo, D. M., Hilt, D. C. (1997) Glial cell line-derived neurotrophic factor attenuates the excitotoxin-induced behavioral and neurochemical deficits in a rodent model of Huntington's disease. *Neuroscience* **81,** 1099–1110.

206. Perez-Navarro, E., Arenas, E., Reiriz, J., Calvo, N., and Alberch, J. (1996) Glial cell line-derived neurotrophic factor protects striatal calbindin immunoreactive neurons from excitotoxic damage. *Neuroscience* **75,** 345–352.

207. Anderson, K. D., Panayotatos, N., Corcoran, T. L., Lindsay, R. M., and Wiegand, S. J. (1996) Ciliary neurotrophic factor protects striatal output neurons in an animal model of Huntington's disease. *Proc. Natl. Acad. Sci. USA* **93,** 7346–7351

208. Emerich, D. F., Lindner, M. D., Winn, S. R., et al. (1996) Implants of encapsulated human CNTF-producing fibroblasts prevent behavioral deficits and striatal degeneration in a rodent model of Huntington's disease. *J. Neurosci.* **16,** 5168–5181.

209. Emerich, D. F., Cain, C. K., Greco, C., Chen, E. Y., Frydel, B. R., and Kordower, J. H. (1997) Cellular delivery of human CNTF prevents motor and cognitive dysfunction in a rodent model of Huntington's disease. *Cell Transplant.* **6,** 249–266.

210. Kordower, J. H., Chu, Y., and Maclennan, A. J. (1997) Ciliary neurotrophic factor receptor α-immunoreactivity in the monkey central nervous system. *J. Comp. Neurol.* **377,** 365–380.

211. Emerich, D. F., Winn, S. R., Hantraye, P. M., Peschanski, M., Chen, E. Y., Chu, Y., McDermott, P., Baetge, E. E., and Kordower, J. H. (1997) Protective effect of encapsulated cells producing neurotrophic factor CNTF in a monkey model of Huntington's disease. *Nature* **386,** 395–399.

212. Gage, F. H., Wolff, J. A., Rosenberg, M. B., Xu, L., Yee, J. K., Shults, C., and Friedmann, T. (1987) Grafting genetically modified cells to the brain: possibilities for the future. *Neuroscience* **23,** 795–807.

213. Snyder, E. Y. and Senut, M. C. (1997) The use of nonneuronal cells for gene delivery. *Neurobiol. Dis.* **4,** 69–102.

214. Kordower, J. H., Chen, E.-Y., Mufson, E. J., Winn, S. R., and Emerich, D. F. (1996) Intrastriatal implants of polymer encapsulated cells genetically modified to secrete human nerve growth factor: trophic effects upon cholinergic and noncholinergic striatal neurons. *Neuroscience* **72,** 63–77.

215. Frim, D. M., Short, M. P., Rosenberg, W. S., Simpson, J., Breakefield, X. O., and Isacson, O. (1993) Local protective effects of nerve growth factor-secreting fibroblasts against excitotoxic lesions in the rat striatum. *J. Neurosurg.* **78,** 267–273.

216. Frim, D. M., Simpson, J., Uhler, T. A., Short, M. P., Bossi, S. R., Breakefield, X. O., and Isacson, O. (1993) Striatal degeneration induced by mitochondrial blockade is prevented by biologically delivered NGF. *J. Neurosci. Res.* **35,** 452–458.

217. Kordower, J. H., Chen, E.-Y., Winkler, C., Fricker, R., Charles, V., Messing, A., Mufson, E. J., Wong, S. C., Rosenstein, J. M., Bjorklund, A., Emerich, D. F., Hammang, J., and Carpenter, M. K. (1997) Grafts of EGF-responsive neural stem cells derived from GFAP-hNGF transgenic mice: trophic and tropic effects in a rodent model of Huntington's disease. *J. Comp. Neurol.* **387,** 96–113.

218. Giordano, M., Hagenmeyer-Houser, S. H., and Sanberg, P. R. (1988) Intraparenchymal fetal striatal transplants and recovery in kainic acid lesioned rats. *Brain Res.* **446,** 183–188.

219. Zhou, F. C. and Buchwald, N. (1989) Connectivities of the striatal grafts in adult rat brain: a rich afference and scant striatonigral efference. *Brain Res.* **504,** 15–30.

220. Xu, Z. C., Wilson, C. J., and Emson, P. C. (1989) Restoration of the corticostriatal projection in rat neostriatal grafts: electron microscopic analysis. *Neuroscience* **29,** 539–550.

221. Dube, L., Smith, A. D., Bolam, J. P. (1988) Identification of synaptic terminals of thalamic or cortical origin in contact with distinct medium-size spiny neurons in the rat neostriatum. *J. Comp. Neurol.* **267,** 455–471.

222. Pritzel, M., Isacson, O., Brundin, P., Wiklund, L., and Bjorklund, A. (1986) Afferent and efferent connections of striatal grafts implanted into the ibotenic acid lesioned neostriatum in adult rats. *Exp. Brain Res.* **65,** 112–126.

223. Roberts, R. C. and DiFiglia, M. (1988) Localization of immunoreactive GABA and enkephalin and NADPH-diaphorase-positive neurons in fetal striatal grafts in the quinolinic-acid-lesioned rat neostriatum. *J. Comp. Neurol.* **274,** 406–421.

224. Wictorin, K., Clarke, D. J., Bolam, J. P., and Bjorklund, A. (1990) Fetal striatal neurons grafted into the ibotenate lesioned adult striatum: efferent projections and synaptic contacts in the host globus pallidus. *Neuroscience* **37,** 301–315.

225. Clarke, D. J. and Dunnett, S. B. (1990) Ultrasructural organization within intrastriatal striatal grafts. *Brain Res.* **82,** 407–415.

226. DiFiglia, M., Schiff, L., and Deckel, A. W. (1988) Neuronal organization of fetal striatal grafts in kainate- and sham-lesioned rat caudate nucleus: light- and electron-microscopic observations. *J. Neurosci.* **8,** 1112–1130.

227. Xu, Z. C., Wilson, C. J., and Emson, P. C. (1991) Synaptic potentials evoked in spiny neurons in rat neostriatal grafts by cortical and thalamic stimulation. *J. Neurophysiol.* **65,** 477–493.

228. Campbell, K., Kalen, P., Wictorin, K., Lundberg, C., Mandel, R. J., and Bjorklund, A. (1993) Characterization of GABA release from intrastriatal striatal transplants: dependence on host-derived afferents. *Neuroscience* **53,** 403–415.

229. Sramka, M., Rattaj, M., Molina, H., Vojtassak, J., Belan, V., and Ruzicky, E. (1992) Stereotactic technique and pathophysiological mechanisms of neurotransplantation in Huntington's chorea. *Stereotact. Funct. Neurosurg.* **58,** 79–83.

230. Madrazo, I., Cuevas, C., Castrejon, H., Guizar-Sahagun, G., Franco-Bourland, R. E., Ostrosky-Solis, F., Aguilera, M., and Magallon, E. (1993) The first homotopic fetal

homograft of the striatum in the treatment of Huntington's disease. *Gac. Med. Mex.* **129,** 109–117.

231. Shannon, K. M. and Kordower. J. H. (1996) Neural transplantation for Huntington's disease; experimental rationale and recommendations for clinical trials. *Cell Transplant.* **5,** 339–352.
232. Peschanski, M., Cesaro, P., and Hantraye, P. (1995) Rationale for intrastriatal grafting of striatal neuroblasts in patients with Huntington's disease. *Neuroscience* **68,** 273–285.
233. Kordower, J. H, Freeman, T. B., Snow, B. J., Vingerhoets, F. J. G., Mufson, E. J., Sanberg, P. R., Hauser, R. A., Smith, D. A., Nauert, G. M., Perl, D. P., and Olanow, C. W. (1995) Neuropathological evidence of graft survival and striatal reinnervation after the transplantation of fetal mesencephalic tissue in a patient with Parkinson's disease. *N. Engl. J. Med.* **332,** 1118–1124.
234. Quinn, N., Brown, R., Craufurd, D., Goldman, S., Hodges, J., Kieburtz, K., Lindvall, O., MacMillan, J., and Roos, R. (1996) Core assessment program for intracerebral transplantation in Huntington's disease (CAPIT-HD). *Move. Disord.* **11,** 143–150.
235. Philpott, L. M., Kopyov, O. V., Lee, A. J., Jacques, S., Duma, C. M., Caine, S., Yang, M., and Eagle, K. S. (1997) Neuropsychological functioning following fetal striatal transplantation in Huntington's chorea: three case presentations. *Cell Transplant.* **6,** 203–212.
236. Annas, G. J. and Elias, S. (1989) The politics of transplantation of human fetal tissue. *N. Engl. J. Med.* **320,** 1079–1082.
237. Isacson, O. and Breakefield, X. O. (1997) Benefits and risks of hosting animal cells in the human brain. *Nature Med.* **3,** 964–970.
238. Hara, K., Uchida, K., Fukunaga, A., Toya, S., and Kawase, T. (1997) Implantation of xenogeneic transgenic neural plate tissues into parkinsonian rat brain. *Cell Transplant.* **6,** 515–519.
239. Duan, W. M., Widner, H., and Brundin, P. (1995) Temporal pattern of host responses against intrastriatal grafts of syngeneic, allogeneic or xenogeneic embryonic neuronal tissue in rats. *Exp. Brain Res.* **104,** 227–242.
240. Honey, C. R., Clarke, D. J., Dalman, M. J., and Charlton, H. M. (1990) Human neural graft function in rats treated with anti interleukin II receptor antibody. *NeuroReport* **1,** 247–249.
241. Pakzaban, P., Deacon, T. W., Burns, L. H., Dinsmore, J., and Isacson, O. (1995) A novel model of immunosuppression of neural xenotransplants: masking of donor major histocompatibility complex class I enhances transplant survival in the central nervous system. *Neuroscience* **65,** 983–996.
242. Wood, M. J., Sloan, D. J., Wood, K. J., and Charlton, H. M. (1996) Indefinite survival of neural xenografts induced with anti-CD4 monoclonal antibodies. *Neuroscience* **70,** 775–789.
243. Okura, Y., Tanaka, R., Ono, K., Yoshida, S., Tanuma, N., and Matsumoto, Y. (1997) Treatment of rat hemiparkinson model with xenogeneic neural transplantation: tolerance induction by anti-T-cell antibodies. *J. Neurosci. Res.* **48,** 385–396.
244. Borlongan, C. V., Stahl, C. E., Cameron, D. F., Saporta, S., Freeman, T. B., Cahill, D. W., and Sanberg, P. R. (1996) CNS immunological modulation of neural graft rejection and survival. *Neurol. Res.* **18,** 297–304.
245. Emerich, D. F., Winn, S. R., Harper, J., Hammang, J. P., Baetge, E. E., and Kordower, J. H. (1994) Implants of polymer-encapsulated human NGF-secreting cells in the nonhuman primate: rescue and sprouting of degenerating cholinergic basal forebrain neurons. *J. Comp. Neurol.* **349,** 148–164.

24

Future Prospects of Gene Therapy for Treating CNS Diseases

Daniel A. Peterson, Jasodhara Ray, and Fred H. Gage

1. INTRODUCTION

In the last several decades, enormous gains have been made in our understanding of the pathogenesis of various human diseases. By understanding the biochemical, molecular, and genetic mechanisms through which a disease state manifests itself, it becomes possible to develop rational strategies for therapeutic intervention. In addition to providing insight into disease mechanisms, studies using molecular biology have provided a tool, gene therapy, through which it is possible to induce a cell to make a specific protein that could interfere with the disease process. Gene therapy provides for a local, regulated delivery of the therapeutic gene product, thereby avoiding some of the systemic side effects of drug therapies. Gene therapy has been investigated with some success in several organ systems and this review discusses efforts to use gene therapy in central nervous system (CNS) diseases. Considerations for using gene therapy and the benefits of various delivery approaches are addressed. The potential efficacy and drawbacks of gene therapy for CNS disease are discussed in light of advances in preclinical research. Finally, the future prospects for the use of gene therapy in the clinic are discussed in terms of safety, ethical considerations, and public opinion.

2. CONSIDERATIONS FOR THE USE OF GENE THERAPY

To apply gene therapy, it is necessary to have a detailed understanding of the molecular and biochemical sequences in the cell biology of both the healthy and the disease state, an understanding of the anatomical framework of the tissue involved, and an understanding of the physiological balance necessary for the health of this tissue and of its function within the context of the whole organism. As a result, the potential utility of gene therapy rests upon a foundation of accumulated knowledge. In the future, additional applications for gene therapy may well be suggested by advances in basic research on healthy and diseased tissue.

Up to the present time, pharmacotherapy has been the primary medical approach for intervention in disease processes. Although this approach has been quite successful and will continue to be the therapeutic approach of choice in most instances, there are

From: Central Nervous System Diseases
Edited by: D. F. Emerich, R. L. Dean, III, and P. R. Sanberg © Humana Press Inc., Totowa, NJ

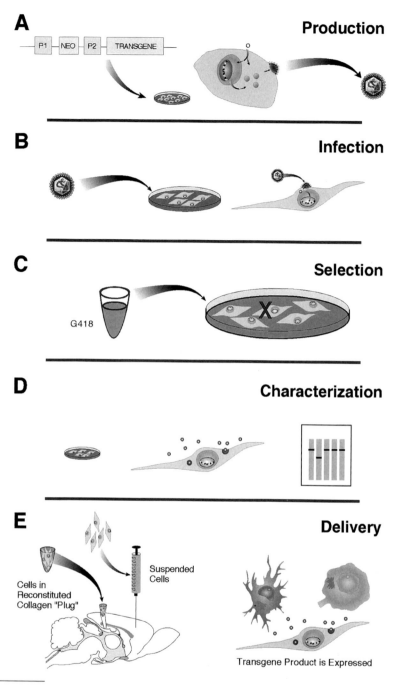

Fig. 1. *(see next page)* Ex vivo gene therapy begins with virus production. **(A)** A construct containing the therapeutic transgene and a selectable antibiotic resistance gene (neomycin resistance in this example) with an appropriate promotor for each (P1 and P2, respectively) is cloned into a viral backbone. The resulting plasmid is introduced by chemical or physical methods into cultured cells that either constituitively express viral packaging proteins or are cotransfected to express these proteins. These cells, called producer cells, generate and package virus which is released into the culture media. **(B)** Media from the producer cell cultures is used to infect the primary cells that will be used for gene delivery. **(C)** To ensure that cells harvested for gene delivery are infected to produce the transgene, cultures are exposed to an antibiotic (G418 in this example) that is lethal (X) to cells not expressing resistance (the neo gene product). **(D)** Following selection, infected cells are characterized by Southern or Western

a number of considerations that make the search for alternative therapies viable. In regard to CNS disease, a major consideration in drug delivery is the size of the compound and its ability to cross the blood–brain barrier. Even if the drug crosses the barrier, it may be necessary to deliver high concentrations of the substance systemically to achieve a suitable dosage in the brain. Depending upon the half-life of the drug, repeated doses may be required to achieve the desired tissue level. Because the drug cannot be directed only to the tissue of interest with systemic administration, all tissues are exposed to the same biological level of the drug achieved in the tissue of interest. As a result, there can be considerable side effects due to both the absolute levels of the drug administered and to the response of other tissues. Finally, with drug delivery there is the danger of sensitization, where increasing levels of the drug are needed to maintain the same level of tissue response. A well known example of these difficulties in the CNS is Parkinson's disease (PD), in which oral administration of the dopamine precursor, levodopa (L-Dopa), which crosses the blood–brain barrier, is given to counteract the reduced striatal dopamine levels that follow the death of substantia nigra dopaminergic neurons. The remarkable recovery of PD patients treated with L-Dopa raised hopes that this drug delivery might be a cure for PD. However, although initially effective, this pharmacotherapy becomes limited after a few years of treatment and also produces side effects *(1)*.

To overcome some of the disadvantages of systemic administration, strategies have been developed to deliver neuroprotective genes directly into the brain in a localized fashion. Experimental gene therapy approaches reviewed in this chapter primarily address somatic gene therapy, where the cell receiving the new transferred gene, called a transgene, is a somatic (nonreproductive) cell. The use of germ line gene therapy is addressed briefly in the last section of this chapter. There are two strategies for transgene delivery: ex vivo and in vivo gene therapy. With ex vivo gene therapy, cultured cells are genetically manipulated to express the corrective transgene and then implanted into an organism (Fig. 1). Depending on the types of cells used, implanted cells reside at the site of implantation and function as "mini-pumps" to deliver the transgene products locally at a even rate or migrate and deliver gene products over a wide area. Alternatively, therapeutic genes can be directly delivered into the organism *in situ* (in vivo gene therapy; Fig. 2). The choice of a particular method depends upon a number of considerations, including the type of disease and how much of and how long the recombinant protein must be delivered.

The identification of a monogenetic, or single gene, defect presents the most straightforward case for the use of gene therapy. Here, the delivery of a single gene and the expression of its product can correct the defect and eliminate the disease. Examples of such diseases where gene therapy approaches are being investigated include the hemophilias *(2,3)* and cystic fibrosis *(4)*. It is more difficult to address disease processes that are polygenetic or whose etiology at the molecular level is not clear. Neuronal

---

Fig. 1. *(continued)* blots or by immunocytochemistry to confirm their expression of the transgene or its product. **(E)** Delivery of transgene product into the brain is accomplished either by direct stereotaxic injection of infected primary cells into the brain parenchyma or by placing a suspension of infected cells in a collagen matrix into a lesion cavity.

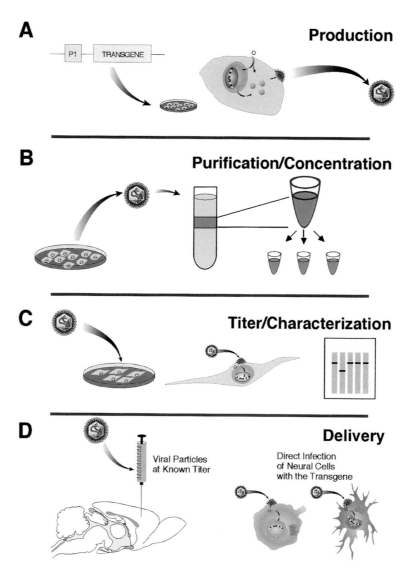

Fig. 2. (A) In vivo gene therapy begins with virus production from producer cells as described in Fig. 1A but without the need to include a selectable marker gene in the construct. **(B)** The virus collected from the producer cell culture media is purified and concentrated by ultracentrifugation and the aliquoted for viral stock. **(C)** Viral titer is determined by infecting a cell line with the incidence of plaque formation used to calculate the number of viral particles per aliquot and viral expression is evaluated by Southern or Western blot or immunocytochemistry. **(D)** The virus is then directly injected into the brain parenchyma where neural cells are directly infected and subsequently produce the gene product.

response to injury and neurodegenerative diseases, such as Alzheimer's and Parkinson's diseases, reflect circumstances where the disease state cannot yet be entirely attributed to a single gene. Gene therapy can still be considered for use with such complex degenerative processes by delivery of a gene whose product is thought to provide trophic support for the injured tissue or to replace the output of the injured tissue. In addition to supplying a corrective gene, a gene can also be delivered whose product selectively inhibits host gene function such as ribozymes *(5,6)*. Gene delivery can also be used to selectively target cells, such as cancer cells, by introducing a gene whose product is either directly lethal or renders the cell susceptible to host immunogenicity or pharmacotherapy *(7,8)*.

In addition to the identification of a suitable transgene and isolating or obtaining its cDNA to clone into a vector, successful gene therapy depends on a number of factors. These include the choice of vector, the elements in the vector that control long-term stable gene expression both in vitro and in vivo, and, in the case of ex vivo gene therapy, the choice of cells for gene delivery. The advantages and weaknesses of the various choices are discussed in the following sections.

3. APPROACHES TO TRANSGENE DELIVERY

3.1. Ex Vivo Gene Delivery

3.1.1. Choice of Vector—Retroviral Vectors

Transgenes are introduced into cells by delivery vehicles or vectors (Fig. 1). Vector constructs contain *cis*-acting elements (e.g. promoters/enhancers, polyadenylation signals, splice signals) that are involved in determining when and in which tissue or cell type a gene will be expressed. There are two classes of vectors: nonviral and viral. Transgenes can be introduced into cells by nonviral vectors using a number of chemical (calcium phosphate, lipofection) or physical methods (electroporation, microinjection) *(9)*. However, gene transfer by nonviral vectors is largely inefficient *(10)*. To date, most gene therapy approaches use viral vectors that have been modified to disable any pathogenic effects. At present, retroviral vectors are the most commonly used viral vectors for ex vivo gene transfer.

Retroviruses are a group of viruses whose RNA genome is converted into DNA in the infected cells. Retroviral vectors are plasmid DNAs derived from Moloney murine leukemia virus (MoMLV) *(10,11)*. In the vector, the viral genes are replaced with transgene sequences and placed under the control of viral long terminal repeats (LTR) or internal promoter/enhancers. The vector also contains a psi packaging sequence required for the encapsidation of genomic-length RNA into virions. Recombinant vector is transduced into packaging cells that contain a provirus lacking packaging signals but provides the viral structural genes in *trans* for the production of transmissible viral particles. The culture medium from the packaging cells containing replication-defective infectious virus particles is then used to infect the target cells.

Despite their high infection efficiency, a critical limitation of retroviral vectors is their inability to infect nondividing cells such as brain cells *(11–13)*. As a result they can be successfully used for ex vivo gene therapy only using specific cell types that show good growth characteristics in vitro. Another disadvantage of retroviral vectors is the rare possibility of their random insertion, causing the disruption of cellular genes, or the insertion of regulatory sequences (promoter or enhancers) near a cellular gene leading to inappropriate or uncontrollable expression of this gene.

Stability of transgene expression from retroviral vectors is still problematic *(12,14,15)*. Although growing cells can stably express transduced genes, the expression decreases considerably in quiescent cells *(16)*. Within 2 mo after grafting to the brain, there is a gradual down-regulation of transgene expression *(17,18)*. Although the mechanism of down-regulation is unclear, the nature of the transgene, the promoters or enhancers, and the cell type used may be some of the factors involved. The appropriate choice of promoter/enhancers can stabilize transgene expression; however, the search for such critical factors for sustained transgene expression would require time-consuming trial and error. These difficulties have prompted an exploration of ways to control transgene expression using regulatable vectors, such as the tetracycline-controlled retrovector system *(19)*, where genes can be turned off or on by exogenous agents.

3.1.2. Choice of Cell Type

The choice of cells for ex vivo gene therapy depends on a number of considerations. First, the cells should withstand the genetic manipulation process and express transgenes stably in vitro. Second, the cells should survive in the brain noninvasively and express the transgene at a physiologically significant level for a long period of time. Early work focused on the use of immortalized cell lines for genetic manipulation, as the robust proliferation of these cells in vitro made them easy to grow and to transduce efficiently with retroviral vectors. However, many of these cell lines continued to proliferate after grafting in the brain and were lethal to the experimental animals, limiting their usefulness for long-term therapeutic applications *(14, 20, 21)*. Other work focused on primary cells, such as skin fibroblasts, as targets for gene transfer as such cells show good but nonpathological survival within the CNS *(14,20–22)*. One advantage of genetically modified fibroblasts cultured from skin biopsies is that they can be used for autologous grafting to reduce immune rejection. One of the disadvantages of fibroblasts is that, being nonneuronal, they will not make functional connections with the host brain, limiting their biological effect to passive transgene expression *(14,20–22)*. Genetically modified fibroblasts from different species grafted in both peripheral (PNS) and central nervous system (CNS) survive in a nonmitotic state for up to 2 yr and produce and release transgene products in vivo that affect the functional and anatomical properties of the host brain *(14,21,23,24)*.

Primary astrocytes also have been used as cellular vehicles for gene delivery in the CNS, as their rapid proliferative capacity in vitro makes them amenable to gene transfer by retroviral vectors. Astrocytes grafted into the brain migrate from the site of implantation and survive for 2–3 mo *(25–27)*. The good survival and transgene expression of astrocytes postgrafting *(14)* have made them useful for gene delivery in vivo.

Genetically modified primary cells function as a "mini-pump" to supply factors locally, and are useful for delivery of factors near compromised neurons to prevent their further damage and halt the progression of neurodegeneration. However, the engraftment of these cells within the brain parenchyma may itself displace and damage neurons or disrupt connections. Furthermore, larger brain structures, such as the striatum, may require multiple deposits of modified primary cells, magnifying this concern. A final consideration for ex vivo gene therapy is that, although the brain is immunoprivileged, grafted cells can trigger immune reaction leading to adverse effects.

3.2. In Vivo Gene Therapy

In recent years, research has focused on the development of vectors that can be used to directly introduce the transgene into nondividing cells in the brain to effect their

Table 1
Characteristics of Viral Vectors Currently in Use for Experimental
In Vivo Gene Delivery

	Integration into host genome?	Transgene expression	Immunological response	Safety considerations
Adenovirus	No	Short-term	Cellular and humoral	Cytotoxicity, inflammation
Adeno-associated virus	May integrate	Long-term	Not established	Cytotoxicity, inflammation, purification from helper virus
Herpes simplex virus-I	Yes	Short-term	Minimal	Cytotoxicity, spread of possible reverted wild-type virus
Lentivirus	Yes	Long-term	Minimal	Possible insertional mutagensis

endogenous production of the gene product (Fig. 2). Furthermore, the vector systems described in the following subheadings can also be used successfully to deliver genes to cells in culture that can be subsequently used for ex vivo gene therapy. The potential advantages of in vivo gene therapy is that the intraparenchymal displacement and immunological stimulation of grafting modified cells into the brain can be minimized and that the host cells would themselves express the transgene product. In contrast to ex vivo gene therapy, this would permit modification of intracellular pathways without requiring a secreted gene product to retain its biological activity in the extracellular environment and then be taken up by or activate receptors on the host cell. The vectors described below and some of their properties are summarized in Table 1.

3.2.1. Adenoviral Vectors (ADVs)

The recombinant ADVs are commonly derived from type 5 adenovirus, which can infect both dividing and nondividing cells and is not associated with human malignancies. Replication-defective ADV can be generated by replacing the immediate early E1 gene, which is essential for replication, with a promoter/enhancer and the transgene *(13,15,28,29)*. The recombinant ADV is then replicated in cells that express the products of the E1 gene. However, the transgene does not integrate into the host genome and expression lasts only a short time. When ADVs expressing the marker gene *LacZ* from different viral promoter/enhancers were injected into different regions of the brain, a large number of neurons, astrocytes, and microglia were found to express β-gal as early as 24 h post-inoculation but the expression was substantially reduced within 1–2 mo *(30–33)*. Although these results indicate that adenoviral vectors can be used for in vivo gene transfer, transgene expression within these vectors is susceptible to adverse regulatory influences. The down-regulation of transgene expression has been reported to be associated with both cellular and humoral immune responses to the viral proteins (34). Modified ADVs have been generated that lack all of the viral structural genes that are provided in trans (35). The transgene expression from these vectors lasted longer than the first generation vectors in which only the E1 gene was deleted, but the titers were low. Presently, ADVs may be useful where short-term transgene expression is required, such as in tumor cell killing.

3.2.2. Adeno-Associated Vectors (AAVs)

AAV is a defective parvovirus that can infect various types of mammalian cells but is not apparently associated with any human disease. AAV depends on the coinfection of ADV to complete its life cycle (13,36). Unlike ADV, AAV can, but does not always, integrate into the host cell genome even when the cells are in a quiescent state. Without a helper virus, AAV establishes latency by integrating into a single locus in human chromosome 19 (37). All of the AAV sequences except inverted terminal repeats (ITRs), which define the beginning and the end of the virus, can be removed to accommodate foreign DNA. Recombinant AAV (rAAV) is produced from a plasmid vector by stimulating lytic infection with helper ADV while supplying the internal coding sequences in trans (38). Although the virus can be produced at a high titer (10^{11}–10^{12} virus particles/mL), the preparation of virus is laborious and no packaging cell line is currently available that can stably provide all the proteins. The requirement for proteins of ADV to be present for active replication of AAV means that this system suffers from some of the same disadvantages as ADV, including the possibility that the transfer of replication-competent ADV may lead to an immunological response.

The usefulness of AAV-based vectors for long-term gene transfer into the brain has been reported. AAV vectors expressing the reporter gene LacZ or tyrosine hydroxylase (TH) injected directly in different regions of the brain showed a transduction efficiency of approximately 10% (39).

3.2.3. Herpes Simplex Virus-I Vectors (HSV-I)

The herpes simplex virus type I (HSV-I)-derived vectors have also been used for gene delivery into postmitotic cells (40,41). Transgenes can be expressed from herpes immediate early (IE) 4/5 promoter or other strong viral promoters. Defective amplicon HSV-I vectors can infect both neurons and glia but show greater efficiency for neurons (41). HSV-I vectors have been used to infect neurons in vitro (42) and for in vivo gene delivery in mouse and rat brain (43–45). One disadvantage of HSV-I is its cytotoxicity that may be helper virus mediated (41). Another disadvantage is that transgene expression is generally of a modest duration, apparently due either to cell death or down-regulation of transgene expression, although low level expression has been reported in vivo for up to 1 yr (46).

3.2.4. Lentiviral Vectors (HIV)

Lentiviruses (e.g., human immunodeficiency virus [HIV]) belong to the retrovirus family but can infect both dividing and nondividing cells. By utilizing the host cell's nuclear import machinery, the lentiviral construct and its transgene are actively transported through the nuclear pore, permitting stable integration. Like retroviruses,

HIV-based vectors have been developed by disabling the virus and cloning the transgene between the LTRs and the packaging signals *(47)*. Replication incompetent lentiviral vectors injected into the brain showed sustained transgene expression up to at least 6 mo at the site of injection by transduced, terminally differentiated neurons *(48)*. There is apparently little host response to lentivirus as injection of a high concentration of the virus showed very little toxicity and did not provoke an immune response *(48,49)*. Thus, this integrating recombinant virus may provide an efficient vehicle for sustained transgene delivery into differentiated nondividing cells in vivo.

3.3. Specificity and Efficiency of Vectors

The vectors available at present for in vivo gene delivery can infect both dividing and nondividing cells within the CNS. However, the specificity with which different vectors infect these categories of cells varies. Indeed, vectors differ in their proclivity to infect different phenotypes of nondividing cells, exhibiting a (nonabsolute) specificity for infecting neurons or glia. Lentivirus- and HSV-I-derived vectors show a propensity for differentiated neurons, whereas more glial cells are infected by ADV and AAV.

The efficiency of gene transfer in the brain by ADV, AAV and HSV-I is about 10%. The viruses can be produced at high titers (ranging from 10^9 virus particle/mL for lentivirus to 10^{12} virus particle/mL for AAV). ADV can infect both neurons and glia in vivo, causing them to express high levels of transgene. However, ADV remains episomal (unintegrated) in the infected cells. Overall down-regulation of transgene expression can result from infected cell loss following host immune response. Also, if infected cells are dividing, then ADV will dilute out with each cell division. For this reason, ADV vectors are presently best used for therapies requiring short-term gene delivery, for example, in the treatment of cancer. HSV-I-derived vectors are cytotoxic to cells and the expression of transgene is short either due to cell death or transgene expression shut-off *(41)*. AAV can integrate into the infected cell chromosome and thus offers the advantage that the newly introduced gene can be stably acquired by the cells and passed onto the progeny upon cell division. Transgene expression from AAV vectors can be sustained for a long period of time. Lentivirus-based vectors showed high transduction efficiency, stable integration in host chromosomes, and long-term gene expression (up to 6 mo; *50*). Although there is no perfect vector, some vectors, such as AAV and HIV, provide a usable balance between advantages and disadvantages. Continued work will, no doubt, provide improvements in the current vectors and introduce superior vector designs not yet envisioned.

4. PRECLINICAL RESEARCH IN CNS DISEASE MODELS

4.1. Ex Vivo or In Vivo Gene Therapy—Which Approach to Use?

The method of choice for gene therapy in the CNS is directed by the nature of the disease, which determines how much and where the gene should be delivered. Ex vivo gene therapy makes a deposit of cells that function as mini-pumps to deliver factors locally. However, the grafted cells occupy space and may perturb endogenous cells and connections. On the other hand, if placed into a region disrupted by injury, the graft may provide a physical bridge for reestablishing connections in addition to supplying the transgene product *(23,51)*. Although in vivo gene therapy can be used under the same situations as ex vivo gene therapy, it can also be used to alter the physiology of endogenous cells by genetically modifying them *in situ*. Furthermore, in vivo gene therapy is the method of choice when it is necessary to deliver factors over a wide area, as virus can be easily injected in multiple areas of the CNS.

4.2. Delivery of Trophic Factor Genes

Numerous lines of evidence suggest that some neurons in the adult CNS are sensitive to endogenously produced proteins that provide a degree of metabolic support. Application of these trophic factors to fetal neurons in vitro can promote their survival, phenotypic maturation, and neurite outgrowth. The in vivo delivery of appropriate exogenous trophic factors can be neuroprotective to some damaged neurons *(52)*. However, the development of pharmacotherapeutic administration of trophic factors has been hampered by the high cost of purified or recombinant trophic factors, concerns about their ability to cross the blood–brain barrier, and their biological effect on peripheral tissues. Even intracerebroventricular infusion of trophic factor proteins can lead to unanticipated and possibly deleterious side effects *(53,54)*.

Ex vivo gene therapy provides an approach to deliver trophic factors locally and in a regulated manner. Ex vivo gene delivery of nerve growth factor (NGF) can rescue cholinergic projection neurons of the basal forebrain *(55–57)* and can stimulate fiber outgrowth through the NGF-secreting graft from basal forebrain neurons *(23,51)* and from other NGF-sensitive central neurons *(58)*. Regenerative outgrowth in a spinal cord injury model can also be elicited by primary fibroblast grafts secreting NGF *(59–61)* or other trophic factors *(62,63)*.

The entorhinal cortex is a region that shows an early vulnerability in Alzheimer's disease *(64)*. Neuroprotection by ex vivo gene delivery has been demonstrated in vulnerable entorhinal cortical neurons by secretion of basic fibroblast growth factor (FGF-2; Fig. 3; *65*). Similarly, grafts of cells producing brain-derived neurotrophic factor (BDNF; *66–68*) have been evaluated for neuroprotection in an animal model of Parkinson's disease.

While these studies using inbred laboratory animal strains found little immunological response to the grafting of autologous primary cells carrying the transgene, concern about immunological response by human patients prompted an alternative approach to ex vivo gene delivery. By placing the modified cells within a polymer capsule, it is possible to introduce an immunologically inert container within the CNS that remains viable through exchange with the host interstitial fluid and whose transgene product reaches the surrounding tissue through the same exchange. Encapsulated ex vivo gene therapy has been investigated for secretion of glial cell line-derived neurotrophic factor (GDNF) and ciliary neurotrophic factor (CNTF) in animal models of Parkinson's and Huntington's diseases and for amyotrophic lateral sclerosis *(69–71)*. Although encapsulated grafts minimize the host immunological response, they are still a space-occupying disruption to the neural parenchyma and are susceptible to the same limitations on transgene expression duration as nonencapsulated cells.

Successful delivery of trophic factor genes using direct in vivo gene therapy has recently been demonstrated. Direct injection in the basal forebrain with a NGF-lentiviral vector infected primarily neurons and protected these neurons from subsequent injury *(Fig. 4; 72)*. Midbrain dopaminergic neurons, directly infected with GDNF

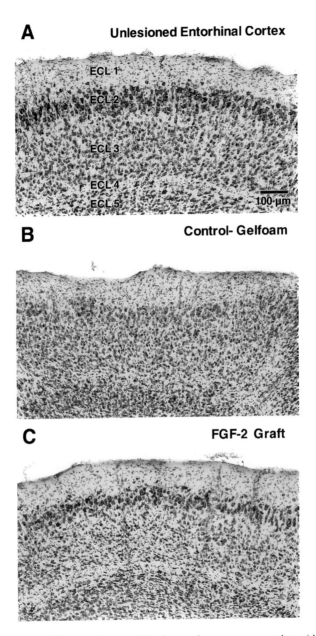

Fig. 3. Ex vivo gene delivery of a trophic factor for neuroprotection. (**A**) The projection neurons of entorhinal cortex occupy a well-defined layer known as layer II (ECL2). (**B**) Within 2 wk following lesion of the perforant pathway there is substantial loss of ECL2 neurons. (**C**) Primary fibroblasts infected to express the growth factor FGF-2 grafted into the lesion cavity protect these neurons from lesion-induced death. (Modified from *ref. 65* with permission.)

by ADV *(73)* or AAV *(74)*, survived subsequent lesion in a Parkinson's disease model. Interestingly, this protection was also achieved when the deposition of a GDNF-adenoviral vector was in these neurons' terminal field in the striatum, suggesting efficient retrograde transport of the vector *(75)*. In contrast, injection of a GDNF-AAV vector into the striatum did not protect midbrain dopaminergic neurons *(76)*. It is difficult to know if the lack of protection with the GDNF-AAV construct was due to decreased efficiency relative to GDNF-ADV or whether AAV was retrogradely transported with less efficiency because the AAV study *(76)* injected vector one week following 6-OHDA lesion whereas the other studies injected prior to lesion.

4.3. Delivery of Neurotransmitter/Precursor Genes

Some neurodegenerative diseases, such as Alzheimer's and Parkinson's diseases, are characterized by a reduction of neurotransmitter levels (acetylcholine and dopamine, respectively). Reestablishment of neurotransmitter levels has been investigated as one therapeutic approach to restore function. For example, fibroblasts modified to express acetylcholine have successfully restored behavioral performance on learning and memory tasks following engraftment to the lesioned cerebral cortex *(77)* or hippocampus *(78)*.

Replacement of dopamine in an animal model of Parkinson's disease has been intensively investigated by grafting cells modified to express L-Dopa *(17,79–81)*. Although these results are promising as a therapeutic model for Parkinson's disease, volumetric constraints become an issue as the size of the human striatum relative to the sphere of influence for each deposit of grafted cells is considered *(82)*.

Because multiple injections of viral vector are less invasive than the equivalent number of grafted cell deposits, the use of direct in vivo gene therapy holds considerable promise as a therapeutic approach to restore dopamine levels throughout the striatum in Parkinson's disease. Several studies have introduced the gene for tyrosine hydroxylase (TH) into 6-hydroxydopamine (6-OHDA)-lesioned rats as a model of Parkinson's disease. Two weeks following intrastriatal injection of ADV vectors expressing TH, a large number of cells (mostly reactive astrocytes) expressed TH and the animals showed modification of apomorphine-induced turning behavior *(29)*. When rAAV vector expressing human TH was injected into caudate nucleus, the transgene was expressed for 4 mo at the site of injection. However, the number of positive cells decreased with time. Injection of the vector into the denervated striatum showed the expression of TH in the striatal neurons and glia for up to 4 mo, which was accompanied by significant behavioral recovery in these animals compared to rats that received AAV expressing *LacZ* gene *(39)*. Intrastriatal delivery of TH by HSV-I-derived vectors resulted in increased TH enzyme activity, increased striatal dopamine levels, and behavioral recovery *(46)*. Although the behavioral recovery lasted for 1 yr in this study, few expressing cells were found upon histological examination, suggesting that even low levels of endogenous production may be adequate for functional recovery.

4.4. Delivery of Anti-Apoptotic Genes

One advantage of in vivo gene therapy over ex vivo gene therapy is the ability to directly infect endogenous cells to express the transgene. It therefore becomes possible to deliver a transgene whose product intervenes in the host cell's intracellular processes. Several studies have sought to protect neurons from apoptotic death by introducing transgenes from the Bcl family, whose products are believed to be anti-apoptotic *(83)*. Injection of HSV-Bcl-2 vector prior to ischemic insult successfully infected neurons within the injection region and increased the amount of viable cortical tissue within the infected tissue volume in a stroke model *(84)*. Likewise, HSV-Bcl-2 vector also infected hippocampal neurons and protected them from subsequent excitotoxic and ischemic injury *(85)*. Using the fimbria–fornix lesion model, Blömer et al. *(72)* have recently reported that prior infection of the basal forebrain with a Lentiviral-Bcl-xL vector infected primarily neurons and significantly increased their survival following lesion (Fig. 4). These studies suggest that in vivo gene therapy delivery of anti-apoptotic genes may be a promising therapeutic strategy for protecting discrete populations of neurons.

4.5. Modification of Neuronal Progenitors

In addition to using primary cells for ex vivo gene therapy, the use of progenitor cells from the fetal and adult CNS for gene delivery has recently generated considerable interest. These cells can be isolated in vitro due to establishment of culture conditions using mitogenic growth factors, such as epidermal growth factor (EGF) or basic fibroblast growth factor (FGF-2) *(86–88)*. During expansion in culture, these cells can be genetically modified using retroviral vectors as previously described. Cell lines of embryonic neuronal progenitors can be established, genetically modified, and subsequently grafted into the brain using the ex vivo gene therapy approach *(89)*.

Neuronal progenitors from the adult CNS also can be isolated, genetically modified in culture, and grafted back to the adult CNS (Fig. 5). Upon grafting in the adult rat hippocampus or rostral migratory pathway leading to the olfactory bulb, the adult neuronal progenitors migrate and differentiate in a site-specific manner *(86,90)*. Implanted neuronal progenitors continued to express the transgene for an extended period of time (up to 6 wk; *91*). The ability to modify these adult neuronal progenitors prior to their return to the adult CNS suggests that it may be possible to have them endogenously express putative migration and differentiation cues that may no longer be present in the adult CNS, thus giving these cells a wider range of potential than has so far been observed. Another possibility is that endogenous neuronal progenitors within the adult CNS may be modified by in vivo gene therapy during cell division and recruited for specific fates. Further understanding regarding the regulation of neuronal differentiation in the adult CNS will be required before such potential could be realized. However, the realization of such potential may permit therapeutic repopulation strategies for regions of the CNS where neuronal loss has already occurred.

5. FUTURE PROSPECTS FOR HUMAN GENE THERAPY

The preclinical research reviewed in the preceding sections suggests that gene therapy approaches can efficiently deliver functional genes resulting in neuronal protection and functional recovery. While additional work will continue in basic science research using gene therapy approaches to address fundamental questions in neurobiology, there are a number of issues that need to be addressed to enable gene therapy to move from the laboratory to the clinic.

5.1. Safety of Vectors

Because highly efficient viral machinery is used to introduce the transgenes, one of the foremost concerns is in the overall safety of the viral vectors being used. Ex vivo

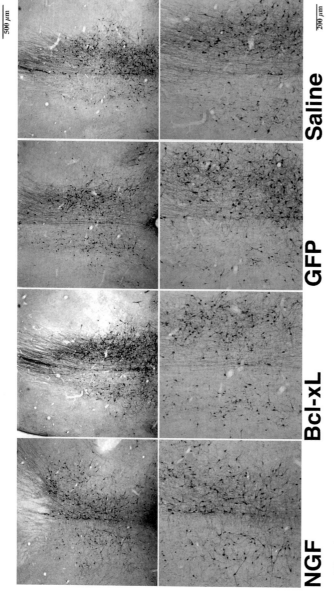

Fig. 4. Direct in vivo gene delivery of a trophic factor for neuroprotection. Unilateral lesion of the fimbria-fornix pathway results in loss within a population of cholinergic neurons in the medial septum ipsilateral to the lesion. The loss of ChAT-positive neurons is seen on the left side of the medial septum at low magnification (*upper panels*) and at higher magnification (*lower panels*) in both saline injected animals and animals receiving a lentiviral vector containing the marker gene for green fluorescent protein (GFP). The lesion-induced loss of ChAT-positive neurons was significantly prevented by injection of lentiviral vectors containing genes for the trophic factor NGF or the anti-apoptotic gene Bcl-xL. (Reprinted with permission from *ref. 72*.)

gene therapy does not present safety concerns with regard to the vectors themselves, as exposure to vectors occurs prior to grafting. The concern with in vivo gene therapy falls into two categories: the continued, unregulated replication of the virus and the toxicity of and immune response to viral proteins.

Vectors developed for in vivo gene therapy are made replication-defective by removing the genes essential for viral replication. Adenoviral vectors used for gene therapy have the E1 gene removed, which renders them replication incompetent. AAV requires additional genes for replication provided generally by wild-type adenovirus. The contaminating wild-type adenoviruses need to be carefully separated from AAV preparations to avoid problems. HSV-I vectors produce wild-type virus at a low reversion frequency during packaging of the viral DNA and this virus may spread in vivo after gene delivery from the injected virus. The HIV-based lentiviral vectors are "first generation" vectors and little is known about their ability for recombination, although the experiments so far have shown them to be helper virus free.

Mammalian and human immune systems are designed to protect against viral infection and the delivery of therapeutic genes by viral vectors can be hampered by host responses. The adenoviruses produce benign respiratory infections in humans and elicit both cellular and humoral immune responses. In the humoral response, antibodies to viral proteins are made, thus preventing subsequent infection from a second exposure to the recombinant virus. Since most of the human populations have antibodies to naturally occurring virus, ADV vectors may not be very useful as gene transfer vectors in these individuals. AAV is a nonpathogenic virus. However, HSV is cytopathic and it will be necessary to remove viral-induced cytotoxic functions before HSV can be considered for widespread use. Because HIV is highly pathogenic in humans, special precautions are presently taken, although recent work suggests that there is minimal immune response *(49)*. For use in humans, it may be prudent to develop vectors with nonhuman lentiviruses such as equine infectious anemia virus, bovine immunodeficiency virus, or simian immunodeficiency virus.

5.2 Ethical Considerations

This chapter has thus far dealt only with technical issues concerning the validity of gene delivery as a therapeutic procedure for CNS disease. However, in addressing the future prospects of CNS gene therapy for the clinic, it is necessary to introduce some ethical considerations. Clearly these ethical issues can be dealt with only superficially here and other reviews specifically addressing gene therapy should be consulted *(92–94)*.

The first distinction to be made is between ex vivo somatic gene therapy and in vivo gene therapy. While this distinction has already been made procedurally (Figs. 1 and 2), there is also a conceptual distinction. Ex vivo gene therapy essentially modifies current drug delivery approaches by creating a biological "pump" to distribute the therapeutic gene product. As a result, the ethical framework for this procedure can be addressed by well established ethical precepts for pharmacotherapy.

In contrast, in vivo gene therapy is essentially genetic engineering, where the endogenous cell, and possibly its progeny, is directly modified. As a result, in vivo gene therapy is placed within a different ethical framework *(95)*. A logical extension to the direct modification of somatic cells is the modification of germ line cells. Human germ line gene therapy could be a powerful tool, as any change to the genome would be part of the genetic composition of future generations. For example, if a genetic alteration that would prevent the onset of Alzheimer's disease were to be identified, would it be desirable to alter the genetic makeup of the germ line to eliminate the incidence of Alzheimer's disease? Or, if not for the general population, could germ line gene therapy be considered for the subpopulation at risk for familial Alzheimer's disease? Such prospects challenge social, ethical, and religious constraints and would require the achievement of a consensus viewpoint before they could be contemplated *(96–99)*.

While falling short of the greater ethical implications of human germ line gene therapy, the CNS in particular could be considered for the application of fetal gene therapy *(100,101)*. Our present technical capabilities permit us to envision injecting a viral vector containing a therapeutic transgene into the developing brain *in utero* at a time when a certain population of neurons is being generated. These modified neurons would subsequently express the transgene as they migrated to their final distribution. In this way, entire neuronal populations within an individual could be engineered to alter their phenotype, connectivity, or ability to survive neurodegenerative diseases. Alternatively, transgenes could be introduced to the placenta, permitting the gene product to cross to and influence the fetus *(102)*.

Beyond the issues raised by the technological advances in gene therapy, the choice of the transgene could have profound ethical implications. The discussion in this chapter has been of therapeutic transgenes, whose product is identified as combating some disease process. However, gene therapy in the CNS provides a framework for broadening this concept to social behavior therapies where, for instance, the ability to reduce or eliminate addictive behavior could be desired if a suitable genetic modification were identified. Even more difficult is the potential use of gene therapy to introduce a transgene to confer an enhancement of neural or brain functioning. Although enhancement gene therapy generates concerns about eugenics, the potential advantages conferred upon the individual could be a strong driving force against ethical constraints *(103,104)*.

5.3. Public Opinion Regarding Human Gene Therapy

Despite any procedural advances in gene therapy, the readiness of the public to accept its use with humans will be largely responsible for any widespread adoption of gene therapy. Although an earlier study in the United States found most of those surveyed knew little or nothing about gene therapy *(105)*, a more recent measurement of international perceptions has revealed that the majority of the populace has some recognition of the concept of gene therapy and appears to favor its use for humans *(106)*. In addition to the suggestion that there have been improvements in education about gene therapy, Macer et al. *(106)* found that most respondents would accept therapeutic benefits from gene therapy for themselves and especially for their children. While there was limited enthusiasm for potential social enhancements, such as appearance or intelligence, respondents supported direct therapeutic (i.e., disease curing) as well as health-enhancing gene delivery. There was little concern over a eugenic application of gene therapy, but this may have reflected the respondents' minimal distinction between somatic and germ line gene therapy. Most of the minority who rejected the use of gene therapy cited it as being "unnatural," "playing God," or producing "unpredictable" results. These concerns should probably not be dismissed as "bio-Luddism," given that

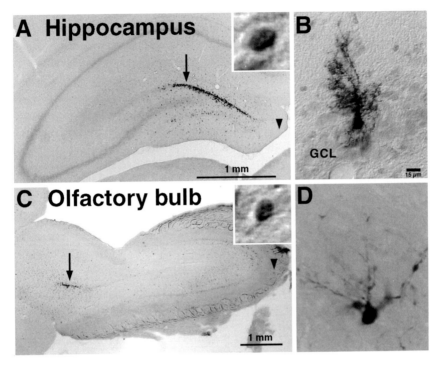

Fig. 5. Genetically modified adult neuronal progenitors engrafted into the adult CNS. **(A)** Adult hippocampal progenitors labeled with bromodeoxyuridine (BrdU) survive engraftment (*arrow*) and disperse throughout the adult hippocampus (cell at *arrowhead* shown in *inset*). **(B)** Differentiation of appropriate granule cell morphology of grafted adult hippocampal progenitors is revealed through expression of the *LacZ* gene product 8 wk after engraftment. **(C)** BrdU-labeled adult hippocampal progenitors grafted into the olfactory bulb/rostral migratory pathway (*arrow*) migrate throughout the extent of the olfactory bulb (cell at *arrowhead* shown in *inset*). **(D)** Progenitors from the adult hippocampus can differentiate into morphologically appropriate olfactory glomerular neurons (location approximately at the *arrowhead* in C) as revealed through expression of the *LacZ* gene product 8 wk after engraftment. (A, C, and D modified from *ref. 90* with permission. B modified from *ref. 86* with permission.)

human efforts to control our environment and ecosystems are replete with examples of miscalculation.

Public opinion will also play a role in influencing the direction of governmental advisory and/or regulatory bodies and in turn influencing the implementation of human gene therapy. In the United States, the Recombinant DNA Advisory Committee to the National Institutes of Health presently restricts enhancing somatic gene therapy and prohibits investigation into human germ line gene therapy, although a dialogue on this topic is emerging *(107,108)*. However, the UNESCO International Bioethics Committee proposed guidelines, approved by the UN General Assembly in 1998, that would permit enhancing somatic gene therapy and permit human germ line gene therapy while outlawing only germ line gene enhancement *(109)*. The resolution of international and domestic guidelines for the use of human gene therapy may take some time to be achieved and will require the concerted effort of scientists, clinicians, and the public.

6. CONCLUSIONS

Delivery of functional genes to replace defective genes or to produce a necessary protein has been shown to be an effective therapeutic strategy in animal models of CNS injury and disease. Recent advances in viral vector design to permit direct in vivo gene delivery have expanded the possible therapeutic applications. However, further work is needed to improve vector design to increase efficiency of infection, maintain stable expression, and minimize host responsiveness. Further work is also needed to identify candidate transgenes that will be most effective for counteracting CNS injury and disease in the various animal models. To improve the prospects for CNS gene therapy in human patients, continued work is required in animal models to elucidate the optimal delivery strategies while simultaneously efforts must be made by clinicians and scientists to educate the public concerning the benefits and risks of human gene therapy and to articulate these points to regulatory bodies.

ACKNOWLEDGMENTS

The authors thank M. L. Gage for editorial review of the manuscript. The work conducted in the authors' laboratory was supported by NINDS P01-NS28121 and by a T. L. L. Temple Foundation/Alzheimer's Association Award to D. A. P.

REFERENCES

1. Hurtig, H. I. (1997) Problems with current pharmacologic treatment of Parkinson's disease. *Exp. Neurol.* **144,** 10–116.
2. Dai, Y., Roman, M., Naviaux, R. K., and Verma, I. M. (1992) Gene therapy via primary myoblasts: long-term expression of factor IX protein following transplantation in vivo. *Proc. Natl. Acad. Sci. USA* **89,** 10,892–10,895.
3. Fallaux, F. J. and Hoeben, R. C. (1996) Gene therapy for the hemophilias. *Curr. Opin. Hematol.* **3,** 385–389.
4. Wagner, J. A. and Gardner, P. (1997) Toward cystic fibrosis gene therapy. *Annu. Rev. Med.* **48,** 203–216.
5. Couture, L. A. and Stinchcomb, D. T. (1996) Anti-gene therapy: the use of ribozymes to inhibit gene function. *Trends Genet.* **12,** 510–515.
6. Looney, D. and Yu, M. (1997) Clinical aspects of ribozymes as therapeutics in gene therapy. *Methods Mol. Biol.* **74,** 469–486.
7. Herrmann, F. (1995) Cancer gene therapy: principles, problems, and perspectives. *J. Mol. Med.* **73,** 157–163.
8. Dachs, G. U., Dougherty, G. J., Stratford, I. J., and Chaplin, D. J. (1997) Targeting gene therapy to cancer: a review. *Oncol. Res.* **9,** 313–325.
9. Ray, J. and Gage, F. H. (1992) Gene transfer into established and primary fibroblast cell lines: comparison of transfection methods and promoters. *Biotechniques* **13,** 598–603.
10. Felgner, P. L. (1997) Nonviral strategies for gene therapy. *Sci. Am.* **276,** 102–106.
11. Verma, I. M. (1990) Gene therapy. *Sci. Am.* **263,** 68–72, 81–84.
12. Mulligan, R. C. (1993) The basic science of gene therapy. *Science* **260,** 926–932.
13. Gunzburg, W. H. and Salmons, B. (1995) Virus vector design in gene therapy. *Mol. Med. Today* **1,** 410–417.
14. Fisher, L. J. (1995) Engineered cells: a promising therapeutic approach for neural disease. *Res. Neurol. Neurosci.* **8,** 49–57.
15. Verma, I. M., and Somia, N. (1997) Gene therapy—promises, problems and prospects. *Nature* **389,** 239–42.

16. Schinstine, M., Rosenberg, M. B., Routledge-Ward, C., Friedmann, T., and Gage, F. H. (1992) Effects of choline and quiescence on *Drosophila* choline acetyltransferase expression and acetylcholine production by transduced rat fibroblasts. *J. Neurochem.* **58,** 2019–2029.

17. Fisher, L. J., Jinnah, H. A., Kale, L. C., Higgins, G. A., and Gage, F. H. (1991) Survival and function of intrastriatally grafted primary fibroblasts genetically modified to produce L-Dopa. *Neuron* **6,** 371–380.

18. Doering, L.C. and Chang, P. L. (1991) Expression of a novel gene product by transplants of genetically modified primary fibroblasts in the central nervous system. *J. Neurosci. Res.* **29,** 292–298.

19. Reeves, S. A. (1998) Retrovirus vectors and regulatable promoters, in *Gene Therapy for Neurological Disorders and Brain Tumors* (Chiocca, E.A. and Breakfield, X.O., eds.), Humana Press, Totowa, NJ, pp. 7–38.

20. Kawaja, M. D., Fisher, L. J., Schinstine, M., Jinnah, H. A., Ray, J., Chen, L. S., and Gage, F. H. (1992) Grafting genetically modified cells within the rat central nervous system: methodological considerations, in *Neural Transplantation: A Practical Approach* (Dunnet, S. D. and Björklund, A., eds.), Oxford University Press, New York.

21. Ray, J., Fisher, L. J., and Gage, F. H. (1994) Implantation of genetically modified cells in the brain, in *Somatic Gene Therapy* (Chang, P. L., ed.), CRC Press, Boca Raton, FL, pp. 161–182.

22. Gage, F. H., Kawaja, M. D., and Fisher, L. J. (1991) Genetically modified cells: applications for intracerebral grafting. *Trends Neurosci.* **14,** 328–333.

23. Kawaja, M. D., Rosenberg, M. B., Yoshida, K., and Gage, F. H. (1992) Somatic gene transfer of nerve growth factor promotes the survival of axotomized septal neurons and the regeneration of their axons in adult rats. *J. Neurosci.* **12,** 2849–2864.

24. Fisher, L. J. and Gage, F. H. (1993) Grafting in the mammalian central nervous system. *Physiol. Rev.* **73,** 583–616.

25. Ignacio, V., Collins, V. P., Suard, I. M., and Jacque, C. M. (1989) Survival of astroglial cell lineage from adult brain transplant. *Dev. Neurosci.* **11,** 175–178.

26. Cunningham, L. A., Hansen, J. T., Short, P., and Bohn, M. C. (1991) The use of genetically altered astrocytes to provide nerve growth factor to adrenal chromaffin cells grafted into the striatum. *Brain Res.* **561,** 192–202.

27. Cunningham, L. A., Short, M. P., Breakefield, X. O., and Bohn, M. C. (1994) Nerve growth factor released by transgenic astrocytes enhances the function of adrenal chromaffin cell grafts in a rat model of Parkinson's disease. *Brain Res.* **658,** 219–231.

28. Graham, F. L. and Prevec, L. (1991) Manipulation of adenovirus vectors. *Methods Mol. Biol.* **7,** 109–127.

29. Horellou, P., Revah, F., Sabate, O., Buc-Caron, M-H., Robert, J-J., and Malet, J. (1996) Adenovirus: a new tool to transfer genes into the central nervous system for treatment of neurodegenerative disorders, in *Genetic Manipulation of the Nervous System* (Latchman, D. S., ed.) Academic Press, San Diego.

30. Akli, S., Caillaud, C., Vigne, E., Stratford-Perricaudet, L. D., Poenaru, L., Perricaudet, M., Kahn, A., and Peschanski, M. R. (1993) Transfer of a foreign gene into the brain using adenovirus vectors. *Nat. Genet.* **3,** 224–228.

31. Bajocchi, G., Feldman, S. H.; Crystal, R. G., and Mastrangeli, A. (1993) Direct in vivo gene transfer to ependymal cells in the central nervous system using recombinant adenovirus vectors. *Nat. Genet.* **3,** 229–234.

32. Davidson, B. L., Allen, E. D., Kozarsky, K. F., Wilson, J. M., and Roessler, B. J. (1993) A model system for in vivo gene transfer into the central nervous system using an adenoviral vector. *Nat. Genet.* **3,** 219–223.

33. Le Gal La Salle, G., Robert, J. J., Berrard, S., Ridoux, V., Stratford-Perricaudet, L. D., Perricaudet, M., and Mallet, J. (1993) An adenovirus vector for gene transfer into neurons and glia in the brain. *Science* **259,** 988–990.

34. Yang, Y., Nunes, F. A., Berencsi, K., Gonczol, E., Engelhardt, J. F., and Wilson, J. M. (1994) Inactivation of E2a in recombinant adenoviruses improves the prospect for gene therapy in cystic fibrosis. *Nat. Genet.* **7**, 362–369.

35. Mitani, K., Graham, F. L., Caskey, C. T., and Kochanek, S. (1995) Rescue, propagation, and partial purification of a helper virus-dependent adenovirus vector. *Proc. Natl. Acad. Sci. USA* **92**, 3854–3858.

36. Tenenbaum, L. and Hooghe-Peters, E. L. (1996) Gene delivery using adeno-associated virus, in *Genetic Manipulation of the Nervous System* (Latchman, D. S., ed.) Academic Press, San Diego.

37. Kotin, R. M., Menninger, J. C., Ward, D. C., and Berns, K. I. (1991) Mapping and direct visualization of a region-specific viral DNA integration site on chromosome 19q13–qter. *Genomics* **10**, 831–834.

38. Samulski, R. J., Chang, L. S., and Shenk, T. (1989) Helper-free stocks of recombinant adeno-associated viruses: normal integration does not require viral gene expression. *J. Virol.* **63**, 3822–3828.

39. Kaplitt, M. G., Leone, P., Samulski, R. J., Xiao, X., Pfaff, D. W., O'Malley, K. L., and During, M. J. (1994) Long-term gene expression and phenotypic correction using adeno-associated virus vectors in the mammalian brain. *Nat. Genet.* **8**, 148–154.

40. Freese, A., Geller, A. I., and Neve, R. (1990) HSV-1 vector mediated neuronal gene delivery. Strategies for molecular neuroscience and neurology. *Biochem. Pharmacol.* **40**, 2189–2199.

41. O'Malley, K. L. and Geller, A. I. (1996) Gene delivery using herpes simplex virus type I plasmid vectors, in *Genetic Manipulation of the Nervous System* (Latchman, D. S., ed.) Academic Press, San Diego.

42. Andersen, J. K., Garber, D. A., Meaney, C. A., and Breakefield, X. O. (1992) Gene transfer into mammalian central nervous system using herpes virus vectors: extended expression of bacterial lacZ in neurons using the neuron-specific enolase promoter. *Hum. Gene Ther.* **3**, 487–499.

43. Palella, T. D., Hidaka, Y., Silverman, L. J., Levine, M., Glorioso, J., and Kelley, W. N. (1989) Expression of human HPRT mRNA in brains of mice infected with a recombinant herpes simplex virus-1 vector. *Gene* **80**, 137–144.

44. Fink, D. J., Sternberg, L. R., Weber, P. C., Mata, M., Goins, W. F., and Glorioso, J. C. (1992) In vivo expression of beta-galactosidase in hippocampal neurons by HSV-mediated gene transfer. *Hum. Gene Ther.* **3**, 11–19.

45. Geller, A. I. (1997) Herpes simplex virus-1 plasmid vectors for gene transfer into neurons. *Adv. Neurol.* **72**, 143–148.

46. During, M. J., Naegele, J. R., O'Malley, K. L., and Geller, A. I. (1994) Long-term behavioral recovery in parkinsonian rats by an HSV vector expressing tyrosine hydroxylase. *Science* **266**, 1399–1403.

47. Naldini, L., Blömer, U., Gallay, P., Ory, D., Mulligan, R., Gage, F. H., Verma, I. M., and Trono, D. (1996) In vivo gene delivery and stable transduction of nondividing cells by a lentiviral vector. *Science* **272**, 263–267.

48. Blömer, U., Naldini, L., Kafri, T., Trono, D., Verma, I. M., and Gage, F. H. (1997) Highly efficient and sustained gene transfer in adult neurons with a lentiviral vector. *J. Virol.* **71**, 6641–6649.

49. Kafri, T., Blomer, U., Peterson, D. A., Gage, F. H., and Verma, I. M. (1997) Sustained expression of genes delivered directly into liver and muscle by lentiviral vectors. *Nat. Genet.* **17**, 314–317.

50. Blömer, U., Naldini, L., Verma, I. M., Trono, D., Gage, F. H. (1996) Applications of gene therapy to the CNS. *Hum. Mol. Genet.* **5**, 1397–1404.

51. Eagle, K. S., Chalmers, G. R., Clary, D. O., and Gage, F. H. (1995) Axonal regeneration

and limited functional recovery following hippocampal deafferentation. *J. Comp. Neurol.* **363,** 377–388.

52. Peterson, D. A. and Gage, F. H. (1999) Trophic factors in experimental models of adult central nervous system injury, in *Cerebral Cortex,* Vol. 14. *Neurodegeneration and Age-Related Changes in Structure and Function of Cerebral Cortex* (Jones, E. G., Peters, A., and Morrison, J. H., eds.), Kluwer Academic/Plenum Publishers, New York, NY.

53. Kuhn, H. G., Winkler, J., Kempermann, G., Thal, L. J., and Gage, F. H. (1997) Epidermal growth factor and fibroblast growth factor-2 have different effects on neural progenitors in the adult rat brain. *J. Neurosci.* **17,** 5820–5829.

54. Winkler, J., Ramirez, G. A., Kuhn, H. G., Peterson, D. A., Day-Lollini, P. A., Stewart, G. R., Tuszynski, M. H., Gage, F. H., and Thal, L. J. (1997) Reversible Schwann cell hyperplasia and sprouting of sensory and sympathetic neurites after intraventricular administration of nerve growth factor. *Ann. Neurol.* **41,** 82–93.

55. Rosenberg, M. B., Friedmann, T., Robertson, R. C., Tuszynski, M., Wolff, J. A., Breakefield, X. O., and Gage, F. H. (1988) Grafting of genetically modified cells to the damaged brain: restorative effects of NGF gene expression. *Science* **242,** 1575–1578.

56. Strömberg, I., Wetmore, C. J., Ebendal, T., Ernfors, P., Persson, H., and Olson, L. (1990) Rescue of basal forebrain cholinergic neurons after implantation of genetically modified cells producing recombinant NGF. *J. Neurosci. Res.* **25,** 405–411.

57. Tuszynski, M. H., Roberts, J., Senut, M. C., U, H. S., and Gage, F. H. (1996) Gene therapy in the adult primate brain: intraparenchymal grafts of cells genetically modified to produce nerve growth factor prevent cholinergic neuronal degeneration. *Gene Ther.* **3,** 305–314.

58. Chalmers, G. R., Peterson, D. A., and Gage, F. H. (1996) Sprouting adult CNS cholinergic axons express NILE and associate with astrocytic surfaces expressing neural cell adhesion molecule. *J. Comp. Neurol.* **371,** 287–299.

59. Tuszynski, M. H., Peterson, D. A., Ray, J., Baird, A., Nakahara, Y., and Gage, F. H. (1994) Fibroblasts genetically modified to produce nerve growth factor induce robust neuritic ingrowth after grafting to the spinal cord. *Exp. Neurol.* **126,** 1–14.

60. Tuszynski, M. H., Gabriel, K., Gage, F. H., Suhr, S., Meyer, S., and Rosetti, A. (1996) Nerve growth factor delivery by gene transfer induces differential outgrowth of sensory, motor, and noradrenergic neurites after adult spinal cord injury. *Exp. Neurol.* **137,** 157–173.

61. Blesch, A. and Tuszynski, M. H. (1997) Robust growth of chronically injured spinal cord axons induced by grafts of genetically modified NGF-secreting cells. *Exp. Neurol.* **148,** 444–452.

62. Nakahara, Y., Gage, F. H., and Tuszynski, M. H. (1996) Grafts of fibroblasts genetically modified to secrete NGF, BDNF, NT-3, or basic FGF elicit differential responses in the adult spinal cord. *Cell Transplant.* **5,** 191–204.

63. Grill, R., Murai, K., Blesch, A., Gage, F. H., and Tuszynski, M. H. (1997) Cellular delivery of neurotrophin-3 promotes corticospinal axonal growth and partial functional recovery after spinal cord injury. *J. Neurosci.* **17,** 5560–72.

64. Gomez-Isla, T., Price, J. L., McKeel, D. W. Jr., Morris, J. C., Growdon, J. H., and Hyman, B. T. (1996) Profound loss of layer II entorhinal cortex neurons occurs in very mild Alzheimer's disease. *J. Neurosci.* **16,** 4491–4500.

65. Peterson, D. A., Lucidi-Phillipi, C. A., Murphy, D. P., Ray, J., and Gage, F. H. (1996) Fibroblast growth factor-2 protects entorhinal layer II glutamatergic neurons from axotomy-induced death. *J. Neurosci.* **16,** 886–898.

66. Frim, D. M., Uhler, T. A., Galpern, W. R., Beal, M. F., Breakefield, X. O., and Isacson, O. (1994) Implanted fibroblasts genetically engineered to produce brain-derived neurotrophic factor prevent 1-methyl-4-phenylpyridinium toxicity to dopaminergic neurons in the rat. *Proc. Natl. Acad. Sci. USA* **91,** 5104–5108.

67. Levivier, M., Przedborski, S., Bencsics, C., and Kang, U. J. (1995) Intrastriatal implanta-

tion of fibroblasts genetically engineered to produce brain-derived neurotrophic factor prevents degeneration of dopaminergic neurons in a rat model of Parkinson's disease. *J. Neurosci.* **15,** 7810–7820.

68. Yoshimoto, Y., Lin, Q., Collier, T. J., Frim, D. M., Breakefield, X. O., and Bohn, M. C. (1995) Astrocytes retrovirally transduced with BDNF elicit behavioral improvement in a rat model of Parkinson's disease. *Brain Res.* **691,** 25–36.

69. Emerich, D. F., Plone, M., Francis, J., Frydel, B. R., Winn, S. R., and Lindner, M. D. (1996) Alleviation of behavioral deficits in aged rodents following implantation of encapsulated GDNF-producing fibroblasts. *Brain Res.* **736,** 99–110.

70. Emerich, D. F., Winn, S. R., Hantraye, P. M., Peschanski, M., Chen, E. Y., Chu, Y., McDermott, P., Baetge, E. E., and Kordower, J. H. (1997) Protective effect of encapsulated cells producing neurotrophic factor CNTF in a monkey model of Huntington's disease. *Nature* **386,** 395–399.

71. Aebischer, P., Schluep, M., Deglon, N., Joseph, J. M., Hirt, L., Heyd, B., Goddard, M., Hammang, J. P., Zurn, A. D., Kato, A. C., Regli, F., and Baetge, E. E. (1996) Intrathecal delivery of CNTF using encapsulated genetically modified xenogeneic cells in amyotrophic lateral sclerosis patients. *Nat. Med.* **2,** 696–699.

72. Blömer, U., Kafri, T., Randolph-Moore, L., Verma, I. M., and Gage, F. H. (1998) Bcl-xL protects adult septal cholinergic neurons from axotomized cell death. *Proc. Natl. Acad. Sci. USA* **95,** 2603–2608.

73. Choi-Lundberg, D. L., Lin, Q., Chang, Y. N., Chiang, Y. L., Hay, C. M., Mohajeri, H., Davidson, B. L., and Bohn, M. C. (1997) Dopaminergic neurons protected from degeneration by GDNF gene therapy. *Science* **275,** 838–841.

74. Mandel, R. J., Spratt, S. K., Snyder, R. O., and Leff, S. E. (1997) Midbrain injection of recombinant adeno-associated virus encoding rat glial cell line-derived neurotrophic factor protects nigral neurons in a progressive 6-hydroxydopamine-induced degeneration model of Parkinson's disease in rats. *Proc. Natl. Acad. Sci. USA* **94,** 14,083–14,088.

75. Bilang-Bleuel, A., Revah, F., Colin, P., Locquet, I., Robert, J. J., Mallet, J., and Horellou, P. (1997) Intrastriatal injection of an adenoviral vector expressing glial-cell-line-derived neurotrophic factor prevents dopaminergic neuron degeneration and behavioral impairment in a rat model of Parkinson disease. *Proc. Natl. Acad. Sci. USA* **94,** 8818–8823.

76. During, M. J. and Leone, P. (1997) Targets for gene therapy of Parkinson's disease: growth factors, signal transduction, and promoters. *Exp. Neurol.* **144,** 74–81.

77. Winkler, J., Suhr, S. T., Gage, F. H., Thal, L. J., and Fisher, L. J. (1995) Essential role of neocortical acetylcholine in spatial memory. *Nature* **375,** 484–487.

78. Dickinson-Anson, H., Aubert, I., Gage, F. H., and Fisher, L. J. (1998) Hippocampal grafts of acetylcholine-producing cells are sufficient to improve behavioural performance following a unilateral fimbria–fornix lesion. *Neuroscience* **84,** 771–781.

79. Wolff, J. A., Fisher, L. J., Xy, L., Jinnah, H. A., Langlais, P. J., Iuvone, P. M., O'Malley, K. L., Rosenberg, M. B., Shimohama, S., Friedmann, T., and Gage, F. H. (1989) Grafting fibroblasts genetically modified to produce L-dopa in a rat model of Parkinson disease. *Proc. Natl. Acad. Sci. USA* **86,** 9011–9014.

80. Horellou, P., Brundin, P., Kalen, P., Mallet, J., and Björklund, A. (1990) In vivo release of DOPA and dopamine from genetically engineered cells grafted to the denervated rat striatum. *Neuron* **5,** 393–402.

81. Lundberg, C., Horellou, P., Mallet, J., and Bjorklund, A. (1996) Generation of DOPA-producing astrocytes by retroviral transduction of the human tyrosine hydroxylase gene: in vitro characterization and in vivo effects in the rat Parkinson model. *Exp. Neurol.* **139,** 39–53.

82. Freeman, T. B. (1997) From transplants to gene therapy for Parkinson's disease. *Exp. Neurol.* **144,** 47–50.

83. Reed, J. C. (1994) Bcl-2 and the regulation of programmed cell death. *J. Cell Biol.* **124,** 1–6.

84. Linnik, M. D., Zahos, P., Geschwind, M. D., and Federoff, H. J. (1995) Expression of bcl-2 from a defective herpes simplex virus-1 vector limits neuronal death in focal cerebral ischemia. *Stroke* **26,** 1670–1674.

85. Lawrence, M. S., Ho, D. Y., Sun, G. H., Steinberg, G. K., and Sapolsky, R. M. (1996) Overexpression of Bcl-2 with herpes simplex virus vectors protects CNS neurons against neurological insults in vitro and in vivo. *J. Neurosci.* **16,** 486–496.

86. Gage, F. H., Coates, P. W., Palmer, T. D., Kuhn, H. G., Fisher, L. J., Suhonen, J. O., Peterson, D. A., Suhr, S. T., and Ray, J. (1995) Survival and differentiation of adult neuronal progenitor cells transplanted to the adult brain. *Proc. Natl. Acad. Sci. USA* **92,** 11,879–11,883.

87. Kilpatrick, T. J., Richards, L. J., and Bartlett, P. F. (1995) The regulation of neural precursor cells within the mammalian brain. *Mol. Cell Neurosci.* **6,** 2–15.

88. Ray, J., Palmer, T. D., Suhonen, J., Takahashi, J., and and Gage, F. H. (1997) Neurogenesis in the adult brain: lessons learned from the studies of progenitor cells from the embryonic and adult central nervous systems, in *Isolation, Characterization and Utilization of CNS Stem Cells* (Gage, F. H. and Christen, Y., eds.), Springer-Verlag, Berlin, pp. 129–149.

89. Martinez-Serrano, A. and Bjorklund, A. (1997) Immortalized neural progenitor cells for CNS gene transfer and repair. *Trends Neurosci.* **20,** 530–538.

90. Suhonen, J. O., Peterson, D. A., Ray, J., and Gage, F. H. (1996) Differentiation of adult hippocampus-derived progenitors into olfactory neurons in vivo. *Nature* **383,** 624–627.

91. Gage, F. H., Ray, J., and Fisher, L. J. (1995) Isolation, characterization, and use of stem cells from the CNS. *Annu. Rev. Neurosci.* **18,** 159–192.

92. Anderson, W. F. and Fletcher, J. C. (1980) Sounding boards. Gene therapy in human beings: when is it ethical to begin? *N. Engl. J. Med.* **303,** 1293–1297.

93. Fletcher, J. C. (1985) Ethical issues in and beyond prospective clinical trials of human gene therapy. *J. Med. Philos.* **10,** 293–309.

94. Juengst, E. and Walters, L. (1995) *Gene Therapy: Ethical and Social Issues. Encyclopedia of Bioethics*, Macmillan, New York.

95. Dyer, A. R. (1997) The ethics of human genetic intervention: a postmodern perspective. *Exp. Neurol.* **144,** 168–172.

96. Danks, D. M. (1994) Germ-line gene therapy: no place in treatment of genetic disease. *Hum. Gene Ther.* **5,** 151–152.

97. Resnik, D. (1994) Debunking the slippery slope argument against human germ-line gene therapy. *J. Med. Philos.* **19,** 23–40.

98. McDonough, P. G. (1997) The ethics of somatic and germline gene therapy. *Ann. NY Acad. Sci.* **816,** 378–382.

99. McGleenan, T. (1995) Human gene therapy and slippery slope arguments. *J. Med. Ethics* **21,** 350–355.

100. Coutelle, C., Douar, A. M., Colledge, W. H., and Froster, U. (1995) The challenge of fetal gene therapy. *Nat. Med.* **1,** 864–866.

101. Fletcher, J. C. and Richter, G. (1996) Human fetal gene therapy: moral and ethical questions. *Hum. Gene Ther.* **7,** 1605–1614.

102. Senut, M.-C., Suhr, S. T., and Gage, F. H. (1998) Gene transfer to the rodent placenta in situ. A new strategy for delivering gene products to the fetus. *J. Clin. Invest.* **101,** 1565–1571.

103. Gardner, W. (1995) Can human genetic enhancement be prohibited? *J. Med. Philos.* **20,** 65–84.

104. Juengst, E. T. (1997) Can enhancement be distinguished from prevention in genetic medicine? *J. Med. Philos.* **22,** 125–142.

105. March of Dimes (1992) *Genetic Testing and Gene Therapy*, National Survey Findings.

106. Macer, D. R., Akiyama, S., Alora, A. T., Asada, Y., Azariah, J., Azariah, H., Boost, M. V., Chatwachirawong, P., Kato, Y., Kaushik, V., Leavitt, F. J., Macer, N. Y., Ong, C. C., Srinives, P., and Tsuzuki, M. (1995) International perceptions and approval of gene therapy. *Hum. Gene Ther.* **6,** 791–803.

107. Juengst, E. T. (1990) The NIH "Points to Consider" and the limits of human gene therapy. *Hum. Gene Ther.* **1,** 425–433.

108. Carmen, I. H. (1993) Human gene therapy: a biopolitical overview and analysis. *Hum. Gene Ther.* **4,** 187–193.

109. Butler, D. (1994) Ethics treaty to target genome implications. *Nature* **371,** 369.

Index